Understanding MOTOR CONTROLS

THIRD EDITION

Stephen L. Herman

Australia • Brazil • Mexico • Singapore • United Kingdom • United States

Understanding Motor Controls, Third Edition
Stephen L. Herman

SVP, GM Skills & Global Product Management: Dawn Gerrain

Product Team Manager: James DeVoe

Senior Director, Development: Marah Bellegarde

Senior Product Development Manager: Larry Main

Senior Content Developer: John Fisher

Product Assistant: Andrew Ouimet

Vice President, Marketing Services: Jennifer Ann Baker

Marketing Manager: Scott Chrysler

Senior Production Director: Wendy Troeger

Production Director: Andrew Crouth

Content Project Management and Art Direction: Lumina Datamatics, Inc.

Technology Project Manager: Joe Pliss

Cover image(s): © Eugene Sergeev; © terekhov igor

Library of Congress Control Number: 2015947066

ISBN-13: 978-1-3054-9812-9

Cengage Learning
20 Channel Center Street
Boston, MA 02210
USA

Cengage Learning is a leading provider of customized learning solutions with office locations around the globe, including Singapore, the United Kingdom, Australia, Mexico, Brazil, and Japan. Locate your local office at: **www.cengage.com/global**

Cengage Learning products are represented in Canada by Nelson Education, Ltd.

To learn more about Cengage Learning, visit **www.cengage.com**

Purchase any of our products at your local college store or at our preferred online store **www.cengagebrain.com**

Printed in the United States of America
Print Number: 04 Print Year: 2019

TABLE OF CONTENTS

PREFACE

A Note from the Author

I have taught the subject of motor control for over 30 years. I have tried different methods and found that some are more successful than others. *Understanding Motor Controls* is the accumulation of this knowledge. I am sure other methods may work equally well, but the methods and information presented in this textbook have worked the best for me. My goal in writing this textbook is to present the subject of motor control in a way that the average student can understand. I have three main objectives:

- Teach the student how to interpret the logic of a schematic diagram.
- Teach the student how to properly connect a circuit using a schematic diagram.
- Teach the student how to troubleshoot a control circuit.

Understanding Motor Controls assumes that the student has no knowledge of motor controls. The student is expected to have knowledge of basic Ohm's law and basic circuits, such as series, parallel, and combination. The book begins with an overview of safety. A discussion of schematics (ladder diagrams) and wiring diagrams is presented early. The discussion of schematics and wiring diagrams is intended to help students understand the written language of motor controls. Standard NEMA symbols are discussed and employed throughout the book when possible. The operation of common control devices is presented to help students understand how these components function and how they are used in motor control circuits. Basic control circuits are presented in a manner that allows students to begin with simple circuit concepts and progress to more complicated circuits.

The textbook contains examples of how a schematic or ladder diagram is converted into a wiring diagram. A basic numbering system is explained and employed to aid students in making this conversion. This is the most effective method I have found of teaching a student how to make the transition from a circuit drawn on paper to properly connecting components in the field.

Understanding Motor Controls also covers solid-state controls for both DC and AC motors. Variable frequency drives and programmable logic controllers are covered in detail. I explain how to convert a ladder diagram into a program that can be loaded into a PLC. The book contains many troubleshooting problems that help the student understand the logic of a control system. Circuit design is also used to help the student develop the concepts of circuit logic.

Understanding Motor Controls contains 16 hands-on laboratory exercises that are designed to use off-the-shelf motor control components. A list of materials and suggested vendors is given for the components used in the exercises. The laboratory exercises begin with very basic concepts and connections and progress through more complicated circuits.

Supplements

An online Instructor Companion website contains an Instructor Guide with answers to end of chapter review questions, test banks, and Chapter presentations done in PowerPoint, and testing powered by Cognero.

Cengage Learning Testing Powered by Cognero is a flexible, online system that allows you to:

- author, edit, and manage test bank content from multiple Cengage Learning solutions
- create multiple test versions in an instant
- deliver tests from your LMS, your classroom, or wherever you want

Contact Cengage Learning or your local sales representative to obtain an instructor account.

Accessing an Instructor Companion Website from SSO Front Door

1. Go to: http://login.cengage.com and login using the Instructor email address and password.
2. Enter author, title or ISBN in the **Add a title to your bookshelf** search box, click on **Search** button.
3. Click **Add to My Bookshelf** to add Instructor Resources.
4. At the Product page click on the **Instructor Companion site** link.

New Users

If you're new to Cengage.com and do not have a password, contact your sales representative.

Acknowledgments

Wes Mozley, Albuquerque Tech
Ralph Potter, Bowling Green Technical College
Richard Schell, Luzerne County Community College
Terry Snarr, Idaho State University
Ron Stadtherr, Ridgewater College
William Quimby, Shelton State Community College
Stephen Vossler, Lansing Community College
Keith Dinwiddie, Ozarks Community College

New for the Third Edition

Updated Illustrations
Extended coverage of control components.
Comparison of NEMA symbols and IEC symbols.
Additional information concerning pressure switches.
Extended coverage of troubleshooting.
Coverage of relays with mercury wetted contacts.
Code references have been updated to the 2014 NEC.
Added information concerning troubleshooting motors.

SAFETY OVERVIEW

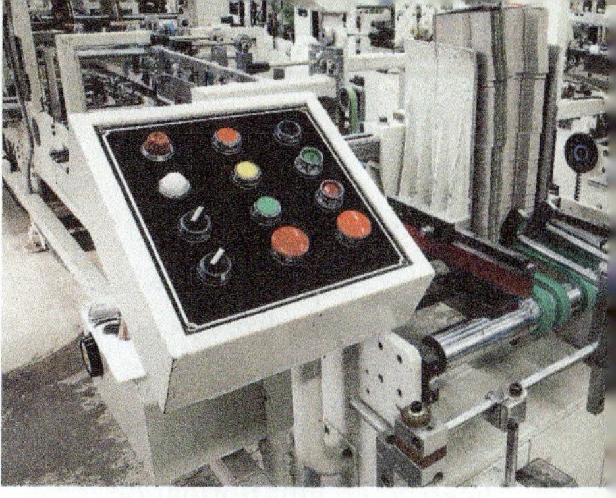

SAFETY OVERVIEW

Objectives

After studying this chapter the student will be able to:

- State basic safety rules.
- Describe the effects of electric current on the body.
- Discuss the origin and responsibilities of OSHA.
- Discuss material safety data sheets.
- Discuss lockout and tagout procedures.
- Discuss types of protective clothing.
- Explain how to properly place a straight ladder against a structure.
- Discuss different types of scaffolds.
- Discuss classes of fires.
- Discuss ground-fault circuit interrupters.
- Discuss the importance of grounding.

Safety is the job of each individual. You should be concerned not only with your own safety but with the safety of others around you. This is especially true for persons employed in the electrical field. Some general rules should be followed when working with electric equipment or circuits.

General Safety Rules

Never Work on an Energized Circuit If the Power Can Be Disconnected

When possible, use the following three-step check to make certain that power is turned off.

1. Test the **meter** on a known live circuit to make sure the meter is operating.
2. Test the circuit that is to become the **de-energized circuit** with the meter.
3. Test the meter on the known live circuit again to make certain the meter is still operating.

Install a warning tag at the point of **disconnection** so people will not restore power to the circuit. If possible, use a lock to prevent anyone from turning the power back on.

Think

Of all the rules concerning safety, this one is probably the most important. No amount of safeguarding or **idiot proofing** a piece of equipment can protect a person as well as taking time to think before acting. Many technicians have been killed by supposedly "dead" circuits. Do not depend on circuit breakers, fuses, or someone else to open a circuit. Test it yourself before you touch it. If you are working on high-voltage equipment, use insulated gloves and meter probes to measure the voltage being tested. *Think* before you touch something that could cost you your life.

Avoid Horseplay

Jokes and **horseplay** have a time and place but not when someone is working on an electric circuit or a piece of moving machinery. Do not be the cause of someone's being injured or killed and do not let someone else be the cause of your being injured or killed.

Do Not Work Alone

This is especially true when working in a hazardous location or on a live circuit. Have someone with you who can turn off the power or give **artificial respiration** and/or **cardiopulmonary resuscitation (CPR).** Several electric shocks can cause breathing difficulties and can cause the heart to go into fibrillation.

Work with One Hand When Possible

The worst kind of electric shock occurs when the current path is from one hand to the other, which permits the current to pass directly through the heart. A person can survive a severe shock between the hand and foot but it would cause death if the current path was from one hand to the other.

Learn First Aid

Anyone working on electric equipment, especially those working with voltages greater than 50 volts, should make an effort to learn first aid. A knowledge of first aid, especially CPR, may save your own or someone else's life.

Avoid Alcohol and Drugs

The use of alcohol and drugs has no place on a work site. Alcohol and drugs are not only dangerous to users and those who work around them; they also cost industry millions of dollars a year. Alcohol and drug abusers kill thousands of people on the highways each year and are just as dangerous on a work site as they are behind the wheel of a vehicle. Many industries have instituted testing policies to screen for alcohol and drugs. A person who tests positive generally receives a warning the first time and is fired the second time.

Effects of Electric Current on the Body

Most people have heard that it is not the voltage that kills but the current. This is true, but do not be misled into thinking that voltage cannot harm you. Voltage is the force that pushes the current though the circuit. It can be compared to the pressure that pushes water through a pipe. The more pressure available, the greater the volume of water flowing through the pipe. Students often ask how much current will flow through the body at a particular voltage. There is no easy answer to this question. The amount of current that can flow at a particular voltage is determined by the resistance of the current path. Different people have different resistances. A body has less resistance on a hot day when sweating, because salt water is a very good conductor. What one eats and drinks for lunch can have an effect on the body's resistance as can the length of the current path. Is the current path between two hands or from one hand to one foot? All of these factors affect body resistance.

Figure S–1 illustrates the effects of different amounts of current on the body. This chart is general— some people may have less tolerance to electricity and others may have a greater tolerance.

A current of 2 to 3 **milliamperes (mA)** (0.002 to 0.003 amperes) usually causes a slight tingling sensation, which increases as current increases and becomes very noticeable at about 10 milliamperes (0.010 amperes). The

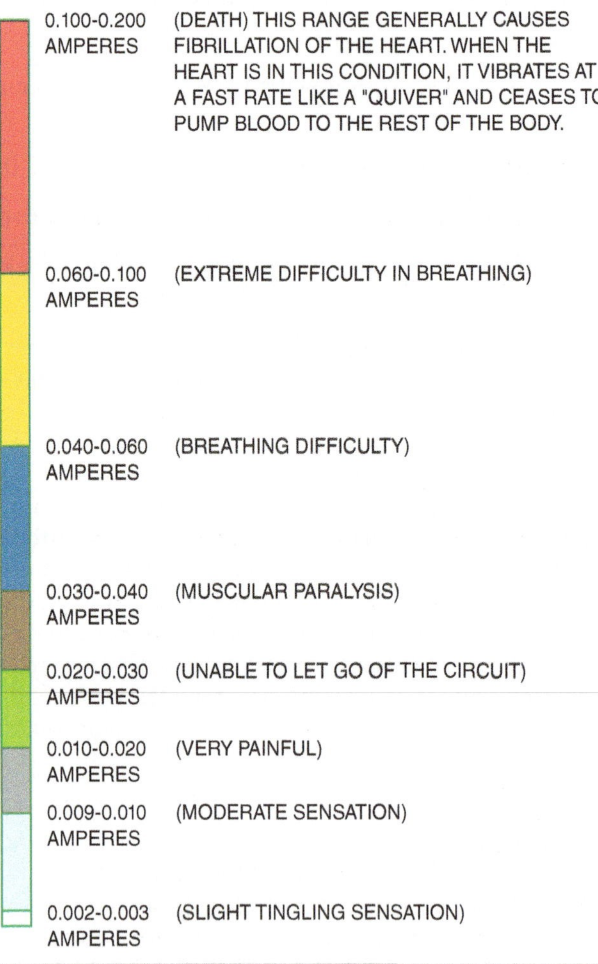

Figure S–1 The effects of electric current on the body.

tingling sensation is very painful at about 20 milliamperes. Currents between 20 and 30 milliamperes cause a person to seize the line and be unable to let go of the circuit. Currents between 30 and 40 milliamperes cause muscular paralysis, and those between 40 and 60 milliamperes cause breathing difficulty. When the current increases to about 100 milliamperes, breathing is extremely difficult. Currents from 100 to 200 milliamperes generally cause death because the heart usually goes into **fibrillation,** a condition in which the heart begins to "quiver" and the pumping action stops. Currents above 200 milliamperes cause the heart to squeeze shut. When the current is removed, the heart usually returns to a normal pumping action. This is the operating principle of a defibrillator. The voltage considered to be the most dangerous to work with is 120 volts, because that generally causes a current flow of between 100 and 200 milliamperes through most people's bodies. Large amounts of current can cause severe electric burns that are often very serious because they occur on the inside of the body. The exterior of the body may not look seriously burned, but the inside may be severely burned.

On the Job

OSHA

OSHA is an acronym for Occupational Safety and Health Administration, U.S. Department of Labor. Created by congress in 1971, its mission is to ensure safe and healthful workplaces in the United States. Since its creation, workplace fatalities have been cut in half, and occupational injury and illness rates have declined by 40%. Enforcement of OHSA regulations is the responsibility of the Secretary of Labor.

OSHA standards cover many areas, such as the handling of hazardous materials, fall protection, protective clothing, and hearing and eye protection. Part 1910 Subpart S deals mainly with the regulations concerning electrical safety. These regulations are available in books and can be accessed at the OSHA website on the Internet at www.osha.org.

Hazardous Materials

It may become necessary to deal with some type of hazardous material. A hazardous material or substance is any substance that if exposed to may result in adverse effects on the health or safety of employees. Hazardous materials may be chemical, biological, or nuclear. OSHA sets standards for dealing with many types of hazardous materials. The required response is determined by the type of hazard associated with the material. Hazardous materials are required to be listed as such. Much information concerning hazardous materials is generally found on **Material Safety Data Sheets (MSDS).** (A sample MSDS is included at the end of the unit.) If you are working in an area that contains hazardous substances, always read any information concerning the handling of the material and any safety precautions that should be observed. After a problem exists is not the time to start looking for information on what to do.

Some hazardous materials require a Hazardous Materials Response Team (HAZMAT) to handle any problems. A HAZMAT is any group of employees designated by the employer that are expected to handle and control an actual or potential leak or spill of a hazardous material. They are expected to work in close proximity to the material. A HAZMAT is not always a fire brigade, and a fire brigade may not necessarily have a HAZMAT. On the other hand, HAZMAT may be part of a fire brigade or fire department.

Employer Responsibilities

Section 5(a)1 of the Occupational Safety and Health Act basically states that employers must furnish each of their employees a place of employment that is free of recognized hazards that are likely to cause death or serious injury. This places the responsibility for compliance on employers. Employers must identify hazards or potential hazards within the work site and eliminate them, control them, or provide employees with suitable protection from them. It is the employee's responsibility to follow the safety procedures set up by the employer.

To help facilitate these safety standards and procedures, OSHA requires that an employer have a competent person oversee implementation and enforcement of these standards and procedures. This person must be able to recognize unsafe or dangerous conditions and have the authority to correct or eliminate them. This person also has the authority to stop work or shut down a work site until safety regulations are met.

MSDS

MSDS stands for material safety data sheets, which are provided with many products. They generally warn users of any hazards associated with the product. They outline the physical and chemical properties of the product; list precautions that should be taken when using the product; and list any potential health hazards, storage consideration, flammability, reactivity, and, in some instances, radioactivity. They sometimes list the name, address, and telephone number of the manufacturer; the MSDS date and emergency telephone numbers; and, usually, information on first aid procedures to use if the product is swallowed or comes in contact with the skin. Safety data sheets can be found on many home products such as cleaning products, insecticides, and flammable liquids.

Trenches

It is often necessary to dig trenches to bury conduit. Under some conditions, these trenches can be deep enough to bury a person if a cave-in occurs. Safety regulations for the shoring of trenches is found in OSHA Standard 1926 Subpart P App C titled "Timber Shoring for Trenches." These procedures and regulations are federally mandated and must be followed. Some general safety rules should be followed, such as:

1. Do not walk close to trenches unless it is necessary. This can cause the dirt to loosen and increase the possibility of a cave-in.
2. Do not jump over trenches if it is possible to walk around them.
3. Place barricades around trenches (Figure S–2).
4. Use ladders to enter and exit trenches.

Confined Spaces

Confined spaces have a limited means of entrance or exit (Figure S–3). They can be very hazardous workplaces, often containing atmospheres that are extremely harmful or deadly. Confined spaces are very difficult to ventilate because of their limited openings. It is often necessary for a worker to wear special clothing and use a separate air supply. OSHA Section 12: "Confined Space Hazards," lists rules and regulations for working in a confined space. In addition, many industries have written procedures that must be followed when working in confined spaces. Some general rules include the following:

Figure S–2 Place a barricade around a trench and use a ladder to enter and exit the trench.

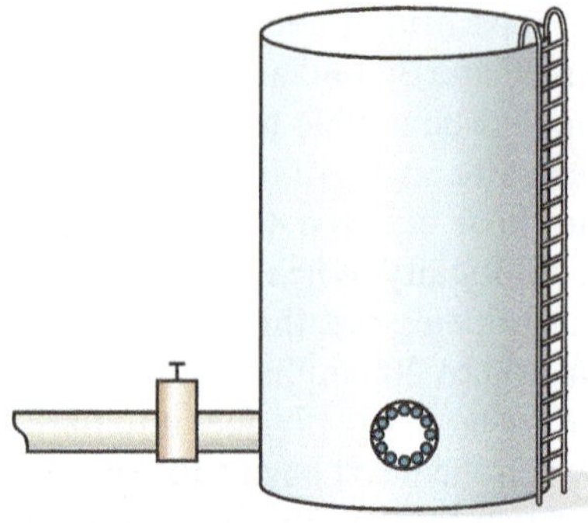

Figure S–3 A confined space is any space having a limited means of entrance or exit.

1. Have a person stationed outside the confined space to watch the person or persons working inside. The outside person should stay in voice or visual contact with the inside workers at all times. He or she should check air sample readings and monitor oxygen and explosive gas levels.
2. The outside person should never enter the space, even in an emergency, but should contact the proper emergency personnel. If he or she enters the space and become incapacitated, no one would be available to call for help.
3. Use only electric equipment and tools that are approved for the atmosphere found inside the confined area. It may be necessary to obtain a burning permit to operate tools that have open brushes and that spark when they are operated.
4. As a general rule, a person working in a confined space should wear a harness with a lanyard that extends to the outside person, so the outside person could pull him or her to safety if necessary.

Lockout and Tagout Procedures

Lockout and tagout procedures are generally employed to prevent someone from energizing a piece of equipment by mistake. This could apply to switches, circuit breakers, or valves. Most industries have their own internal policies and procedures. Some require that a tag similar to the one shown in Figure S–4 be placed on the piece

Figure S–4 Safety tag used to tagout equipment.

of equipment being serviced; some also require that the equipment be locked out with a padlock. The person performing the work places the lock on the equipment and keeps the key in his or her possession. A device that permits the use of multiple padlocks and a safety tag is shown in Figure S–5. This is used when more than one person is working on the same piece of equipment. Violating lockout and tagout procedures is considered an extremely serious offense in most industries and often results in immediate termination of employment. As a general rule, there are no first-time warnings.

After locking out and tagging a piece of equipment, it should be tested to make certain that it is truly de-energized before working on it. A simple three-step procedure is generally recommended for making certain that a piece of electric equipment is de-energized. A voltage tester or voltmeter that has a high enough range to safely test the voltage is employed. The procedure is as follows:

Figure S–5 The equipment can be locked out by several different people.

1. Test the voltage tester or voltmeter on a known **energized circuit** to make certain the tester is working properly.
2. Test the circuit you intend to work on with the voltage tester or voltmeter to make sure that it is truly de-energized.
3. Test the voltage tester or voltmeter on a known energized circuit to make sure that the tester is still working properly.

This simple procedure helps to eliminate the possibility of a faulty piece of equipment indicating that a circuit is de-energized when it is not.

Protective Clothing

Maintenance and construction workers alike are usually required to wear certain articles of protective clothing, dictated by the environment of the work area and the job being performed.

Head Protection

Some type of head protection is required on almost any work site. A typical electrician's hard hat, made of nonconductive plastic, is shown in Figure S–6. It has a pair of safety goggles attached that can be used when desired or necessary.

Eye Protection

Eye protection is another piece of safety gear required on almost all work sites. Eye protection can come in different forms, ranging from the goggles shown in Figure S–6 to the safety glasses with side shields shown in Figure S–7. Common safety glasses may or may not be prescription glasses, but almost all provide side protection (Figure S–7). Sometimes a full face shield may be required.

Hearing Protection

Section III, Chapter 5 of the OSHA Technical Manual includes requirements concerning hearing protection.

Figure S–6 Typical electrician's hard hat with attached safety goggles.

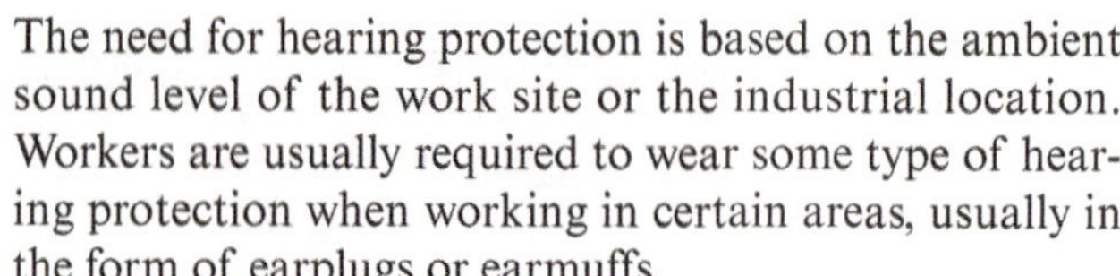

The need for hearing protection is based on the ambient sound level of the work site or the industrial location. Workers are usually required to wear some type of hearing protection when working in certain areas, usually in the form of earplugs or earmuffs.

Fire-Retardant Clothing

Special clothing made of fire-retardant material is required in some areas, generally certain industries as opposed to all work sites. **Fire-retardant clothing** is often required for maintenance personnel who work with high-power sources such as transformer installations and motor-control centers. An arc flash in a motor-control center can easily catch a person's clothes on fire. The typical motor-control center can produce enough energy during an arc flash to kill a person 30 feet away.

Gloves

Another common article of safety clothing is gloves. Electricians often wear leather gloves with rubber inserts when it is necessary to work on energized circuits (Figure S–8). These gloves are usually rated for a certain amount of voltage. They should be inspected for holes or tears before they are used. Kevlar gloves (Figure S–9) help protect against cuts when stripping cable with a sharp blade.

Safety Harness

Safety harnesses provide protection from falling. They buckle around the upper body with leg, shoulder, and chest straps; and the back has a heavy metal D-ring (Figure S–10). A section of rope approximately 6 feet

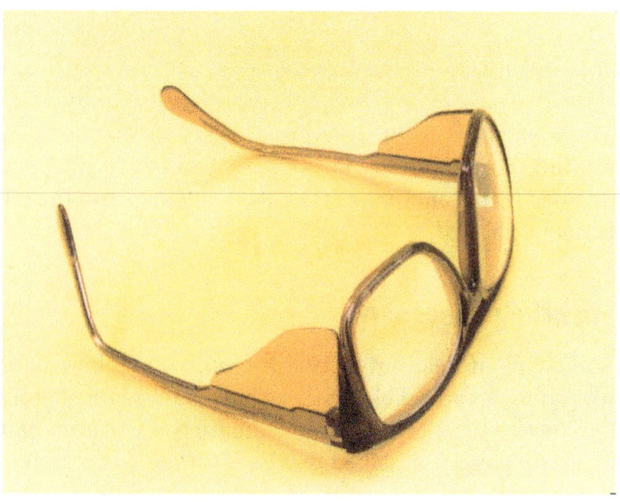

Figure S–7 Safety glasses provide side protection.

Figure S–8 Leather gloves with rubber inserts.

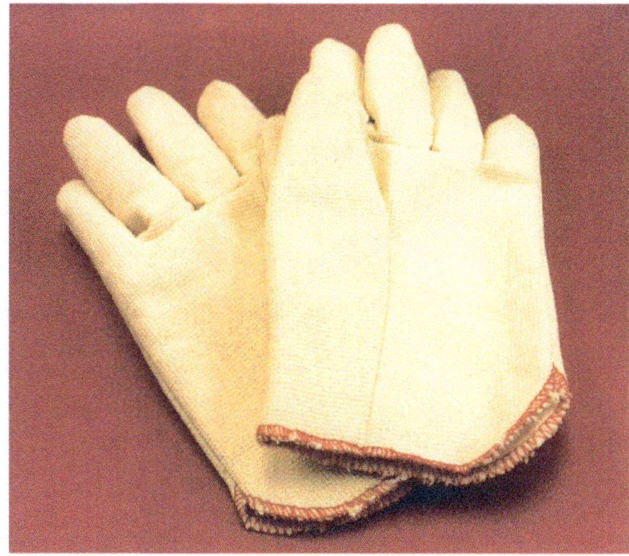

Figure S–9 Kevlar gloves protect against cuts.

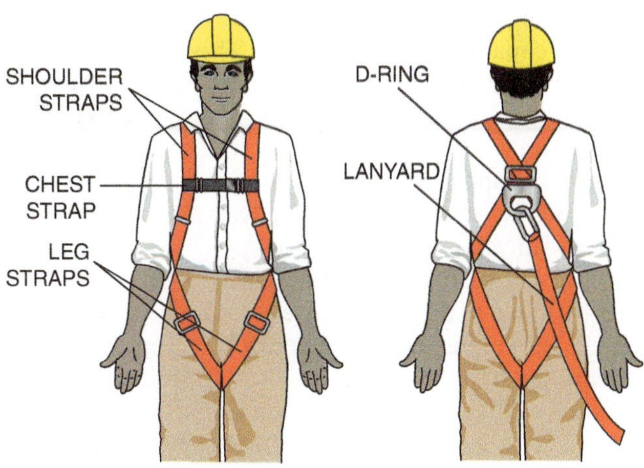

Figure S–10 Typical safety harness.

in length, called a lanyard, is attached to the D-ring and secured to a stable structure above the worker. If the worker falls, the lanyard limits the distance he or she can drop. A safety harness should be worn:

1. When working more than 6 feet above the ground or floor
2. When working near a hole or drop-off
3. When working on high scaffolding

A safety harness is shown in Figure S–11.

Figure S–11 Safety harness.

Ladders and Scaffolds

It is often necessary to work in an elevated location. When this is the case, ladders or scaffolds are employed. **Scaffolds** generally provide the safest elevated working platforms. They are commonly assembled on the work site from standard sections (Figure S–12). The bottom sections usually contain adjustable feet that can be used to level the sections. Two end sections are connected by X braces that form a rigid work platform (Figure S–13). Sections of scaffolding are stacked on top of each other to reach the desired height.

Rolling Scaffolds

Rolling scaffolds are used in areas that contain level floors, such as inside a building. The major difference between a rolling scaffold and those discussed previously is that it is equipped with wheels on the bottom section

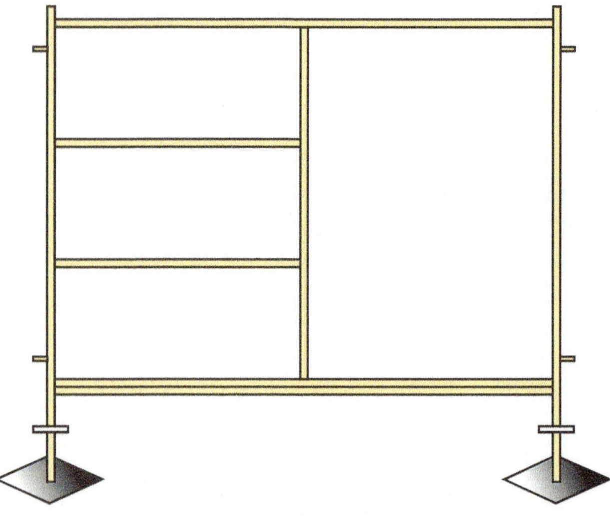

Figure S–12 Typical section of scaffolding.

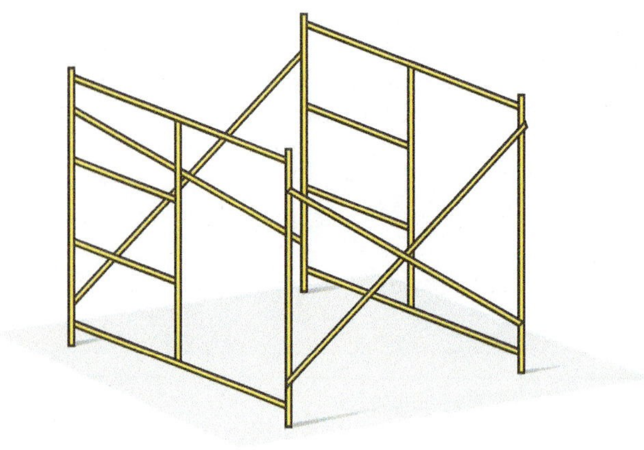

Figure S–13 X braces connect scaffolding sections together.

that permit it to be moved from one position to another. The wheels usually contain a mechanism that permits them to be locked after the scaffold is rolled to the desired location.

Hanging or Suspended Scaffolds

Hanging or suspended scaffolds are suspended by cables from a support structure. They are generally used on the sides of buildings to raise and lower workers by using hand cranks or electric motors.

Straight Ladders

Ladders can be divided into two main types, straight and step. Straight ladders are constructed by placing rungs between two parallel rails (Figure S–14). They generally

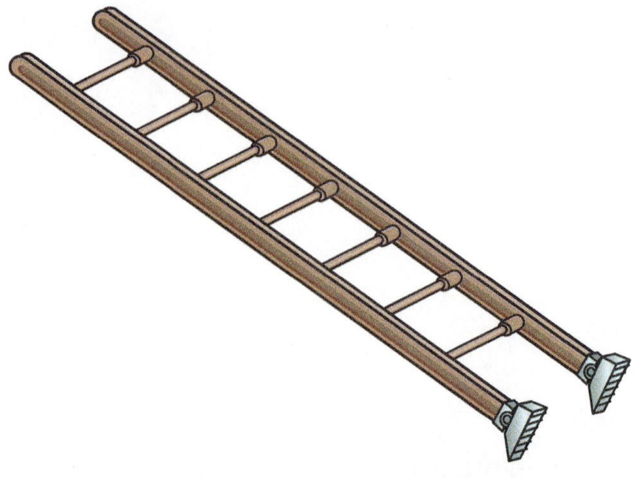

Figure S–14 Straight ladder.

contain safety feet on one end that help prevent the ladder from slipping. Ladders used for electrical work are usually wood or fiberglass; aluminum ladders are avoided because they conduct electricity. Regardless of the type of ladder used, you should check its load capacity before using it. This information is found on the side of the ladder. Load capacities of 200 pounds, 250 pounds, and 300 pounds are common. Do not use a ladder that does not have enough load capacity to support your weight plus the weight of your tools and the weight of any object you are taking up the ladder with you.

Straight ladders should be placed against the side of a building or other structure at an angle of approximately 76° (Figure S–15). This can be accomplished by moving the base of the ladder away from the structure a distance equal to one fourth the height of the ladder. If the ladder is 20 feet high, it should be placed 5 feet from the base of the structure. If the ladder is to provide access to the top of the structure, it should extend 3 feet above the structure.

Step Ladders

Step ladders are self-supporting, constructed of two sections hinged at the top (Figure S–16). The front section has two rails and steps, the rear portion two rails and braces. Like straight ladders, step ladders are designed to withstand a certain load capacity. Always check the load capacity before using a ladder. As a general rule, ladder manufacturers recommend that the top step not be used because of the danger of becoming unbalanced and falling. Many people mistakenly think the top step is the top of the ladder, but it is actually the last step before the ladder top.

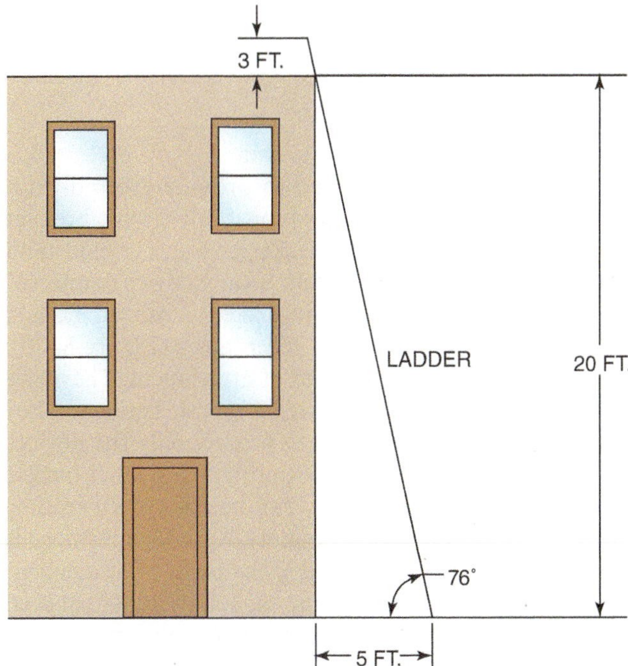

Figure S–15 A ladder should be placed at an angle of approximately 76°.

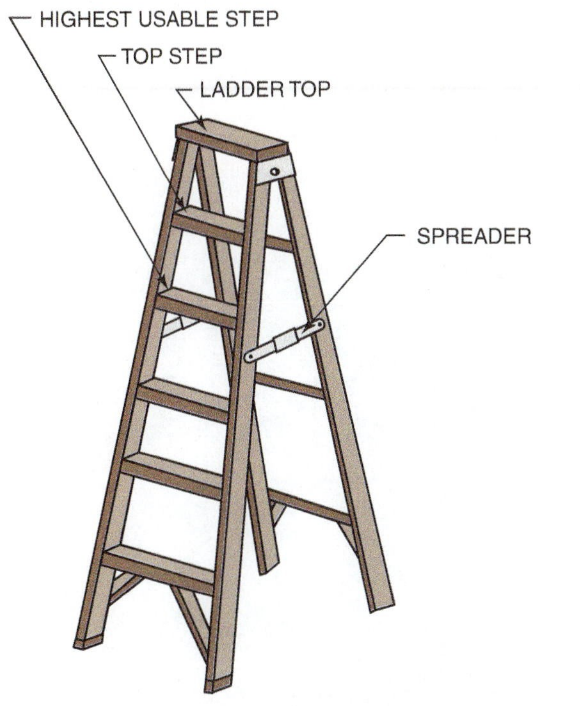

Figure S–16 Typical step ladder.

 Fires

For a fire to burn, it must have three things: fuel, heat, and oxygen. Fuel is anything that can burn, including materials such as wood, paper, cloth, combustible dusts, and even some metals. Different materials require different amounts of heat for combustion to take place. If the temperature of any material is below its combustion temperature, it will not burn. Oxygen must be present for combustion to take place. If a fire is denied oxygen, it will extinguish.

Fires are divided into four classes: A, B, C, and D. Class A fires involve common combustible materials such as wood or paper. They are often extinguished by lowering the temperature of the fuel below the combustion temperature. Class A fire extinguishers often use water to extinguish a fire. A fire extinguisher listed as Class A only should never be used on an electrical fire.

Class B fires involve fuels such as grease, combustible liquids, or gases. A Class B fire extinguisher generally employs carbon dioxide (CO_2), which greatly lowers the temperature of the fuel and deprives the fire of oxygen. Carbon dioxide extinguishers are often used on electrical fires, because they do not destroy surrounding equipment by coating it with a dry powder.

Class C fires involve energized electric equipment. A Class C fire extinguisher usually uses a dry powder to smother the fire. Many fire extinguishers can be used on multiple types of fires; for example, an extinguisher labeled ABC could be used on any of the three classes of fire. The important thing to remember is never to use an extinguisher on a fire for which it is not rated. Using a Class A extinguisher filled with water on an electrical fire could be fatal.

Class D fires consist of burning metal. Spraying water on some burning metals actually can cause the fire to increase. Class D extinguishers place a powder on top of the burning metal that forms a crust to cut off the oxygen supply to the metal. Some metals cannot be extinguished by placing powder on them, in which case the powder should be used to help prevent the fire from spreading to other combustible materials.

Ground-Fault Circuit Interrupters

Ground-fault circuit interrupters (GFCI) are used to prevent people from being electrocuted. They work by sensing the amount of current flow on both the ungrounded (hot) and grounded (neutral) conductors supplying power to a device. In theory, the amount of current in both conductors should be equal but opposite in polarity (Figure S–17). In this example, a current of 10 amperes flows in both the hot and neutral conductors.

A ground fault occurs when a path to ground other than the intended path is established (Figure S–18).

Assume that a person comes in contact with a defective electric appliance. If the person is grounded, a current path can be established through the person's body. In the example shown in Figure S–18, it is assumed that a current of 0.1 ampere is flowing through the person. This means that the hot conductor now has a current of 10.1 amperes but the neutral conductor has a current of only 10 amperes. The GFCI is designed to detect this current difference to protect personnel by opening the circuit when it detects a current difference of approximately 5 milliamperes (0.005 ampere). The *National Electrical Code® (NEC®) 210.8* lists places where ground-fault protection is required in dwellings. The *National Electrical Code®* and *NEC®* are registered trademarks of the National Fire Protection Association, Quincy, MA.

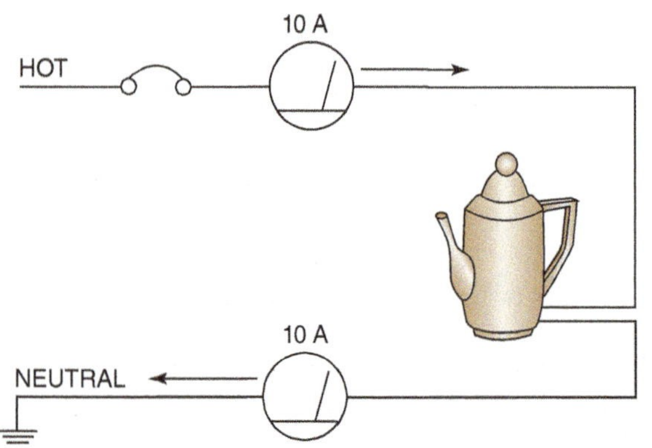

Figure S–17 The current in both the hot and neutral conductors should be the same but flowing in opposite directions.

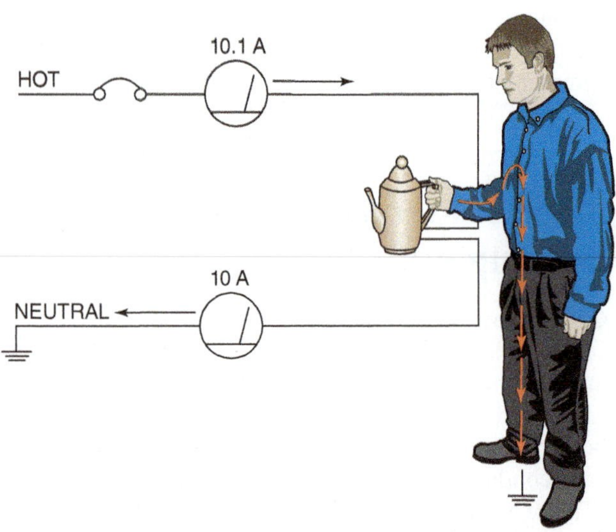

Figure S–18 A ground fault occurs when a path to ground other than the intended path is established.

GFCI Devices

Several devices can be used to provide ground-fault protection, including the ground-fault circuit breaker (Figure S–19). The circuit breaker provides ground-fault protection for an entire circuit, so any device connected to the circuit is ground-fault protected. A second method of protection, ground-fault receptacles (Figure S–20), provide protection at the point of attachment. They have some advantages over the GFCI circuit breaker. They can be connected so that they protect only the devices connected to them and do not protect any other outlets on the same circuit, or they can be connected so they provide protection to other outlets. Another advantage is that, because they are located at the point of attachment for the device, there is no stray capacitance loss between the panel box and the equipment is being protected. Long wire runs often cause nuisance tripping of GFCI circuit breakers. A third ground-fault protective device is the GFCI extension cord (Figure S–21). It can be connected into any standard electric outlet, and any devices connected to it are then ground-fault protected.

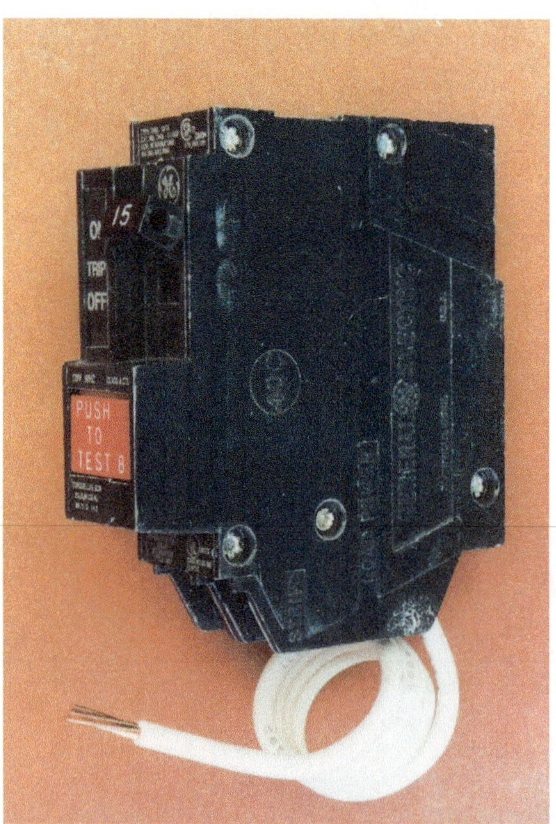

Figure S–19 Ground-fault circuit breaker.

Figure S–20 Ground-fault receptacle.

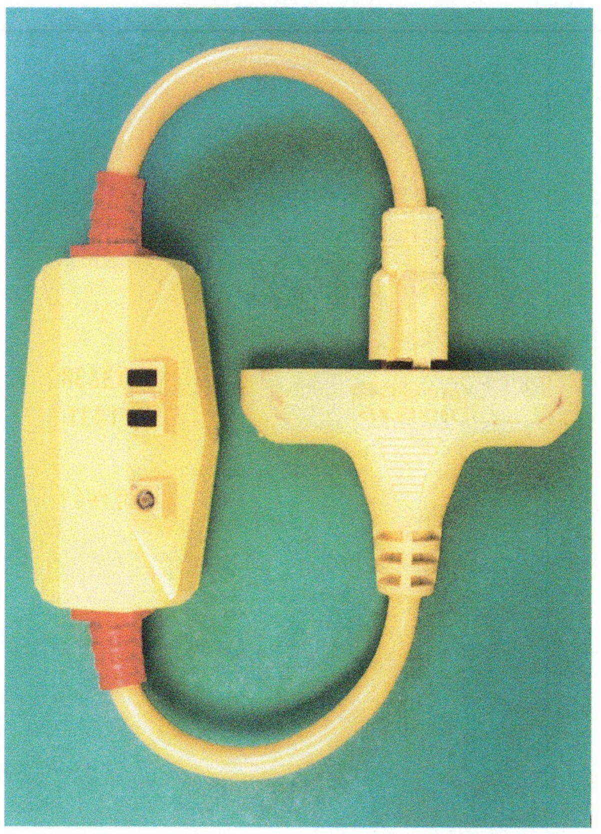

Figure S–21 Ground-fault extension.

Grounding

Grounding is one of the most important safety considerations in the electrical field. Grounding provides a low resistance path to ground to prevent conductive objects from existing at a high potential. Many electric appliances are provided with a three-wire cord. The third prong is connected to the case of the appliance and forces the case to exist at ground potential. If an ungrounded conductor comes in contact with the case, the grounding conductors conduct the current directly to ground. The third prong on a plug should never be cut off or defeated. Grounding requirements are far too numerous to list in this chapter, but *NEC® 250* covers the requirements for the grounding of electrical systems.

Review Questions

1. What is the most important rule of electrical safety?
2. Why should a person work with only one hand when possible?
3. What range of electric current generally causes death?
4. What is fibrillation of the heart?
5. What is the operating principle of a defibrillator?
6. Who is responsible for enforcing OSHA regulations?
7. What is the mission of OSHA?
8. What is an MSDS?
9. A padlock is used to lock out a piece of equipment. Who should have the key?
10. A ladder is used to reach the top of a building 16 feet tall. What distance should the bottom of the ladder be placed from the side of the building?
11. What is a ground fault?
12. What is the approximate current at which a ground-fault detector will open the circuit?
13. Name three devices used to provide ground-fault protection.
14. What type of fire is Class B?
15. What section of the NEC® covers grounding?

Table S–1 Heavy Duty Clear LO-VAC PVC Cement.

Section 1	**Identity of Material**
Trade Name	OATEY HEAVY DUTY CLEAR LO-VOC PVC CEMENT
Product Numbers	31850, 31851, 31853, 31854
Formula	PVC Resin in Solvent Solution
Synonyms	PVC Plastic Pipe Cement
Firm Name & Mailing Address	OATEY CO., 4700 West 160th Street, P.O. Box 35906 Cleveland, Ohio 44135, U.S.A. http://www.oatey.com
Oatey Phone Number	1-216-267-7100
Emergency Phone Numbers	For Emergency First Aid call 1-303-623-5716 COLLECT. For chemical transportation emergencies ONLY, call Chemtrec at 1-800-424-9300
Prepared By	Charles N. Bush, Ph.D.

Section 2	**Hazardous Ingredients**		
Ingredients	**%**	**Cas Number**	**Sec 313**
Acetone	0–5%	67-64-1	No
Amorphous Fumed Silica (Non-Hazardous)	1–3%	112945-52-5	No
Proprietary (Nonhazardous)	5–15%	N/A	No
PVC Resin (Nonhazardous)	10–16%	9002-86-2	No
Cyclohexanone	5–15%	108-94-1	No
Tetrahydrofuran	30–50%	109-99-9	No
Methyl Ethyl Ketone	20–35%	78-93-3	Yes

Section 3	**Known Hazards Under U.S. 29 CFR 1910.1200**				
Hazards	**Yes**	**No**	**Hazards**	**Yes**	**No**
Combustible Liquid		x	Skin Hazard	x	
Flammable Liquid	x		Eye Hazard	x	
Pyrophoric Material		x	Toxic Agent	x	
Explosive Material		x	Highly Toxic Agent		x
Unstable Material		x	Sensitizer		x
Water Reactive Material		x	Kidney Toxin	x	
Oxidizer		x	Reproductive Toxin	x	
Organic Peroxide		x	Blood Toxin		x
Corrosive Material		x	Nervous System Toxin	x	
Compressed Gas		x	Lung Toxin	x	
Irritant	x		Liver Toxin	x	
Carcinogen NTP/IARC/OSHA		x			

Table S–1 Continued

Section 4	Emergency and First Aid Procedures—Call 1-303-623-5716 Collect
Skin	If irritation arises, wash thoroughly with soap and water. Seek medical attention if irritation persists. Remove dried cement with Oatey Plumber's Hand Cleaner or baby oil.
Eyes	If material gets into eyes or if fumes cause irritation, immediately flush eyes with water for 15 minutes. If irritation persists, seek medical attention.
Inhalation	Move to fresh air. If breathing is difficult, give oxygen. If not breathing, give artificial respiration. Keep victim quiet and warm. Call a poison control center or physician immediately. If respiratory irritation occurs and does not go away, seek medical attention.
Ingestion	**DO NOT INDUCE VOMITING.** This product may be aspirated into the lungs and cause chemical pneumonitis, a potentially fatal condition. Drink water and call a poison control center or physician immediately. Avoid alcoholic beverages. Never give anything by mouth to an unconscious person.
Section 5	**Fire Fighting Measures**
Precautions	Do not use or store near heat, sparks, or flames. Do not smoke when using. Vapors may accumulate in low places and may cause flash fires.
Special Fire Fighting Procedures	For Small Fires: Use dry chemical, CO_2, water, or foam extinguisher. For Large Fires: Evacuate area and call Fire Department immediately.
Section 6	**Accidental Release Measures**
Spill or Leak Procedures	Remove all sources of ignition and ventilate area. Stop leak if it can be done without risk. Personnel cleaning up the spill should wear appropriate personal protective equipment, including respirators if vapor concentrations are high. Soak up spill with absorbent material such as sand, earth, or other noncombusting material. Put absorbent material in covered, labeled metal containers. Contaminated absorbent material may pose the same hazards as the spilled product.
Section 7	**Handling and Storage**
Precautions	**HANDLING & STORAGE:** Keep away from heat, sparks, and flames; store in cool, dry place. **OTHER:** Containers, even empties, will retain residue and flammable vapors.
Section 8	**Exposure Controls/Personal Protection**
Protective Equipment Types	**EYES:** Safety glasses with side shields. **RESPIRATORY:** NIOSH-approved canister respirator in absence of adequate ventilation. **GLOVES:** Rubber gloves are suitable for normal use of the product. For long exposures to pure solvents chemical resistant gloves may be required. OTHER: Eye wash and safety shower should be available.
Ventilation	**LOCAL EXHAUST:** Open doors and windows. Exhaust ventilation capable of maintaining emissions at the point of use below PEL. If used in enclosed area, use exhaust fans. Exhaust fans should be explosion-proof or set up in a way that flammable concentrations of solvent vapors are not exposed to electrical fixtures or hot surfaces.

Table S–1 Continued

Section 9	Physical and Chemical Properties			
NFPA Hazard Signal	Health 2	Stability 1	Flammability 3	Special None
HMIS Hazard Signal	Health 3	Stability 1	Flammability 4	Special None
Boiling Point	151 Degrees F/66 C			
Melting Point	N/A			
Vapor Pressure	145 mmHg @ 20 Degrees C			
Vapor Density (Air = 1)	2.5			
Volatile Components	70–80%			
Solubility In Water	Negligible			
PH	N/A			
Specific Gravity	0.95 ± 0.015			
Evaporation Rate	(BUAC = 1) = 5.5 − 8.0			
Appearance	Clear Liquid			
Odor	Ether-Like			
Will Dissolve In	Tetrahydrofuran			
Material Is	Liquid			

Section 1

BASIC CONTROL CIRCUITS AND COMPONENTS

Chapter 1

GENERAL PRINCIPLES OF MOTOR CONTROL

Objectives

After studying this chapter the student will be able to:

» State the purpose and general principles of motor control.

» Discuss the differences between manual and automatic motor control.

» Discuss considerations when installing motors or control equipment.

» Discuss the basic functions of a control system.

» Discuss surge protection for control systems.

Several factors should be considered when selecting the motor needed to perform a specific task and the control components that govern the operation of the motor. An electrician should not only be capable of properly installing a motor, but he or she should also be capable of maintaining the equipment and troubleshooting a control circuit when necessary. This textbook is designed to give students the skills they need to succeed in an industrial environment.

Many years ago, machines were operated by a line shaft (Figure 1–1). A central prime mover, whether a steam engine, electric motor, or water wheel, powered all the machines by connecting them to the line shaft with belts. Although this concept worked, the number of machines that could be operated at the same time was limited and control of the machine processes was very difficult. Some applications not only called for the machine to start and stop, but also to reverse direction, increase or decrease speed, and brake to a stop. The advent of connecting individual power sources to each machine changed the world of motor control forever.

Installation of Motors and Control Equipment

When installing electric motors and equipment, several factors should be considered. When a machine is installed, the motor, machine, and controls are all interrelated and must be considered as a unit. Some machines have the motor or motors and control equipment mounted on the machine itself when it is delivered from the manufacturer, and the electrician's job is generally to make a simple power connection to the machine. A machine of this type is shown in Figure 1–2. Other types of machines require separately mounted motors that are connected by belts, gears, or chains. Some

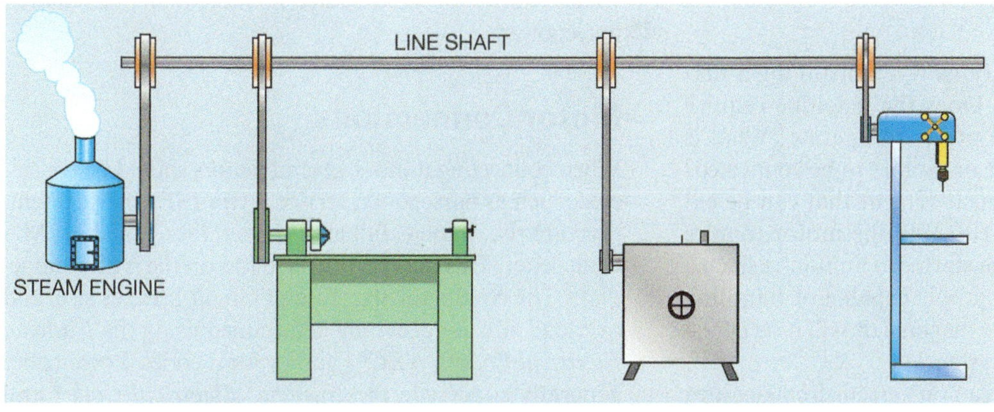

Figure 1–1 Power machines driven by a line shaft.

Figure 1–2 Machine was delivered with self-contained motors and controls.

machines also require the connection of pilot-sensing devices such as photo switches, limit switches, or pressure switches. Regardless of how easy or complex the connection is, several factors must be considered.

Power Source

One of the main considerations when installing a machine is the power source. Does the machine require single phase or three phase power to operate? What is the horsepower of the motor or motors to be connected? What is the amount of in-rush current that can be expected when the motor starts? Will the motor require some type of reduced voltage starter to limit in-rush current? Is the existing power supply capable of handling the power requirement of the machine or will it be necessary to install a new power system?

The availability of power can vary greatly from one area of the country to another. Companies that supply power to heavily industrialized areas can generally permit larger motors to be started across-the-line than companies that supply power to areas that have light industrial needs. In some areas the power company may permit a motor of several thousand horsepower to be started across-the-line, and in other areas the power company may require a reduced voltage starter for motors rated no more than one hundred horsepower.

Motor Connections

When connecting motors, several factors should be considered, such as **horsepower, service factor (SF),** marked temperature rise, **voltage,** full load current rating, and NEMA Code letter. This information is found on the motor nameplate. The conductor size, fuse or circuit breaker size, and overload size are generally determined using the *National Electrical Code® (NEC®)* and/or local codes. Local codes generally supersede the *National Electrical Code®* and should be followed when they apply. Motor installation based on the *NEC®* will be covered in this textbook.

Controller Type

Different operating conditions require different types of control. Some machines simply require the motor to start and stop. Some machines require a soft start, which means bringing the motor up to speed over a period of time instead of all at once. This is especially true of gear-driven machines or motors that must start heavy inertia loads, such as flywheels or centrifuges. Other machines may require variable speed or the application of a brake when the motor is stopped. Inching and jogging may also be a consideration. Regardless of the specific conditions, all control systems should be able to start and stop the motor, and also provide overload protection for the motor and short-circuit protection for the circuit.

Environment

Another consideration is the type of environment in which the motor and control system operates. Can the controls be housed in a general purpose enclosure similar to the one shown in Figure 1–3, or is the system subject to moisture or dust? Are the motor and controls to be operated in a hazardous area that requires explosion-proof enclosures similar to that shown in Figure 1–4? Some locations may contain corrosive vapor or liquid, or extremes of temperature. All of these conditions should be considered when selecting motors and control components.

Codes and Standards

Another very important consideration is the safety of the operator and persons that work around the machine. In 1970, the Occupational Safety and Health Act (OSHA) was established. In general, OSHA requires employers to provide an environment that is free of recognized hazards that are likely to cause serious injury.

Another organization that exhibits much influence on the electrical field is Underwriters Laboratories (UL). Insurance companies established Underwriters

Courtesy Schneider Electric USA, Inc.

Figure 1–3 Manual starter in a general purpose enclosure.

Figure 1–4 Magnetic starter in an explosion-proof enclosure.

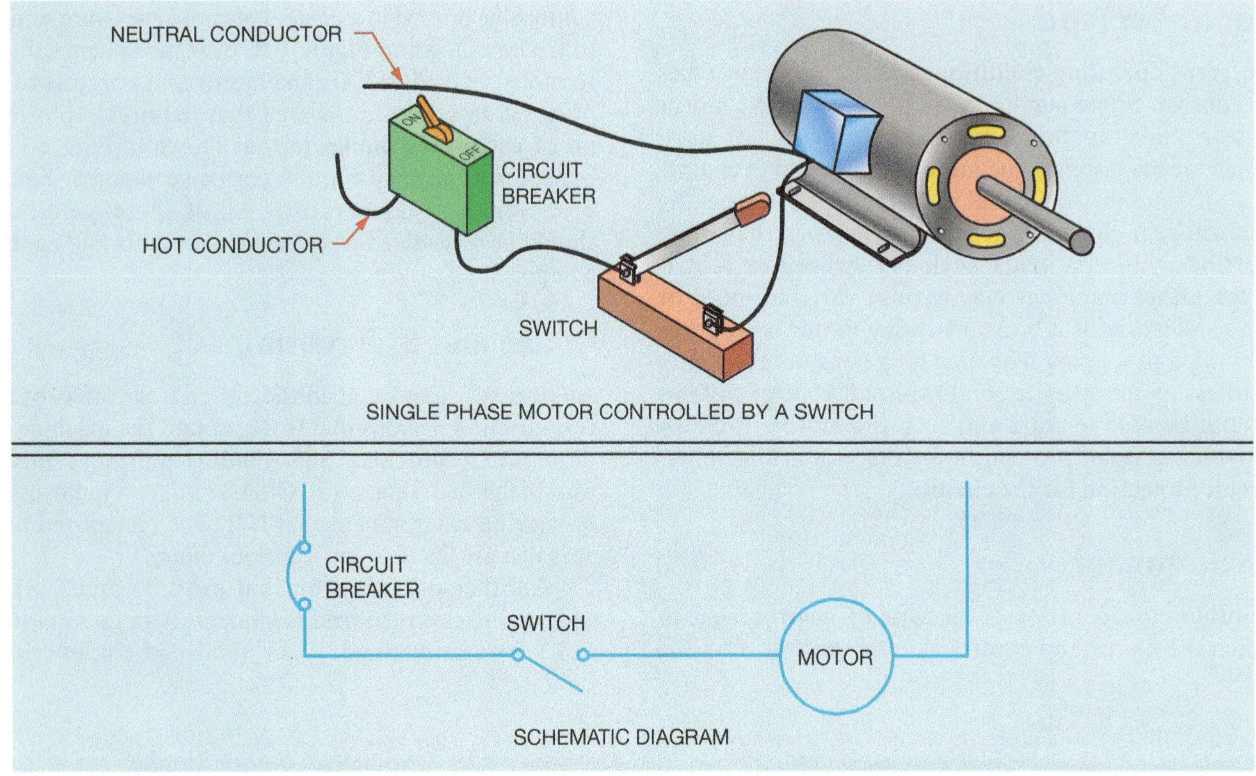

Figure 1–5 Pictorial and schematic diagram of a single phase motor controlled by a switch.

Laboratories in an effort to reduce the number of fires caused by electrical equipment. UL tests equipment to determine if it is safe under different conditions. Approved equipment is listed in its annual publication, which is kept current with bimonthly supplements.

A previously mentioned document is the ***National Electrical Code®***. The ***NEC®*** is published by the National Fire Protection Association. The *NEC®* establishes rules and specifications for the installation of electrical equipment. The *National Electrical Code®* is not a law unless it is made law by a local authority.

Two other organizations that have a great influence on control equipment are the **National Electrical Manufacturers Association** (**NEMA**) and the **International Electrotechnical Commission** (**IEC**). Both of these organizations will be discussed later in the textbook.

Types of Control Systems

Motor control systems can be divided into three major types: manual, semiautomatic, and automatic. Manual controls are characterized by the fact that the operator must go to the location of the controller to initiate any change in the state of the control system. **Manual controllers** are generally very simple devices that connect the motor directly to the line. They may or may not provide overload protection or low voltage release. Manual control may be accomplished by simply connecting a switch in series with the motor (Figure 1–5).

Semiautomatic control is characterized by the use of push buttons, limit switches, pressure switches, and other sensing devices to control the operation of a magnetic contactor or starter. The starter actually connects the motor to the line and the push buttons and other pilot devices control the coil of the starter. This permits the actual control panel to be located away from the motor or starter. The operator must still initiate certain actions, such as starting and stopping, but he or she does not have to go to the location of the motor or starter to perform the action. A typical control panel is shown in Figure 1–6. A schematic and wiring diagram of a START–STOP push button station is shown in Figure 1–7. A **schematic diagram** shows components in their electrical sequence without regard for physical location. A **wiring diagram** is basically a pictorial representation of the control components with connecting wires. Although the

Figure 1–6 Typical operator's control panel.

two circuits shown in Figure 1–7 look different, electrically they are the same.

Automatic control is very similar to semiautomatic control, in that pilot-sensing devices are employed to operate a magnetic contactor or starter that actually controls the motor. With automatic control, however, an operator does not have to initiate certain actions. Once the control conditions have been set, the system continues to operate on its own. A good example of an automatic control system is the heating and cooling system found in many homes. Once the thermostat has been set to the desired temperature, the heating or cooling system operates without further attention of the home owner. The control circuit contains sensing devices that automatically shut the system down in the event of an unsafe condition such as motor overload, excessive current, or no pilot light or ignition in gas heating systems.

Functions of Motor Control

Motor control systems perform some basic functions. The ones listed are by no means the only ones, but are very common. These basic functions will be discussed in greater detail in this book. It is important to not only understand these basic functions of a control system but also to know how control components are employed to achieve the desired circuit logic.

Starting

Starting the motor is one of the main purposes of a motor control circuit. Several methods can be employed depending on the requirements of the circuit. The simplest method is **across-the-line** starting. This is accomplished by connecting the motor directly to the

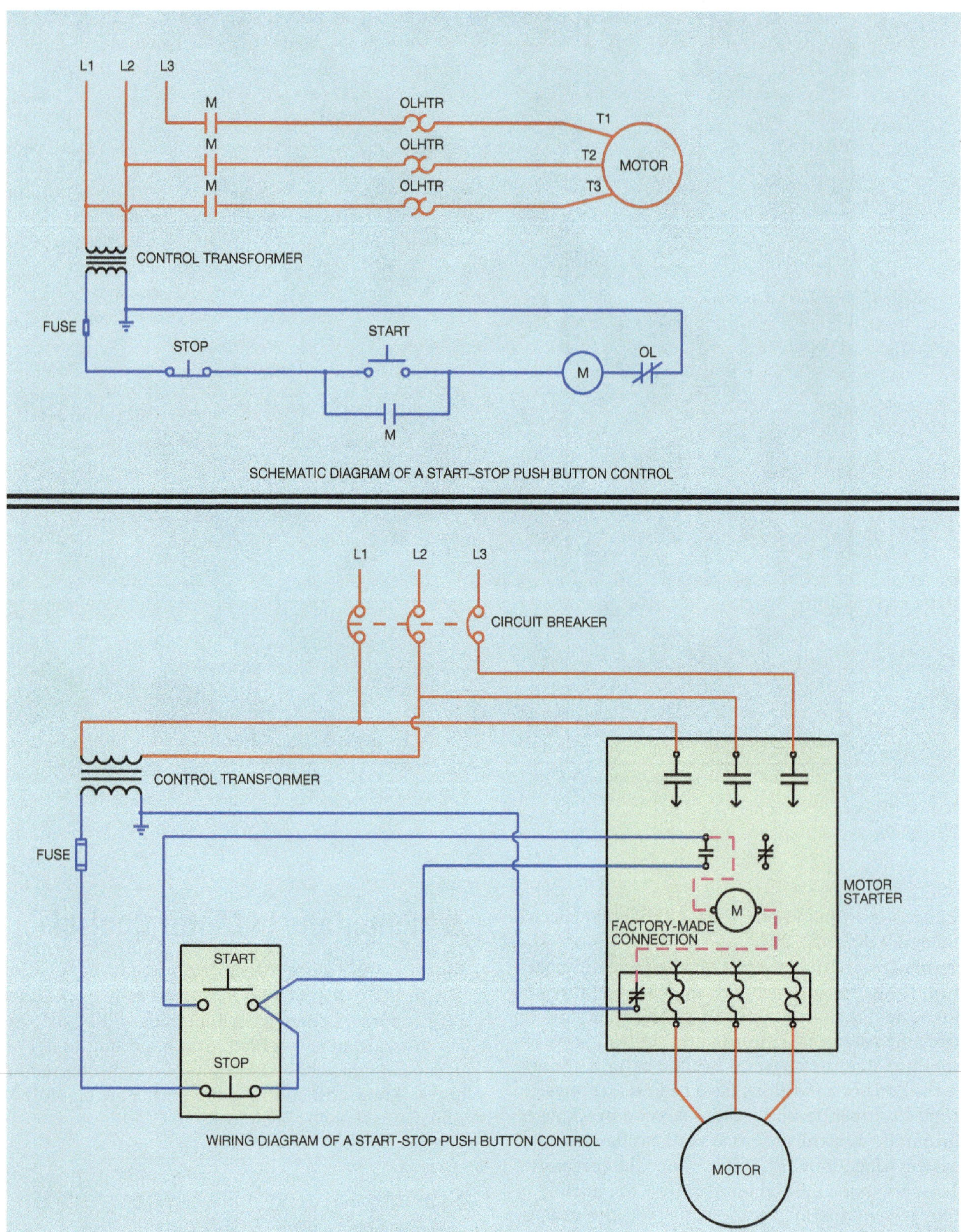

Figure 1–7 Schematic and wiring diagram of a start-stop push button control.

power line. Some situations, however, may require the motor to start at a low speed and accelerate to full speed over some period of time. This situation is often referred to as **ramping.** In other situations it may be necessary to limit the amount of current or torque during starting. Some of these methods will be discussed later in the textbook.

Stopping

Another function of the control system is to stop the motor. The simplest method is to disconnect the motor from the power line and permit it to coast to a stop. Some conditions, however, may require that the motor be stopped more quickly, or that a brake hold a load when the motor is stopped.

Jogging or Inching

Jogging and **inching** are methods employed to move a motor with short jabs of power—generally to move a motor or load into some desired position. The difference between jogging and inching is that jogging is accomplished by momentarily connecting the motor to full line voltage and inching is accomplished by momentarily connecting the motor to reduced voltage.

Speed Control

Some control systems require variable speed. There are several ways to accomplish variable speed. One of the most common is with variable frequency control for alternating current motors, or by controlling the voltage applied to the armature and fields of a direct current motor. Another method may involve the use of a direct current clutch. These methods will be discussed in more detail later in this textbook.

Motor and Circuit Protection

One of the major functions of most control systems is to provide protection for both the circuit components and the motor. Fuses and circuit breakers are generally employed for circuit protection, and overload relays are used to protect the motor. The different types of overload relays will be discussed later.

Surge Protection

Another concern in many control circuits is the voltage spikes or surges produced by collapsing magnetic fields when power to the coil of a relay or contactor is turned off. These collapsing magnetic fields can induce voltage spikes that are hundreds of volts (Figure 1–8). These high voltage surges can damage electronic components connected to the power line. Voltage spikes are of greatest concern in control systems that employ computer controlled devices, such as programmable logic controllers and measuring instruments used to sense temperature, pressure, and so on. Coils connected to alternating current often have a metal oxide varistor (**MOV**) connected across the coil (Figure 1–9). Metal oxide varistors are voltage-sensitive resistors. They have the ability to change their resistance value in accord with the amount of voltage applied to them. The MOV will have a voltage rating greater than that of the coil they are connected across. An MOV connected across a coil intended to operate on 120 volts, for example, will have a rating of about 140 volts. As long as the voltage applied to the MOV is below its voltage rating, it will exhibit an extremely high amount of resistance, generally several million ohms. The current flow through the MOV is called **leakage current** and is so small that it does not affect the operation of the circuit.

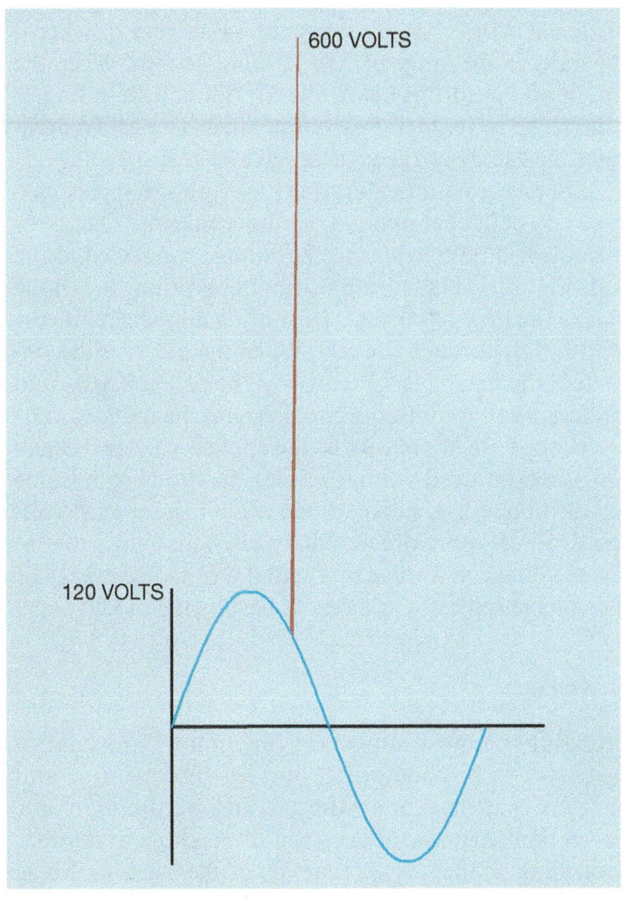

Figure 1–8 Voltage spikes can be hundreds of volts.

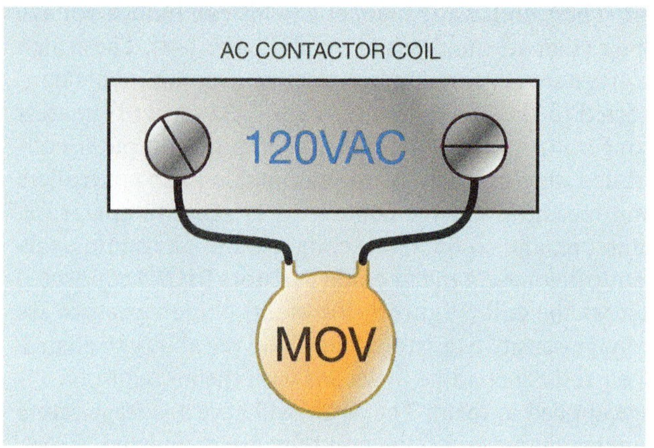

Figure 1–9 A metal oxide varistor is used to eliminate voltage spikes on alternating current coils.

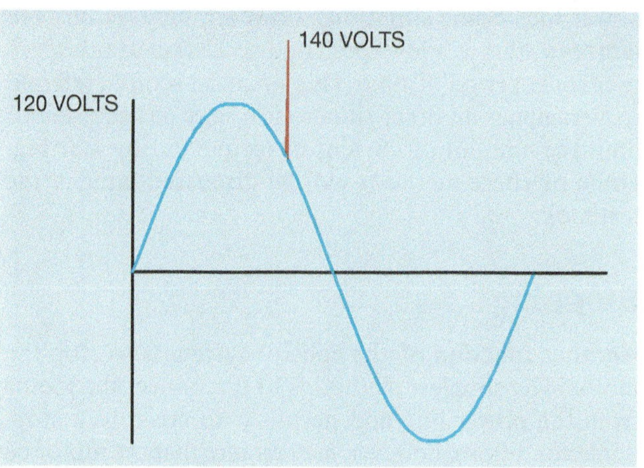

Figure 1–10 The metal oxide varistor limits the voltage spike to 140 volts.

If the voltage across the coil becomes greater than the voltage rating of the MOV, the resistance of the MOV suddenly changes to a very low value, generally in the range of 2 to 3 ohms. This effectively short-circuits the coil and prevents the voltage from becoming any higher than the voltage rating of the MOV (Figure 1–10). Metal oxide varistors change resistance value very quickly—generally in the range of 3 to 10 nanoseconds. When the circuit voltage drops below the MOV's voltage rating, it will return to its high resistance value. The MOV dissipates the energy of the voltage spike as heat.

Diodes are used to suppress the voltage spikes produced by coils that operate on direct current. The diode is connected reverse bias to the voltage connected to the coil, Figure 1–11. During normal operation, the diode blocks the flow of current, permitting all the circuit current to flow through the coil. When the power is disconnected, the magnetic field around the coil collapses and induces a voltage into the coil. Because the induced voltage is opposite in polarity to the applied voltage, (**Lenz's Law**), the induced voltage causes the diode to become forward biased. A silicon diode exhibits a forward voltage drop of approximately 0.7 volt. This limits the induced voltage to a value of about 0.7 volt. The energy of the voltage spike is dissipated as heat by the diode.

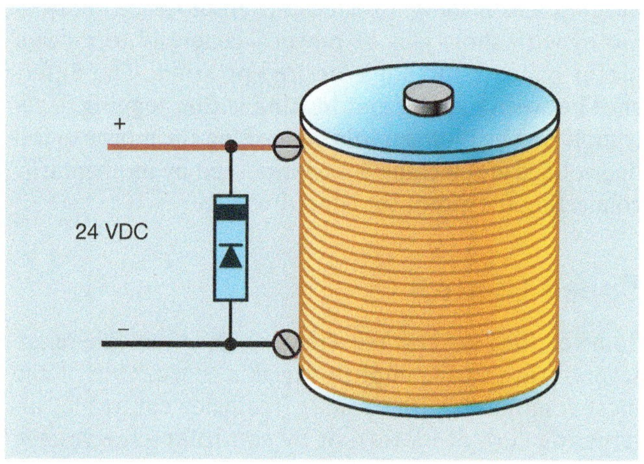

Figure 1–11 A diode is used to prevent voltage spikes on direct current coils.

Safety

Probably the most important function of any control system is to provide protection for the operator and persons that may be in the vicinity of the machine. Protection varies from one type of machine to another depending on the specific function of the machine. Many machines are provided with both mechanical and electrical safeguards.

Review Questions

1. When installing a motor control system, list four major factors to consider concerning the power system.
2. Where is the best place to look to find specific information about a motor, such as horsepower, voltage, full load current, service factor, and full load speed?
3. Is the *National Electrical Code®* a law?
4. Explain the difference between manual control, semiautomatic control, and automatic control.

5. What is the simplest of all starting methods for a motor?
6. Explain the difference between jogging and inching.
7. What is the most common method of controlling the speed of an alternating current motor?
8. What agency requires employers to provide a workplace free of recognized hazards for its employees?
9. What is meant by the term *ramping*?
10. What is the most important function of any control system?

Chapter 2

SYMBOLS AND SCHEMATIC DIAGRAMS

Objectives

After studying this chapter the student will be able to:

›› Discuss symbols used in the drawing of schematic diagrams.

›› Determine the difference between switches that are drawn normally open, normally closed, normally open held closed, and normally closed held open.

›› Draw standard NEMA control symbols.

›› State rules that apply to schematic or ladder diagrams.

›› Interpret the logic of simple ladder diagrams.

When you learned to read, you were first taught a set of symbols that represented different sounds. This set of symbols is called the alphabet. Schematics and wiring diagrams are the written language of motor controls. Before you can learn to properly determine the logic of a control circuit, it is necessary to first learn the written language. Unfortunately, there is no actual standard used for motor control symbols. Different manufacturers and companies often use their own set of symbols for their in-house schematics. Also, schematics drawn in other countries may use an entirely different set of symbols to represent different control components. European schematics often contain symbols adopted by the International Electrotechnical Commission (IEC). Although symbols can vary from one manufacturer to another, or from one country to another, once you have learned to interpret circuit logic, it is generally possible to determine what the different symbols represent by the way they are used in the schematic. The most standardized set of symbols in the United States is provided by the National Electrical Manufacturer's Association (NEMA). It is these symbols that will be discussed in this chapter.

Push Buttons

One of the most commonly used symbols in control schematics is the push button. Push buttons can be shown as normally closed or normally open. Most are momentary contact devices in that they make or break connection as long as pressure is applied to them. When pressure is removed, they return to their normal position. Push buttons contain both movable and stationary contacts. The stationary contacts are connected to terminal screws. The normally open push button symbol is characterized by drawing the movable contact above and not touching the stationary contacts (Figure 2–1). Because the movable contact is not touching the stationary contacts the circuit is open and current cannot flow from one stationary contact to the other. The normally closed push button symbol is characterized by drawing the movable contact below and touching the two stationary contacts as shown in Figure 2–1. Because the movable contact is touching the two stationary contacts there is a complete circuit and current can flow from one stationary contact to the other.

Normally Closed Push Buttons

The movable contact of the normally closed push button makes contact with the two stationary contacts when no pressure is applied to the button as shown in Figure 2–2. Because the movable contact touches the two stationary contacts a complete circuit exists and current can flow from one stationary contact to the other. If pressure is applied to the button, the movable contact moves away from the two stationary contacts and opens the circuit. When pressure is removed from the button, a spring causes the movable contact to return and bridge the two stationary contacts.

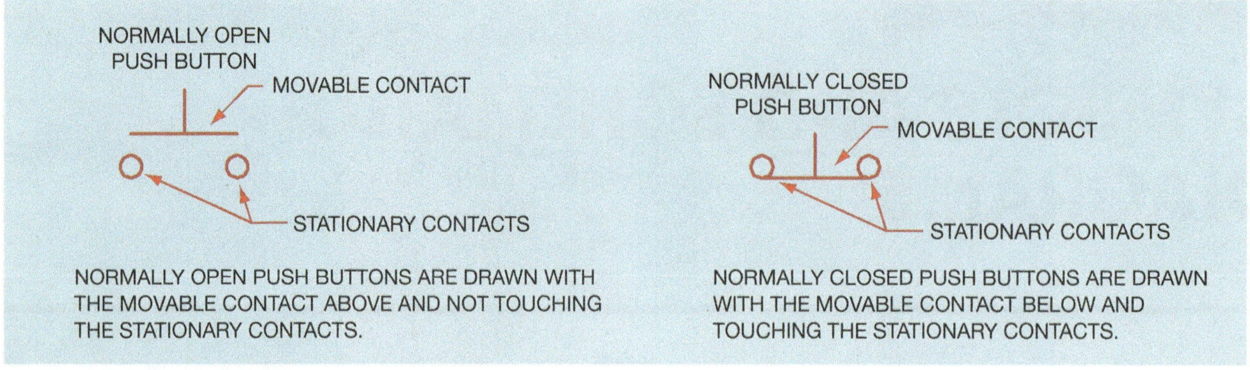

Figure 2–1 NEMA standard push button symbols.

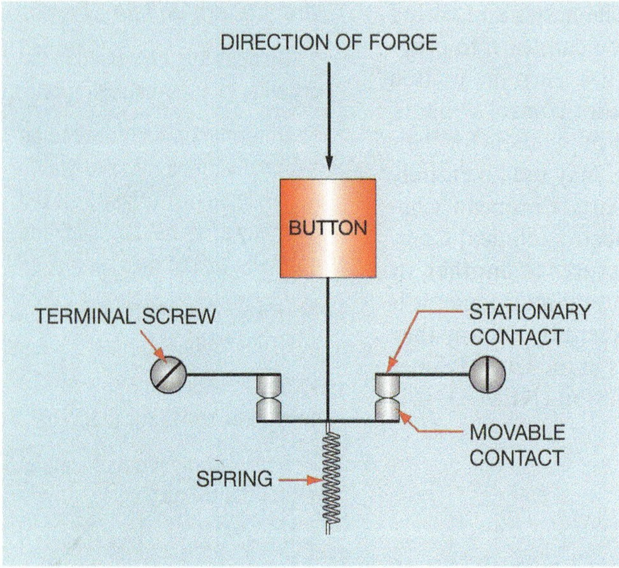

Figure 2–2 The movable contact bridges the two stationary contacts.

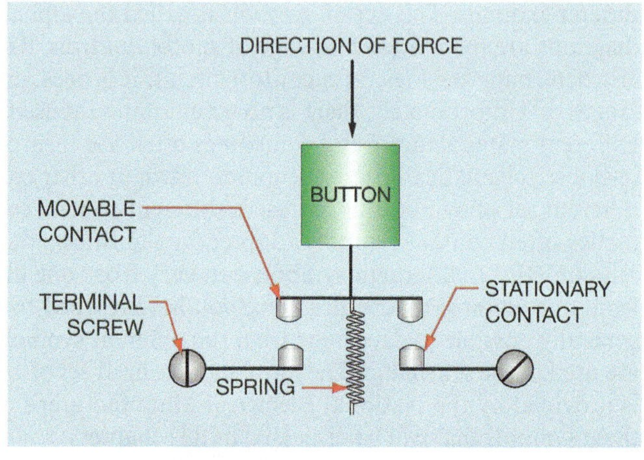

Figure 2–3 The movable contact does not bridge the two stationary contacts.

The normally closed push button symbol shown in Figure 2–1 is characterized by drawing the movable contact below and touching the two stationary contacts.

Normally Open Push Button

The normally open push button is similar to the normally closed except that the movable contact does not make connection with the two stationary contacts in its normal position (Figure 2–3). The normally open push button symbol is characterized by drawing the movable contact above and not touching the two stationary contacts as shown in Figure 2–1. When the button is pressed, the movable contact moves down and bridges the two stationary contacts to form a complete circuit. When pressure is removed from the button, a spring returns the movable contact to its original position and a circuit no longer exists between the two stationary contacts.

Double Acting Push Buttons

Another very common push button found throughout industry is the double acting push button. Double acting push buttons contain both normally open and normally closed contacts (Figure 2–4). When pressure is applied to the button, the movable contacts break connection with the two normally closed stationary contacts, creating an open circuit. The movable contacts then bridge the two normally open stationary contacts, creating a complete circuit. When pressure is removed from the button, a spring causes the movable contacts to return to their normal position. Double acting push buttons contain four terminal screws, two for the normally closed contacts and two for the normally open contacts. When connecting these push buttons, it is important to make certain that the wires are connected to the correct set of contacts. A schematic symbol for a double acting push button is shown in Figure 2–5. The symbol for double acting push buttons can actually be drawn in several ways (Figure 2–6). The symbol on the left is drawn with two movable contacts connected by a common shaft. When

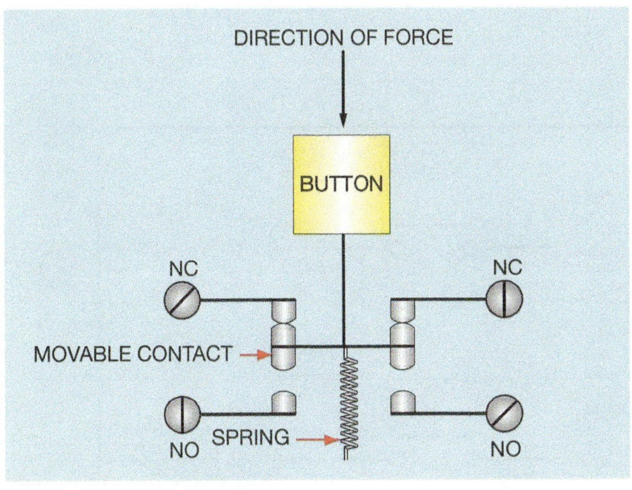

Figure 2–4 Double acting push buttons contain both normally open and normally closed contacts.

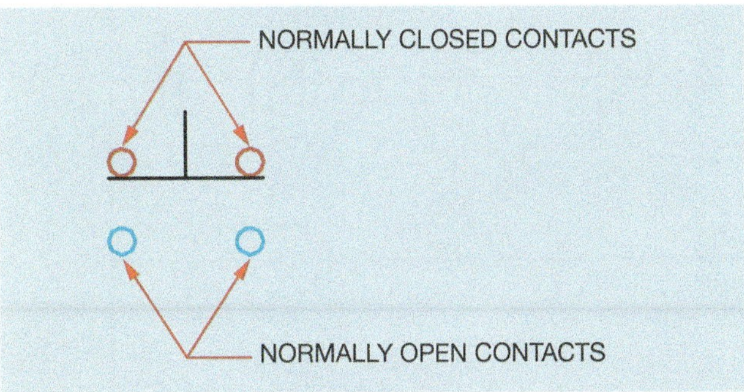

Figure 2–5 Double-acting push button.

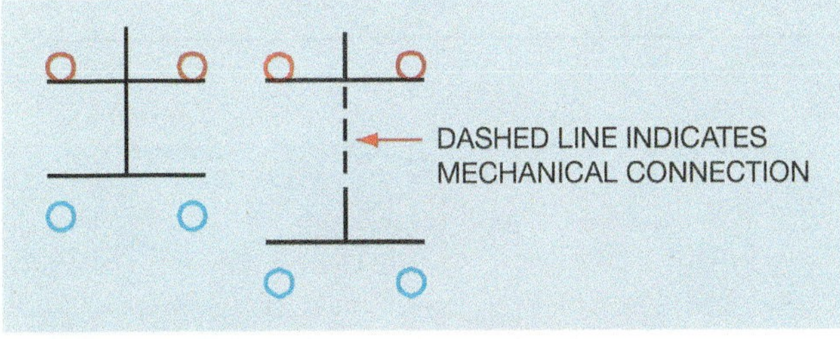

Figure 2–6 Other symbols used to represent double-acting push buttons.

the button is pressed, the top movable contact breaks away from the two stationary contacts at the top to open the circuit, and the bottom movable contact bridges to two bottom stationary contacts to complete the circuit.

The symbol on the right is very similar to the one on the left in that it shows two separate movable contacts. The right hand symbol, however, connects the two push button symbols together with a dashed line. When components are connected together with a dashed line in a schematic diagram, it indicates that the components are mechanically connected together. If one is pressed, all that are connected

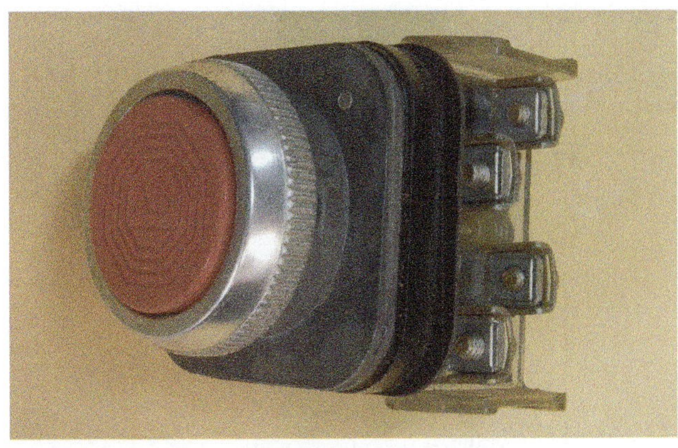

Figure 2–7 Double acting push buttons have four terminal screws.

by the dashed line are pressed at the same time. A typical double acting push button is shown in Figure 2–7.

Stacked Push Buttons

A circuit employing the use of multiple push buttons is shown in Figure 2–8. This circuit illustrates the control of three separate motors. An emergency stop button can be used to de-energize all three motors at the same time. When the emergency stop button is pressed, three normally closed push buttons open to disconnect power to all three motors, and a normally open push button energizes control relay CR. The normally closed CR contact opens and disconnects power to the main control circuit. A normally open CR contact is used as a holding contact to maintain connection to CR coil after the emergency stop button is released, and a separate normally open CR contact closes to turn on a red indicator light. The red light indicates that the emergency stop was activated. The circuit will remain in this condition until the reset button is pressed. In this circuit, four separate push buttons are controlled at the same time. Push buttons that contain multiple contacts are often called stacked push buttons. *Stacked push buttons* are made by connecting multiple contact units together, and controlling them with a single push button (Figure 2–9).

Mushroom Head Push Buttons

The button portion of most push buttons generally exhibits a small surface area and some are slightly recessed to help prevent accidental activation as shown in Figure 2–7. Mushroom head push buttons, however, have a button with a large surface area that extends above the button to make them very accessible. They are often used as an emergency stop button. A mushroom head push button is shown in Figure 2–10. The schematic symbol for a mushroom head push button is shown in Figure 2–11.

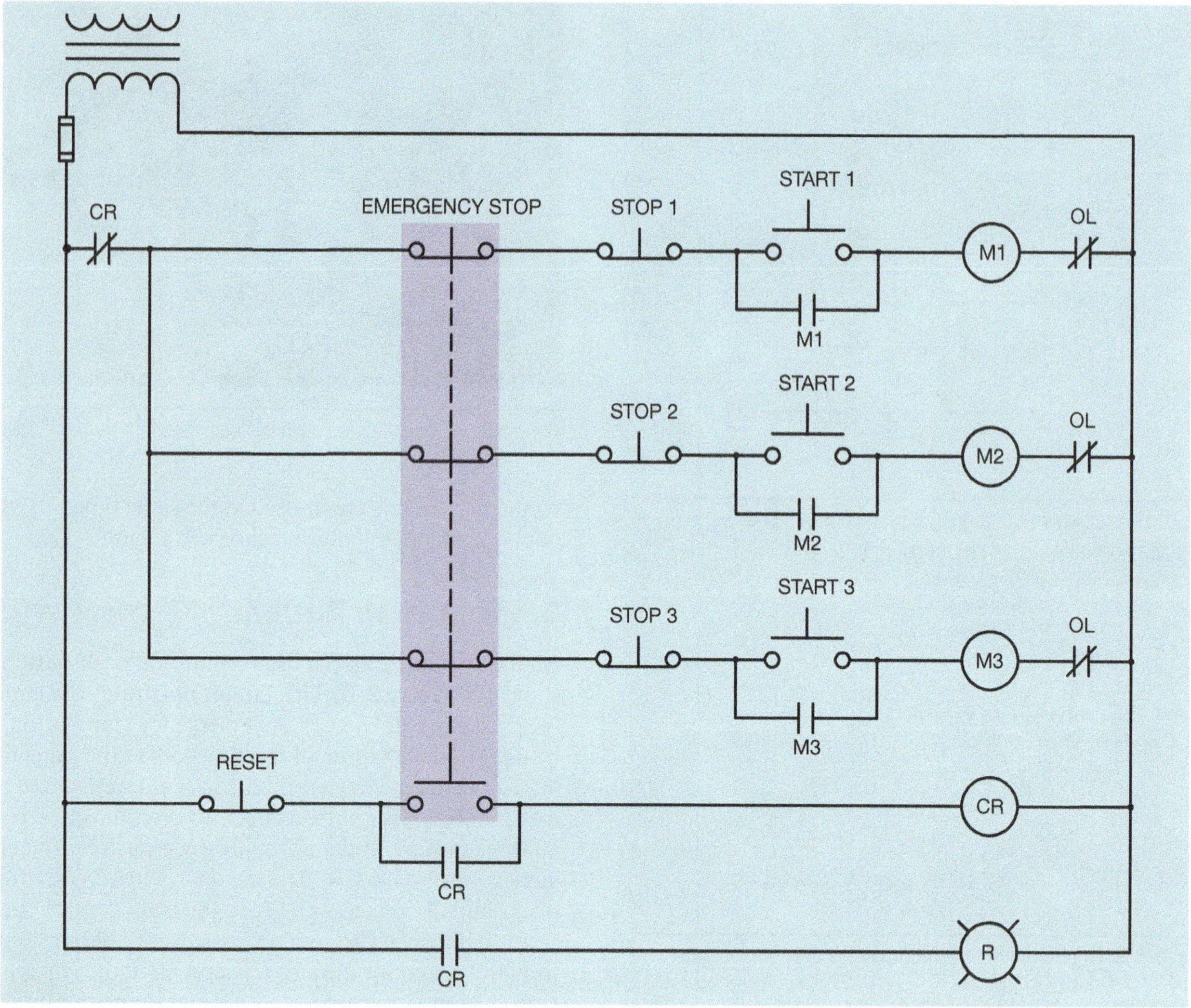

Figure 2–8 An emergency stop button stops all three motors.

Figure 2–9 Stacked push buttons are made by connecting multiple contact sets together. Components Courtesy of Wholesale Electric Supply.

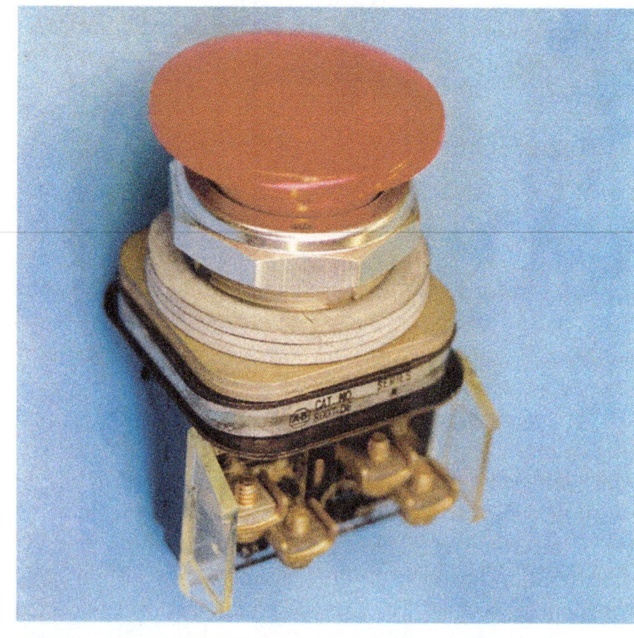

Figure 2–10 Mushroom head push button.

Figure 2–11 Schematic symbol for a mushroom head push button.

Figure 2–12 Push-pull button.

Push-Pull Buttons

Another push button that has found wide use is the push-pull button (Figure 2–12). Some push-pull buttons contain both normally open and normally closed contacts similar to double acting push buttons, but the contact arrangement is different. Push-pull button can provide both the start and stop function in one push button, eliminating the space needed for a second push button. The symbol for a push-pull button of this type is shown in Figure 2–13. A circuit employing a push-pull button in a start-stop control circuit is shown in Figure 2–14. When the button is pulled, the normally closed contact

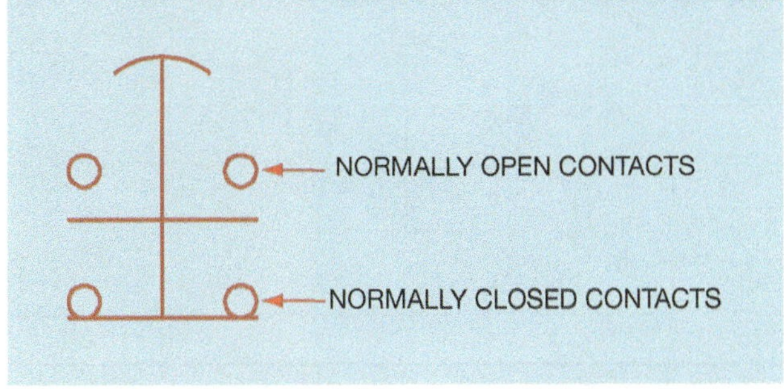

Figure 2–13 The push-pull button contains one set of normally closed contacts and one set of normally open contacts.

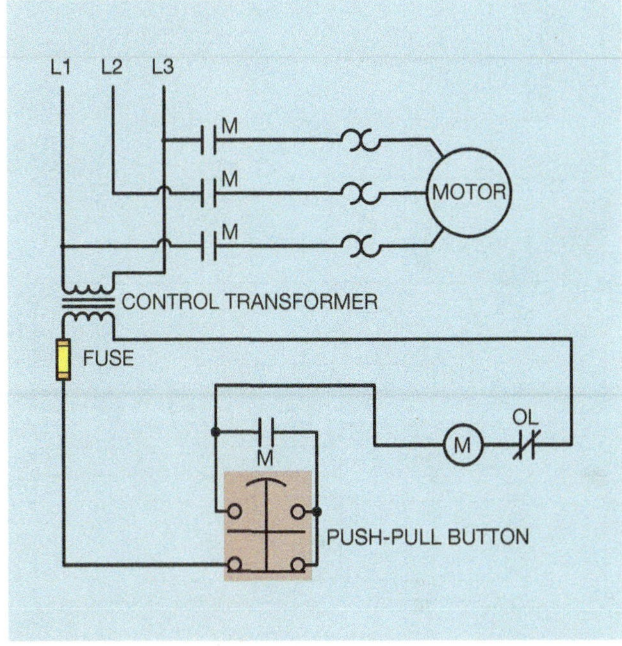

Figure 2–14 A push-pull button is used in a start-stop control circuit.

remains closed, and the normally open contact bridges the two stationary contacts to complete a circuit to the coil of M starter. A set of normally open M contacts close to maintain the circuit when the button is released. When the button is pushed, the normally closed contacts open and open the circuit to the coil of M starter.

Push-pull button that contain two normally open contacts or two normally closed contacts are also available. The symbol for a push-pull button that contains two normally open contacts is shown in Figure 2–15. Push-pull buttons of this type are often used for run-jog controls. A run-jog circuit using a push-pull button is shown in Figure 2–16. To make the motor run, pull the button. When the button is pulled, a circuit is completed to control relay CR, causing both CR contacts to close. One CR contact maintains the circuit around the normally open contacts of the push-pull button, and the

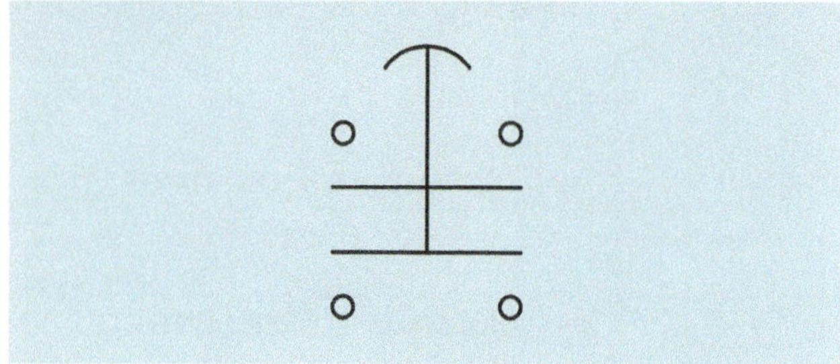

Figure 2–15 Push-pull buttons also contain two normally open contacts.

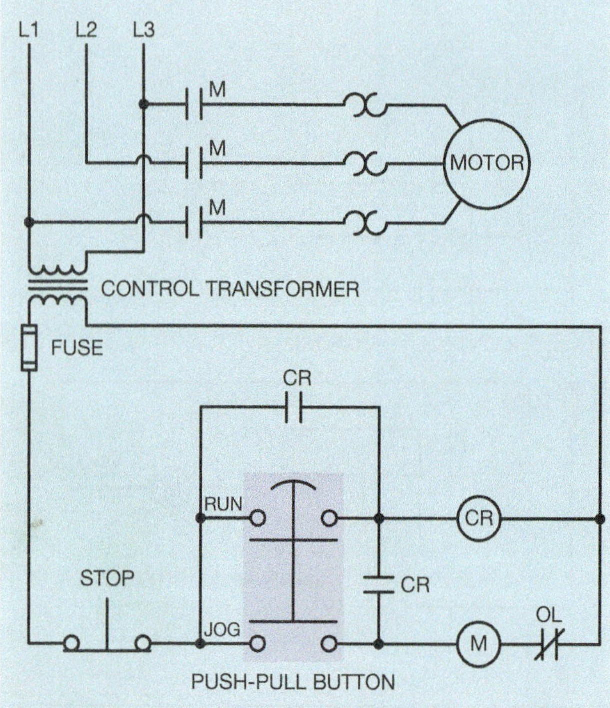

Figure 2–16 A push-pull button with two normally open contacts is used to control a run-jog circuit.

other supplies power to the coil of M starter. A separate stop button is used to stop the motor. If the push-pull button is pressed, a circuit is completed directly to M starter coil. Because there are no M contacts to hold the circuit, when the button is released the circuit to M coil is open and the motor stops running.

Illuminated Push Buttons

Illuminated push buttons are another example of providing a second function in a single space (Figure 2–17). They are often used to indicate that a motor is running, stopped, or tripped on overload. Most illuminated push buttons are equipped with a small transformer to reduce the control voltage to a much lower value (Figure 2–18). Lens caps of different colors are available. The schematic symbol for an illuminated push button is shown in Figure 2–19.

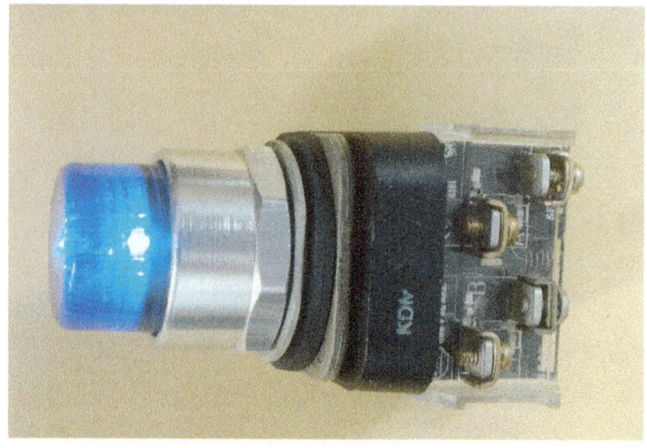

Figure 2–17 Illuminated push button.

Figure 2–18 Illuminated push buttons generally employ a small transformer to reduce the control voltage to a lower value.

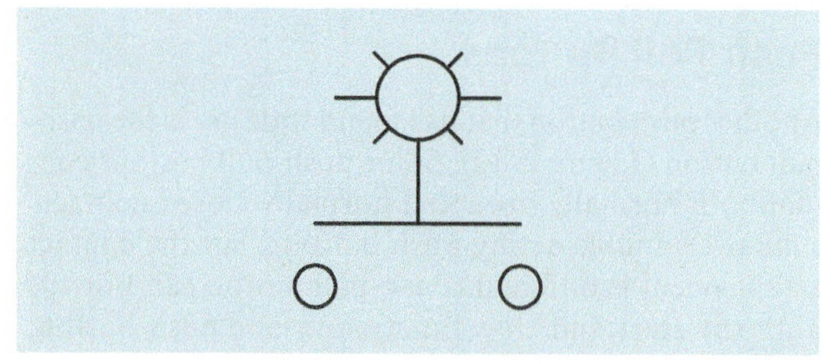

Figure 2–19 Schematic symbol for an illuminated push button.

Hand-Off-Automatic Switches

Hand-off-automatic (HOA) switches are generally used to select the function of a motor controller either manually or automatically. They can be a stand-alone control as shown in Figure 2–20, or incorporated into a start-stop push button station as shown in Figure 2–21. A single-break HOA switch used to control the coil of a motor starter is shown in Figure 2–22. If the HOA switch is set in the off position, no current can flow to the coil of the starter. If set in the hand position, the coil is connected directly to the power line and the motor will run continuously. If the HOA switch is set in the auto position, the starter coil is controlled by a float switch.

Figure 2–20 Hand-off-automatic switch.

Hand-off-automatic or selector switches often contain double break contacts as shown in Figure 2–23. Switches of this type are generally provided with a chart showing the contact connections. The X indicates that the contacts are closed and the O indicates that they

Figure 2–21 HOA switch incorporated with a start-stop push button control.

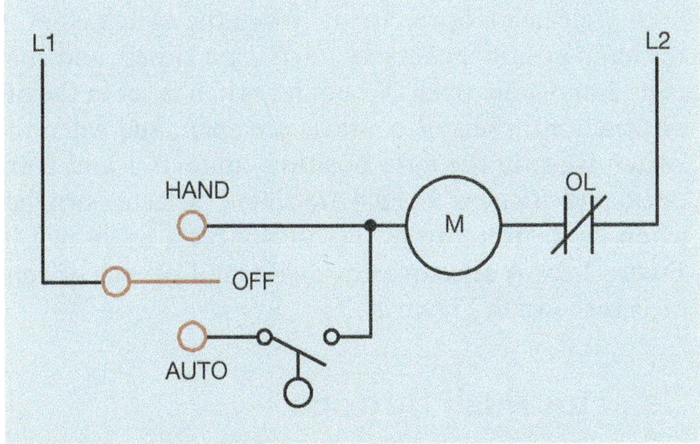

Figure 2–22 A single-break HOA switch controls the coil of a motor starter.

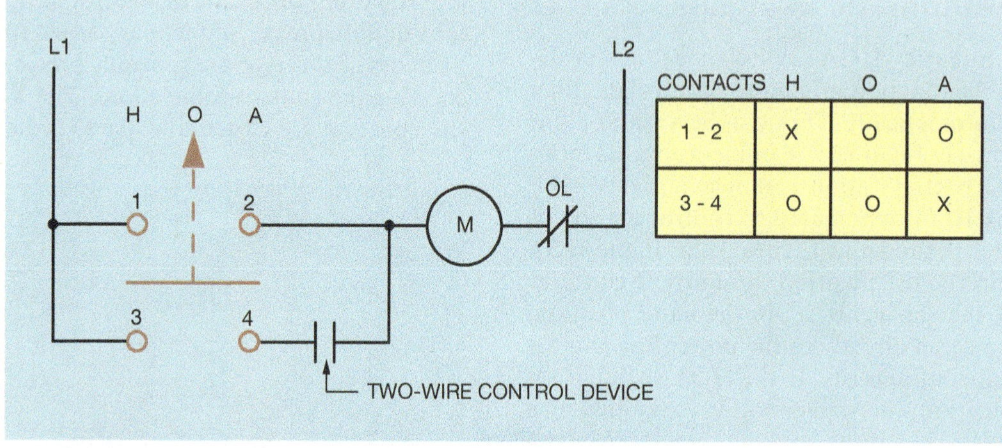

CONTACTS	H	O	A
1 - 2	X	O	O
3 - 4	O	O	X

Figure 2–23 Selector switches often contain double break contacts.

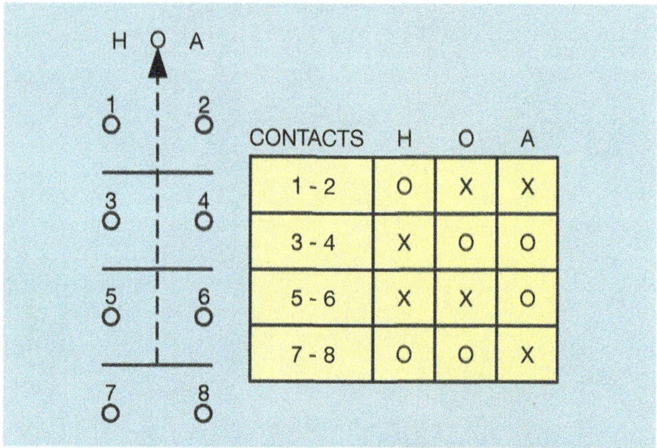

CONTACTS	H	O	A
1 - 2	O	X	X
3 - 4	X	O	O
5 - 6	X	X	O
7 - 8	O	O	X

Figure 2–24 Selector switch with multiple double break contacts.

Figure 2–25 Selector switch with multiple contacts.

are open. Some contact connection charts use an X to indicate a complete circuit and a blank space instead of O to indicate an open circuit. When the switch is set in the hand position, contacts 1 and 2 are closed, and contacts 3 and 4 are open. When the switch is set in the off position, both sets of contacts are open, and when the switch is set in the auto position contacts 1 and 2 are open, and contacts 3 and 4 are closed. Selector switches often contain multiple sets of contacts as shown in Figure 2–24. A selector switch with multiple sets of contacts is shown in Figure 2–25.

Selector Push Buttons

Selector push buttons combine the operation of a selector switch and push button in the same unit (Figure 2–26). The selector switch is controlled by turning the push button sleeve. Some selector push buttons permit the sleeve to be set in any of three positions and others permit the sleeve to be set in only two positions. The push button sleeve

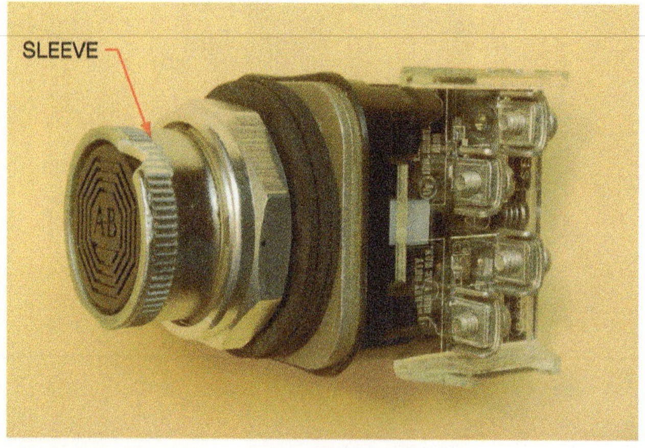

Figure 2–26 Selector push button.

shown in Figure 2–26 may be set in either of two positions. Different contact blocks are available that permit different contact settings. The contact block shown in Figure 2–27 contains two sets of bridge type contacts designated as A and B. A chart indicating contact connections for different conditions is shown in Figure 2–28. To better understand the chart and how this selector push button works, refer to Figures 2–29 A, B, C, and D.

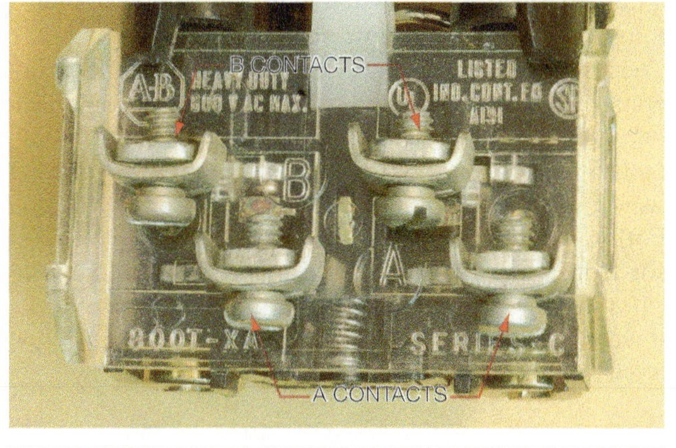

Figure 2–27 Contact block of selector push button.

	SLEEVE LEFT		SLEEVE RIGHT	
	PF	PD	PF	PD
B	O	O	X	O
A	O	X	O	X

PF = PUSH BUTTON FREE
PD = PUSH BUTTON DEPRESSED
X = CONTACTS CLOSED
O = CONTACTS OPEN

Figure 2–28 Contact chart for selector push button.

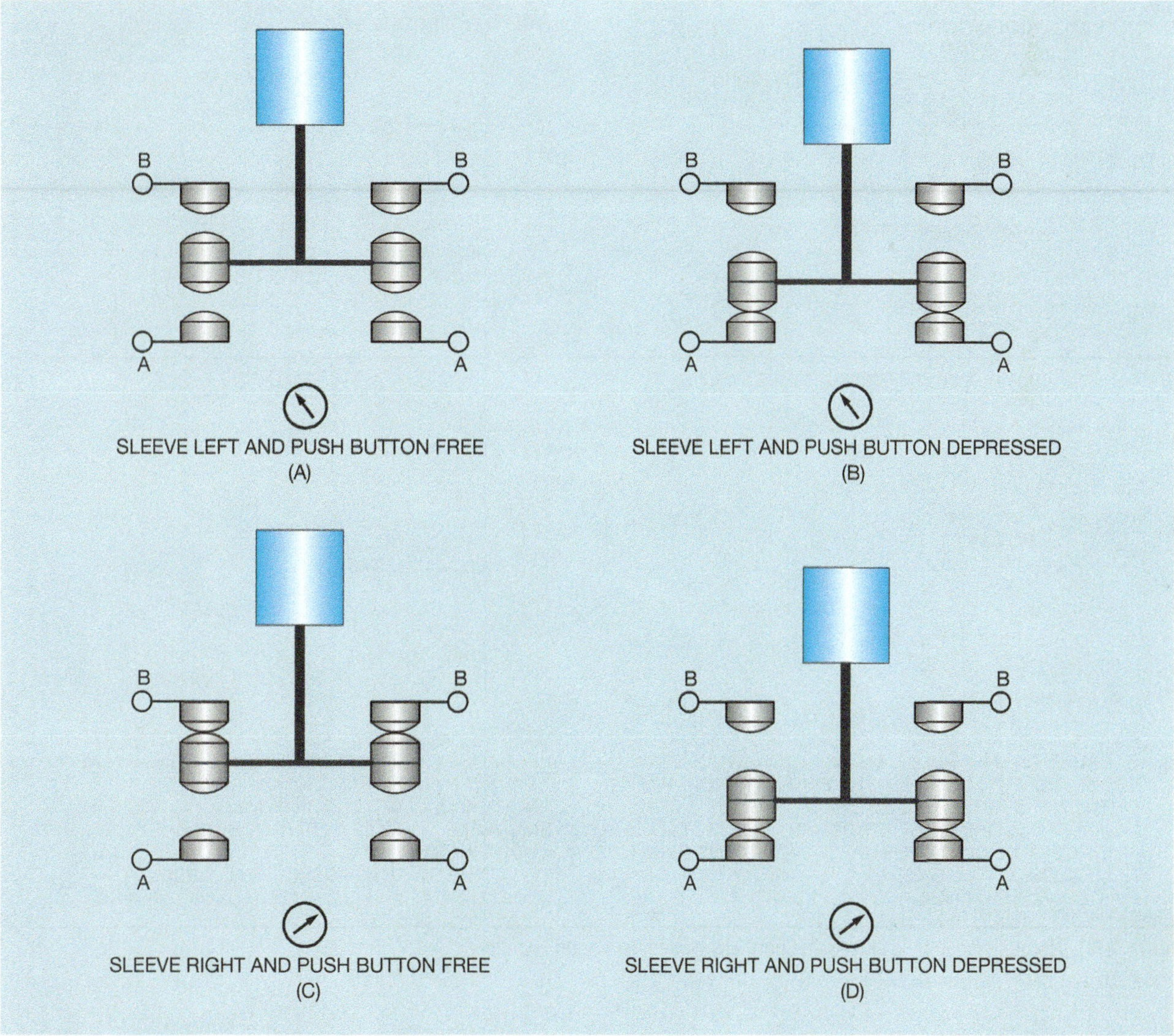

B B A A
SLEEVE LEFT AND PUSH BUTTON FREE
(A)

B B A A
SLEEVE LEFT AND PUSH BUTTON DEPRESSED
(B)

B B A A
SLEEVE RIGHT AND PUSH BUTTON FREE
(C)

B B A A
SLEEVE RIGHT AND PUSH BUTTON DEPRESSED
(D)

Figure 2–29 Contact connections for different settings of the selector push button.

Switch Symbols

Switch symbols are employed to represent many common control sensing devices. There are four basic symbols: **normally open (NO)**, **normally closed (NC)**, **normally open held closed (NOHC)**, and **normally closed held open (NCHO)**. To understand how these switches are drawn, it is necessary to begin with how normally open and normally closed switches are drawn (Figure 2–30). Normally open switches are drawn with the movable contact *below and not touching* the stationary contact. Normally closed switches are drawn with the movable contact *above and touching* the stationary contact.

The normally open held closed and normally closed held open switches are shown in Figure 2–31. Note that the movable contact of the normally open held closed switch is drawn below the stationary contact. The fact that the movable contact is drawn *below* the stationary contact indicates that the switch is normally open. Because the movable contact is touching the stationary contact, however, a complete circuit does exist because something is holding the contact closed. A very good example of this type of switch is the low pressure switch found in many air conditioning circuits (Figure 2–32). The low pressure switch is being held closed by the refrigerant in the sealed system. If the refrigerant should leak out, the pressure will drop low enough to permit the contact to return to its normal open position. This contact would open the circuit and de-energize coil C, causing both C contacts to open and disconnect the compressor from the power line. Although the schematic indicates that the switch is closed during normal operation, it would have to be connected as an open switch when it is wired into the circuit.

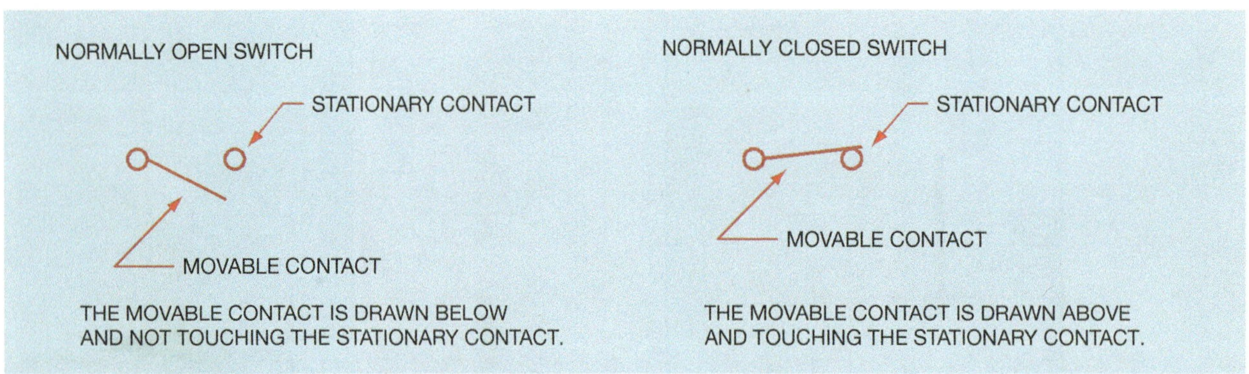

Figure 2–30 Normally open and normally closed switches.

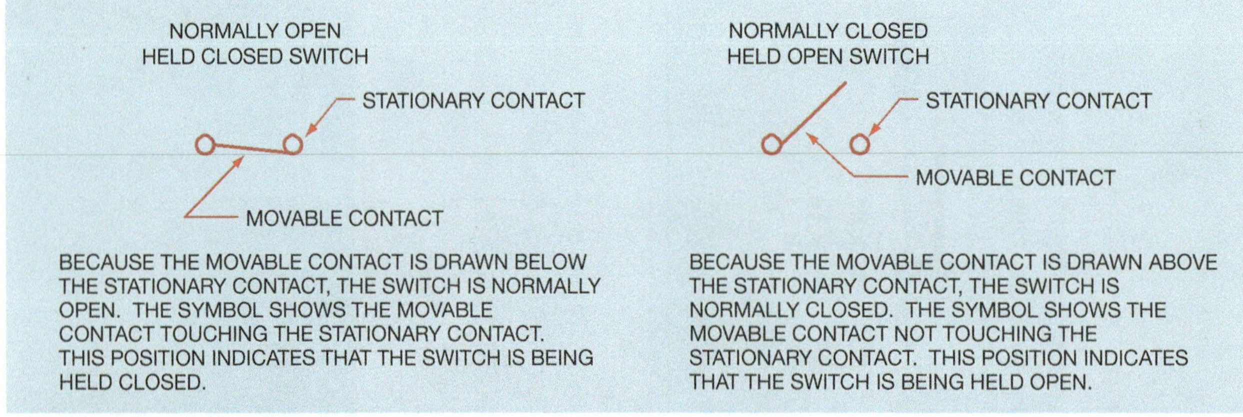

Figure 2–31 Normally open held closed and normally closed held open switches.

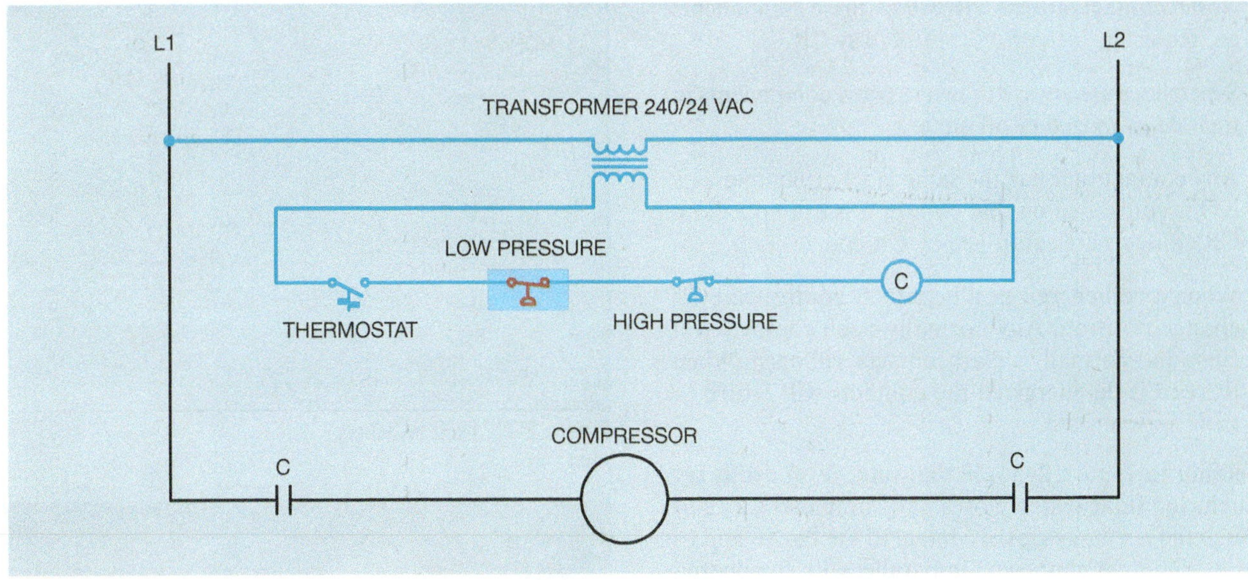

Figure 2–32 If system pressure drops below a certain value, the normally open held closed low pressure switch opens and de-energizes coil C.

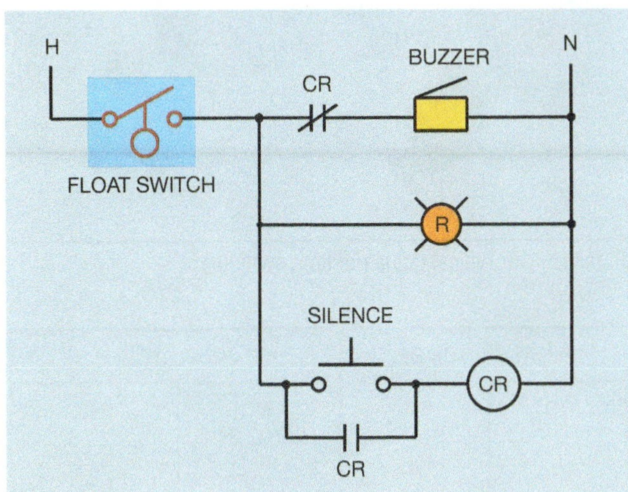

Figure 2–33 Low water warning circuit.

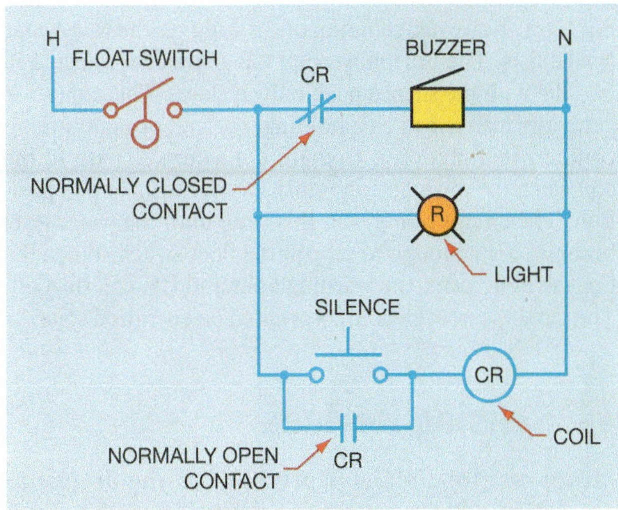

Figure 2–34 Circuit with labeled components.

The normally closed held open switch is shown open in Figure 2–31. Although the switch is shown open, it is actually a normally closed switch because the movable contact is drawn *above* the stationary contact, indicating that something is holding the switch open. A good example of how this type of switch can be used is shown in Figure 2–33. This circuit is a low water warning circuit for a steam boiler. The float switch is held open by the water in the boiler. If the water level should drop sufficiently, the contacts will close and energize a buzzer and warning light.

Basic Schematics

To understand the operation of the circuit shown in Figure 2–33, you must understand some basic rules concerning schematic or ladder diagrams.

1. Schematics show components in their electrical sequence without regard for physical location. The schematic in Figure 2–33 has been redrawn in Figure 2–34. Labels have been added and show a coil labeled CR and one normally open and one normally

closed contact labeled CR. All of these components are physically located on control relay CR.

2. Schematics are always drawn to show components in their de-energized or off state.

3. Any contact that has the same label or number as a coil is controlled by that coil. In this example, both CR contacts are controlled by CR coil.

4. When a coil energizes, all contacts controlled by it change position. Any normally open contacts will close and normally closed contacts will open. When the coil is de-energized the contacts will return to their normal state.

Referring to Figure 2–34, if the water level drops far enough, the float switch closes and completes a circuit through the normally closed contact to the buzzer and to the warning light connected in parallel with the buzzer. At this time both the buzzer and warning light are turned on. If the silence push button is pressed, coil CR will energize and both CR contacts change position. The normally closed contact opens and turns off the buzzer. The warning light, however, remains on as long as the low water level exists. The normally open CR contact connected in parallel with the silence push button closes. This contact is generally referred to as a holding, sealing, or maintaining contact. Its function is to maintain a current path to the coil when the push button returns to its normal open position. The circuit remains in this state until the water level becomes high enough to reopen the float switch. When the float switch opens, the warning light and CR coil turn off. The circuit is now back in its original de-energized state.

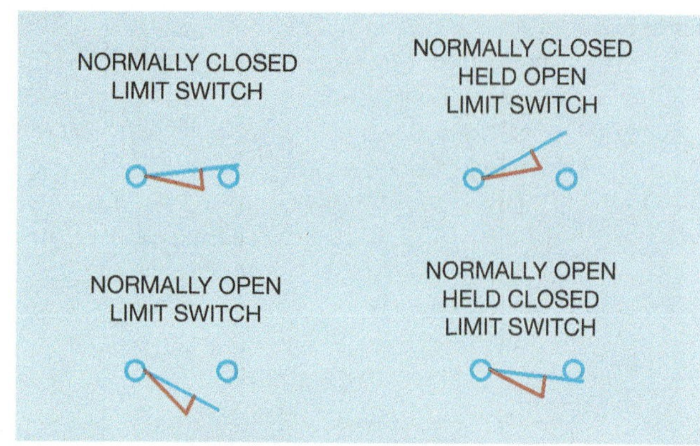

Figure 2–35 Limit switches.

Courtesy of Honeywell International, Inc.

Figure 2–36 Typical industrial limit switches.

FLOAT SWITCHES		FLOW SWITCHES	
NO	NC	NO	NC

PRESSURE SWITCHES		TEMPERATURE SWITCHES	
NO	NC	NO	NC

Figure 2–37 Schematic symbols for sensing switches.

Sensing Devices

Motor control circuits depend on sensing devices to determine what conditions are occurring. They act very much like the senses of the body. The brain is the control center of the body. It depends on input information such as sight, touch, smell, and hearing to determine what is happening around it. Control systems are very similar in that they depend on such devices as temperature switches, float switches, limit switches, flow switches, etc, to know the conditions that exist in the circuit. These sensing devices are covered in greater detail later in the textbook. The four basic types of switches are used in conjunction with other symbols to represent some of these different kinds of sensing switches.

Limit Switches

Limit switches are drawn by adding a wedge to one of the four basic switches (Figure 2–35). The wedge represents the bumper arm. Common industrial limit switches are shown in Figure 2–36.

Float, Pressure, Flow, and Temperature Switches

The symbol for a float switch illustrates a ball float. It is drawn by adding a circle to a line (Figure 2–37). The flag symbol of the flow switch represents the paddle that senses movement. The flow switch symbol is used for both liquid and air flow switches. The symbol for a pressure switch is a half circle connected to a line. The flat part of the semicircle represents a diaphragm. The symbol for a temperature switch represents a bimetal helix. The helix contracts and expands with a change of temperature. Any of these symbols can be used with any of the four basic switches.

There are many other types of sensing switches that do not have a standard symbol. Some of these are photo switches, proximity switches, sonic switches, Hall effect switches, and others. Some manufacturers employ a special type of symbol and label the symbol to indicate the type of switch. An example of this is shown in Figure 2–38.

Coils

The most common coil symbol used in schematic diagrams is the circle. The reason is that letters and/or numbers may be written in the circle to identify the coil. Contacts controlled by the coil are given the same number. Several standard coil symbols are shown in Figure 2–39.

Timed Contacts

Timed contacts are either normally open or normally closed. They are not drawn as normally open held closed or normally closed held open. The two basic types of timers are **on delay** and **off delay**. Timed contact symbols use an arrow to point in the direction that the contact will move at the end of the time cycle. Timers will be discussed in detail in a later chapter. Standard timed contact symbols are shown in Figure 2–40.

Contact Symbols

Another very common symbol used on control schematics is the contact symbol. The symbol is two parallel lines connected by wires (Figure 2–41). The normally open contacts are drawn to represent an open connection. The normally closed contact symbol is the same as the normally open symbol with the exception that a diagonal line is drawn through the contacts. The diagonal line indicates that a complete current path exists.

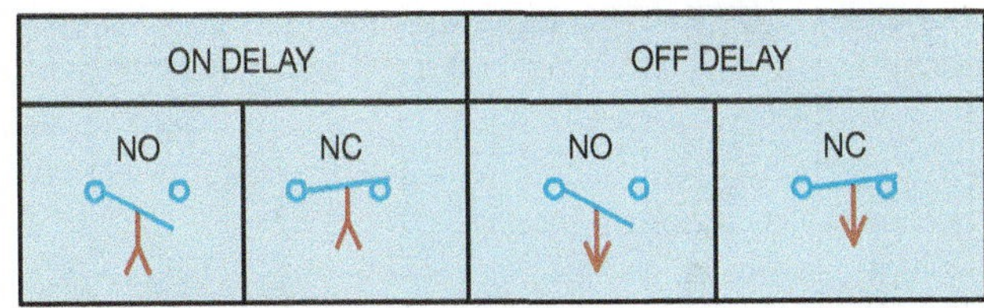

Figure 2–40 Timed contact symbols.

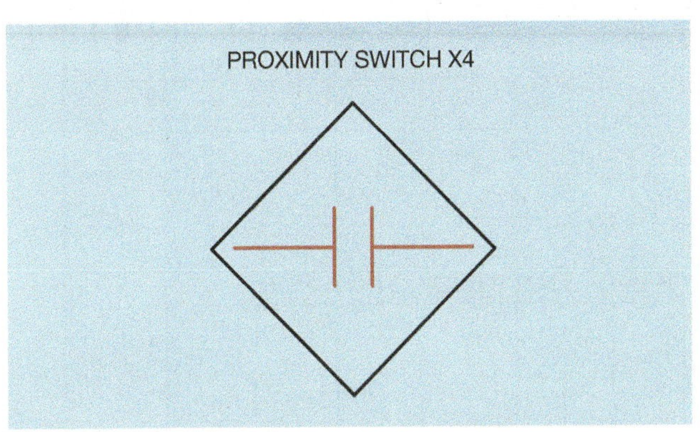

Figure 2–38 Special symbols are often used for sensing devices that do not have a standard symbol.

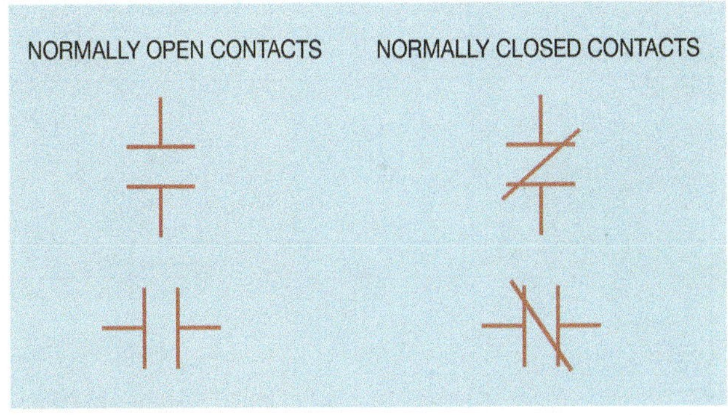

Figure 2–41 Normally open and normally closed contact symbols.

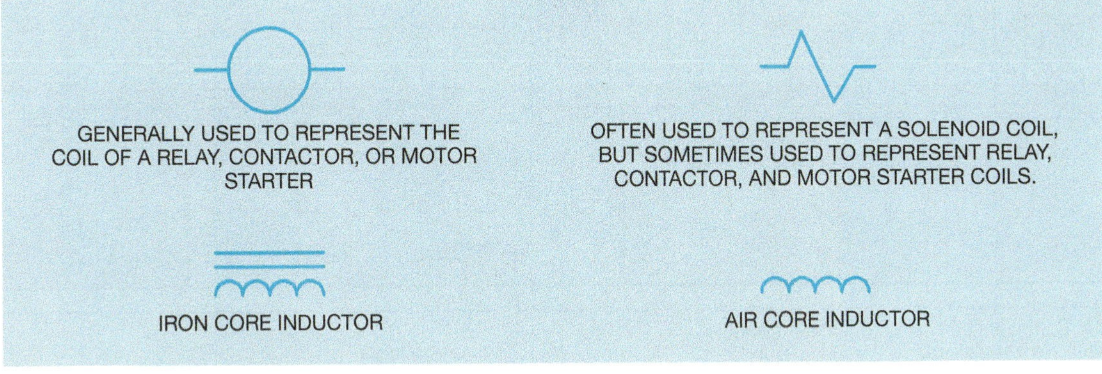

Figure 2–39 Common coil symbols.

Other Symbols

Not only does NEMA have standard symbols for coils and contacts, but there are also symbols for transformers, motors, capacitors, and special types of switches. Figure 2–42 shows both common control and electrical symbols in a chart.

IEC Symbols

Many schematic diagrams provided by European companies employ the use of symbols adopted by the International Electrotechnical Commission (IEC). These Symbols can be confusing to electricians working in the United States and Canada. Table 2–1 provides a comparison of NEMA symbols and IEC symbols.

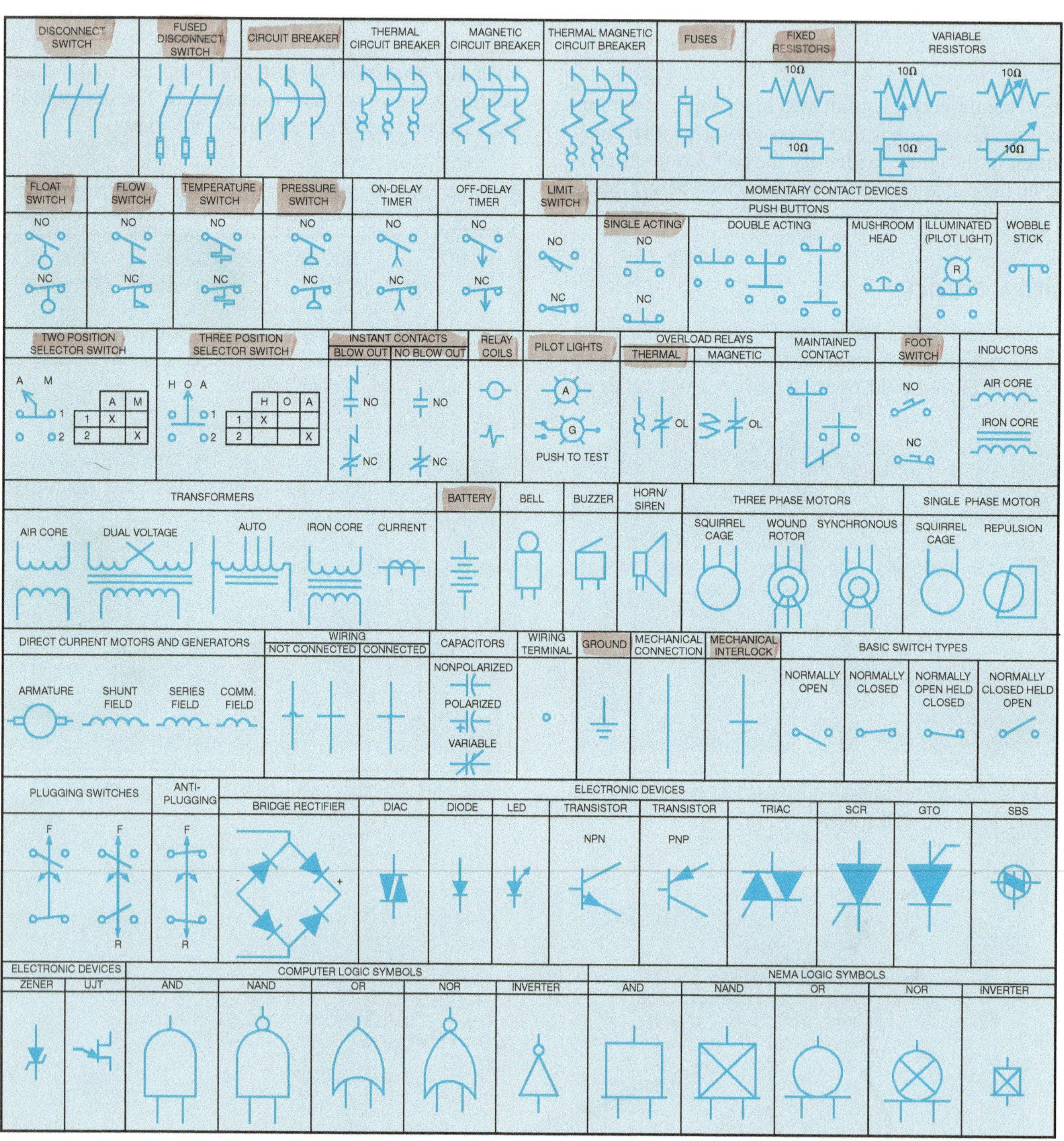

Figure 2–42 Common control and electrical symbols.

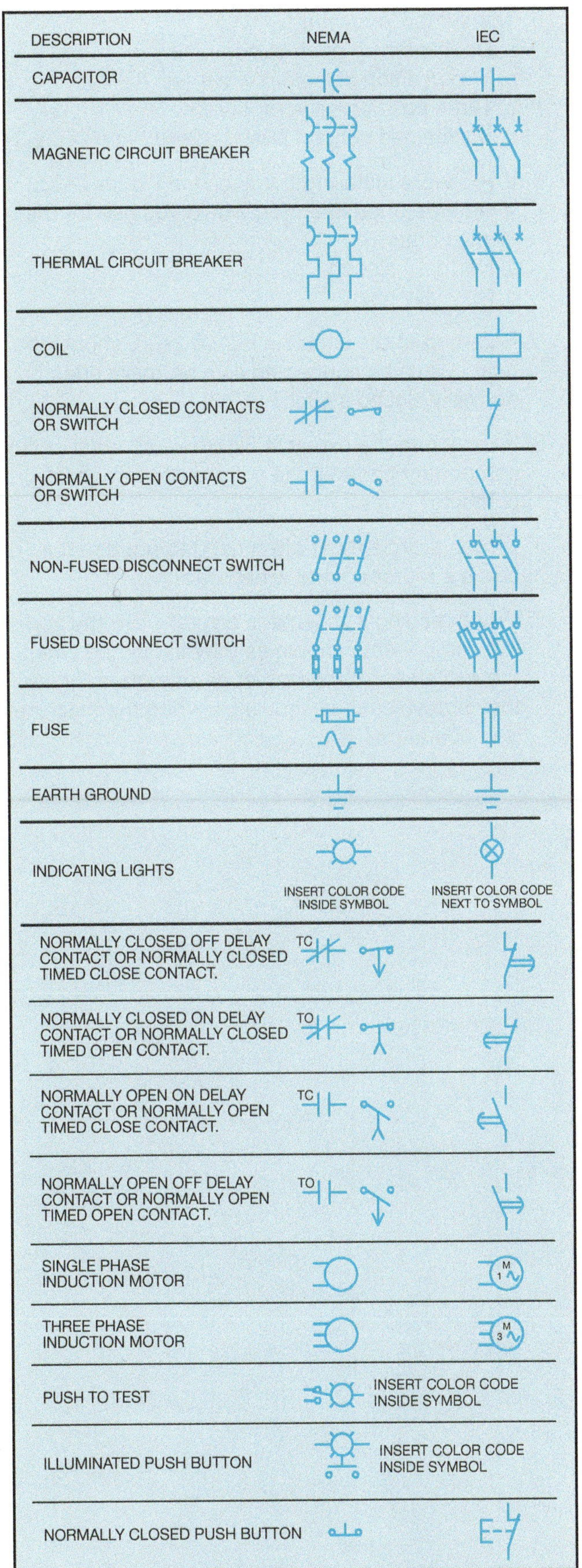

DESCRIPTION	NEMA	IEC
CAPACITOR		
MAGNETIC CIRCUIT BREAKER		
THERMAL CIRCUIT BREAKER		
COIL		
NORMALLY CLOSED CONTACTS OR SWITCH		
NORMALLY OPEN CONTACTS OR SWITCH		
NON-FUSED DISCONNECT SWITCH		
FUSED DISCONNECT SWITCH		
FUSE		
EARTH GROUND		
INDICATING LIGHTS	INSERT COLOR CODE INSIDE SYMBOL	INSERT COLOR CODE NEXT TO SYMBOL
NORMALLY CLOSED OFF DELAY CONTACT OR NORMALLY CLOSED TIMED CLOSE CONTACT.	TC	
NORMALLY CLOSED ON DELAY CONTACT OR NORMALLY CLOSED TIMED OPEN CONTACT.	TO	
NORMALLY OPEN ON DELAY CONTACT OR NORMALLY OPEN TIMED CLOSE CONTACT.	TC	
NORMALLY OPEN OFF DELAY CONTACT OR NORMALLY OPEN TIMED OPEN CONTACT.	TO	
SINGLE PHASE INDUCTION MOTOR		M 1
THREE PHASE INDUCTION MOTOR		M 3
PUSH TO TEST	INSERT COLOR CODE INSIDE SYMBOL	
ILLUMINATED PUSH BUTTON	INSERT COLOR CODE INSIDE SYMBOL	
NORMALLY CLOSED PUSH BUTTON		

DESCRIPTION	NEMA	IEC
NORMALLY OPEN PUSH BUTTON		
MUSHROOM PUSH BUTTON (N.C.)		
MUSHROOM PUSH BUTTON (N.O.)		
RESISTOR	OR	
NORMALLY CLOSED FLOAT SWITCH		
NORMALLY OPEN FLOAT SWITCH		
NORMALLY CLOSED FLOW SWITCH		
NORMALLY OPEN FLOW SWITCH		
NORMALLY CLOSED FOOT SWITCH		
NORMALLY OPEN FOOT SWITCH		
NORMALLY CLOSED LIMIT SWITCH		
NORMALLY OPEN LIMIT SWITCH		
NORMALLY CLOSED PRESSURE SWITCH		P
NORMALLY OPEN PRESSURE SWITCH		P
NORMALLY CLOSED TEMPERATURE SWITCH		
NORMALLY OPEN TEMPERATURE SWITCH		
TWO POSITION SELECTOR SWITCH	1 2 B A; LETTER / POSITION 1 2; A X; B X	1 2
THREE POSITION SELECTOR SWITCH	1 2 3 B A; LETTER / POSITION 1 2 3; A X; B X	1 2 3
CURRENT TRANSFORMER		
ISOLATION OR VOLTAGE TRANSFORMER		
THERMAL OVERLOAD ELEMENT	OR	
MAGNETIC OVERLOAD ELEMENT		

Table 2–1 NEMA symbols as compared to IEC symbols.

Review Questions

1. The symbol shown is:
   a. Polarized capacitor
   b. Normally closed switch
   c. Normally open held closed switch
   d. Normally open contact
2. The symbol shown is:
   a. Normally closed float switch
   b. Normally open held closed float switch
   c. Normally open float switch
   d. Normally closed held open float switch
3. The symbol shown is:
   a. Iron core transformer
   b. Auto transformer
   c. Current transformer
   d. Air core transformer
4. The symbol shown is:
   a. Normally open pressure switch
   b. Normally open flow switch
   c. Normally open float switch
   d. Normally open temperature switch
5. The symbol shown is:

   a. Double-acting push button
   b. Two position selector switch
   c. Three position selector switch
   d. Maintained contact push button
6. If you were installing the circuit in Figure 2–33, what type of push button would you use for the silence button?
   a. Normally closed
   b. Normally open
7. Referring to the circuit in Figure 2–33, should the float switch be connected as a normally open or normally closed switch?
8. Referring to the circuit in Figure 2–33, what circuit component controls the actions of the two CR contacts?
9. Why is a circle most often used to represent a coil in a motor control schematic?
10. When reading a schematic diagram, are the control components shown as they should be when the machine is turned off or de-energized, or are they shown as they should be when the machine is in operation?

Chapter 3

MANUAL STARTERS

Objectives

After studying this chapter the student will be able to:

- Discuss the operation of manual motor starters.
- Discuss low voltage release.
- Connect a manual motor starter.

Manual starters are characterized by the fact that the operator must go to the location of the starter to initiate any change of action. There are several different types of manual starters. Some look like a simple toggle switch with the addition of an overload heater. Others are operated by push buttons and may or may not be capable of providing low voltage protection.

Fractional Horsepower Single Phase Starters

One of the simplest manual motor starters resembles a simple toggle switch with the addition of an overload heater (Figure 3–1). The toggle switch lever is mounted on the front of the starter and is used to control the on and off operation of the motor. In addition to being an on and off switch, it also provides overload protection for the motor. An overload heater symbol has been added to the photograph to indicate where the overload heater should be connected. A schematic diagram showing the overload heater connected in series with the switch is shown in Figure 3–2. When current flows, the heater produces heat in proportion to the amount of motor current. If the heater is sized correctly, it will never get hot enough to open the circuit under normal operating conditions. If the motor should become overloaded, however, current increases causing a corresponding increase in the heat production by the heater. If the heat becomes great enough, it causes a mechanical mechanism to trip and open the switch contacts and disconnect the motor from the power line. If the starter trips on overload, the switch lever moves to a center position. The starter must be reset before the motor can be restarted by moving the lever to the full OFF position. This action is basically the same as resetting a tripped circuit breaker. The starter shown in this example has only one line contact and is generally used to protect motors intended to operate on 120 volts.

Starters that are intended to protect motors that operate on 240 volts should contain two load contacts (Figure 3–3). Although a starter that contains only one contact would control the operation of a 240-volt motor, it could create a hazardous situation. If the motor were switched off and an electrician tried to disconnect the motor, one power line would still be connected directly to the motor. *Section 430.103* of the *National Electrical Code® (NEC®)* requires that a disconnecting means open all ungrounded supply conductors to a motor.

Manual starters of this type are intended to control fractional horsepower motors only. Motors of 1 horsepower or less are considered fractional horsepower. Starters of this type are across-the-line starters. This means that they connect the motor directly to the power line. Some motors can draw up to 600 percent of rated full load current during starting. These starters generally do not contain large enough contacts to handle the current surge of multi-horsepower motors.

Figure 3–1 Single phase manual motor starter.

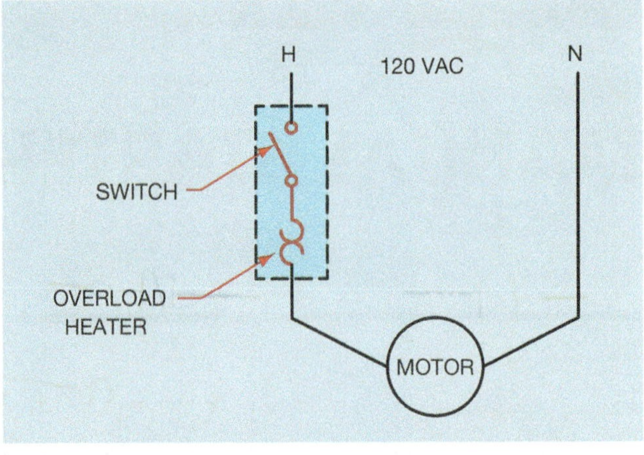

Figure 3–2 Schematic diagram of a single pole manual starter.

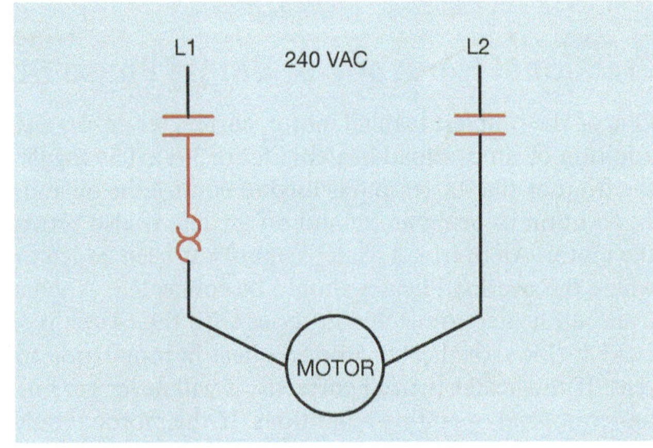

Figure 3–3 Schematic diagram of a two pole manual starter.

Another factor to consider when using a starter of this type is that it does not provide low voltage release. Most manual starters are strictly mechanical devices and do not contain an electrical coil. The contacts are mechanically opened and closed. This simply means that if the motor is in operation and the power fails, the motor will restart when the power is restored. This can be an advantage in some situations where the starter controls unattended devices such as pumps, fans, blowers, air conditioning, and refrigeration equipment. This feature saves the maintenance electrician from having to go around the plant and restart all the motors when power returns after a power failure.

This automatic restart feature can also be a disadvantage on equipment such as lathes, milling machines, saws, drill presses, and any other type of machine that may have an operator present. The unexpected and sudden restart of a piece of equipment could cause injury.

Mounting

Mounting this type of starter is generally very simple because it requires very little space. The compact design of this starter permits it to be mounted in a single gang switch or conduit box or directly on a piece of machinery. The open type starter can be mounted in the wall and covered with a single gang switch cover plate. The ON and OFF markings on the switch lever make it appear to be a simple toggle switch.

Like larger starters, fractional horsepower starters can be obtained in different enclosures. Some are simple sheet metal and are intended to be mounted on the surface or on a piece of machinery. If the starter is to be mounted in an area containing hazardous vapors or gasses, it may require an explosion-proof enclosure (Figure 3–4). Other areas that are subject to high moisture may require a waterproof enclosure (Figure 3–5). For areas that have a high concentration of flammable dust, the starter may be housed in a dust-proof enclosure similar to the one shown in Figure 3–6.

Automatic Operation

It is sometimes necessary to combine the manual starter with other sensing devices to obtain the proper control desired. When using some type of sensing pilot device to

Figure 3–4 Explosion-proof enclosure.

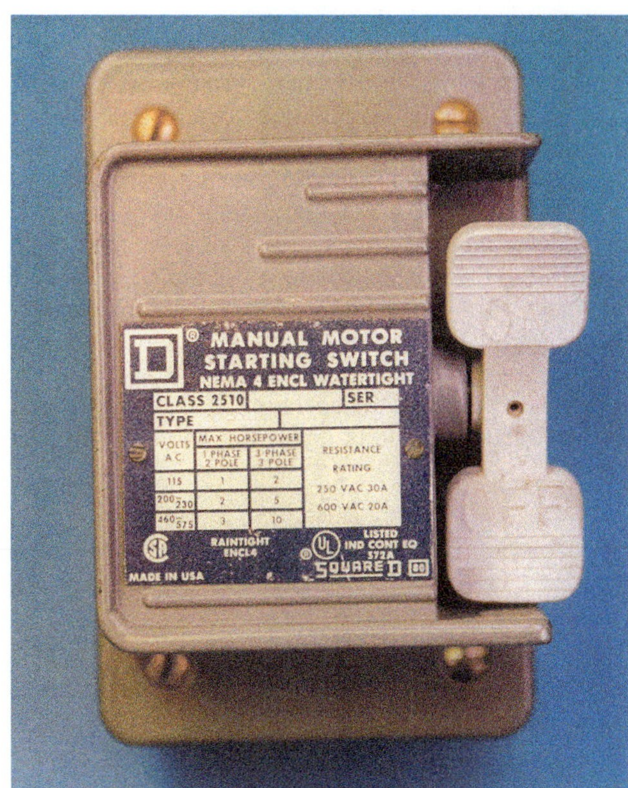

Figure 3–5 Waterproof enclosure.

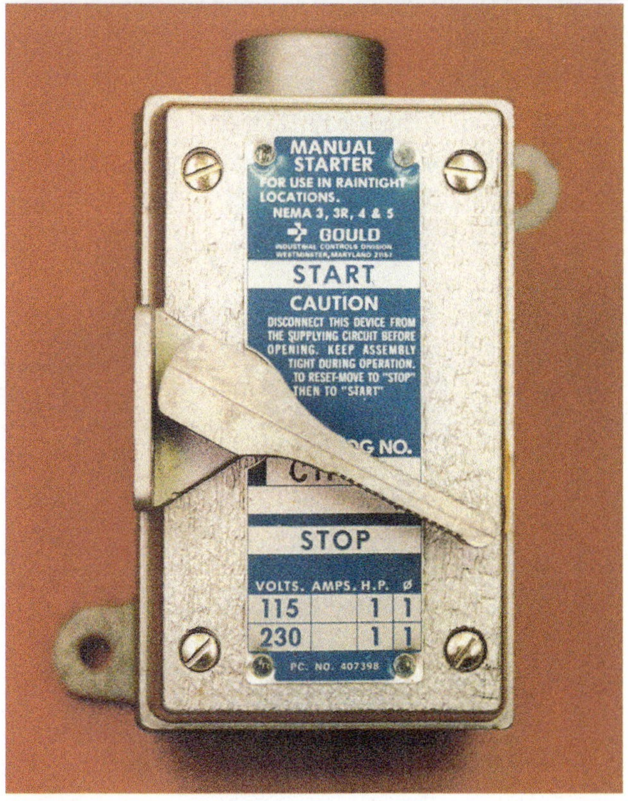

Figure 3–6 Dust-proof enclosure.

directly control the operation of a motor, you must make sure that the pilot device is equipped with contacts that can handle the rated current of the motor. These devices are generally referred to as line voltage devices. Line voltage devices have larger contacts than sensing pilot devices intended for use in a motor control circuit that employs a magnetic motor starter. The smaller pilot devices intended for use with magnetic motor starters have contacts that are typically rated from 1 to 3 amperes. Line voltage devices may have contacts rated for 15 to 20 amperes. A good example of how a line voltage sensing device can be used in conjunction with the manual starter is shown in Figure 3–7. In this circuit, a line voltage thermostat is used to control the operation of a blower motor. When the temperature rises to a sufficient level, the thermostat contacts close, connecting the motor directly to the power line if the manual starter contacts are closed. When the temperature drops, the thermostat contact opens and turns off the motor. A line voltage thermostat is shown in Figure 3–8.

Another circuit that permits the motor to be controlled either manually or automatically is shown in Figure 3–9. In this circuit a manual/automatic switch is used to select either manual or automatic operation of a pump. The pump is used to fill a tank when the water falls to a certain level. The schematic is drawn to assume that the tank is full of water during normal operation.

In the manual position, the pump is controlled by turning the starter on or off. An amber pilot light indicates when the manual starter contacts are closed or turned on. If the manual/automatic switch is moved to the automatic position, as in Figure 3–10, a line voltage float switch controls the operation of the pump motor. When water in the tank drops low enough, the float switch contact closes and starts the pump motor. When water rises to a high enough level, the float switch contact opens and disconnects the pump motor from the line.

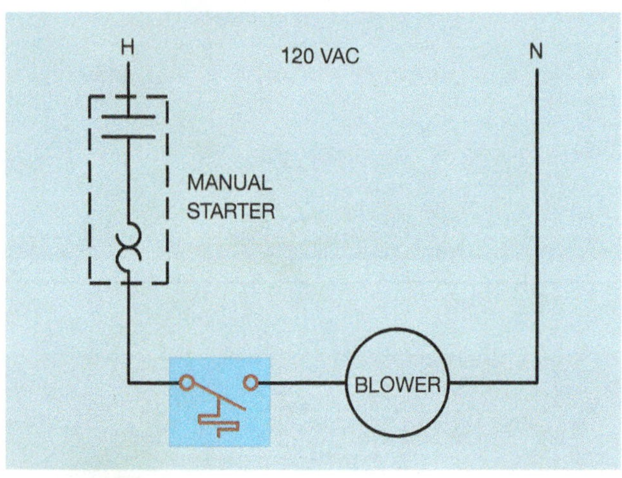

Figure 3–7 A line voltage thermostat controls the operation of a blower motor.

Figure 3–8 Line voltage thermostat.

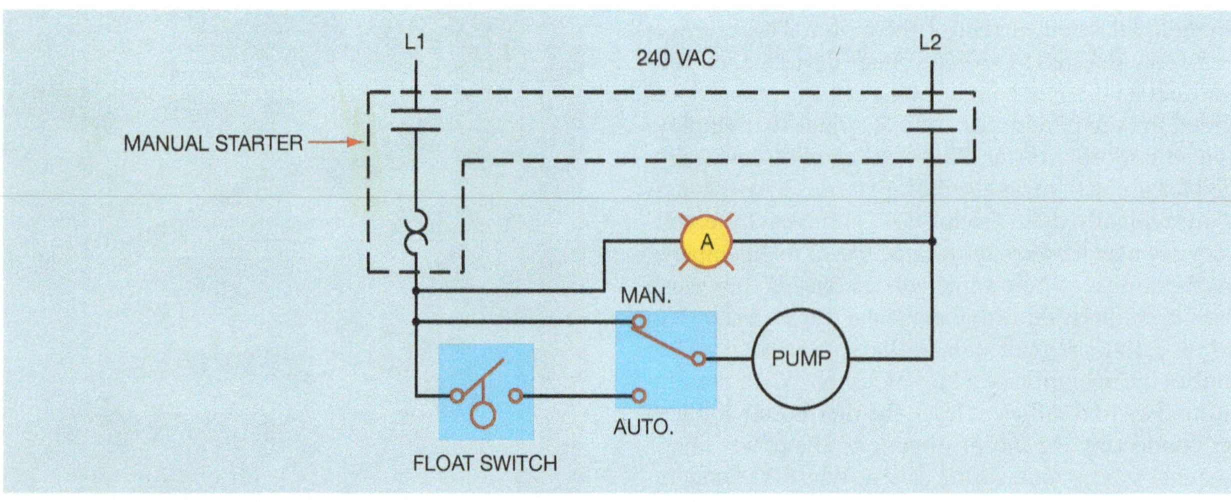

Figure 3–9 Pump can be controlled either manually or automatically.

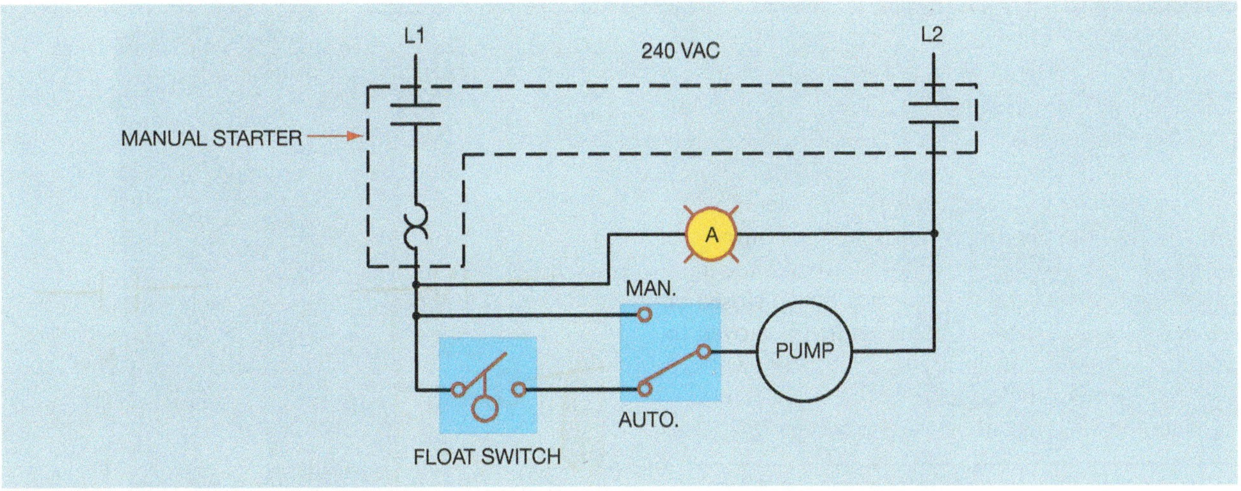

Figure 3–10 Moving the switch to the automatic position permits the float switch to control the pump.

Manual Push Button Starters

Manual push button line voltage starters are manufactured with two or three line voltage contacts. The two contact models are intended to control single phase motors that operate on 240 volts, or direct current motors. The starters that contain three contacts are intended to control three phase motors. Push button type manual starters are integral, not fractional, horsepower starters. Generally, they can control single phase motors rated up to 5 horsepower, direct current motors up to 2 horsepower, and three phase motors up to 10 horsepower. A typical three contact manual push button starter is shown in Figure 3–11. A schematic diagram for this type of starter is shown in Figure 3–12.

If any one of the overloads should trip, a mechanical mechanism opens the load contacts and disconnects the motor from the line. Once the starter has tripped on overload, it must be reset before the motor can be restarted. After allowing enough time for the overload heaters to cool, resetting the starter is accomplished by pressing the STOP push button with more than normal pressure. This extra pressure causes the mechanical mechanism to reset so that the motor can restart when the START push button is pressed. These starters are economical and are generally used with loads that are not started or stopped at frequent intervals. Although this type of starter provides overload protection, it does not provide low voltage release. If the power should fail and then be restored, the motor this starter controls will restart without warning.

Figure 3–11 Three phase line voltage manual starter.

Manual Starter with Low Voltage Release

Integral horsepower manual starters with low voltage release will not restart after a power failure without being reset. This is accomplished by connecting a

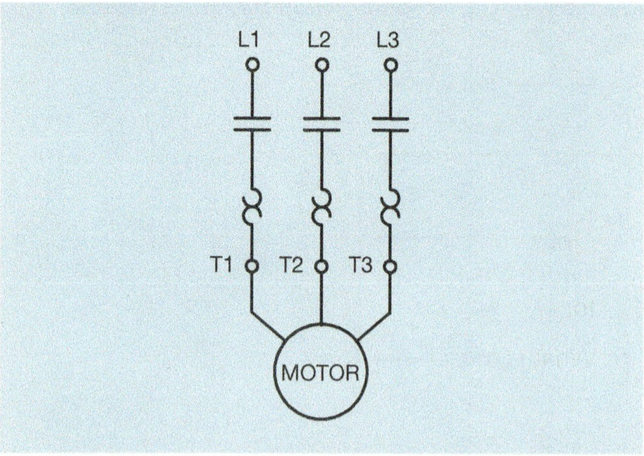

Figure 3–12 Schematic diagram for a three pole line voltage manual starter.

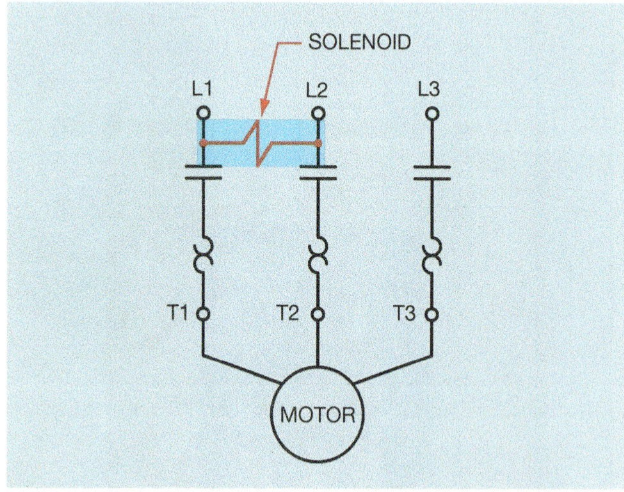

Figure 3–13 Solenoid provides low voltage release for the manual starter.

solenoid across the incoming power lines (Figure 3–13). As long as power is supplied to the starter, the solenoid holds a spring-loaded mechanism in place. As long as the mechanism is held in place, the load contacts can be closed when the START button is pressed. If the power is interrupted, the spring-loaded mechanism mechanically opens the contacts and prevents them from being reclosed until the starter has been manually reset. This starter will not operate unless power is present at the line terminal. This starter should not be confused with magnetic starters controlled by a coil. Magnetic type starters are designed to be used with other pilot control devices that control the operation of the starter. A manual starter with low voltage release is shown in Figure 3–14.

Courtesy Schneider Electric USA, Inc.

Figure 3–14 Manual starter with low voltage release.

Troubleshooting

Anytime a motor has tripped on overload, the electrician should check the motor and circuit to determine why the overload tripped. The first step is generally to determine whether the motor is actually overloaded. Some common causes of motor overloads are bad bearings in either the motor or the load the motor operates. Shorted windings in the motor can cause the motor to draw excessive current without being severe enough to blow a fuse or trip a circuit breaker. The simplest way to determine if the motor is overloaded is to find the motor full load current on the nameplate and then check the running current with an ammeter (Figure 3–15). If checking a single phase motor, it is necessary to check only one of the incoming lines. If checking a three phase motor, each line should be checked individually. The current flow in each line of a three phase motor should be relatively the same. A small amount of variation is not uncommon, but if the current is significantly different in any of the lines, it is an indication of internally shorted windings. Overloads are generally set to trip at 115 percent to 125 percent of motor full load current, depending on the motor. If the ammeter reveals that the motor is drawing excessive current, the reason must be determined before the motor can be put back into operation.

Excessive current is not the only cause for an overload trip. Thermal overloads react to heat. Any heat source can cause an overload to trip. If the motor is not drawing an excessive amount of current, the electrician should determine any other sources of heat. Loose

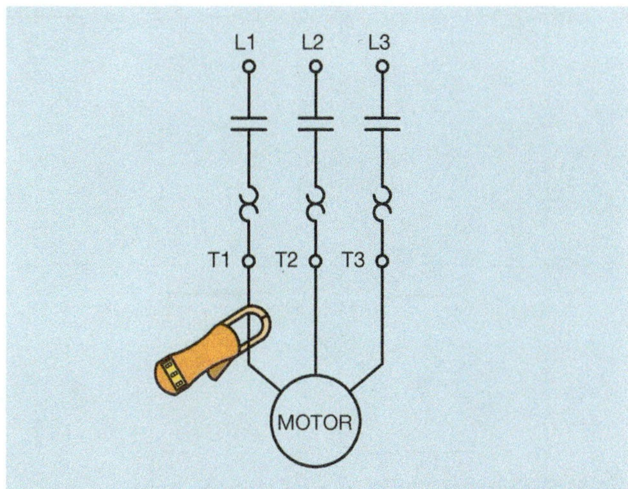

Figure 3–15 Checking motor current.

connections are one of the greatest sources of heat. Check the wires for insulation that has been overheated close to terminal screws. Any loose connection on the starter can cause an overload trip. Make sure that all connections are tight. Another source of heat is ambient or surrounding air temperature. In hot climates, the surrounding air temperature combined with the heat caused by motor current can be enough to cause the overload to trip. It may be necessary to set a fan that blows on the starter to help remove excess heat. Manual starters that are installed in a switchbox inside a wall are especially susceptible to ambient temperature problems. In this case it may be necessary to install some type of vented cover plate.

Review Questions

1. A manual motor starter controls a single phase 120 volt motor. The motor is not running, and the switch handle on the starter is found to be in the center position. What does this indicate?
2. Referring to the above question, what action is necessary to restart the motor and how is it accomplished?
3. A single phase motor operates on 240 volts. Why should a starter that contains two load contacts be used to control this motor?
4. A push button manual starter has tripped on overload. Explain how to reset the starter so the motor can be restarted.
5. What is meant by the term *line voltage* on some pilot sensing devices?
6. Explain the difference between manual motor starters that provide low voltage release and those that do not.
7. What is the simplest way to determine if a motor is overloaded?
8. Refer to the circuit shown in Figure 3–7. What type of switch is connected in series with the motor and is the switch normally open, normally closed, normally open held closed, or normally closed held open?
9. Refer to the circuit shown in Figure 3–10. When would the amber pilot light be turned on?
   a. When the manual/automatic switch is set in the manual position.
   b. When the float switch contacts are closed.
   c. Anytime the manual starter is turned on.
   d. Only when the manual/automatic switch is set in the manual position.
10. Refer to the circuit shown in Figure 3–10. Is the float switch normally open, normally closed, normally open held closed, or normally closed held open?

Chapter 4

OVERLOAD RELAYS

Objectives

After studying this chapter the student will be able to:

» Discuss differences between fuses and overloads.

» List different types of overload relays.

» Describe how thermal overload relays operate.

» Describe how magnetic overload relays operate.

» Describe how dashpot overload relays operate.

Overloads should not be confused with fuses or circuit breakers. **Fuses** and **circuit breakers** are designed to protect the circuit from a direct ground or short-circuit condition. Overloads are designed to protect the motor from an overload condition. Assume, for example, that a motor has a full load current rating of 10 amperes. Also assume that the motor is connected to a circuit that is protected by a 20 ampere circuit breaker (Figure 4–1). Now assume that the motor becomes overloaded and has a current draw of 15 amperes. The motor is drawing 150 percent of full load current. This much of an overload will overheat the motor and damage the windings. Because the current is only 15 amperes, the 20 ampere circuit breaker will not open the circuit to protect the motor. Overload relays are designed to open the circuit when the current becomes 115 percent to 125 percent of the motor full load current. The setting of the overload depends on the properties of the motor that is to be protected.

Overload Properties

All **overload relays** must possess certain properties in order to protect a motor.

1. *They must have some means of sensing motor current.* Some overload relays do this by converting motor current into a proportionate amount of heat and others sense motor current by the strength of a magnetic field.
2. *They must have some type of time delay.* Motors typically have a current draw of 300 percent to 800 percent of motor full load current when they start. Motor starting current is referred to as **locked rotor current**. Because overload relays are generally set to trip at 115 percent to 125 percent of full load motor current, the motor could never start if the overload relay tripped instantaneously.
3. *Overload relays are divided into two separate sections: the current sensing section and the contact section.* The current sensing section is connected in series with the motor and senses the amount of motor current. This section is typically connected to voltages that range from 120 volts to 600 volts. The contact section is part of the control circuit and operates at the control circuit voltage. Control circuit voltages generally range from 24 volts to 120 volts.

Dual Element Fuses

Some fuses are intended to provide both short circuit protection and overload protection. These fuses are called dual element time delay fuses. They contain two sections (Figure 4–2). The fuse link is designed to open quickly under a large amount of excessive current. This link protects the circuit against direct grounds and short circuits. The second contains a solder link that is connected to a spring. The solder

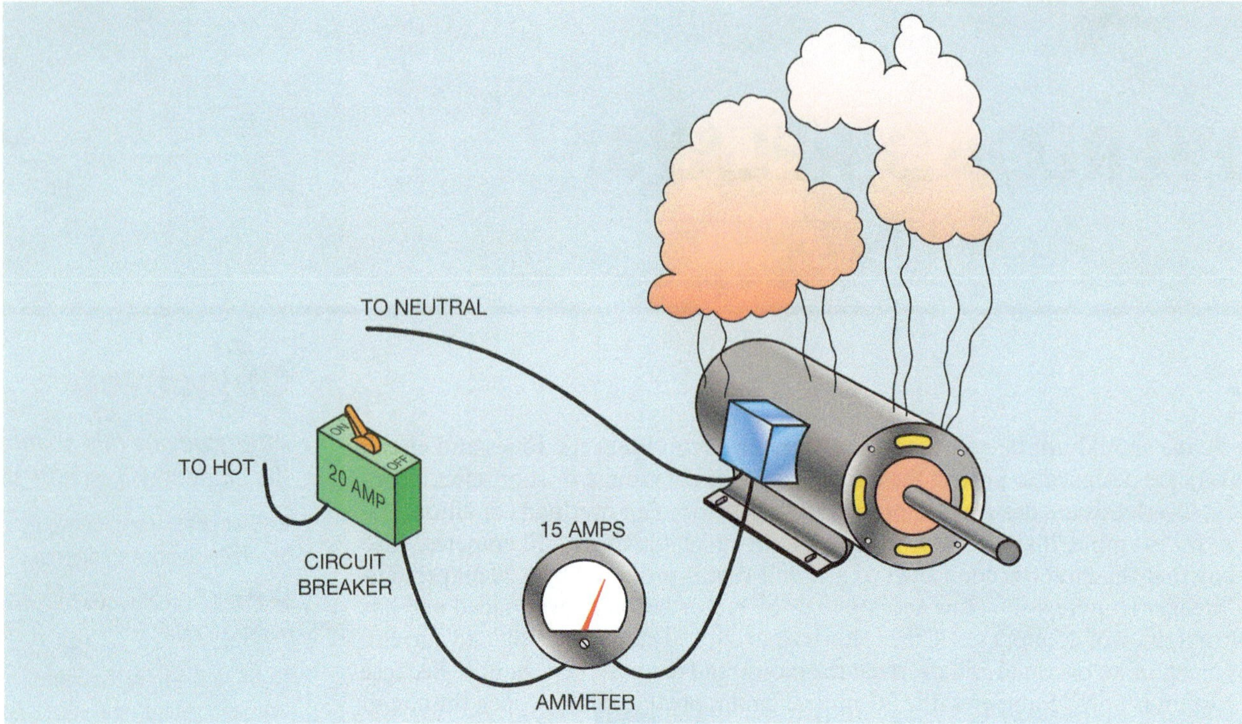

Figure 4–1 The circuit breaker does not protect the motor from an overload.

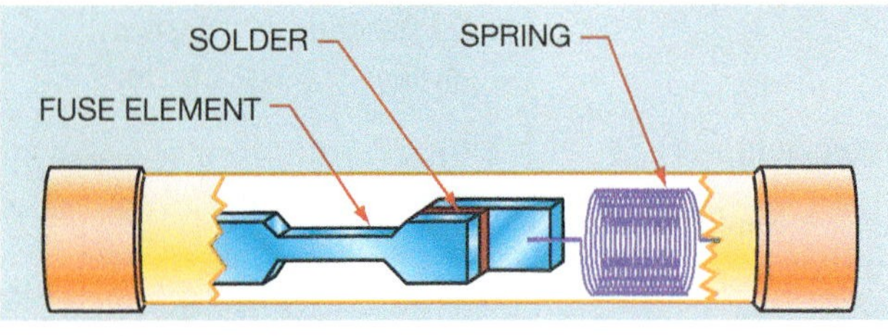

Figure 4–2 Dual element time delay fuse.

is a highly controlled alloy designed to melt at a particular temperature. If motor current becomes excessive, the solder melts and the spring pulls the link apart. The necessary time delay is achieved because of the time it takes for the solder to melt, even under a large amount of current. If motor current returns to normal after starting, the solder will not get hot enough to melt.

Thermal Overload Relays

There are two major types of overload relays: thermal and magnetic. Thermal overloads operate by connecting a heater in series with the motor. The amount of heat produced depends on motor current. Thermal overloads can be divided into two types: solder melting type or **solder pot**, and **bimetal strip**. Because thermal overload relays operate on the principle of heat, they are sensitive to ambient (surrounding air) temperature. They trip faster when located in a warm area than in a cool area.

Solder Melting Type

Solder melting type overloads are often called *solder pot overloads*. A brass shaft is placed inside a brass tube. A serrated wheel is connected to one end of the brass shaft. A special alloy solder that melts at a very specific temperature keeps the brass shaft mechanically connected to the brass tube (Figure 4–3). The serrated wheel keeps a set of spring-loaded contacts closed (Figure 4–4). An electric heater is placed around or close to the brass tube. The heater is connected in series with the motor. Motor current causes the heater to produce heat. If the current is great enough for a long enough period of time, the solder melts and permits the brass shaft to turn inside the tube, causing the contact to open. The fact that time must elapse before the solder can become hot enough to melt provides the delay for this overload relay. A large overload

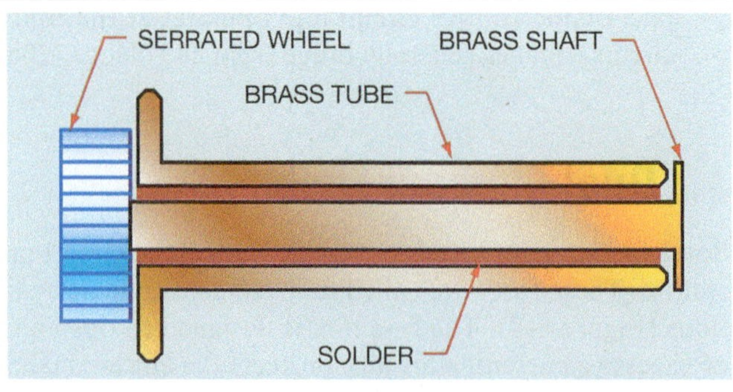

Figure 4–3 Construction of a typical solder pot overload.

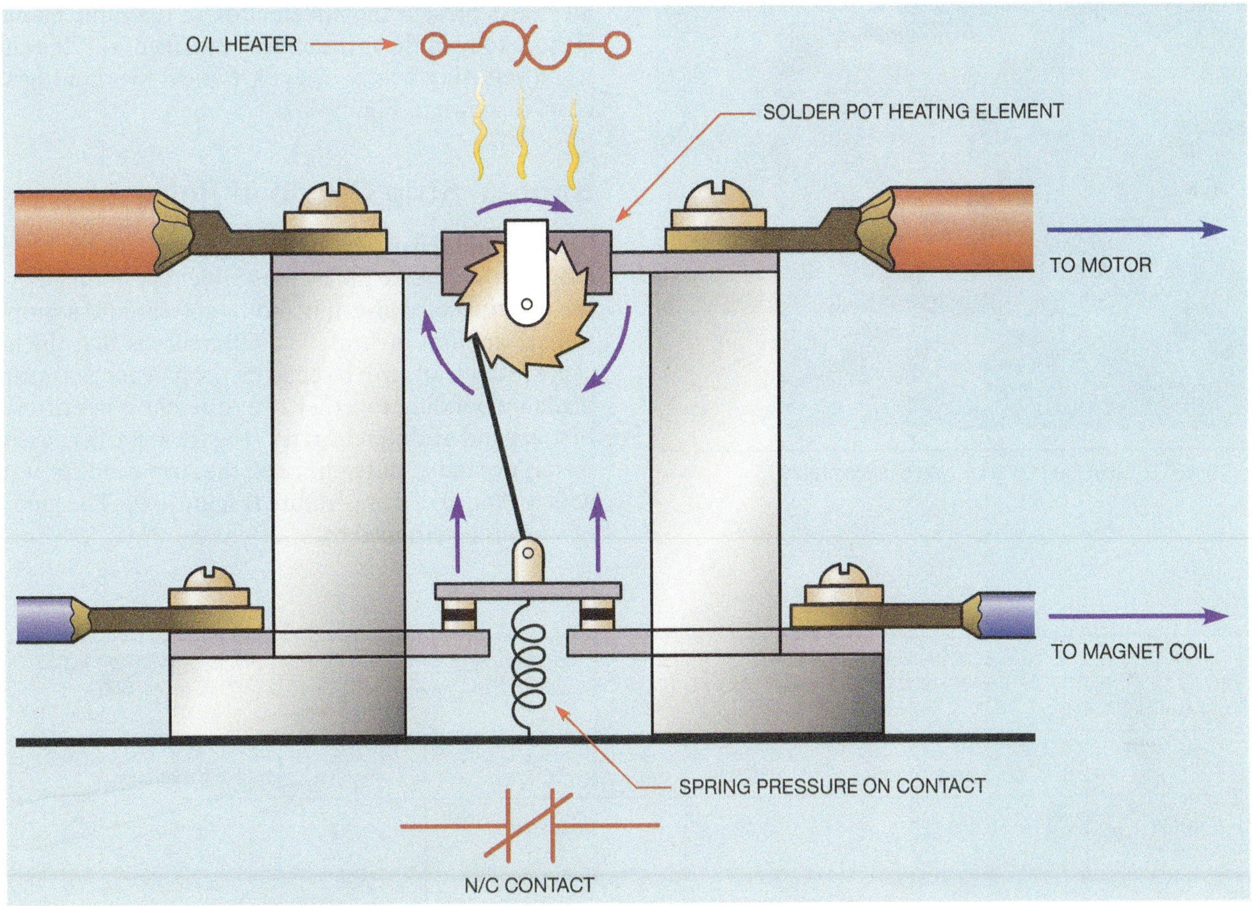

Figure 4–4 Melting alloy thermal overload relay. Spring pushes contact open as heat melts alloy allowing ratchet wheel to turn freely. Note electrical symbols for heater and normally closed contact.

causes the solder to melt faster and causes the contacts to open quicker than a smaller amount of overload current.

Manufacturers construct overload heaters differently, but all work on the same principle. Two different types of melting alloy heater assemblies are shown in Figures 4–5 A and B. A typical melting alloy type overload relay is shown in Figure 4–6. After the overload relay has tripped, it is necessary to allow the relay to cool for 2 or 3 minutes before it can be reset. This cool-down time is necessary to permit the solder to become hard again after it has melted.

The current setting can be changed by changing the heater. Manufacturers provide charts that indicate what size heater should be installed for different amounts of motor current. It is necessary to use the chart that corresponds to the particular type of overload relay. Not

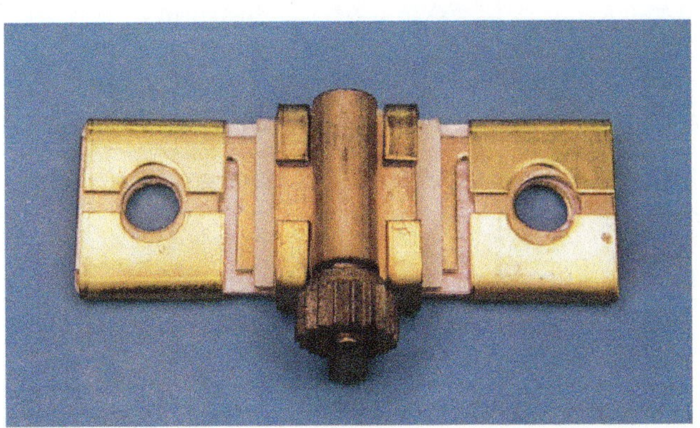

Figure 4–5A Melting alloy type overload heater.

Figure 4–5B Melting alloy type overload heater.

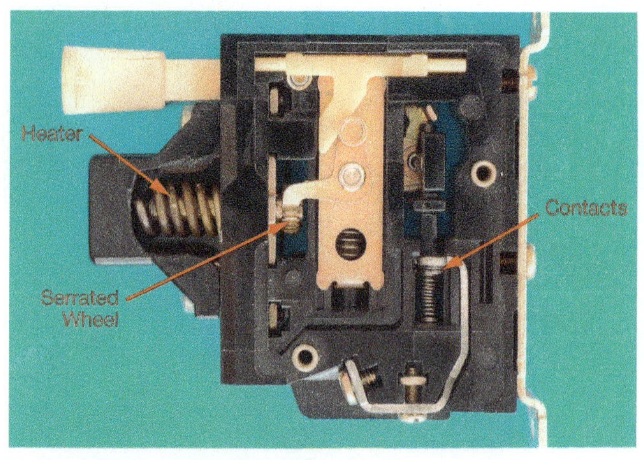

Figure 4–6 Typical melting alloy type overload relay.

all charts present the information in the same manner. Be sure to read the instructions contained with the chart when selecting heater sizes. A typical overload heater chart is shown in Figure 4–7.

Bimetal Strip Overload Relay

The second type of thermal overload relay is the bimetal strip overload. Like the melting alloy type, it operates on the principle of converting motor current into a proportionate amount of heat. The difference is that the heat causes a bimetal strip to bend or warp. A bimetal strip is made by bonding together two different types of metal that expand at different rates (Figure 4–8). Because the metals expand at different rates, the strip bends or warps with a change of temperature (Figure 4–9). The amount of warp is determined by:

OVERLOAD HEATER SELECTION FOR NEMA STARTER SIZES 00–1. HEATERS ARE CALIBRATED FOR 115% OF MOTOR FULL LOAD CURRENT. FOR HEATERS THAT CORRESPOND TO 125% OF MOTOR FULL LOAD CURRENT USE THE NEXT SIZE LARGER HEATER.					
HEATER CODE	MOTOR FULL LOAD CURRENT	HEATER CODE	MOTOR FULL LOAD CURRENT	HEATER CODE	MOTOR FULL LOAD CURRENT
XX01	0.25–0.27	XX18	1.35–1.47	XX35	6.5–7.1
XX02	0.28–0.31	XX19	1.48–1.62	XX36	7.2–7.8
XX03	0.32–0.34	XX20	1.63–1.78	XX37	7.9–8.5
XX04	0.35–0.38	XX21	1.79–1.95	XX38	8.6–9.4
XX05	0.39–0.42	XX22	1.96–2.15	XX39	9.5–10.3
XX06	0.43–0.46	XX23	2.16–2.35	XX40	10.4–11.3
XX07	0.47–0.50	XX24	2.36–2.58	XX41	11.4–12.4
XX08	0.51–0.55	XX25	2.59–2.83	XX42	12.5–13.5
XX09	0.56–0.62	XX26	2.84–3.11	XX43	13.6–14.9
XX10	0.63–0.68	XX27	3.12–3.42	XX44	15.0–16.3
XX11	0.69–0.75	XX28	3.43–3.73	XX45	16.4–18.0
XX12	0.76–0.83	XX29	3.74–4.07	XX46	18.1–19.8
XX13	0.84–0.91	XX30	4.08–4.39	XX47	19.9–21.7
XX14	0.92–1.00	XX31	4.40–4.87	XX48	21.8–23.9
XX15	1.01–1.11	XX32	4.88–5.3	XX49	24.0–26.2
XX16	1.12–1.22	XX33	5.4–5.9		
XX17	1.23–1.34	XX34	6.0–6.4		

Figure 4–7 Typical overload heater chart.

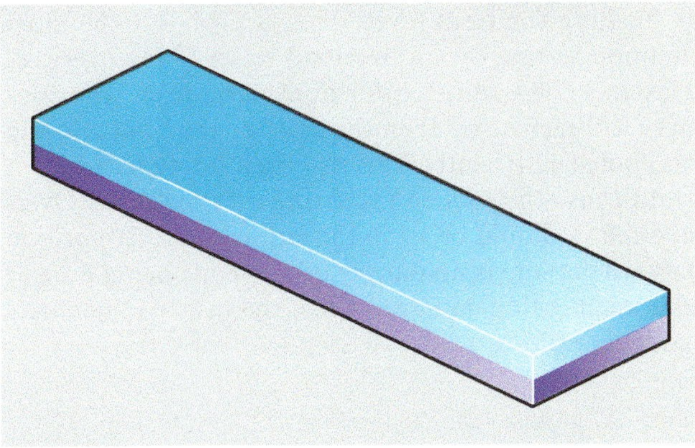

Figure 4–8 A bimetal strip is constructed by bonding two different metals together.

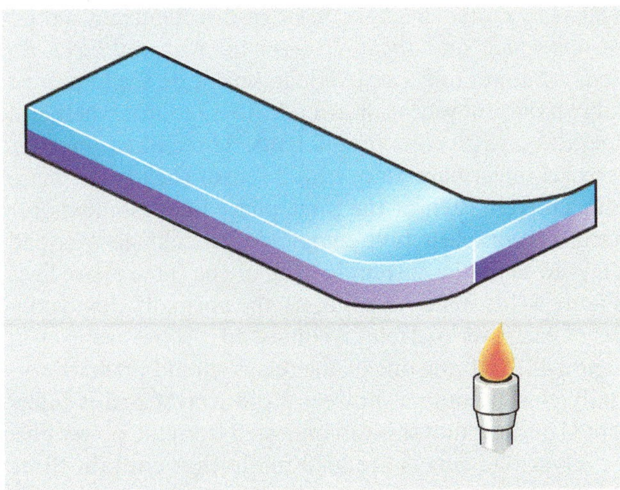

Figure 4–9 A bimetal strip warps with a change of temperature.

1. The type of metals used to construct the bimetal strip.
2. The difference in temperature between the two ends of the strip.
3. The length of the strip.

The overload heater heats the bimetal strip when motor current flows through it. The heat causes the bimetal strip to warp. If the bimetal strip becomes hot enough, it causes a set of contacts to open (Figure 4–10). Once the overload contact has opened, about 2 minutes of cool-down time is needed to permit the bimetal strip to return to a position that permits the contacts to reclose. The time delay factor for this overload relay is the time required for the bimetal strip to warp a sufficient amount to open the normally closed contact. A large amount of overload current causes the bimetal strip to warp at a faster rate and open the contact sooner.

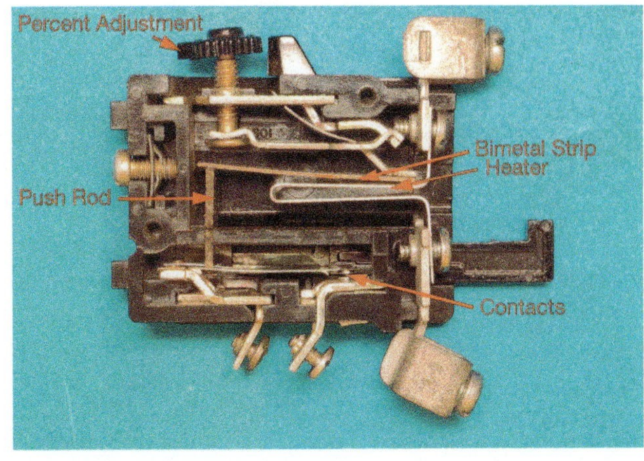

Figure 4–10 Bimetal strip type overload relay.

Most bimetal strip type overload relays have a couple of features that are not available with solder melting type overload relays. As a general rule, the trip range can be adjusted by turning the knob shown in Figure 4–10. This knob adjusts the distance the bimetal strip must warp before opening contacts. This adjustment permits the sensitivity to be changed due to changes in ambient air temperature. If the knob is set in the 100 percent position (Figure 4–11), the

Figure 4–11 An adjustment knob permits the full load motor current to be adjusted between 85 percent and 115 percent.

Figure 4–12 Many bimetal strip type overload relays can be set for manual or automatic reset.

overload operates at the full load current rating as determined by the size overload heater installed. In cold winter months, this setting may be too high to protect the motor. The knob can be adjusted to operate at any point from 100 percent to 85 percent of the motor full load current. In hot summer months, the motor may nuisance trip due to high ambient temperatures. The adjustment knob permits the overload relay to also be adjusted between 100 percent and 115 percent of motor full load current.

Another difference is that many bimetal strip type overload relays can be set for either manual or automatic reset. A spring located on the side of the overload relay permits this setting (Figure 4–12). When set in the manual position, the contacts must be reset manually by pushing the reset lever. This is probably the most common setting for an overload relay. If the overload relay has been adjusted for automatic reset, the contacts will reclose by themselves after the bimetal strip has cooled sufficiently. This may be a safety hazard if it could cause the sudden restarting of a machine. Overload relays should be set in the automatic reset position only when there is no danger of someone being hurt or equipment being damaged when the overload contacts suddenly reclose.

Three Phase Overloads

The overload relays discussed so far are intended to detect the current of a single conductor supplying power to a motor (Figure 4–13). An application for this type of overload relay is to protect a single phase or direct current motor. *Section 430.37* and *Table 430.37* of the *National Electrical Code*® requires only one overload sensor device to protect a direct current motor or a single phase motor whether it operates on 120 volts or 240 volts. Three phase motors, however, must have an overload sensor (heaters or magnetic coils) in each of the three phase lines. Some motor starters accomplish this by employing three single overload relays to sense the current in each of the three phase lines (Figure 4–14). When this is done, the normally closed contact of each overload relay is connected in series as shown in Figure 4–15. If any one of the relays should open its normally closed contact, power to the starter coil is interrupted and the motor is disconnected from the power line.

Overload relays are also made that contain three overload heaters and one set of normally closed contacts (Figure 4–16). These relays are generally used

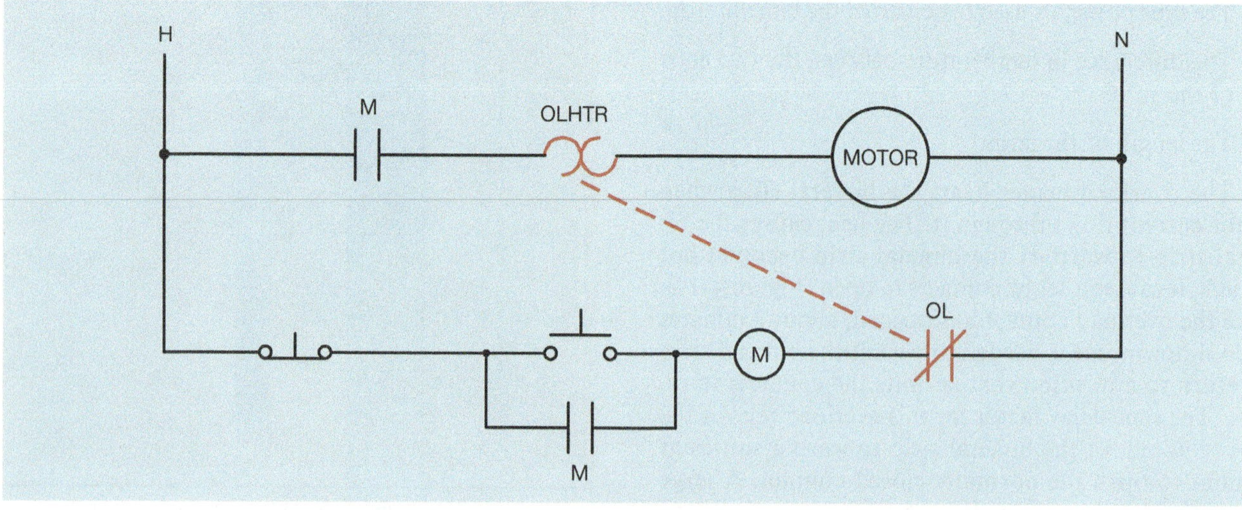

Figure 4–13 A single overload relay is used to protect a single phase motor.

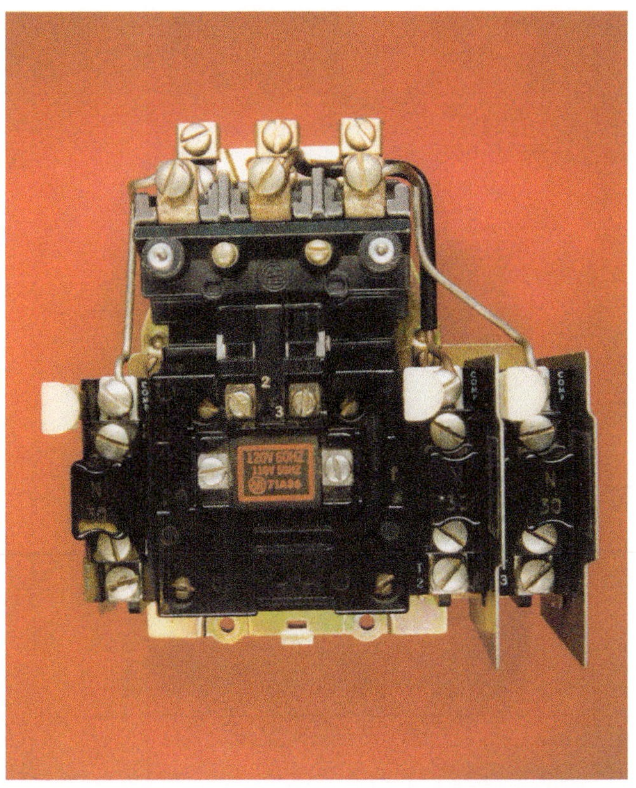

Figure 4–14 Three single overload relays are used to sense the current in each of the three phase lines.

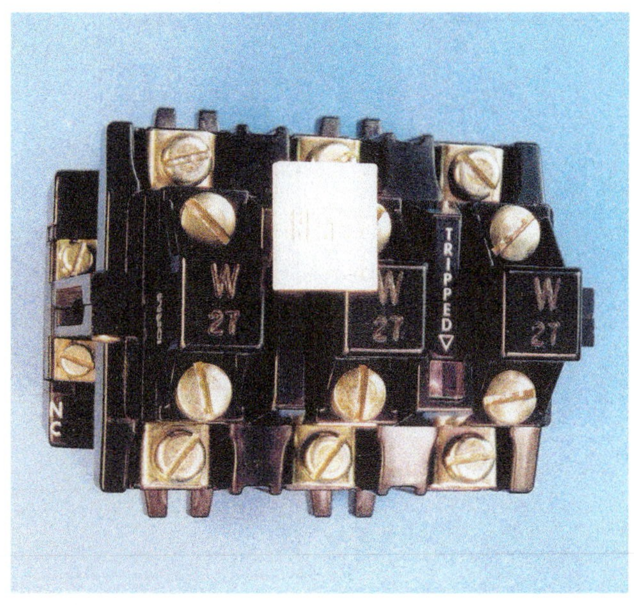

Figure 4–16 Three phase thermal overload relay.

to protect three phase motors. Although there is only one set of normally closed contacts, if an overload occurs on any one of the three heaters, it causes the contacts to open and disconnect the coil of the motor starter (Figure 4–17).

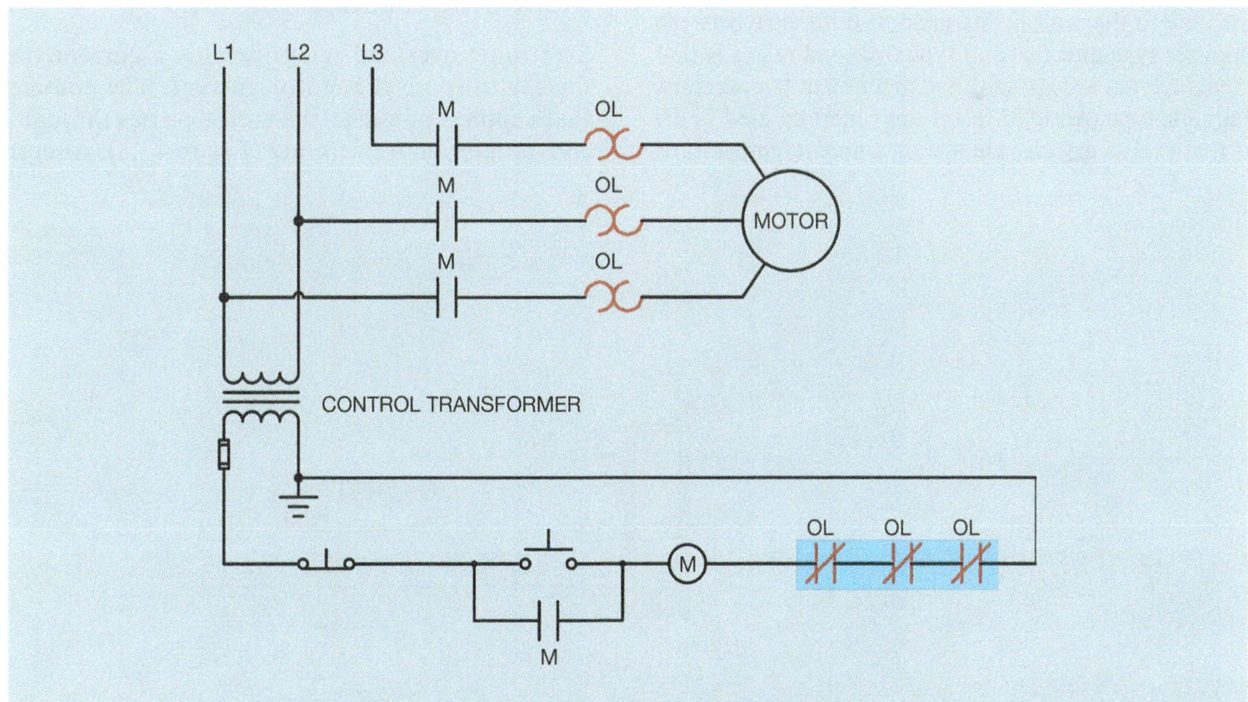

Figure 4–15 When three single overload relays are employed to protect a three phase motor, all normally closed overload contacts are connected in series.

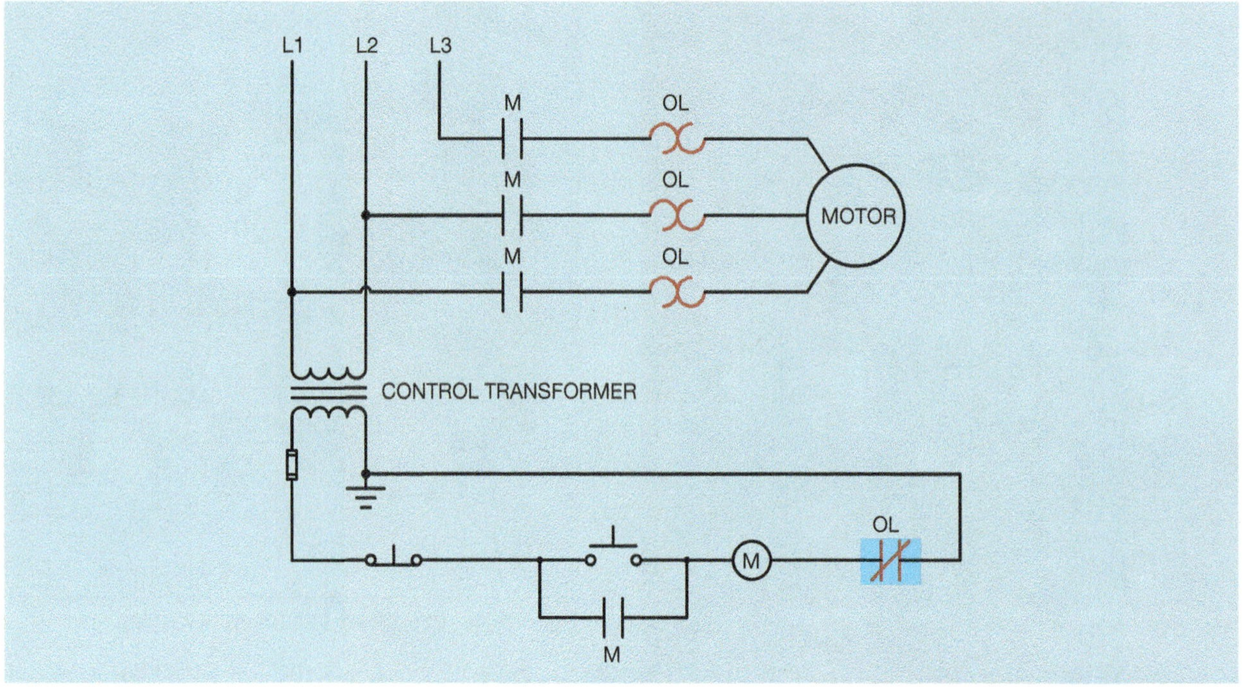

Figure 4–17 A three phase overload relay contains three heaters but only one set of normally closed contacts.

Magnetic Overload Relays

Magnetic type overload relays operate by sensing the strength of the magnetic field produced by the current flow to the motor. The greatest difference between magnetic type and thermal type overload relays is that magnetic types are *not* sensitive to ambient temperature. Magnetic type overload relays are generally used in areas that exhibit extreme changes in ambient temperature.

Magnetic overload relays can be divided into two major types: electronic and **dashpot**.

Electronic Overload Relays

Electronic overload relays employ a current transformer to sense the motor current. The conductor that supplies power to the motor passes through the core of a toroid transformer (Figure 4–18). As current

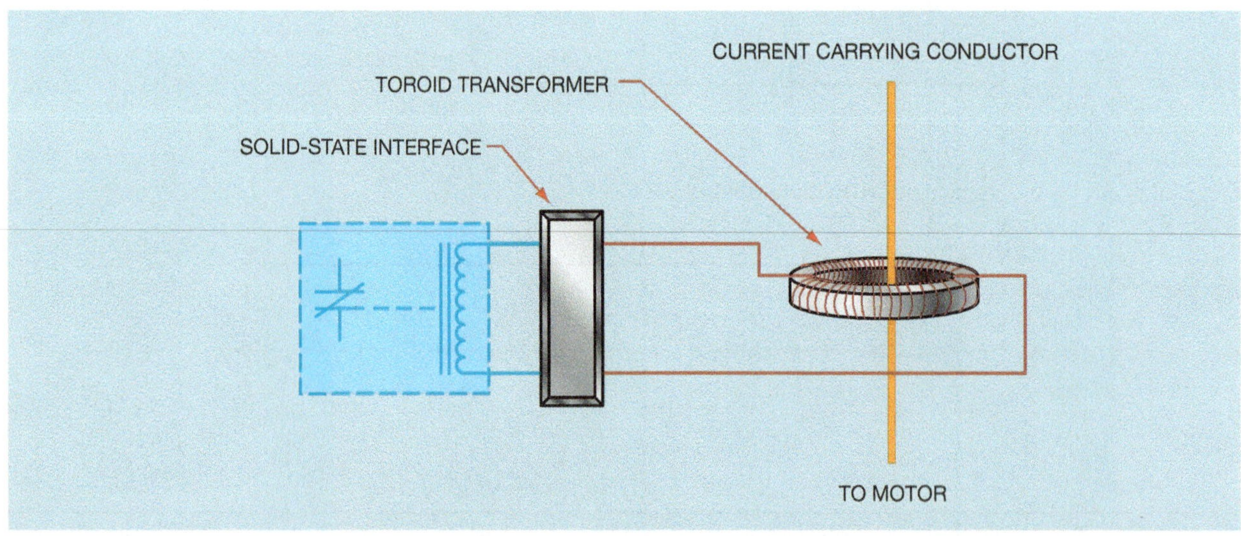

Figure 4–18 Electronic overloads sense motor current by measuring the strength of a magnetic field.

Courtesy Schneider Electric USA, Inc.

Figure 4–19 Three phase electronic overload relay.

flows through the conductor, the alternating magnetic field around the conductor induces a voltage into the toroid transformer. The amount of induced voltage is proportional to the amount of current flowing through the conductor. The same basic principle of operation is employed by most clamp-on type ammeters. The voltage induced into the toroid transformer is connected to an electronic interface that provides the time delay necessary to permit the motor to start. Many electronic type overload relays are programmable and can be set for the amount of full load motor current, maximum and minimum voltage levels, percent of overload, and other factors. A three phase electronic overload relay is shown in Figure 4–19.

Dashpot Overload Relays

Dashpot overload relays receive their name from the method used to accomplish the time delay that permits the motor to start. A dashpot timer is used to provide this time delay. A dashpot timer is basically a container, a piston, and a shaft (Figure 4–20). The piston is placed inside the container and the container is filled with a special type of oil called *dashpot oil* (Figure 4–21). Dashpot oil maintains a constant viscosity over a wide range of temperatures. The type and viscosity of oil used is one of the factors that determine the amount of time delay for the timer. The other factor is setting the opening of the orifices in the piston (Figure 4–22). Orifices permit the oil to flow through the piston as it rises through the oil. The opening of the orifices can be set by adjusting a sliding valve on the piston.

Figure 4–20 A dashpot timer is a container, piston, and shaft.

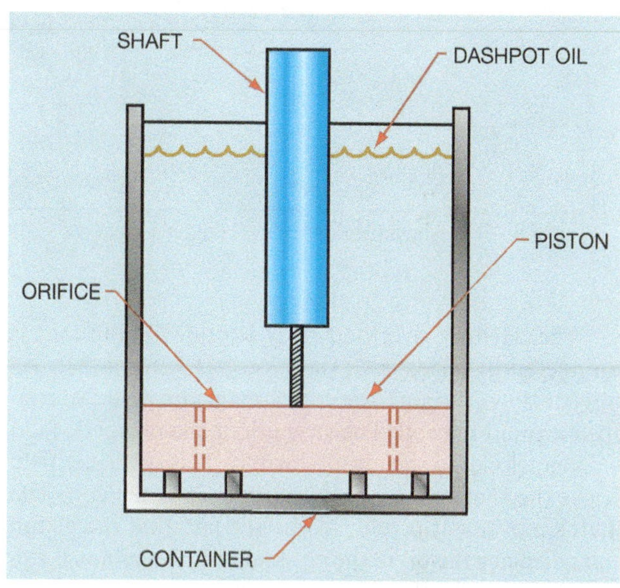

Figure 4–21 Basic construction of a dashpot timer.

Figure 4–22 Setting the opening of the orifices affects the tide delay of the dashpot timer.

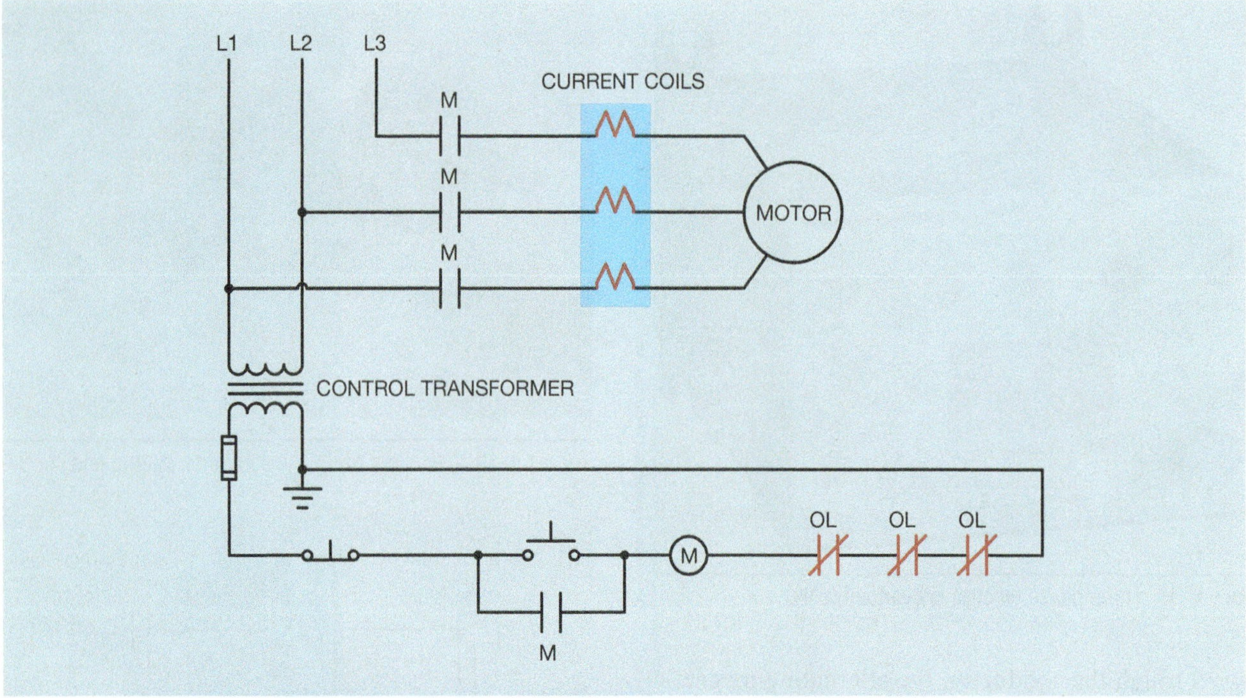

Figure 4–23 Dashpot overload relays contain current coils connected in series with the motor.

The dashpot overload relay contains a coil that is connected in series with the motor (Figure 4–23). As current flows through the coil, a magnetic field is developed around the coil. The strength of the magnetic field is proportional to the motor current. This magnetic field draws the shaft of the dashpot timer into the coil. The shaft's movement is retarded by the fact that the piston must displace the oil in the container. If the motor is operating normally, the motor current drops to a safe level before the shaft is drawn far enough into the coil to open the normally closed contact (Figure 4–24). If the motor is overloaded, however, the magnetic field will be strong enough to continue drawing the shaft into the coil until it opens the overload contact. When power is disconnected from the motor, the magnetic field collapses and the piston returns to the bottom of the container. Check valves permit the piston to return to the bottom of the container almost immediately when motor current ceases.

Dashpot overloads generally provide some method that permits the relay to be adjusted for different full load current values. To make this adjustment, the shaft is connected to a threaded rod (Figure 4–25). This permits the shaft to be lengthened or shortened inside the coil—the greater the length of the shaft the less current required to draw the shaft into the coil far enough to open the contacts. A nameplate on the coil lists the different current settings for a particular overload relay (Figure 4–26). The adjustment is made by moving the

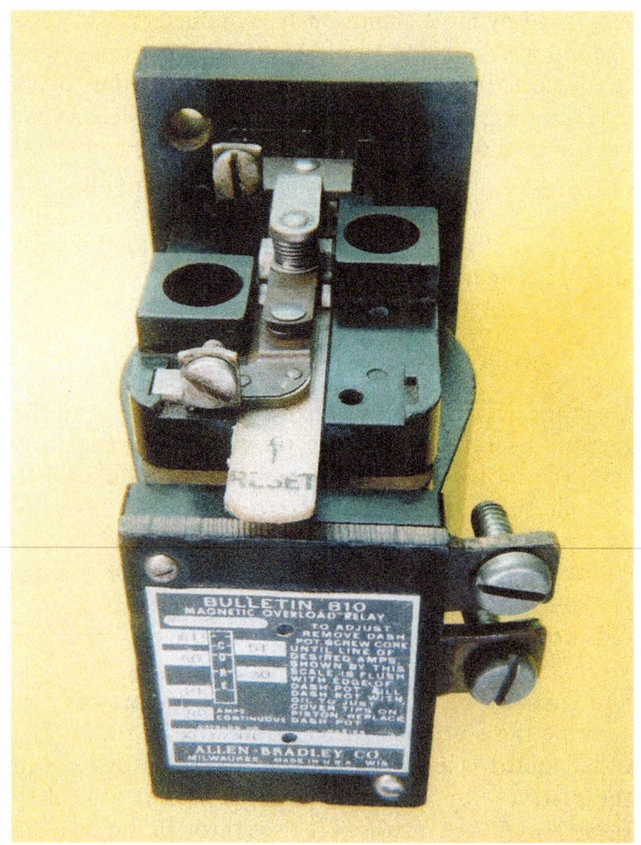

Figure 4–24 Normally closed contact of a dashpot timer.

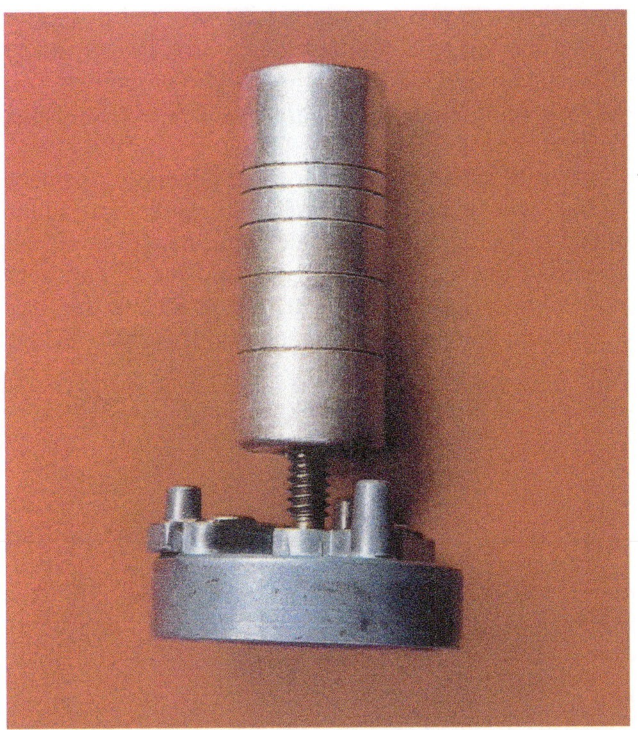

Figure 4–25 The length of the shaft can be adjusted for different values of current.

Figure 4–26 The nameplate lists different current settings.

Figure 4–27 The line on the shaft that represents the desired amount of current is set flush with the top of the dashpot container.

shaft until the line on the shaft that represents the desired current is flush with the top of the dashpot container (Figure 4–27). A dashpot overload relay is shown in Figure 4–28.

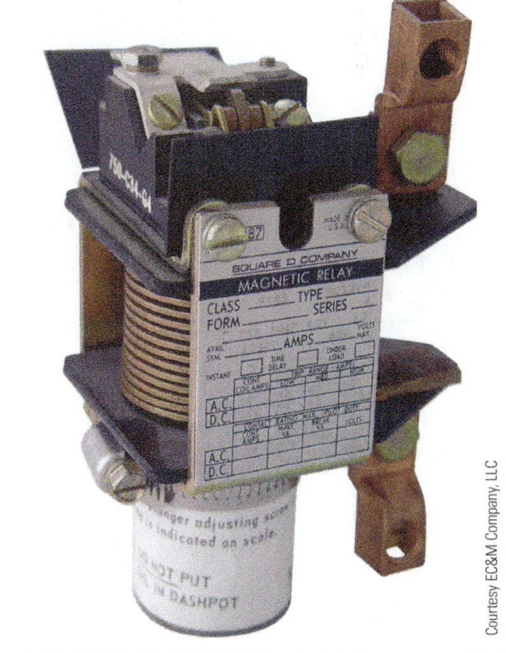

Courtesy EC&M Company, LLC

Figure 4–28 Dashpot overload relay.

Overload Contacts

Although all overload relays contain a set of normally closed contacts, some manufacturers also add a set of normally open contacts as well. These two sets of contacts are either in the form of a single pole double throw switch or as two separate contacts. The single pole double throw switch arrangement contains a common terminal (C), a normally closed terminal (NC), and a normally open terminal (NO). There are several reasons for adding the normally open set of contacts. The starter shown in Figure 4–29 uses the normally closed section to disconnect the motor starter in the event of an overload, and the normally open section to turn on a light to indicate that the overload has tripped.

Another common use for the normally open set of contacts on an overload relay is to provide an input signal to a programmable logic controller (PLC). If the overload trips, the normally closed set of contacts opens and disconnects the starter coil from the line. The normally open set of contacts closes and provides

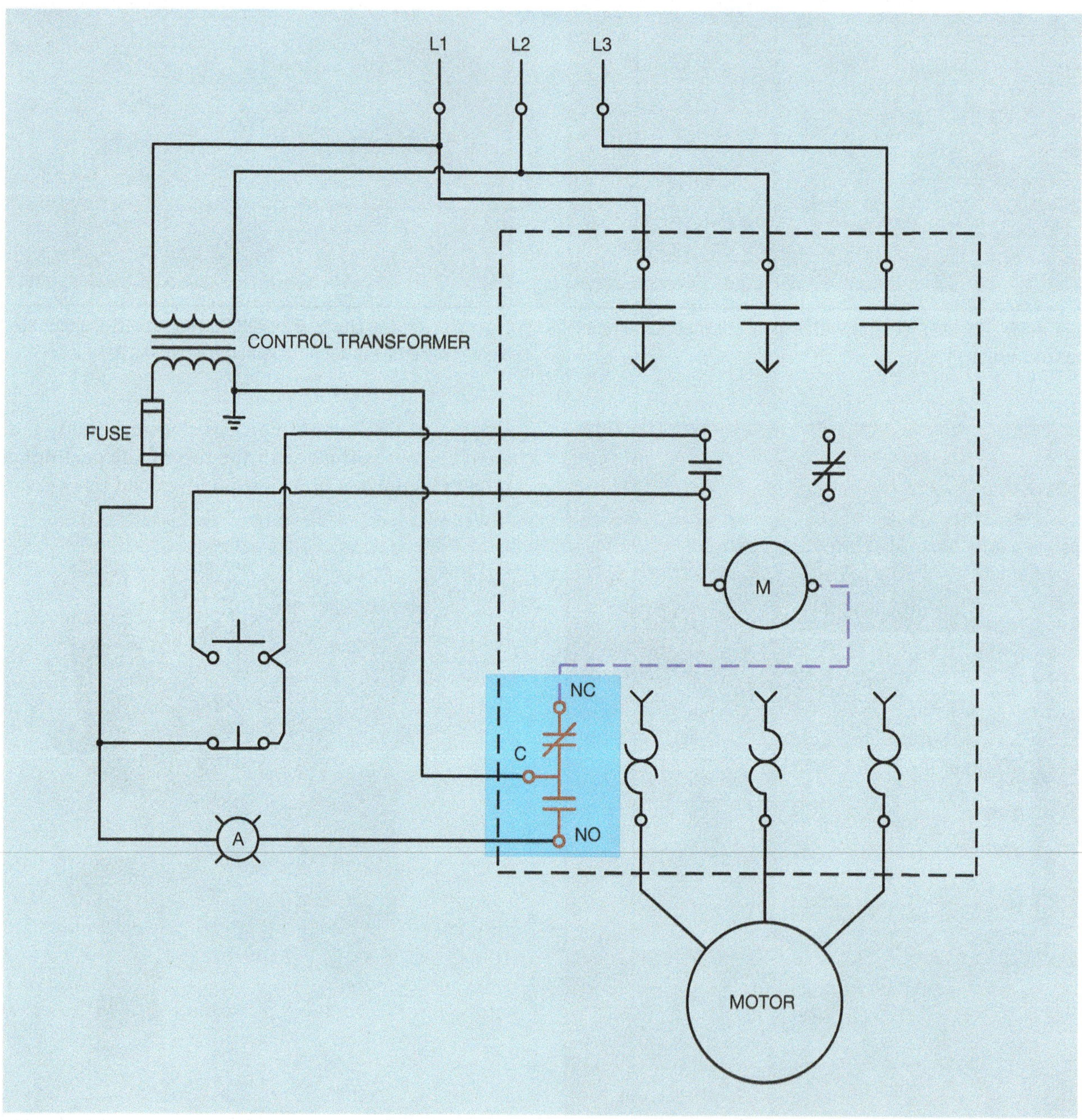

Figure 4–29 The overload relay contains a single pole double throw set of contacts.

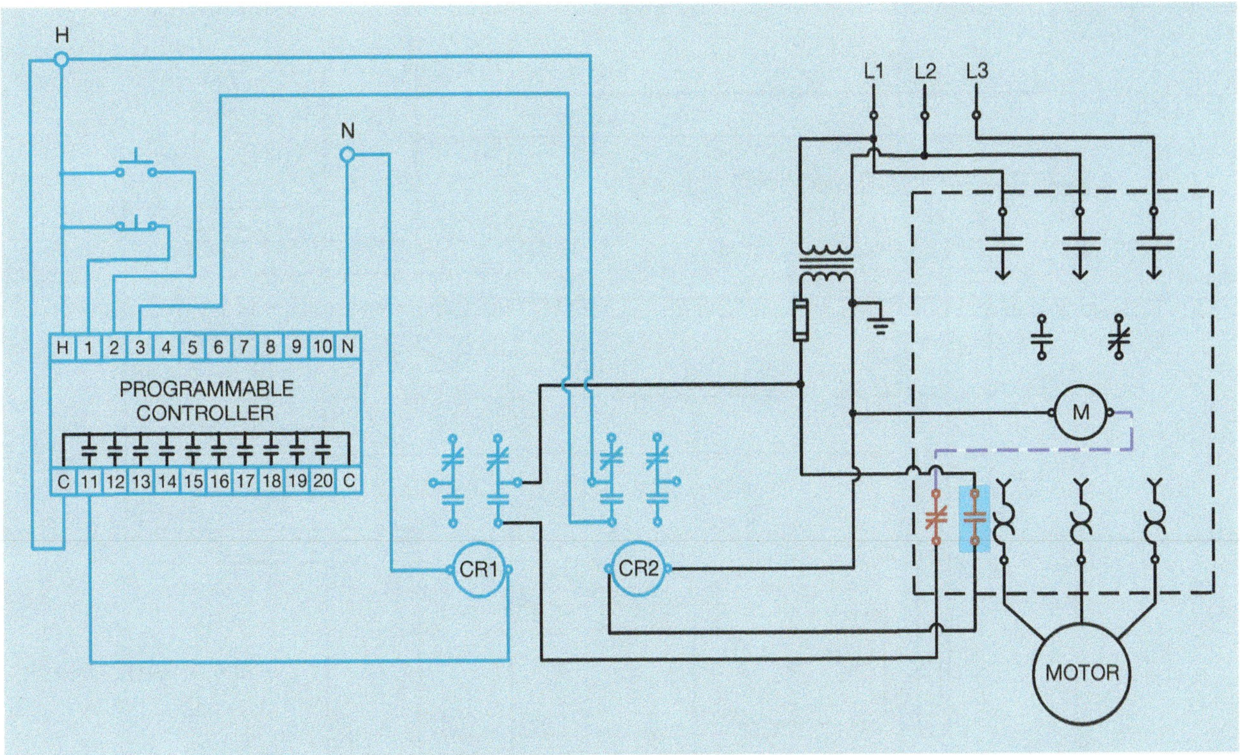

Figure 4–30 The normally open overload contact provides a signal to the input of a programmable logic controller.

a signal to the input of the PLC (Figure 4–30). Notice that two interposing relays, CR1 and CR2, are used to separate the programmable logic controller and the motor starter. This separation is often done for safety reasons. The control relays prevent more than one source of power from entering the starter or programmable controller. Note that the starter and programmable controller each have a separate power source. If the power was disconnected from the starter, for example, it could cause an injury if the power from the programmable controller was connected to any part of the starter.

Protecting Large Horsepower Motors

Large horsepower motors often have current draws of several hundred amperes, making the sizing of overload heaters difficult. When this is the case, it is common practice to use current transformers to reduce the amount of current to the overload heaters (Figure 4–31). The current transformers shown in Figure 4–31 have ratios of 150:5. This ratio means that when 150 amperes of current flows through the primary, which is the line connected to the motor, the transformer secondary produces a current of 5 amperes when the secondary terminals are shorted together. The secondaries of the current transformers are connected to the overload heaters to provide protection for the motor (Figure 4–32). Assume that the motor connected to the current transformers in Figure 4–32 has a full load current of 136 amperes. A simple calculation reveals that current transformers with a

Figure 4–31 Current transformers reduce overload current.

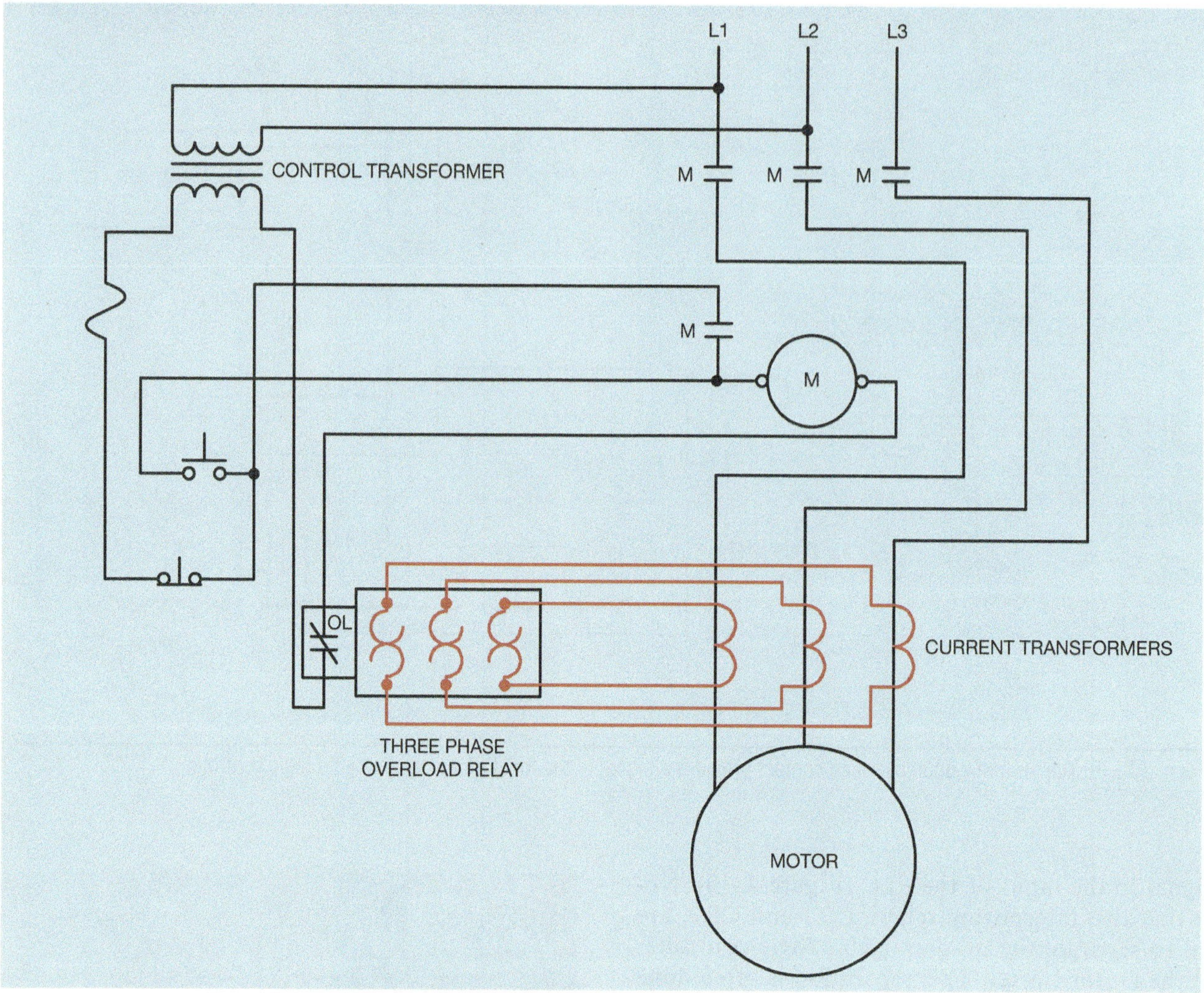

Figure 4–32 Current transformers reduce the current to the overload heaters.

ratio of 150:5 would produce a secondary current of 4.533 amperes when 136 amperes flow through the primary.

$$\frac{150}{5} = \frac{136}{X}$$

$$150X = 680$$

$$X = \frac{680}{150}$$

$$X = 4.533$$

The overload heaters would actually be sized for a motor with a full load current of 4.533 amperes. The typical overload heater chart shown in Figure 4–7 indicates that a XX31 heater would be used if the overload is sized at 115 percent of motor full load current and a XX32 heater would be used if the overload is sized at 125 percent of motor full load current.

Large size starters often contain current transformers as an integral part of the starter. The size 6 starter illustrated in Figure 4–33 contains current transformers

Figure 4–33 Large starters are often provided with current transformers.

with a ratio of 300:5 on each of the input connections. The secondary or outputs of these transformers are connected directly to the overload heater elements.

Review Questions

1. What are the two basic types of overload relays?
2. What is the major difference in characteristics between thermal type and magnetic type overload relays?
3. What are the two major types of thermal overload relays?
4. What type of thermal overload relay can generally be set for manual or automatic operation?
5. Why is it necessary to permit a solder melting type of overload relay to cool for 2 to 3 minutes after it has tripped?
6. All overload relays are divided into two sections. What are these two sections?
7. What device is used to sense the amount of motor current in an electronic overload relay?
8. What two factors determine the time setting for a dashpot timer?
9. How many overload sensors are required by the *National Electrical Code*® to protect a direct current motor?
10. A large motor has a full load current rating of 425 amperes. Current transformers with a ratio of 600:5 are used to reduce the current to the overload heaters. What should be the full load current rating of the overload heaters?

Chapter 5

RELAYS, CONTACTORS, AND MOTOR STARTERS

Objectives

After studying this chapter the student will be able to:

- Discuss the operation of magnetic type relay devices.
- Explain the difference between relays, contactors, and motor starters.
- Connect a relay in a circuit.
- Identify the pins of 8 and 11 pin relays.
- Discuss the differences between DC and AC type relays and contactors.
- Discuss difference between NEMA- and IEC-rated starters.

Relays and **contactors** are electromechanical switches. They operate on the **solenoid** principle. A coil of wire is connected to an electric current. The magnetic field developed by the current is concentrated in an iron pole piece. The electromagnet attracts a metal armature. Contacts are connected to the metal armature. When the coil is energized, the contacts open or close. There are two basic methods of constructing a relay or contactor. The clapper type uses one movable contact to make connection with a stationary contact. The bridge type uses a movable contact to make connection between two stationary contacts.

Relays

Relays are electromechanical switches that contain auxiliary contacts. Auxiliary contacts are small and are intended to be used for control applications. As a general rule, they are not intended to control large amounts of current. Current ratings for most relays can vary from 1 to 10 amperes depending on the manufacturer and type of relay. A clapper type relay is illustrated in Figure 5–1. When the

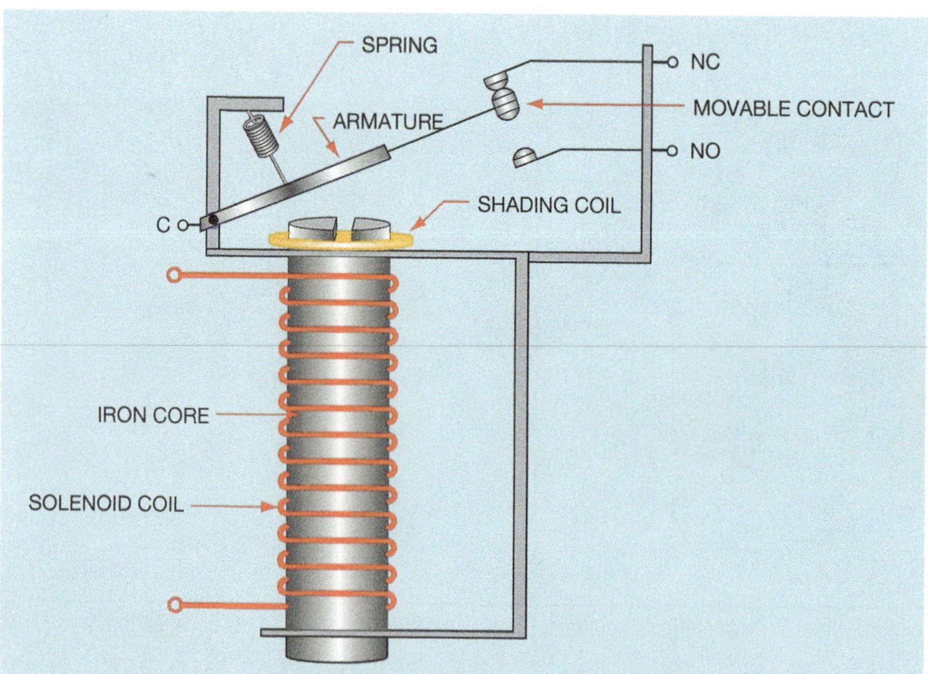

Figure 5–1 A magnetic relay is basically a solenoid with movable contacts attached.

coil is energized, the armature is attracted to the iron core inside the coil. This attraction causes the movable contact to break away from one stationary contact and make connection with another. The common terminal is connected to the armature, which is the movable part of the relay. The movable contact is attached to the armature. The two stationary contacts form the normally closed and normally open contacts. A spring returns the armature to the normally closed position when power is removed from the coil. The shading coil is necessary to prevent the contacts from chattering. All solenoids that operate on alternating current must have a shading coil. Relays that operate on direct current do not require them. A clapper type relay is shown in Figure 5–2.

Bridge Type Relay

A bridge type relay operates by drawing a piece of metal or plunger inside a coil, Figure 5–3. The plunger is connected to a bar that contains movable contacts. The movable contacts are mounted on springs and are insulated from the bar. The plunger and bar assembly are called the armature because they are the moving part of the relay. Bridge contacts receive their name because when the solenoid coil is energized and the plunger is drawn inside the coil, the movable contacts bridge across the two stationary contacts. Bridge contacts can control more voltage than clapper types because they break connection at two places instead of one. When power is removed from the coil, the force of gravity or a spring returns the movable contacts to their original position. A relay with bridge type contacts is shown in Figure 5–4.

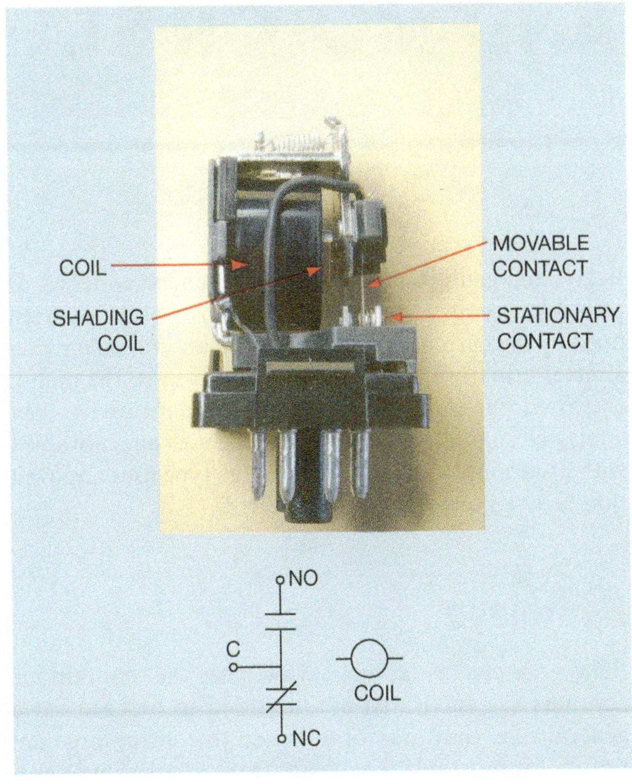

Figure 5–2 Single pole double throw clapper type relay.

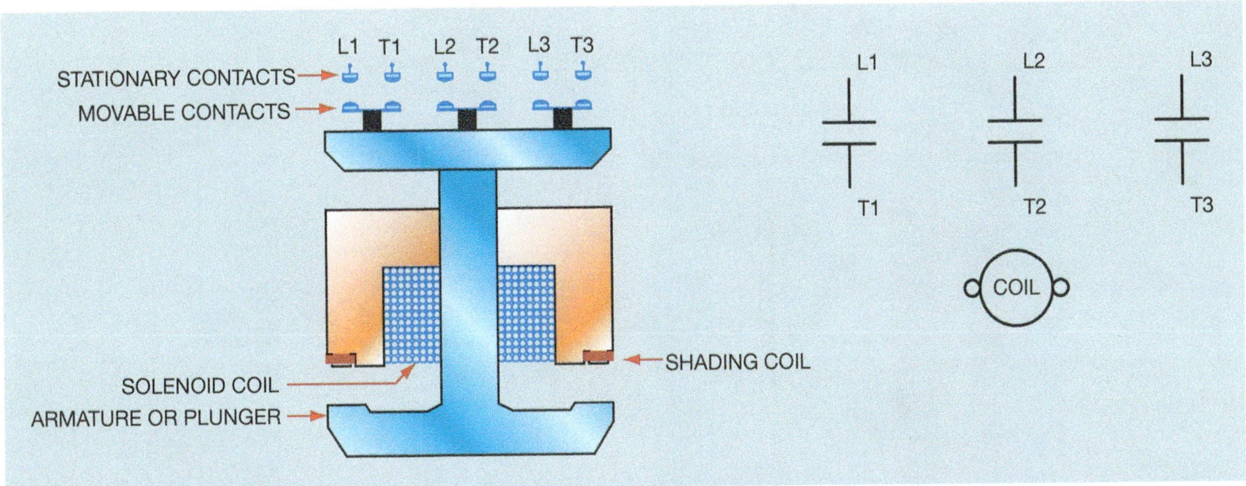

Figure 5–3 Bridge type contacts use one movable and two stationary contacts. They can control higher voltages because they break connection in two places instead of one.

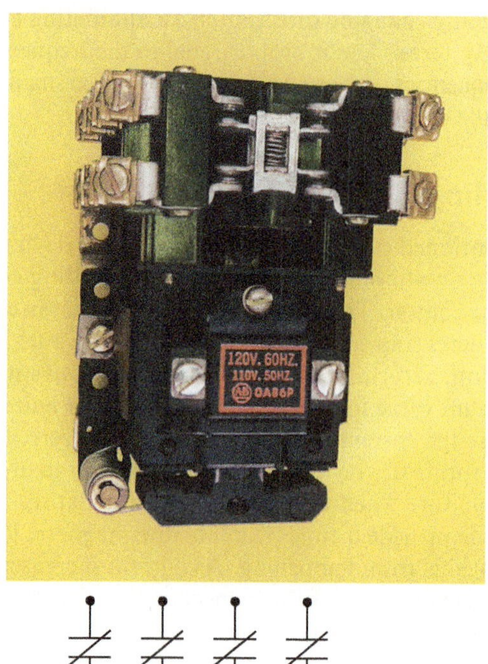

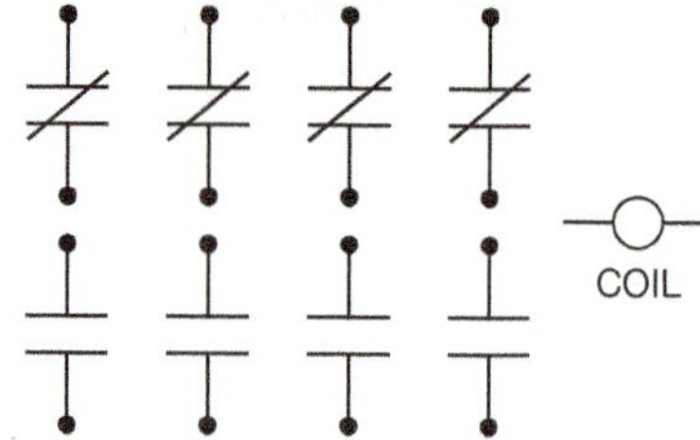

Figure 5–4 A relay with bridge type contacts.

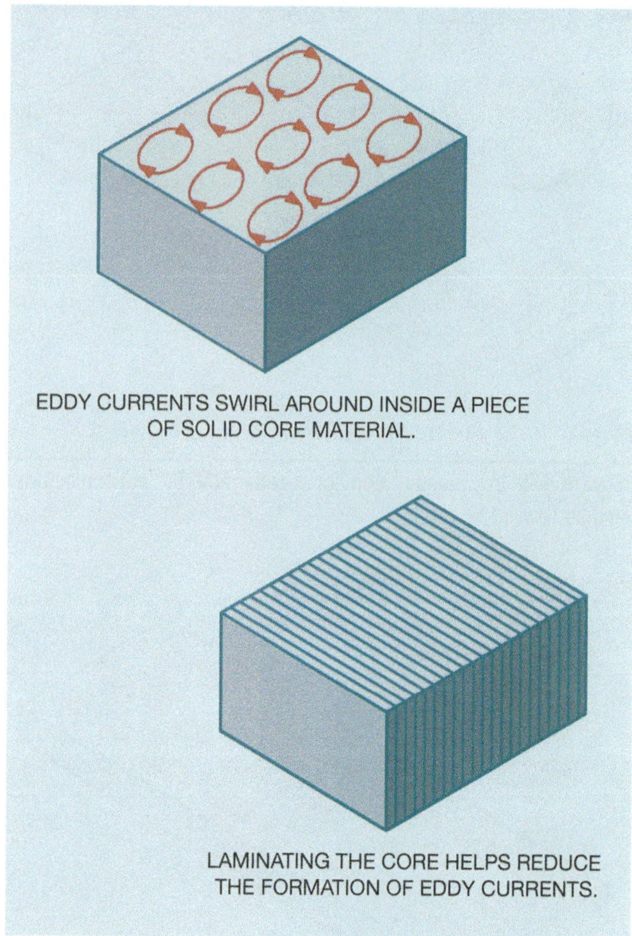

Figure 5–5 Eddy currents are induced into the metal core and produce power loss in the form of heat.

Electromagnet Construction

The construction of the electromagnetic part of a relay or contactor greatly depends on whether it is to be operated by direct or alternating current. Relays and contactors that are operated by direct current generally contain solid core materials, while those intended for use with alternating current contain laminated cores. The main reason is the core losses associated with alternating current caused by the continuous changing of the electromagnetic field.

Core Losses

This continuous change of both amplitude and polarity of the magnetic field causes currents to be induced into the metal core material. These currents are called **eddy currents** because they are similar to eddies (swirling currents) found in rivers. Eddy currents tend to swirl around inside the core material producing heat (Figure 5–5). Laminated cores are constructed with thin sheets of metal stacked together. A thin layer of oxide forms between the laminations. This oxide is an insulator and helps reduce the formation of eddy currents.

Another type of core loss associated with alternating current devices is called **hysteresis loss**. Hysteresis loss is caused by the molecules inside magnetic materials changing direction. Magnetic materials such as iron or soft steel contain magnetic domains or magnetic molecules. In an unmagnetized piece of material, these magnetic domains are not aligned in any particular order (Figure 5–6). If the metal becomes magnetized, the magnetic molecules or domains align themselves in an orderly fashion (Figure 5–7). If the polarity of the magnetic field is reversed, the molecules realign themselves to the new polarity (Figure 5–8). Although the domains realign to correspond to a change of polarity, they resist the realignment. The power required to cause them to change polarity is a power loss in the form of heat. Hysteresis loss is often referred to as **molecular friction** because the molecules are

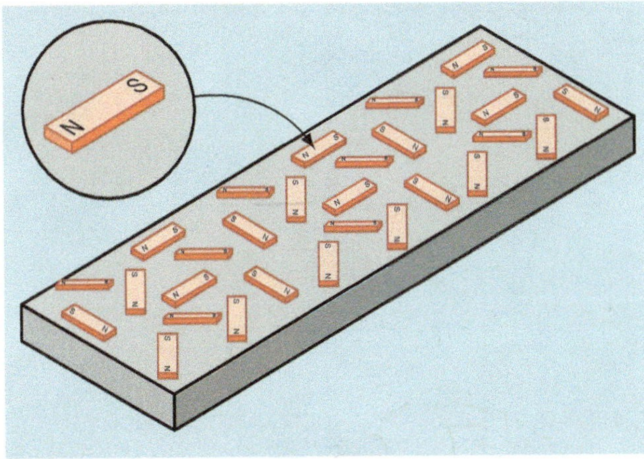

Figure 5–6 In a piece of unmagnetized metal, the molecules are in disarray.

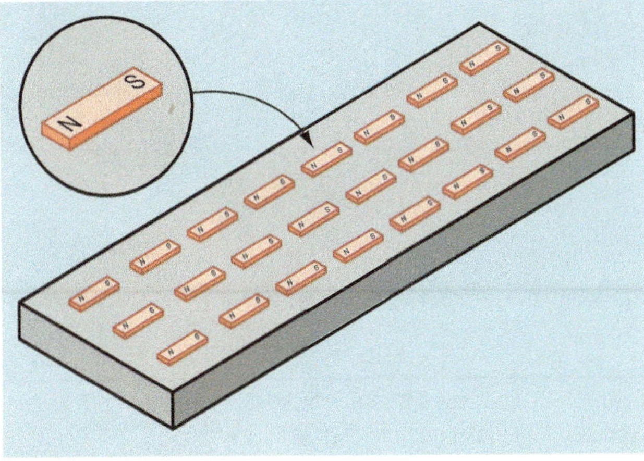

Figure 5–7 In a piece of magnetized material, the magnetic molecules are aligned.

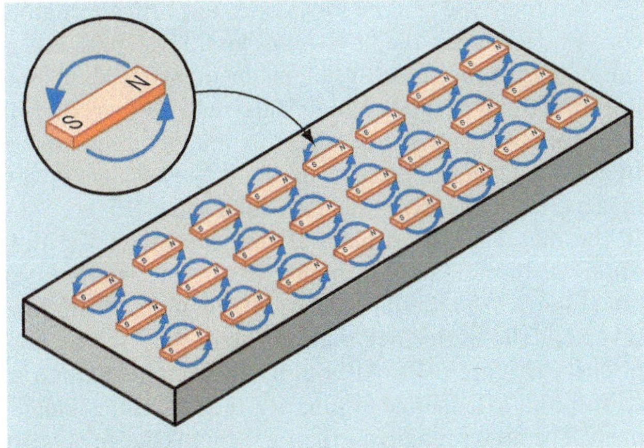

Figure 5–8 When the magnetic polarity changes, all the molecules change position.

continually changing direction in an alternating current field. Hysteresis loss is proportional to the frequency. At low frequencies such as 60 Hz, it is generally so small that it is of little concern.

Shading Coils

As mentioned previously, all solenoid-type devices that operate on alternating current contain shading coils to prevent chatter. The current in an AC circuit is continually increasing from zero to a maximum value in one direction, returning to zero, and then increasing to a maximum value in the opposite direction (Figure 5–9). Because the current is continually falling to zero, the solenoid spring or gravity would continually try to drop out the armature when the magnetic field collapses. Shading coils provide a time delay for the magnetic field to prevent this from happening. As current increases from zero, magnetic lines of flux concentrate in the metal pole piece (Figure 5–10). This increasing magnetic field cuts the shading coil and induces a voltage into it. Because the shading coil or loop is a piece of heavy copper, it has a very low resistance. A very small induced voltage can cause a large amount of current to flow in the loop. The current flow in the shading coil causes a magnetic field to be developed around the shading coil also. This magnetic field acts in opposition to the magnetic field in the pole piece and causes it to bend away from the shading coil (Figure 5–11). As long as the AC current is changing in amplitude, a voltage will be induced in the shading loop.

When the current reaches its maximum or peak value, the magnetic field is no longer changing and no voltage is induced in the shading coil. Because the shading coil

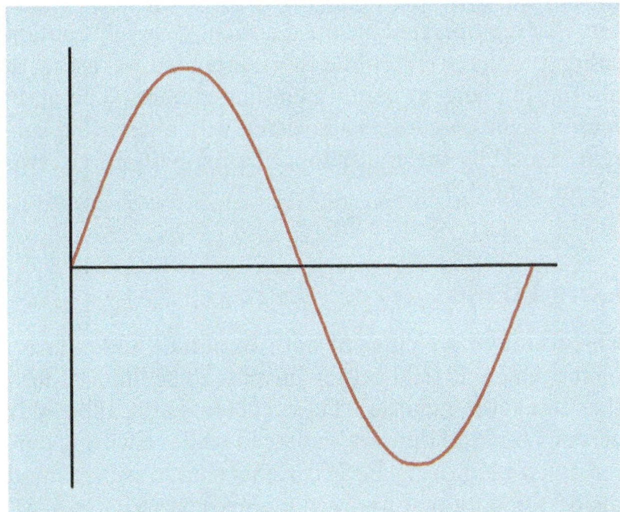

Figure 5–9 The current in an AC circuit continually changes amplitude and direction.

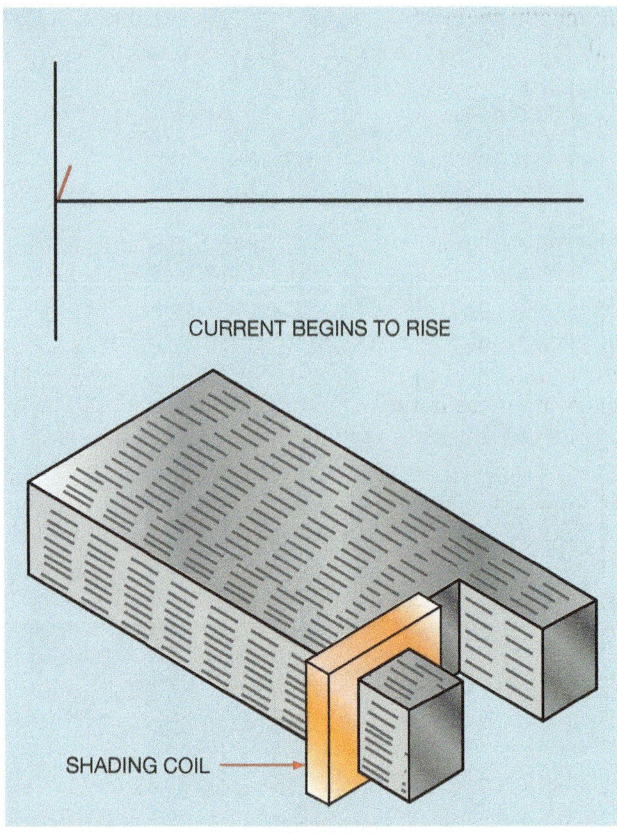

Figure 5–10 As current begins to rise, a magnetic field is concentrated in the pole piece.

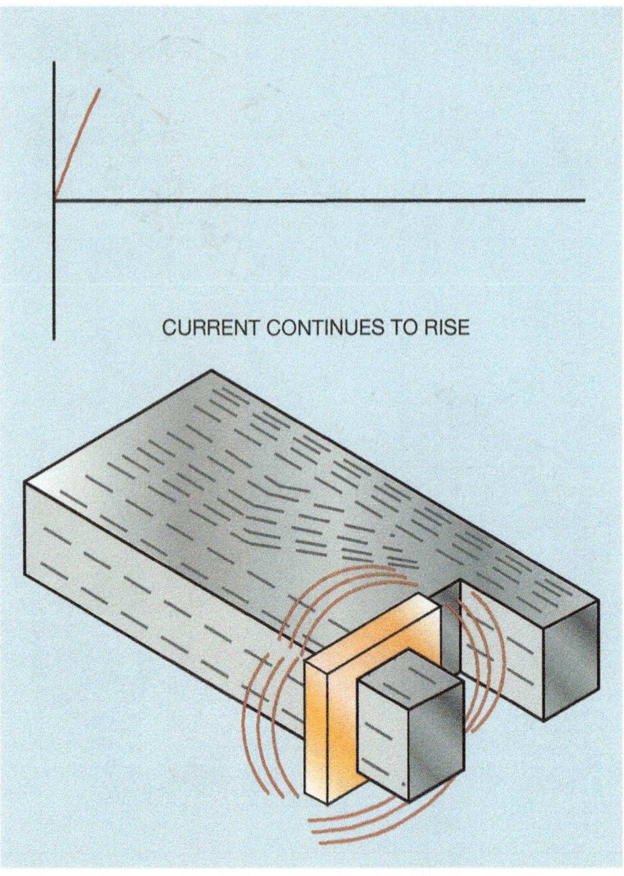

Figure 5–11 The magnetic field of the shading coil causes the magnetic field of the pole piece to bend away and concentrate in the unshaded portion of the pole piece.

has no current flow, there is no magnetic field to oppose the magnetic field of the pole piece (Figure 5–12).

When the current begins to decrease, the magnetic field of the pole piece begins to collapse. The collapsing magnetic field again induces a voltage into the shading coil. Because the collapsing magnetic field is moving in the opposite direction, the voltage induced in the shading coil causes current to flow in the opposite direction producing a magnetic field of the opposite polarity around the shading coil. The magnetic field of the shading coil now tries to maintain the collapsing magnetic field of the pole piece (Figure 5–13). This collapse causes the magnetic flux lines of the pole piece to concentrate in the shaded part of the pole piece. The shading coil provides a continuous magnetic field to the pole piece, preventing the armature from dropping out. A laminated pole piece with shading coils is shown in Figure 5–14.

Control Relay Types

Control relays can be obtained in a variety of styles and types (Figure 5–15). Most have multiple sets of contacts and some are constructed in such a manner that their contacts can be set as either normally open or normally closed. This flexibility can be a great advantage. When a control circuit is being constructed, one relay may require three normally open contacts and one normally closed, while another may need two normally open and two normally closed contacts.

Relays that are designed to plug into 8 or 11 pin tube or relay sockets are very popular for many applications (Figure 5–16). These sockets are often referred to as tube sockets because they were originally used for making connection to vacuum tubes. Relays of this type are relatively inexpensive and replacement is fast and simple in the event of failure. Because the relays plug into a socket, the wiring is connected to the socket, not the relay. Replacement is a matter of removing the defective relay and plugging in a new one. Both 8 and 11 pin tube or relay sockets are shown in Figure 5–17. Both 8 and 11 pin relays can be obtained with different coil voltages. Coil voltages of 12 VDC, 24 VDC, 24 VAC, and 120 VAC are common. Their contact ratings generally range from 5 to 10 amperes depending on

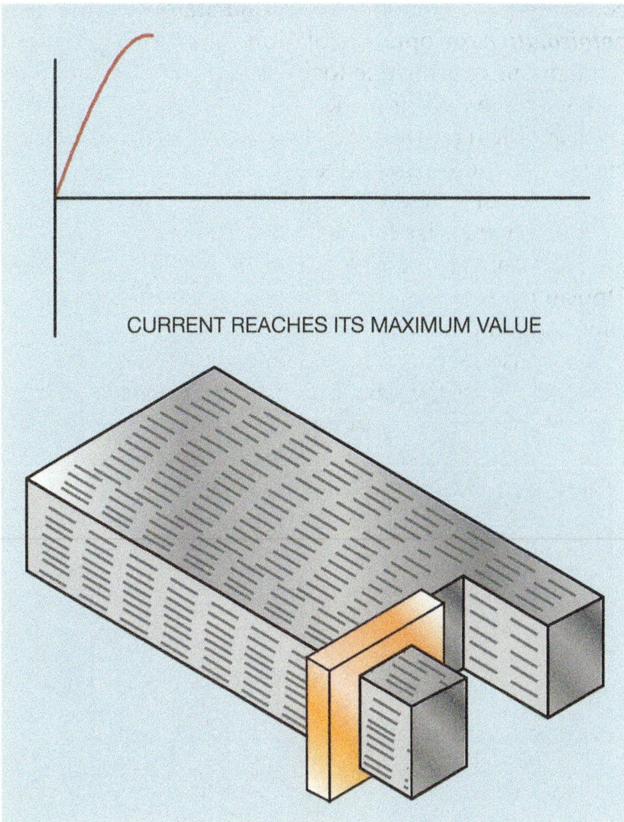

Figure 5–12 When the current reaches its maximum value, the magnetic field is no longer changing and the shading coil offers no resistance to the magnetic field of the pole piece.

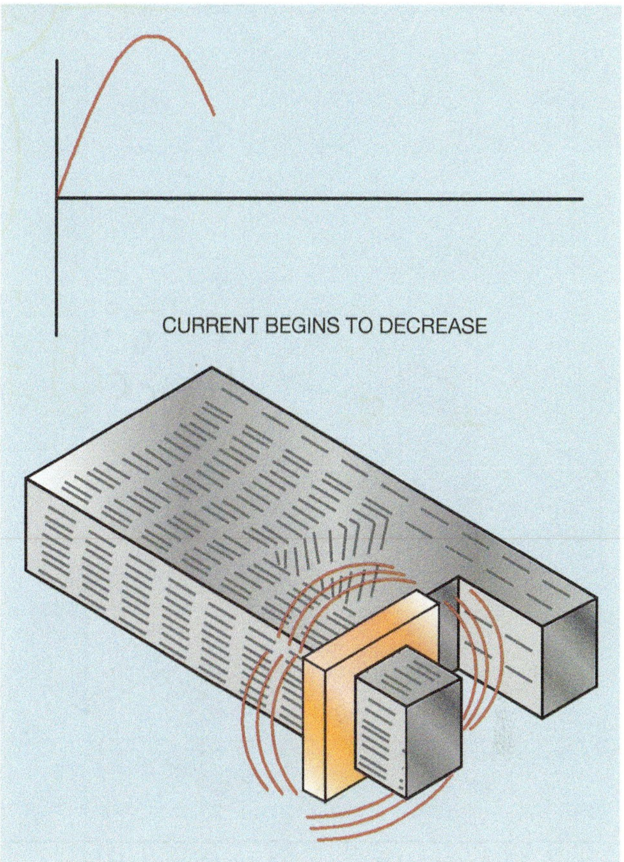

Figure 5–13 As current decreases, the collapsing magnetic field again induces a voltage into the shading coil. The shading coil now aids the magnetic field of the pole piece and flux lines are concentrated in the shaded section of the pole piece.

relay type and manufacturer. The connection diagram for 8 and 11 pin relays is shown in Figure 5–18. The pin numbers for 8 and 11 pin relays can be determined by holding the relay with the bottom facing you. Hold the relay so that the key is facing down. The pins are numbered as shown in Figure 5–18. The 11 pin relay contains three separate single pole double throw contacts. Pins 1 and 4, 6 and 5, and 11 and 8 are normally closed contacts. Pins 1 and 3, 6 and 7, and 11 and 9 are normally open contacts. The coil is connected to pins 2 and 10.

The 8 pin relay contains two separate single pole double throw contacts. Pins 1 and 4, and 8 and 5 are normally closed. Pins 1 and 3, and 8 and 6 are normally open. The coil is connected across pins 2 and 7.

Figure 5–14 Laminated pole piece with shading coils.

Solid-State Relays

Another type of relay that is found in many applications is the solid-state relay. Solid-state relays employ the use of **solid-state devices** to connect the load to the line instead of mechanical contacts. Solid-state relays that are intended to connect alternating current loads to the line use a device called a **triac**. The triac is bidirectional device, which means it permits current to flow through it in either direction. A couple of methods are used to control when the triac turns on or off. One method employs a small relay device that controls the gate of the triac (Figure 5–19). The relay can be controlled by a low voltage source. When energized, the relay contact closes and supplies power to the gate of the triac, which connects the load to the line. Another common method for

Figure 5–15 Control relays can be obtained in a variety of case styles.

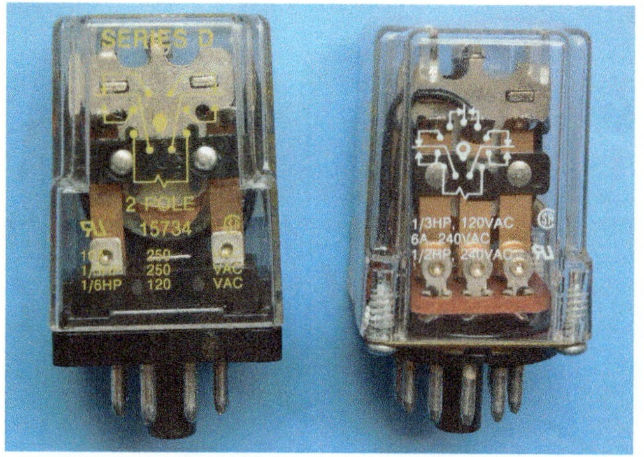

Figure 5–16 Relays designed to plug into 8 and 11 pin tube sockets.

controlling the operation of a solid-state relay is called *optoisolation*, or optical isolation. This method is used by many programmable logic controllers to communicate with the output device. Optoisolation is achieved by using the light from a light-emitting diode (LED) to energize a photo triac (Figure 5–20). The arrows pointing away from the diode symbol indicate that it emits light when energized. The arrows pointing toward the triac symbol indicate that it must receive light to turn on. Optical isolation is very popular with electronic devices such as computers and programmable logic controllers because there are no moving contacts to wear and the load side of the relay is electrically isolated from the control side. This isolation prevents any electrical noise generated on the load side from being transferred to the control side.

Solid-state relays can also be obtained that control loads connected to direct current circuits (Figure 5–21).

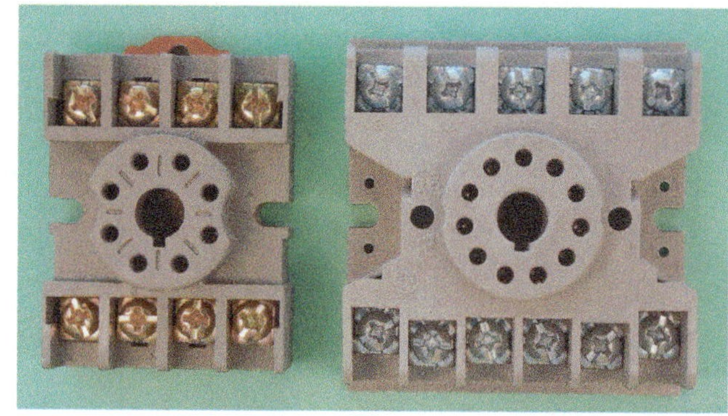

Figure 5–17 8 and 11 pin sockets.

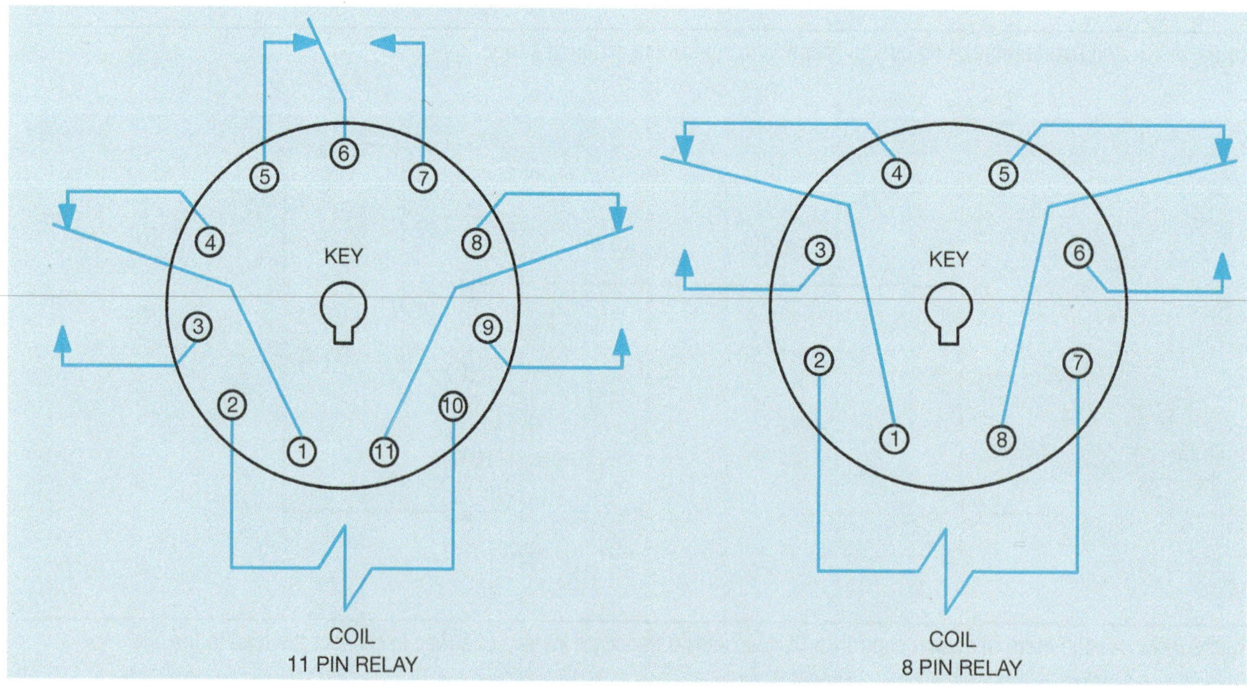

Figure 5–18 Connection diagrams for 8 and 11 pin relays.

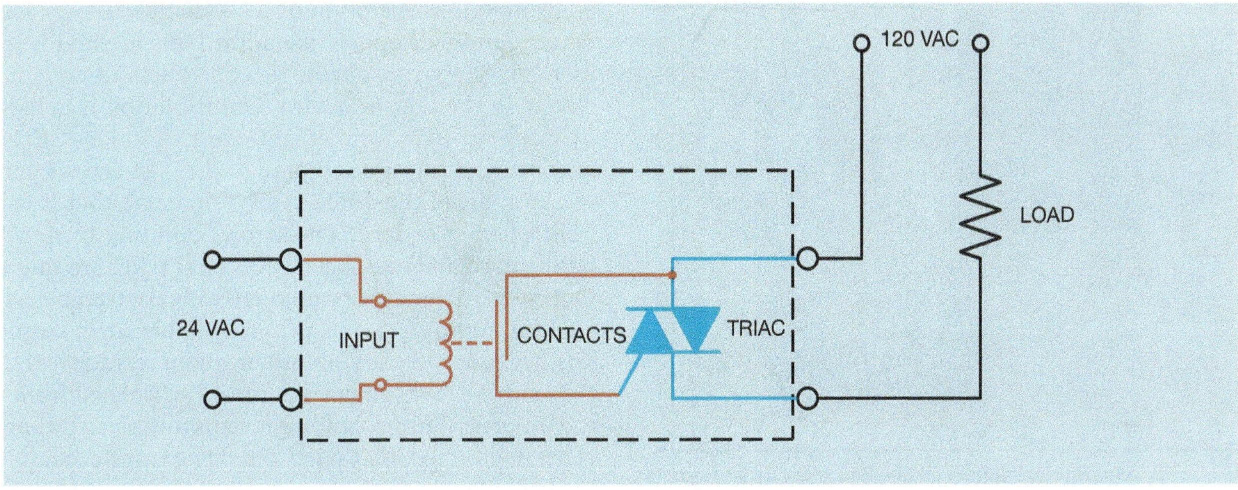

Figure 5–19 Solid-state relay using a relay to control the action of a triac.

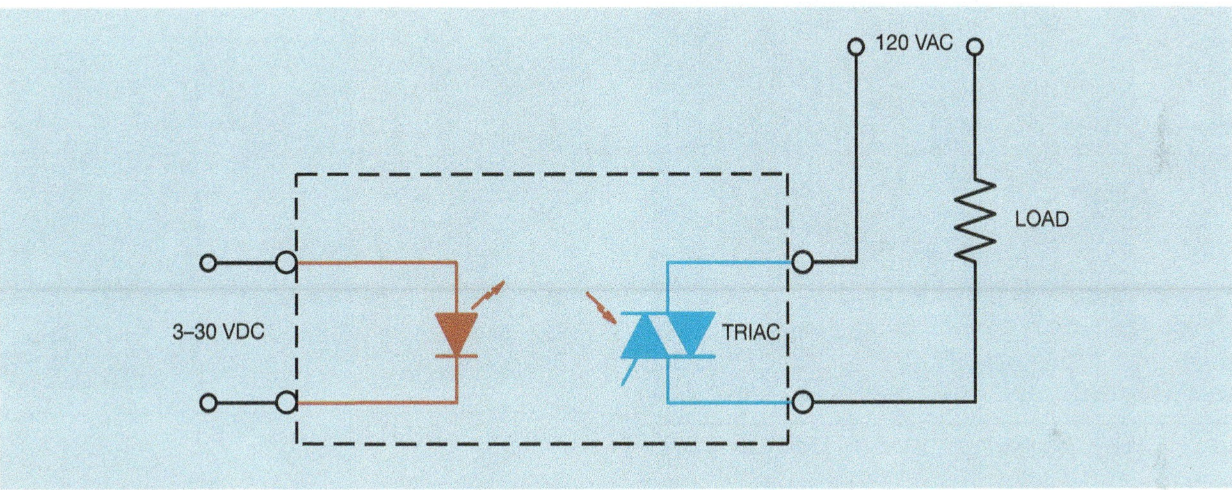

Figure 5–20 Solid-state relay using optical isolation to control the action of a triac.

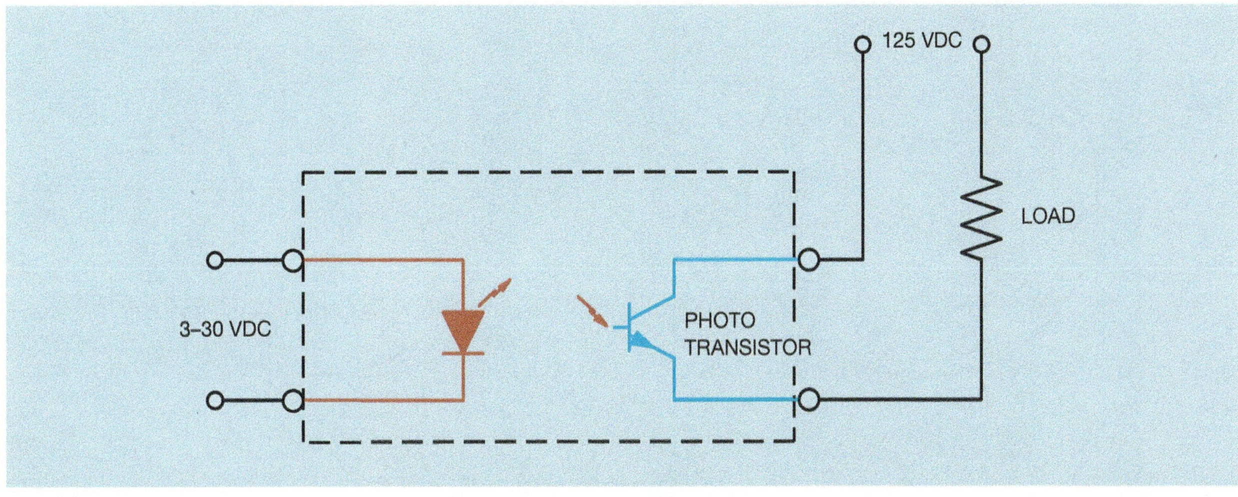

Figure 5–21 A solid-state relay that controls a DC load uses a transistor instead of a triac to connect the load to the line.

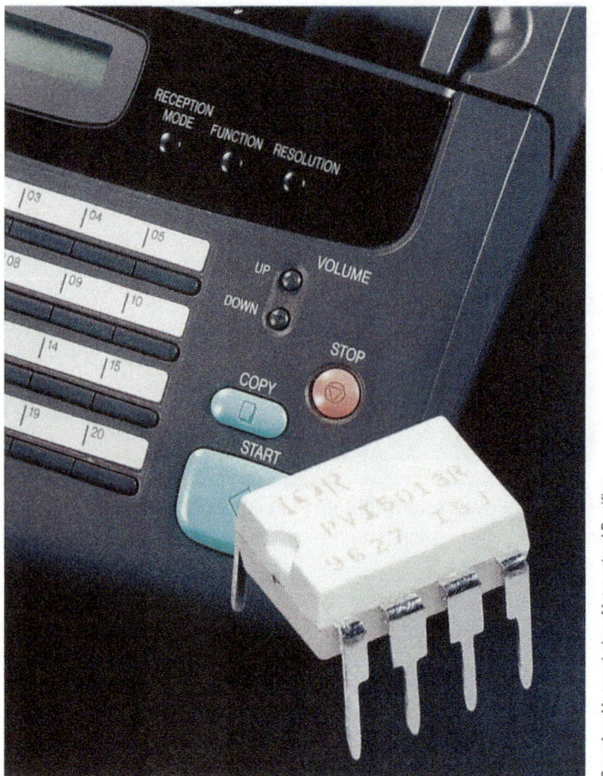

Figure 5–22 An 8 pin integrated circuit containing two low power solid-state relays.

These relays use a **transistor** to connect the load to the line instead of a triac.

Solid-state relays can be obtained in a variety of case styles and ratings. Some have voltage ratings that range from about 3 to 30 volts and can control only a small amount of current, while others can control hundreds of volts and several amperes. The 8 pin IC (integrated circuit) shown in Figure 5–22 contains two solid-state relays that are intended for low power applications. The solid-state relay shown in Figure 5–23 is rated to control a load of 8 amperes connected to a 240 volt AC circuit. For this solid-state relay to be capable of controlling that amount of power, it must be mounted on a heat sink to increase its ability to dissipate heat. Although this relay is rated 240 volts, it can control devices at a lower voltage also.

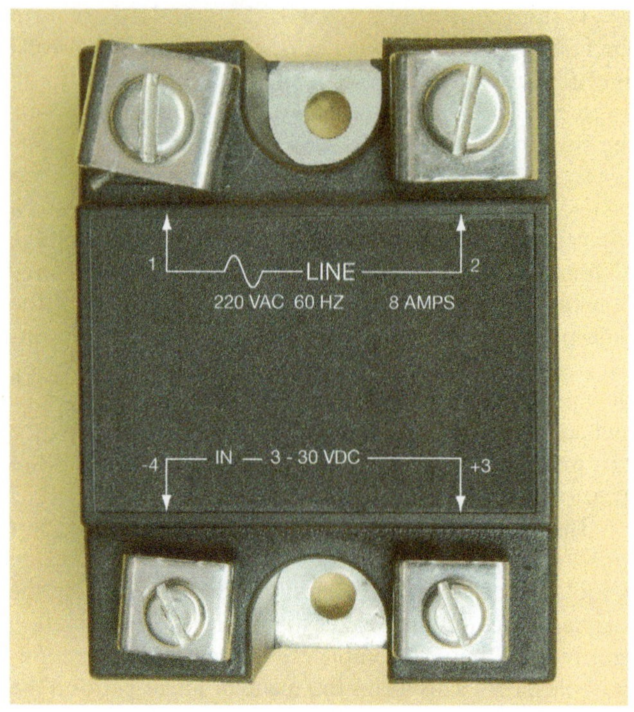

Figure 5–23 Solid-state relay that can control 8 amperes at 240 volts.

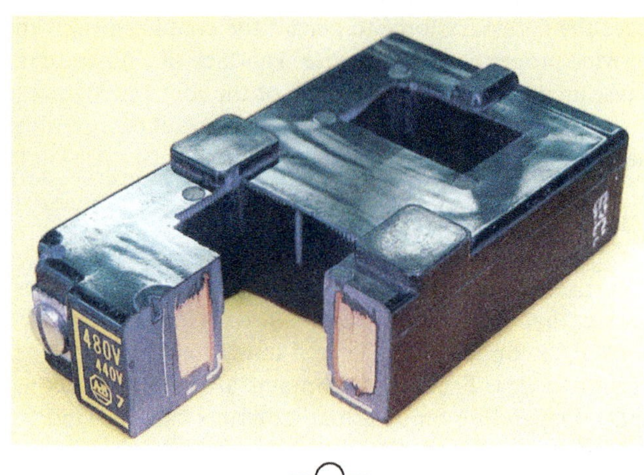

ELECTRICAL SYMBOL FOR COIL

Figure 5–24 Magnetic coil cut away to show insulated copper wire wound on a spool and protected by a molding.

Contactors

Contactors are very similar to relays in that they are electromechanical devices. Contactors can be obtained with coils designed for use on higher voltages than most relays. Most relay coils are intended to operate on voltages that range from 5 to 120 volts AC or DC. Contactors can be obtained with coils that have voltage ranges from 24 volts to 600 volts. Although these higher voltage coils can be obtained, for safety reasons, most contactors operate on voltages that generally do not exceed 120 volts. Contactors can be made to operate on different control circuit voltages by changing the coil. Manufacturers make coils to interchange with specific types of contactors. Most contain many turns of wire and are mounted in some type of molded case that can be replaced by dissembling the contactor (Figure 5–24).

It should be noted that NEMA standards require the magnetic switch device to operate properly on voltages that range from 85 percent to 110 percent of the rated coil voltage. Voltages can vary from one part of the country to another, as well as the variation of voltage that often occurs inside a plant. If coil voltage is excessive, it draws too much current causing the insulation to overheat and eventually burn out. Excessive voltage also causes the armature to slam into the stationary pole pieces with a force that can cause rapid wear of the pole pieces and shorten the life to the contactor. Another effect of too much voltage is the wear caused by the movable contacts slamming into the stationary contacts causing excessive contact bounce. Contact bounce can produce arcing that creates more heat and more wear on the contacts.

Insufficient coil voltage can produce as much, if not more, damage than excessive voltage. If the coil voltage is too low, the coil will have less current flow causing the magnetic circuit to be weaker than normal. The armature may pick up, but not completely seal against, the stationary pole pieces. This forms too much of an air gap and the coil current does not drop to its sealed value, causing excessive coil current, overheating, and coil burn out. A weak magnetic circuit can cause the movable contacts to touch the stationary contacts and provide a connection, but not have the necessary force to permit the contact springs to provide proper contact pressure. This lack of pressure can cause arcing and possible welding of the contacts. Without proper contact pressure, high currents produce excessive heat and greatly shorten the life of the contacts.

Load Contacts

The greatest difference between relays and contactors is that contactors are equipped with contacts intended to connect high current loads to the power line (Figure 5–25). These large contacts are called *load* contacts. Depending on size, load contacts can be rated to control several hundred amperes. Most will contain some type of arcing chamber to help extinguish the arc that is produced when heavy current loads are disconnected from the power line. Arcing chambers can be seen in Figure 5–25.

Other contacts may contain arc chutes that lengthen the path of the arc to help extinguish it. When the contacts open, the established arc rises because of the heat the arc produces (Figure 5–26). The horn of the arc chute pulls the arc farther and farther apart until it can no longer sustain itself. Another device that operates on a similar principle is the **blowout coil.** Blowout coils are connected in series with the load (Figure 5–27). When the contact opens, the arc is attracted to the magnetic field and rises at a rapid rate. This same basic action causes the armature of a direct current motor to turn. Because the arc is actually a flow or current, a magnetic field exists around the arc. The arc's magnetic field is

Figure 5–25 Contactors contain load contacts designed to connect high current loads to the power line.

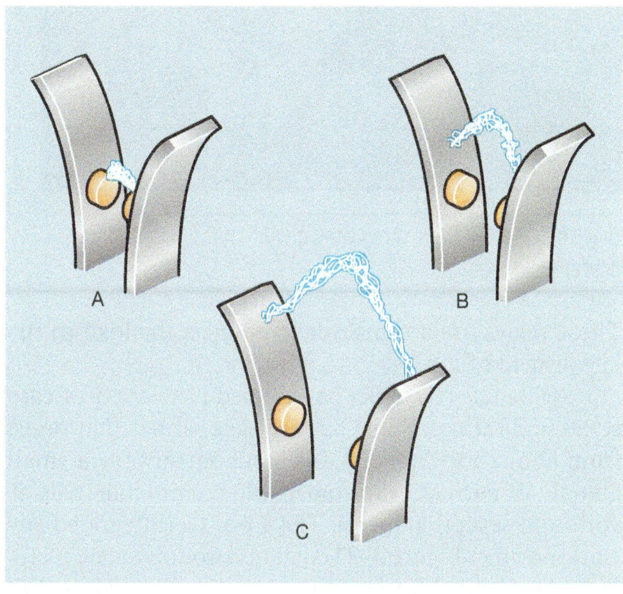

Figure 5–26 The arc rises between the arc chutes because of heat.

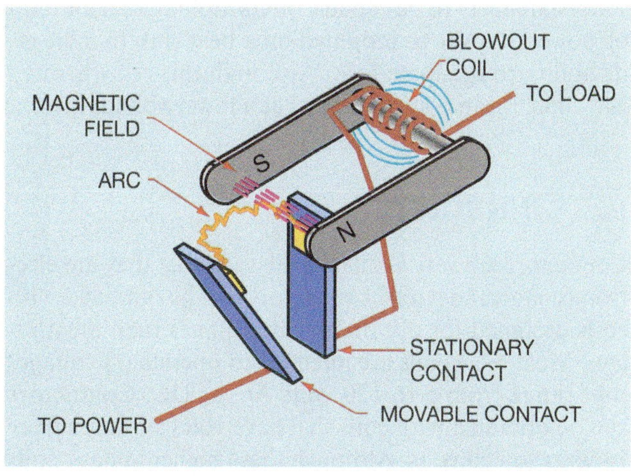

Figure 5–27 Magnetic blowout coils are connected in series with the load to establish a magnetic field.

attracted to the magnetic field produced by the blowout coil causing it to move upward. The arc is extinguished at a faster rate than is possible with an arc chute that depends on heat to draw the arc upward. Blowout coils are sometimes used on contactors that control large amounts of alternating current, but they are most often employed with contactors that control direct current loads. Alternating current turns off each half cycle when the waveform passes through zero, which helps to extinguish arcs in alternating current circuit. Direct current, however, does not turn off at periodic intervals. Once a DC arc is established, it is much more difficult to extinguish. Blowout coils are an effective means of extinguishing these arcs. A contactor with a blowout coil is shown in Figure 5–28.

Most contactors contain auxiliary contacts as well as load contacts. The auxiliary contacts can be used in the control circuit if required. The circuit shown in Figure 5–29 uses a three pole contactor to connect a bank of three phase heaters to the power line. Note that a normally open auxiliary contact is used to control an amber pilot light that indicates that the heaters are turned on, and a normally closed contact controls a red pilot light that indicates that the heaters are turned off. A thermostat controls the action of HR contactor coil. In the normal de-energized state, the normally closed HR auxiliary contact provides power to the red pilot light. When the thermostat contact closes, coil HR energizes and all HR contacts change position. The three load contacts close and connect the heaters to the line. The normally closed HR auxiliary contact opens and turns off the red pilot light, and the normally open HR auxiliary contact closes and turns on the amber pilot light. A size 1 contactor with auxiliary contacts is shown in Figure 5–30.

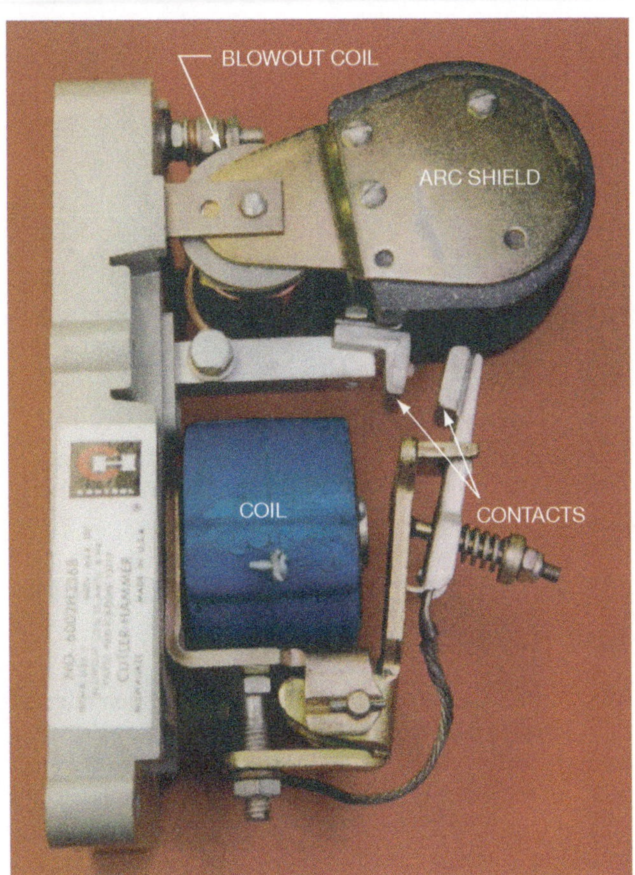

Figure 5–28 Clapper type contactor with blowout coil.

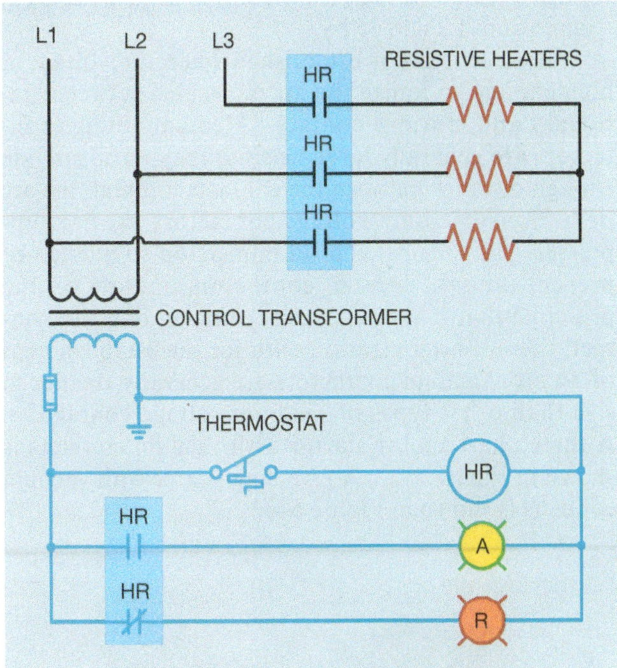

Figure 5–29 The contactor contains both load and auxiliary contacts.

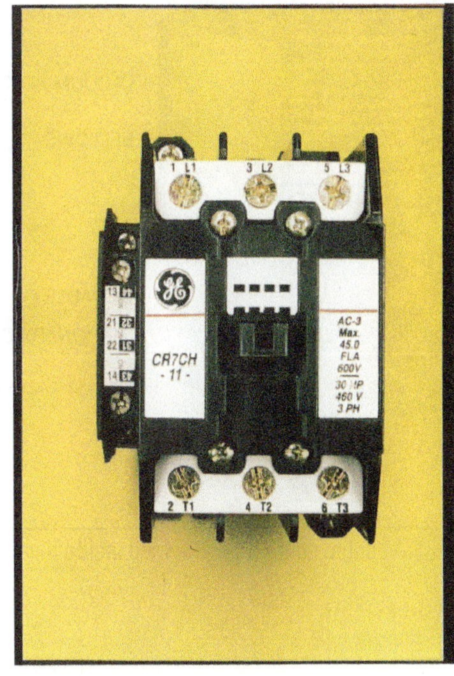

Figure 5–30 Size 1 contactor with auxiliary contacts.

Vacuum Contactors

Vacuum contactors enclose their load contacts in a sealed vacuum chamber. A metal bellows connected to the movable contact permits it to move without breaking the seal (Figure 5–31). Sealing contacts inside a vacuum chamber permits them to switch higher voltages with a relative narrow space between the contacts without establishing an arc. Vacuum contactors are generally employed for controlling devices connected to medium voltage. Medium voltage is generally considered to be in a range from 1 kV to 35 kV.

An electric arc is established when the voltage is high enough to ionize the air molecules between stationary and movable contacts. Medium voltage contactors are generally large because they must provide enough distance between the contacts to break the arc path. Some medium voltage contactors use arc suppressers, arc shields, and oil immersion to quench or prevent an arc. Vacuum contactors operate on the principle that if there is no air surrounding the contact, there is no ionization path for the establishment of an arc. Vacuum contactors are generally smaller in size than other types of medium voltage contactors. A three phase motor starter with vacuum contacts is shown in Figure 5–32. A reversing starter with vacuum contacts is shown in Figure 5–33.

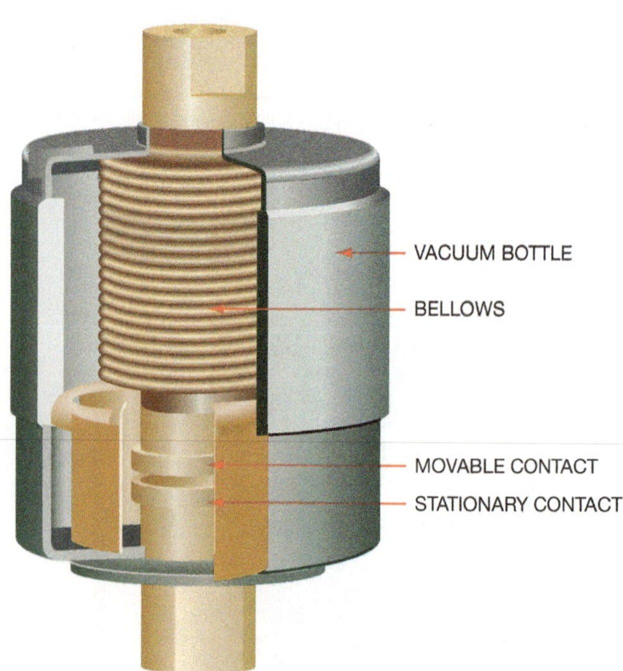

Figure 5–31 Vacuum contacts are sealed inside a vacuum chamber.

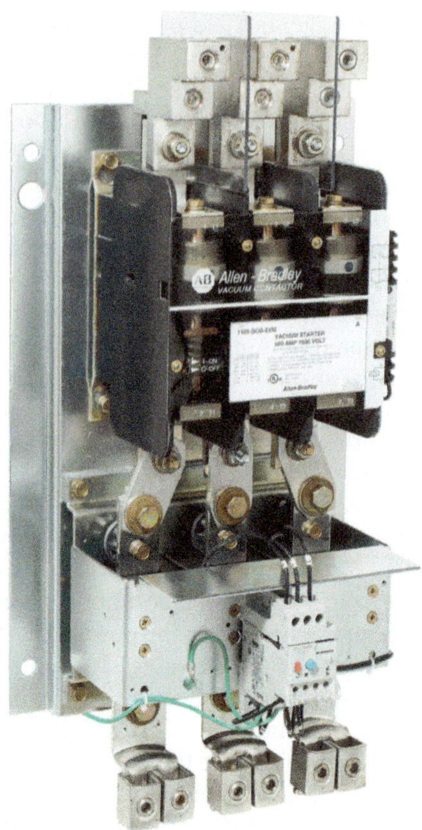

Figure 5–32 Three phase motor starter with vacuum contacts.

Courtesy of Rockwell Automation, Inc.

Figure 5–33 Reversing starter with vacuum contacts.

Mechanically Held Contactors and Relays

Mechanically held contactors and relays are often referred to as **latching** contactors or relays. They employ two electromagnets to operate. One coil is generally called the *latch* coil and the other is called the *unlatch* coil. The latch coil causes the contacts to change position and mechanically hold in position after power is removed from the latch coil. To return the contacts to their normal de-energized position, the unlatch coil must be energized. Power to both coils is momentary. The coils of most mechanically held contactors and relays are intended for momentary use and continuous power often cause burnout.

Unlike common magnetic contactors or relays, the contacts of latching relays and contacts do not return to a normal position if power is interrupted. They should be used only where there is danger of persons or equipment being harmed if power is suddenly restored after a power failure.

Sequence of Operation

Many latching type relays and contactors contain contacts that are used to prevent continuous power being supplied to the coil after it has been energized. These contacts are generally called **coil clearing contacts**. In Figure 5–34, coil L is the latching coil and coil U is the unlatch coil. When the ON push button is pressed, current can flow to coil L, through normally closed contact L to neutral. When the relay changes to the latch position, the normally closed L contact connected in series with L coil opens and disconnects power to L coil. This disconnection prevents further power from being supplied to coil L. At the same time, the open U contact connected in series with coil U closes to permit operation of coil U when the OFF push button is pressed. When coil L energized, it also closed the L load contacts, energizing a bank of lamps. The lamps can be turned off by pressing the OFF push button and energizing coil U. This process causes the relay to return to the normal position. Notice that the coil clearing contacts prevent power from being supplied continuously to the coils of the mechanically held relay.

An 11 pin latching relay is shown in Figure 5–35. The connection diagram for this relay is shown in Figure 5–36.

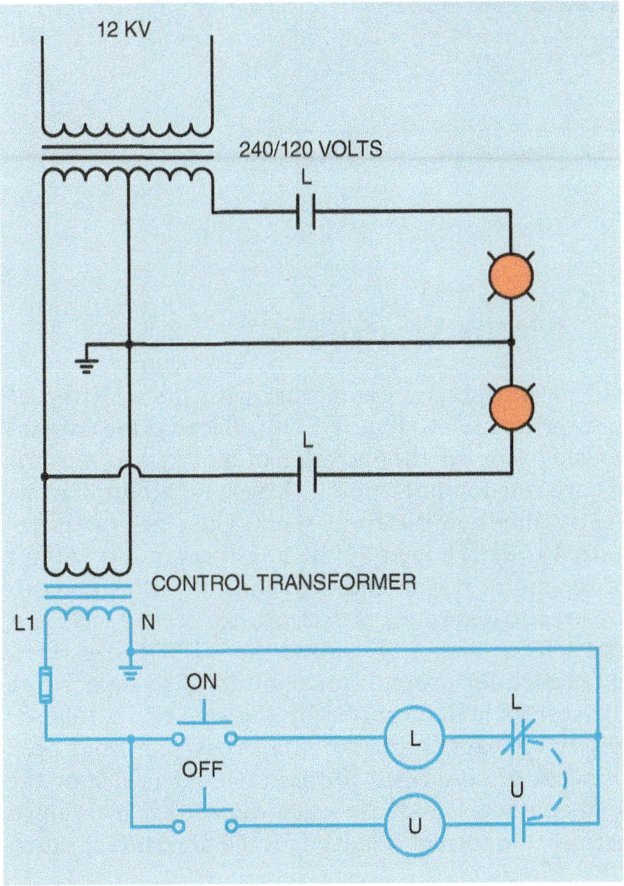

Figure 5–34 Latching type relays and contactors contain a latch and unlatch coil.

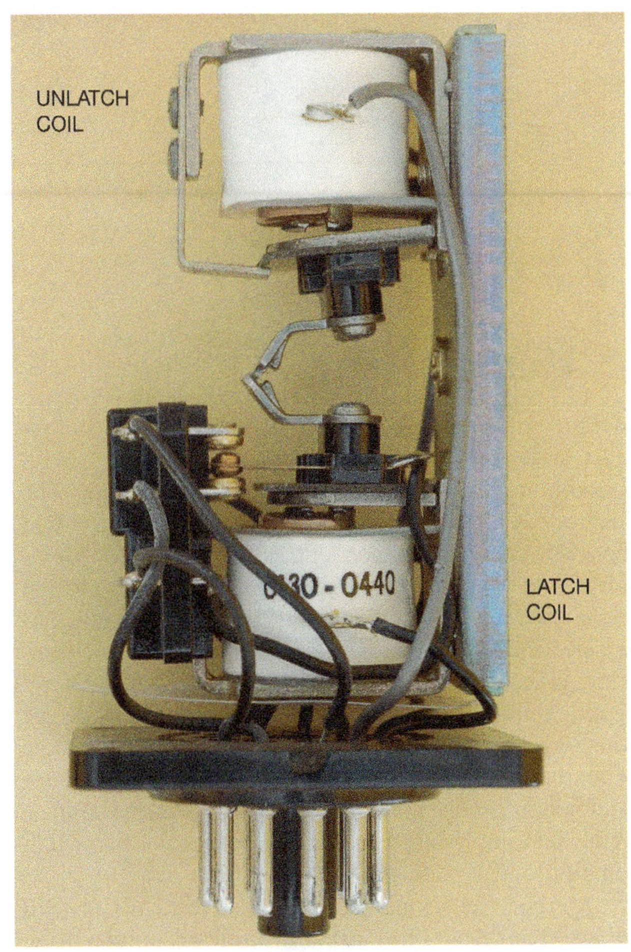

Figure 5–35 An 11 pin latching relay.

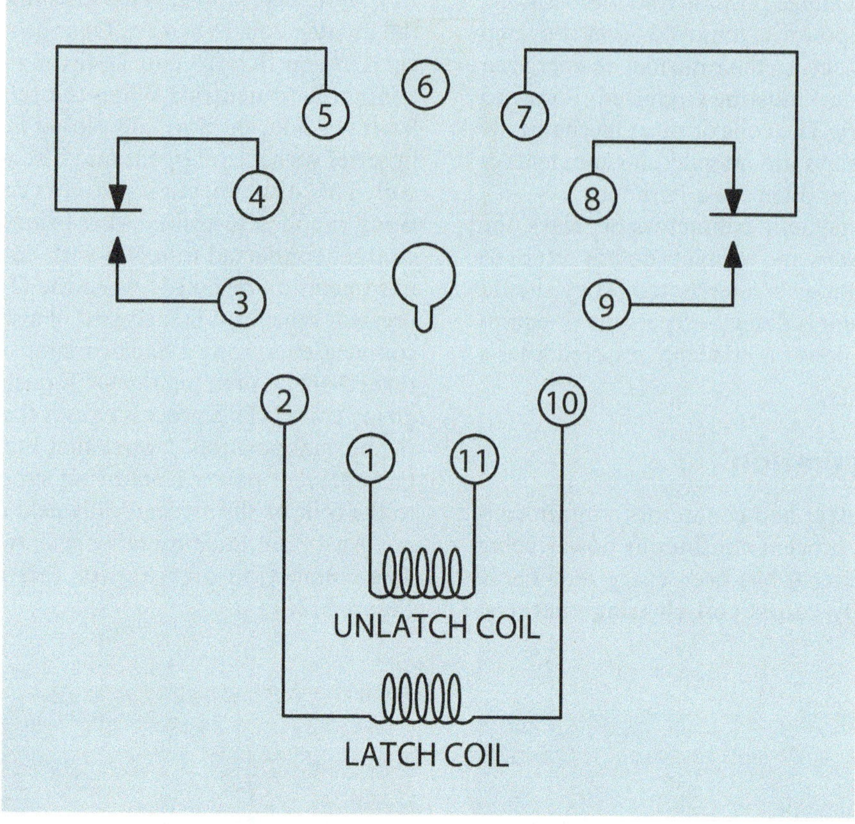

Figure 5–36 Connection diagram for an 11 pin latching relay.

Mercury Relays

Mercury relays employ the use of mercury-wetted contacts instead of mechanical contacts. Mercury relays contain one stationary contact called the electrode. The electrode is located inside the electrode chamber. When the coil is energized, a magnetic sleeve is pulled down inside a pool of liquid mercury, causing the mercury to rise in the chamber and make connection with the stationary electrode (Figure 5–37). The advantage of mercury relays is that each time the relay is used the contact is renewed, eliminating burning and pitting caused by an arc when connection is made or broken. The disadvantage of mercury relays is that they contain mercury. Mercury is a toxic substance that has been shown to cause damage to the nervous system and kidneys. Mercury is banned in some European countries.

Mercury relays must be mounted vertically instead of horizontally. They are available in single-pole, double-pole, and three-pole configurations. A single-pole mercury relay is shown in Figure 5–38.

Motor Starters

Motor starters are contactors with the addition of an overload relay (Figure 5–39). Because they are intended to control the operation of motors, motor starters are rated in horsepower. Magnetic motor starters are available in different sizes. The size starter required is determined by the horsepower and voltage of the motor it is intended to control. Two standards are used to determine the size starter needed—NEMA and IEC. Figure 5–40 shows the NEMA-size starters needed for normal starting duty. The capacity of the starter is determined by the size of its load or power contacts and the wire cross-sectional area that can be connected to the starter. The size of the load contacts is reduced when the voltage is doubled because the current is halved for the same power rating ($P = E \times I$).

The number of **poles** refers to the load contacts and does not include the number of control or auxiliary contacts. Three pole starters are used to control

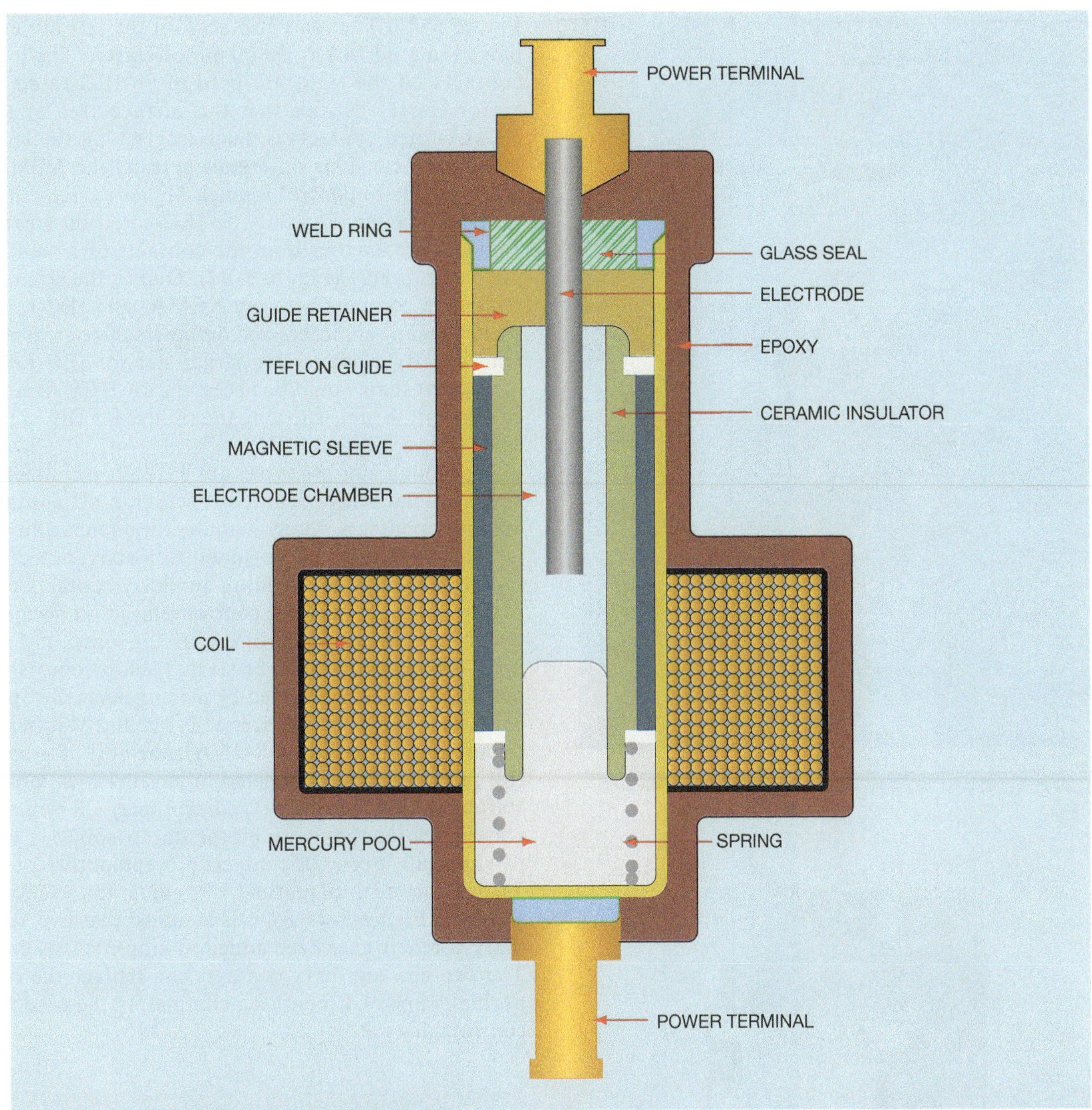

Figure 5–37 Diagram of a mercury relay.

three phase motors, and two pole starters are used for single phase motors.

NEMA and IEC

The IEC establishes standards and ratings for different types of equipment just as NEMA does. The IEC, however, is more widely used throughout Europe than in the United States. Many equipment manufacturers are now beginning to specify IEC standards for their products produced in the United States. The main reason is that much of the equipment produced in the United States is also marketed in Europe. Many European companies will not purchase equipment that is not designed with IEC standard equipment.

Although the IEC uses some of the same ratings as similar NEMA-rated equipment, there is often a vast difference in the physical characteristics of the two. Two sets of load contacts are shown in

Figure 5–38 Single-pole mercury relay.

Courtesy Schneider Electric USA, Inc.

Figure 5–39 A motor starter is a contactor combined with an overload relay. Reproduced by permission of International Rectifier.

Figure 5–41. The load contacts on the left are employed in a NEMA-rated 00 motor starter. The load contacts on the right are used in an IEC-rated 00 motor starter. Notice that the surface area of the NEMA-rated contacts is much larger than the IEC-rated contacts. This difference permits the NEMA-rated starter to control a much higher current than the IEC starter. In fact, the IEC-rated 00 starter contacts are smaller than the contacts of a small 8 pin control relay (Figure 5–42). Due to the size difference in contacts between NEMA- and IEC-rated starters, many engineers and designers of control systems specify a one to two larger size for IEC-rated equipment than would be necessary for NEMA-rated equipment. A table showing the ratings for IEC starters is shown in Figure 5–43.

Although motor starters are basically a contactor and overload relay mounted together, most contain auxiliary contacts. Many manufacturers make auxiliary contacts that can be added to a starter or contactor (Figure 5–44). Adding auxiliary contacts can often reduce the need for control relays that perform part of the circuit logic. In the circuit shown in Figure 5–45, motor #1 must be started before motors #2 or #3. This is accomplished by placing normally open contacts in series with starter coils M2 and M3. In the circuit shown in Figure 5–45(A), the coil of a control relay has been connected in parallel with motor starter coil M1. In this way, control relay CR will operate in conjunction with motor starter coil M1. The two normally open CR contacts prevent motors 2 and 3 from starting until motor 1 is running. In the circuit shown in Figure 5–45(B), it is assumed that two auxiliary contacts have been added to motor starter M1. The two new auxiliary contacts can replace the two normally open CR contacts, eliminating the need for control relay CR.

CAUTION: By necessity, motor control centers have very low impedance and can produce extremely large fault currents. It is estimated that the typical MCC can deliver enough energy in an arc-fault condition to kill a person 30 feet away. For this reason, many industries now require electricians to wear full protection (flame-retardant clothing, face shield, ear plugs, and hard hat) when opening the door on a combination starter or energizing the unit. When energizing the starter, always stand to the side of the unit and not directly in front of it. In a direct short condition, it is possible for the door to be blown off or open.

Maximum Horsepower Rating—Nonplugging and Nonjogging Duty				Maximum Horsepower Rating—Nonplugging and Nonjogging Duty			
NEMA Size	**Load Volts**	**Single Phase**	**Poly Phase**	**NEMA Size**	**Load Volts**	**Single Phase**	**Poly Phase**
	115	½	. . .		115	7½	. . .
	200	. . .	1½		200	. . .	25
	230	1	1½		230	15	30
00	380	. . .	1½	3	380	. . .	50
	460	. . .	2		460	. . .	50
	575	. . .	2		575	. . .	50
	115	1	. . .		200	. . .	40
	200	. . .	3		230	. . .	50
	230	2	3		380	. . .	75
0	380	. . .	5	4	460	. . .	100
	460	. . .	5		575	. . .	100
	575	. . .	5				
	115	2	. . .		200	. . .	75
	200	. . .	7½		230	. . .	100
	230	3	7½		380	. . .	150
1	380	. . .	10	5	460	. . .	200
	460	. . .	10		575	. . .	200
	575	. . .	10				
					200	. . .	150
	115	3	. . .		230	. . .	200
*1P	230	5	. . .	6	380	. . .	300
					460	. . .	400
					575	. . .	400
	115	3	. . .		230	. . .	300
	200	. . .	10	7	460	. . .	600
	230	7½	15		575	. . .	600
2	380	. . .	25		230	. . .	450
	460	. . .	25	8	460	. . .	900
	575	. . .	25		575	. . .	900

Tables are taken from NEMA Standards.

*1¾, 10 hp is available.

Figure 5–40 Motor starter sizes and ratings.

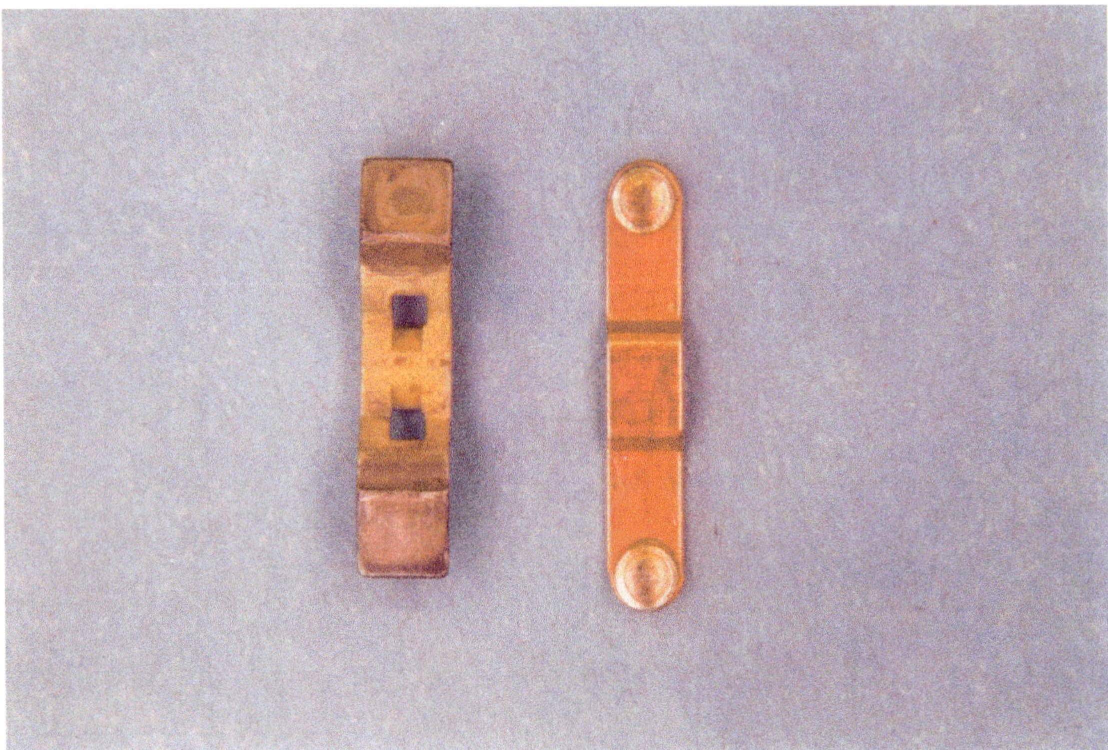

Figure 5–41 The load contacts on the left are NEMA size 00. The load contacts on the right are IEC size 00.

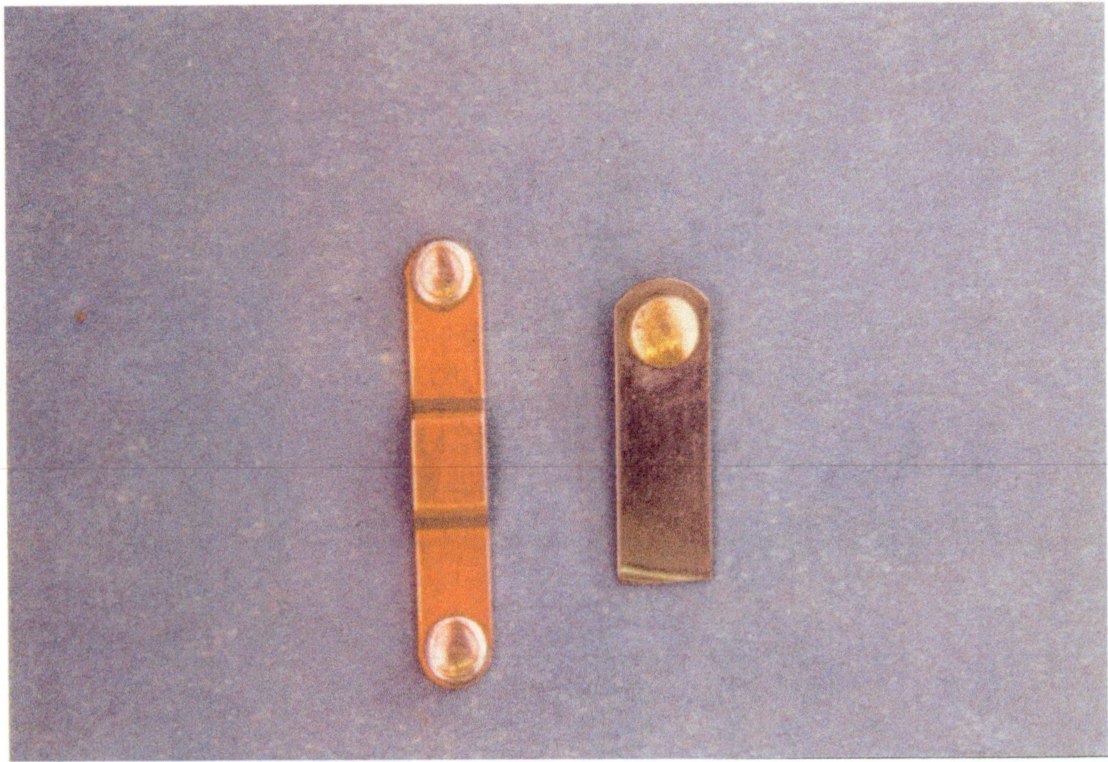

Figure 5–42 The load contacts of an IEC 00 starter shown on the left are smaller than the auxiliary contacts of an 8 pin control relay shown on the right.

IEC MOTOR STARTERS (60 HZ)

SIZE	MAX AMPS	MOTOR VOLTAGE	MAX. HORSEPOWER	
			SINGLE PHASE	THREE PHASE
A	7	115	1/4	
		200		1 1/2
		230	1/2	1 1/2
		460		3
		575		5
B	10	115	1/2	
		200		2
		230	1	2
		460		5
		575		7 1/2
C	12	115	1/2	
		200		3
		230	2	3
		460		7 1/2
		575		10
D	18	115	1	
		200		5
		230	3	5
		460		10
		575		15
E	25	115	2	
		200		5
		230	3	7 1/2
		460		15
		575		20
F	32	115	2	
		200		7 1/2
		230	5	10
		460		20
		575		25
G	37	115	3	
		200		7 1/2
		230	5	10
		460		25
		575		30
H	44	115	3	
		200		10
		230	7 1/2	15
		460		30
		575		40
J	60	115	5	
		200		15
		230	10	20
		460		40
		575		40
K	73	115	5	
		200		20
		230	10	25
		460		50
		575		50
L	85	115	7 1/2	
		200		25
		230	10	30
		460		60
		575		75
M	105	115	10	
		200		30
		230	10	40
		460		75
		575		100
N	140	115	10	
		200		40
		230	10	50
		460		100
		575		125
P	170	115		
		200		50
		230		60
		460		125
		575		125
R	200	115		
		200		60
		230		75
		460		150
		575		150
S	300	115		
		200		75
		230		100
		460		200
		575		200
T	420	115		
		200		125
		230		125
		460		250
		575		250
U	520	115		
		200		150
		230		150
		460		350
		575		250
V	550	115		
		200		150
		230		200
		460		400
		575		400
W	700	115		
		200		200
		230		250
		460		500
		575		500
X	810	115		
		200		250
		230		300
		460		600
		575		600
Z	1215	115		
		200		450
		230		450
		460		900
		575		900

Figure 5–43 IEC motor starters rated by size, horsepower, and voltage for 60 Hz circuits.

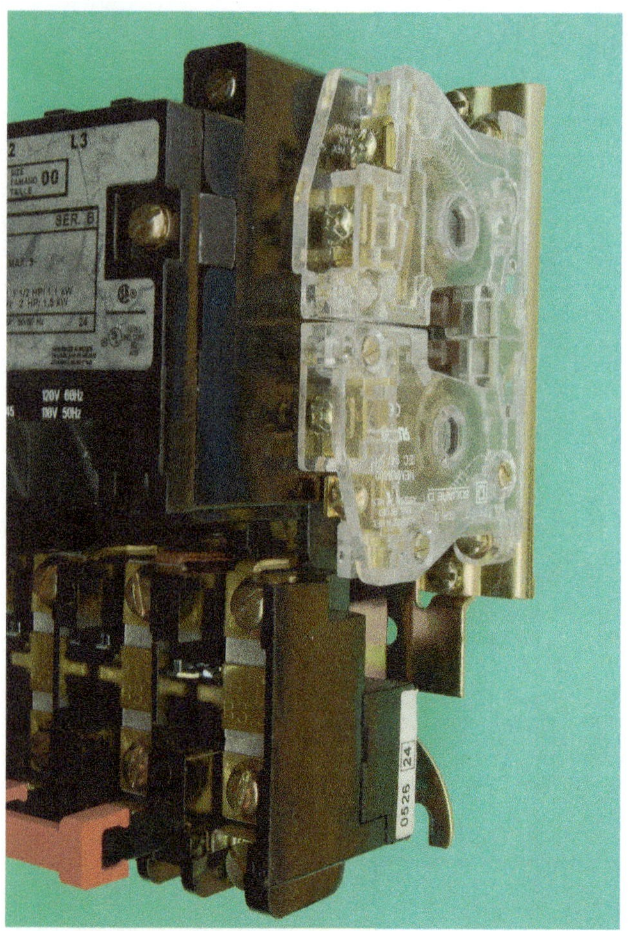

Figure 5–44 Motor starter with additional auxiliary contacts.

Motor Control Centers

Motor starters are often grouped with other devices, such as circuit breakers, fuses, disconnects, and control transformers, and referred to as a combination starter. These components are often contained inside one enclosure (Figure 5–46).

Motor control centers employ the use of combination starters that are mounted in special enclosures designed to plug into central buss bars that supply power for several motors. The enclosure for this type of combination starter is often referred to as a module, cubical, or can (Figure 5–47). They are designed to be inserted into a motor control center (MCC) (Figure 5–48). Connection to individual modules is generally made with terminal strips located inside the module. Most manufacturers provide some means of removing the entire terminal strip without having to remove each individual wire. If a starter should fail, this permits rapid installation of a new starter. The defective starter can then be serviced at a later time.

Current Requirements

When the coil of an alternating current relay or contactor is energized, it will require more current to pull the armature in than to hold it in. The reason is the change of inductive reactance caused by the air gap (Figure 5–49). When the relay is turned off, a large air gap exists between the metal of the stationary pole piece and the armature. This air gap causes a poor

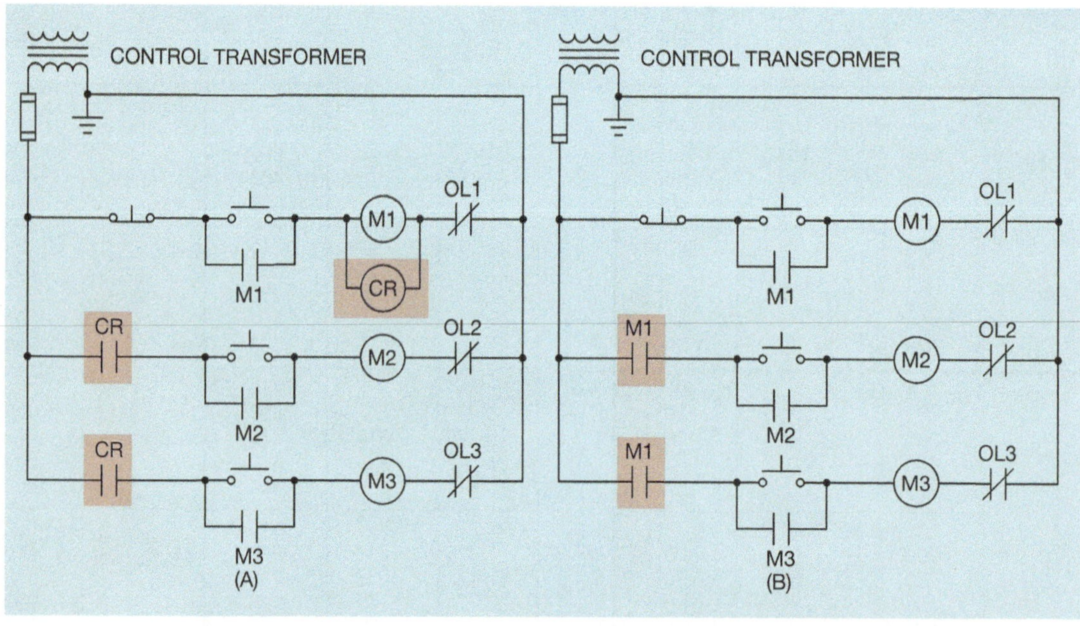

Figure 5–45 Control relays can sometimes be eliminated by adding auxiliary contacts to a motor starter.

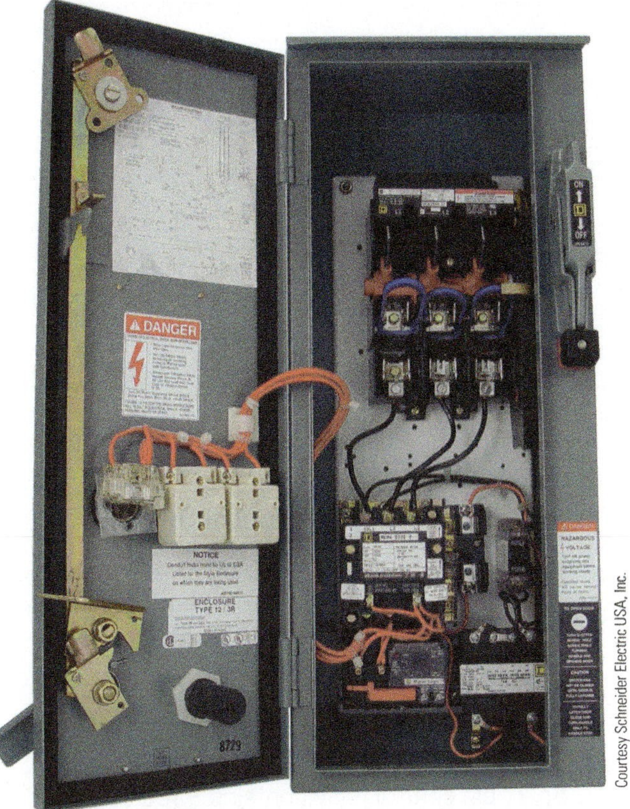

Courtesy Schneider Electric USA, Inc.

Figure 5–46 A combination starter with fused disconnect, control transformer, push buttons, and motor starter.

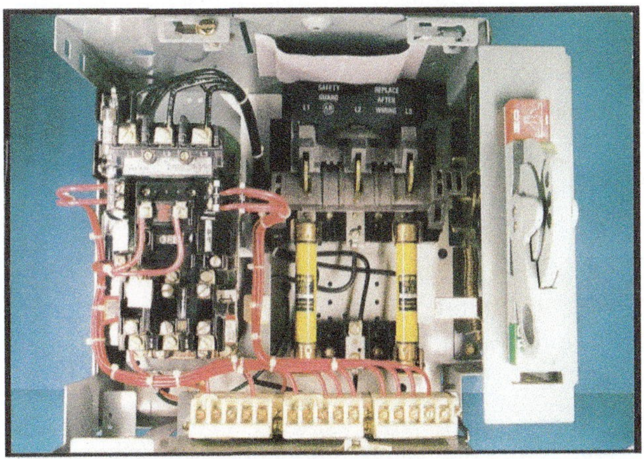

Figure 5–47 Combination starter with fused disconnect intended for use in a motor control center (MCC). Note that only two fuses are used in this module. Delta connected power systems with one phase grounded do not require a fuse in the grounded conductor.

magnetic circuit and the inductive reactance X_L has a low ohmic value. Although the wire used to make the coil does have some resistance, the main current limiting factor of an inductor is inductive reactance. After

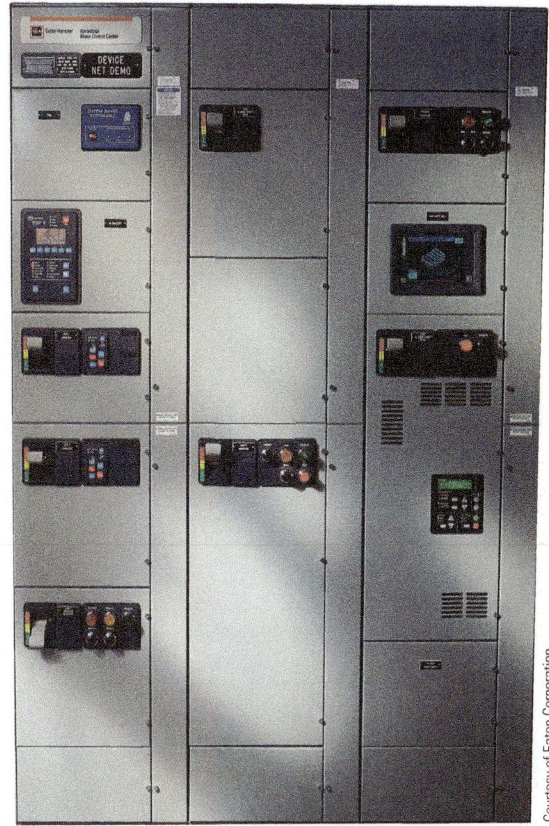

Courtesy of Eaton Corporation

Figure 5–48 Motor control center.

the coil is energized and the armature makes contact with the stationary pole piece, there is a very small air gap between the armature and pole piece. This small air gap permits a better magnetic circuit, which increases the inductive reactance causing the current to decrease. If dirt or some other foreign matter should prevent the armature from making a seal with the stationary pole piece, the coil current will remain higher than normal, which can cause overheating and eventual coil burn out.

Direct current relays and contactors depend on the resistance of the wire used to construct the coil to limit current flow. For this reason, the coils of DC relays and contactors exhibit a higher resistance than coils of AC relays. Large direct current contactors are often equipped with two coils instead of one (Figure 5–50). When the contactor is energized, both coils are connected in parallel to produce a strong magnetic field in the pole piece. A strong field is required to provide the attraction needed to attract the armature. Once the armature has been attracted, a much weaker magnetic field can hold the armature in place. When the armature closes, a switch disconnects one of the coils reducing the current to the contactor.

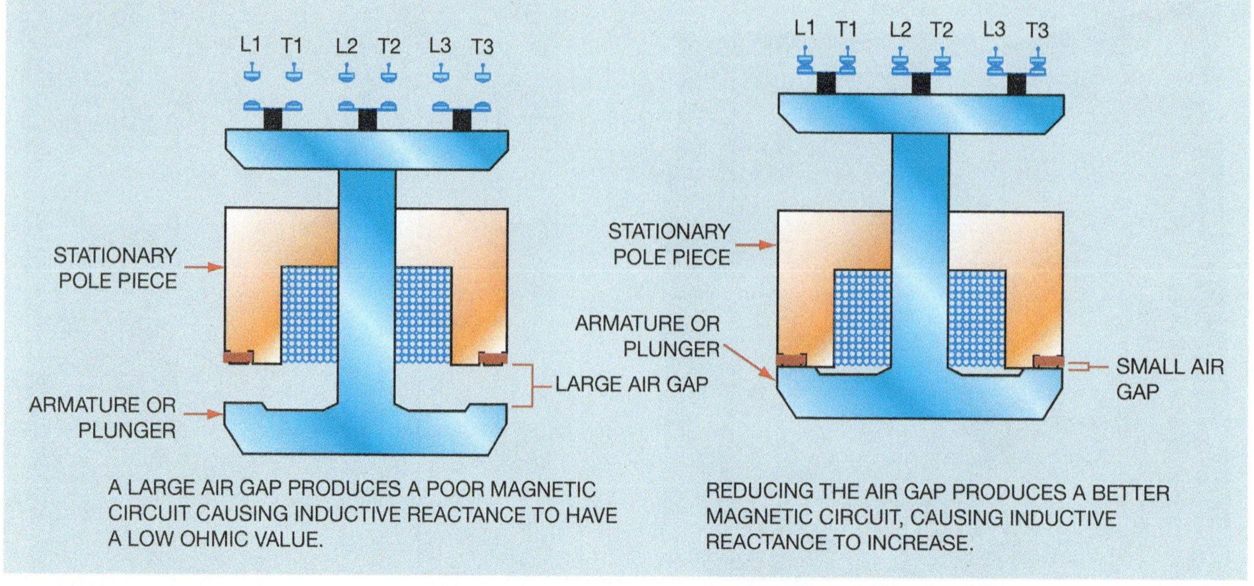

Figure 5–49 The air gap determines the inductive reactance of the solenoid.

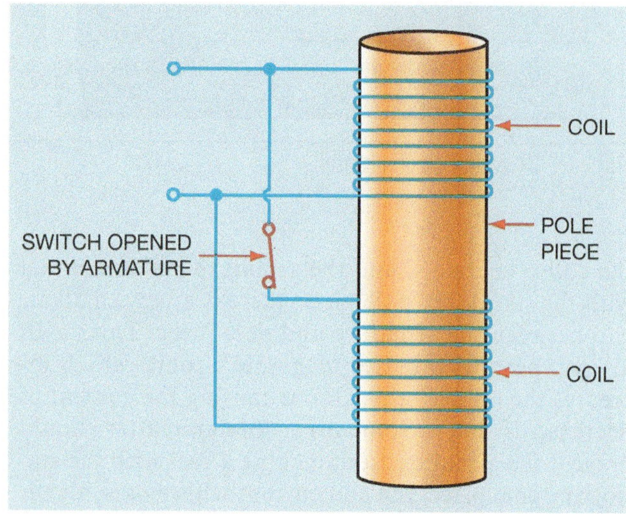

Figure 5–50 Direct current contactors often contain two coils.

Review Questions

1. Explain the difference between clapper type contacts and bridge type contacts.
2. What is the advantage of bridge type contacts over clapper type contacts?
3. Explain the difference between auxiliary contacts and load contacts.
4. What type of electronic device is used to connect the load to the line in a solid-state relay used to control an alternating current load?
5. What is optoisolation and what is its main advantage?
6. What pin numbers are connected to the coil of an 8 pin control relay?
7. An 11 pin control relay contains three sets of single pole double throw contacts. List the pin numbers by pairs that can be used as normally open contacts.
8. What is the purpose of the shading coil?
9. Refer to the circuit shown in Figure 5–29. Is the thermostat contact normally open, normally closed, normally closed held open, or normally open held closed?
10. What is the difference between a motor starter and a contactor?
11. A 150-horsepower motor is to be installed on a 480-volt three phase line. What is the minimum size NEMA starter that should be used for this installation?
12. What is the minimum size IEC starter rated for the motor described in question 11?
13. When energizing or de-energizing a combination starter what safety precaution should always be taken?
14. What is the purpose of coil clearing contacts?
15. Refer to the circuit shown in Figure 5–29. In this circuit, the HR contactor is equipped with five contacts. Three are load contacts and two are auxiliary contacts. From looking at the schematic diagram, how is it possible to identify which contacts are the load contacts and which are the auxiliary contacts?

Chapter 6

THE CONTROL TRANSFORMER

Objectives

After studying this chapter the student will be able to:

- Discuss the use of control transformers in a control circuit.
- Connect a control transformer for operation on a 240- or 480-volt system.

Most industrial motors operate on voltages that range from 240 to 480 volts. Magnetic control systems, however, generally operate on 120 volts. A control transformer is used to step the 240 or 480 volts down to 120 volts to operate the control system. There is really nothing special about a control transformer except that most of them are made with two primary windings and one secondary winding. Each primary winding is rated at 240 volts and the secondary winding is rated at 120 volts. This means there is a turns ratio of 2:1 (2 to 1) between each primary winding and the secondary winding. For example, assume that each primary winding contains 200 turns of wire and the secondary winding contains 100 turns. There are two turns of wire in each primary winding for every one turn of wire in the secondary.

One of the primary windings of the control transformer is labeled H1 and H2. The other primary winding is labeled H3 and H4. The secondary winding is labeled X1 and X2. If the transformer is to be used to step 240 volts down to 120 volts, the two primary windings are connected parallel to each other as shown in Figure 6–1. Notice that in Figure 6–1 the H1 and H3 leads are connected together, and the H2 and H4 leads are connected together. Because the voltage applied to each primary winding is same, the effect is the same as having only one primary winding with 200 turns of wire in it. This means that when the transformer is connected in this manner, the turns ratio is 2:1. When 240 volts are connected to the primary winding, the secondary voltage is 120 volts.

If the transformer is to be used to step 480 volts down to 120 volts, the primary windings are connected in series, as shown in Figure 6–2. With the windings connected in series, the primary winding now has a total of 400 turns of wire, which makes a turns ratio of 4:1. When 480 volts are connected to the primary winding, the secondary winding has an output of 120 volts.

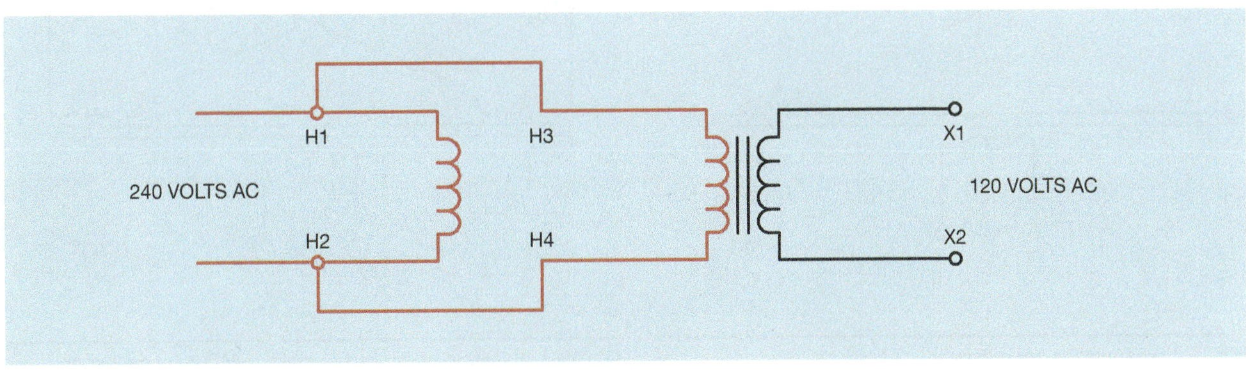

Figure 6–1 Primaries connected in parallel for 240-volt operation.

Control transformers generally have screw terminals connected to the primary and secondary leads. The H2 and H3 leads are crossed to make connection of the primary winding easier, Figure 6–3. For example, if the transformer is to be connected for 240-volt operation, the two primary windings must be connected parallel to each other as shown in Figure 6–1. This connection can be made on the transformer by using one metal link to connect leads H1 and H3, and another metal link to connect H2 and H4 (Figure 6–4).

If the transformer is to be used for 480-volt operation, the primary windings must be connected in series as shown in Figure 6–2. This connection can be made on the control transformer by using a metal link to connect H2 and H3 as shown in Figure 6–5. A typical control transformer is shown in Figure 6–6.

Multi-Tapped Control Transformers

Some control transformers provide connection to different voltages by providing different taps on the high voltage winding, Figure 6–7.

The transformer in this example can be connected to 208, 277, or 380 volts. The secondary or low voltage winding provides an output of 120 volts. The connection diagram for this transformer is shown in Figure 6–8.

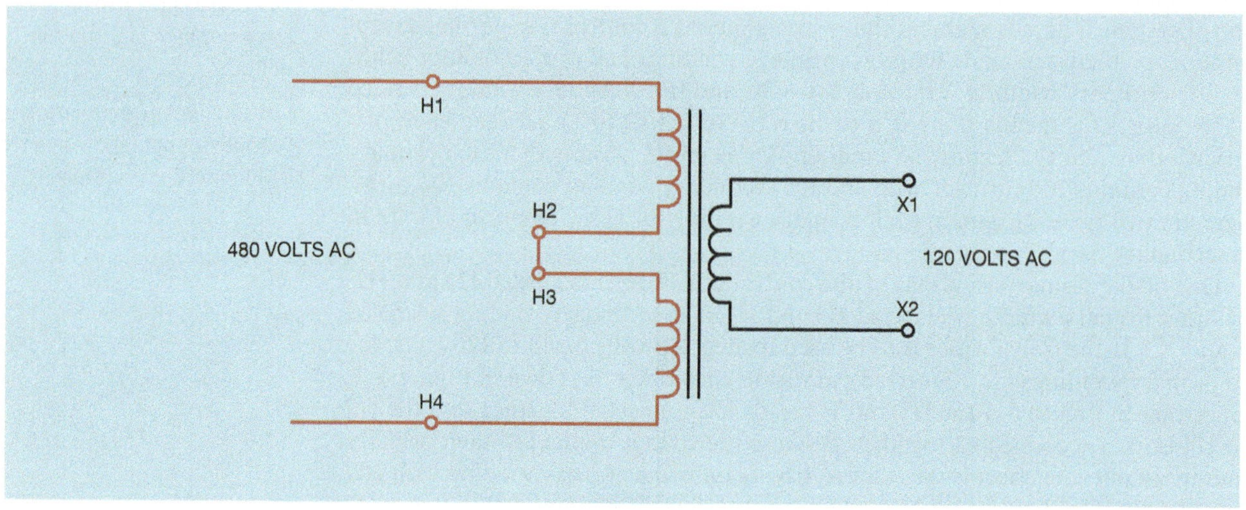

Figure 6–2 Primary windings connected in series for 480-volt operation.

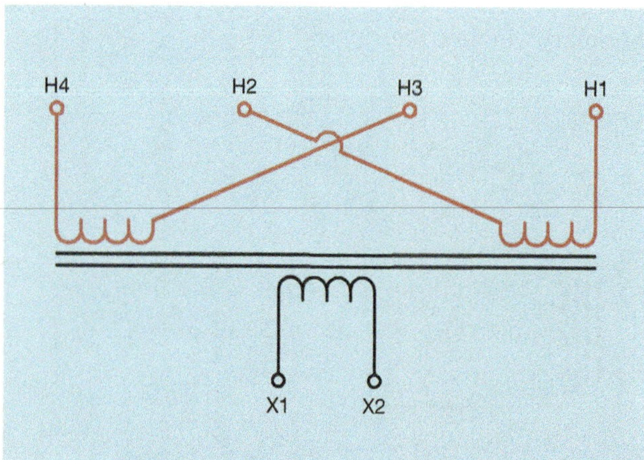

Figure 6–3 Primary leads are crossed.

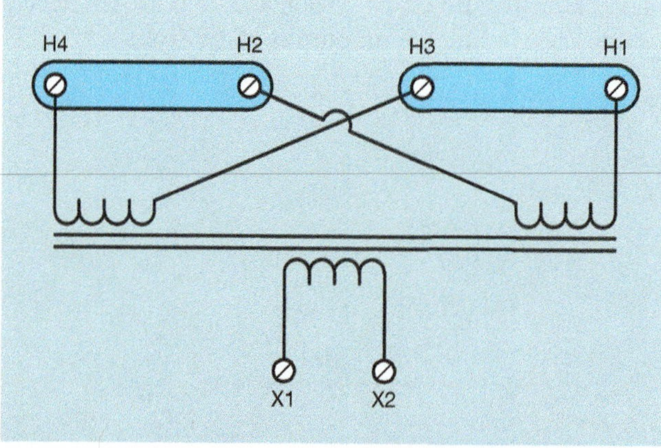

Figure 6–4 Metal links used to make a 240-volt connection.

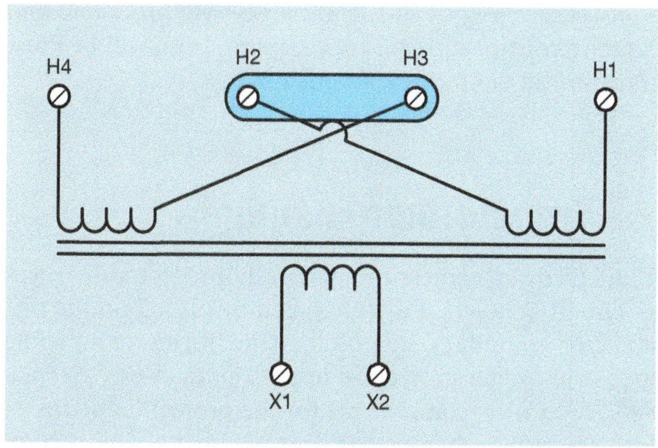

Figure 6–5 Metal link used to make a 480-volt connection.

Figure 6–6 Control transformer.

If the transformer is connected to 208 volts, primary taps H1 and H2 would be used. Taps H3 and H4 are left disconnected. If the transformer is to be connected to 277 volts, taps H1 and H3 would be used. Taps H2 and H4 would be left disconnected. Taps H1 and H4 can be connected to 380 volts, which is a common voltage used in Europe.

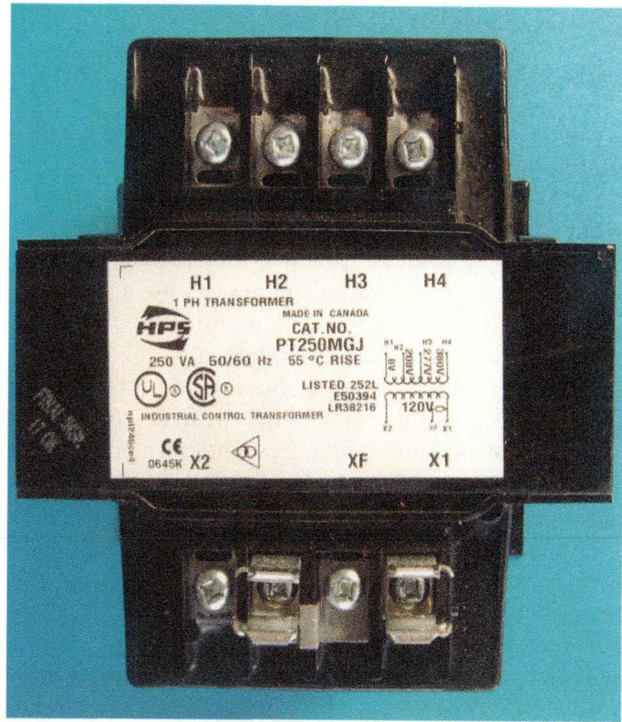

Figure 6-7 A multi-tapped control transformer.

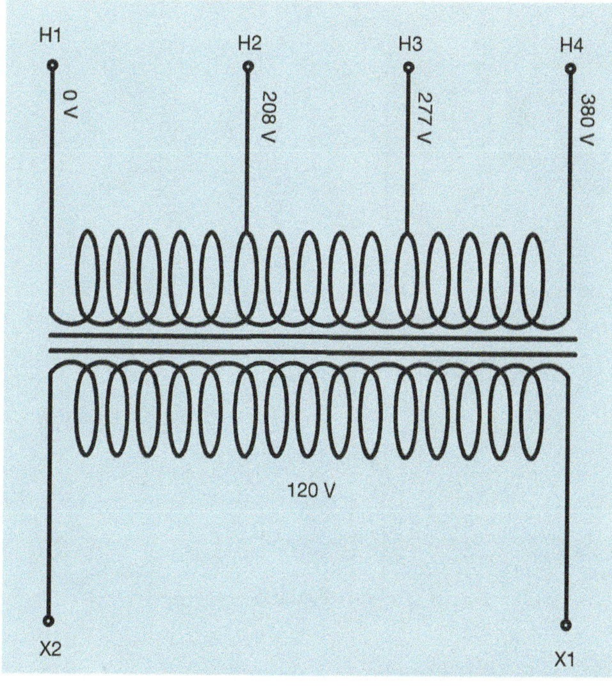

Figure 6-8 Diagram of a multi-tapped control transformer.

Grounded and Floating Control Systems

One side of the secondary winding of a control transformer is often grounded (Figure 6–9). When this is done, the control system is referred to as a **grounded system**. Many industries prefer to ground the control system and it is a very common practice. Some technicians believe that it is an aid when troubleshooting a problem. Grounding one side of the control transformer permits one lead of a voltmeter to be connected to any grounded point and the other voltmeter lead is used to test voltage at various locations throughout the circuit (Figure 6–10).

It is also a common practice to not ground one side of the control transformer. This is generally referred to as a **floating system**. If one voltmeter probe was connected to a grounded point, the meter reading would be erroneous or meaningless because there is not a complete circuit (Figure 6–11). High impedance voltmeters would probably indicate some amount of voltage caused by the capacitance of the ground and induced voltage produced by surrounding magnetic fields. These are generally referred to as **ghost voltages**. A low impedance meter such as a plunger type voltage tester would indicate no voltage. Accurate voltage measurement can be made in a float control system, however, by connecting one voltmeter probe directly to one side of the control transformer (Figure 6–12). Because both grounded and floating control systems are common, both will be illustrated throughout this textbook.

Transformer Fusing

Control transformers are generally protected by fuses or circuit breakers. Protection can be placed on the primary or secondary side of the transformer, and some industries prefer protection in both sides. *NEC® Section 430.72(C)* lists requirements for the protection of transformers employed in motor control circuits. This section basically states that control transformers that have a primary current of less than 2 amperes shall be protected by an overcurrent device set at not more than 500 percent of the rated primary current. This large percentage is necessary because of the high in-rush current associated with transformers. To determine the rated current of the transformer, divide the volt-ampere rating of the transformer by the primary voltage.

The secondary fuse size can be set at a lower percentage of the rated current because the secondary does not experience the high in-rush current of the primary. Because primary and secondary fuse protection is common throughout industry, control circuits presented in this textbook illustrate both.

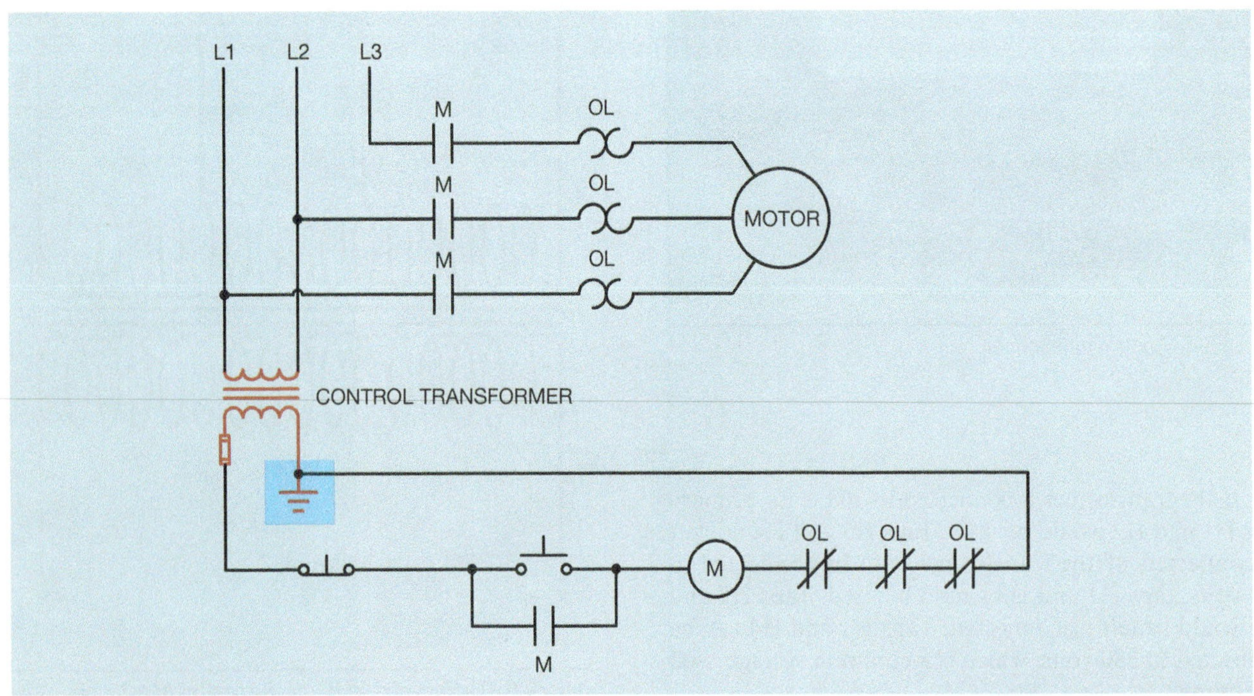

Figure 6–9 One side of the transformer has been grounded.

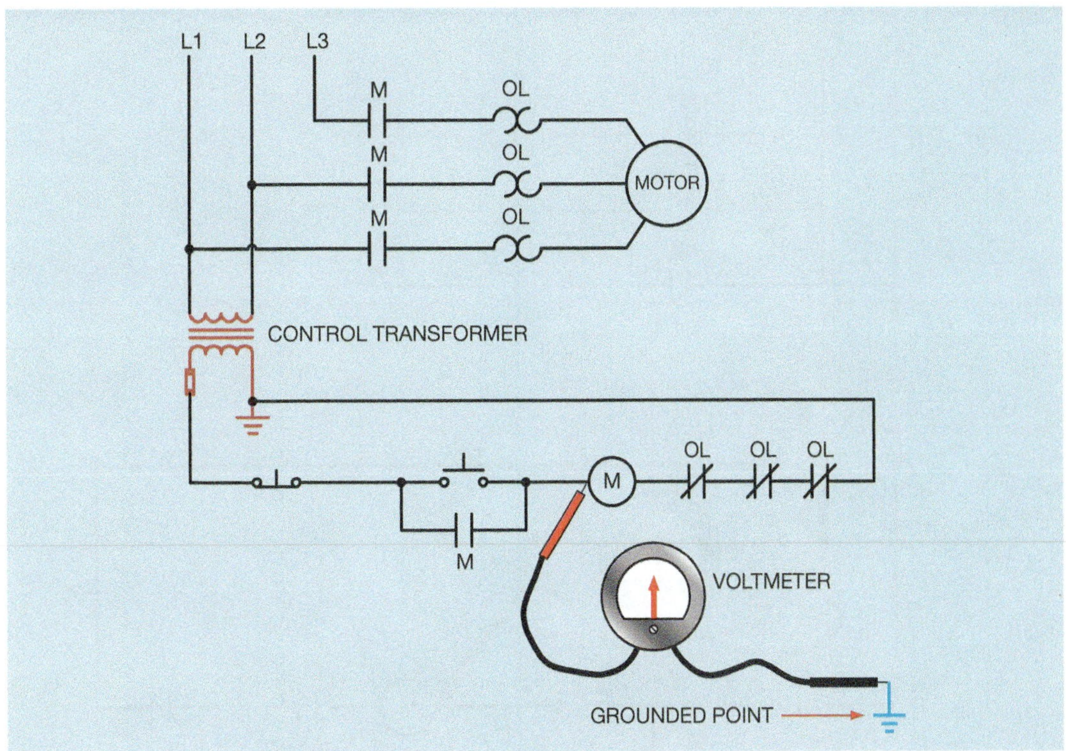

Figure 6–10 Voltage can be measured by connecting one meter probe to any grounded point.

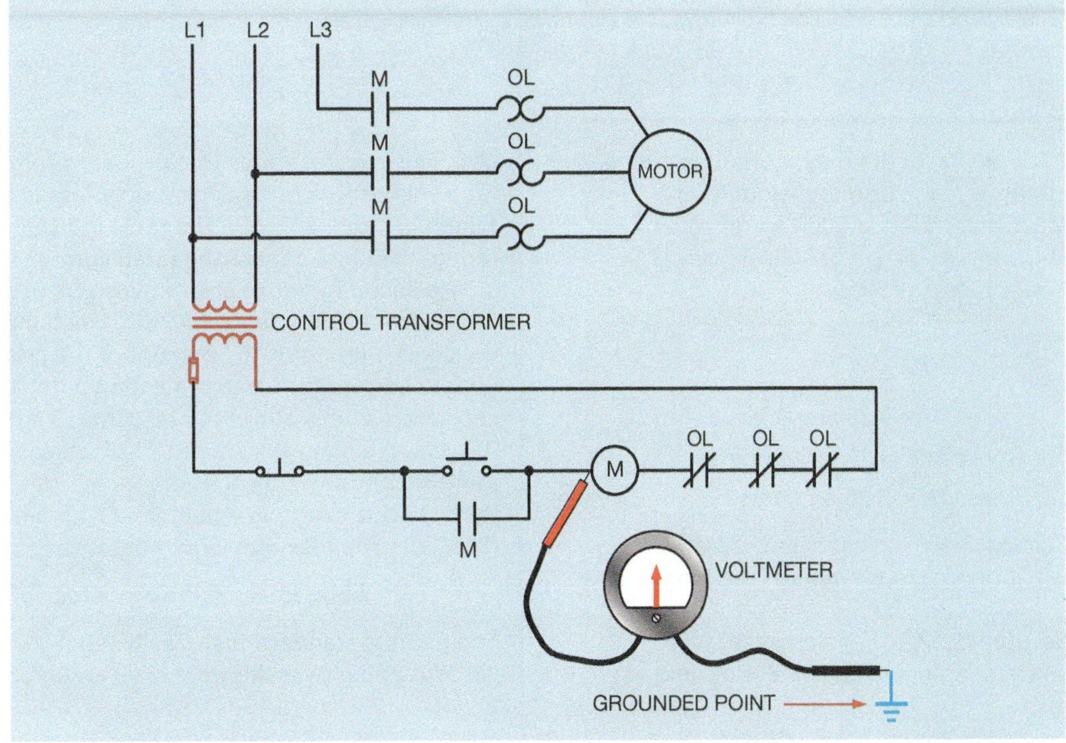

Figure 6–11 Floating control systems do not ground one side of the transformer. Connecting a voltmeter probe to a grounded point would provide meaningless values of voltage because a complete circuit does not exist.

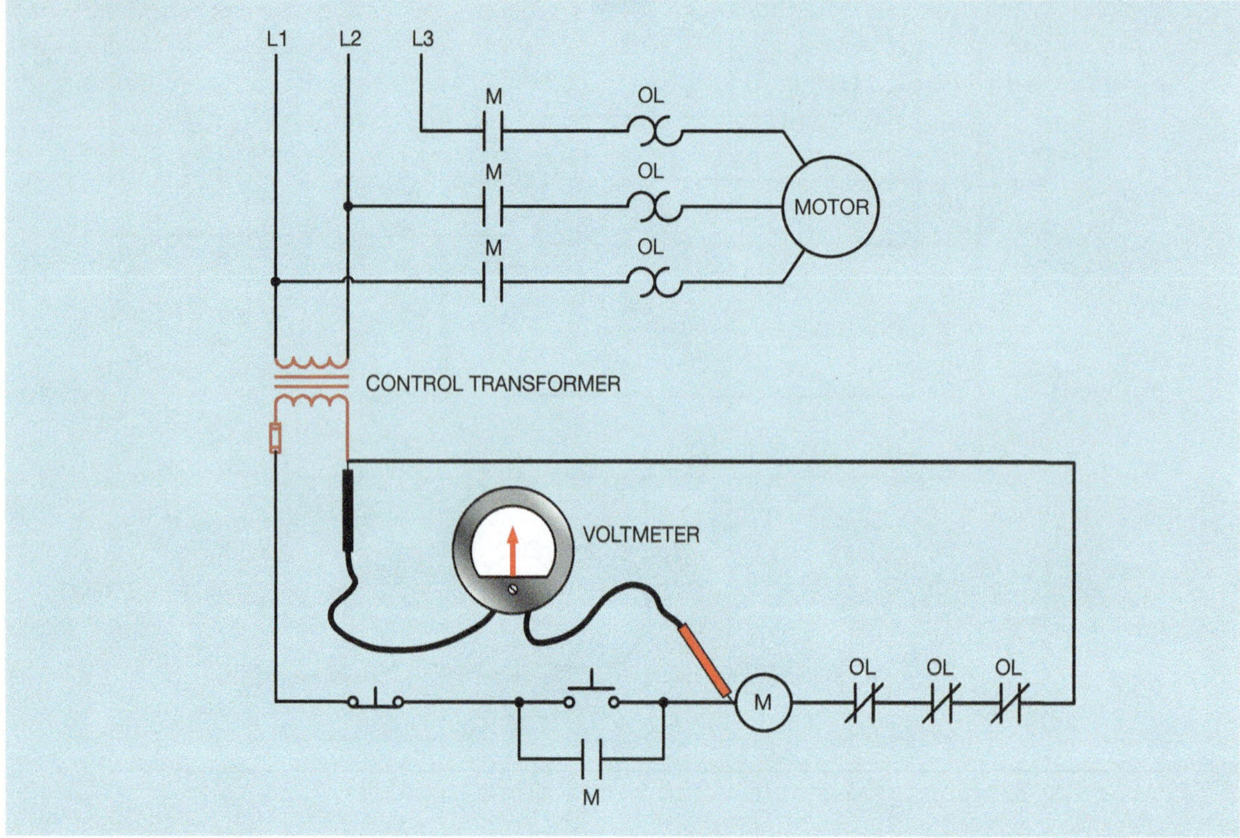

Figure 6–12 One meter probe is connected directly to one side of the control transformer.

EXAMPLE: What is the maximum fuse size permitted to protect the primary winding of a control transformer rated at 300 VA and connected to 240 volts?

$$I = \frac{VA}{E}$$

$$I = \frac{300}{240}$$

$$I = 1.25 \text{ amperes}$$

$$\textit{Fuse size} = 1.25 \times 5 \text{ amperes}$$

$$\textit{Fuse size} = 6.25 \text{ amperes}$$

NEC® Section 240.6 indicates that a standard fuse size is 6 amperes. A 6-ampere fuse would be used.

NEC® Section 430.72(C)(2) states that fuse protection in accordance with 450.3 is permitted also. This section states that primary protection for transformers rated 600 volts or less is determined in *Table 430.3(B)*. The table indicates a rating of 300 percent of the rated current.

The secondary fuse size can also be determined from *NEC® Table 450.3(B)*. The table indicates a rating of 167 percent of the rated secondary current for fuses protecting a transformer secondary with a current of less than 9 amperes. Assuming a control voltage of 120 volts, the rated secondary current of the transformer in the previous example would be 2.5 amperes (300/120). The fuse size would be:

$$2.5 \times 1.67 = 4.175 \text{ amperes}$$

The nearest standard fuse size listed in 240.6 without going over this value is 3 amperes.

Review Questions

1. What is the operating voltage of most magnetic control systems?
2. How many primary windings do control transformers have?
3. How are the primary windings connected when the transformer is to be operated on a 240-volt system?
4. How are the primary windings connected when the transformer is to be operated on a 480-volt system?
5. Why are two of the primary leads crossed on a control transformer?

Section 2

BASIC CONTROL CIRCUITS

Chapter 7

START–STOP PUSH BUTTON CONTROL

Objectives

After studying this chapter the student will be able to:

» Explain the operation of a START–STOP push button control.

» Place wire numbers on a schematic diagram.

» Place corresponding numbers on control components.

» Draw a wiring diagram from a schematic diagram.

» Define the difference between a schematic or ladder diagram and a wiring diagram.

» Connect a START–STOP push button control circuit.

The START–STOP push button circuit is the basis for many other control circuits. In this chapter you will learn the difference between schematic or ladder diagrams and wiring diagrams, and how to convert a ladder diagram into a wiring diagram. The laboratory exercise section provides instructions in step-by-step order for connecting a START–STOP push button circuit.

Schematics and Wiring Diagrams

Schematic or ladder diagrams show electrical components in their electrical sequence without regard for physical location. Several basic rules apply to how schematics are drawn:

1. **Schematic or ladder diagrams are read like a book, from left to right and from top to bottom.**
2. **Electrical components are always shown in their de-energized or off state.**
3. **Any contact that is marked with the same letter or number as a coil is controlled by that coil regardless of where it is located in the schematic.**
4. **Dashed lines between components indicate that the components are mechanically connected together.**

Wiring diagrams show a pictorial illustration of electrical components with connecting wires. Beginning electricians often prefer wiring diagrams when connecting a circuit because they show precise placement of the wires, but with time electricians discover that it is much easier to troubleshoot circuits and make connections using a schematic diagram.

Circuit Operation

The schematic of a START–STOP push button circuit is shown in Figure 7–1. The operation of the circuit is as follows:

- When the START push button is pressed, current can flow to the coil of M motor starter.
- When M motor starter energizes, all M contacts change position (Figure 7–2).
- The three M load contacts close and connect the motor to the power line.
- The M auxiliary contact connected in parallel with the START button closes. This contact is generally referred to as a **holding contact**, **sealing contact**, or **maintaining contact**, because its function is to maintain or hold a current

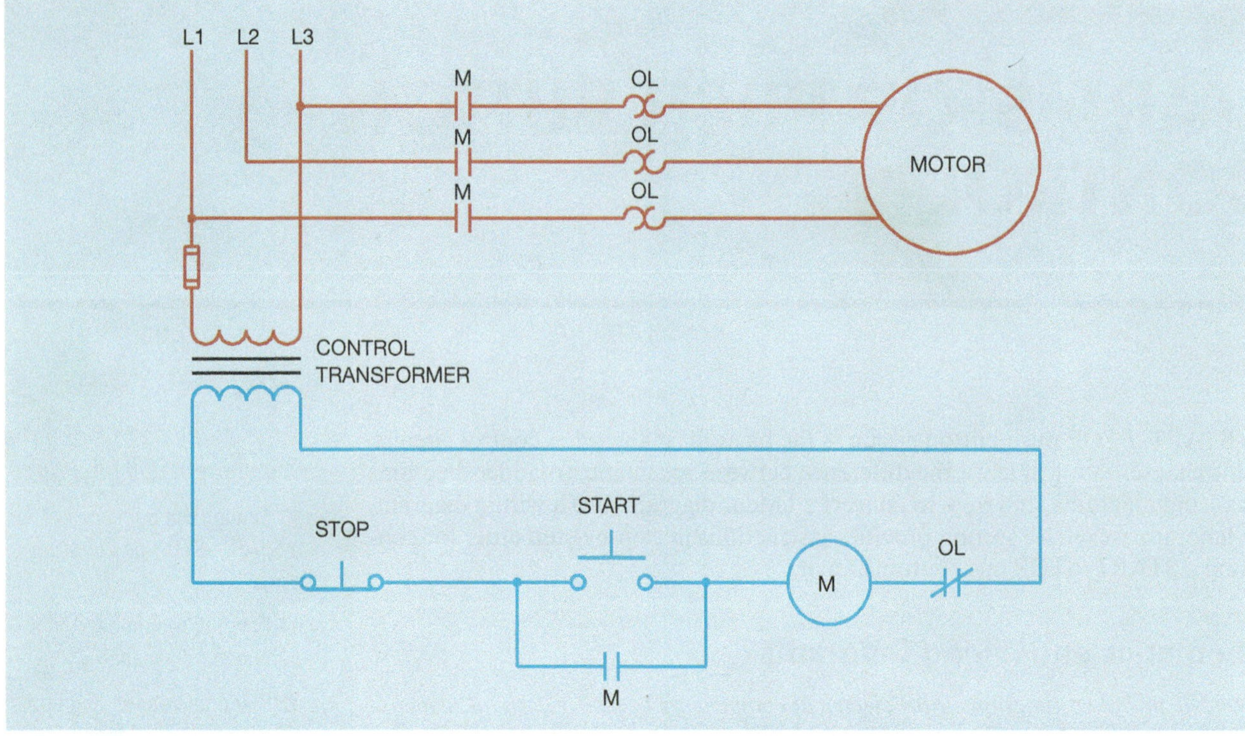

Figure 7–1 Schematic diagram of a basic START–STOP push button control circuit.

path to the coil when the push button is released (Figure 7–3).

- When the STOP button is pressed, current flow to the coil is interrupted and all M contacts return to their original position, (Figure 7–4).
- If an overload should occur, the normally closed overload contact will open. This opening has the same effect as pressing the STOP button.

Converting a Schematic Diagram into a Wiring Diagram

The pictorial representation of the components is shown in Figure 7–5.

To simplify the task of converting the schematic diagram into a wiring diagram, wire numbers are added to the schematic diagram. These numbers are then transferred to the control components shown in Figure 7–5. The rules for numbering a schematic diagram are as follows:

1. A set of numbers can be used only once.
2. Each time you go through a component the number set must change.
3. All components that are connected together have the same number.

To begin the numbering procedure, begin at Line 1 (L1) with the number 1 and place a number 1 beside each component that is connected to L1 (Figure 7–6). The number 2 is placed beside each component connected to L2 (Figure 7–7), and a 3 is placed beside each component connected to L3 (Figure 7–8 on page 106). The number 4 will be placed on the other side of the M load contact that already has a number 1 on one side and on one side of the overload heater (Figure 7–9 on page 107). Number 5 is placed on the other side of the M load contact, which has one side number with a 2 and a 5 will be placed beside the second overload heater. The other side of the M load contact that has been numbered with a 3 will be numbered with a 6, and one side of the third overload heater will be labeled with a 6. Numbers 7, 8, and 9 are placed between the other side of the overload heaters and the motor T leads.

The number 10 will begin at one side of the control transformer secondary and go to one side of the normally closed STOP push button. The number 11 is placed on the other side of the STOP button and on one side of the normally open START push button and normally open M auxiliary contact. A number 12 is placed on the other

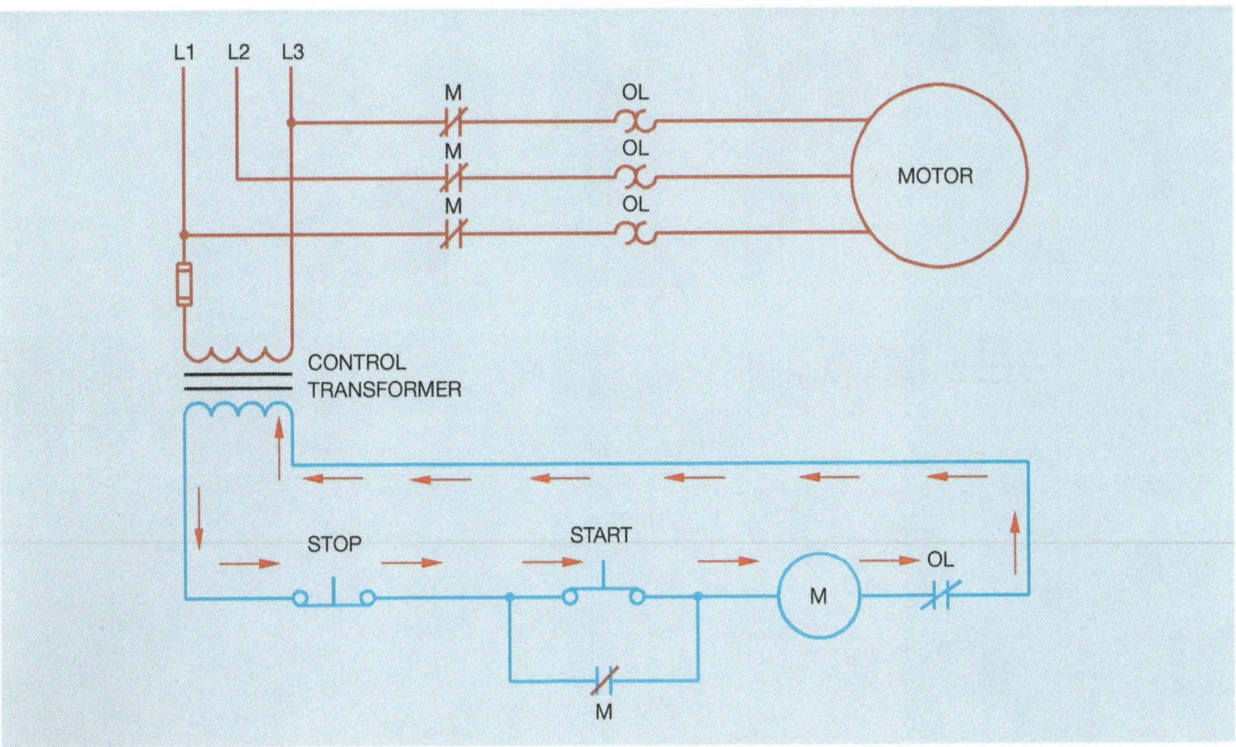

Figure 7–2 When the START button is pressed, current flows through M coil.

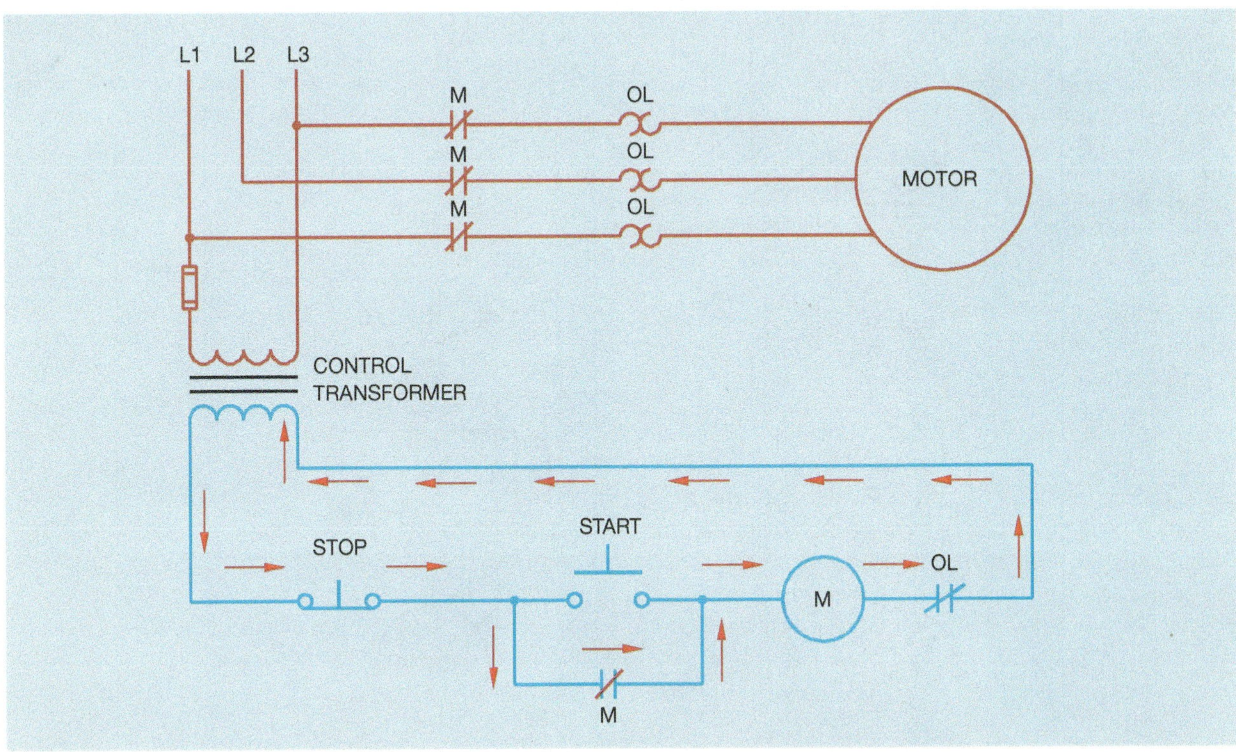

Figure 7–3 The auxiliary contact maintains the current path to the coil after the START button is released.

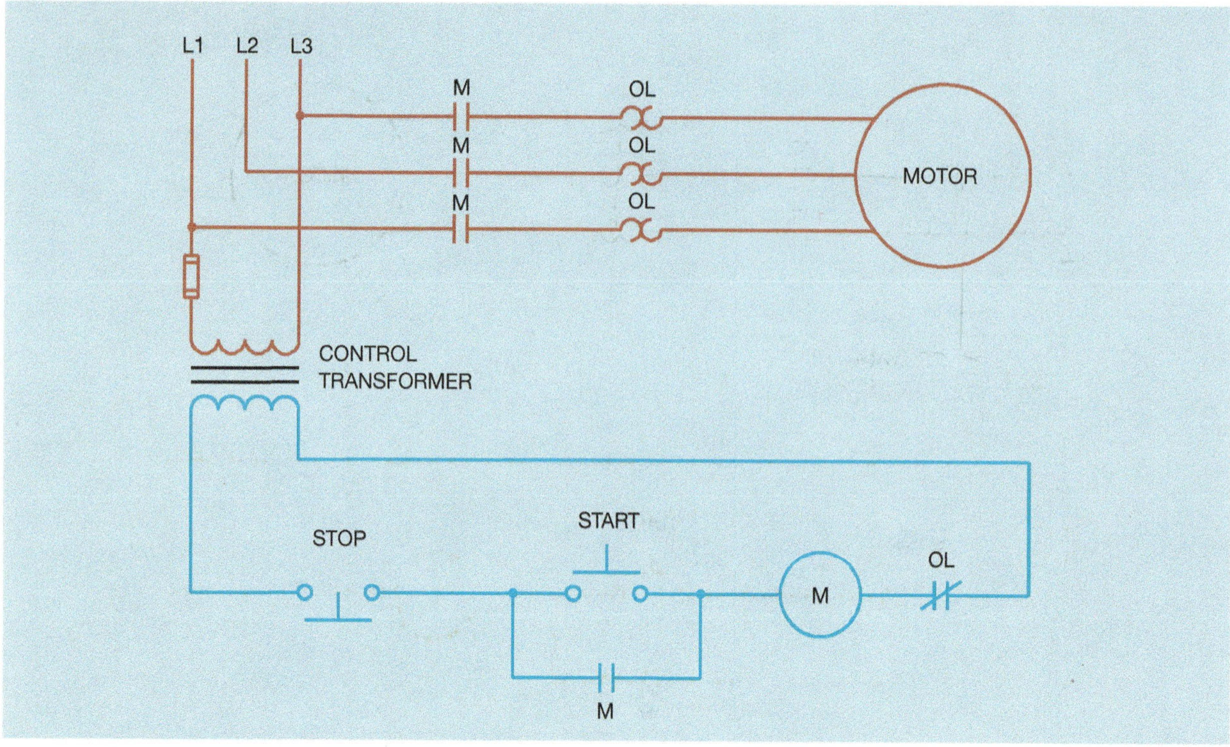

Figure 7–4 Pressing the STOP button de-energizes the coil and all M contacts return to their normal position.

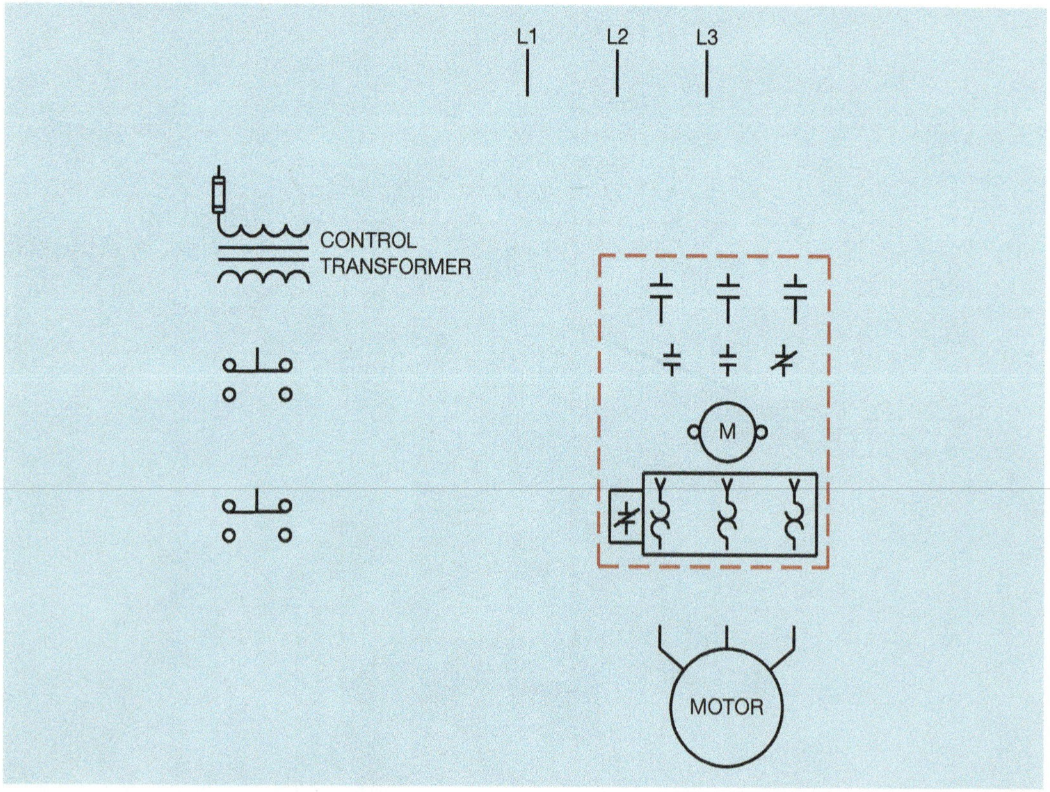

Figure 7–5 Components of the basic START–STOP control circuit.

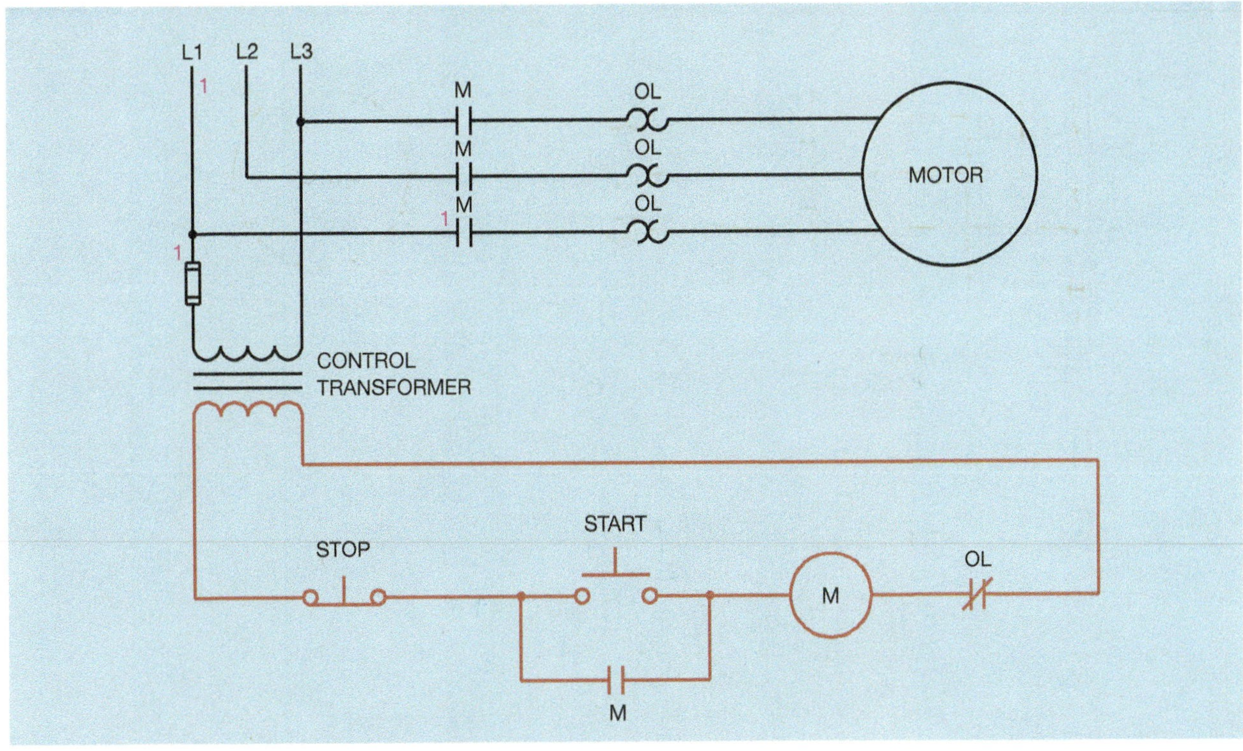

Figure 7–6 The number 1 is placed beside each component connected to Line 1.

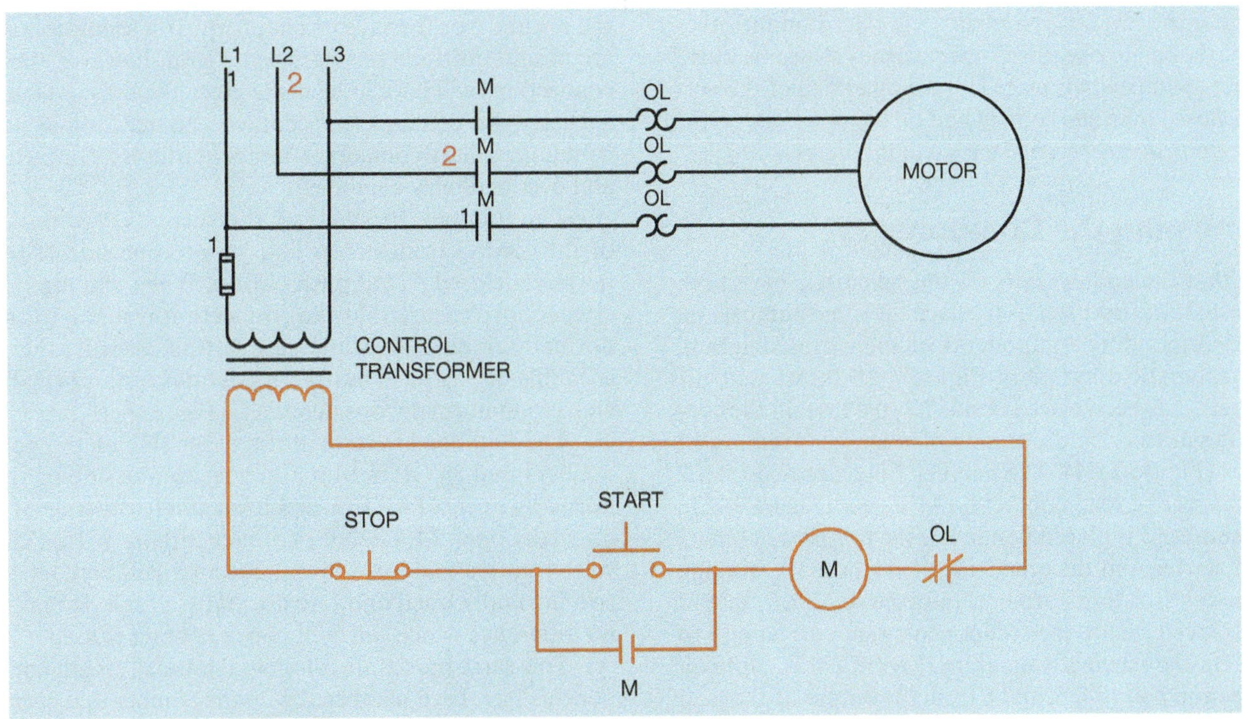

Figure 7–7 A number 2 is placed beside each component connected to Line 2.

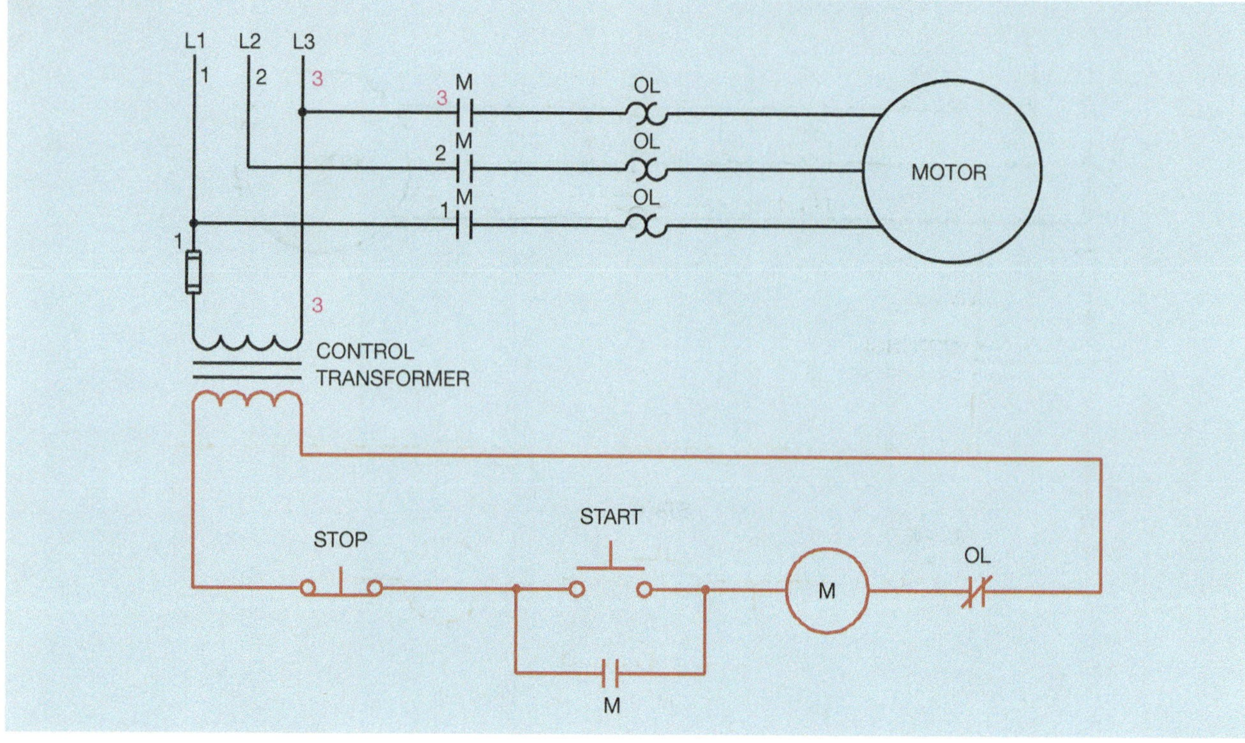

Figure 7–8 A number 3 is placed beside each component connected to Line 3.

side of the START button and M auxiliary contact and on one side of M coil. Number 13 is placed on the other side of the coil to one side of the normally closed overload contact. Number 14 is placed on the other side of the normally closed overload contact and on the other side of the control transformer secondary winding (Figure 7–10).

Numbering the Components

Now that the components on the schematic have been numbered, the next step is to place the same numbers on the corresponding components of the wiring diagram. The schematic diagram in Figure 7–10 shows that the number 1 has been placed beside L1, the fuse on the control transformer, and one side of a load contact on M starter (Figure 7–11). The number 2 is placed beside L2 and the second load contact on M starter (Figure 7–12). The number 3 is placed beside L3, the third load contact on M starter, and the other side of the primary winding on the control transformer. Numbers 4, 5, 6, 7, 8, and 9 are placed beside the components that correspond to those on the schematic diagram (Figure 7–13). Note on connection points 4, 5, and 6 from the output of the load contacts to the overload heaters, that these connections are factory made on a motor starter and do not have to be made in the field. These connections are not shown in the diagram for the sake of simplicity. If a separate contactor and overload relay are being used, however, these connections will have to be made. Recall that a contactor is a relay that contains *load* contacts and may or may not contain auxiliary contacts. A motor starter is a contactor and overload relay combined.

The number 10 starts at the secondary winding of the control transformer and goes to one side of the normally closed STOP push button. When making this connection, ensure that the connection is made to the normally closed side of the push button. Because this is a double-acting push button, it contains both normally closed and normally open contacts (Figure 7–14).

The number 11 starts at the other side of the normally closed STOP button and goes to one side of the normally open START push button and to one side of a normally open M auxiliary contact (Figure 7–15). The starter in this example shows three auxiliary contacts—two normally open and one normally closed. It makes no difference which normally open contact is used.

This same procedure is followed until all circuit components have been numbered with the number that corresponds to the same component on the schematic diagram (Figure 7–16 on page 111).

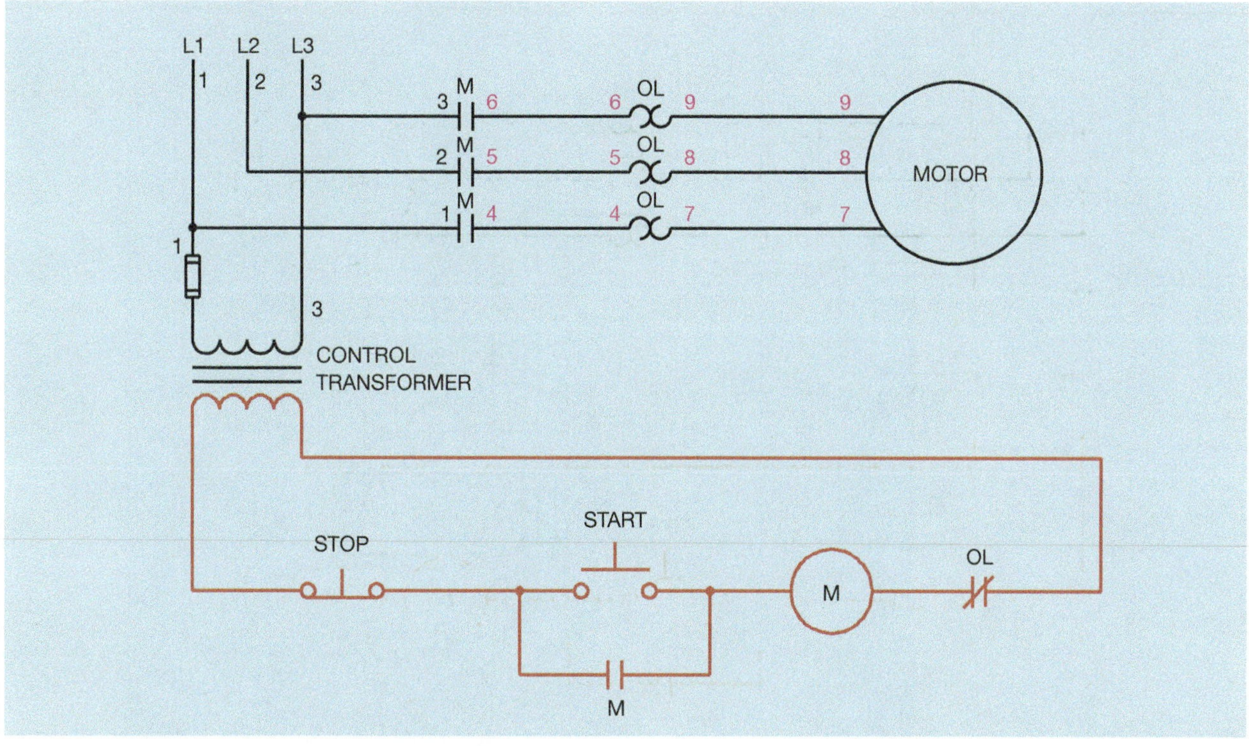

Figure 7–9 The number changes each time you proceed across a component.

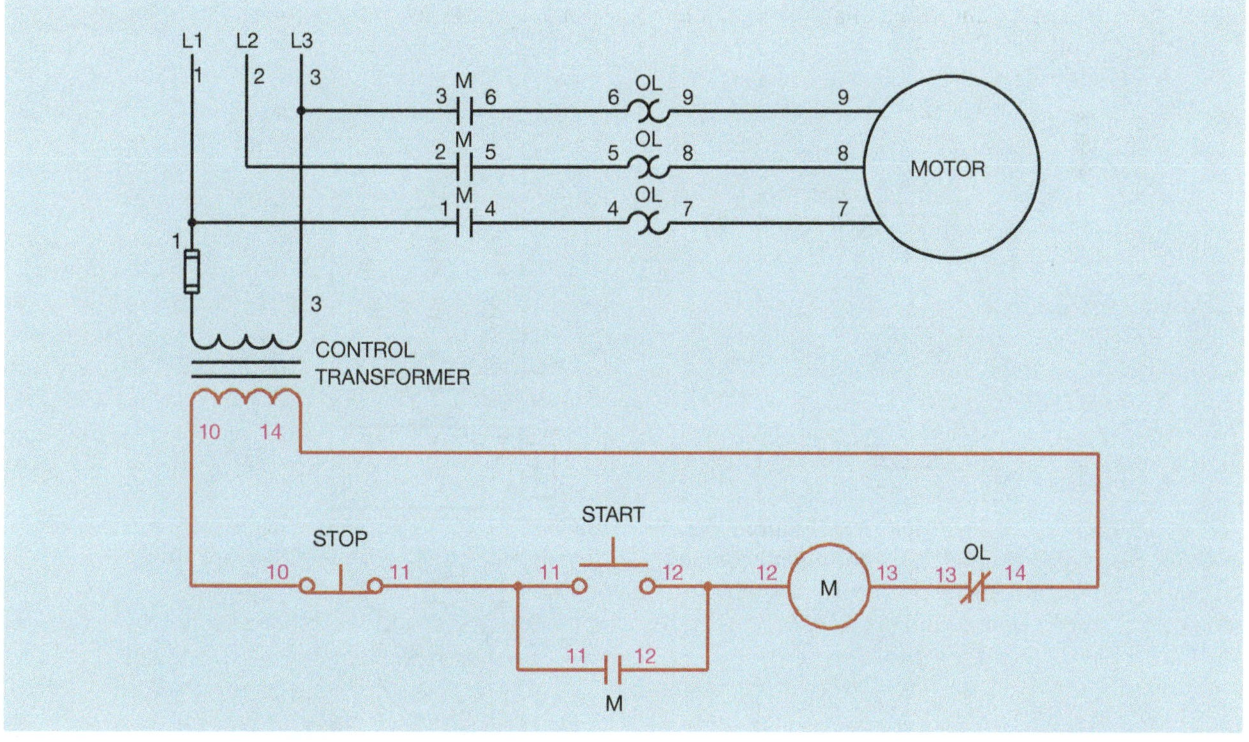

Figure 7–10 Numbers are placed beside all components.

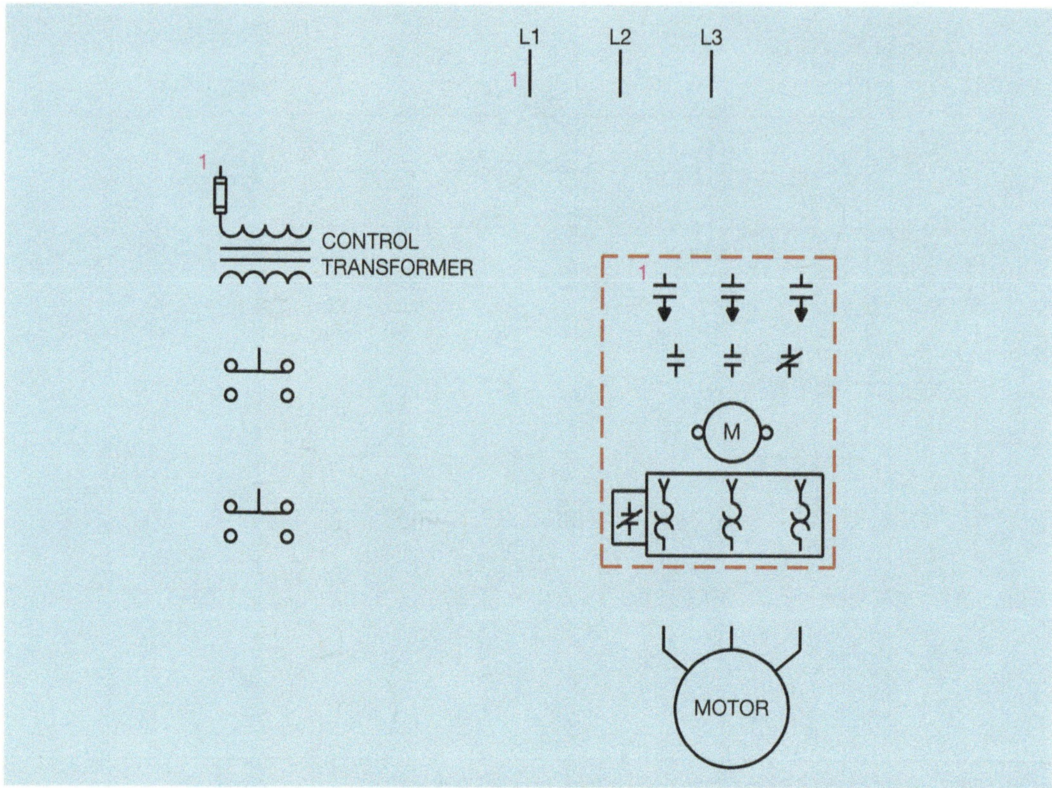

Figure 7–11 A number 1 is placed beside Line 1, the control transformer fuse, and M load contact.

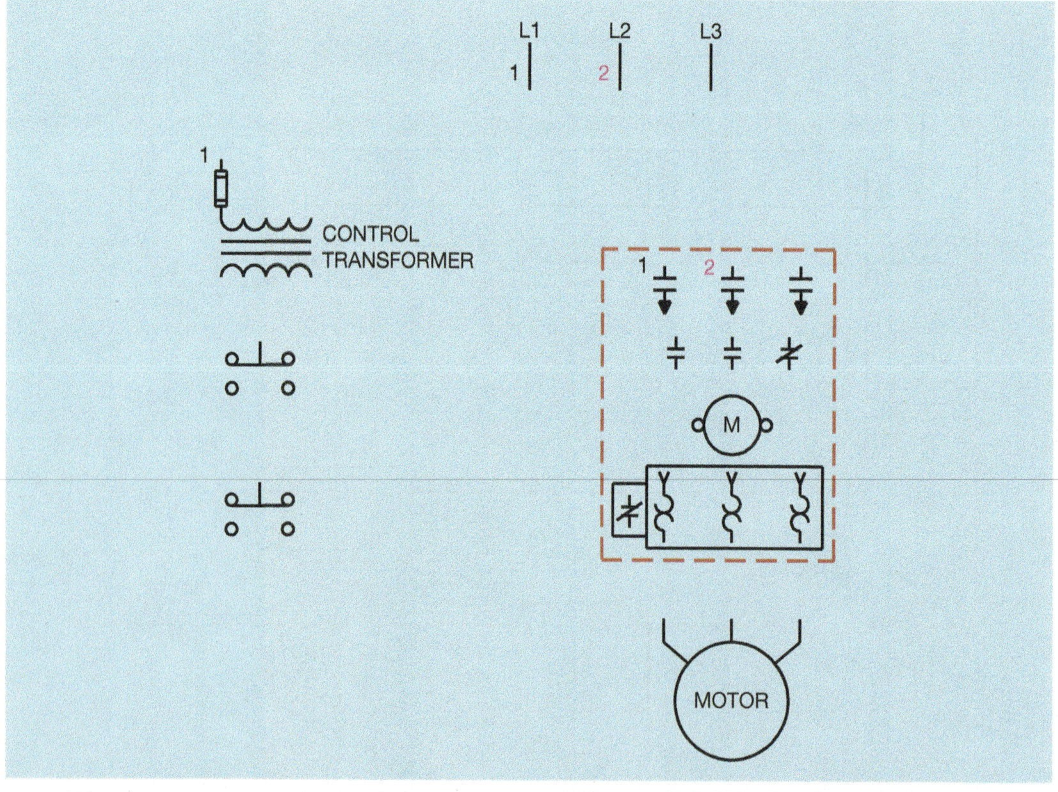

Figure 7–12 The number 2 is placed beside Line 2 and the second load contact on M starter.

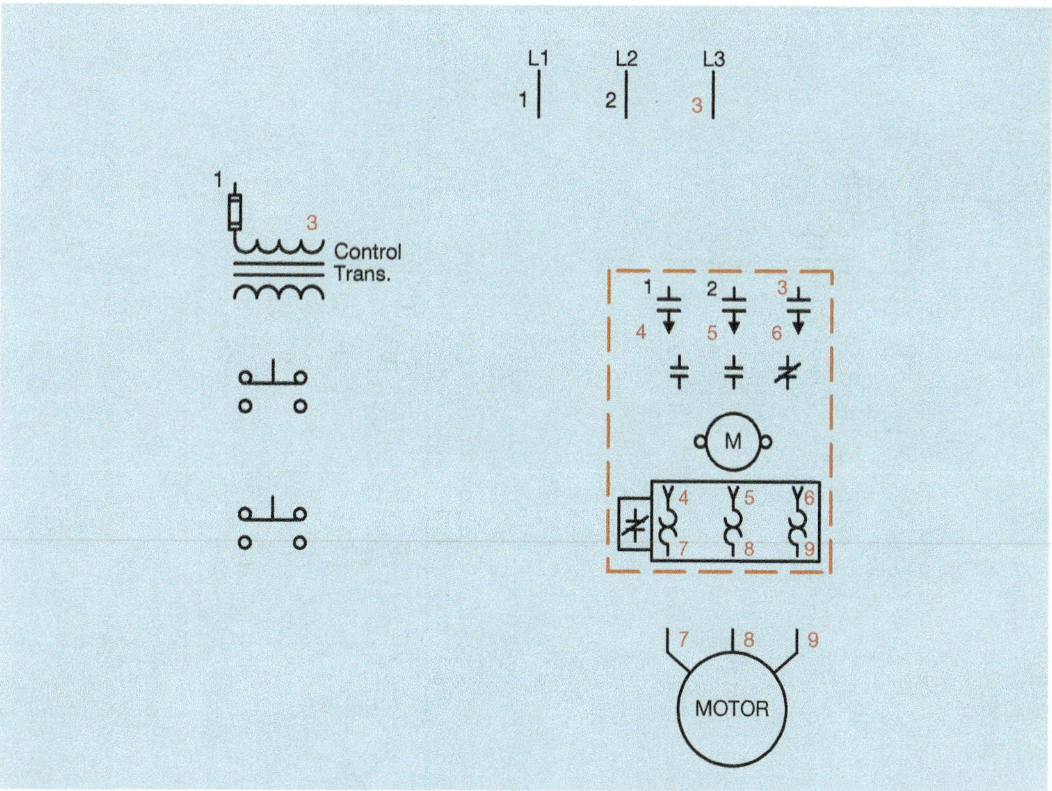

Figure 7–13 Numbers 3, 4, 5, 6, 7, 8, and 9 are placed beside the proper components.

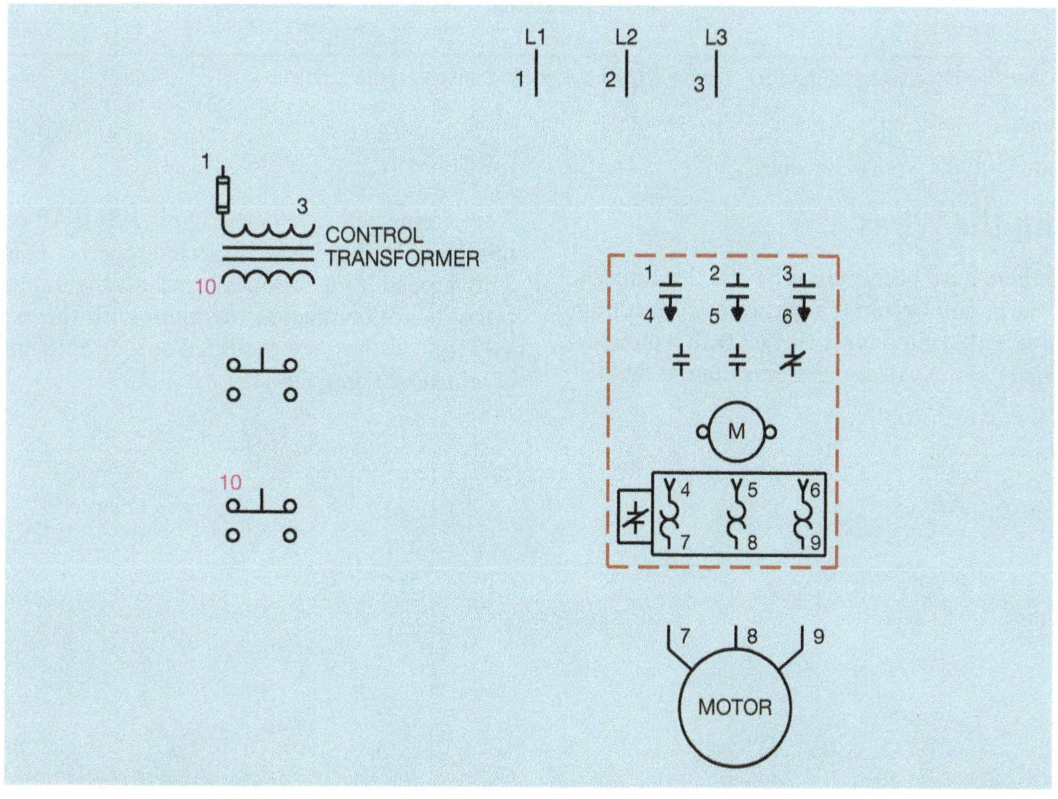

Figure 7–14 Wire number 10 connects from the transformer secondary to the STOP button.

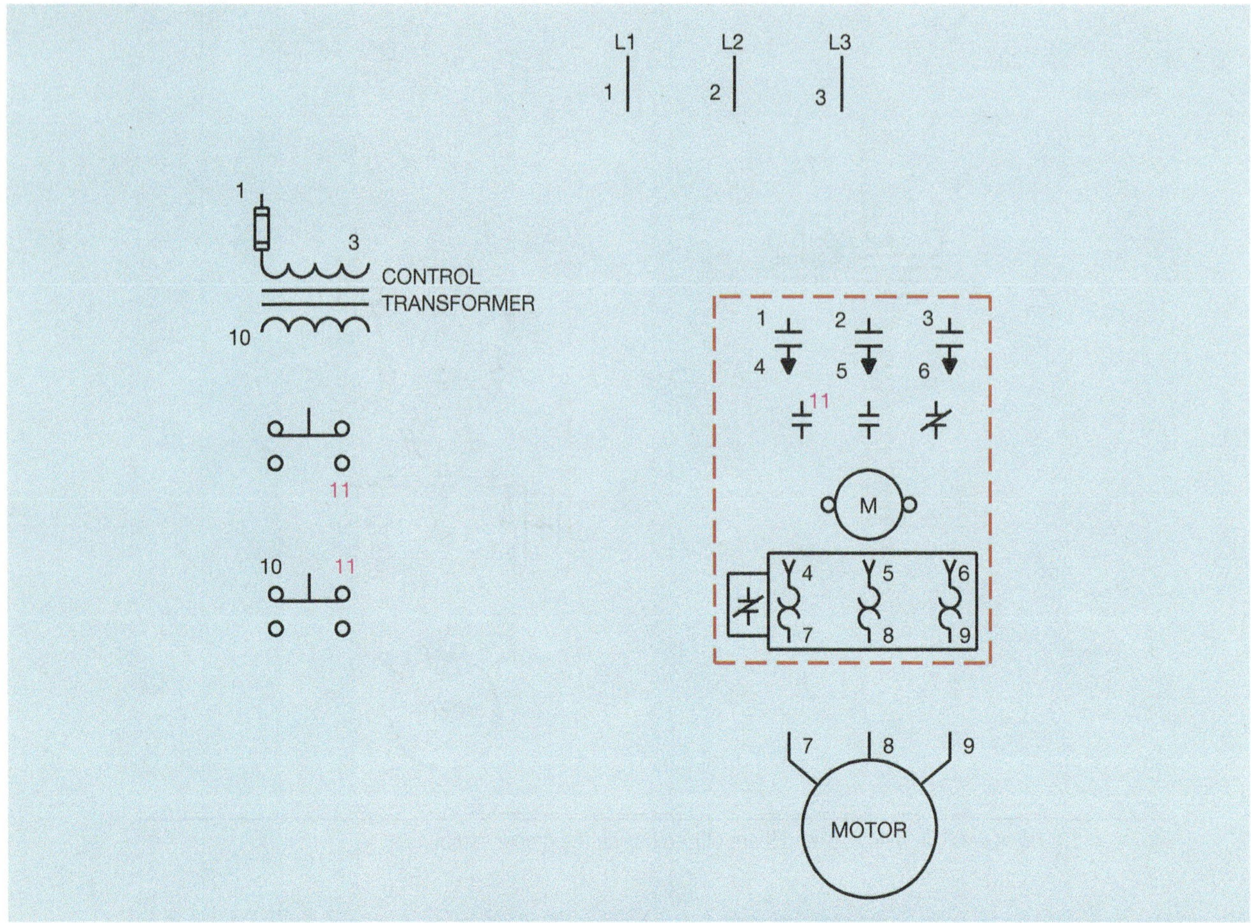

Figure 7–15 Number 11 connects to the STOP button, START button, and holding contact.

Connecting the Wires

Now that numbers have been placed beside the components, wiring the circuit becomes a matter of connecting numbers. Connect all components labeled with a number 1 together (Figure 7–17). All components number with a 2 are connected together (Figure 7–18). All components numbered with a 3 are connected together (Figure 7–19). This procedure is followed until all the numbered components are connected together, with the exception of 4, 5, and 6 that are assumed to be factory connected (Figure 7–20 on page 113).

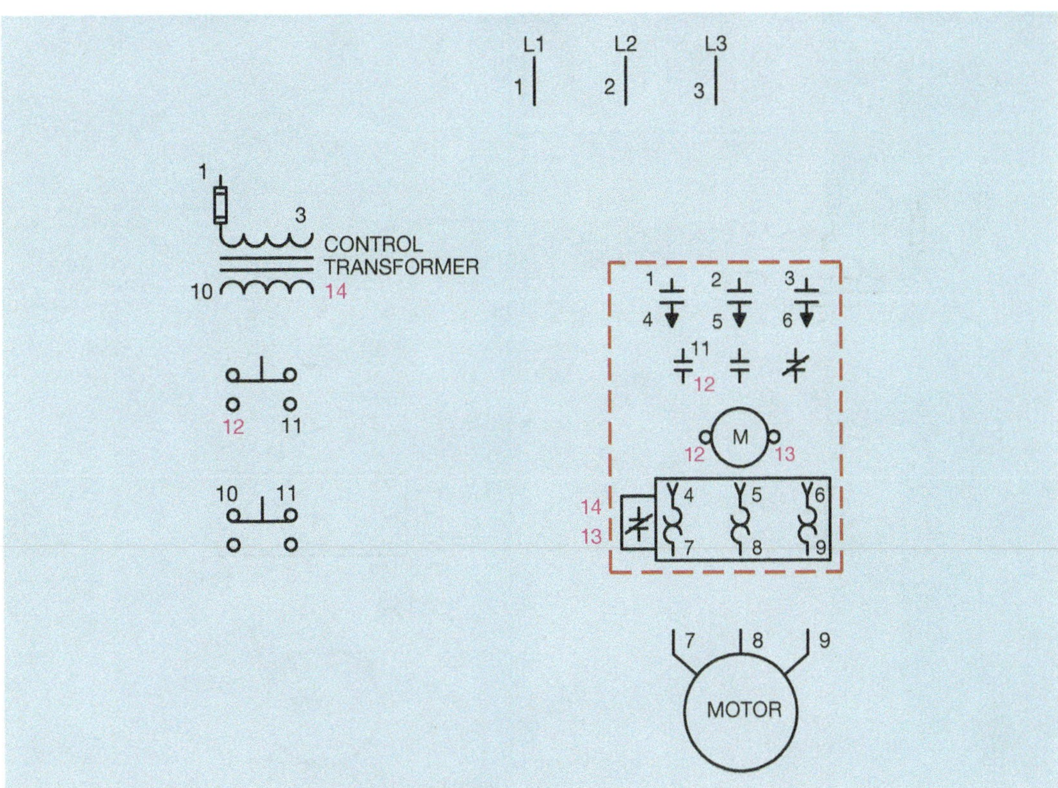

Figure 7–16 All components have been numbered.

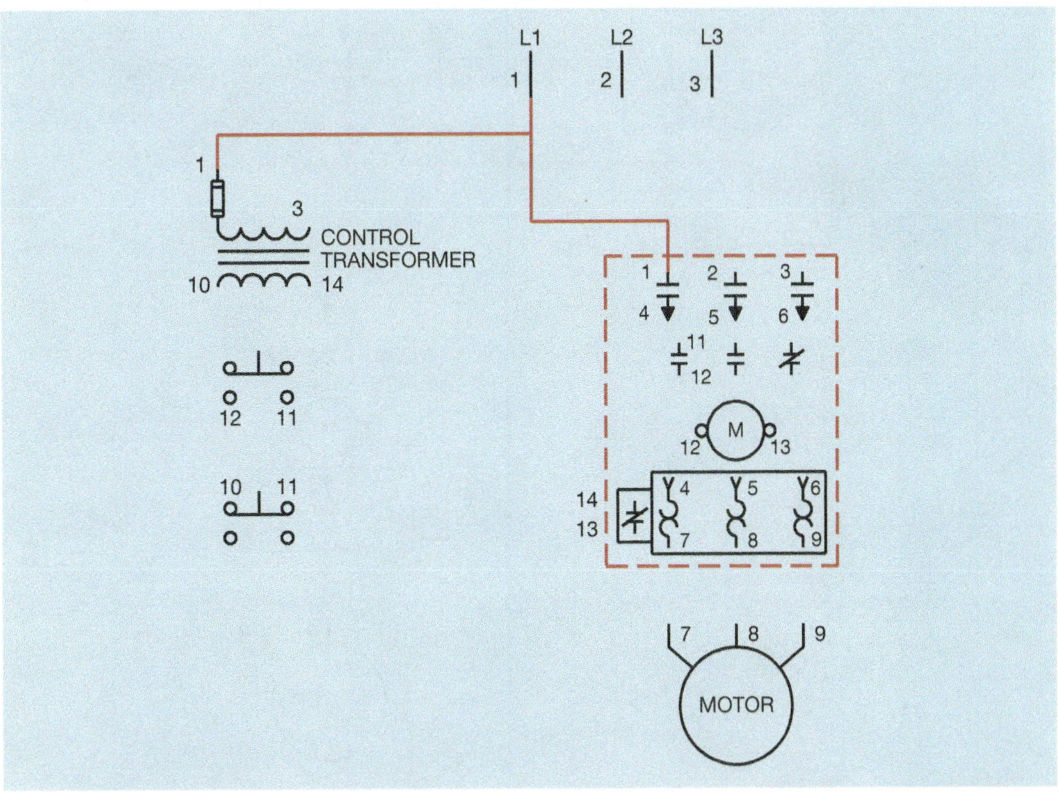

Figure 7–17 All the components with a number 1 are connected together.

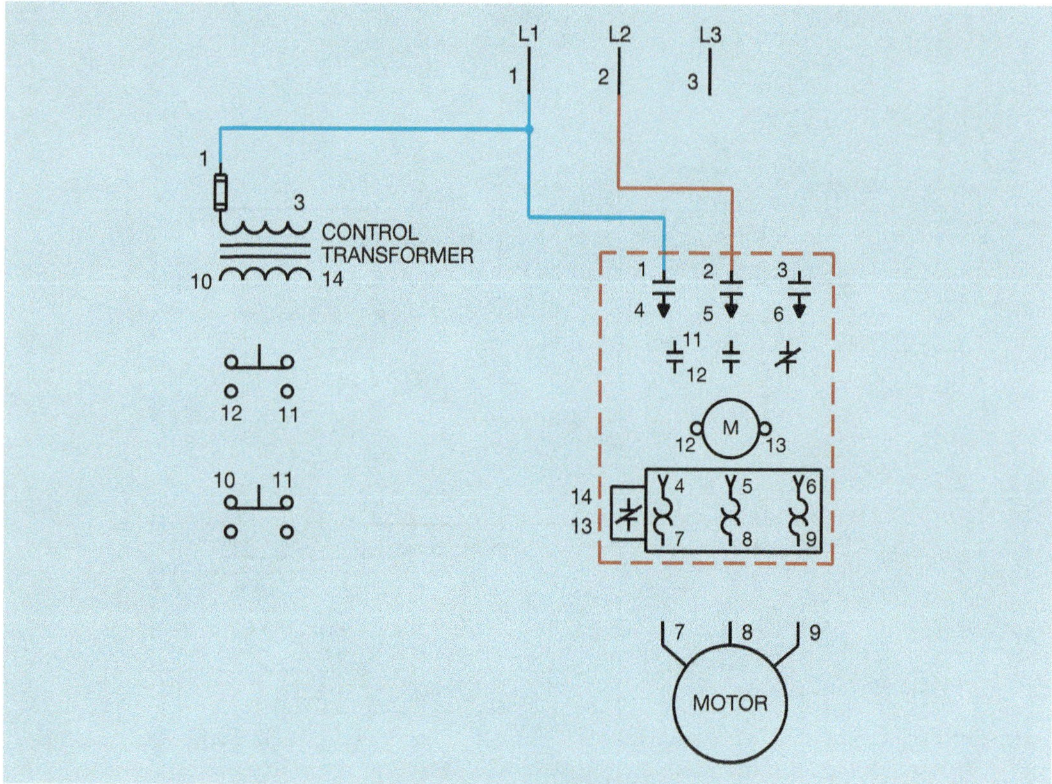

Figure 7–18 All components with a number 2 are connected together.

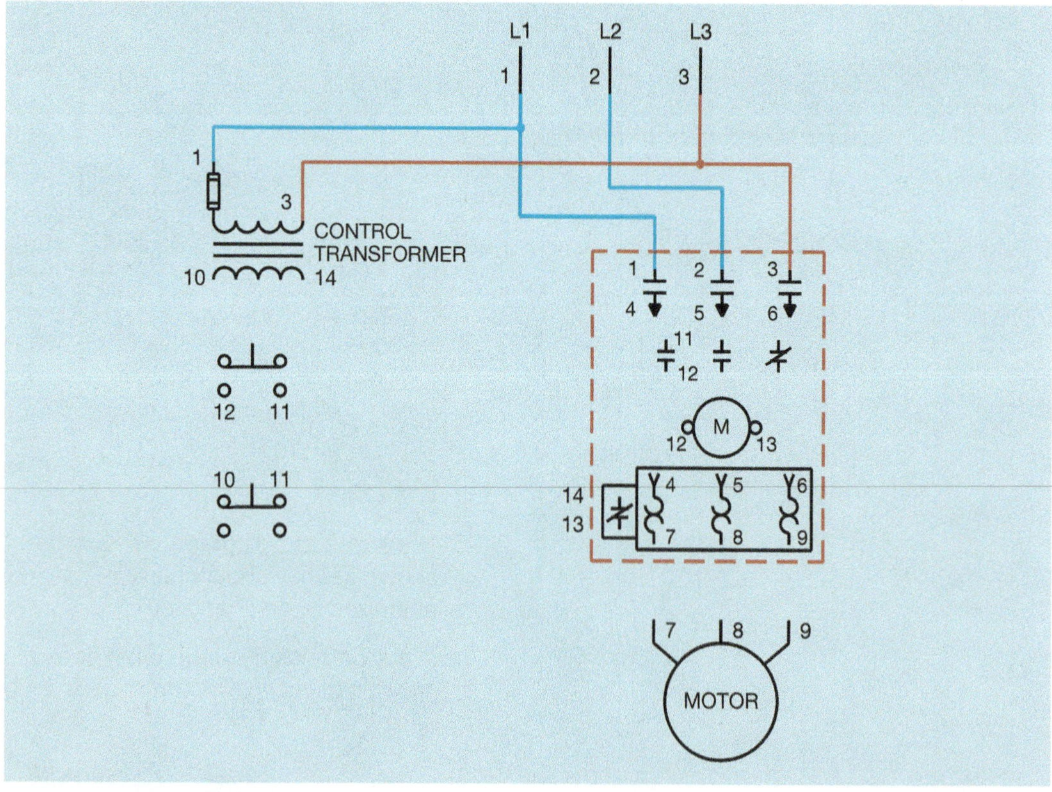

Figure 7–19 All the components with a number 3 are connected together.

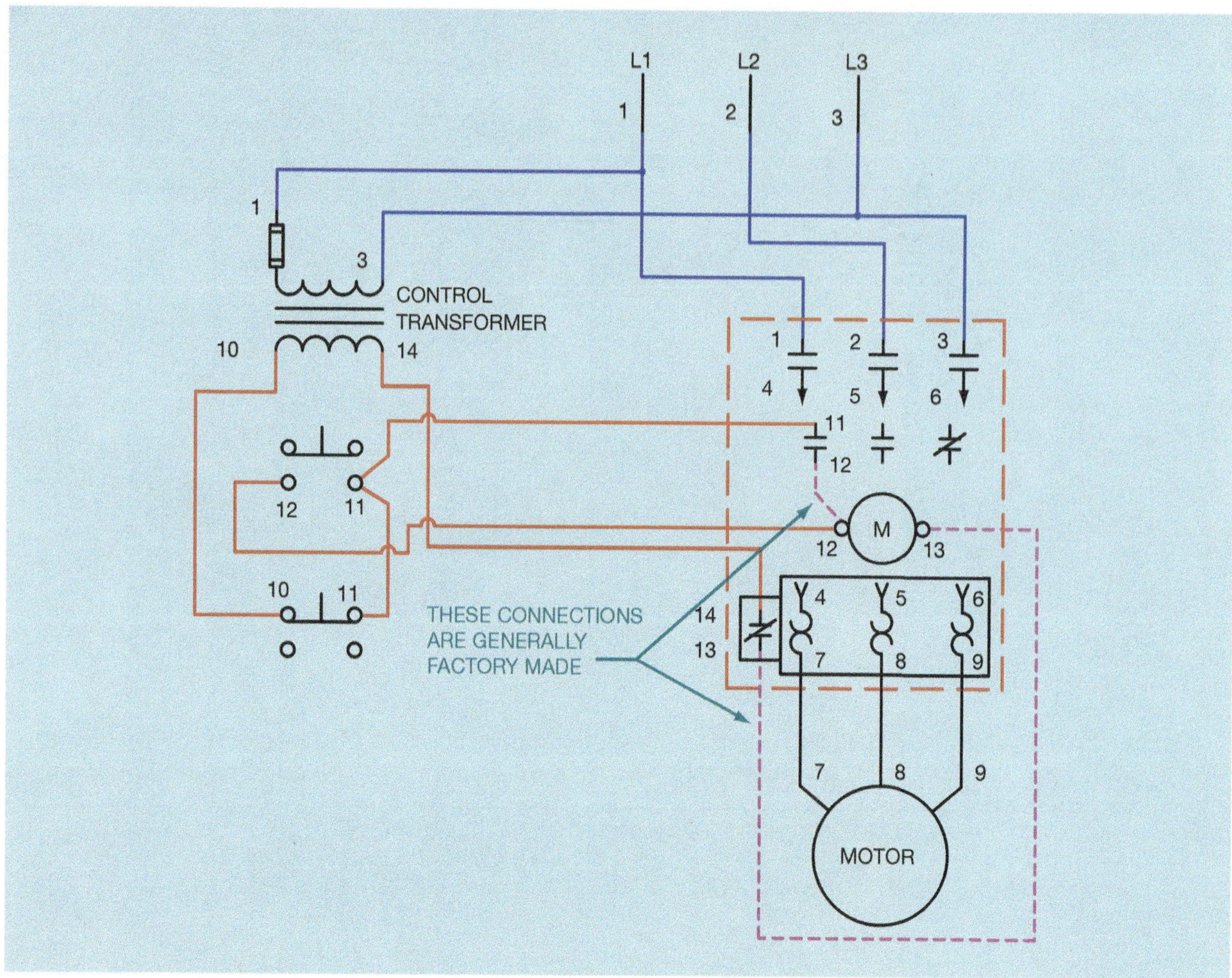

Figure 7–20 Completing the wiring diagram.

Review Questions

1. Refer to the circuit shown in Figure 7–10. If wire number 11 were disconnected at the normally open auxiliary M contact, how would the circuit operate?

2. Assume that when the START button is pressed M starter does not energize. List seven possible causes for this problem.

   a. ____________________

   b. ____________________

   c. ____________________

   d. ____________________

   e. ____________________

   f. ____________________

   g. ____________________

3. Explain the difference between a motor starter and a contactor.

4. Refer to the schematic in Figure 7–10. Assume that when the START button is pressed, the control transformer fuse blows. What is the most like cause of this trouble?

5. Explain the difference between load and auxiliary contacts.

6. In a schematic diagram, are the components shown as they should be when the circuit is energized or de-energized?

7. In a schematic diagram, what does a dashed line drawn between two components indicate?

Chapter 8

MULTIPLE PUSH BUTTON STATIONS

Objectives

After studying this chapter the student will be able to:

- Place wire numbers on a schematic diagram.
- Place corresponding numbers on control components.
- Draw a wiring diagram from a schematic diagram.
- Connect a control circuit using two STOP and two START push buttons.
- Discuss how components are to be connected to perform the functions of START or STOP for a control circuit.

There may be times when it is desirable to have more than one START–STOP push button station to control a motor. In this chapter, the basic START–STOP push button control circuit discussed in Chapter 7 will be modified to include a second STOP and START push button.

When a component is used to perform the function of *stop* in a control circuit, it will generally be a normally closed component and be connected in series with the motor starter coil. In this example, a second STOP push button is to be added to an existing START–STOP control circuit. The second push button will be added to the control circuit by connecting it in series with the existing STOP push button (Figure 8–1).

When a component is used to perform the function of *start* it is generally normally open and connected in parallel with the existing START button (Figure 8–2). If either START button is pressed, a circuit will be completed to M coil. When M coil energizes all M contacts change position. The three load contacts connected between the three phase power line and the motor close to connect the motor to the line. The normally open auxiliary contact connected in parallel with the two START buttons close to maintain the circuit to M coil when the START button is released.

Developing a Wiring Diagram

Now that the circuit logic has been developed in the form of a schematic diagram, a wiring diagram will be drawn from the schematic. The components needed to connect this circuit are shown in Figure 8–3. Following the same procedure discussed in Chapter 7, wire numbers will be placed on the schematic diagram (Figure 8–4). After wire numbers are placed on the schematic, corresponding numbers will be placed on the control components (Figure 8–5).

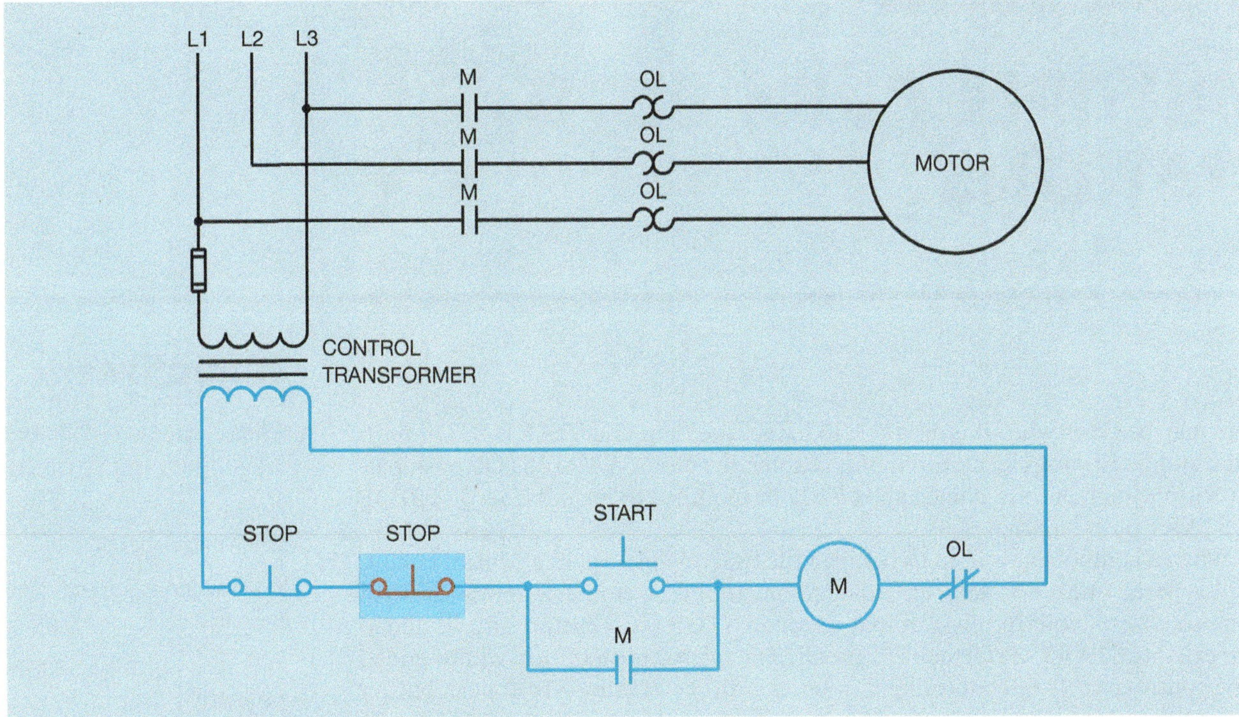

Figure 8–1 Adding a STOP button to the circuit.

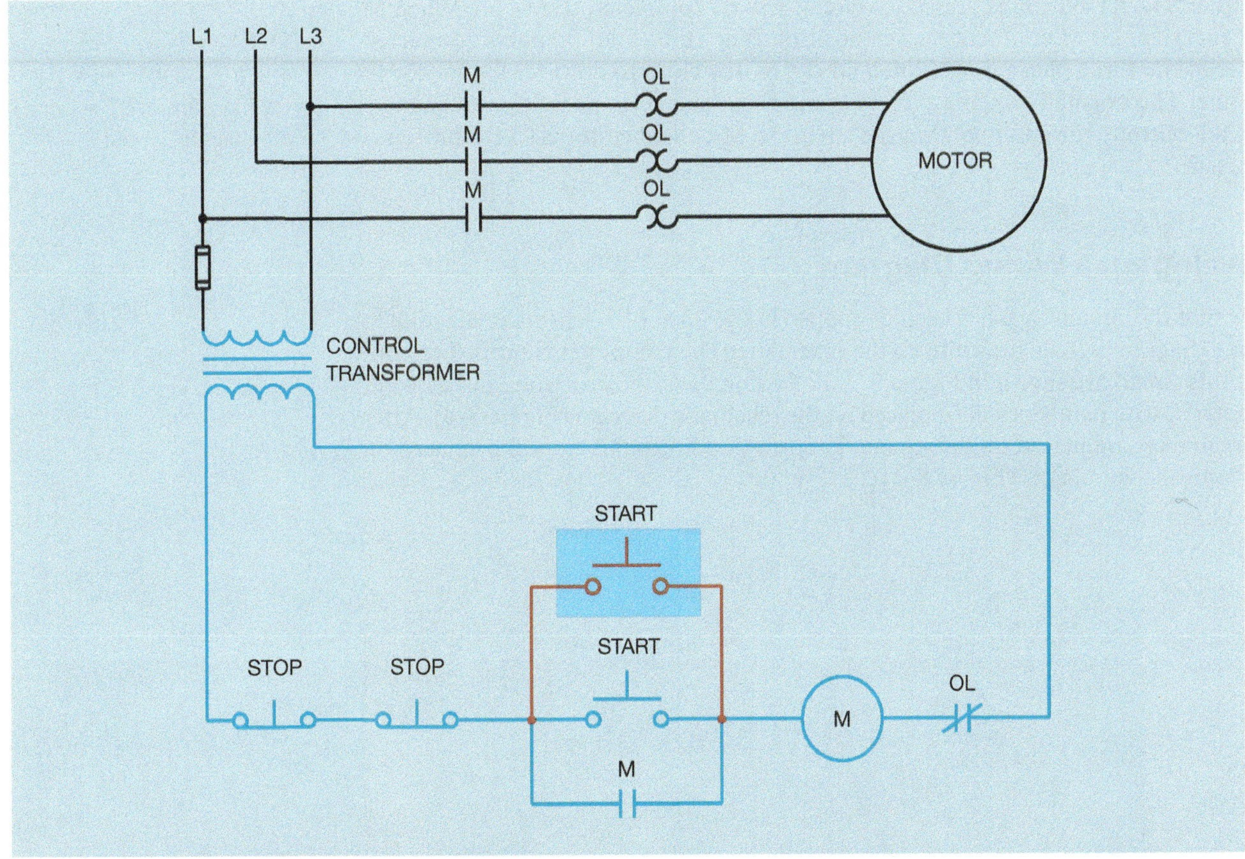

Figure 8–2 A second START button is added to the circuit.

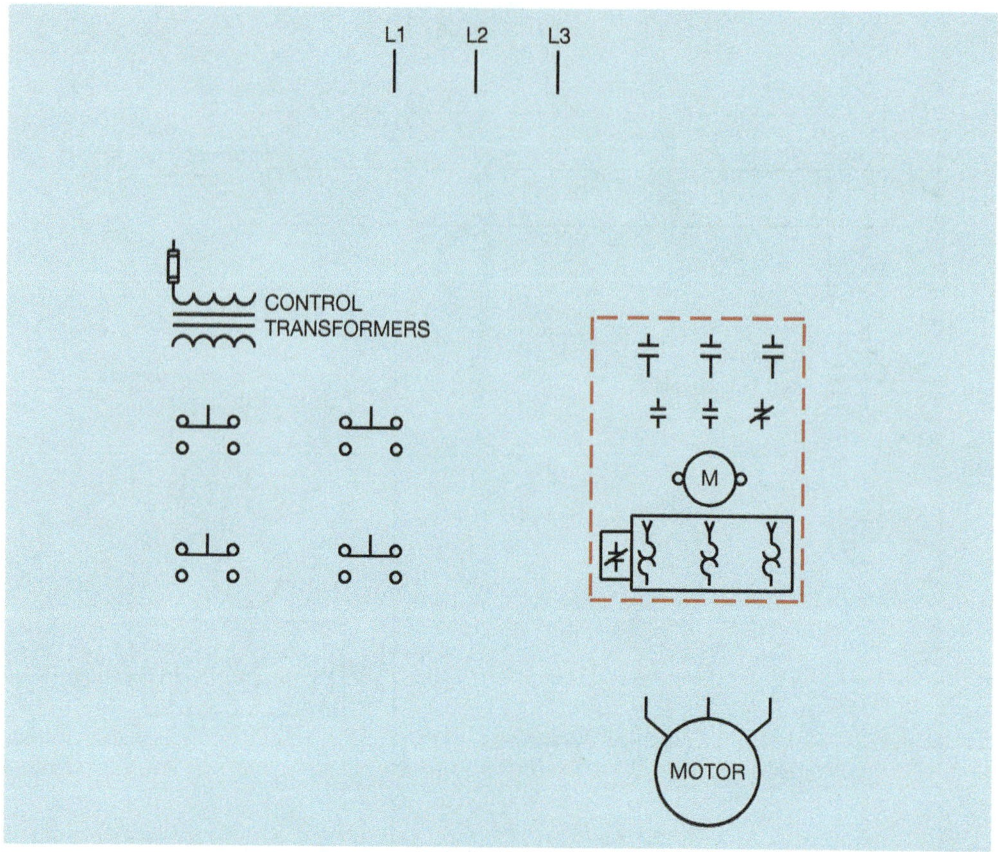

Figure 8–3 Components needed to produce a wiring diagram.

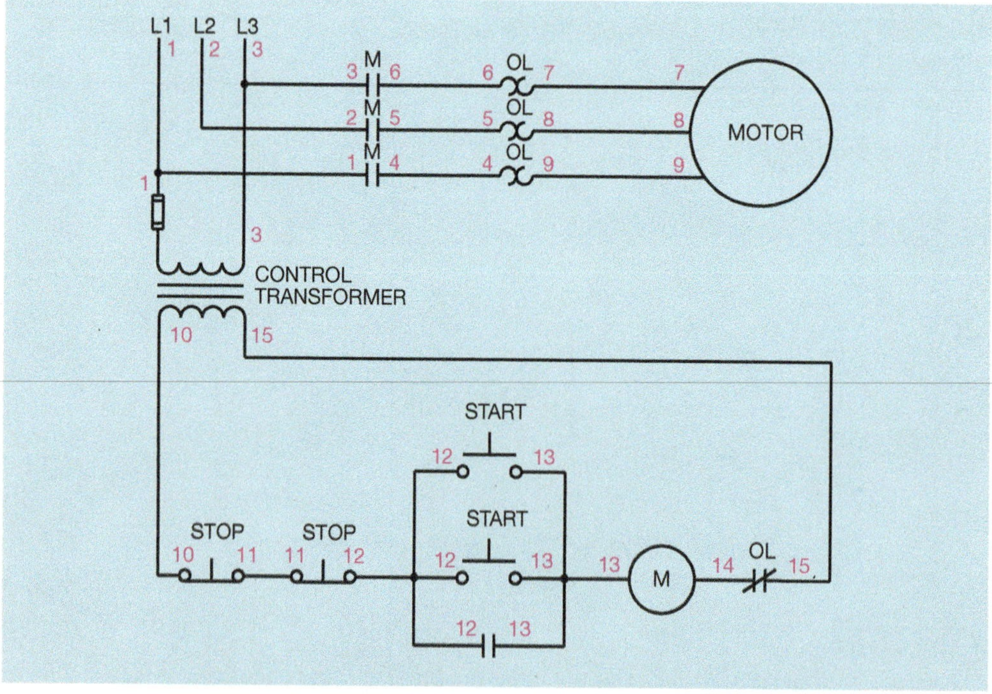

Figure 8–4 Numbering the schematic diagram.

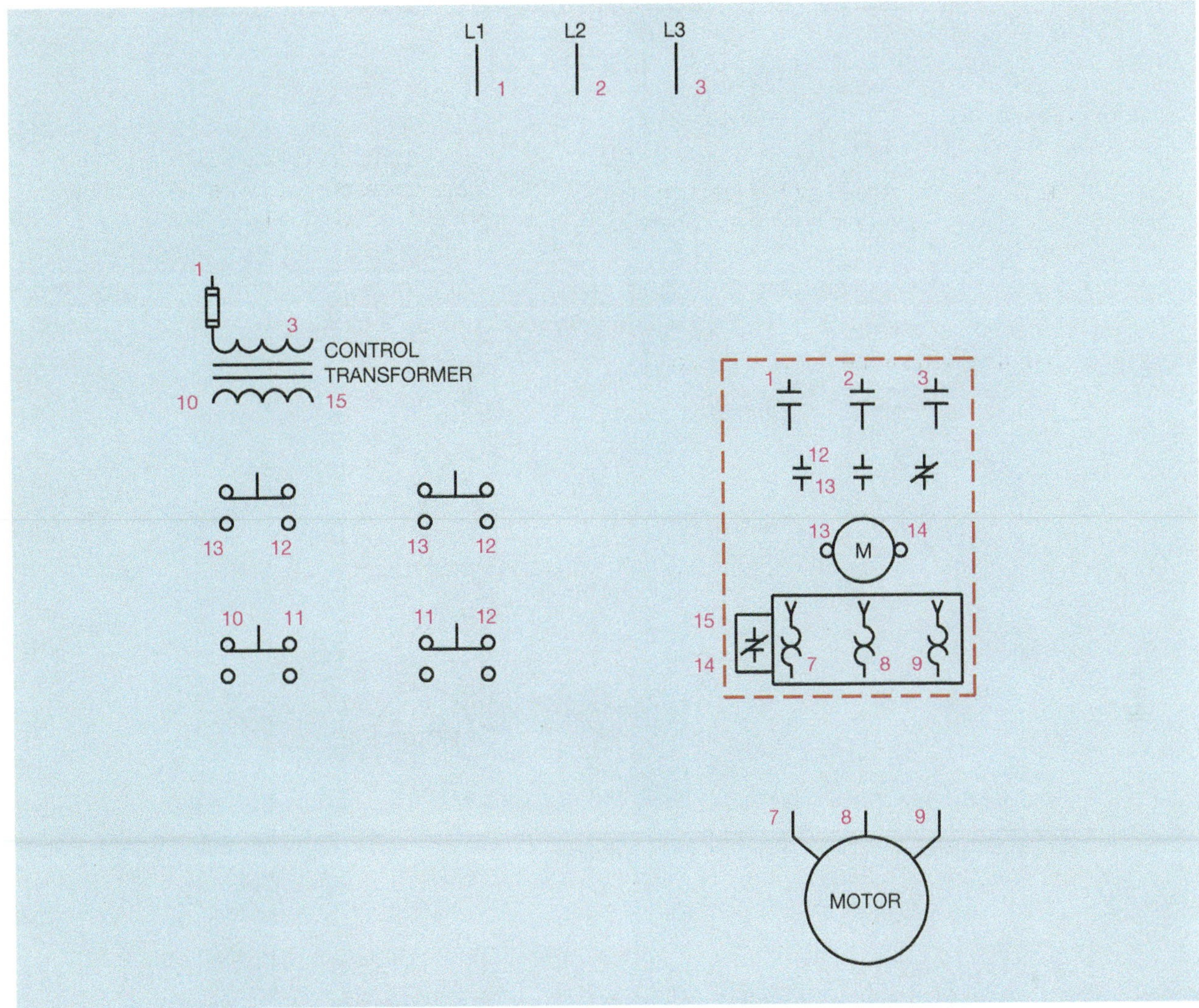

Figure 8–5 Numbering the components.

Review Questions

1. When a component is to be used for the function of start, is the component generally normally open or normally closed?
2. When a component is to be used for the function of stop, is the component generally normally open or normally closed?
3. The two STOP push buttons in Figure 8–2 are connected in series with each other. What would be the action of the circuit if they were to be connected in parallel as shown in Figure 8–6?
4. What would be the action of the circuit if both START buttons were to be connected in series as shown in Figure 8–7?
5. Following the procedure discussed in Chapter 7, place wire numbers on the schematic in Figure 8–7. Place corresponding wire numbers on the components shown in Figure 8–8.

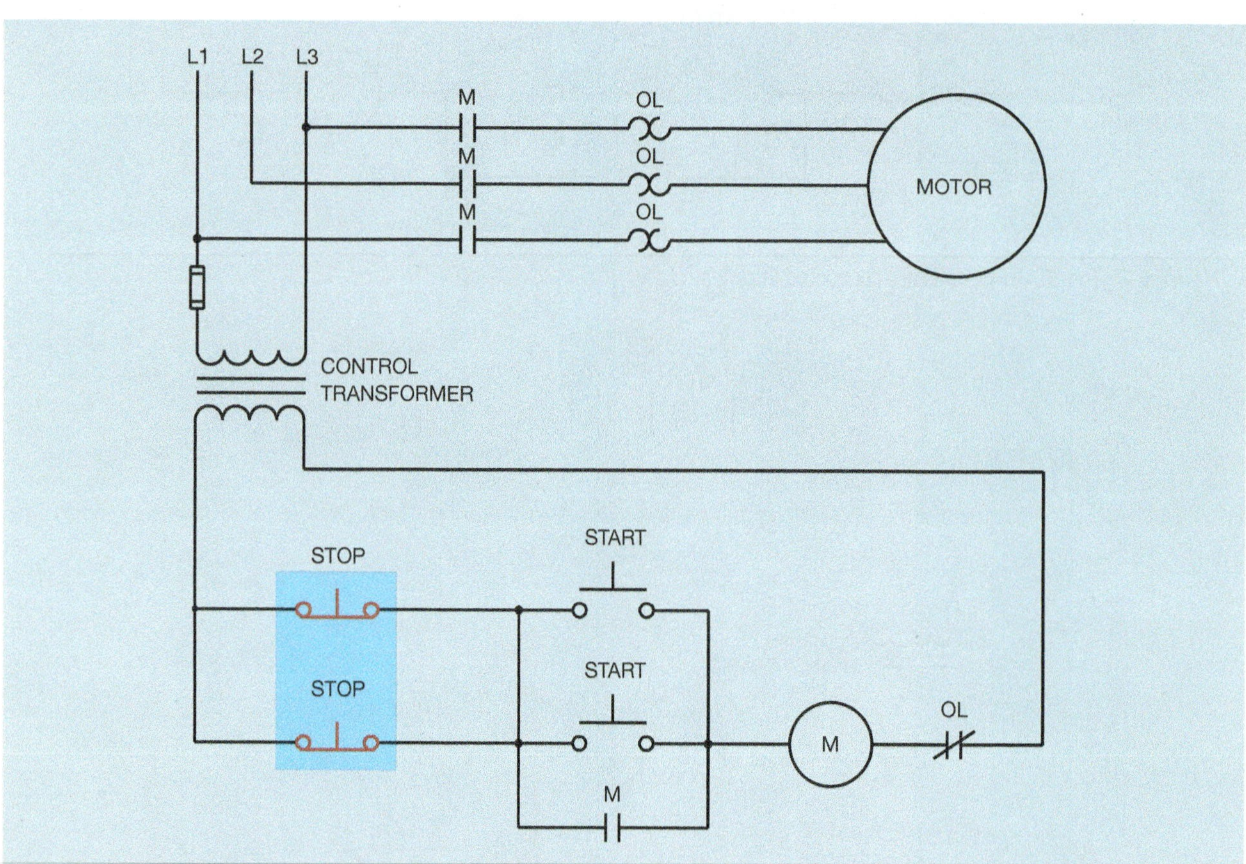

Figure 8–6 The STOP buttons have been connected in parallel.

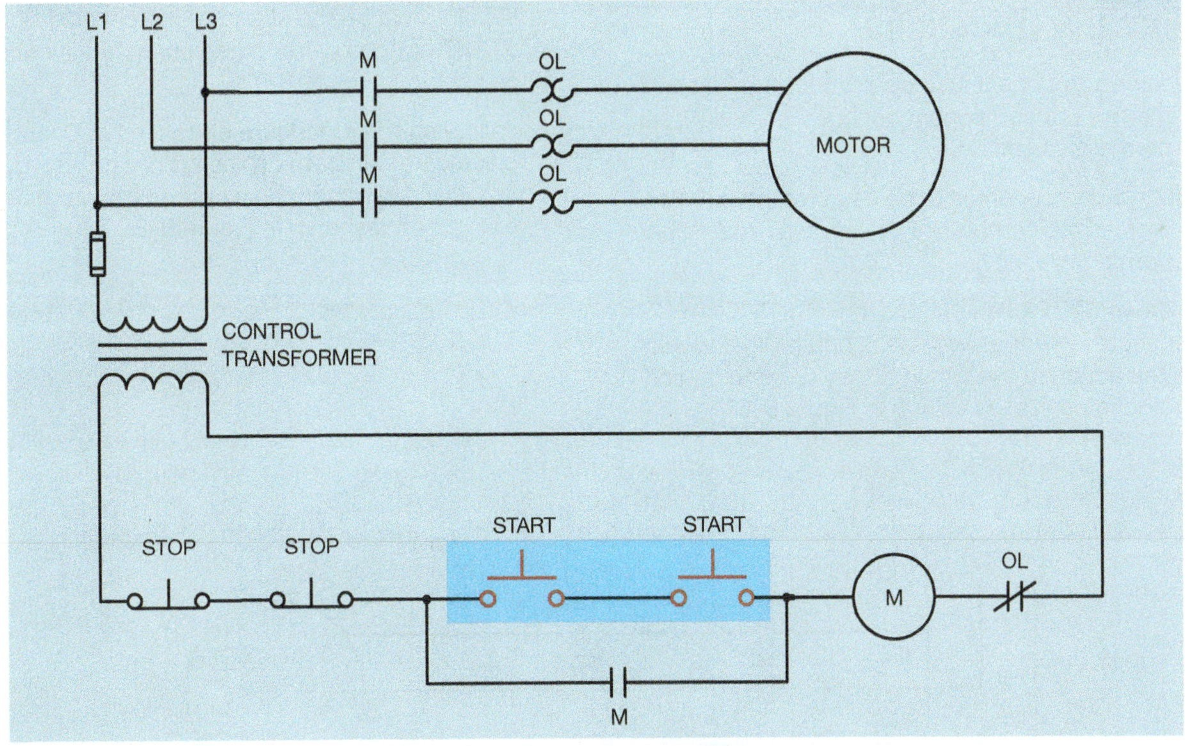

Figure 8–7 The START buttons are connected in series.

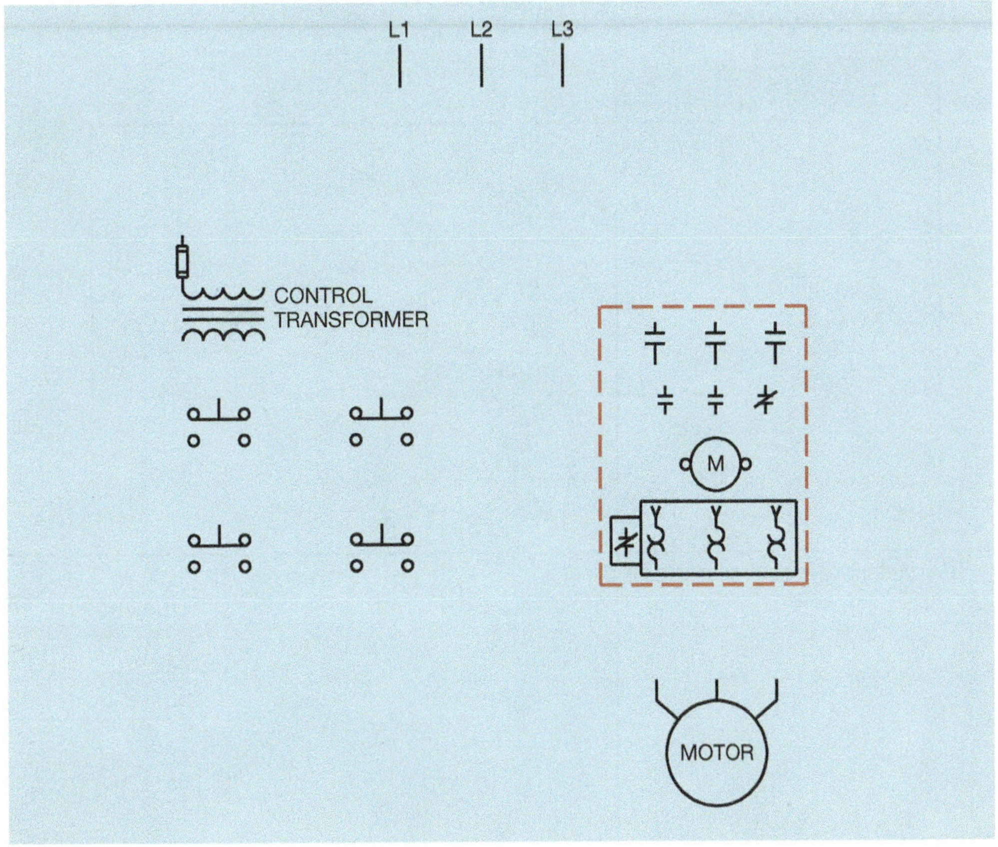

Figure 8–8 Add wire numbers to these components.

Chapter 9

FORWARD-REVERSE CONTROL

Objectives

After studying this chapter the student will be able to:

- Discuss cautions that must be observed in reversing circuits.
- Explain how to reverse a three phase motor.
- Discuss interlocking methods.
- Connect a forward–reverse motor control circuit.

The direction of rotation of any three phase motor can be reversed by changing any two motor T leads (Figure 9–1). Because the motor is connected to the power line regardless of which direction it operates, a separate contactor is needed for each direction. If the reversing starters adhere to NEMA standards, T leads 1 and 3 will be changed (Figure 9–2). Because only one motor is in operation, however, only one overload relay is needed to protect the motor. True reversing controllers contain two separate contactors and one overload relay built into one unit. A vertical reversing starter with overload relay is shown in Figure 9–3, and a horizontal reversing starter without overload relay is shown in Figure 9–4.

Interlocking

Interlocking prevents some action from taking place until some other action has been performed. In the case of reversing starters, interlocking is used to prevent both contactors from being energized at the same time. Having both contactors energized would result in two of the three phase lines being shorted together. Interlocking forces one contactor to be de-energized before the other one can be energized. Three methods can be employed to assure interlocking. Many reversing controls use all three.

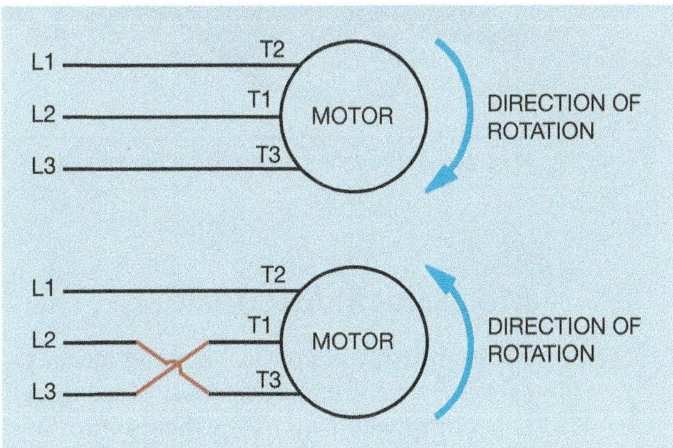

Figure 9–1 The direction of rotation of any three phase motor can be changed by reversing connection to any two motor T leads.

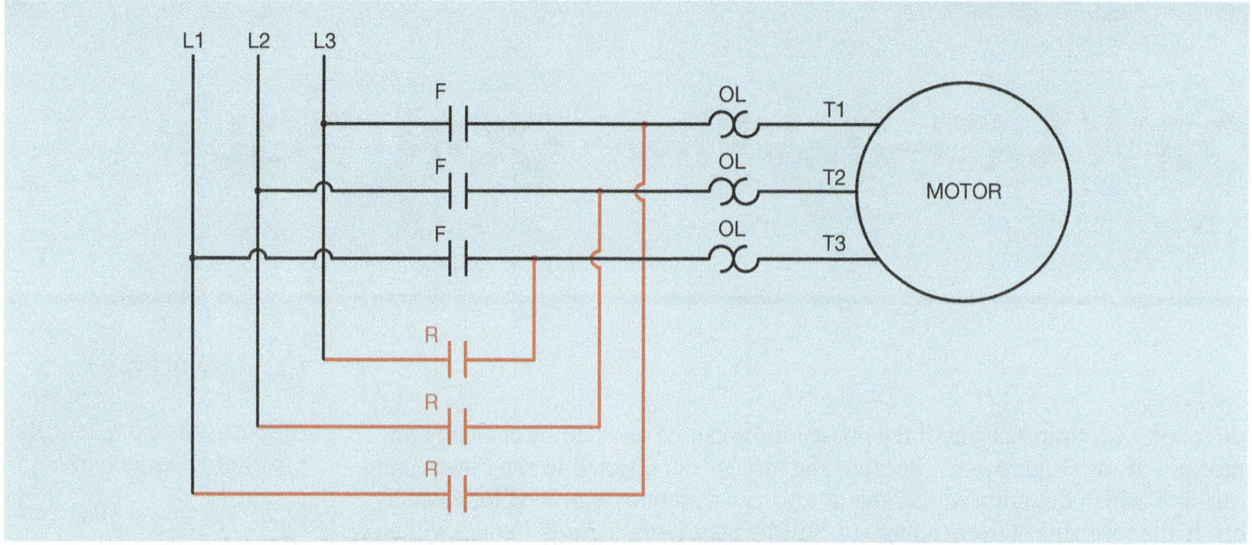

Figure 9–2 Magnetic reversing starters generally change T leads 1 and 3 to reverse the motor.

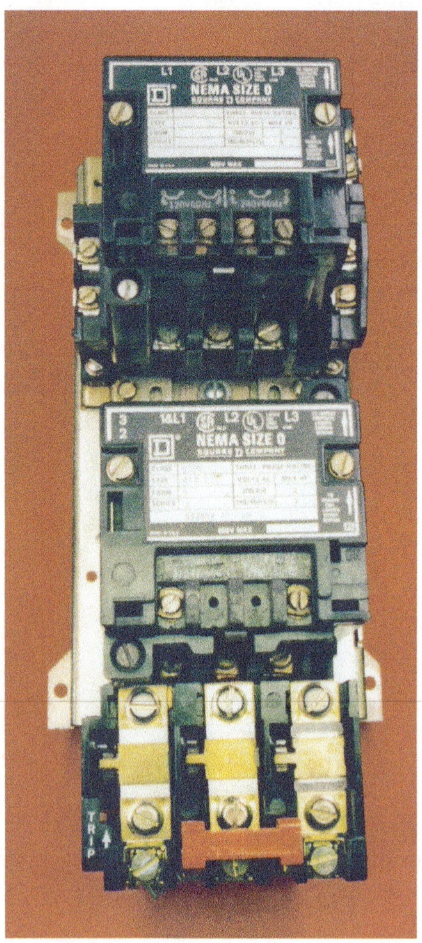

Figure 9–3 Vertical reversing starter with overload relay.

Courtesy Schneider Electric USA, Inc.

Figure 9–4 Horizontal reversing starter without overload relay.

Mechanical Interlocking

Most reversing controllers contain mechanical interlocks as well as electrical interlocks. Mechanical interlocking is accomplished by using the contactors to operate a mechanical lever that prevents the other contactor from closing while one is energized. Mechanical interlocks are supplied by the manufacturer and are built into reversing starters. In a schematic diagram, mechanical interlocks are shown as dashed lines from each coil joining at a solid line (Figure 9–5).

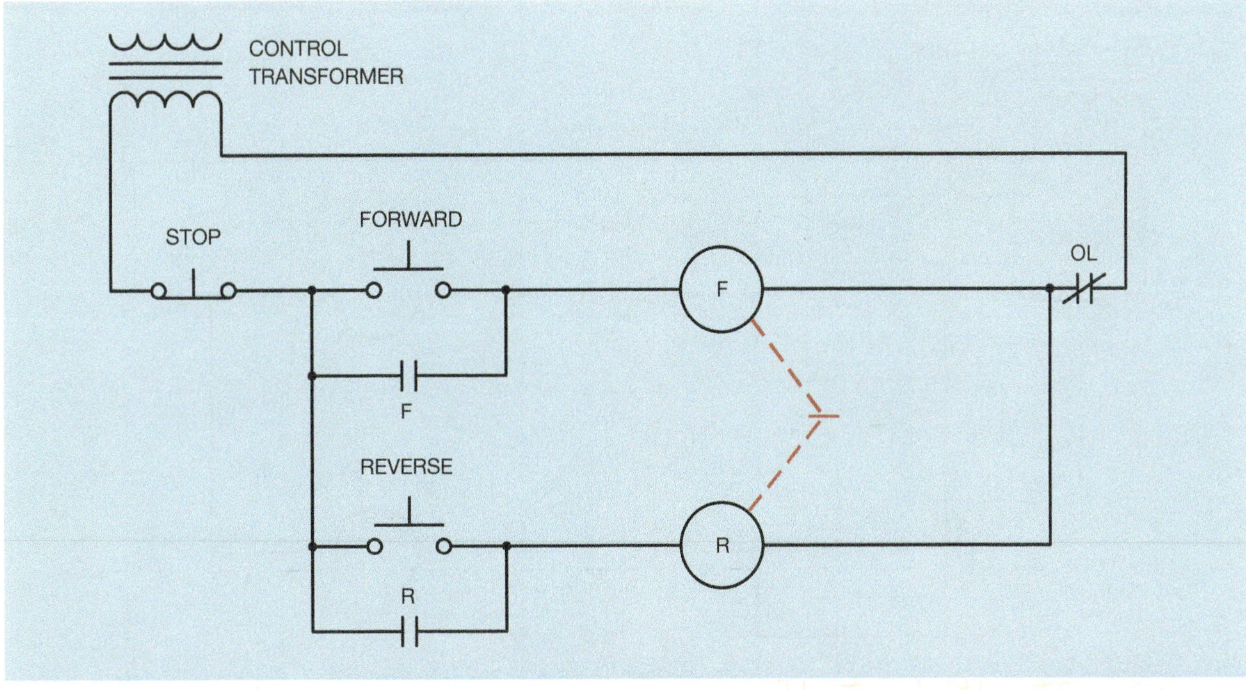

Figure 9–5 Mechanical interlocks are indicated by dashed lines extending from each coil.

Electrical Interlocking

Two methods of electrical interlocking are available. One method is accomplished with the use of double-acting push buttons (Figure 9–6). The dashed lines drawn between the push buttons indicate that they are mechanically connected. Both push buttons will be pushed at the same time. The normally closed part of the forward push button is connected in series with R coil, and the normally closed part of the reverse push button is connected in series with F coil. If the motor is running in the forward direction and the reverse push button is pressed, the normally closed part of the push button opens and disconnects F coil from the line before the normally open part closes to energize R coil. The normally closed section of either push button has the same effect on the circuit as pressing the STOP button.

The second method of electrical interlocking is accomplished by connecting the normally closed auxiliary contacts on one contactor in series with the coil of the other contactor (Figure 9–7). Assume that the forward push button is pressed and F coil energizes. This causes all F contacts to change position. The three F load contacts close and connect the motor to the line. The normally open F auxiliary contact closes to maintain the circuit when the forward push button is released, and the normally closed F auxiliary contact connected in series with R coil opens (Figure 9–8).

If the opposite direction of rotation is desired, the STOP button must be pressed first. If the reverse push button were to be pressed first, the now open F auxiliary contact connected in series with R coil would prevent a complete circuit from being established. Once the STOP button has been pressed, however, F coil de-energizes and all F contacts return to their normal position. The reverse push button can now be pressed to energize R coil (Figure 9–9). When R coil energizes, all R contacts change position. The three R load contacts close and connect the motor to the line. Notice, however, that two of the motor T leads are connected to different lines. The normally closed R auxiliary contact opens to prevent the possibility of F coil being energized until R coil is de-energized.

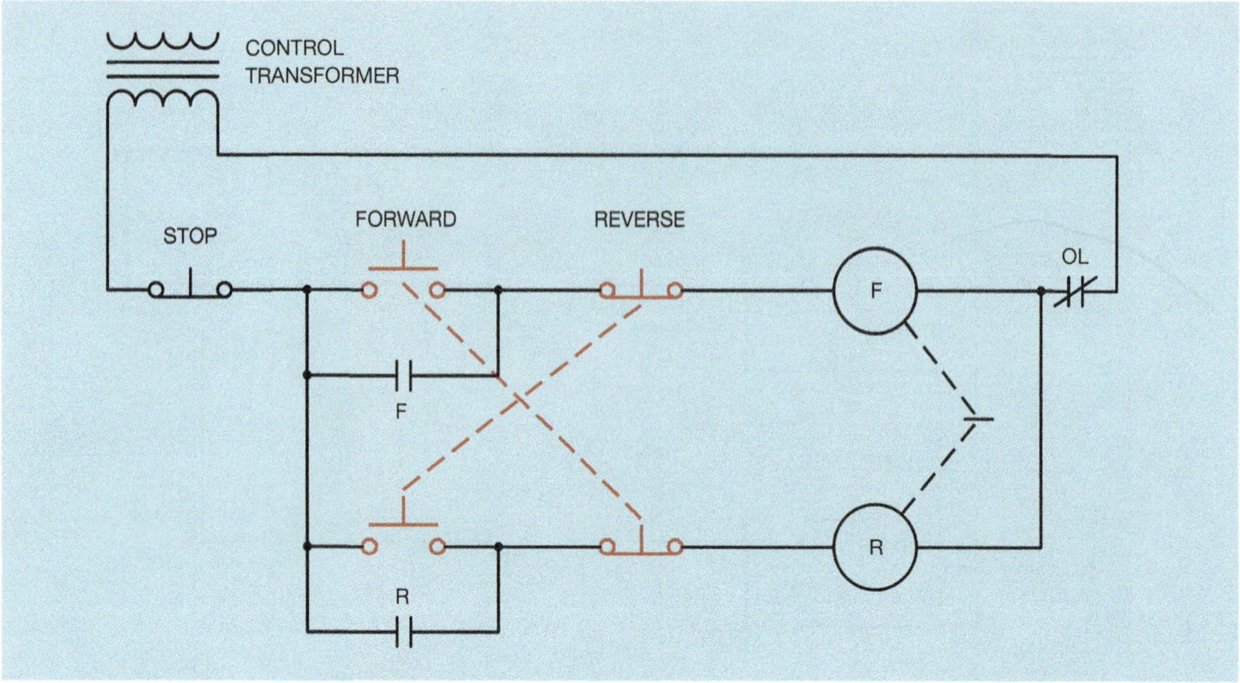

Figure 9–6 Interlocking with double-acting push buttons.

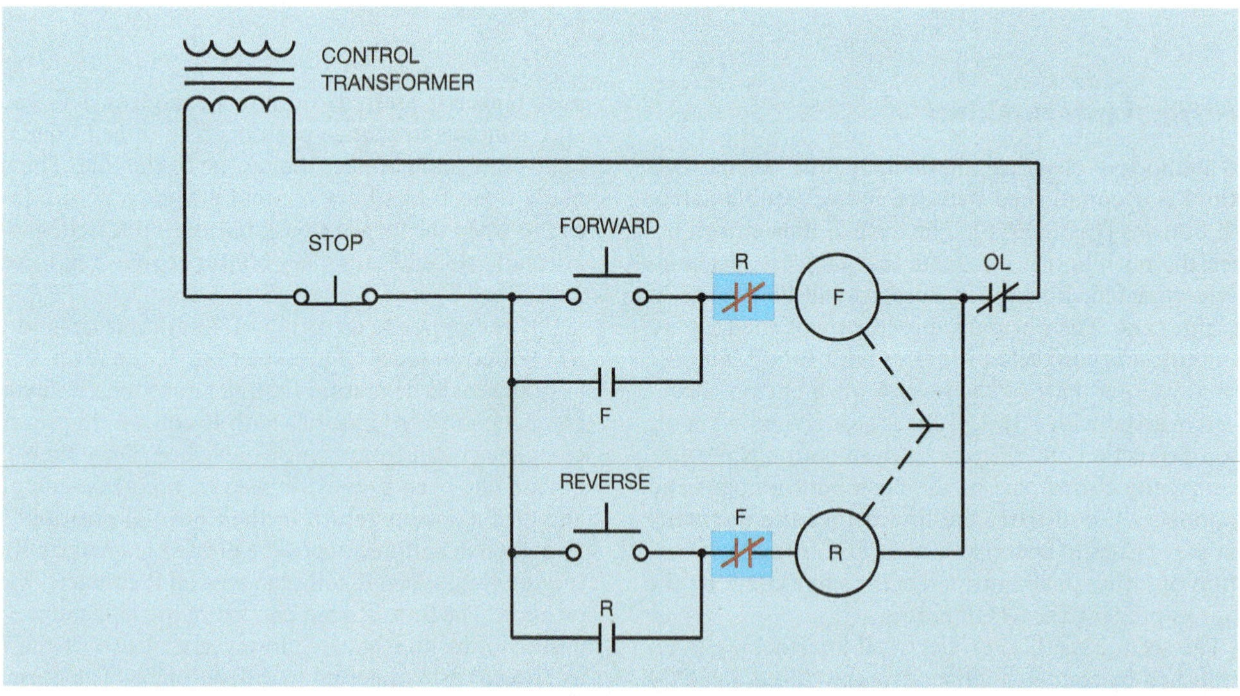

Figure 9–7 Electrical interlocking is also accomplished with normally closed auxiliary contacts.

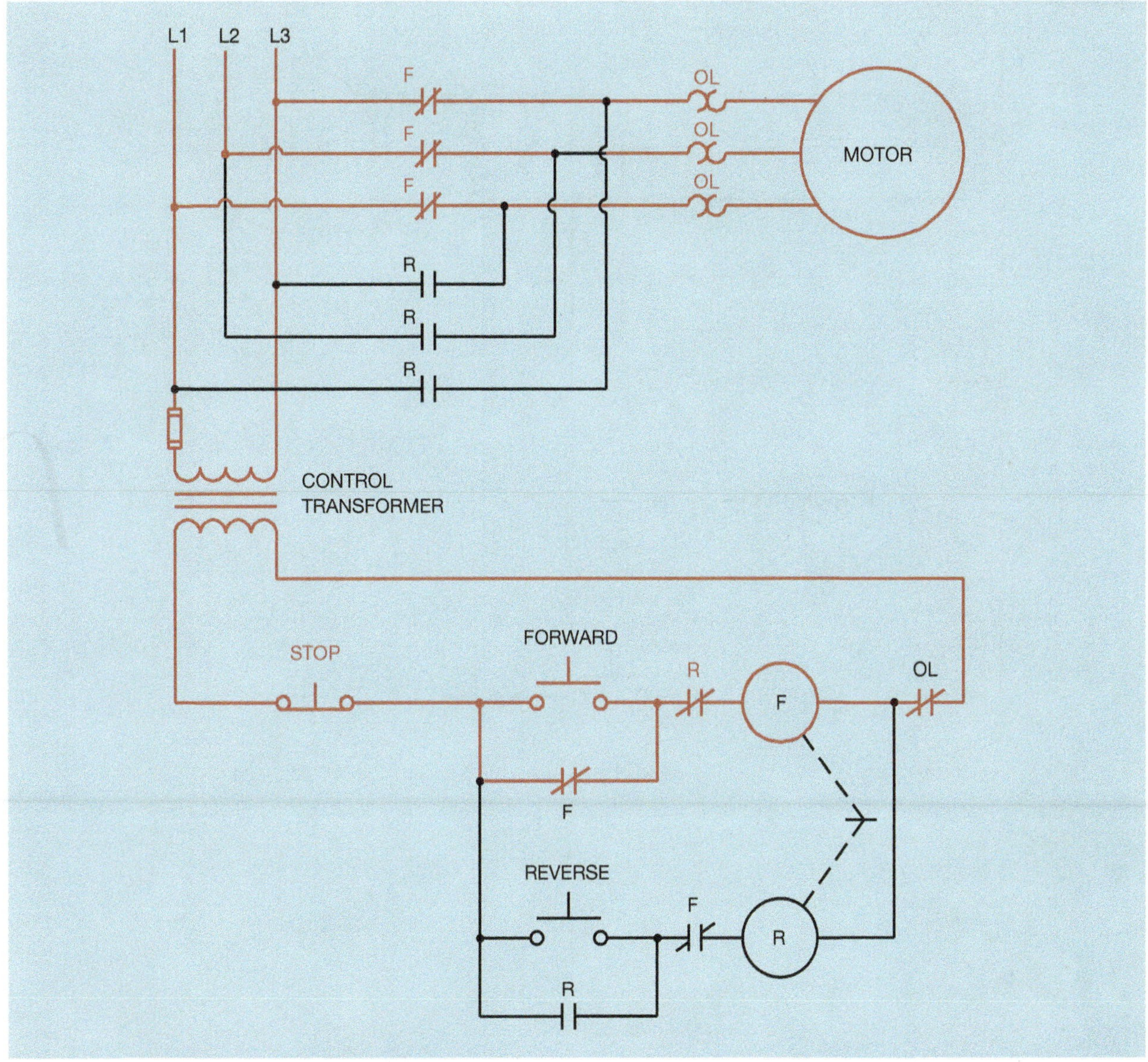

Figure 9–8 Motor operating in the forward direction.

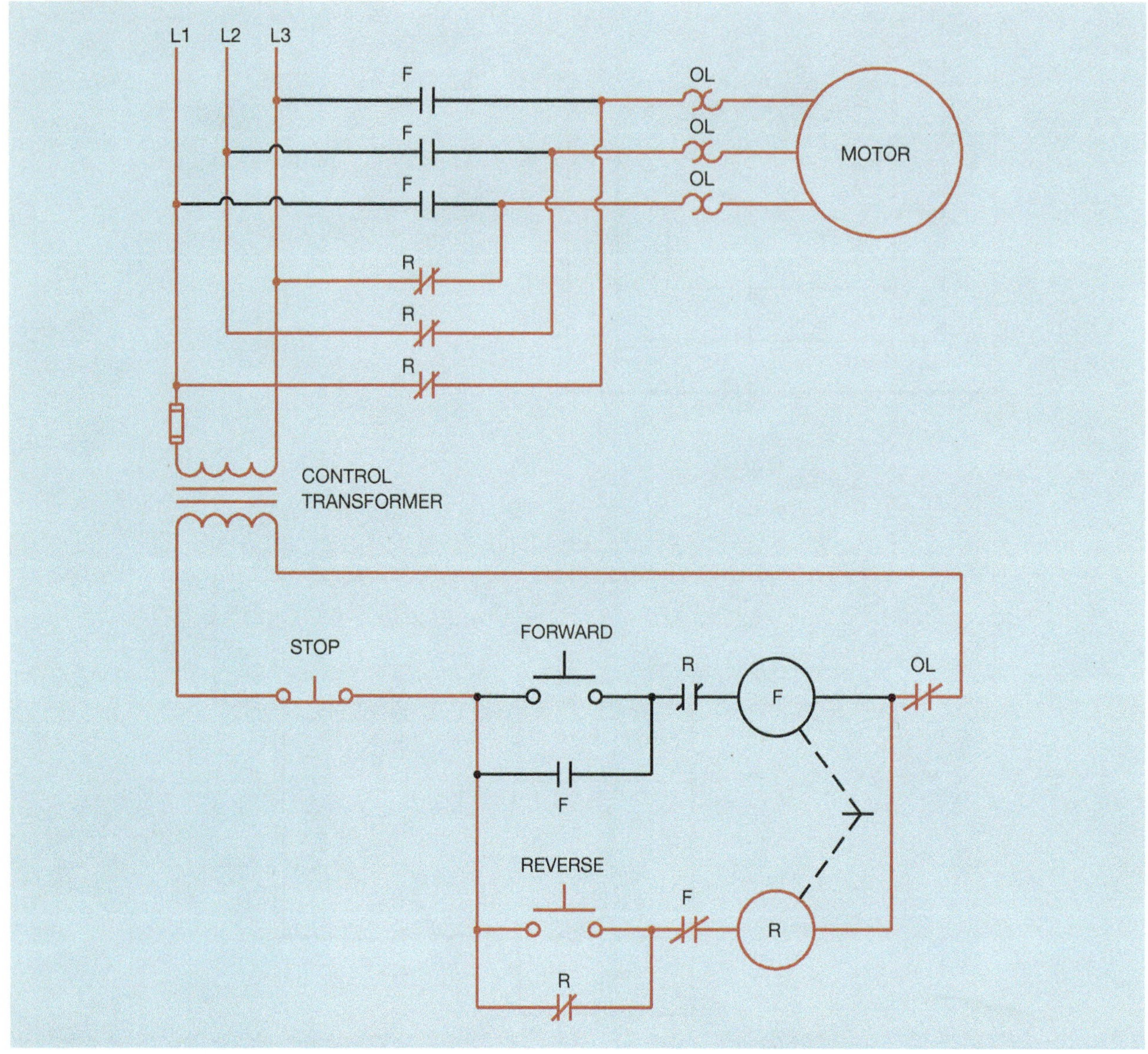

Figure 9–9 Motor operating in the reverse direction.

Developing a Wiring Diagram

The same basic procedure will be used to develop a wiring diagram from the schematic as was followed in the previous chapters. The components needed to construct this circuit are shown in Figure 9–10. In this example it is assumed that two contactors and a separate three phase overload relay will be used.

The first step is to place wire numbers on the schematic diagram. A suggested numbering sequence is shown in Figure 9–11. The next step is to place the wire numbers beside the corresponding components of the wiring diagram as in Figure 9–12. The circuit with connecting wires is shown in Figure 9–13.

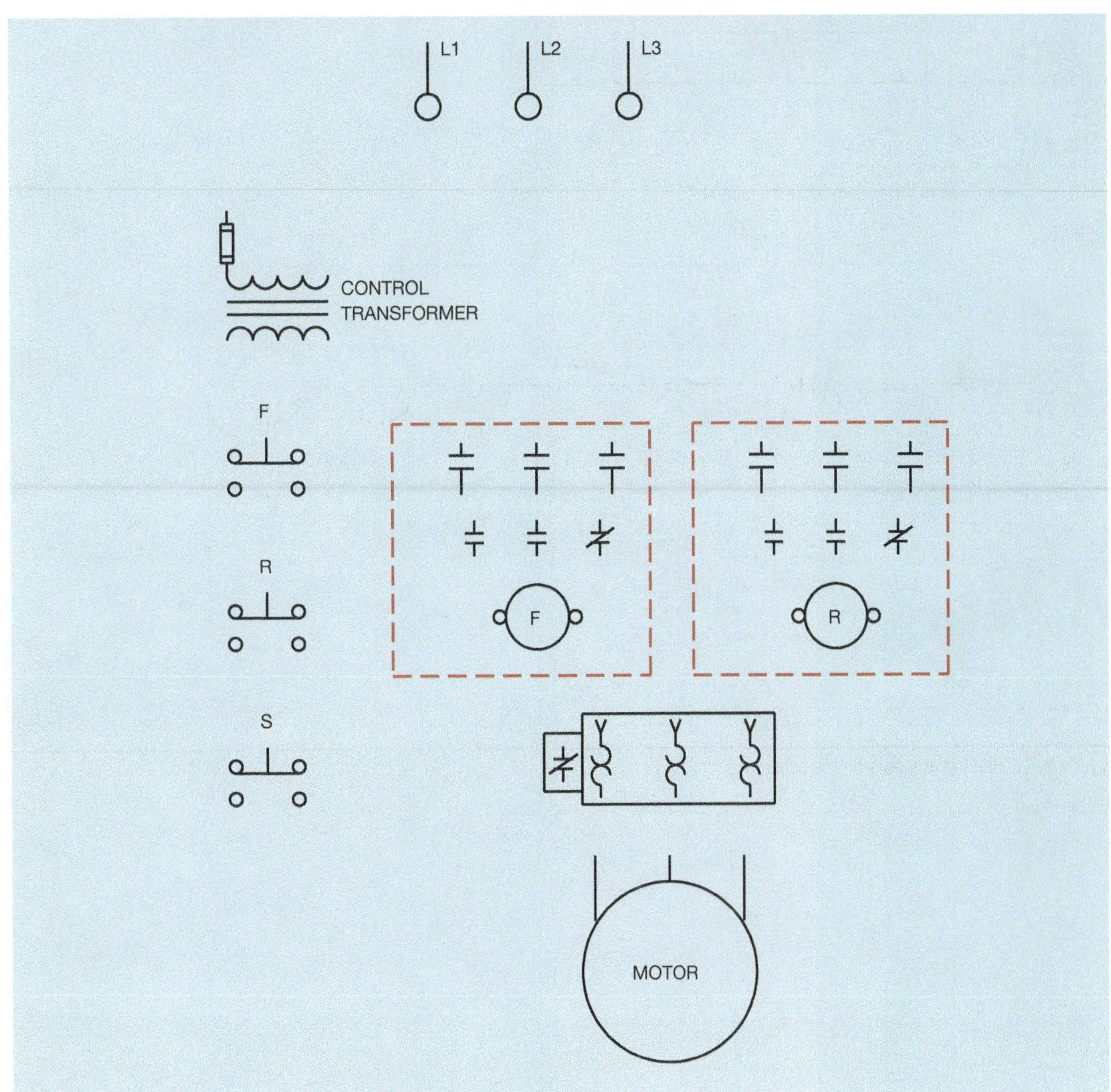

Figure 9–10 Components needed to construct a reversing control.

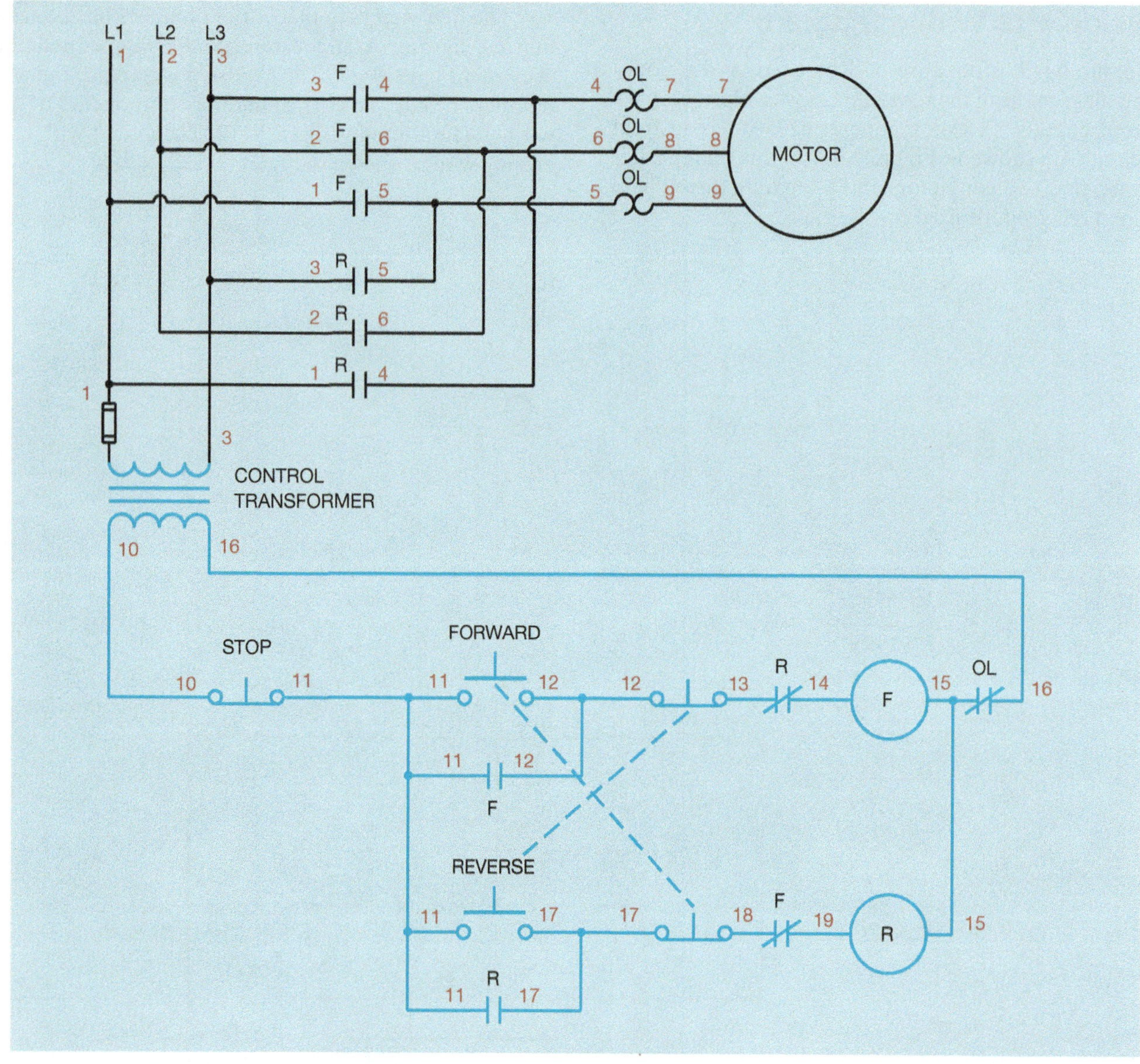

Figure 9–11 Placing numbers on the schematic.

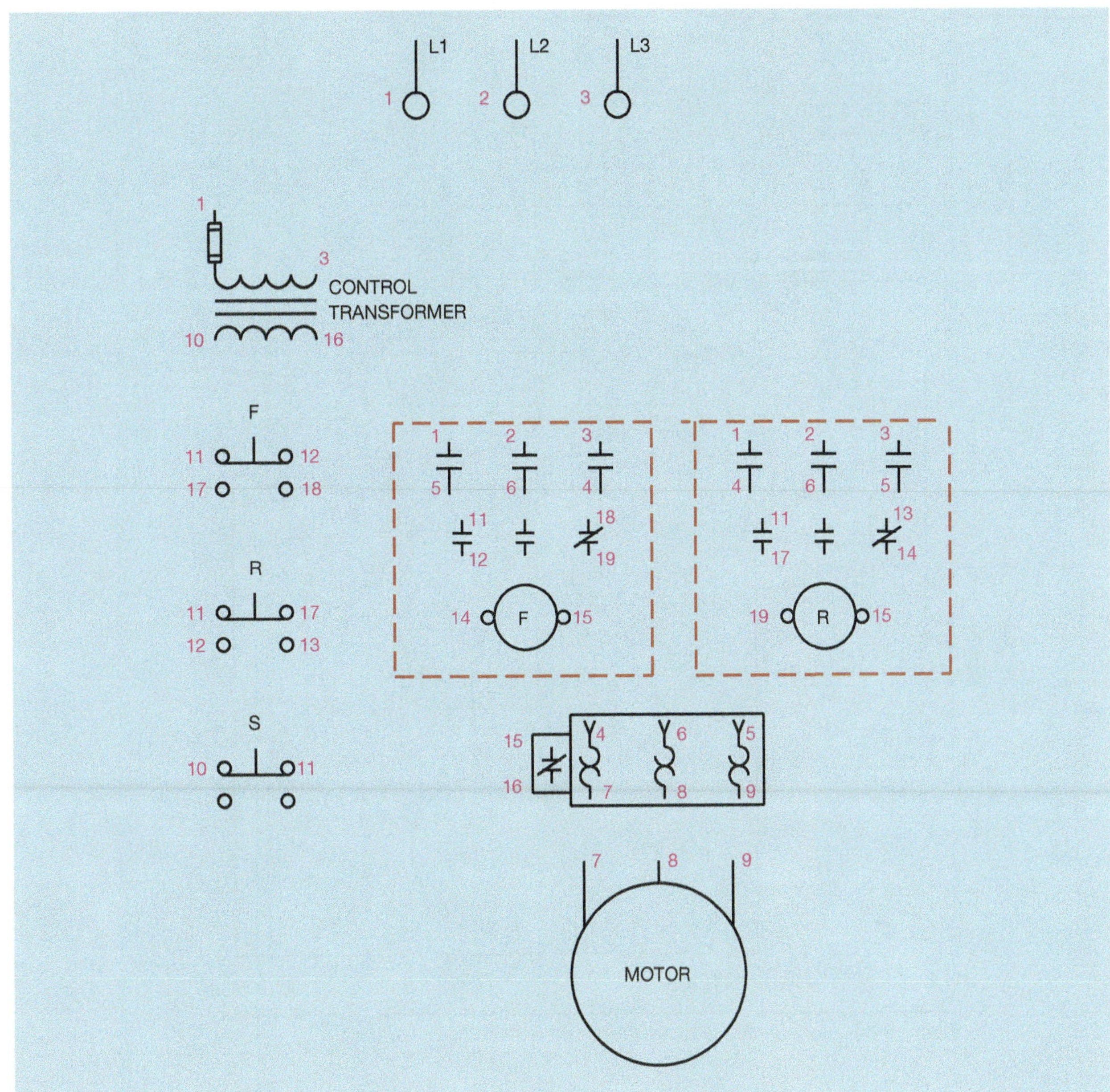

Figure 9–12 Components needed to construct a reversing control circuit.

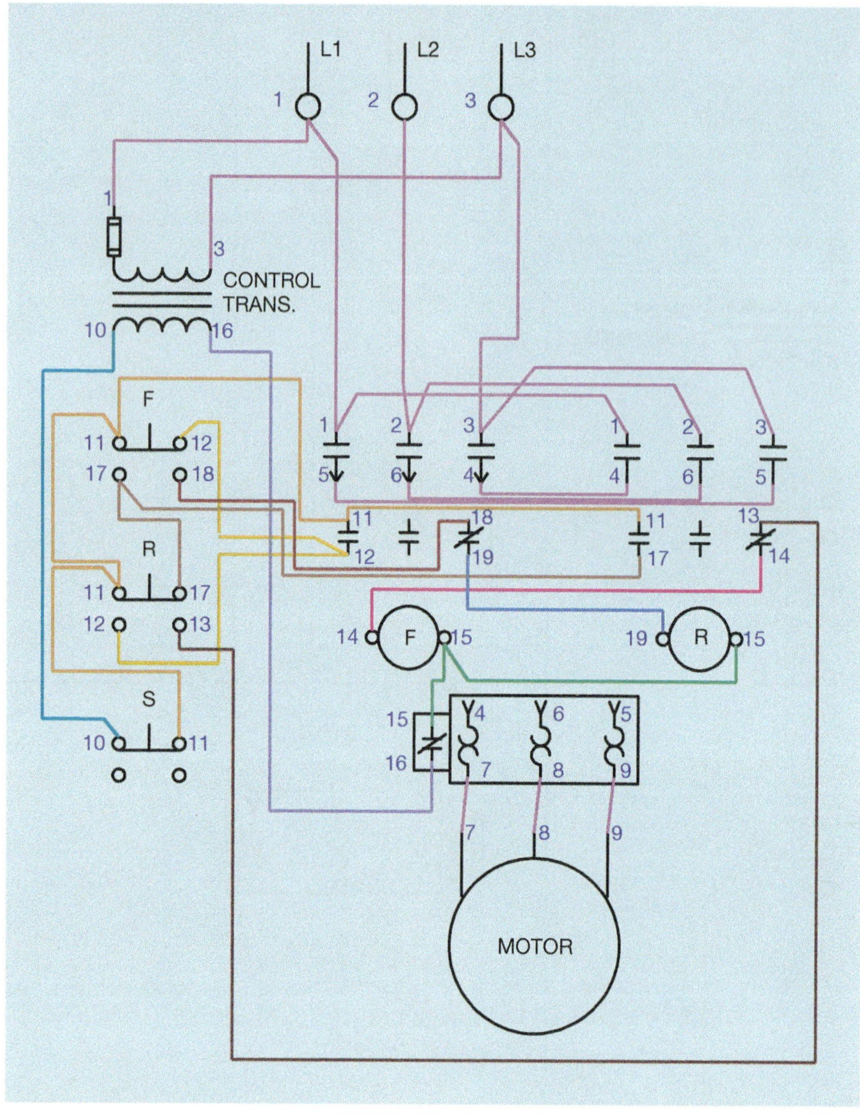

Figure 9–13 Forward–reverse circuit shown with connecting wires.

Review Questions

1. How can the direction of rotation of a three phase motor be changed?
2. What is interlocking?
3. Referring to the schematic shown in Figure 9–7, how would the circuit operate if the normally closed R contact connected in series with F coil were to be connected normally open?
4. What would be the danger, if any, if the circuit were to be wired as stated in question 3?
5. How would the circuit operate if the normally closed auxiliary contacts were to be connected so that F contact was connected in series with F coil and R contact was connected in series with R coil (Figure 9–7)?
6. Assume that the circuit shown in Figure 9–7 were to be connected as shown in Figure 9–14. In what way would the operation of the circuit be different, if at all?

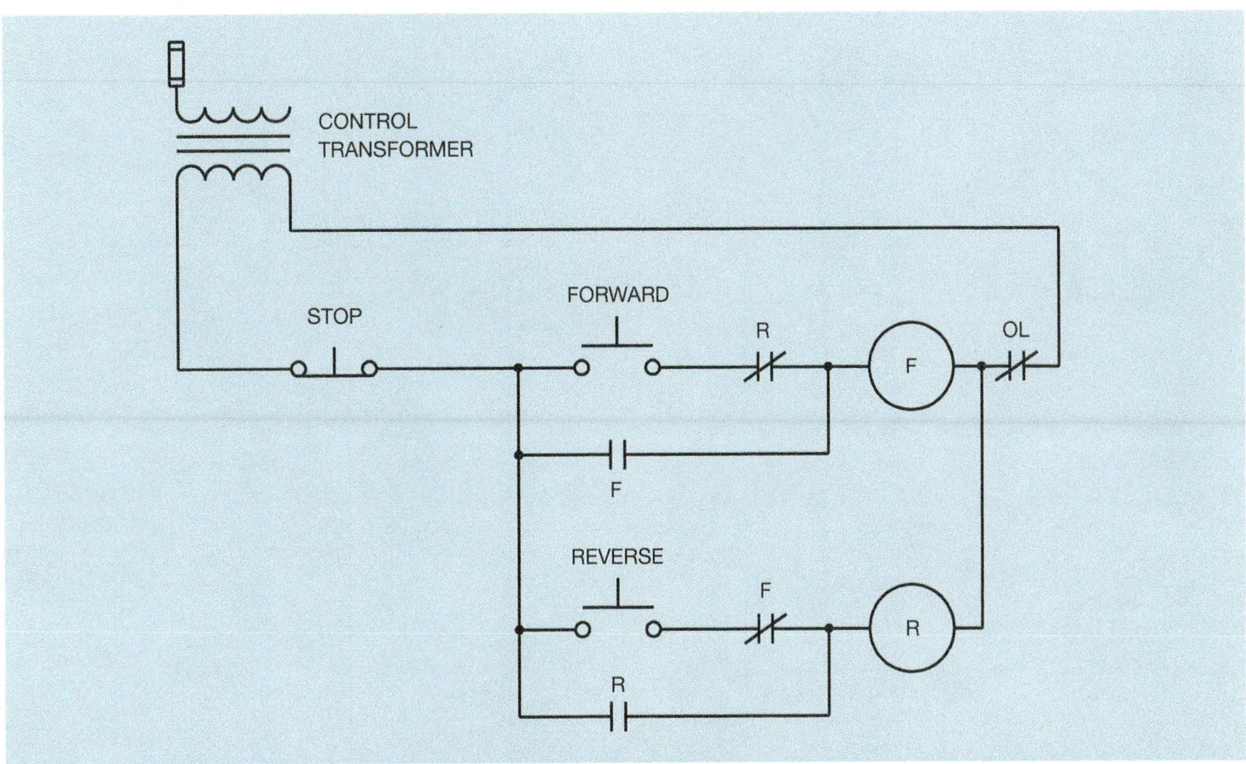

Figure 9–14 The position of the holding contacts has been changed.

Chapter 10

JOGGING AND INCHING

Objectives

After studying this chapter the student will be able to:

- Define the term *jogging.*
- State the purpose of jogging.
- State difference between jogging and inching.
- Describe the operation of a jogging control circuit using control relays.
- Describe the operation of a jogging control circuit using a selector switch.
- Connect a jogging circuit.

The definition of jogging or inching as described by NEMA is *"the quickly repeated closure of a circuit to start a motor from rest for the purpose of accomplishing small movements of the driven machine."* The term *jogging* actually means to start a motor with short jabs of power at full voltage. The term *inching* means to start a motor with short jabs of power at reduced voltage. Although the two terms mean different things, they are often used interchangeably because they are accomplished by preventing a holding circuit.

Jogging Circuits

Various jogging circuits will be presented in this chapter. As with many other types of control circuits, jogging can be accomplished in different ways. Basically, jogging is accomplished by preventing the holding contact from sealing the circuit around the START push button when the motor starter energizes. It should also be noted that jogging circuits require special motor starters rated for jogging duty.

One of the simplest jogging circuits is shown in Figure 10–1. This circuit is basically a START–STOP push button control circuit that has been reconnected so that the START button is in parallel with both the STOP button and holding contact. To jog the motor, simply hold down the STOP button and jog the circuit by pressing the START button. To run the motor, release the STOP button and press the START button. If the motor is in operation, the STOP button breaks the circuit to the holding contact and de-energizes M coil.

Double-Acting Push Buttons

Jogging can also be accomplished using a double-acting push button. Two circuits of that type are shown in Figure 10–2. The normally closed section of the jog push button is connected in such a manner that when the button is pushed, it will defeat the holding contact and prevent it from sealing the circuit. The normally open section of the jog button completes a circuit to energize the coil of the motor starter. When the button is released, the normally open section breaks the circuit to M coil before the normally closed section reconnects to the circuit. This permits the starter to reopen the holding contacts before the normally closed section of the jog button reconnects. Although this circuit is sometimes used for jogging, it does have a severe problem. The action of either of these two circuits depends on the normally open M auxiliary contact (holding contact) used to seal the circuit being open before the normally closed section of the jog button makes connection. Because push buttons employ a spring to return the contacts to their normal position, if a person's finger slips off the jog button, it is possible for the spring to reestablish connection with the normally closed contacts before the holding

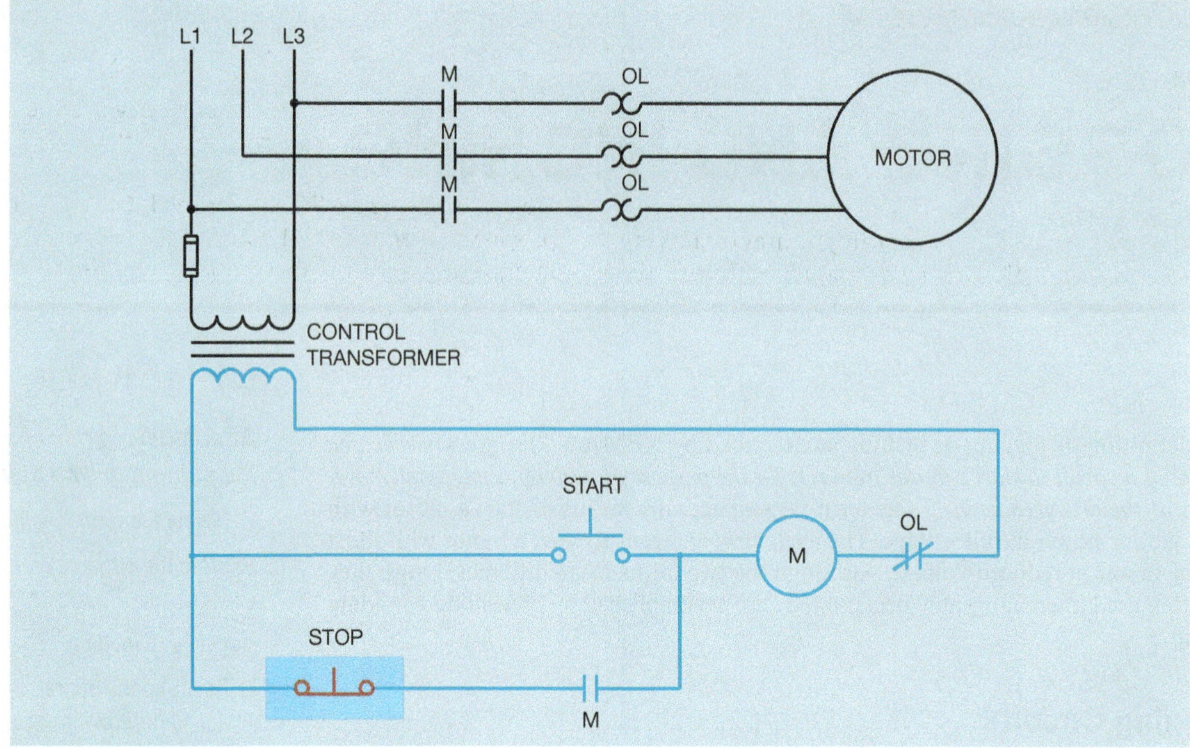

Figure 10–1 The STOP button prevents the holding contact from sealing the circuit.

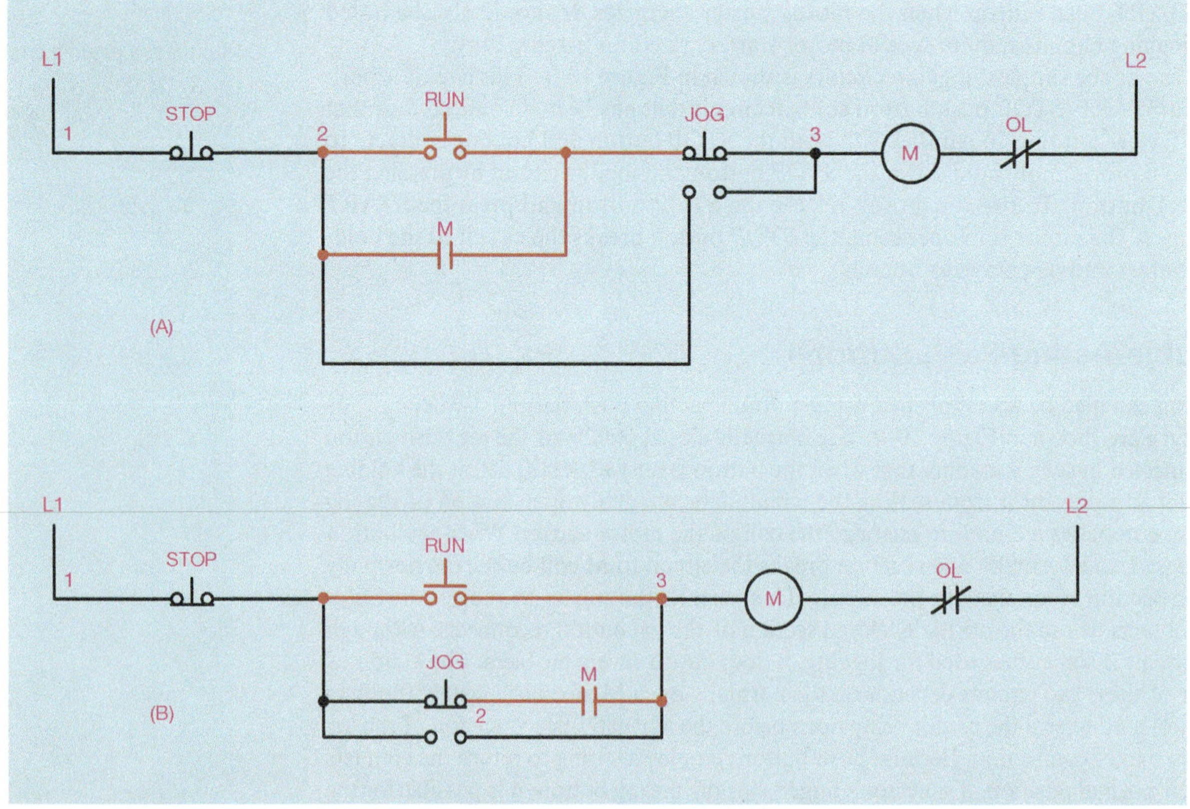

Figure 10–2 Double-acting push buttons are used to provide jogging control.

contact has time to reopen. This would cause the motor to continue running instead of stopping. In some cases, this could become a significant safety hazard.

Using a Control Relay

The addition of a control relay to the jog circuit eliminates the problem of the holding contacts making connection before the normally closed section of the jog push button makes connection. Two circuits that employ a control relay to provide jogging are shown in Figure 10–3. In both of these circuits, the control relay provides the auxiliary holding contacts, not the M starter. The jog push button energizes the coil of M motor starter, but does not energize the coil of control relay CR. The START push button is used to energize the coil of CR relay. When energized, CR relay contacts provide connection to M coil. The use of control relays in a jogging circuit is very popular because of the simplicity and safety offered.

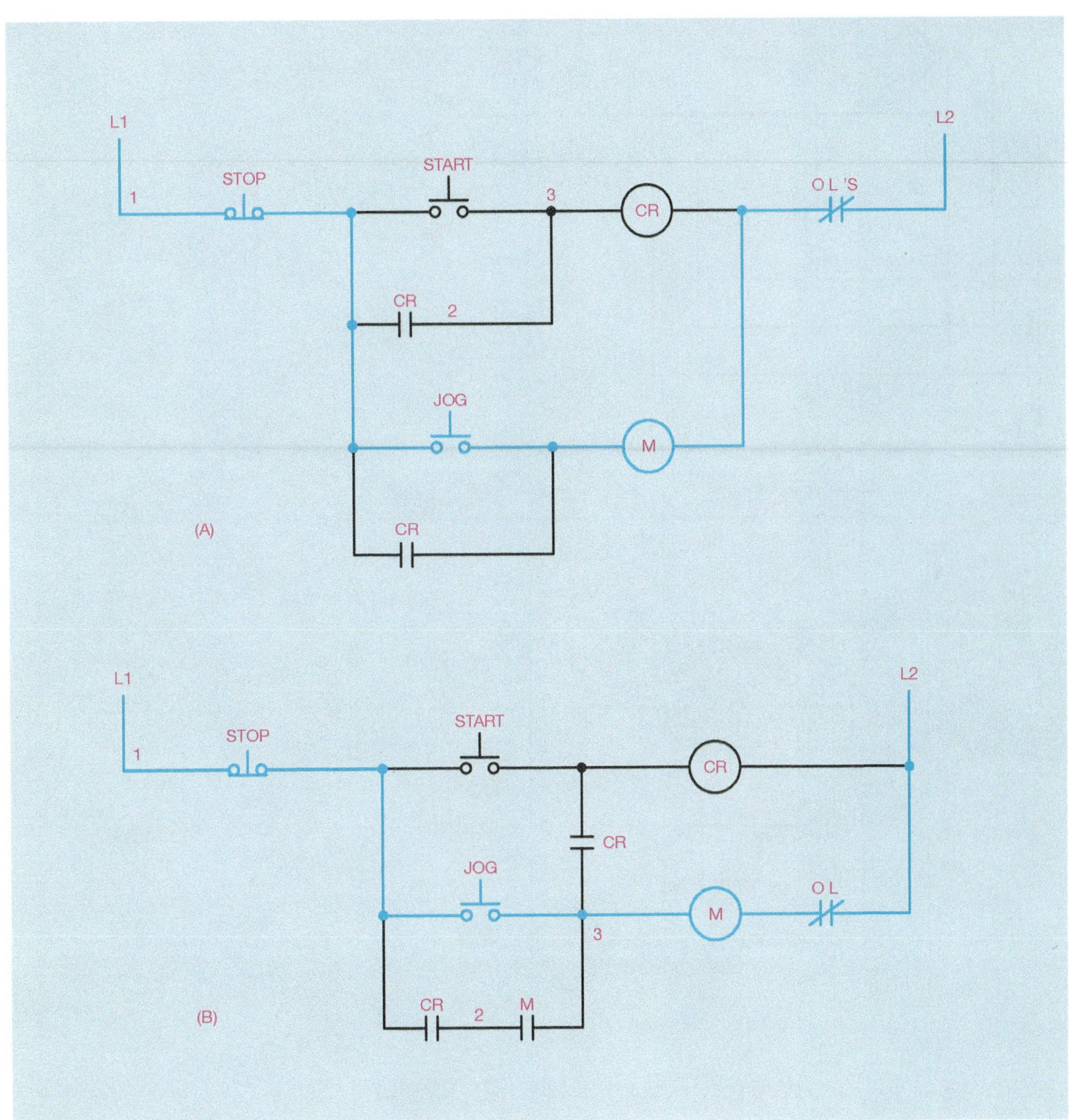

Figure 10–3 Control relays provide jogging control.

A jogging circuit for a forward–reverse control is shown in Figure 10–4. Note that a control relay is used to provide jogging in either direction. When the forward jog push button is pressed, the normally open section makes connection and provides power to F coil. This causes F load contacts to close and connect the motor to the power line. The normally open F auxiliary contact closes, also, but the normally closed section of the forward jog button is now open, preventing coil CR from being energized. Because CR contact remains open, the circuit to F coil cannot be sealed by the normally open F auxiliary contact.

If the forward START button is pressed, a circuit is completed to F coil causing all F contacts to change position. The normally open F auxiliary contact closes and provides a path through the normally closed section

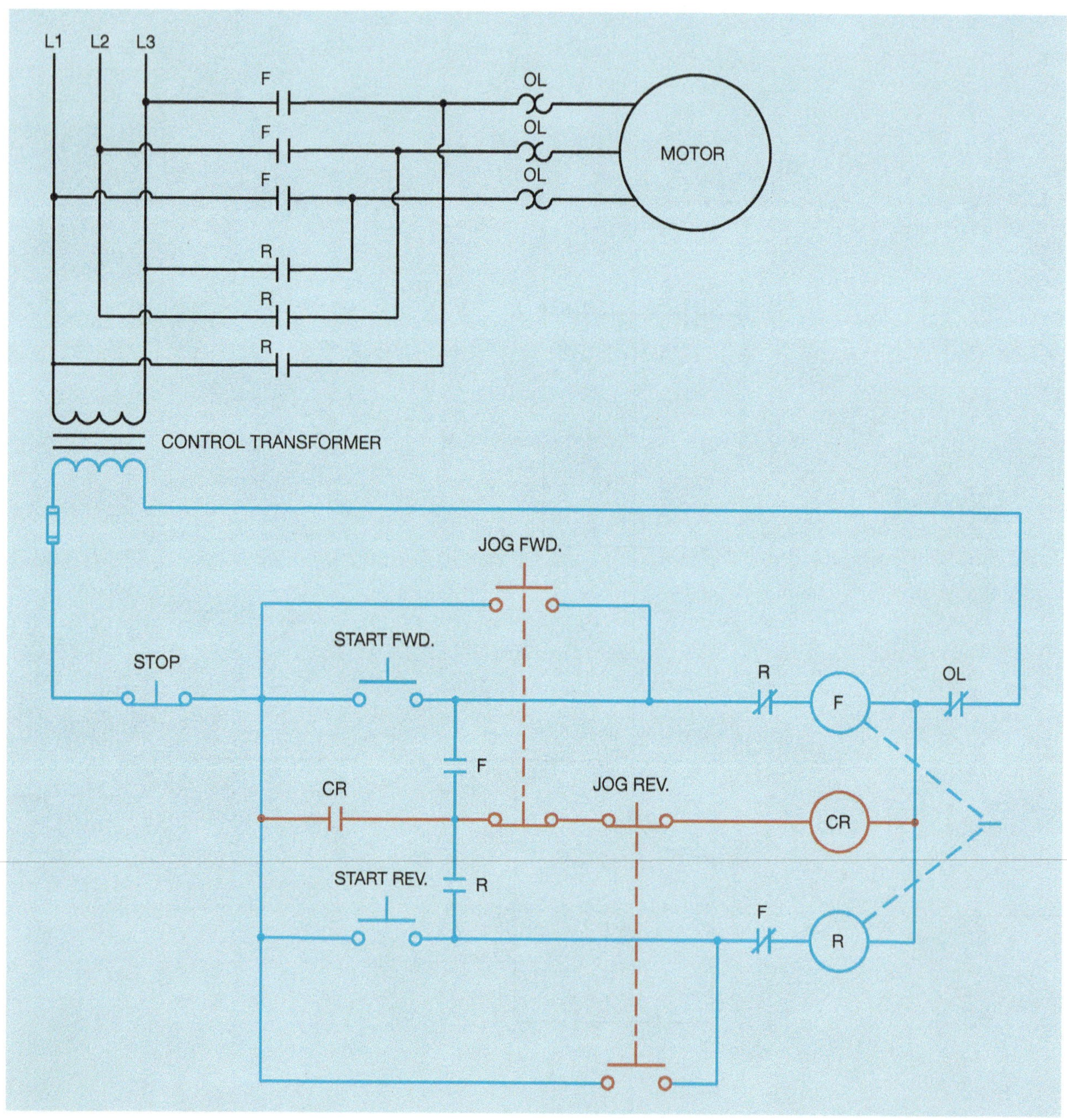

Figure 10–4 Jogging using a control relay on a forward–reverse control.

of both jog buttons to CR coil. This causes CR auxiliary contact to close and provide a current path through the now closed F auxiliary contact to F coil, sealing the circuit when the forward push button is released. The reverse jog button and reverse START button operate the same way. Note also that normally closed F and R auxiliary contacts are used to provide interlocking for the forward–reverse control.

Jogging Controlled by a Selector Switch

Another method of obtaining jogging control is with the use of a selector switch. The switch is used to break the connection to the holding contacts (Figure 10–5). In this circuit a single pole, single throw toggle switch is used. When the switch is in the ON position, connection is made to the holding contacts. If the switch is in the OFF position the holding contacts cannot seal the circuit when the START button is released. Note that the START button acts as both the START and jog button for this circuit. A selector switch can be used to provide the same basic type control (Figure 10–6).

Inching Controls

As stated previously, jogging and inching are very similar in that both are accomplished by providing short jabs of power to a motor to help position certain pieces

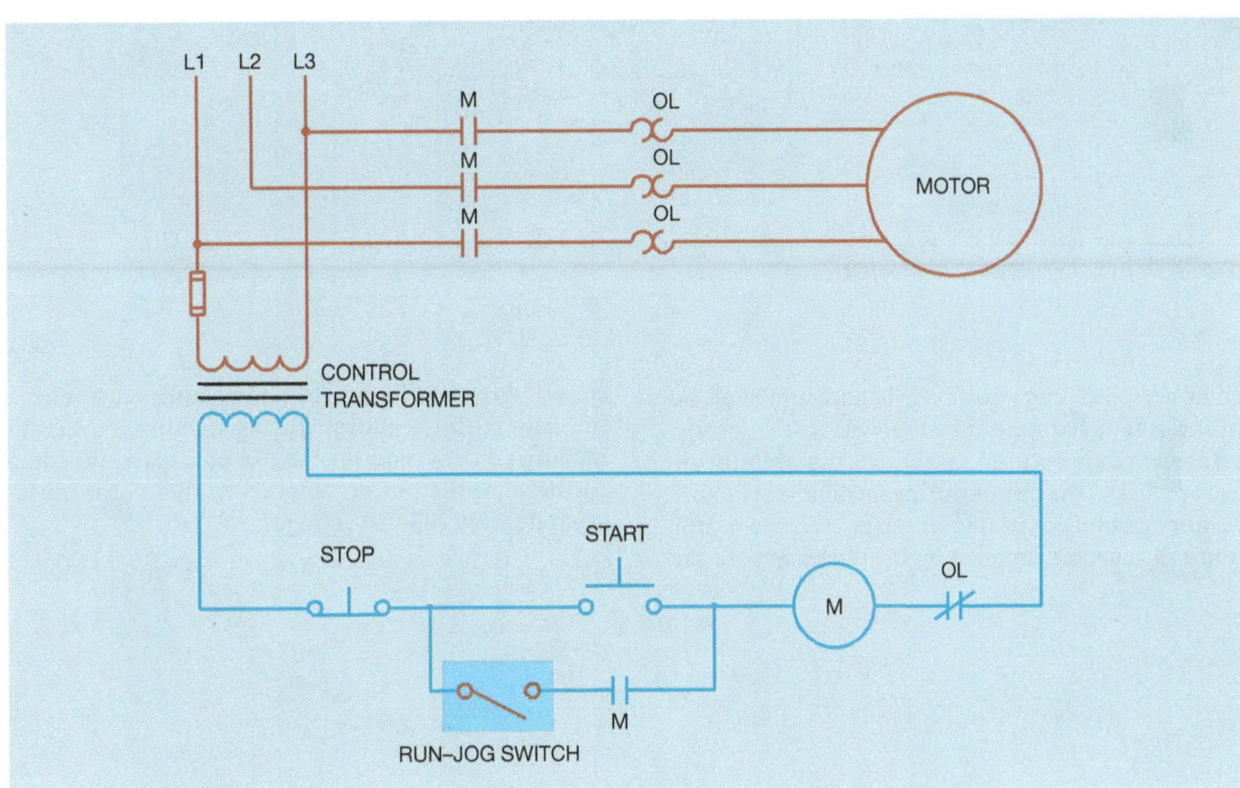

Figure 10–5 A single pole, single throw toggle switch provides jog or run control.

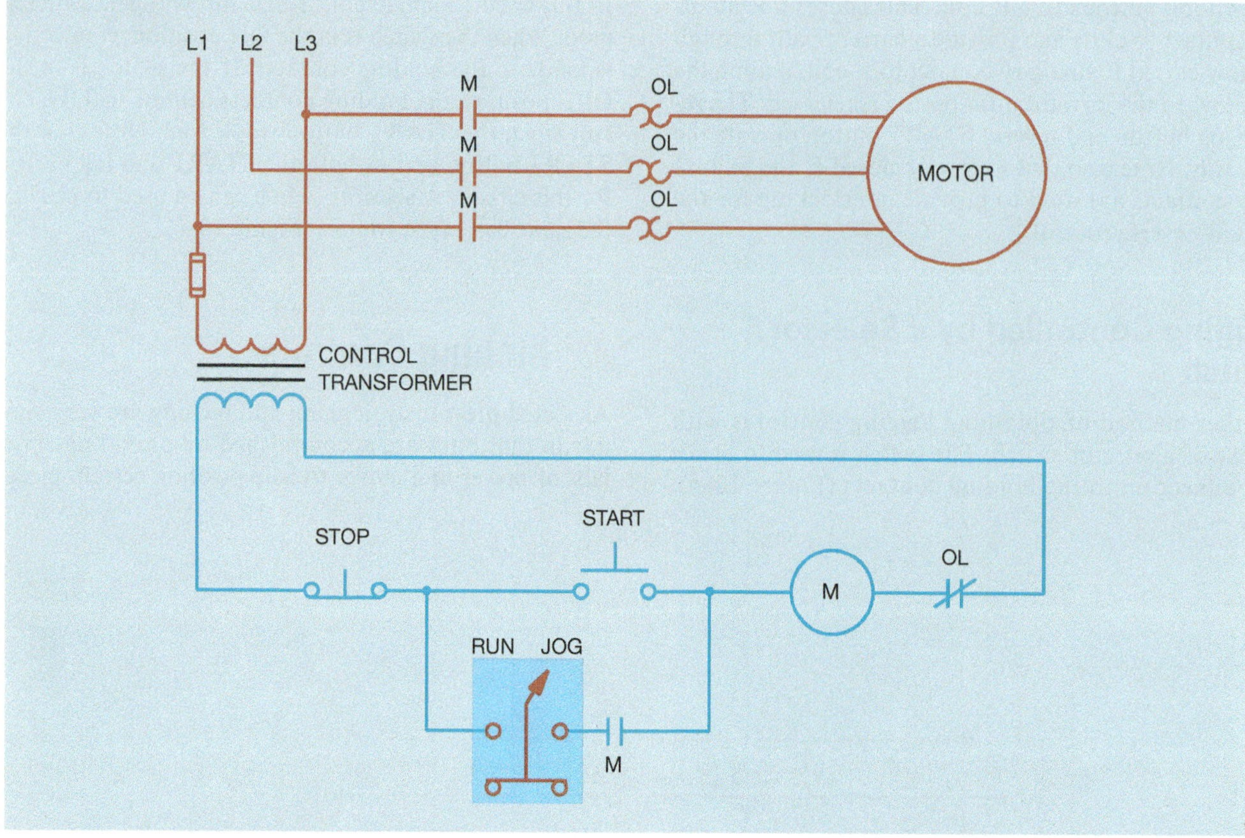

Figure 10–6 A selector switch provides run–jog control.

of machinery. Inching, however, is accomplished by providing a reduced amount of power to the motor. Transformers can be used to reduce the amount of voltage applied to the motor during inching, or reactors or resistors can be connected in series with the motor to reduce the current supplied by the power line. In the circuit shown in Figure 10–7, resistors are connected in series with the motor during inching. Notice that inching control requires the use of a separate contactor because the power supplied to the motor must be separate from full line voltage.

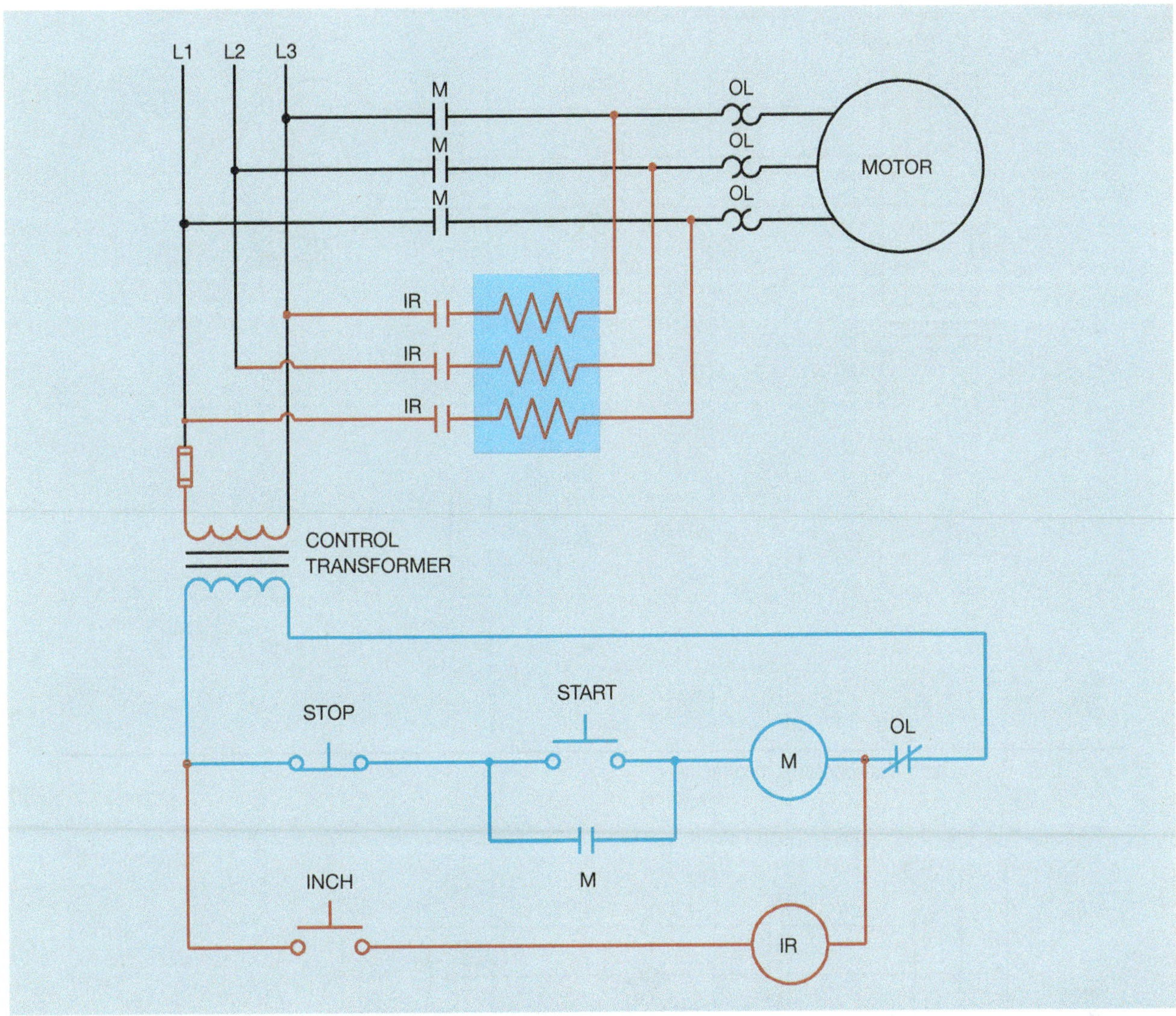

Figure 10–7 Resistors are used to reduce power to the motor.

Review Questions

1. Explain the difference between inching and jogging.
2. What is the main purpose of jogging?
3. Refer to the circuit shown in Figure 10–8. In this circuit, the jog button has been connected incorrectly. The normally closed section has been connected in parallel with the run push button and the normally open section has been connected in series with the holding contacts. Explain how this circuit operates.
4. Refer to the circuit shown in Figure 10–9. In this circuit the jog push button has again been connected incorrectly. The normally closed section of the button has been connected in series with the normally open run push button and the normally open section of the jog button is connecting in parallel with the holding contacts. Explain how this circuit operates.

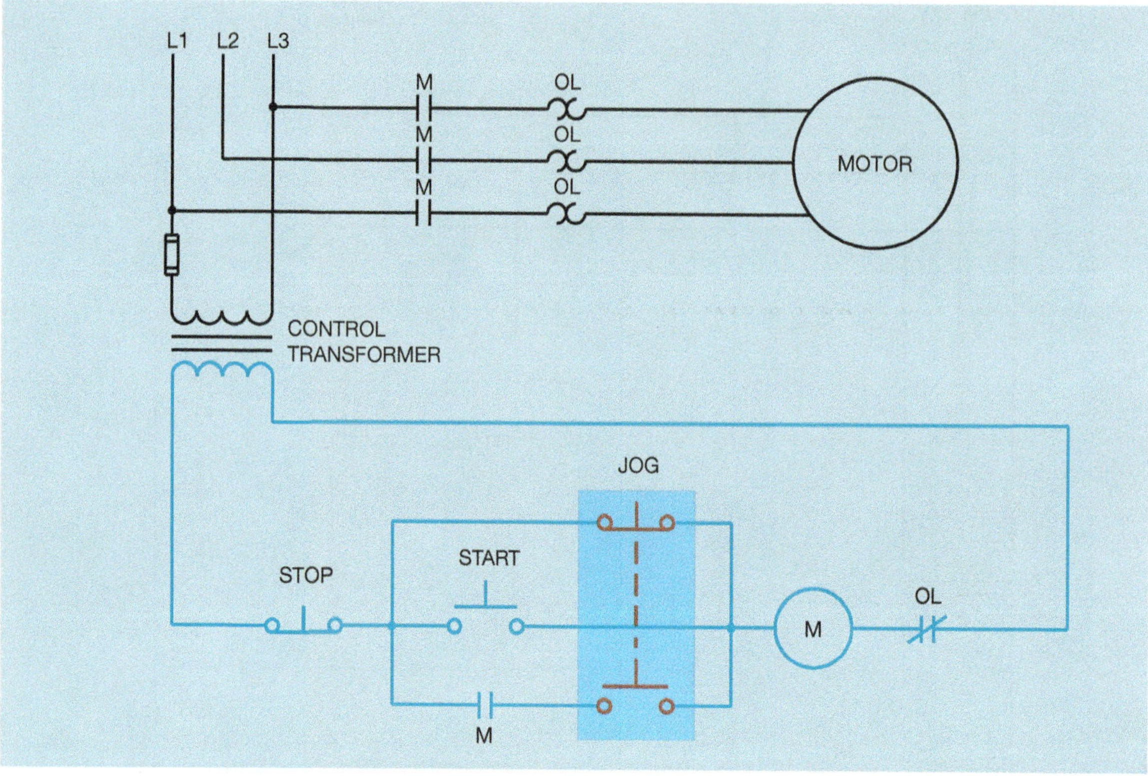

Figure 10–8 The jog button is connected incorrectly.

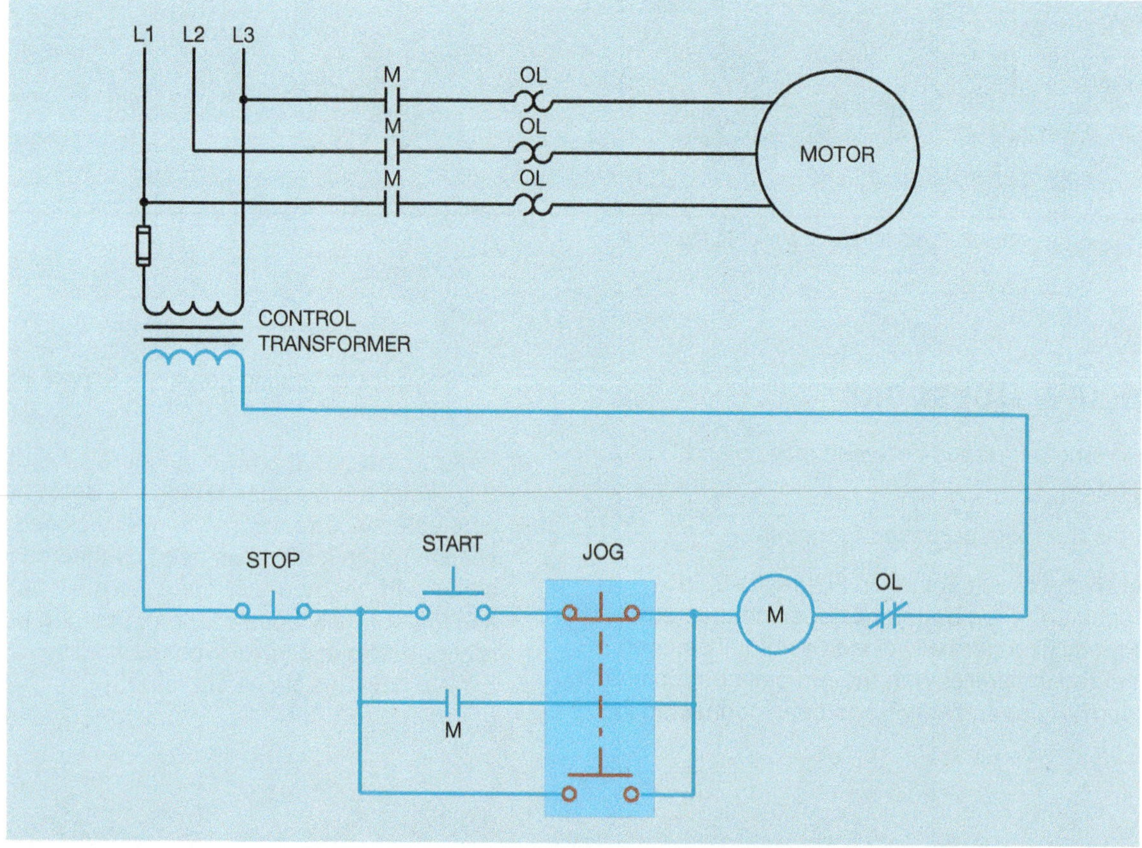

Figure 10–9 Another incorrect connection for the jog button.

Chapter 11

TIMING RELAYS

Objectives

After studying this chapter the student will be able to:

- Identify the primary types of timing relays.
- Explain the basic steps in the operation of the common timing relays.
- List the factors that affect the selection of a timing relay for a particular use.
- List applications of several types of timing relays.
- Draw simple circuit diagrams using timing relays.
- Identify *on-* and *off-*delay timing wiring symbols.

Time delay relays can be divided into two general classifications: the on-delay relay and the off-delay relay. The on-delay relay is often referred to as DOE, which stands for "Delay On Energize." The off-delay relay is often referred to as DODE, which stands for "Delay On De-Energize."

Timer relays are similar to other control relays in that they use a coil to control the operation of some number of contacts. The difference between a control relay and a timer relay is that the contacts of the timer relay delay changing their position when the coil is energized or de-energized. When power is connected to the coil of an on-delay timer, the contacts delay changing position for some period of time. For this example, assume that the timer has been set for a delay of 10 seconds. Also assume that the contact is normally open. When voltage is connected to the coil of the on-delay timer, the contacts remain in the open position for 10 seconds and then close. When voltage is removed and the coil is de-energized, the timed contact immediately changes back to its normally open position. The contact symbols for an on-delay relay are shown in Figure 11–1.

The operation of the off-delay timer is the opposite of the operation of the on-delay timer. For this example, again assume that the timer has been set for a delay of 10 seconds, and also assume that the contact is normally open. When voltage is applied to the coil of the off-delay timer, the contact changes immediately from open to closed. When the coil is de-energized, however, the timed contact remains in the closed position for 10 seconds before it reopens. The contact symbols for an off-delay relay are shown in Figure 11–2. Time delay relays can have normally open, normally closed, or a combination of normally open and normally closed contacts.

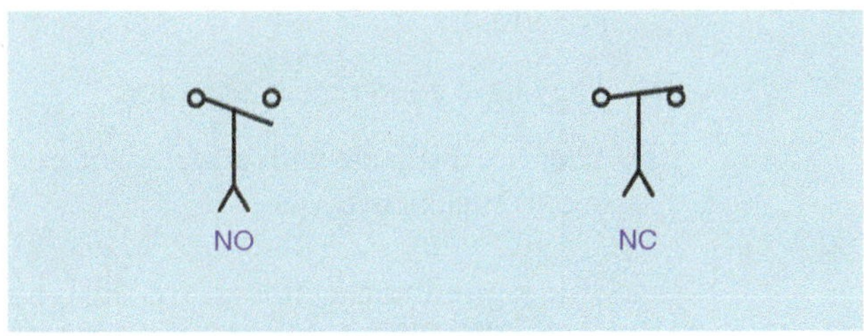

Figure 11–1 On-delay normally open and normally closed contacts.

Although the contact symbols shown in Figures 11–1 and 11–2 are standard NEMA symbols for on-delay and off-delay contacts, some control schematics may use a different method of indicating timed contacts. The abbreviations TO and TC are used with some control schematics to indicate a time-operated contact. *TO stands for time opening, and TC stands for time closing*. If these abbreviations are used with standard contact symbols, their meaning can be confusing. Figure 11–3 shows a standard normally open contact symbol with the abbreviation TC written beneath it. This contact must be connected to an on-delay relay if it is to be time delayed when closing. Figure 11–4 shows the same contact with the abbreviation TO beneath it. If this contact is to be time delayed when opening, it must be operated by an off-delay timer. These abbreviations can also be used with standard NEMA symbols as shown in Figure 11–5.

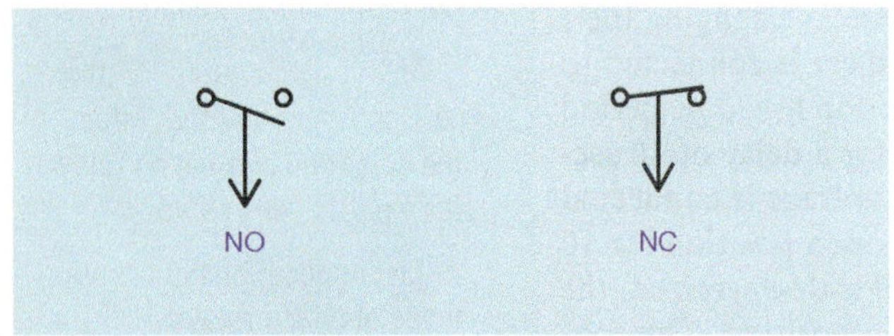

Figure 11–2 Off-delay normally open and normally closed contacts.

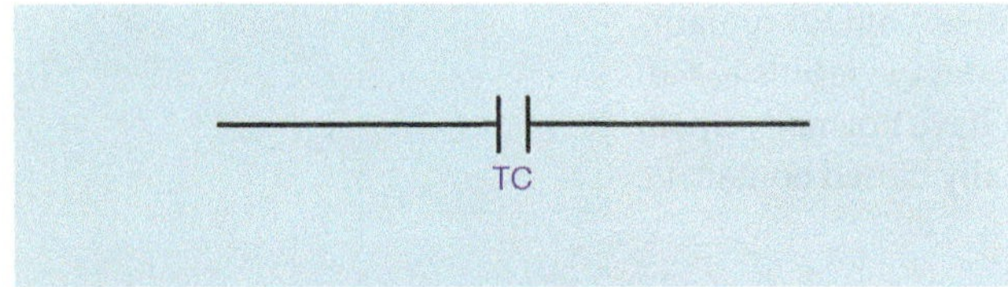

Figure 11–3 Time closing contact.

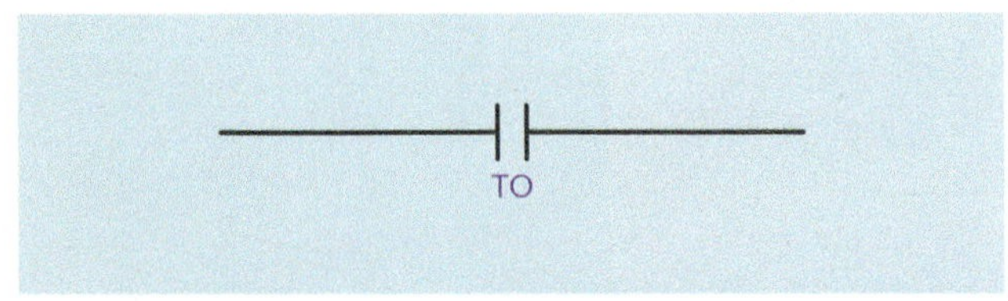

Figure 11–4 Time opening contact.

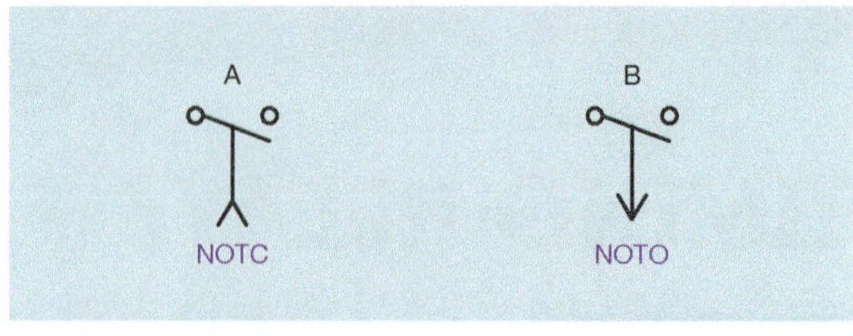

Figure 11–5 Contact A is an on-delay contact with the abbreviation NOTC (normally open time closing). Contact B is an off-delay contact with the abbreviation NOTO (normally open time opening).

Pneumatic Timers

Pneumatic, or air **timers**, operate by restricting the flow of air through an orifice to a rubber bellows or diaphragm. Figure 11–6 illustrates the principle of operation of a simple bellows timer. If rod "A" pushes against the end of the bellows, air is forced out of the bellows through the check valve as the bellows contracts. When the bellows is moved back, contact TR changes from an open to a closed contact. When rod "A" is pulled away from the bellows, the spring tries to return the bellows to its original position. Before the bellows can be returned to its original position, however, air must enter the bellows through the air inlet port. The rate at which the air is permitted to enter the bellows is controlled by the needle valve. When the bellows returns to its original position, contact TR returns to its normally open position.

Pneumatic timers are popular throughout industry because they have the following characteristics:

A. They are unaffected by variations in ambient temperature or atmospheric pressure.

B. They are adjustable over a wide range of time periods.

C. They have good repeat accuracy.

D. They are available with a variety of contact and timing arrangements.

Some pneumatic timers are designed to permit the timer to be changed from on delay to off delay, and the contact arrangement to be changed to normally opened or normally closed (Figure 11–7). This type of flexibility is another reason for the popularity of pneumatic timers.

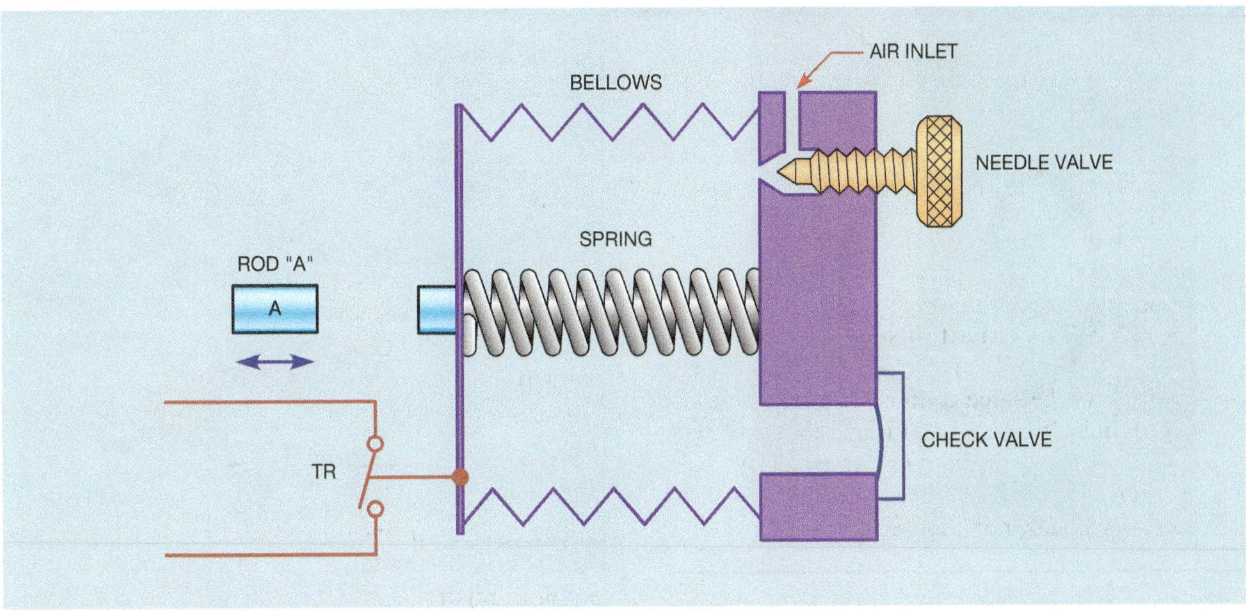

Figure 11–6 Bellows-operated pneumatic timer.

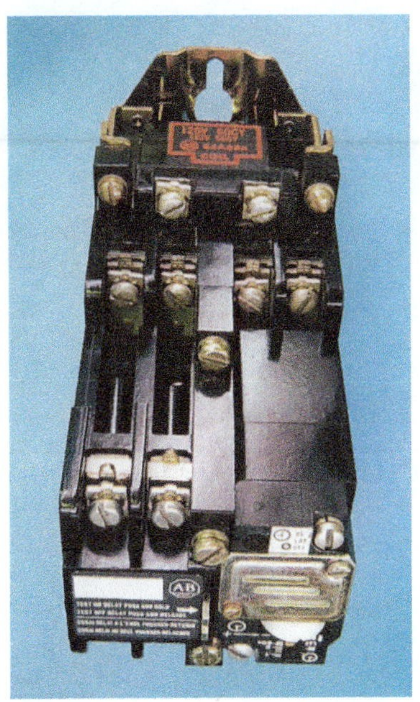

Figure 11–7 Pneumatic timer.

Many timers are made with contacts that operate with the coil as well as time delayed contacts. When these contacts are used, they are generally referred to as *instantaneous contacts* and indicated on a schematic diagram by the abbreviation, inst., printed below the contact (Figure 11–8). These instantaneous contacts change their positions immediately when the coil is energized and change back to their normal positions immediately when the coil is de-energized.

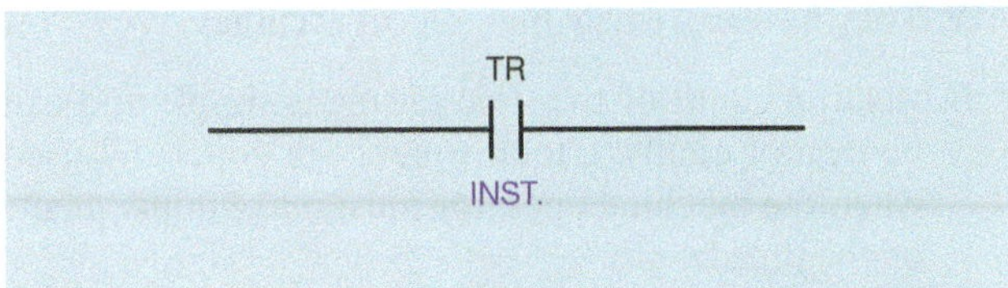

Figure 11–8 Normally open instantaneous contact of a timer relay.

Clock Timers

Another timer frequently used is the **clock timer** (Figure 11–9). Clock timers use a small AC synchronous motor similar to the motor found in a wall clock to provide the time measurement for the timer. The length of time of one clock timer may vary greatly from the length of time of another. For example, one timer may have a full range of 0 to 5 seconds and another timer may have a full range of 0 to 5 hours. The same type of timer motor could be used with both timers. The gear ratio connected to the motor would determine

Figure 11–9 Clock-driven timer.

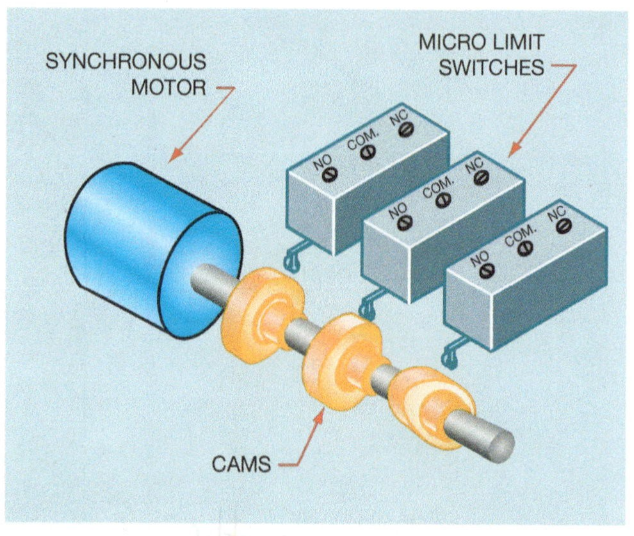

Figure 11–10 Cam switch.

the full range of time for the timer. Some advantages of clock timers are:

A. They have extremely high repeat accuracy.

B. Readjustment of the time setting is simple and can be done quickly. Clock timers are generally used when the machine operator must make adjustments of the time length.

Cam or Sequence Timers

Cam timers are generally used to operate several switches. The operation of the switches is controlled by the action of adjustable cams attached to a common shaft. The shaft is turned by a small synchronous motor (Figure 11–10). Cam timers are also known as sequence timers because they repeat the same action as long as the motor is turning the cams. They are often used to control flashing lights that must turn on or off in a certain sequence, such as the lights that appear to move from one place to another, or lights that spell out a word one letter at a time. A motor-driven cam timer is shown in Figure 11–11.

Programmable logic controllers (PLCs) are most often used to perform the functions of a cam timer. PLCs can be programmed as a sequencer and drive hundreds of outputs. It is much simpler to program the outputs of the PLC than it is to set all the cams on a mechanical timer.

Courtesy Industrial Timer Co.

Figure 11–11 Motor-driven process timer. Often referred to as a cam timer.

Electronic Timers

Electronic timers use solid-state components to provide the time delay desired. Some of these timers use an RC time constant to obtain the time base and others use quartz clocks as the time base (Figure 11–12). RC time constants are inexpensive and have good repeat times. The quartz timers, however, are extremely accurate and can often be set for 0.1 second times. These timers are generally housed in a plastic case and are designed to be plugged into some type of socket. An electronic timer that is designed to be plugged into a standard 8 pin relay socket is shown in Figure 11–13. The length of the time delay can be set by adjusting the control knob shown on top of the timer.

Eight pin electron timers similar to the one shown in Figure 11–13 are intended to be used as on-delay timers only. Many electronic timers are designed to plug into an 11 pin relay or tube socket (Figure 11–14) and are more flexible. Two such timers are shown in Figures 11–15A and 11–15B. Either of these timers can be used as an on-delay

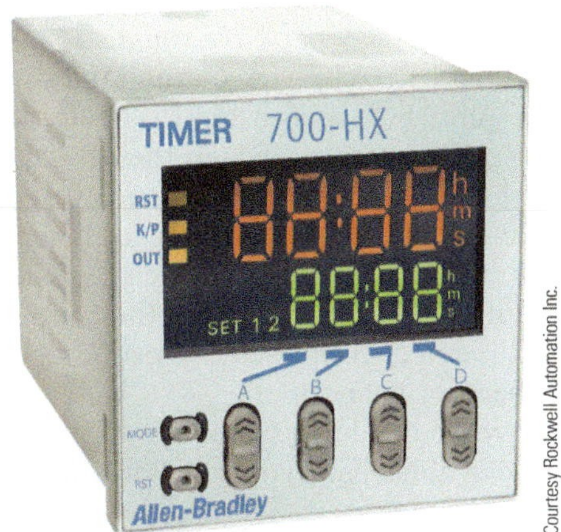

Figure 11–12 Digital clock timer.

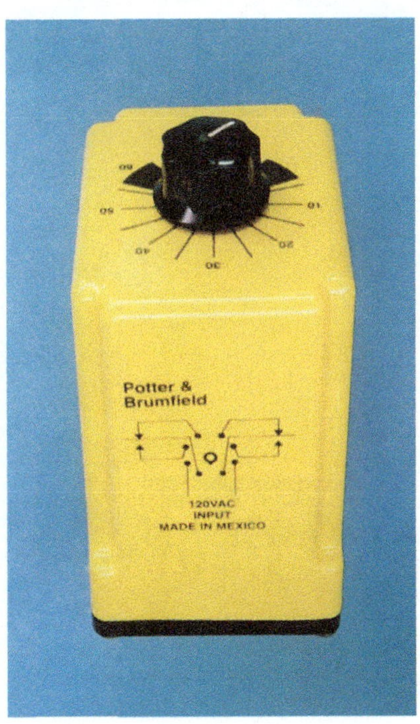

Figure 11–13 Electronic timer.

Figure 11–14 11 pin relay sockets.

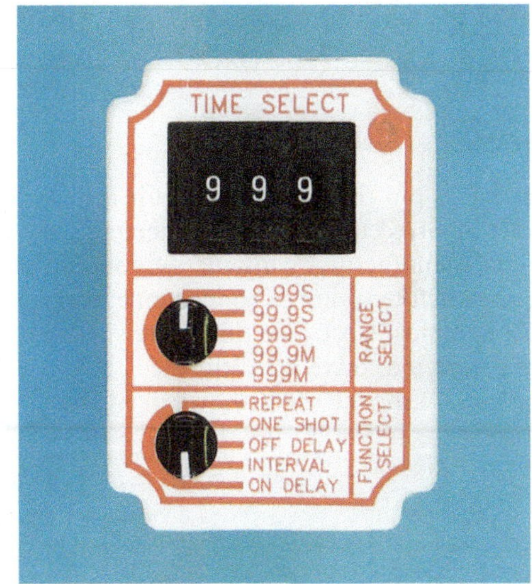

Figure 11–15A Dayton electronic timer.

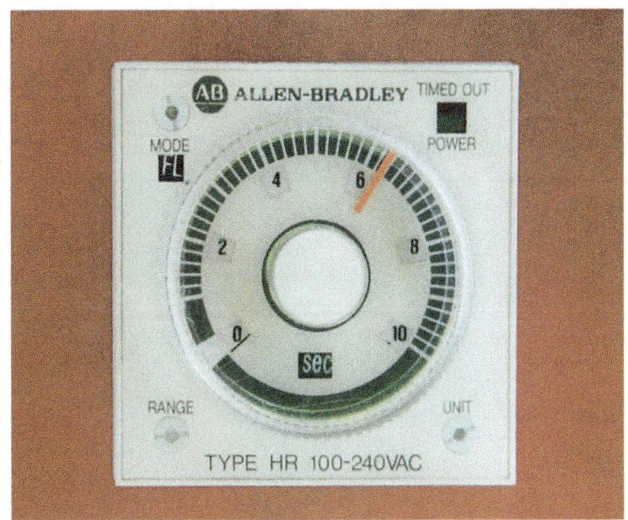

Figure 11–15B Allen-Bradley electronic timer.

timer, an off-delay timer, a pulse timer, or as a one-shot timer. Pulse timers continually turn on and off at regular intervals. A timing period chart for a pulse timer set for a delay of 1 second is shown in Figure 11–16. A one-shot timer operates for one time period only. A timing period chart for a one-shot timer set for 2 seconds is shown in Figure 11–17.

Most electronic timers can be set for a wide range of times. The timer shown in Figure 11–15A uses a thumbwheel switch to enter the timer setting. The top selector switch can be used to set the full range value from 9.99 seconds to 999 minutes. This timer has a range from 0.01 second to 999 minutes (16 hrs. 39 min.). The timer shown in Figure 11–15B can be set for a range of 0.01 second to 100 hours by adjusting the range and units settings on the front of the timer. Most electronic timers have similar capabilities.

Connecting 11 Pin Timers

Connecting 11 pin timers into a circuit is generally a little more involved than simply connecting the coil to power. The manufacturer's instructions should always be consulted before trying to connect one of these timers. Although most electronic timers are similar in how they are connected, there are differences. The pin connection diagram for the timer shown in Figure 11–15A is shown in Figure 11–18. Notice that a normally open push button switch is shown across terminals 5 and 6. This switch is used to start the action of the timer when it is set to function as an off-delay timer or as a one-shot timer. The reason is that when the timer is to function as an off-delay timer, power must be applied to the timer at all times to permit the internal timing circuit to operate. If power is removed, the internal timer cannot function. The START switch is actually used to initiate the operation of the timer when it is set to function in the off-delay mode. Recall the logic of an off-delay timer: *When the coil is energized, the contacts change position immediately. When the coil is de-energized, the contacts delay returning to their normal position.* According to the pin chart shown in Figure 11–18, pins 2 and 10 connect to the coil of the timer. To use this timer in the off-delay mode, power must be connected to pins 2 and 10 at all times. Shorting pins 5 and 6 together causes the timed contacts to change position immediately. When the short circuit between pins 5 and 6 is removed, the time sequence begins. At the end of the preset time period, the contacts return to their normal position.

If electronic off-delay timers are to replace pneumatic off-delay timers in a control circuit, it is generally

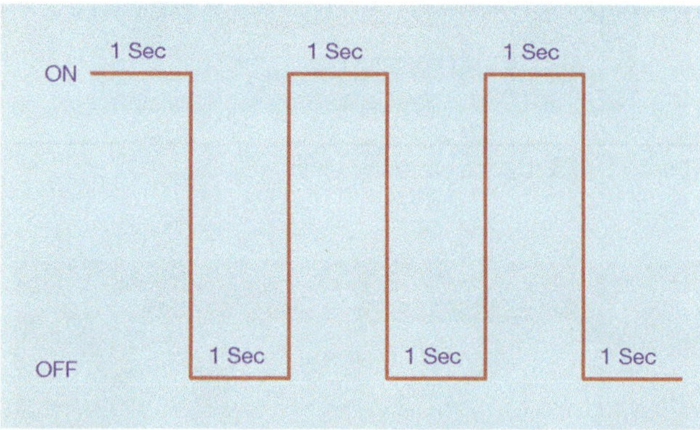

Figure 11–16 Time chart for pulse timer.

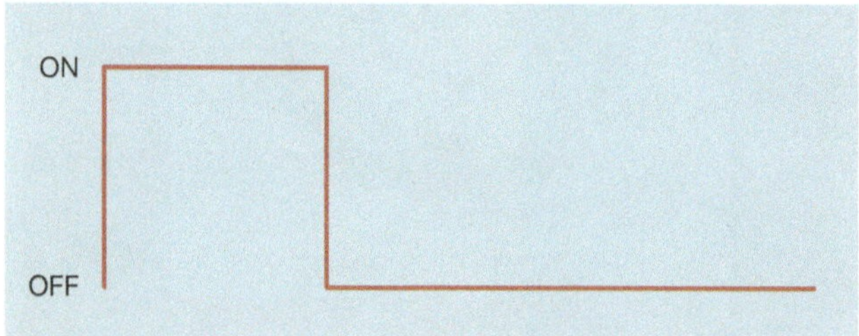

Figure 11–17 Time chart for a one-shot timer.

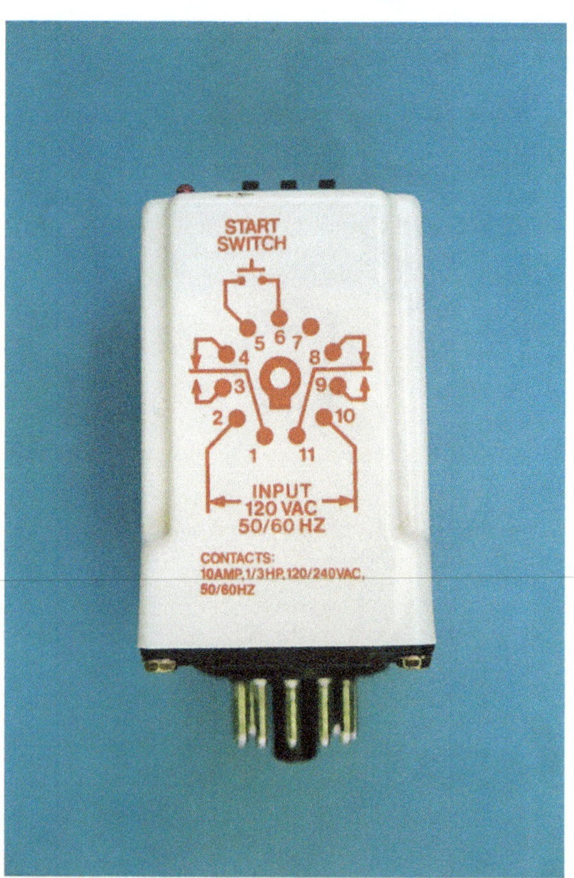

Figure 11–18 Pin connection diagram for Dayton timer.

necessary to modify the circuit. For example, in the circuit shown in Figure 11–19, it is assumed that starters 1M and 2M control the operation of two motors and timer TR is a pneumatic off-delay timer. When the START button is pressed both motors start at the same time. The motors continue to operate until the STOP button is pressed, which causes motor #1 to stop running immediately. Motor #2, however, will continue to run for a period of 5 seconds before stopping.

Now assume that the pneumatic off-delay timer is replaced with an electronic off-delay timer (Figure 11–20). In this circuit, notice that the coil of the timer is connected directly across the incoming power, which permits it to remain energized at all times. In the circuit shown in Figure 11–19, the timer actually operates with starter 1M. When coil 1M energizes, timer TR energizes at the same time. When coil 1M de-energizes, timer TR de-energizes also. For this reason, a normally open auxiliary contact on starter 1M will be used to control the operation of the electronic off-delay timer. In the circuit shown in Figure 11–20 a set of normally open 1M contacts is connected to pins 5 and 6 of the timer. When coil 1M energizes, contact 1M closes and shorts pins 5 and 6, causing the normally open TR contacts to close and energize starter coil 2M. When coil 1M is de-energized, the contacts reopen and timer TR begins timing. After 5 seconds contacts TR reopen and de-energize starter coil 2M.

All electronic timers are similar, but there are generally differences in how they are to be connected. The connection diagram for the timer shown in Figure 11–15B is shown in Figure 11–21. Notice that this timer contains RESET, START, and GATE pins. Connecting pin 2 to pin 5 activates the GATE function, which interrupts or suspends the operation of the internal clock. Connecting pin 2 to pin 6 activates the START function, which operates in the same manner as the timer shown in Figure 11–15A. Connecting pin 2 to pin 7 activates the RESET function, which resets the internal clock to zero. If this timer were to be used in the circuit shown in Figure 11–20, it would have to be modified as shown in Figure 11–22 by connecting the 1M normally open contact to pins 2 and 6 instead of pins 5 and 6.

Construction of a Simple Electronic Timer

The schematic for a simple on-delay timer is shown in Figure 11–23. The timer operates as follows: When switch S1 is closed, current flows through **resistor** RT and begins charging **capacitor** C1. When capacitor C1 has been charged to the trigger value of the **unijunction transistor, the UJT** turns on and discharges capacitor C1 through resistor R2 to ground. The sudden discharge of capacitor C1 causes a spike voltage to appear across resistor R2. This voltage spike travels through capacitor C2 and fires the gate of the SCR. When the SCR turns on, current is provided to the coil of relay K1.

Resistor R1 limits the current flow through the UJT. Resistor R3 is used to keep the SCR turned off until the UJT provides the pulse to fire the gate. Diode D1 is used to protect the circuit from the spike voltage produced by the collapsing magnetic field around coil K1 when the current is turned off.

By adjusting resistor RT, capacitor C1 can be charged at different rates. In this manner, the relay can be adjusted for time. Once the SCR has turned on, it will remain on until switch S1 is opened.

Programmable controllers, which will be discussed in a later chapter, contain "internal" electronic

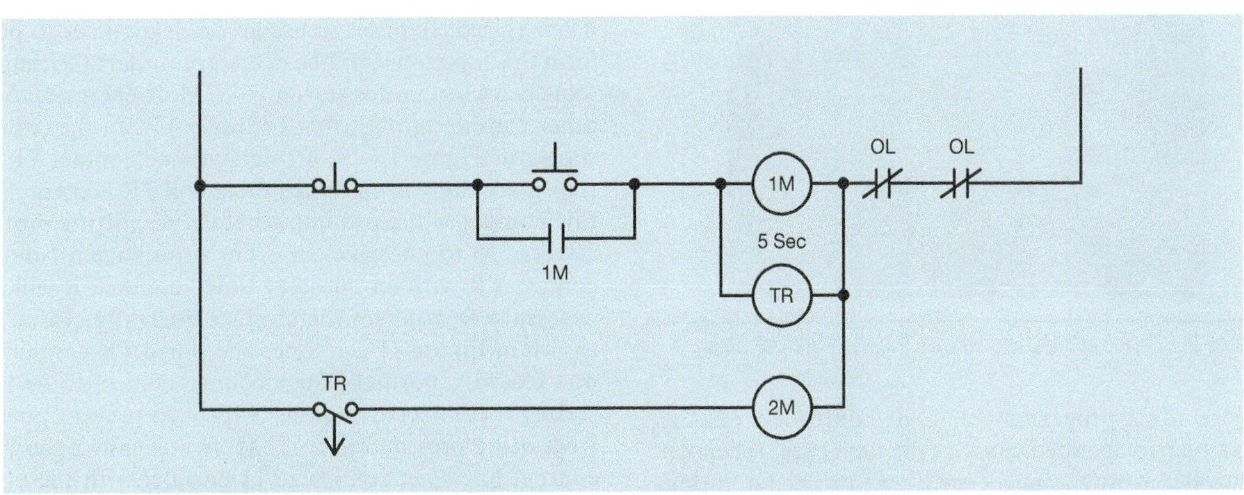

Figure 11–19 Off-delay timer circuit using a pneumatic timer.

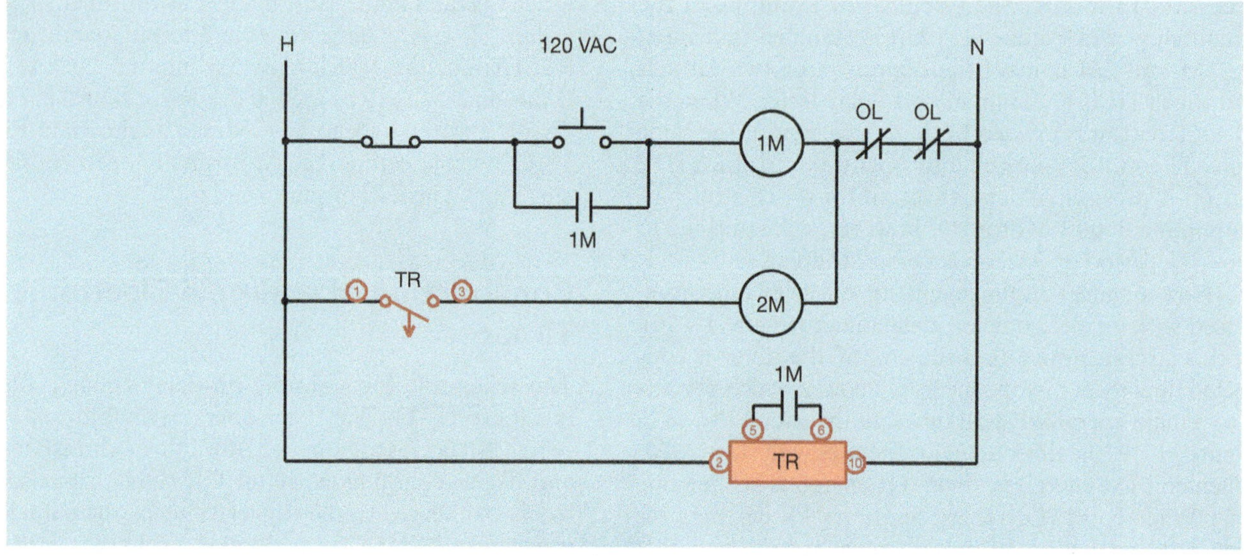

Figure 11–20 Modifying the circuit for an electronic off-delay timer.

Figure 11–21 Pin connection diagram for Allen-Bradley timer.

timers. Most programmable logic controllers (PLCs) use a quartz-operated clock as the time base. When the controller is programmed, the timers can be set in time increments of 0.1 second. This, of course, provides very accurate time delays for the controller.

Making an On-Delay Timer Function as an Off-Delay Timer

On-delay timers can be used to perform the same logic function as off-delay timers. There may be instances that an off-delay timer is needed and only on-delay timers are available. When this is the case, the circuit can be amended to permit the on-delay timer to perform the logic of an off-delay timer. In the circuit shown in Figure 11–19, motors 1 and 2 start at approximately the same time. When the stop button is pressed, motor 1 stops operating immediately but motor 2 continues to operate for an additional 5 seconds. Now assume that off-delay timer TR is to be replaced with an on-delay timer. In the circuit shown in Figure 11–19, off-delay timer TR begins its time sequence when motor starter coil 1M is de-energized. In order to use an on-delay timer to perform this function it will be necessary to start the timing sequence when motor starter coil 1M de-energizes. Another consideration is timer contact TR. In the circuit shown in Figure 11–19, off-delay timed contact TR is shown normally open. When timer coil TR is energized, this contact will close immediately permitting motor starter 2M to energize. In order to replace off-delay contact TR with an on-delay timed contact, it will be necessary to connect the contact normally closed as shown in Figure 11–24. Since the timed TR contact is now closed, a normally open contact controlled by 1M starter is connected in series with it to prevent power from being provided to coil 2M. A normally open 2M contact has been connected in parallel with the normally open 1M contact to maintain a current path when starter 1M is de-energized.

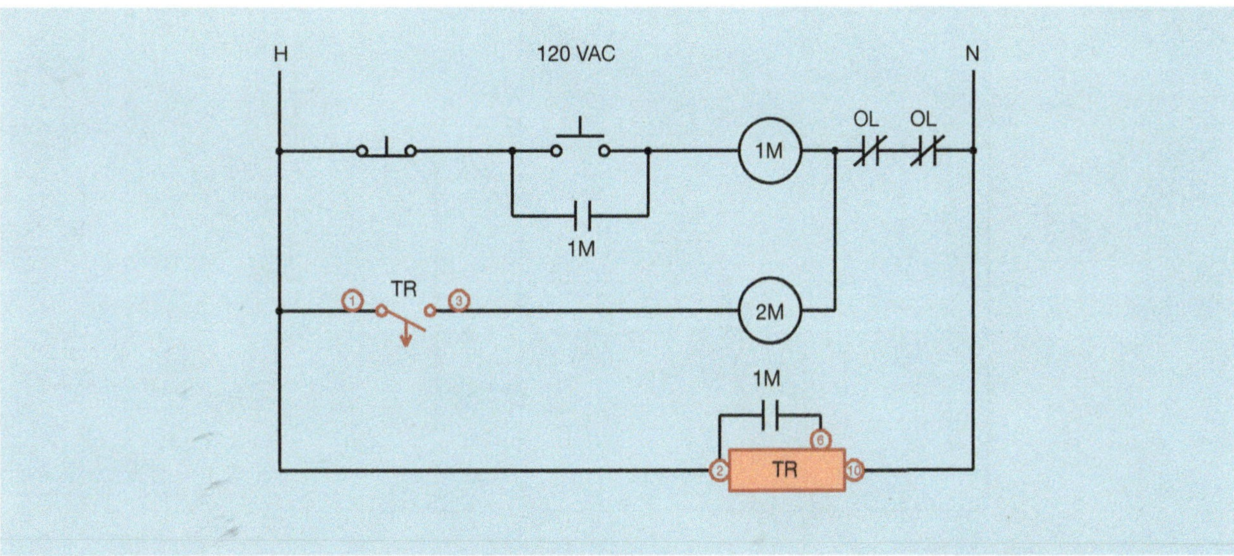

Figure 11–22 Replacing the Dayton timer with the Allen-Bradley timer.

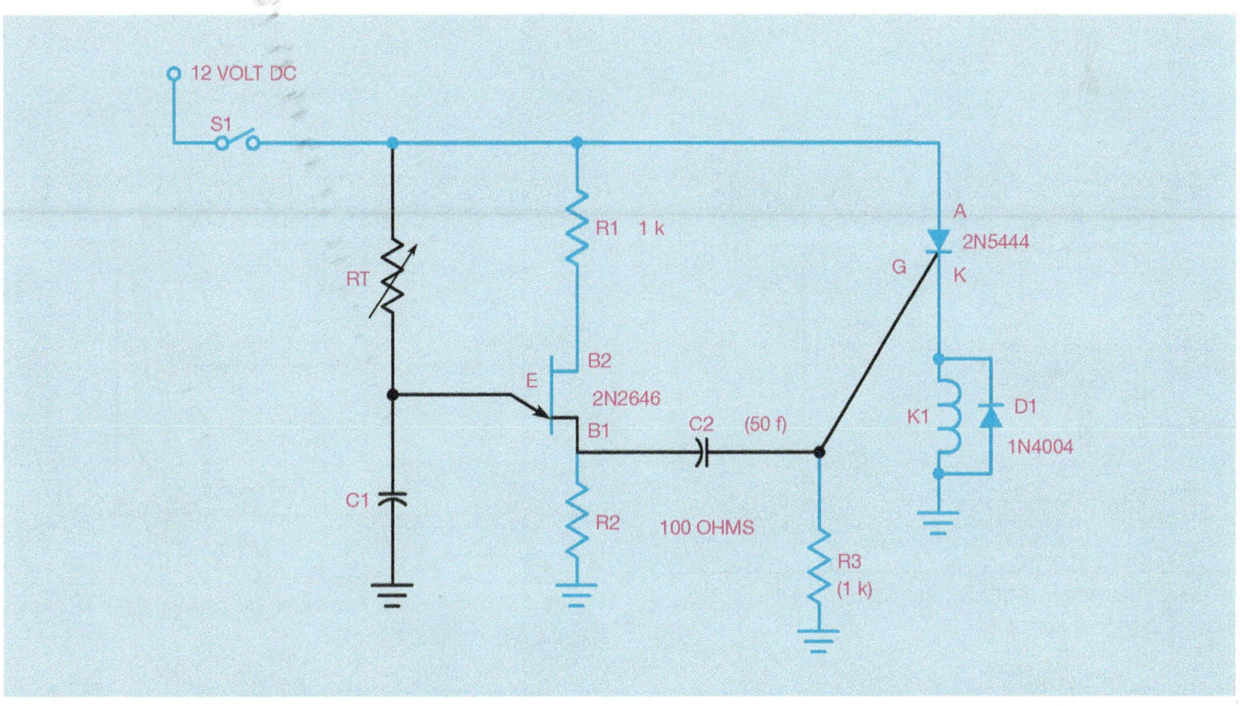

Figure 11–23 Schematic of electronic on-delay timer.

Timer TR must begin its time sequence when coil 1M is de-energized. To accomplish this, a normally closed contact controlled by starter coil 1M is connected in series with the coil of timer TR. Since the 1M contact is normally closed, a normally open contact controlled by starter coil 2M is connected in series to prevent power from being applied to timer coil TR when the circuit is turned off.

The operation of the circuit is as follows:

1. When the start button is pressed, starter 1M energizes and all 1M contacts change position, Figure 11–25.
2. Coil 2M energizes and both normally open 2M auxiliary contacts close. The circuit to timer coil TR is interrupted by the now open normally closed 1M contact.

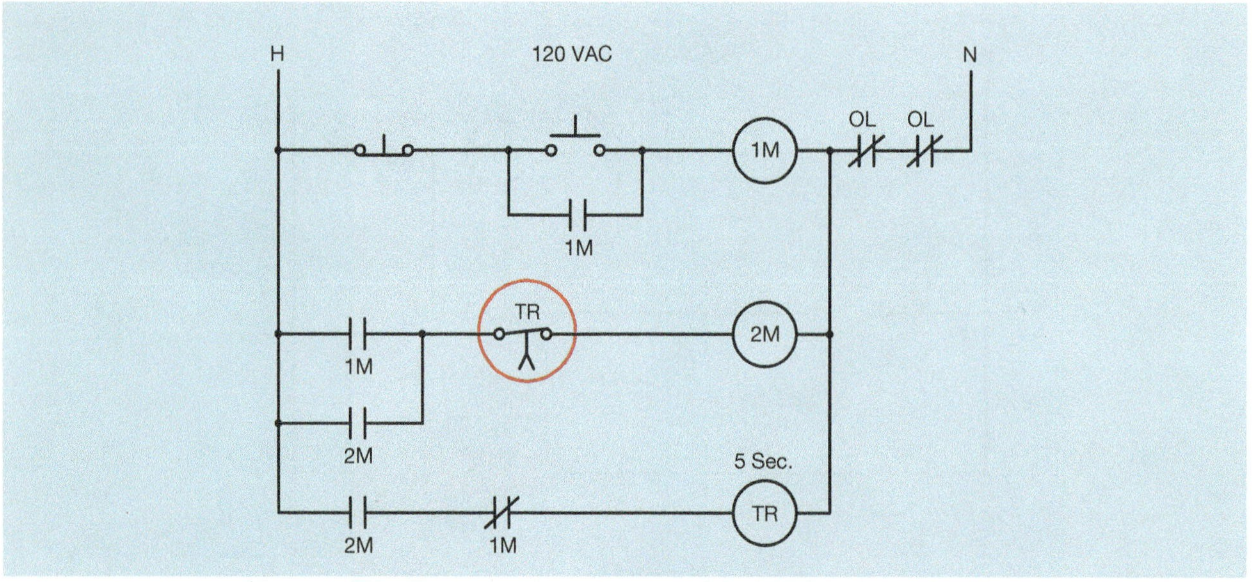

Figure 11–24 Using an on-delay timer to perform the logic of an off-delay timer.

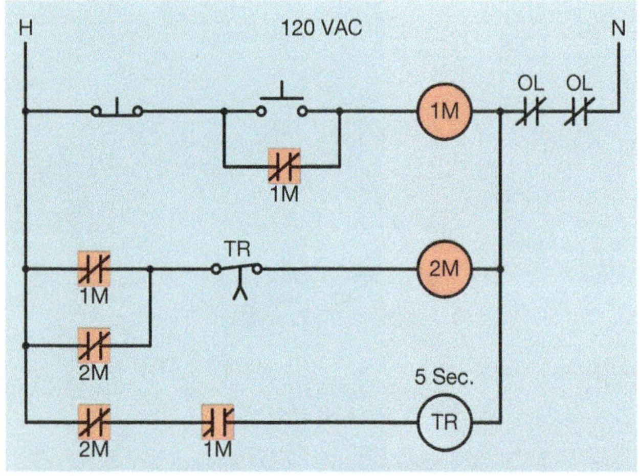

Figure 11–25 Starters 1M and 2M energize causing all 1M and 2M contacts to change position.

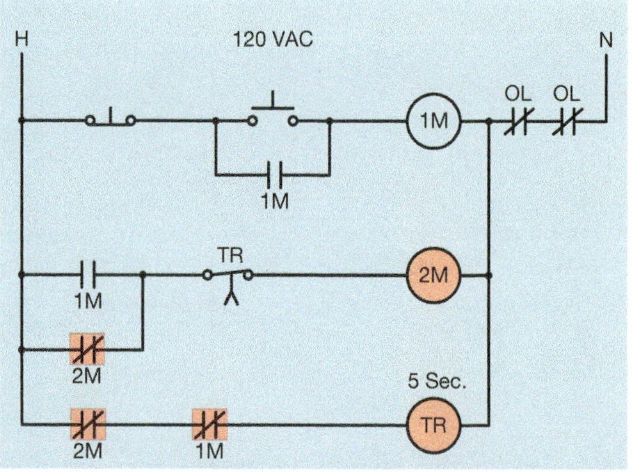

Figure 11–26 Starter 1M de-energizes causing timer TR to begin its time sequence.

3. When the start button is pressed, coil 1M de-energizes and all 1M contacts return to their normal position, Figure 11–26. A circuit path is maintained to starter coil 2M by the closed 2M contact connected in parallel with the now open 1M contact.

4. When the 1M contact connected in series with timer coil TR returned to the closed condition, a circuit was completed, starting the time sequence for timer TR.

5. At the end of the 5-second time period, timed contact TR opens and disconnects power to starter coil 2M, causing all 2M contacts to return to their normal position.

6. When the 2M contact connected in series with coil TR reopens, power is disconnected to timer coil TR, causing timed contact TR to return to its normally closed position. The circuit is now in its original position.

Review Questions

1. What are the two basic classifications of timers?
2. Explain the operation of an on-delay relay.
3. Explain the operation of an off-delay relay.
4. What are instantaneous contacts?
5. How are pneumatic timers adjusted?
6. Name two methods used by electronic timers to obtain their time base.

Chapter 12

SEQUENCE CONTROL

Objectives

After studying this chapter the student will be able to:

» State the purpose for starting motors in a predetermined sequence.

» Read and interpret sequence control schematics.

» Convert a sequence control schematic into a wiring diagram.

» Connect a sequence control circuit.

Sequence control forces motors to start or stop in a predetermined order. One motor cannot start until some other motor is in operation. Sequence control is used by such machines as hydraulic presses that must have a high pressure pump operating before they can be used, or by air conditioning systems that require that the blower be in operation before the compressor starts. Sequence control can be achieved by several methods. One design that meets the requirements is shown in Figure 12–1. In this circuit, push button #1 must be pressed before power can be provided to push button #2. When motor starter #1 energizes, the normally open auxiliary contact 1M closes providing power to coil 1M and to push button #2. Motor starter #2 can now be started by pressing push button #2. Once motor starter #2 energizes, auxiliary contact 2M closes and provides power to coil 2M and push button #3. If the STOP button should be pressed or any overload contact opens, power will be interrupted to all starters.

A Second Circuit for Sequence Control

A second method of providing sequence control is shown in Figure 12–2. Because the motor connections are the same as the previous circuit, only the control part of the schematic is shown. In this circuit, normally open auxiliary contacts located on motor starters 1M and 2M are used to ensure that the three motors start in the proper sequence. A normally open 1M auxiliary contact connected in series with starter coil 2M prevents motor #2 from starting before motor #1, and a normally open 2M auxiliary contact connected in series with coil 3M prevents motor #3 from starting before motor #2. If the STOP button is pressed or if any overload contact opens, power will be interrupted to all starters.

Sequence Control Circuit #3

A third circuit that is almost identical to the previous circuit is shown in Figure 12–3. This circuit also employs the use of normally open auxiliary contacts to prevent motor #2 from starting before motor #1, and motor #3 cannot start before motor #2. These normally open auxiliary contacts that control the starting sequence are often called *permissive* contacts because they permit action to take place. The main difference between the two circuits is that in the circuit shown in Figure 12–2 the

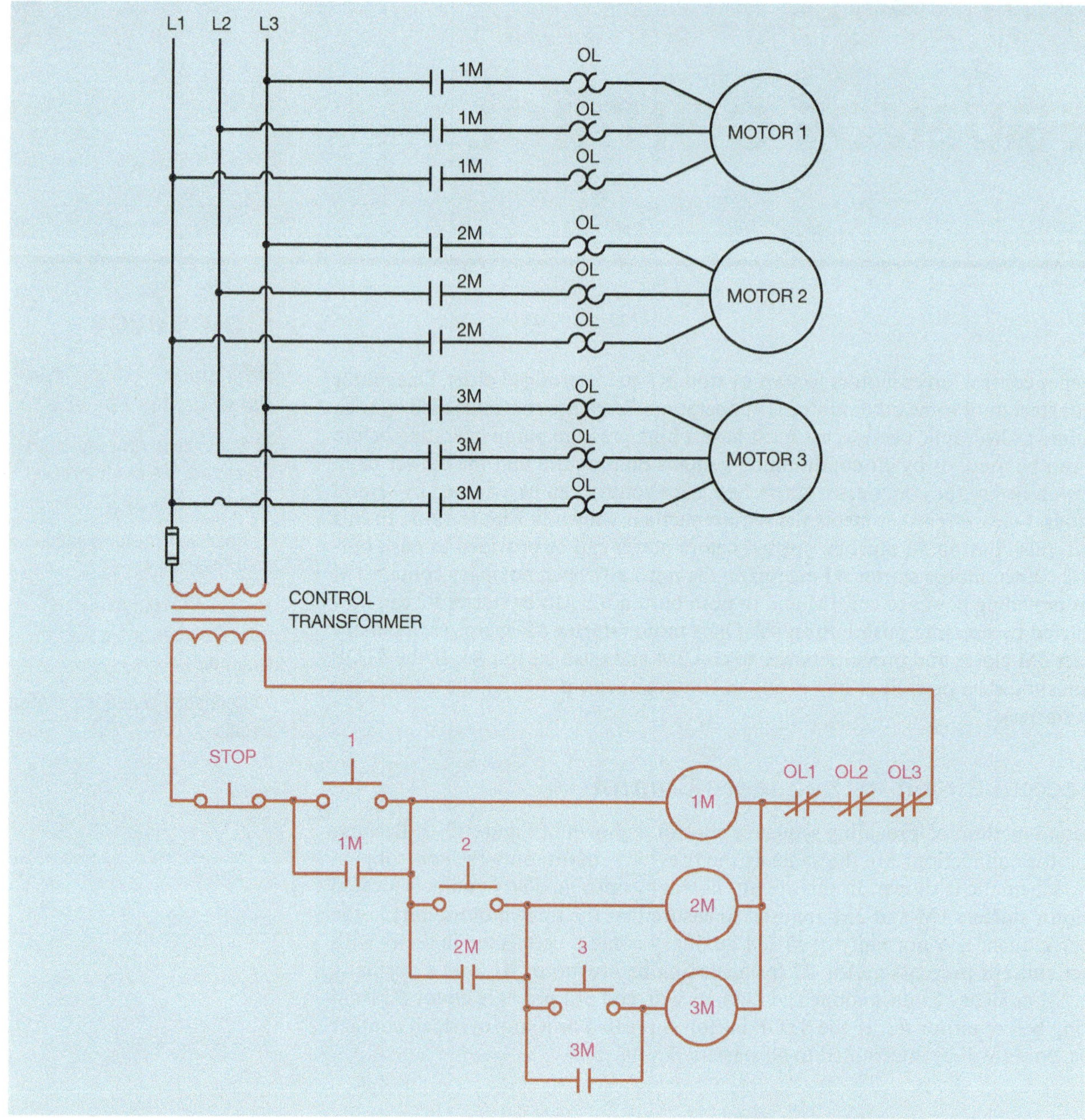

Figure 12–1 One example of a circuit that provides sequence control.

STOP push button interrupts the power to all the motor starters. The circuit in Figure 12–3 depends on the normally open auxiliary contacts reopening to stop motors #2 and #3.

Automatic Sequence Control

Circuits that permit the automatic starting of motors in sequence are common. A number of methods can be employed to determine when the next motor should start. Some sense motor current. When the current of a motor drops to a predetermined level, it permits the next motor to start. Other circuits sense the speed of one motor before permitting the next one to start. One of the most common methods is time delay. The circuit shown in Figure 12–4 permits three motors to start in sequence. Motor #1 starts immediately when the START button is pressed. Motor #2 starts 5 seconds after motor

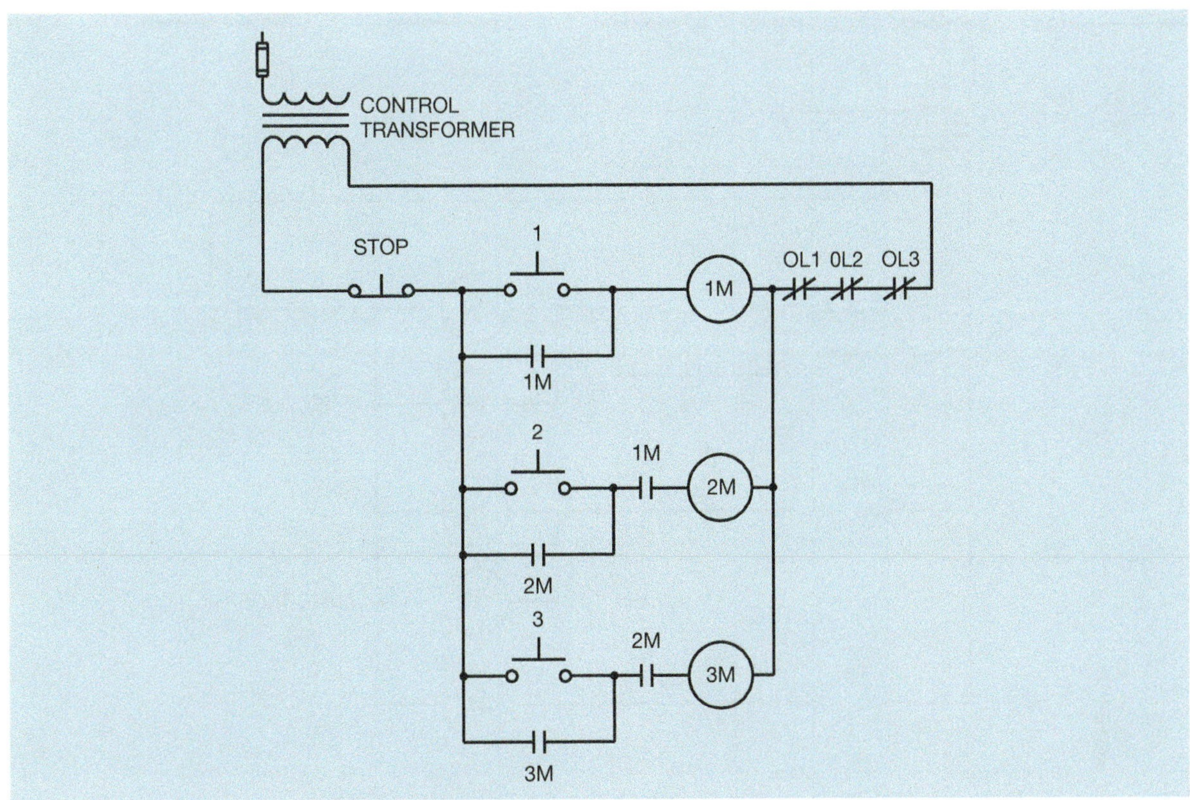

Figure 12–2 A second circuit for sequence control.

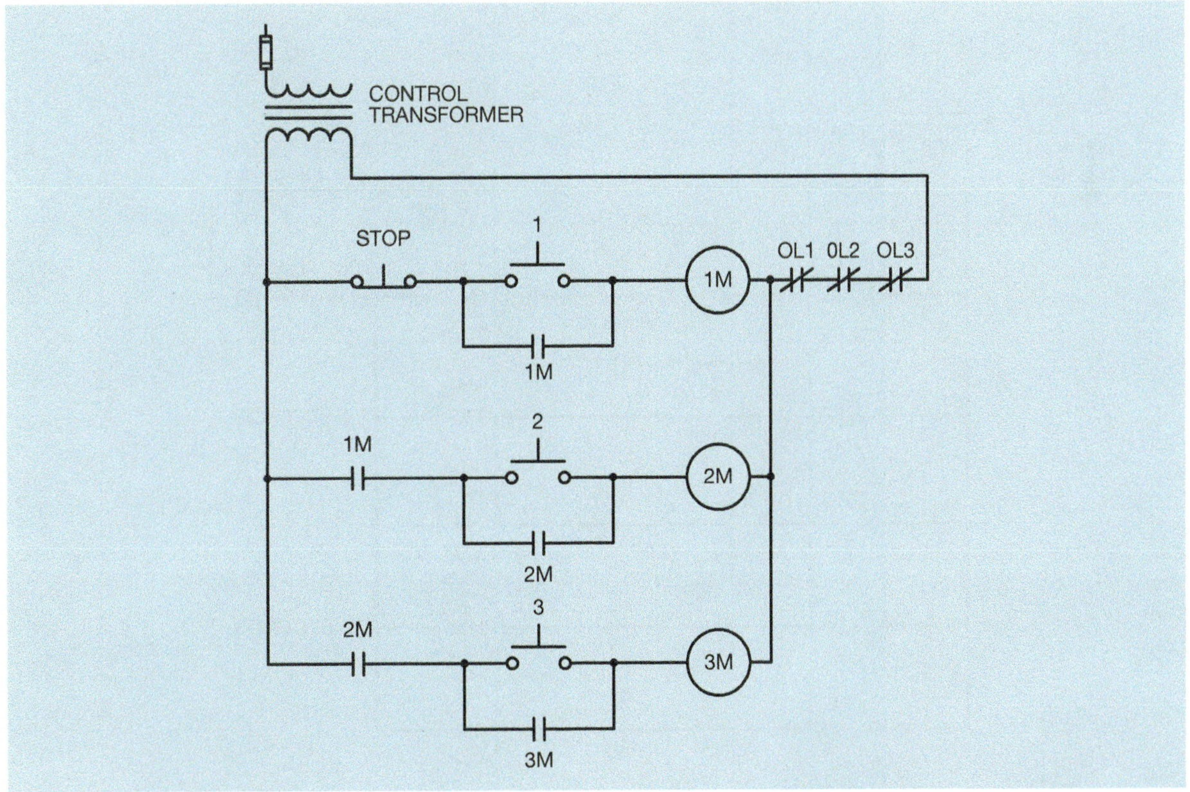

Figure 12–3 A third circuit for sequence control.

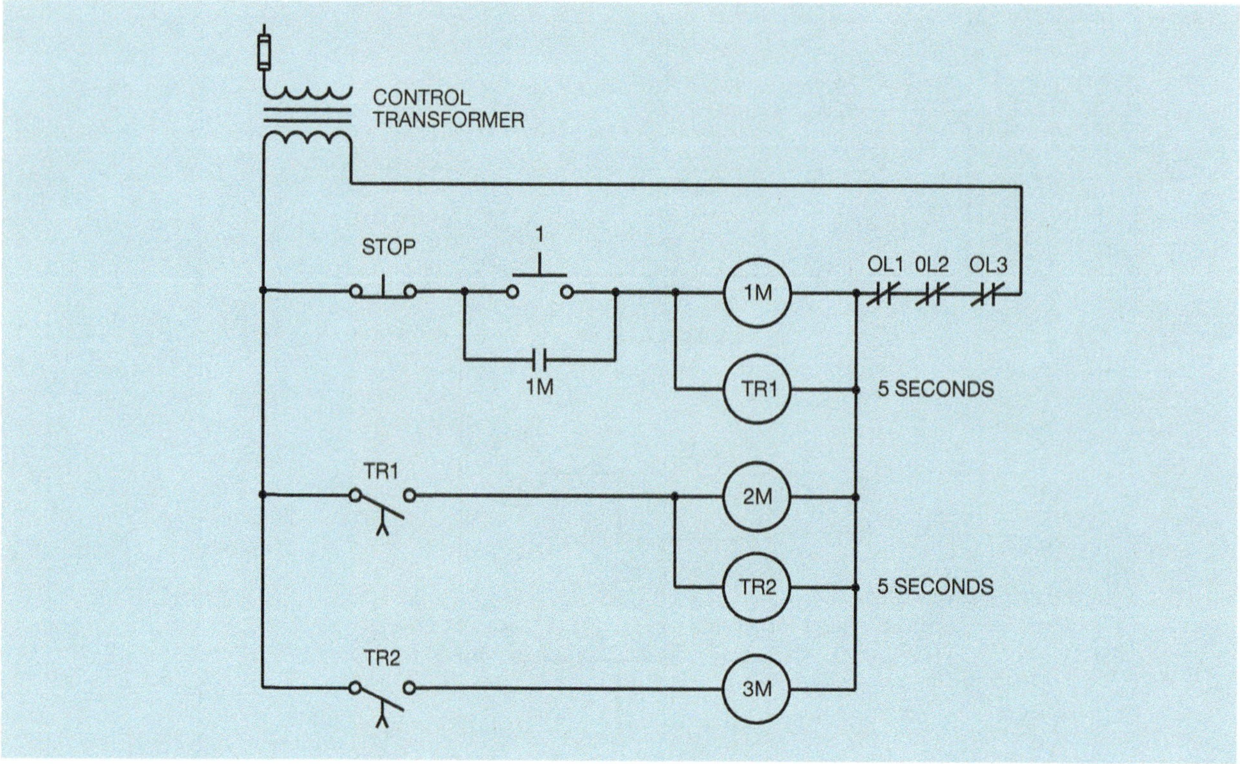

Figure 12–4 Timed starting for three motors.

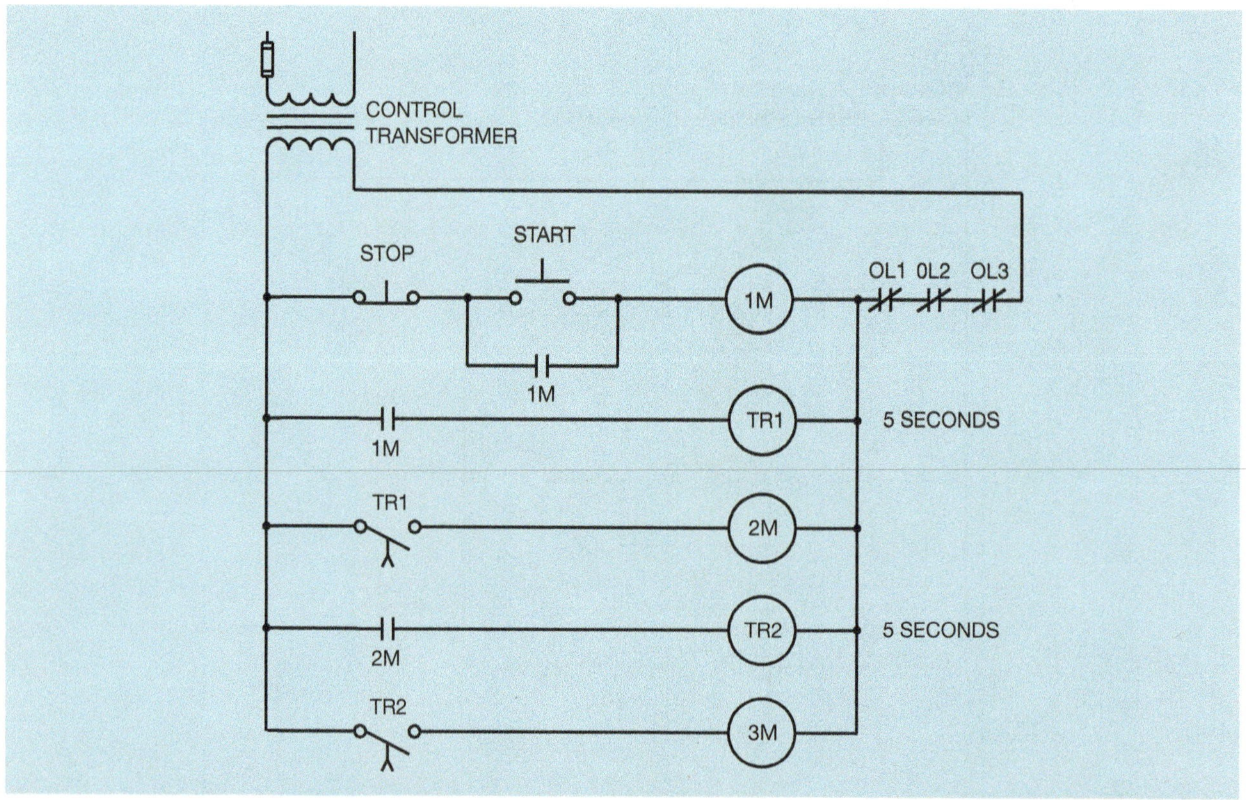

Figure 12–5 Circuit is modified to eliminate parallel coils.

#1 starts, and motor #3 starts 5 seconds after motor #2 starts. Timer coil TR1 is connected in parallel with 1M starter coil. Because they are connected in parallel, they will energize at the same time. After a delay of 5 seconds, TR1 contact closes and energizes coils 2M and TR2. Motor #2 starts immediately, but timed contact TR2 will delay closing for 5 seconds. After the delay period, starter coil 3M will energize and start motor #3. When the STOP button is pushed, all motors stop at virtually the same time.

Although the circuit logic in Figure 12–4 is correct, most ladder diagrams do not show coils connected in parallel. A modification of the circuit is shown in Figure 12–5. In this circuit, auxiliary contacts on the motor starters are used to control the action of the timed relays. Note that the logic of the circuit is identical to that of the circuit in Figure 12–4.

Stopping the Motors in Sequence

Some circuit requirements may demand that the motors turn off in sequence instead of turning on in sequence. This circuit requires the use of off-delay timers. Also, a control relay with four contacts is needed. The circuit shown in Figure 12–6 permits the motors to start in sequence from #1 to #3 when the START button is pressed. Although they start in sequence, the action will be so fast that it will appear they all start at approximately the same time. When the STOP button is pressed, however, they will stop in sequence from #3 to #1 with a time delay of 5 seconds between each motor. Motor #3 will stop immediately. Five seconds later motor #2 will stop, and 5 seconds after motor #2 stops, motor #1 will stop. An overload on any motor will stop all motors immediately.

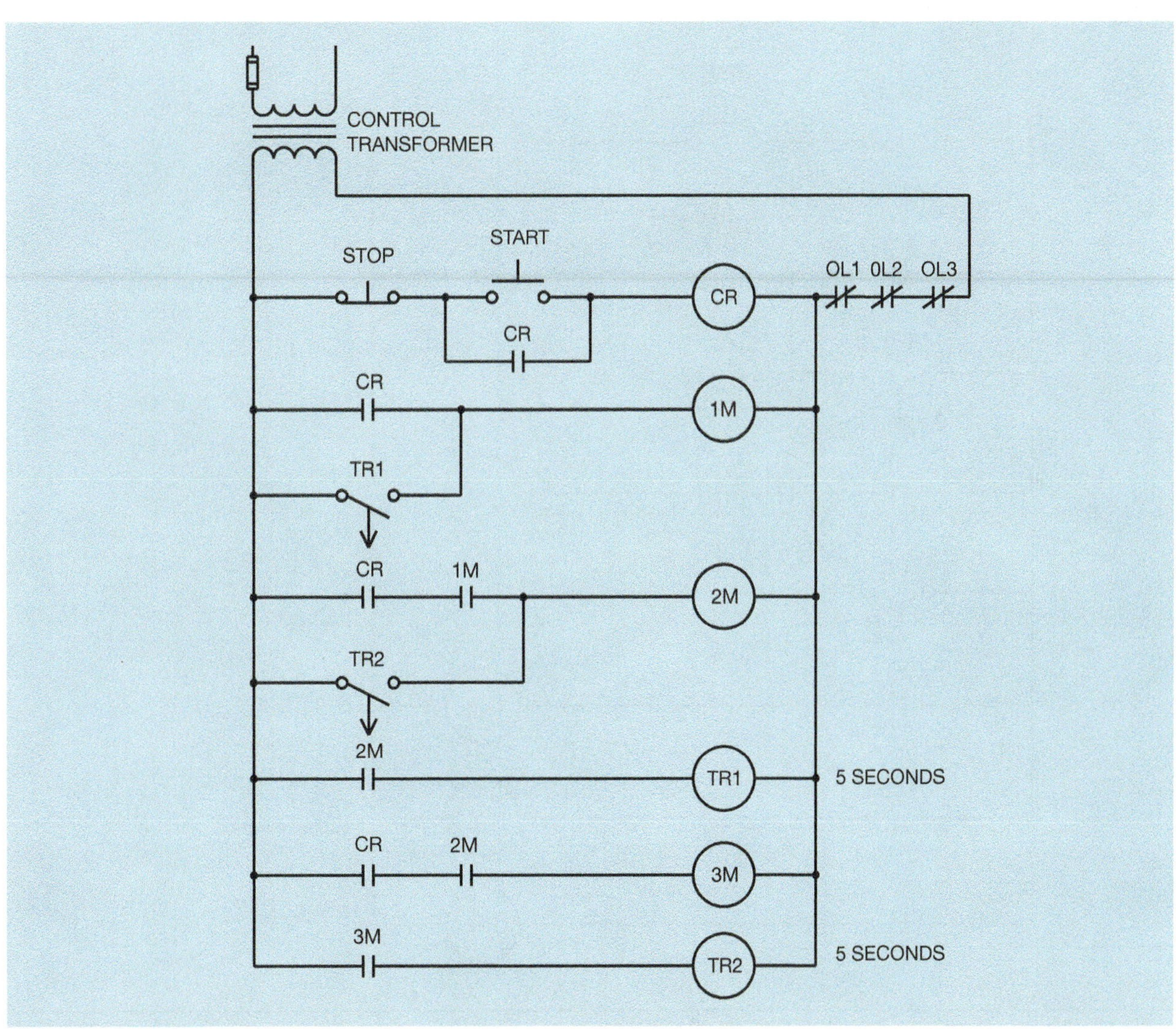

Figure 12–6 Motors start in sequence from 1 to 3 and stop in sequence from 3 to 1 with a delay of 5 seconds between the stopping of each motor.

Circuit Operation

When the START push button is pressed, control relay CR energizes and causes all CR contacts to close (Figure 12–7). Motor starter 2M cannot energize because of the normally open 1M contact connected in series with coil 2M, and motor starter 3M cannot energize because of the normally open 2M contact connected in series with coil 3M. Motor starter 1M does energize, starting motor #1 and closing all 1M contacts (Figure 12–8).

The 1M contact connected in series with coil 2M closes and energizes coil 2M (Figure 12–9). This causes motor #2 to start and all 2M contacts to close. Off-delay timer TR1 energizes and immediately closes the TR1 contact connected in parallel with the CR contact connected in series with coil 1M.

When the 2M contact connected in series with coil 3M closes, starter coil 3M energizes and starts motor #3. The 3M auxiliary contact connected in series with off-delay timer coil TR2 closes and energizes the timer

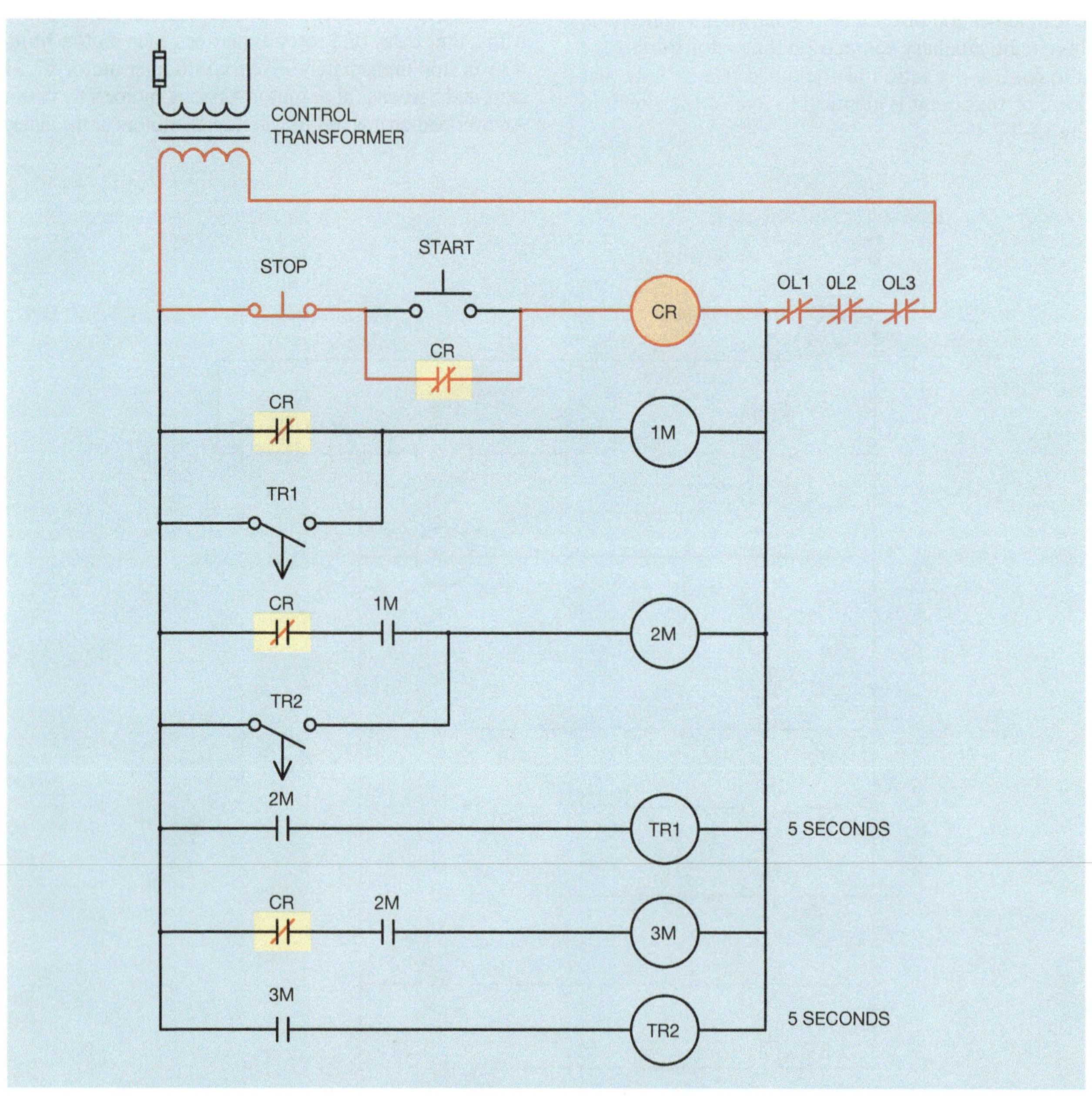

Figure 12–7 Control relay CR energizes and closes all CR contacts.

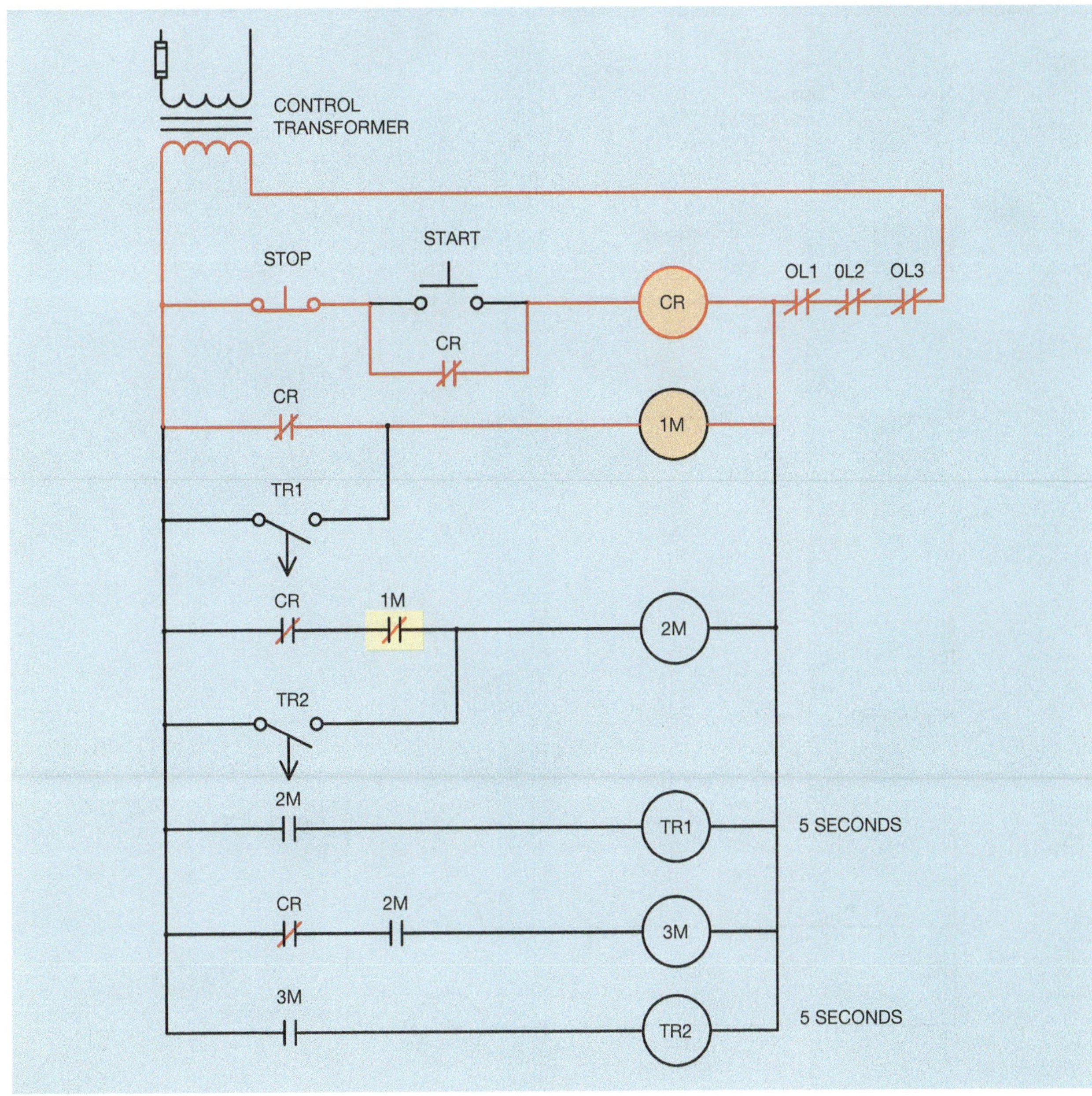

Figure 12–8 Motor #1 starts.

causing timed contacts TR2 to close immediately (Figure 12–10). Although this process seems long when discussed in step-by-step order, it actually takes place almost instantly.

When the STOP button is pressed, all CR contacts open immediately (Figure 12–11 on page 162). Motor #1 continues to run because the now closed TR1 contact maintains a circuit to the coil of 1M starter. Motor #2 continues to run because of the now closed TR2 contact. Motor #3, however, stops immediately when the CR contact connected in series with coil 3M opens. This causes the 3M auxiliary contact connected in series with TR2 coil to open and de-energizes the timer. Because TR2 is an off-delay timer, the timing process starts when the coil is de-energized. TR2 contact remains closed for a period of 5 seconds before it opens.

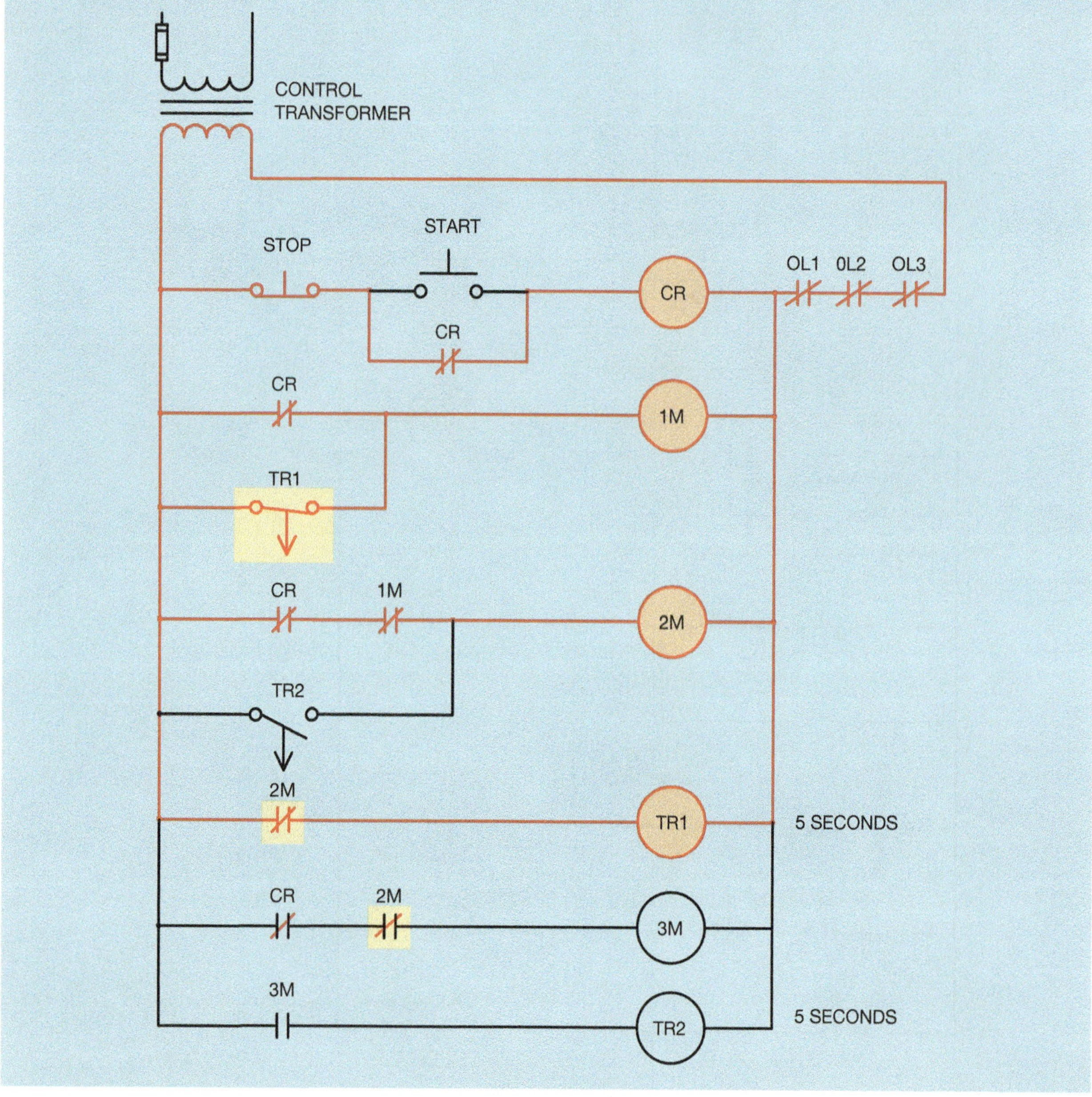

Figure 12–9 Motor #2 starts.

When TR2 contact opens, coil 2M de-energizes and stops motor #2. When the 2M auxiliary contacts open, TR1 coil de-energizes and starts the time delay for contact TR1 (Figure 12–12 on page 163). After a delay of 5 seconds, timed contact TR1 opens and de-energizes coil 1M, stopping motor #1 and opening the 1M auxiliary contact connected in series with coil 2M. The circuit is now back in its normal de-energized state as shown in Figure 12–6.

Timed Starting and Stopping of Three Motors

The addition of two timers makes it possible to start the motors in sequence from #1 to #3 with a time delay between the starting of each motor as well as stopping the motors in sequence from #3 to #1 with a time delay between the stopping of each motor. The circuit shown in Figure 12–13 on page 164 makes this amendment.

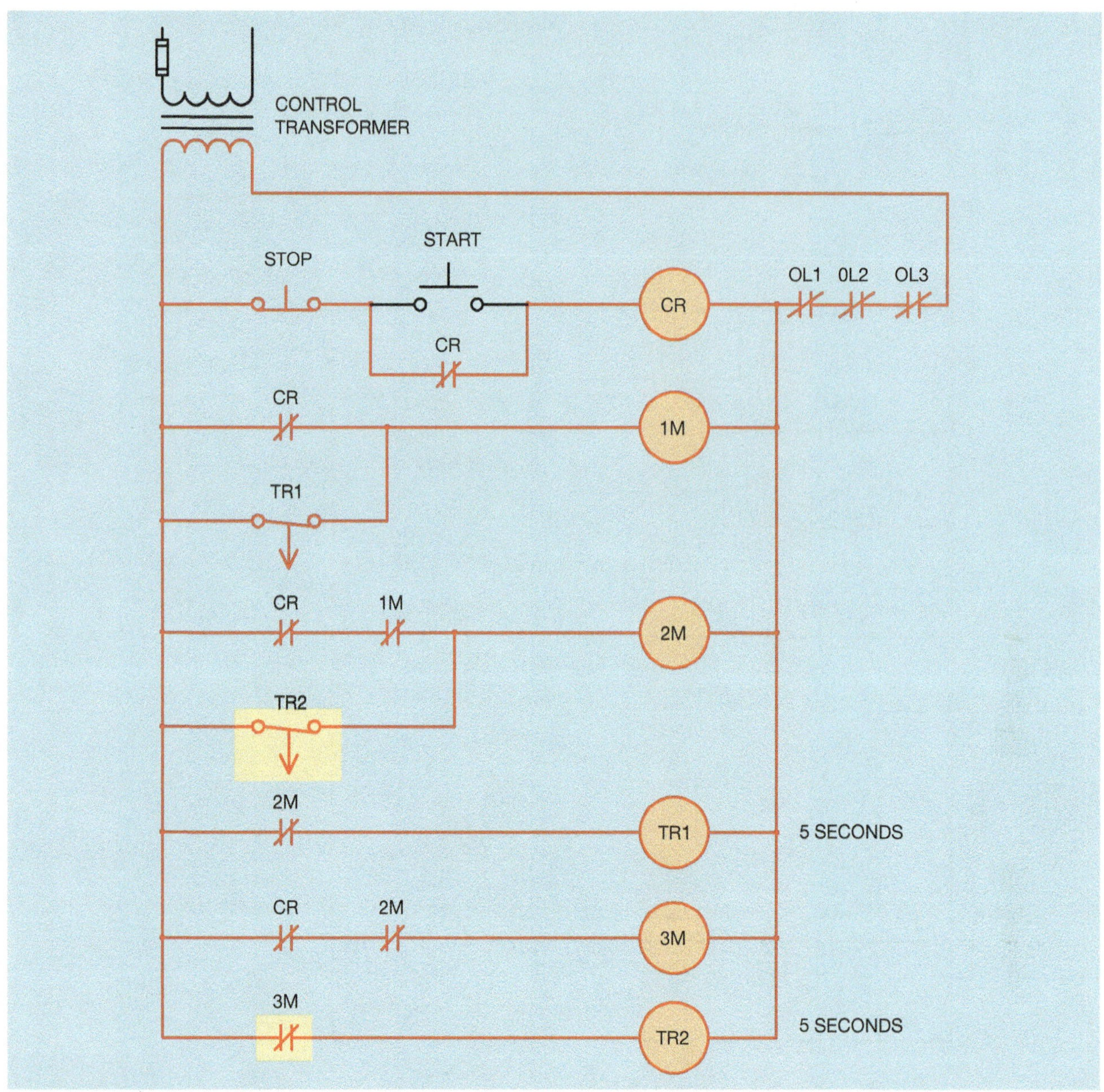

Figure 12–10 Motor #3 starts.

When the START button is pressed, all CR contacts close. Motor #1 starts immediately when starter 1M energizes. The 1M auxiliary contact closes and energizes on-delay timer TR3. After 5 seconds, starter 2M energizes and starts motor #2. The 2M auxiliary contact connected in series with off-delay timer TR1 closes causing timed contact TR1 to close immediately. The second 2M auxiliary contact connected in series with on-delay timer TR4 closes and starts the timing process. After 5 seconds, timed contact TR4 closes and energizes starter coil 3M starting motor #3. The 3M auxiliary contact connected in series with off-delay timer TR2 closes and energizes the timer. Timed contact TR2 closes immediately. All motors are now running.

When the STOP button is pressed, all CR contacts open immediately. This de-energizes starter 3M stopping

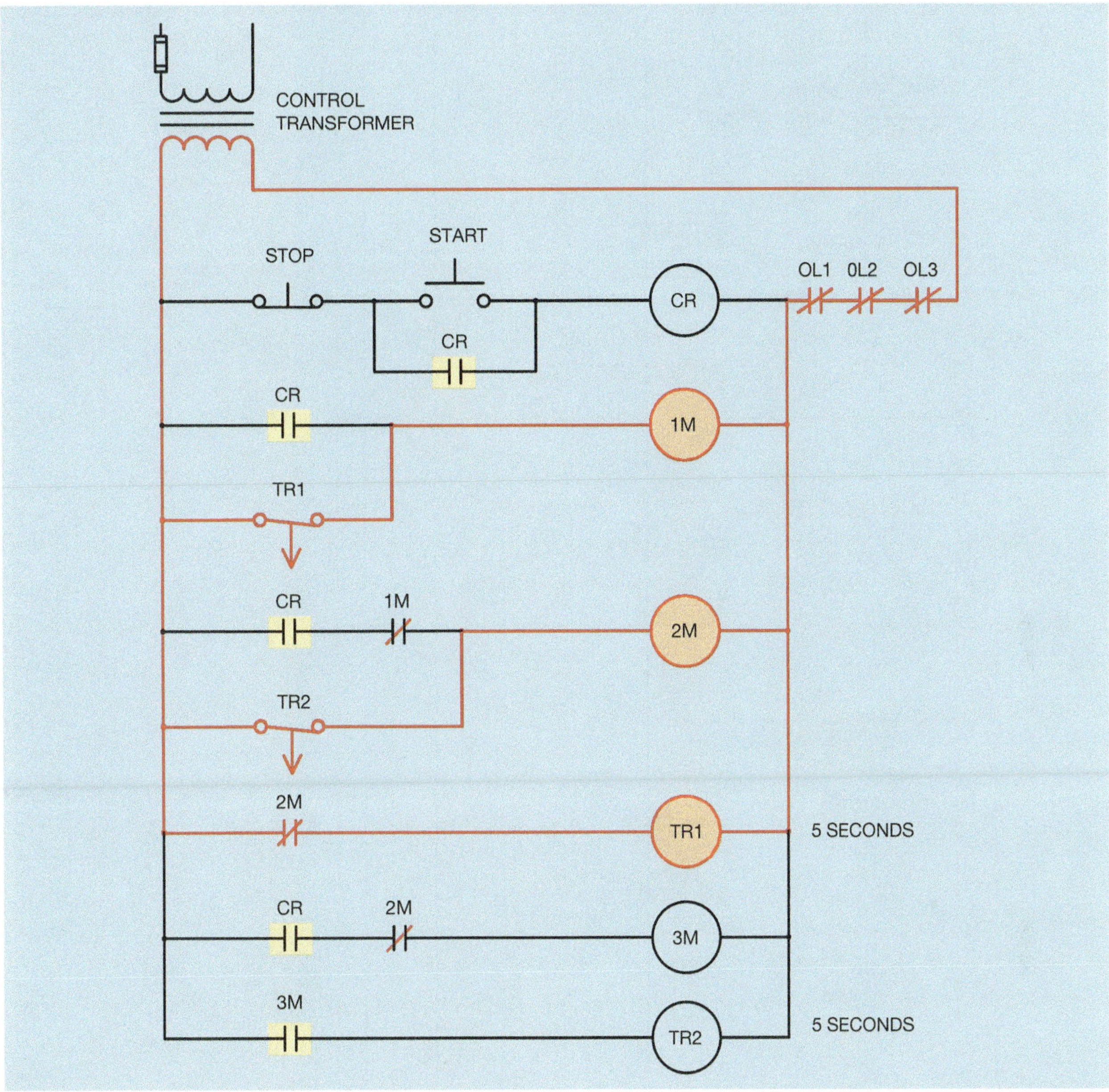

Figure 12–11 Motor #3 stops.

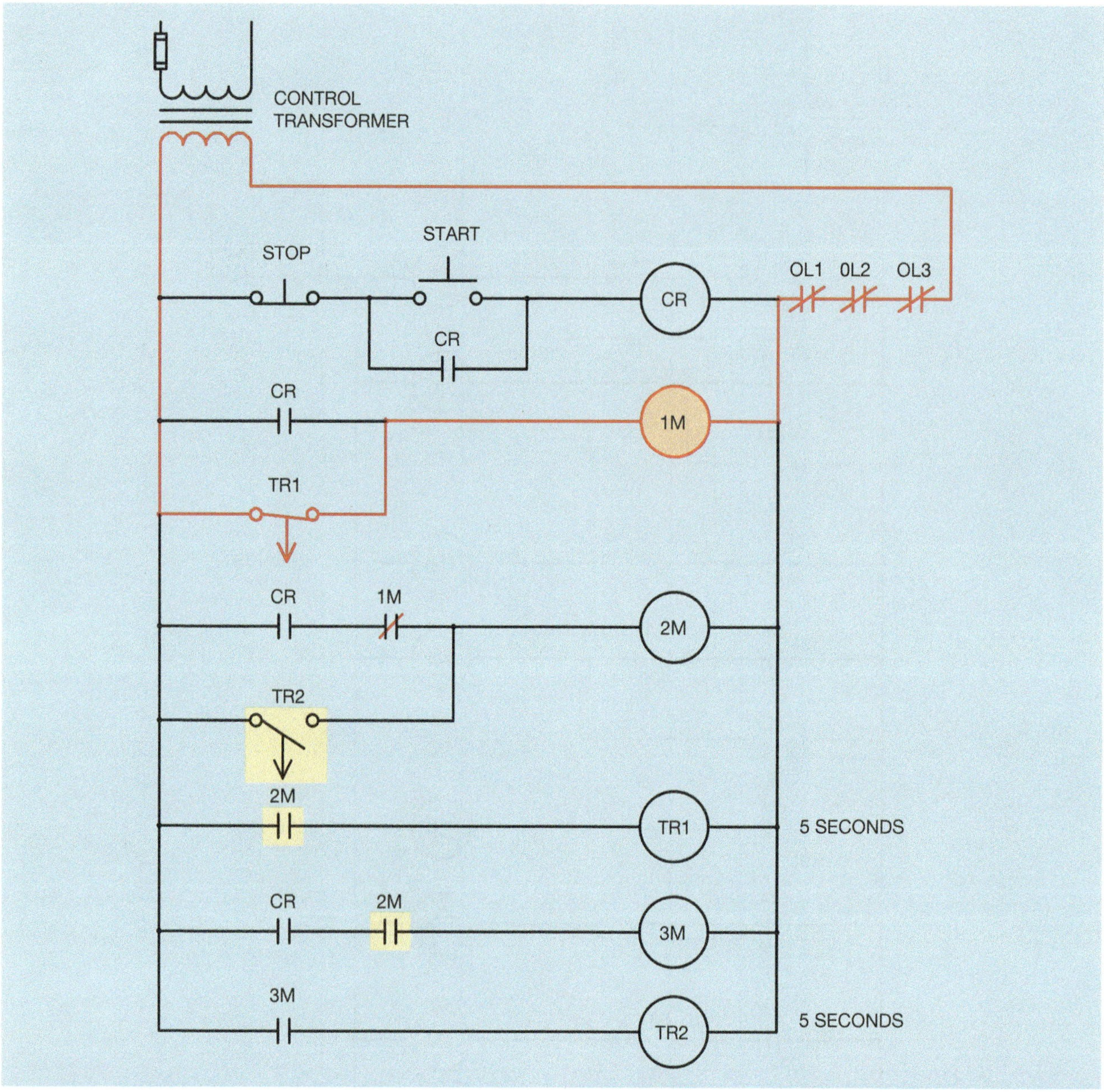

Figure 12–12 Motor #2 stops.

motor #3 and de-energizing off-delay timer coil TR2. After a delay of 5 seconds, timed contact TR2 opens and de-energizes starter 2M. This causes motor #2 to stop, off-delay timer TR1 to de-energize, and on-delay timer TR4 to de-energize. TR4 contact reopens immediately.

After a delay of 5 seconds, timed contact TR1 opens and de-energizes starter coil 1M. This causes motor #1 to stop and on-delay timer TR3 to de-energize. Contact TR3 reopens immediately and the circuit is back in its original de-energized state.

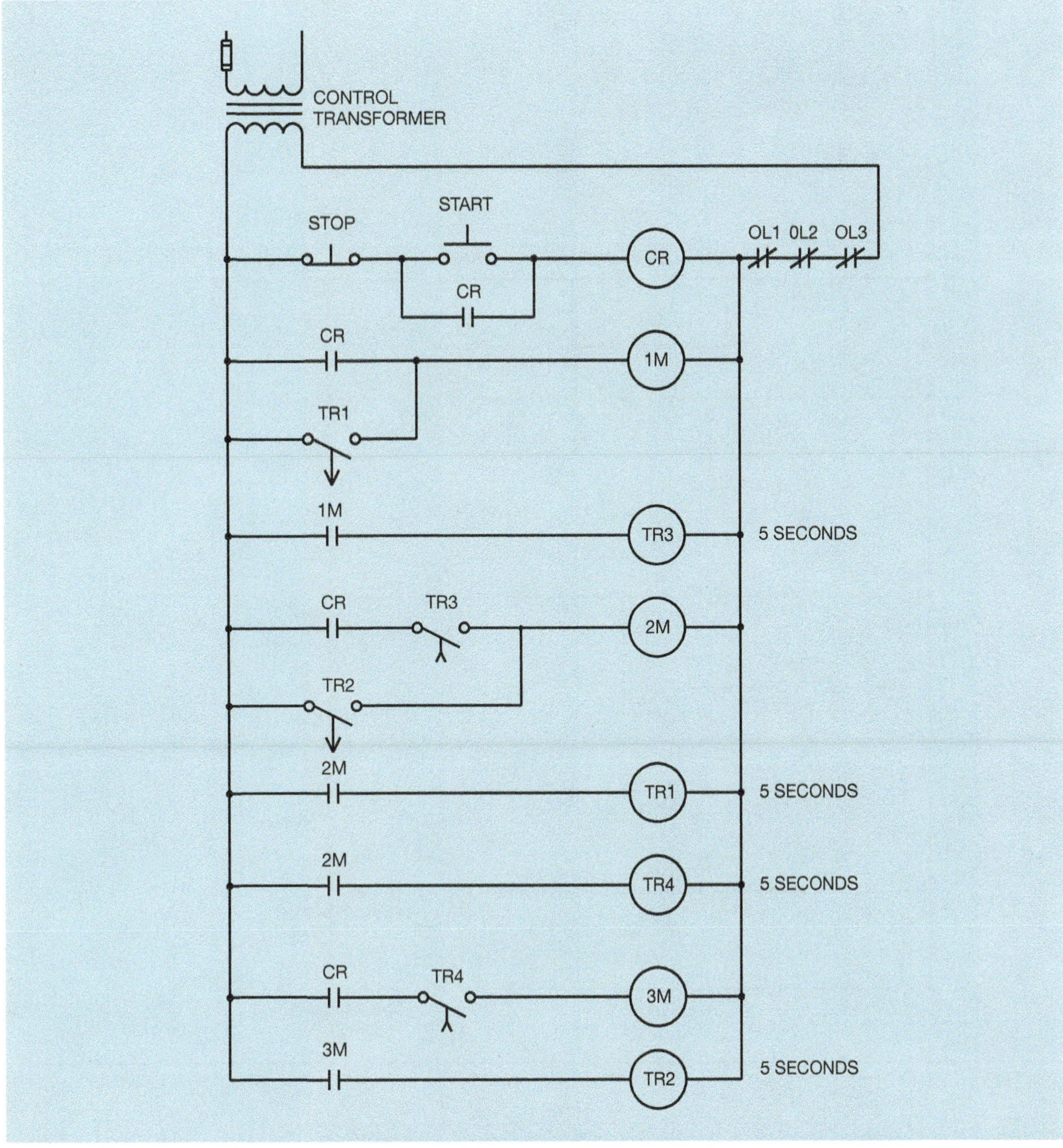

Figure 12–13 Motors start and stop in sequence with a time delay between starting and stopping.

Review Questions

1. What is the purpose of sequence control?
2. Refer to the schematic diagram in Figure 12–14. Assume that the 1M contact located between wire numbers 29 and 30 had been connected normally closed instead of normally open. How would this circuit operate?
3. Assume that all three motors shown in Figure 12–14 are running. Now assume that the STOP button is pressed and motors #1 and #2 stop running, but motor #3 continues to operate. Which of the following could cause this problem?
   a. STOP button is shorted.
   b. The 2M contact between wire numbers 31 and 32 is hung closed.

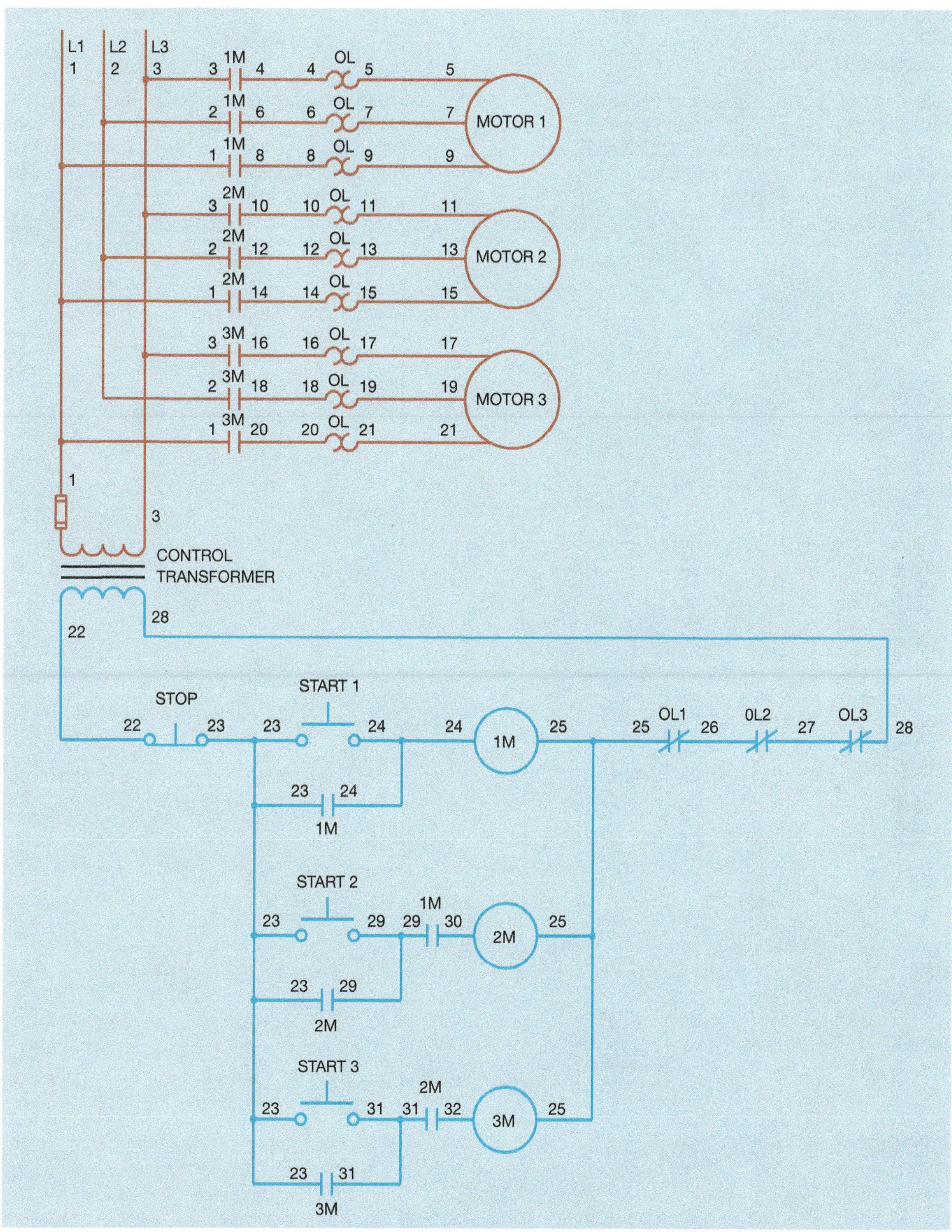

Figure 12–14 Sequence control schematic with wire numbers. Basic control circuits.

c. The 3M load contacts are welded shut.
d. The normally open 3M contact between wire numbers 23 and 31 is hung closed.

4. Referring to Figure 12–14, assume that the normally open 2M contact located between wire numbers 23 and 29 is welded closed. Also assume that none of the motors are running. What would happen if:
   a. The number 2 push button was to be pressed before the number 1 push button?
   b. The number 1 push button was to be pressed first?

5. In the control circuit shown in Figure 12–2, if an overload occurs on any motor, all three motors will stop running. Using a separate sheet of paper, redesign the circuit so that the motors must still start in sequence from #1 to #3, but an overload on any motor will stop only that motor. If an overload should occur on motor #1, for example, motors #2 and #3 would continue to operate.

Section 3

SENSING DEVICES

Chapter 13

PRESSURE SWITCHES AND SENSORS

Objectives

After studying this chapter the student will be able to:

- Describe the operation of high pressure switches.
- Describe the operation of low pressure switches.
- Make connection of a high pressure switch.
- Make connection of a low pressure switch.
- Discuss differential setting of pressure switches.
- Discuss pressure sensors that convert pressure to current for instrumentation purposes.

Pressure switches are found throughout the industry in applications where it is necessary to sense the pressure of pneumatic or hydraulic systems. Pressure switches are available that can sense pressure changes of less than 1 psi (pounds per square inch) or pressures over 15,000 psi. A diaphragm-operated switch can sense small pressure changes at low pressure (Figure 13–1).

A metal bellows type switch can sense pressures up to 2,000 psi. The metal bellows type pressure switch employs a metal bellows that expands with pressure (Figure 13–2). Although this switch can be used to sense a much higher pressure than the diaphragm type, it is not as sensitive in that it takes a greater change in pressure to cause the bellows to expand enough to activate a switch. A piston type pressure switch can be used for pressures up to 15,000 psi (Figure 13–3).

Regardless of the method used to sense pressure, all pressure switches activate a set of contacts. The contacts may be either single pole or double pole depending on the application, and will be designed with some type of snap action mechanism. Contacts cannot be permitted to slowly close or open. This would produce a bad connection and cause burning of the contacts as well as low

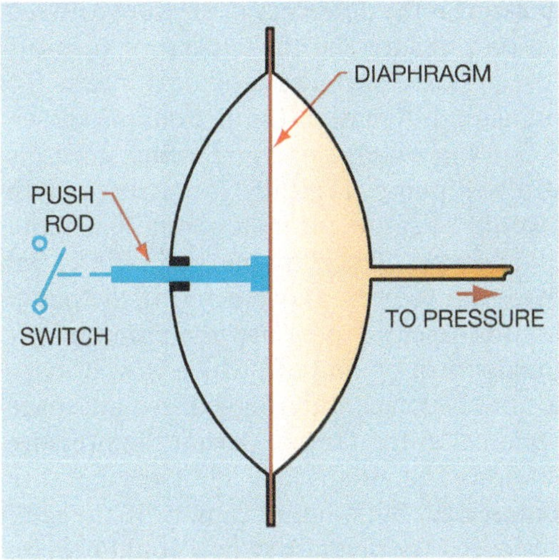

Figure 13–1 A diaphragm type pressure switch can sense small pressure changes at a low pressure.

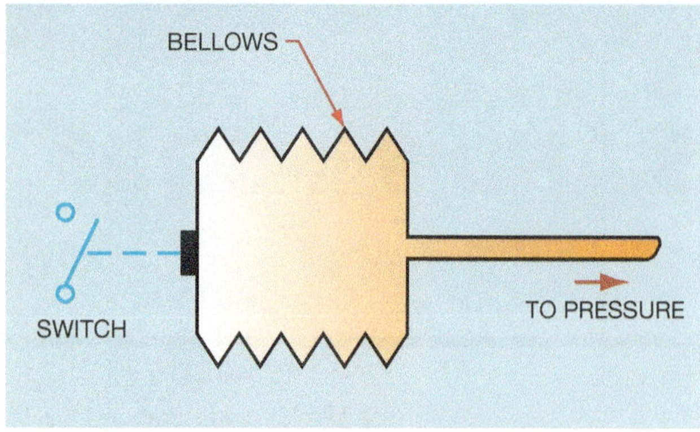

Figure 13–2 A metal bellows type pressure switch can be used for pressures up to 2000 psi.

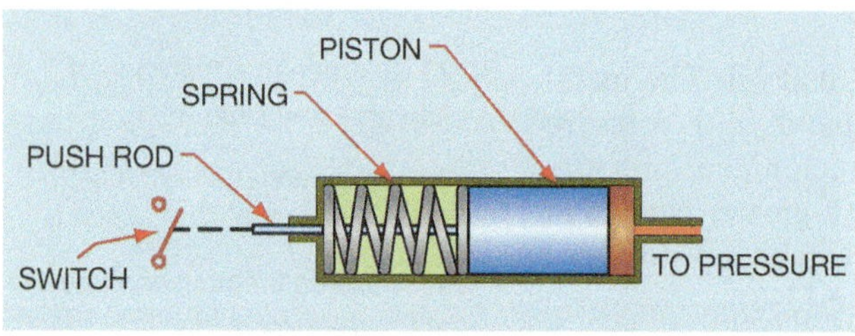

Figure 13–3 Piston type pressure switches can be used for pressures up to 15,000 psi.

Figure 13–4 Line voltage pressure switch.

voltage problems to the equipment they control. Some pressure switches are equipped with contacts large enough to connect a motor directly to the power line, and others are intended to control the operation of a relay coil. A line voltage type pressure switch is shown in Figure 13–4. Pressure switches of this type are often used to control the operation of well pumps and air compressors (Figure 13–5).

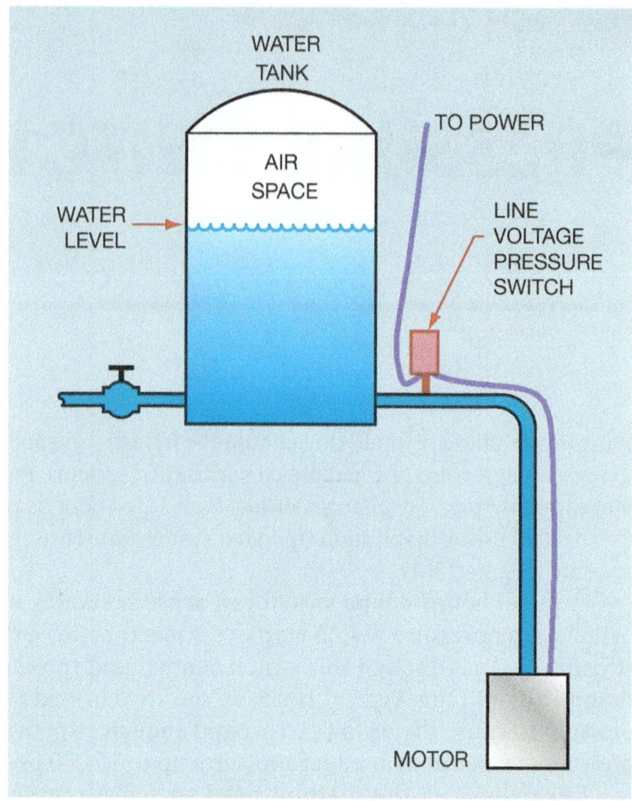

Figure 13–5 A line voltage pressure switch controls the operation of a pump motor.

Differential Pressure

Differential pressure is the difference in pressure between cut-in or turn-on pressure and the cut-out or turn-off pressure. Most pressure switches provide a means for setting the pressure differential. In the example shown in Figure 13–5, a line voltage pressure switch controls the motor of a well pump. Typically, a pressure switch of this type would be set to cut in at about 30 psi and cut out at about 50 psi. The 20 pounds of differential pressure is necessary to prevent overworking the pump motor. Without differential pressure, the pump motor would continually turn on and off, which is what happens when a tank becomes waterlogged. An air space must be maintained in the tank to permit the pressure switch to function. The air space is necessary because air can be compressed, but a liquid cannot. If the tank becomes waterlogged the pressure switch would turn on and off immediately each time a very small amount of water was removed from the tank. Pressure switch symbols are shown in Figure 13–6.

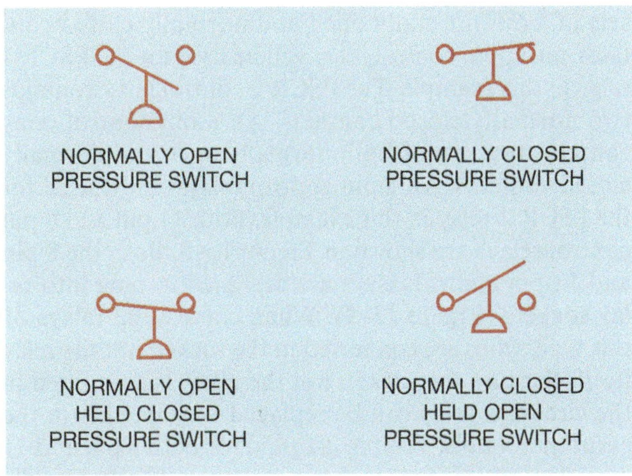

Figure 13–6 Pressure switch symbols.

Typical Application

Pressure switches are used in many common industrial applications. The circuit shown in Figure 13–7 employs a pressure switch to disconnect a motor and turn on two warning lights in the event of a high-pressure condition. The pressure switch contains both normally open and normally closed contacts. The normally closed part of the pressure switch is connected to a control relay labeled PSCR 2 (pressure switch control relay #2), and the normally open part of the switch is connected to a control relay labeled PSCR 1. In order for the control circuit to perform its desired function, the pressure switch must contain five different contacts, three normally closed and two normally open. Because pressure switches do not contain contacts in this arrangement,

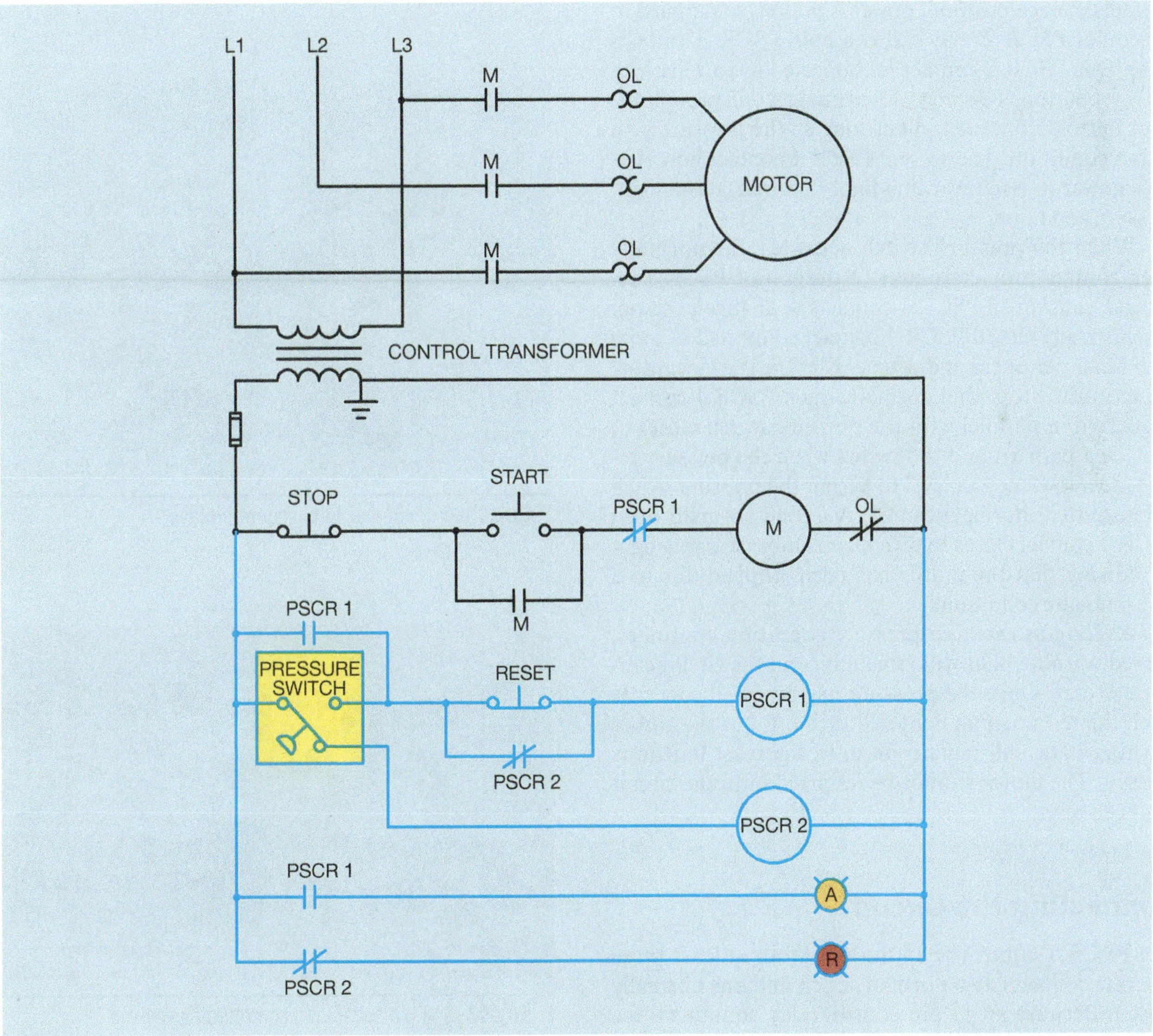

Figure 13–7 High pressure turns off the motor and turns on a warning light.

it is common practice to permit sensing devices such as pressure switches, limit switches, float switches, and others to operate control relays that do contain the necessary contacts.

Circuit Operation

When the pressure is below the value that will cause the pressure switch to activate, the normally closed part of the switch will provide continuous power to the coil of PSCR 2 control relay. Both PSCR 2 normally closed contacts will, therefore, be open anytime that the pressure is below that necessary to activate the pressure switch.

If the pressure should increase to a value that causes the pressure switch to activate, the switch contacts will change from the closed to the open position. When the contacts change position, power is no longer supplied to the coil of PSCR 2 relay coil and both PSCR 2 contacts close. One PSCR 2 contact is connected in parallel with the reset button. This now closed contact will prevent the reset button from working as long as the pressure is in a high condition. The second PSCR 2 contact now supplies power to a red warning light that indicates a high-pressure condition.

When the pressure switch activates, the normally open contact provides power to the coil of PSCR 1 relay coil, causing all PSCR 1 contacts to change position. The normally closed PSCR 1 contact connected in series with M starter opens and de-energizes the starter causing the motor to stop. One normally open PSCR 1 contact connected in parallel with the pressure switch closes to provide a path around the switch when the pressure returns to a low-enough level to permit the pressure switch to return to its normal position. A second normally open PSCR 1 contact closes to turn on an amber warning light to indicate that the motor has been stopped due to a high-pressure condition.

As long as the high-pressure condition continues, the red warning light will remain on and the circuit cannot be reset. Once the pressure has returned to a safe level, the red warning light will turn off, but the amber warning light will remain on until the reset button is pressed. The motor cannot be restarted until the circuit has been reset.

Connecting the Circuit

The PSCR 1 control relay must contain at least three separate contacts, two normally open and one normally closed. Because an 11 pin control relay contains three sets of both normally open and normally closed contacts, an 11 pin control relay will be used for the PSCR 1 relay in this example. The PSCR 2 control relay contains two normally closed contacts. An 8 pin control relay contains two sets of both normally open and normally closed contacts. An 8 pin control relay will be used for the PSCR 2 relay in this example. Both 11 pin and 8 pin control relays are shown in Figure 13–8. Both the 8 pin and 11 pin control relays are designed to plug into relay sockets (Figure 13–9). When connecting relays of this type, wires are connected to the socket, not the relay itself. Because the socket, not the relay, is connected in the circuit, a relay can be replaced very quickly in the event that it fails. The pin diagram for both 8 pin and 11 pin relays of this type is shown in Figure 13–10. The circuit shown in Figure 13–7 is shown in Figure 13–11 with the addition of wire numbers. There are several criteria

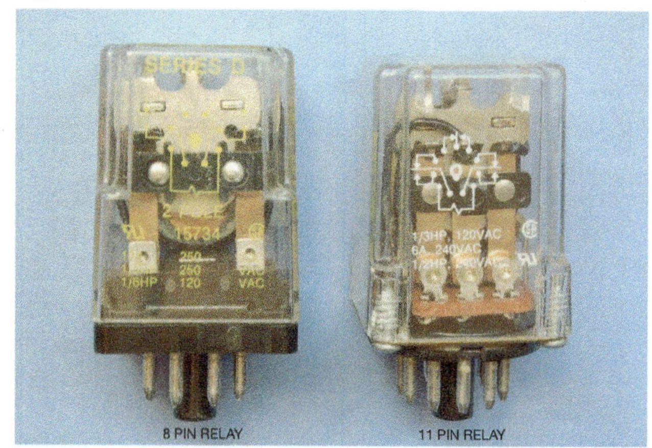

Figure 13–8 8 pin and 11 pin control relays.

Figure 13–9 8 pin and 11 pin control relay sockets.

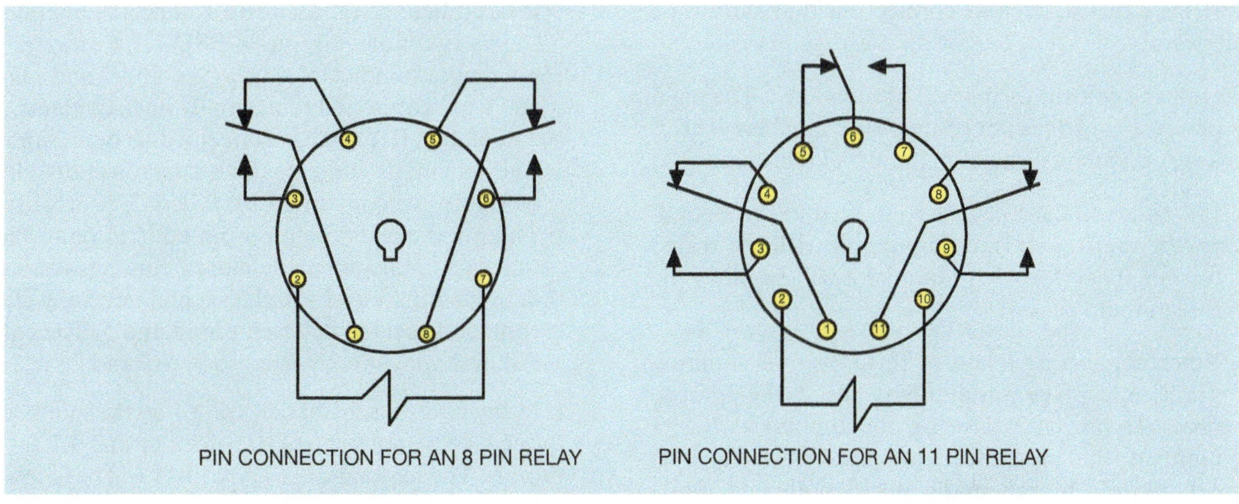

Figure 13–10 Pin connection diagrams for 8 and 11 pin relays.

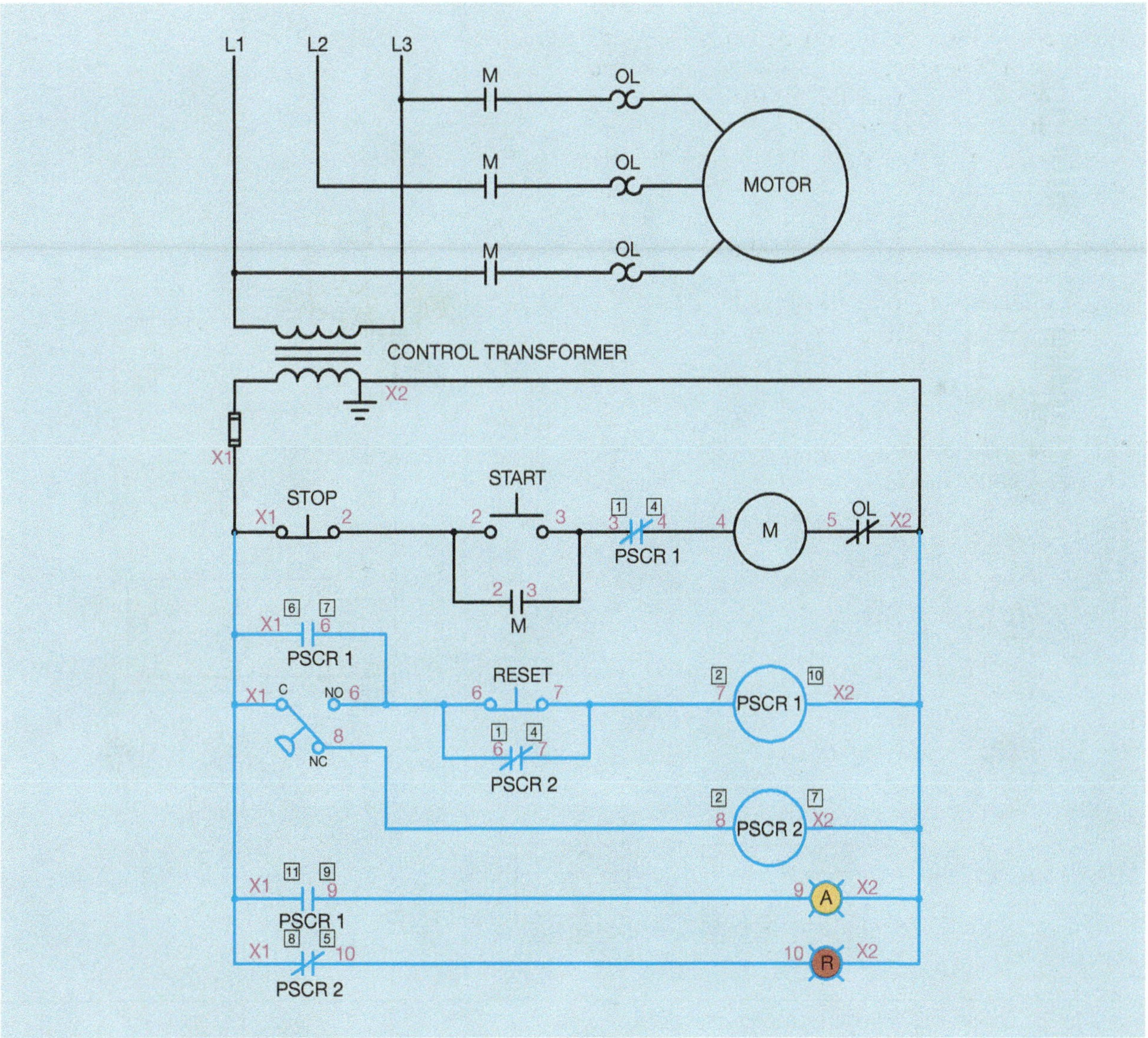

Figure 13–11 Control circuit with wire numbers.

concerning this particular connection that should be understood:

1. Only the control wiring will be numbered. The main power wiring does not contain wire numbers. This is a very common practice in industrial schematics.
2. It is assumed that the fuse connected to the secondary of the transformer is integral with the transformer. Therefore, wire numbers will begin at the fuse output.
3. When connecting relays of this type, it is common practice to place pin numbers beside the components. To prevent confusing pin numbers with wire numbers, the pin numbers are placed inside a circle or square. The schematic shows that a normally closed PSCR 1 contact is connected in series with the coil of M starter. The pin diagram in Figure 13-10 indicates that a normally closed contact is located between pins 1 and 4. The schematic also shows two normally open PSCR 1 contacts. The pin diagram indicates that pins 6 and 7 and pins 11 and 9 are connected to normally open contacts. The coil of PSCR 1 relay is connected to pins 2 and 10. The schematic indicates that two normally closed contacts are controlled by PSCR 2 control relay. The pin diagram for an 8 pin control relay shows that one of the normally closed contacts is located between pins 1 and 4 and a second normally closed contact is located between pins 8 and 5. The coil of PSCR 2 relay is connected to pins 2 and 7.
4. Many control schematics use X1 as the number for one side of the control transformer and X2 for the other. The schematic in Figure 13-11 will be numbered in the manner described by the preceding criteria.

The components with corresponding wire numbers are shown in Figure 13-12, and the circuit with wire

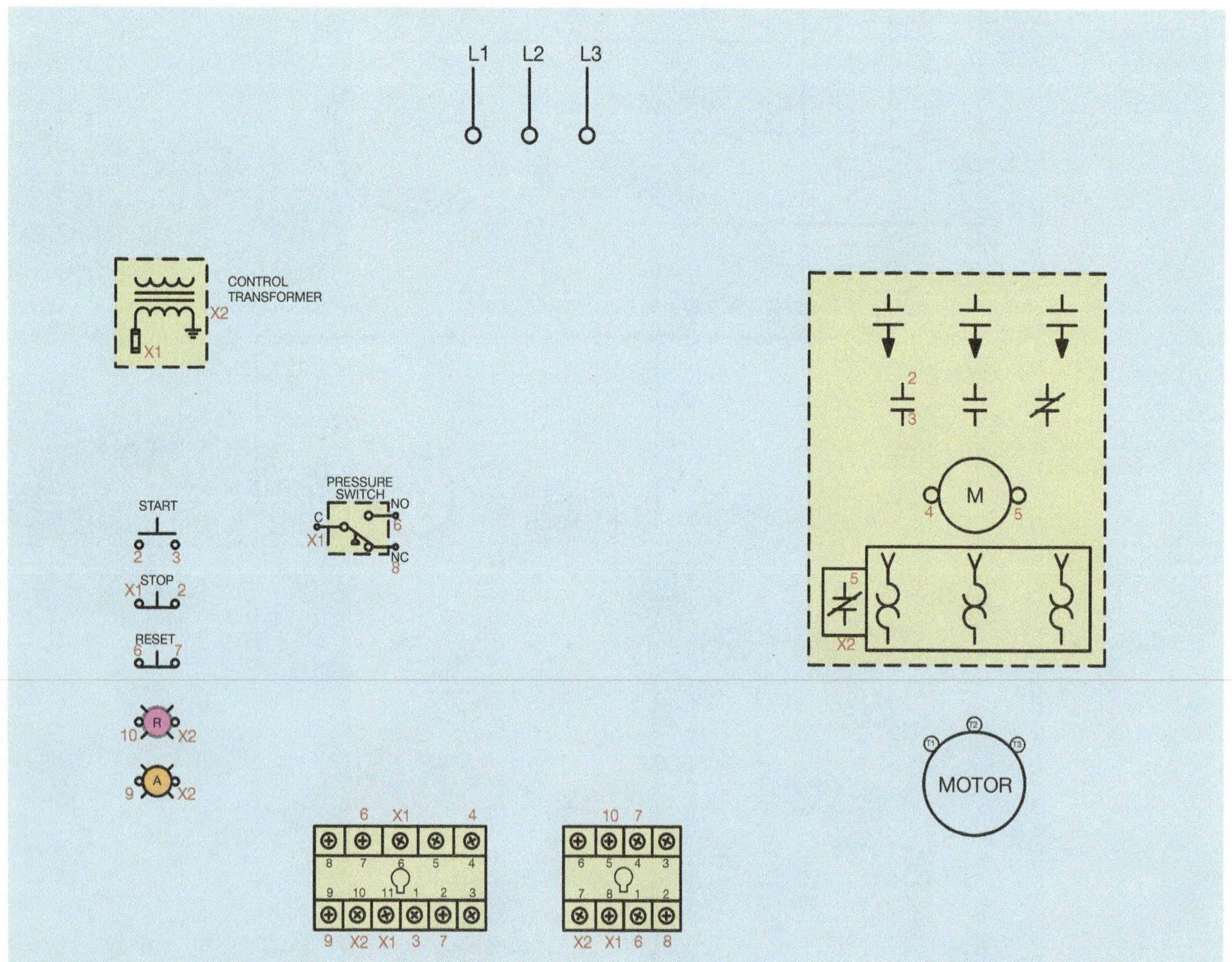

Figure 13–12 Components with corresponding wire numbers.

connections is shown in Figure 13-13. The wire connections are made by connecting all components with the same wire number together.

Pressure Sensors

Pressure switches are not the only pressure sensing devices that an electrician is likely to encounter on the job. This is especially true in an industrial environment. It is often necessary to not only know if the pressure has reached a certain level but also to know the amount of pressure. Although sensors of this type are generally considered to be in the instrumentation field, an electrician should be familiar with some of the various types and how they operate.

Pressure sensors are designed to produce an output voltage or current that is dependent on the amount of pressure being sensed. Piezoresistive sensors are very popular because of their small size, reliability, and accuracy (Figure 13–14). These sensors are available in ranges from 0–1 psi (pounds per square inch) and 0–30 psi. The sensing element is a silicon diaphragm integrated with an integrated circuit chip. The chip contains four implanted piezoresistors connected to form a bridge circuit (Figure 13–15). When pressure is applied to the diaphragm, the resistance of piezoresistors changes proportionally to the applied pressure, which changes the balance of the bridge. The voltage across V0 changes in

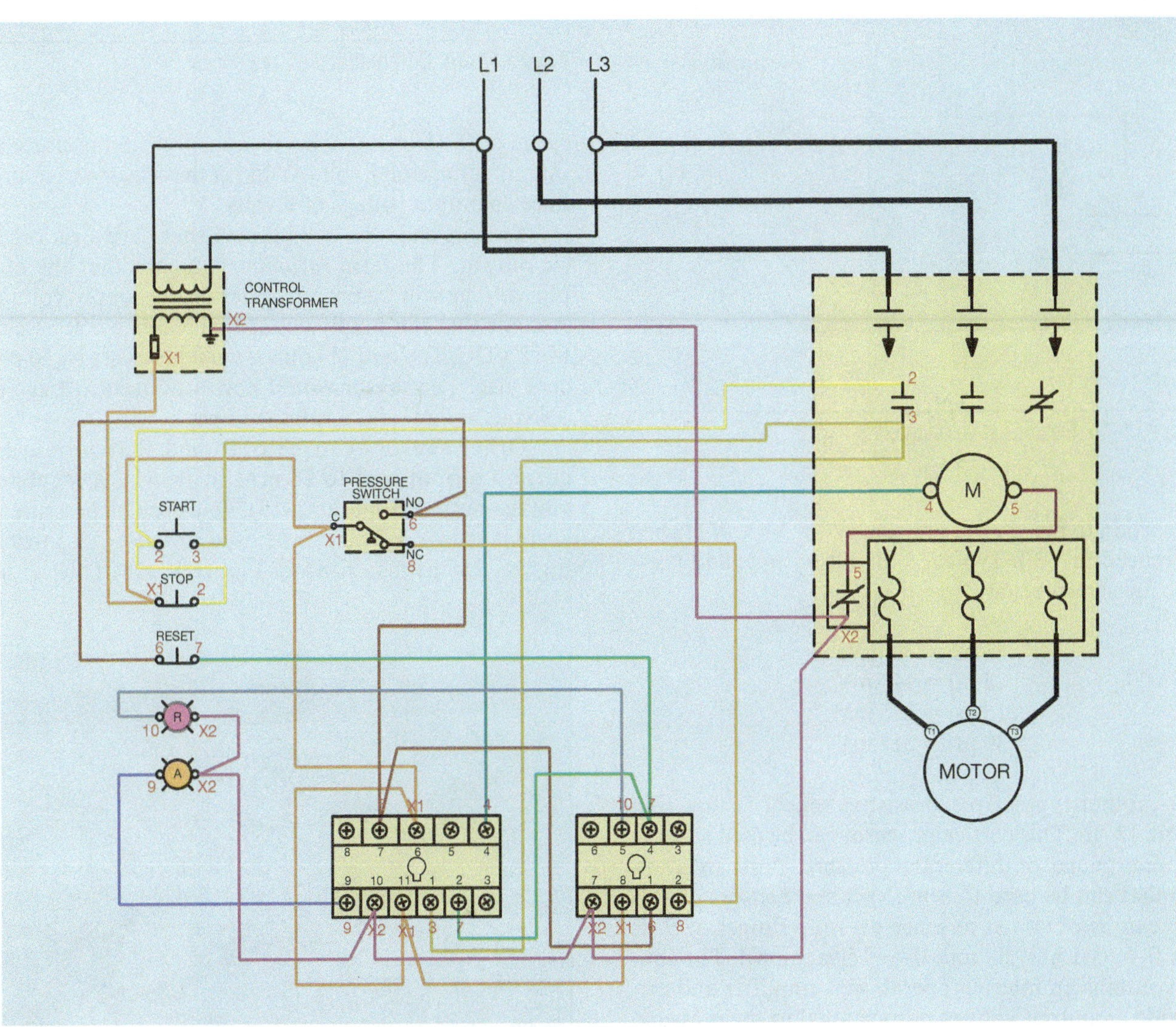

Figure 13–13 Circuit with wire connections.

Figure 13–14 Piezoresistive pressure sensor.

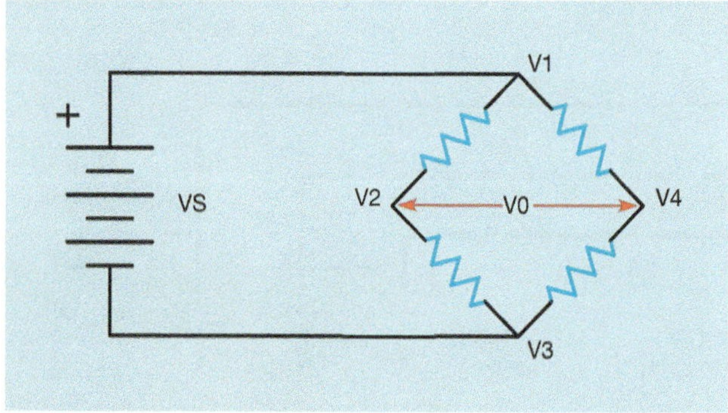

Figure 13–15 Piezoresistive bridge.

proportion to the applied pressure (V0 = V4 − V2 [when referenced to V3]). Typical millivolt outputs and pressures are shown below:

1 psi = 44 mV
5 psi = 115 mV
15 psi = 225 mV
30 psi = 315 mV

Another type of piezoresistive sensor is shown in Figure 13–16. This particular sensor can be used to sense absolute, gauge, or differential pressure. Units are available that can be used to sense vacuum. Sensors of this type can be obtained to sense pressure ranges of 0–1, 0–2, 0–5, 0–15, 0–30, and 0–(−15[vacuum]). The sensor contains an internal operational amplifier and can provide an output voltage proportional to the pressure. Typical supply voltage for this unit is 8 VDC. The *regulated* voltage output for this unit is 1–6 volts. Assume, for example, that the sensor is intended to sense a pressure range of 0–15 psi. At 0 psi the sensor would produce an output voltage of 1 volt. At 15 psi the sensor would produce an output voltage of 6 volts.

Figure 13–16 Differential pressure sensor.

Sensors can also be obtained that have a ratiometric output. The term *ratiometric* means that the output voltage will be proportional to the supply voltage. Assume that the supply voltage increases by 50 percent to 12 VDC. The output voltage would increase by 50 percent also. The sensor would now produce a voltage of 1.5 volts at 0 psi and 9 volts at 15 psi.

Other sensors can be obtained that produce a current output of 4 to 20 mA, instead of a regulated voltage output (Figure 13–17). One type of pressure to current sensor, which can be used to sense pressures as high as 250 psi, is shown in Figure 13–18. This sensor

Figure 13–17 Pressure to current sensor for low pressures.

Courtesy of Honeywell International Inc.

Figure 13–18 Pressure to current sensor for high pressure.

can also be used as a set point detector to provide a normally open or normally closed output. Sensors that produce a proportional output current instead of voltage have fewer problems with induced noise from surrounding magnetic fields, and with voltage drops due to long wire runs.

A flow-through pressure sensor is shown in Figure 13–19. This type of sensor can be placed in line with an existing system. In-line pressure sensors make it easy to add a pressure sensor to an existing system.

Another device that is basically a pressure sensor is the force sensor (Figure 13–20). This sensor uses silicon piezoresistive elements to determine the amount of pressure to the sensing element.

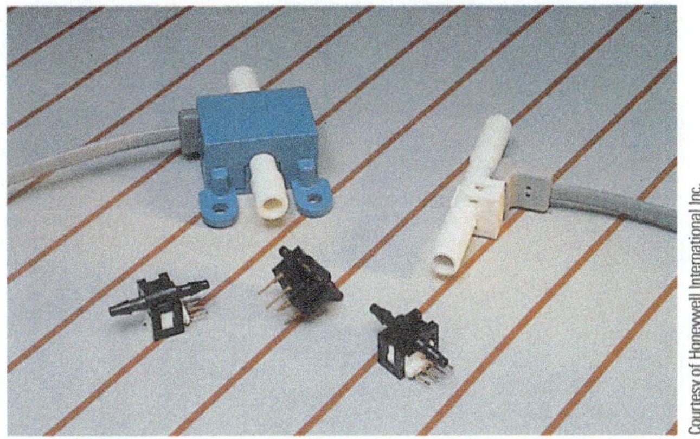

Courtesy of Honeywell International Inc.

Figure 13–19 Flow-through pressure sensor.

Courtesy of Honeywell International Inc.

Figure 13–20 Force sensor.

Review Questions

1. What type of pressure switch is generally used to sense small changes in low pressure systems?
2. A pressure switch is set to cut in at a pressure of 375 psi and cut out at 450 psi. What is the pressure differential for this switch?
3. A pressure switch is to be installed on a system with pressures that can range from 1500 psi to 1800 psi. What type of pressure switch should be used?
4. A pressure switch is to be installed in a circuit that requires it to have three normally open contacts and one normally closed contact. The switch actually has one normally open contact. What must be done to permit this pressure switch to operate in this circuit?
5. What is a piezoresistor?
6. Refer to the circuit shown in Figure 13–7. If the pressure should become high enough for the pressure switch to close and stop the motor, is it possible to restart the motor before the pressure drops to a safe level?
7. Refer to the circuit shown in Figure 13–7. Assume that the motor is running and an overload occurs and causes the overload contact to open and disconnect coil M to stop the motor. What effect does the opening of the overload contact have on the pressure switch circuit?

Chapter 14

FLOAT SWITCHES AND LIQUID LEVER SENSORS

Objectives

After studying this chapter the student will be able to:

- Describe the operation of a float switch.
- Draw schematic symbols for float switches.
- Describe methods of sensing the level of a liquid.

Float switches are used to control the action of a pump in accord with the level of a liquid in a tank or sump. The operation of the float switch is controlled by the upward or downward movement of a float in a tank of liquid. There are several styles of float switches. One employs the use of a rod with a float mounted on one end. Adjustable stops on the rod determine the amount of movement that must take place before a set of contacts is opened or closed (Figure 14–1).

Another common type of float switch is chain operated (Figure 14–2). A float is attached to one end of a chain and a counterweight is connected to the other. The float weighs more than the counterweight, which permits it to control the movement of the chain as the liquid level rises or falls. Some float switches contain large contacts that can be used to connect the motor directly to the line. The contacts may be normally open or normally closed depending on contact arrangement and may not be submerged. Float switches can be used to pump water from a tank or sump, or to fill a tank depending on the requirements of the circuit.

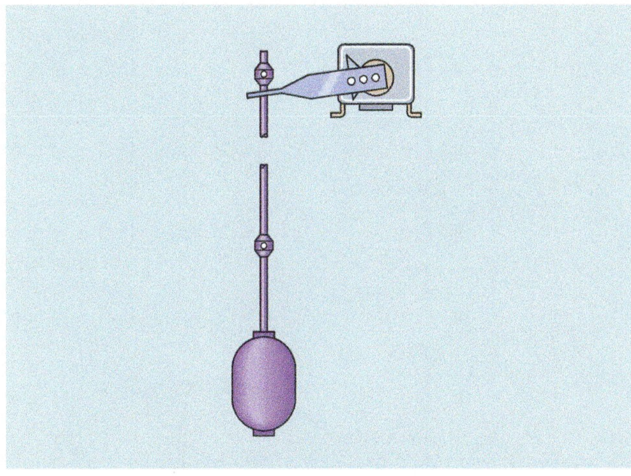

Figure 14–1 Rod-operated float switch.

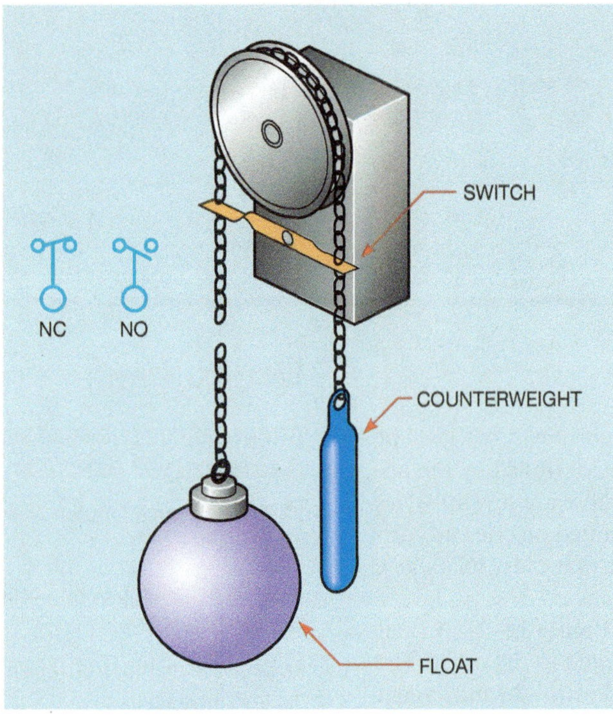

Figure 14–2 Chain-operated float switch with float switch symbols.

Mercury Bulb Float Switch

Another float switch that has become increasingly popular is the mercury bulb type of float switch. This type of float switch does not depend on a float rod or chain to operate. The mercury bulb switch appears to be a rubber bulb connected to a conductor. A set of mercury contacts is located inside the bulb. When the liquid level is below the position of the bulb, it is suspended in a vertical position (Figure 14–3A). When the liquid level rises to the position of the bulb, it changes to a horizontal position (Figure 14–3B). This change of position changes the state of the contacts in the mercury switch.

Because the mercury bulb float switch does not have a differential setting as does the rod or chain type of float switch, it is necessary to use more than one mercury bulb float switch to control a pump motor. The differential level of the liquid is determined by suspending mercury bulb switches at different heights in the tank. Figure 14–4 illustrates the use of four mercury bulb type switches used to operate two pump motors and provide a high liquid level alarm. The control circuit is shown in Figure 14–5. Float switch FS1 detects the lowest point of liquid level in the tank and is used to turn both pump motors off. Float switch FS2 starts the

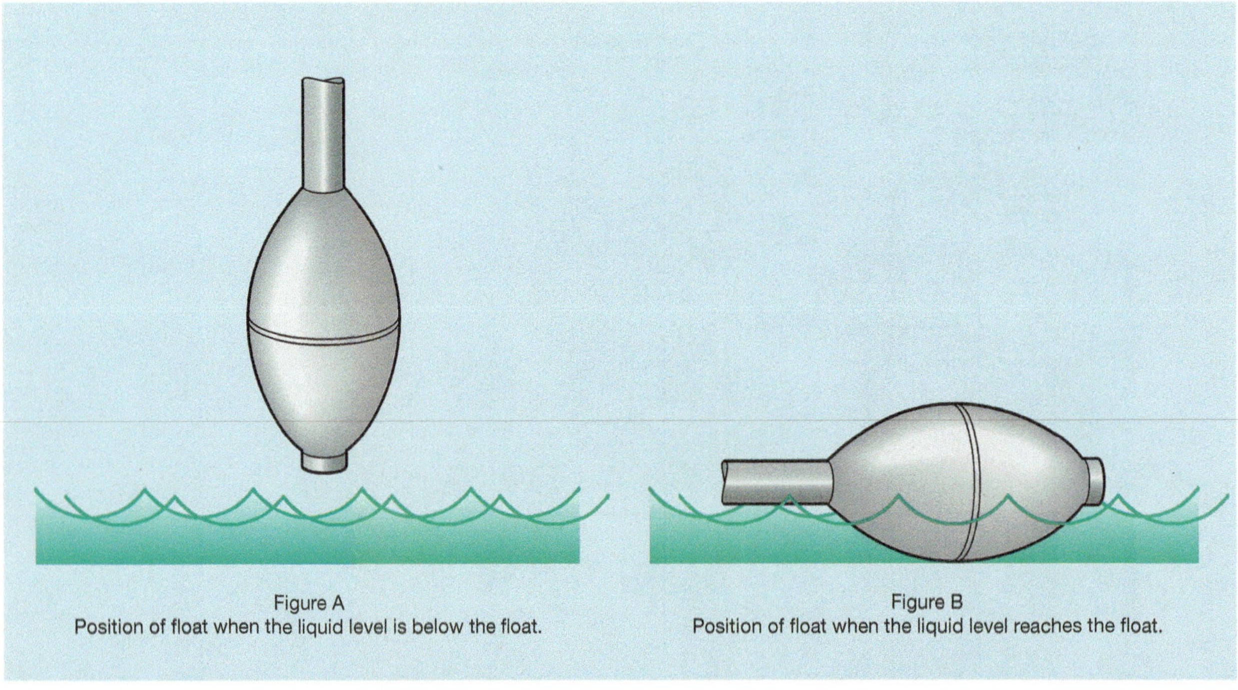

Figure 14–3 Mercury bulb type float switch.

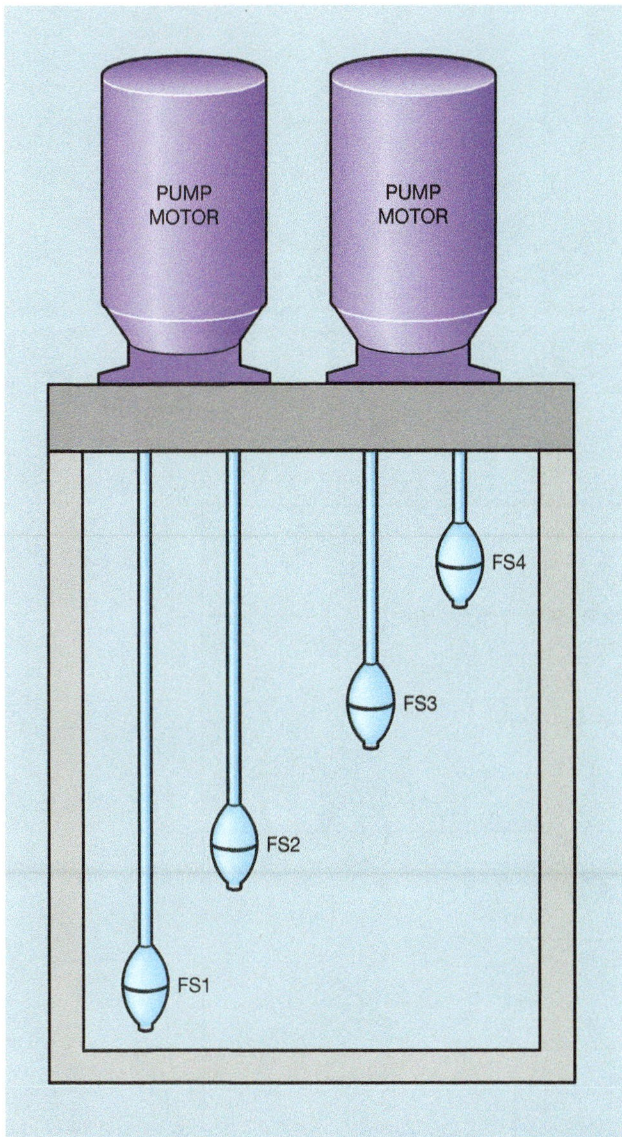

Figure 14–4 Float level is set by the length of the conductor.

first pump when the liquid level reaches that height. If pump #1 is unable to control the level of the tank, float switch FS3 will start pump motor #2 if the liquid level should rise to that height. Float switch FS4 operates a warning light and buzzer to warn that the tank is about to overflow. A reset button can be used to turn off the buzzer, but the warning light remains on until the water level drops below the level of float switch FS4.

The Bubbler System

Another method often used to sense liquid level is the bubbler system. This method does not employ the use of float switches. The liquid level is sensed by pressure switches (Figure 14–6). A great advantage of this system is that the pressure switches are located outside the tank, which makes it unnecessary to open the tank to service the system.

The bubbler system is connected to an air line, which is teed to a manifold and another line that extends down into the tank. A hand valve is used to adjust the maximum air flow. The bubbler system operates on the principle that as the liquid level increases in the tank, it requires more air pressure to blow air through the line in the tank. For example, consider a pipe with an inside area of 1 square inch (1 in^2). Each inch in length would represent a volume of 1 cubic inch (1 $in^2 \times 1$ in = 1 in^3). A cubic inch of water weighs 0.0361 pounds. If a pipe with an inside area of 1 square inch were inside a tank of water 10 feet deep, the weight of the water inside the pipe would be 4.332 pounds.

$$10\,ft \times 12\,in^3\,per\,foot = 120\,in^3$$

$$120\,in^3 \times 0.0361 = 4.332\,lb$$

It would require 4.332 pounds of air pressure to remove the water from inside the pipe. If the water inside the tank were to drop to a depth of 7 feet, it would require 3.032 pounds of air pressure to keep the pipe clear of water.

The bubbler system can be employed to measure the depth of virtually any liquid. The pressure needed would depend on the weight of the liquid. Gasoline, for example, weighs an average of 6.073 pounds per gallon, #2 diesel fuel weighs an average of 7.15 pounds per gallon, and water weighs an average of 8.35 pounds per gallon.

Because the pressure required to bubble air through the pipe is directly proportional to the height of the liquid, the pressure switches provide an accurate measure of the liquid level. The pressure switches shown in Figure 14–6 could be used to control the two pump circuit previously discussed by replacing the float switches with pressure switches in the circuit shown in Figure 14–5.

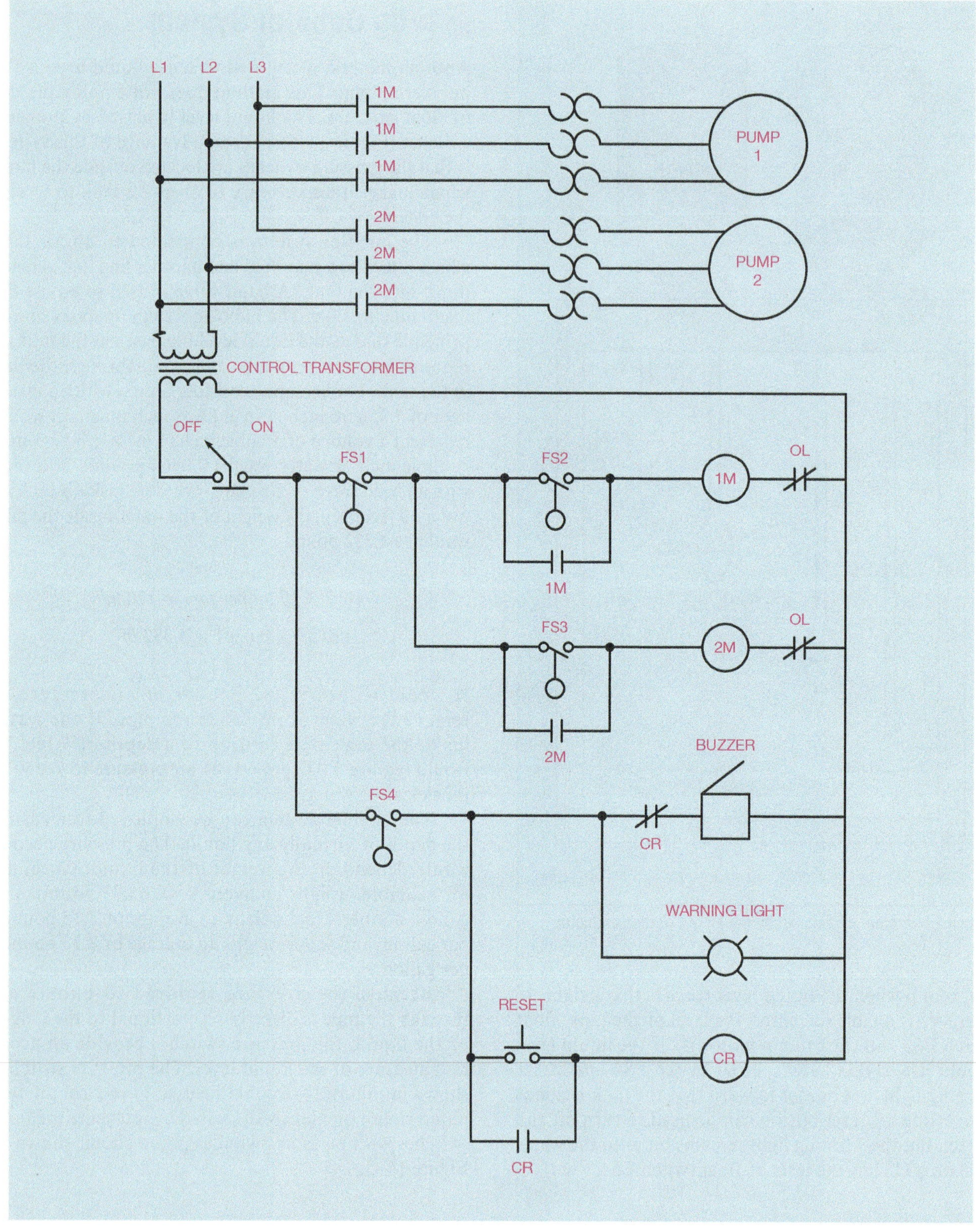

Figure 14–5 Two pump control with high liquid level warning.

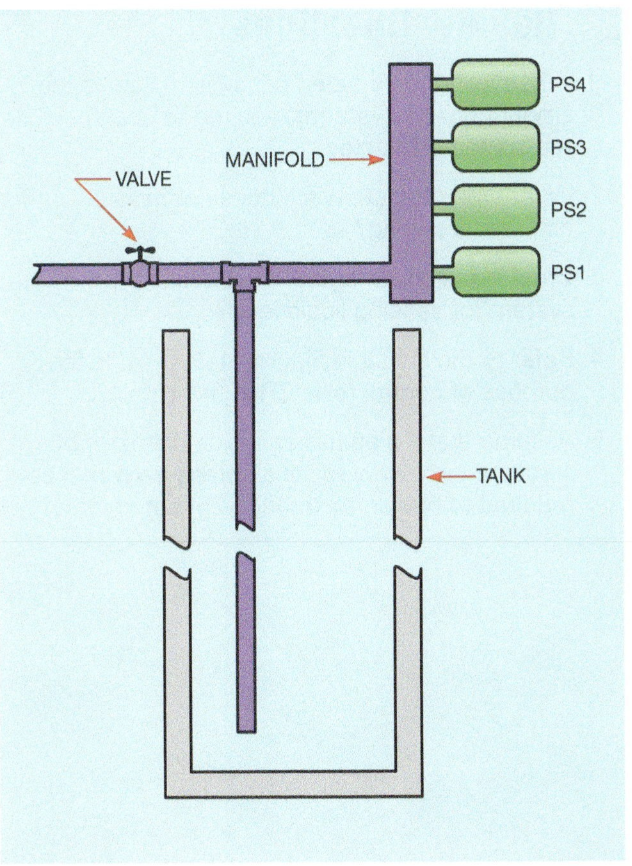

Figure 14–6 Bubbler system for detecting liquid level.

Figure 14–7 Operation of the radar gauge.

Microwave Level Gauge

The microwave level gauge operates by emitting a high frequency signal of approximately 24 GHz into a tank and then measuring the frequency difference of the return signal that bounces off the product (Figure 14–7). A great advantage of the microwave level gauge is that no mechanical object touches or is inserted into the product. The gauge is ideal for measuring the level of turbulent, aerated, solids-laden, viscous, corrosive fluids. It also works well with pastes and slurries. A cut-away view of a microwave level gauge is shown in Figure 14–8.

The gauge shown in Figure 14–9 has a primary 4–20 mA analog signal. The gauge can accept one RTD (resistance temperature detector) input signal. The gauge can be configured to display the level, calculated volume, or standard volume. A microwave level gauge with meter is shown in Figure 14–9.

Figure 14–8 Cut-away view of a microwave level gauge.

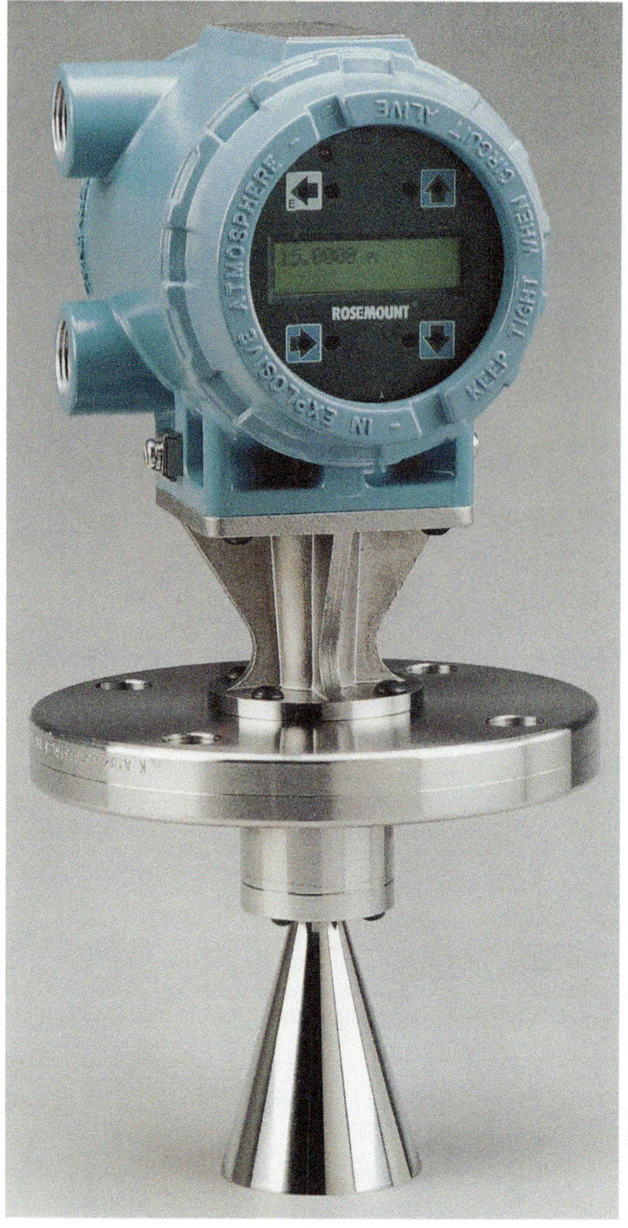

Figure 14–9 Microwave level gauge with meter.

Review Questions

1. When using a rod type float switch, how is the amount float movement required to open or close the contacts adjusted?
2. What type of float switch does not have a differential setting?
3. What is the advantage of the bubbler type system for sensing liquid level?
4. Refer to the circuit in Figure 14–5. What is the purpose of control relay CR in this circuit?
5. Assume that a pipe has an inside diameter of 1 square inch. How much air pressure would be required to bubble air through 25 feet of water?

Chapter 15

FLOW SWITCHES

Objectives

After studying this chapter the student will be able to:

» Describe the operation of flow switches.

» Connect a flow switch in a circuit.

» Draw the NEMA symbols that represent a flow switch in a schematic diagram.

Flow switches are used to detect the movement of air or liquid through a duct or pipe. Air flow switches are often called *sail switches* because the sensor mechanism resembles a sail (Figure 15–1). The air flow switch is constructed from a snap action micro switch. A metal arm is attached to the micro switch. A piece of thin metal or plastic is connected to the metal arm. The thin piece of metal or plastic has a large surface area and offers resistance to the flow of air. When a large amount of air flow passes across the sail, enough force is produced to cause the metal arm to operate the contacts of the switch.

Air flow switches are often used in air conditioning and refrigeration circuits to give a positive indication that the evaporator or condenser fan is operating before the compressor is permitted to start. A circuit of this type is shown in Figure 15–2. When the thermostat contact closes, control relay CR energizes and closes all CR contacts. This energizes both the condenser fan motor (CFM) relay

Courtesy of Honeywell International, Inc.

Figure 15–1 Air flow switch.

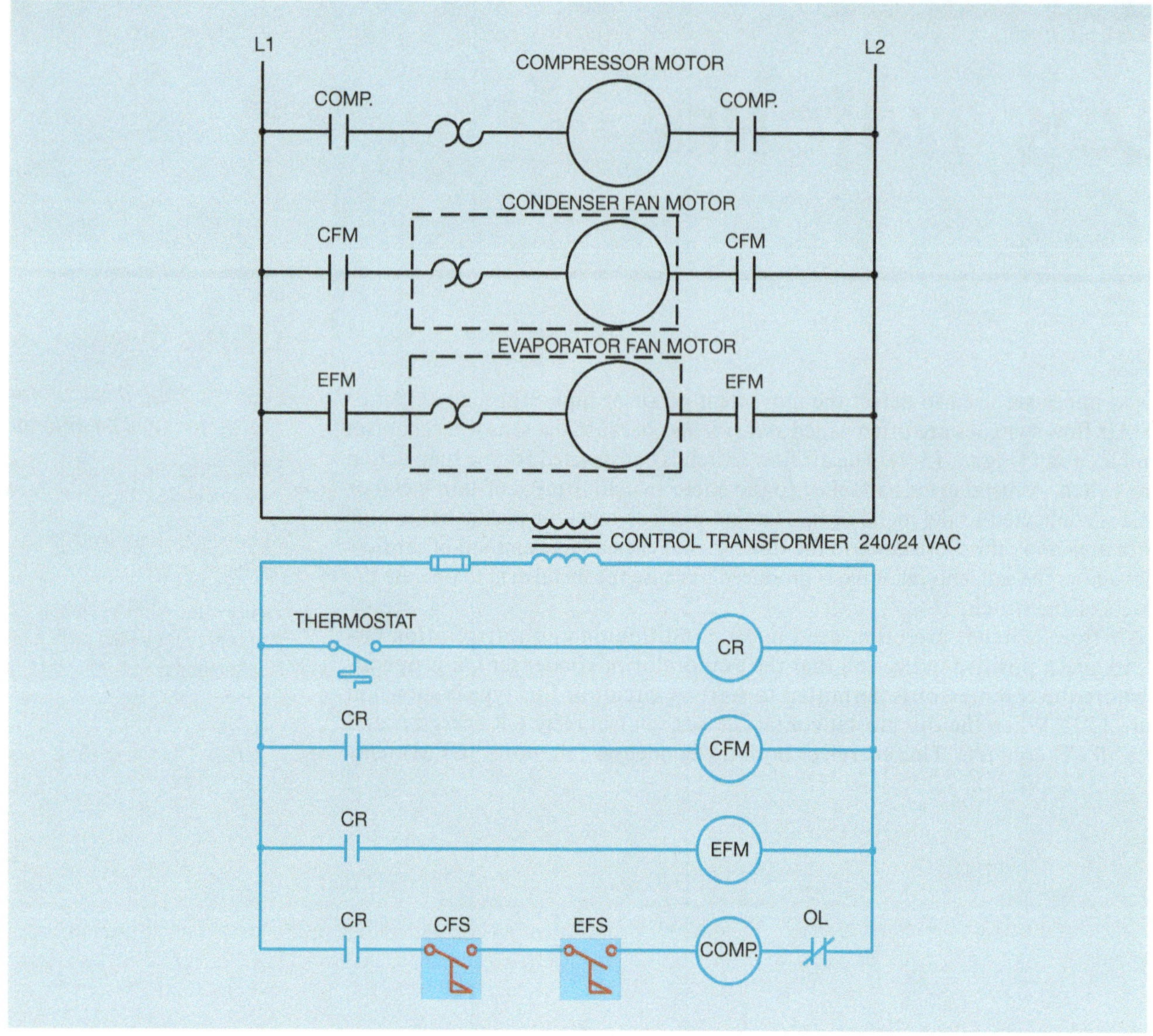

Figure 15–2 Air flow switches indicate a positive movement of air before the compressor can start.

and the evaporator fan motor (EFM) relay. The compressor relay (COMP.) cannot start because of the two normally open air flow switches. If both the condenser fan and evaporator fan start, air movement causes both air flow switches to close and complete a circuit to the compressor relay.

Notice in this circuit that a normally closed overload contact is shown in series with the compressor contactor only. Also notice that a dashed line has been drawn around the condenser fan motor and overload symbol, and around the evaporator fan motor and overload symbol. This indicates that the overload for these motors is located on the motor itself and is not part of the control circuit.

Liquid flow switches are equipped with a paddle that inserts into the pipe (Figure 15–3). A flow switch can be installed by placing a tee in the line as shown in Figure 15–4. When liquid moves through the line, force is exerted against the paddle causing the contacts to change position.

Regardless of the type of flow switch used, they generally contain a single pole double throw micro switch (Figure 15–5). Flow switches are used to control low current loads, such as contactor or relay coils or pilot

Courtesy of Flow Network/Kobold Instruments

Figure 15–3 Liquid flow switch.

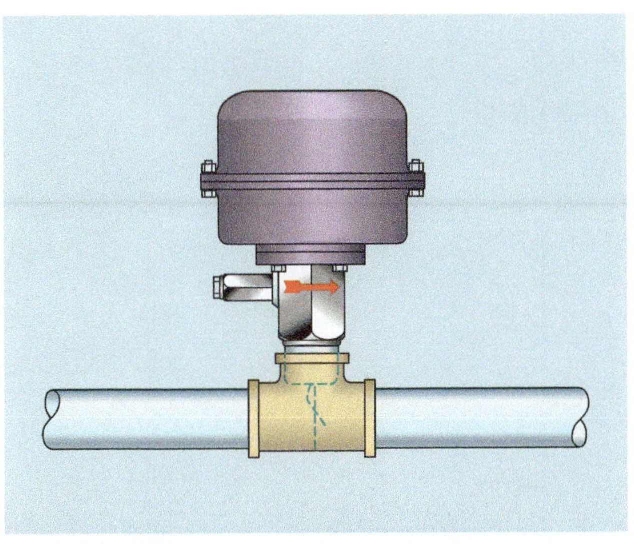

Figure 15–4 Flow switch installed in a tee.

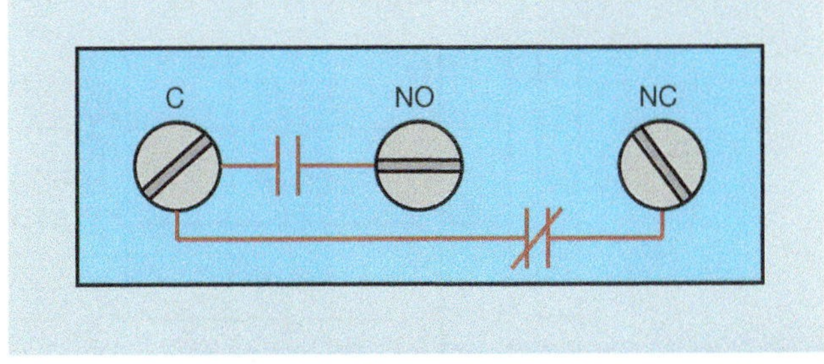

Figure 15–5 Connections of a single pole double throw micro switch.

lights. A circuit that employs both the normally open and normally closed contact of a flow switch is shown in Figure 15–6. The circuit is designed to control the operation of an air compressor. A pressure switch controls the operation of the compressor. In this circuit, a normally open push button is used as a reset button. The control relay must be energized before power can be supplied to the rest of the control circuit. When the pressure switch contact closes, power is supplied to the lube oil pump relay. The flow switch detects the flow of lubricating oil before the compressor is permitted to start. Note that a red warning light indicates when there is no flow of oil. To connect the flow switch in this circuit, power from the control relay contact must be connected to the common terminal of the flow switch so that power can be supplied to both the normally open and normally closed contacts (Figure 15–7). The normally open section of the switch connects to the coil of the compressor contactor, and the normally closed section connects to the red pilot light.

Regardless of whether a flow switch is intended to detect the movement of air or liquid, the NEMA symbol for both is the same. Standard NEMA symbols for flow switches are shown in Figure 15–8.

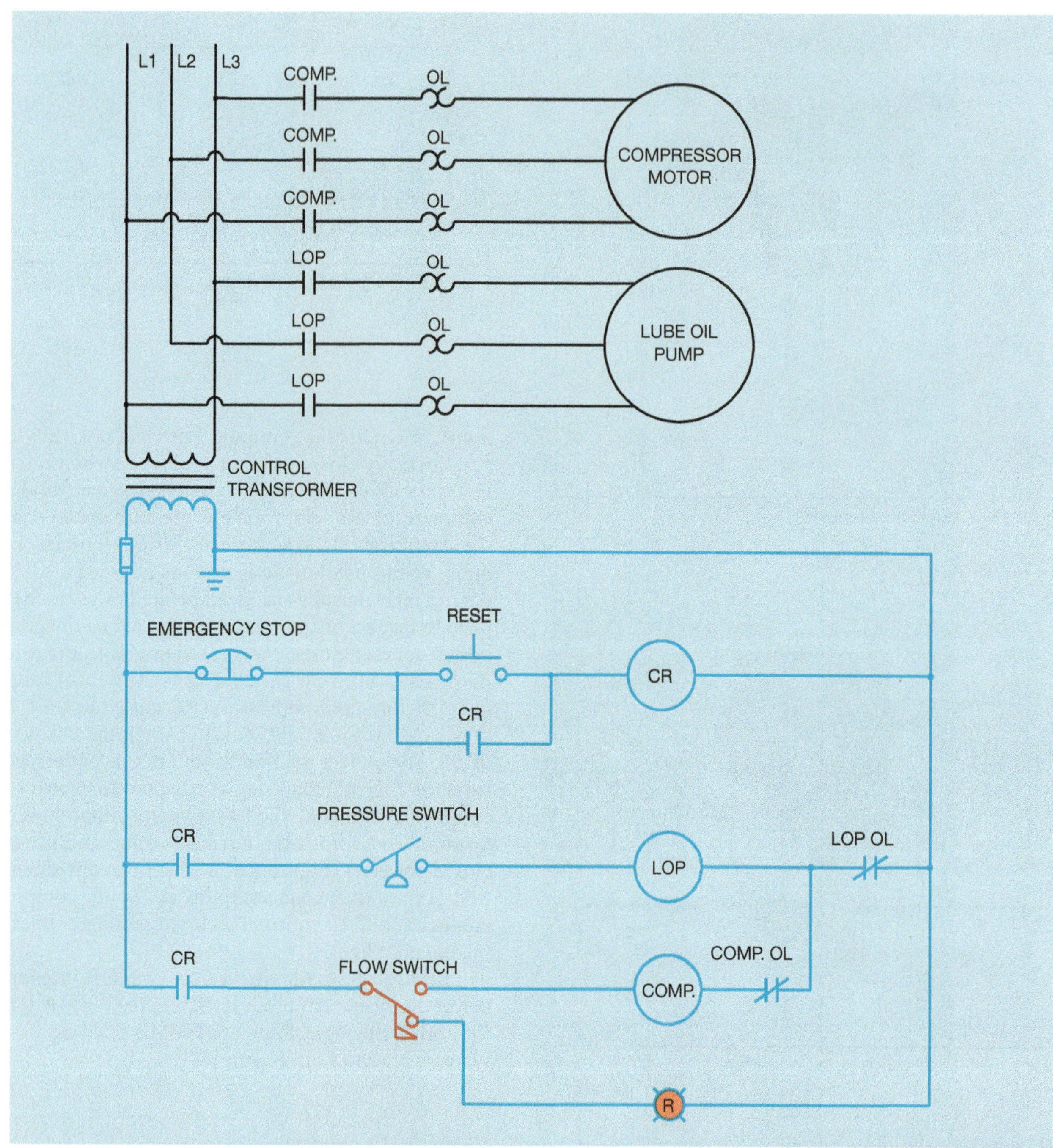

Figure 15–6 A red warning light indicates there is no oil flow.

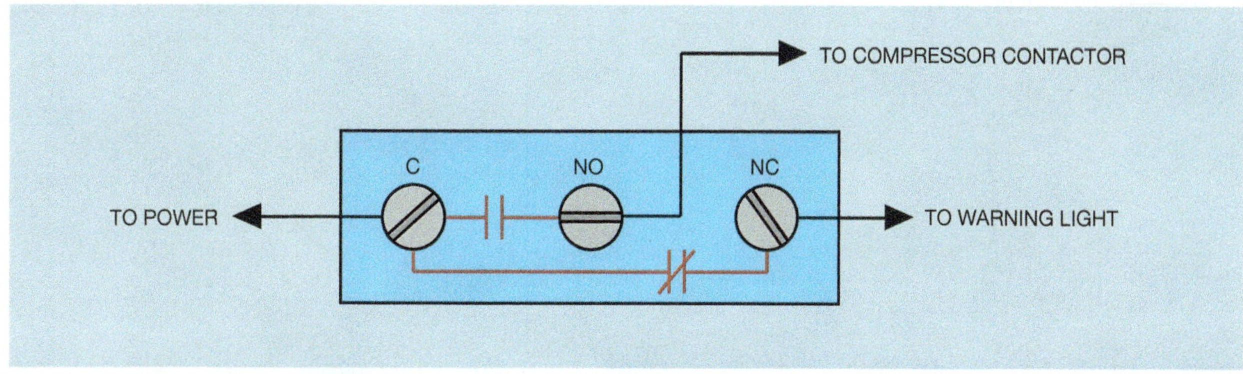

Figure 15–7 Connecting the flow switch.

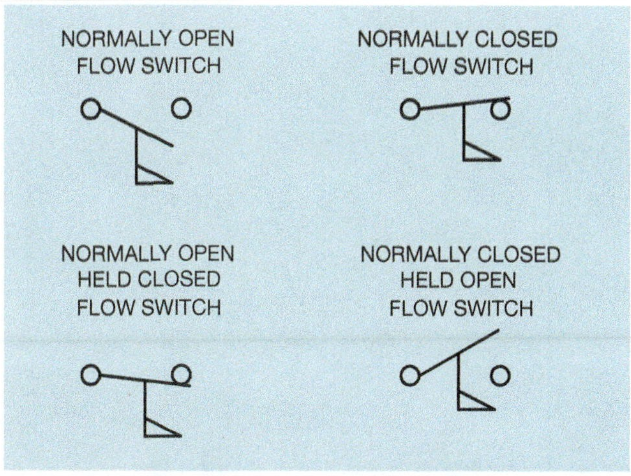

Figure 15–8 NEMA standard flow switch symbols.

Review Questions

1. What is the common name for an air flow switch?
2. What type of switch is contained on most flow switches?
3. Refer to the circuit shown in Figure 15–2. Why is there an overload contact symbol shown in series with the compressor contactor but not the condenser fan contactor or evaporator fan contactor?
4. Refer to the circuit shown in Figure 15–6. The STOP button is shown to be an emergency STOP button. What does the symbol used for the STOP button actually represent?
5. Refer to the circuit shown in Figure 15–6. If an overload should occur on the compressor motor and the overload contact open, would it affect the operation of the lube oil pump?
6. Refer to the circuit shown in Figure 15–6. If the circuit is in operation and an overload should occur on the lube oil pump and the overload contact open, would it affect the operation of the compressor?
7. Refer to the circuit shown in Figure 15–6. The pressure switch is:
   a. normally open
   b. normally closed
   c. normally open held closed
   d. normally closed held open

Chapter 16

LIMIT SWITCHES

Objectives

After studying this chapter the student will be able to:

- Discuss the operation of a limit switch.
- Connect a limit switch in a circuit.
- Recognize limit switch symbols in a ladder diagram.
- Discuss the different types of limit switches.

Limit switches are used to detect when an object is present or absent from a particular location. They can be activated by the motion of a machine or by the presence or absence of a particular object. Limit switches contain some type of bumper arm that is impacted by an object. The type of bumper arm used is determined by the application of the limit switch. When the bumper arm is impacted, it causes the contacts to change position. Figure 16–1 illustrates the use of a limit switch used to detect the position of boxes on a conveyer line. This particular limit switch uses a long metal rod that is free to move in any direction when hit by an object. This type of bumper arm is generally called a wobble stick or wiggle stick. Limit switches with different types of bumper arms are shown in Figure 16–2. They vary in size and contact arrangement depending on the application. Some are constructed of heavy gauge metal and are intended to be struck by moving objects thousands of times. Others are small and designed to fit into constricted spaces. Some contain a single set of contacts and others

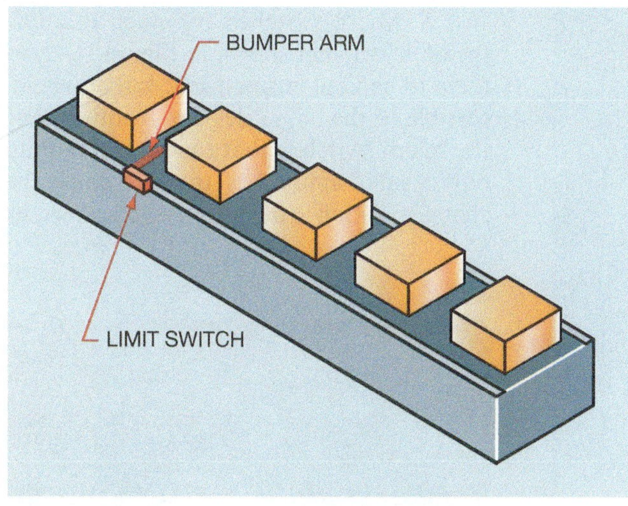

Figure 16–1 Limit switch detects position of boxes on a conveyer line.

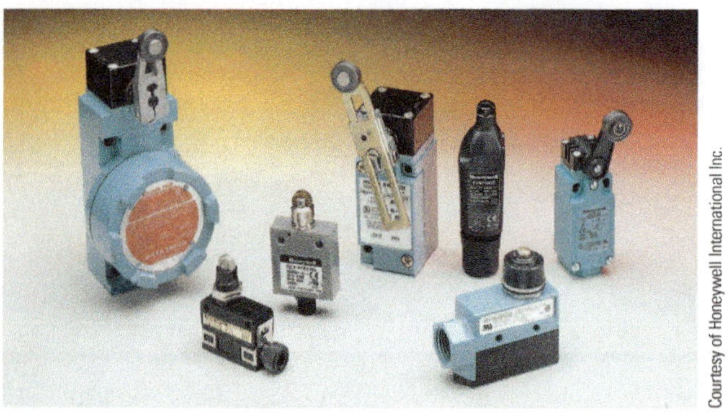

Courtesy of Honeywell International Inc.

Figure 16–2 Limit switches.

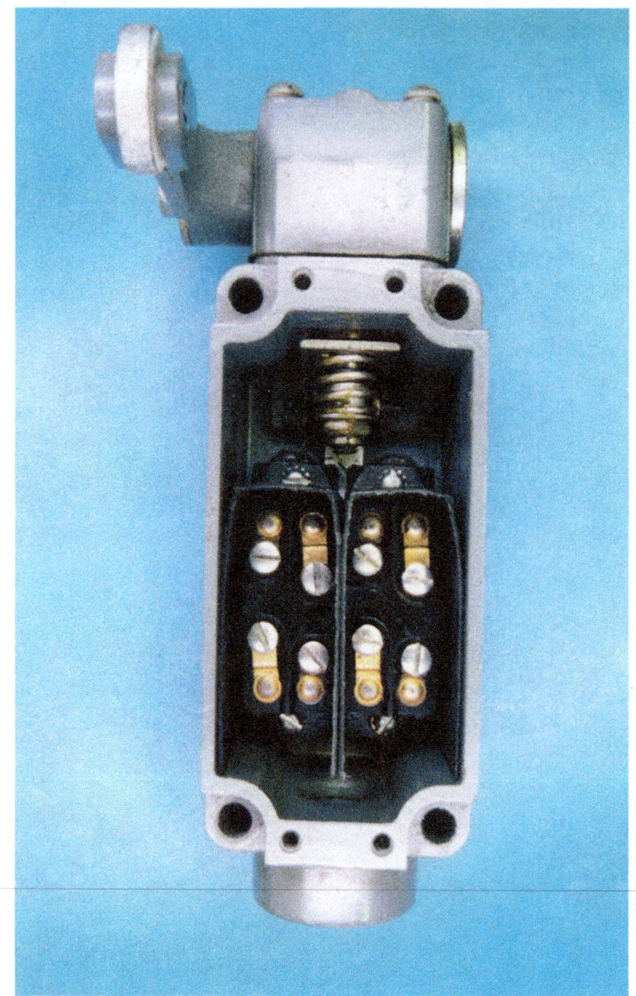

Figure 16–3 Limit switch with cover removed to show multiple contacts.

contain multiple contacts as shown in Figure 16–3. Some limit switches are momentary contact (spring returned) and others are maintained contact.

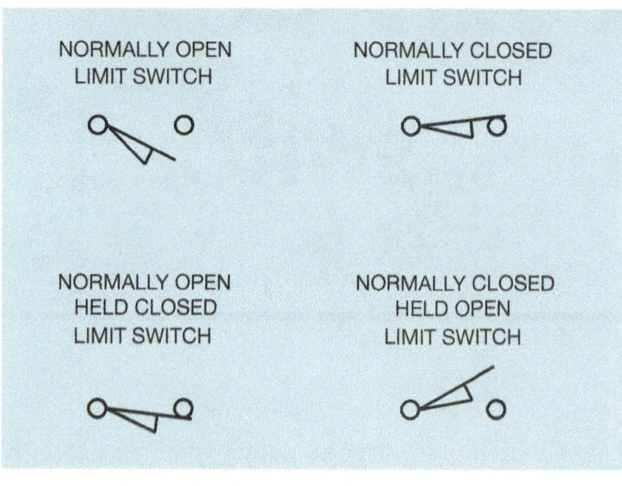

Figure 16–4 NEMA standard symbols for limit switches.

Generally, limit switches are used as pilot devices that control the coil of relays and motor starters in control circuits. The standard NEMA symbols used to indicate limit switches are shown in Figure 16–4. The wedge drawn under the switch symbol represents the bumper arm of the switch.

Micro Limit Switches

Another type of limit switch often used in different types of control circuits is the micro limit switch or *micro switch*. Micro switches are much smaller in size than the limit switches shown in Figure 16–2, which permits them to be used in small spaces that would never be accessible to the larger devices. Another characteristic of the micro switch is that the actuating plunger requires only a small amount of travel to cause the contacts to change position. The micro switch shown in Figure 16–5

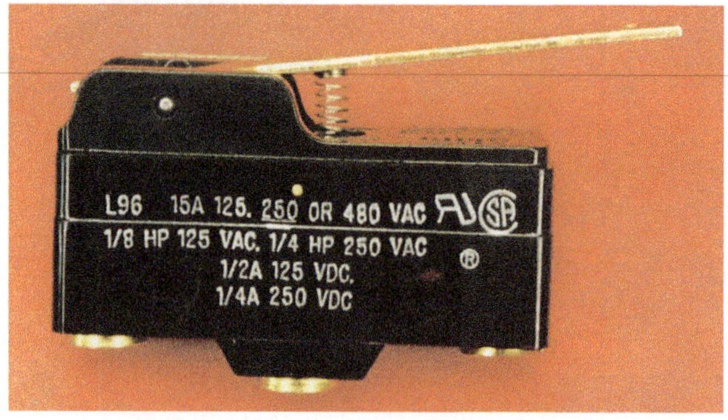

Figure 16–5 Micro limit switch.

has an activating plunger located at the top of the switch. This switch requires that the plunger be depressed approximately 0.015 inch or 0.38 mm. Switching the contact position with this small amount of movement is accomplished by spring loading the contacts as shown in Figure 16–6. A small amount of movement against the spring causes the movable contact to snap from one position to another.

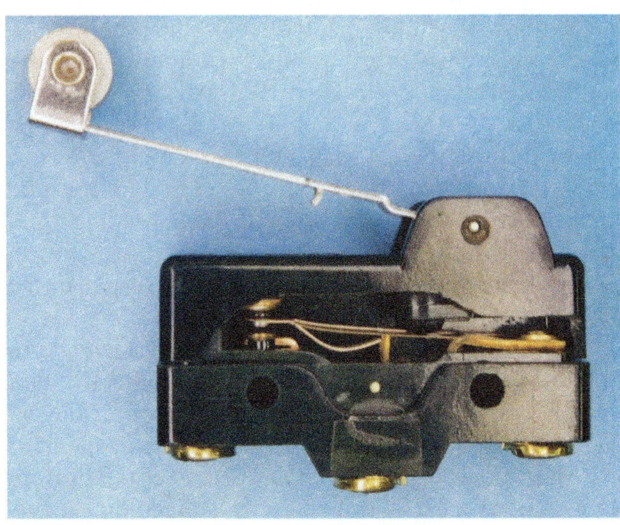

Figure 16–6 Spring-loaded contacts of a basic micro switch.

Electrical ratings for the contacts of the basic micro switch are generally in the range of 250 volts AC and 10 to 15 amps depending on the type of switch. The basic micro switch can be obtained with a variety of different activating arms as shown in Figure 16–7.

Subminiature Micro Switches

The *subminiature micro switch* employs a similar spring contact arrangement as the basic micro switch (Figure 16–8). The switch shown in Figure 16–8 contains a single normally open contact instead of a contact with a common terminal, a normally open contact terminal, and a normally closed contact terminal. The subminiature switches are approximately one-half to one-quarter the size of the basic switch, depending on the model. Due to their reduced size, the contact rating of subminiature switches ranges from about 1 ampere to about 7 amperes depending on the switch

Courtesy of Honeywell International Inc.

Figure 16–7 Micro switches can be obtained with different types of activating arms.

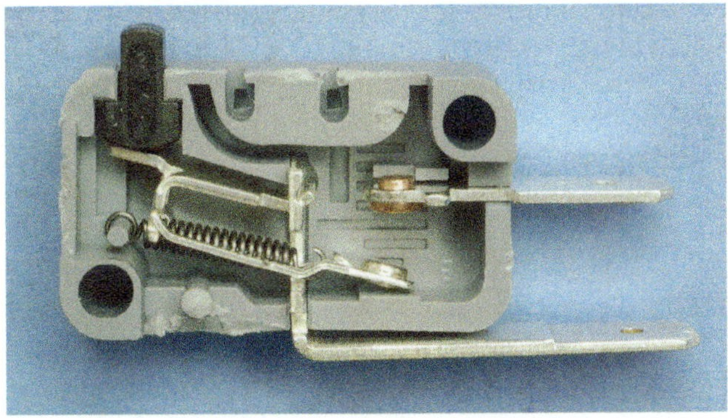

Figure 16–8 Subminiature micro switches employ a similar set of spring-loaded contacts.

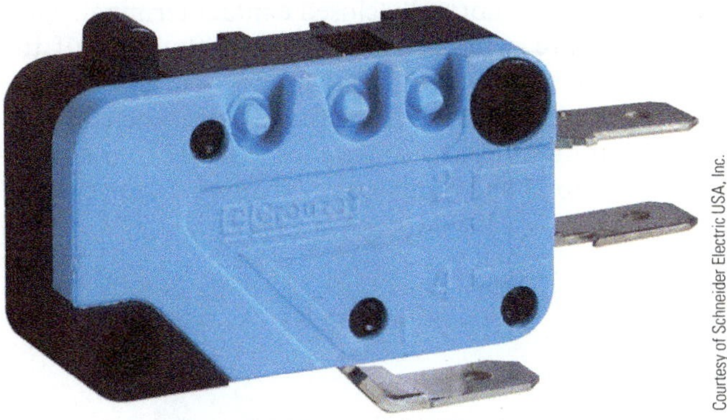

Courtesy of Schneider Electric USA, Inc.

Figure 16–9 Subminiature micro switch with both a normally open and a normally closed contact.

type. A subminiature micro switch containing both a normally open and a normally closed contact is shown in Figure 16–9.

Limit Switch Application

Figure 16–10 illustrates a common use for limit switches. A platform is used to raise material from a bottom floor to an upper floor. A hydraulic cylinder is used to raise the platform. A limit switch located on the bottom floor detects when the platform is in that position, and a second limit switch on the upper floor detects when the platform has reached the upper floor. A hydraulic pump is used to raise the platform. When the platform is to travel from the upper floor to the lower floor, a solenoid valve opens and permits oil to return to a holding tank. It is not necessary to use the pump to lower the platform because the weight of the platform will return it to the lower floor.

The schematic for this control circuit is shown in Figure 16–11. The schematic shows both limit switches to be normally closed. When the platform is at the extent of travel in one direction, however, one of the limit switches will be open. If the platform is at the bottom floor, limit switch LS2 will be open. If the UP push button is pressed, a circuit will be completed to M starter causing the motor to start raising the platform. The M normally closed contact will open to prevent CR from

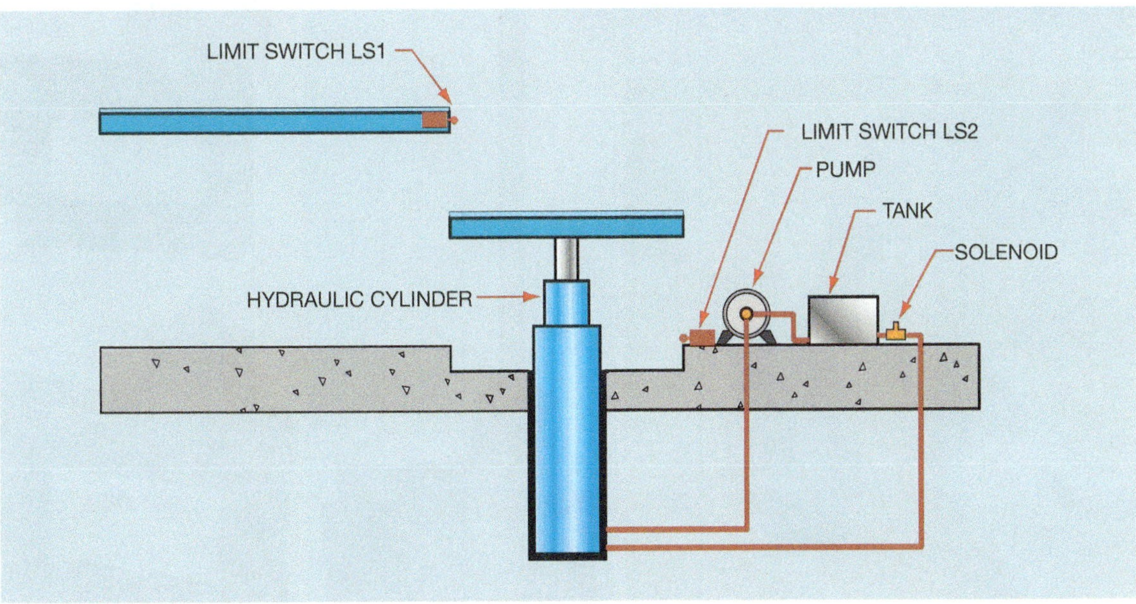

Figure 16–10 Platform rises between floors.

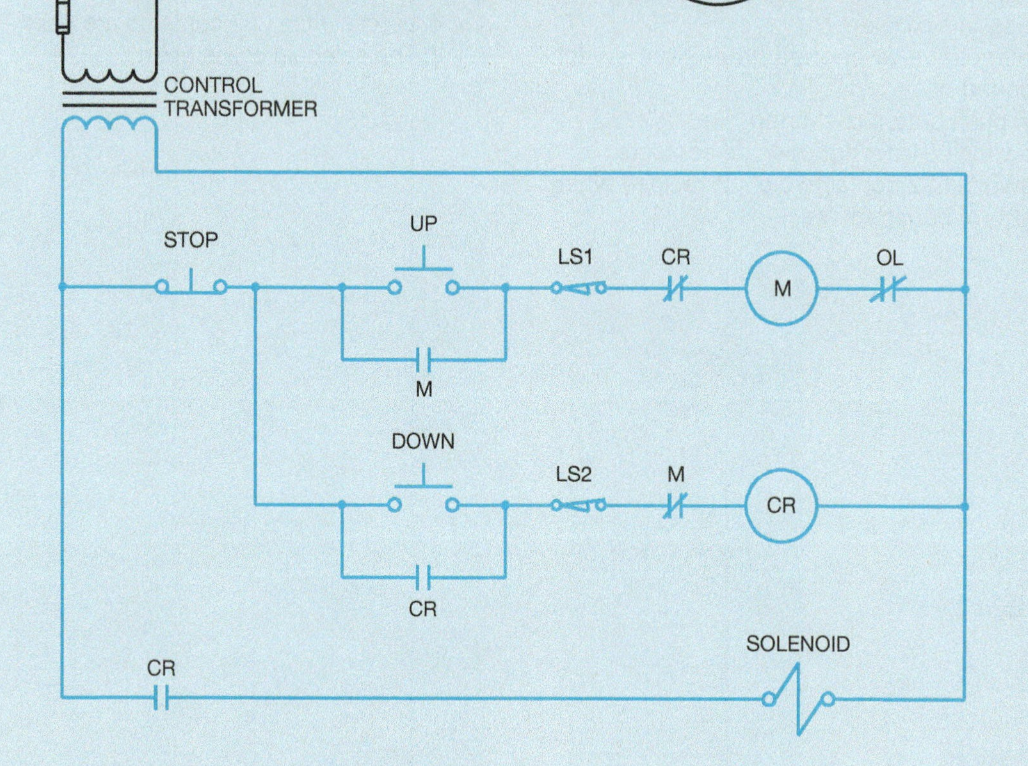

Figure 16–11 Control circuit to raise and lower platform.

being energized at the same time. When the platform begins to rise, limit switch LS2 will close. The platform continues upward until it reaches the top, causing limit switch LS1 to open. This de-energizes M contactor, causing the motor to stop and the normally closed auxiliary contact in series with CR coil to reclose.

When the DOWN push button is pressed, control relay CR energizes. The normally closed CR contact connected in series with M contactor opens to interlock the circuit, and the normally open CR contact connected in series with the solenoid coil closes. When the solenoid coil energizes, the platform starts downward, causing limit switch LS1 to reclose. When the platform reaches the bottom floor, limit switch LS2 opens and de-energizes coil CR.

Review Questions

1. What is the primary use of a limit switch?
2. Why are the contacts of a micro switch spring loaded?
3. Refer to the circuit shown in Figure 16–11. Assume that the platform is located on the bottom floor. When the UP push button is pressed the pump motor does not start. Which of the following could *not* cause this problem?
   a. The contacts of limit switch LS1 are open.
   b. The contacts of limit switch LS2 are open.
   c. Motor starter coil M is open.
   d. The overload contact is open.

4. Refer to the circuit shown in Figure 16–11. Assume that the platform is located on the lower floor. When the UP push button is pressed, the platform rises. When the platform reaches the upper floor, however, the pump does not turn off but continues to run until the overload relay opens the overload contacts. Which of the following could cause this problem?
   a. The solenoid valve opened when limit switch LS1 opened.
   b. The UP push button is shorted.
   c. Limit switch LS1 did not open its contacts.
   d. Limit switch LS2 contacts did not reclose when the platform began to rise.

5. Refer to the circuit shown in Figure 16–11. Assume that the platform is located at the upper floor. When the DOWN push button is pressed, the platform does not begin to lower. Which of the following could *not* cause the problem?
   a. Control relay coil CR is open.
   b. Limit switch LS1 contacts are open.
   c. Limit switch LS2 contacts are open.
   d. The solenoid coil is open.

Chapter 17

TEMPERATURE SENSING DEVICES

Objectives

After studying this chapter the student will be able to:

- Describe different methods for sensing temperature.
- Discuss different devices intended to be operated by a change of temperature.
- List several applications for temperature sensing devices.
- Read and draw the NEMA symbols for temperature switches.

There are many times when the ability to sense temperature is of great importance. The industrial electrician will encounter some devices designed to change a set of contacts with a change of temperature and other devices used to sense the amount of temperature. The method used depends a great deal on the applications of the circuit and the amount of temperature that must be sensed.

Expansion of Metal

A very common and reliable method for sensing temperature is by the expansion of metal. It has long been known that metal expands when heated. The amount of expansion is proportional to two factors:

1. The type of metal used.
2. The amount of temperature.

Consider the metal bar shown in Figure 17–1. When the bar is heated, its length expands. When the metal is permitted to cool, it will contract. Although the amount of movement due to contractions and expansion is small, a simple mechanical principle can be used to increase the amount of movement (Figure 17–2).

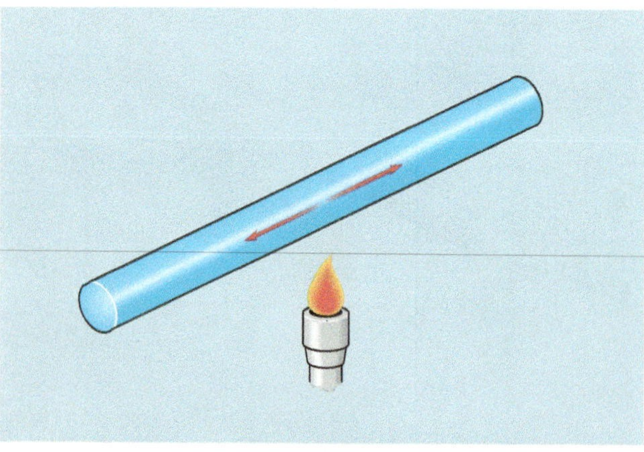

Figure 17–1 Metal expands when heated.

The metal bar is mechanically held at one end. This permits the amount of expansion to be in one direction only. When the metal is heated and the bar expands, it pushes against the mechanical arm. A small movement of the bar causes a great amount of movement in the mechanical arm. This increased movement in the arm can be used to indicate the temperature of the bar by attaching a pointer and scale, or to operate a switch as shown. It should be understood that illustrations are used to convey a principle. In actual practice, the switch shown in Figure 17–2 would be spring loaded to provide a snap action for the contacts. Electrical contacts must never be permitted to open or close slowly. This produces poor contact pressure and will cause the contacts to burn or will cause erratic operation of the equipment they are intended to control.

Hot-Wire Starting Relay

A very common device that uses the principle of expanding metal to operate a set of contacts is the *hot-wire starting relay* found in the refrigeration industry. The hot-wire relay is so named because it uses a length of resistive wire connected in series with the motor to sense motor current. A diagram of this type of relay is shown in Figure 17–3.

When the thermostat contact closes, current can flow from line L1 to terminal L of the relay. Current then flows through the resistive wire, the movable arm, and the normally closed contacts to the run and start windings. When current flows through the resistive wire, its temperature increases. This increase of temperature causes the wire to expand in length. When the length increases, the movable arm is forced downward. This downward pressure produces tension on the springs of both contacts. The relay is so designed that the start contact will snap open first, disconnecting the motor start winding from the circuit. If the motor current is not excessive, the wire will never become hot enough to cause the overload contact to open. If the motor current should become too great, however, the temperature of the resistive wire will become high enough to cause the wire to expand to the point that it will cause the overload contact to snap open and disconnect the motor run winding from the circuit.

The Mercury Thermometer

Another very useful device that works on the principle of contraction and expansion of metal is the *mercury thermometer*. Mercury is a metal that remains in a liquid state at room temperature. If the mercury is confined in a glass tube as shown in Figure 17–4, it rises up the tube as it expands due to an increase in temperature. If the tube is calibrated correctly, it provides an accurate measurement for temperature.

The Bimetal Strip

The bimetal strip is another device that operates by the expansion of metal. It is probably the most common heat sensing device used in the production of

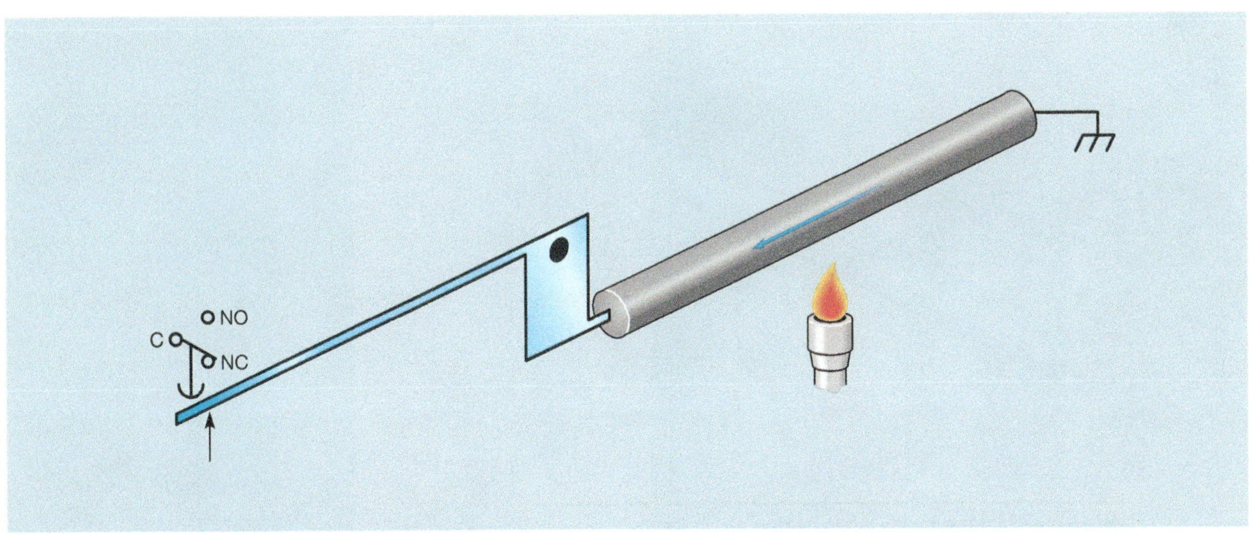

Figure 17–2 Expanding metal operates a set of contacts.

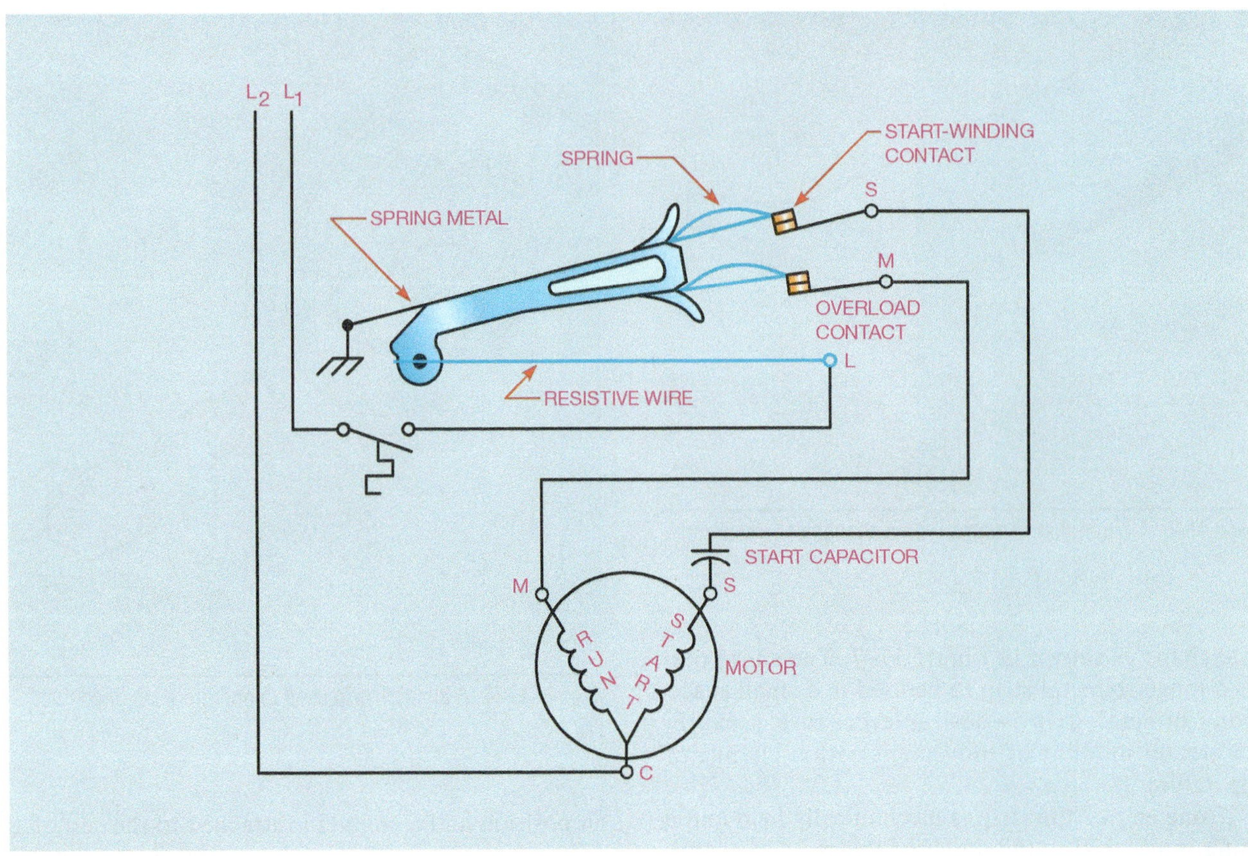

Figure 17–3 Hot-wire relay connection.

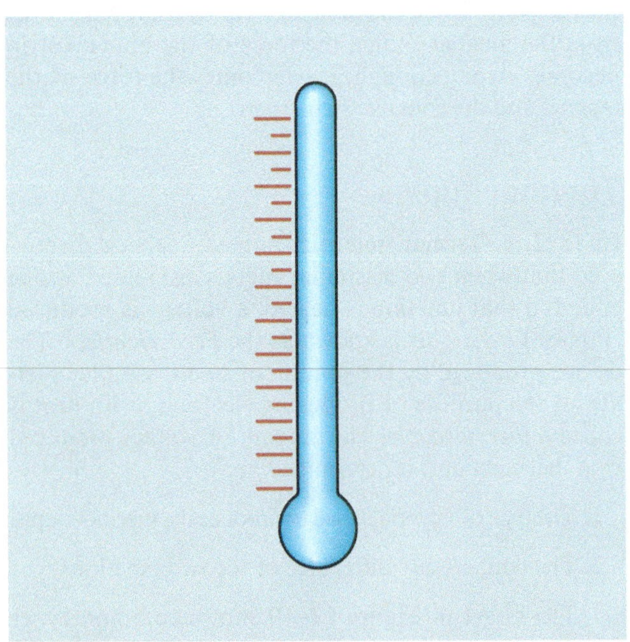

Figure 17–4 A mercury thermometer operates by the expansion of metal.

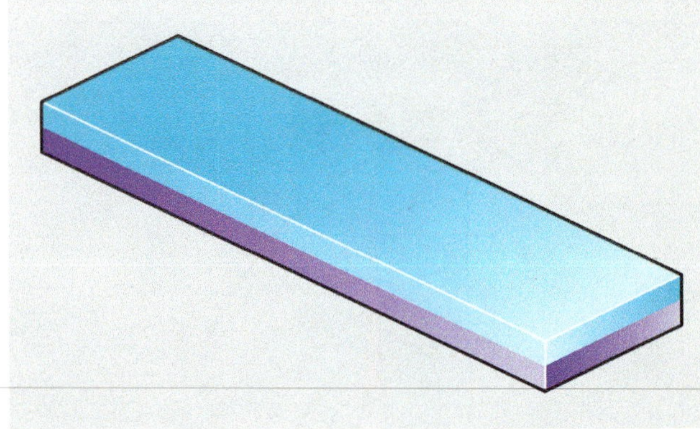

Figure 17–5 A bimetal strip.

room thermostats and thermometers. The bimetal strip is made by bonding two dissimilar types of metal together (Figure 17–5). Because these two metals are not alike, they have different expansion rates. This difference causes the strip to bend or warp when heated (Figure 17–6). A bimetal strip is often formed into a

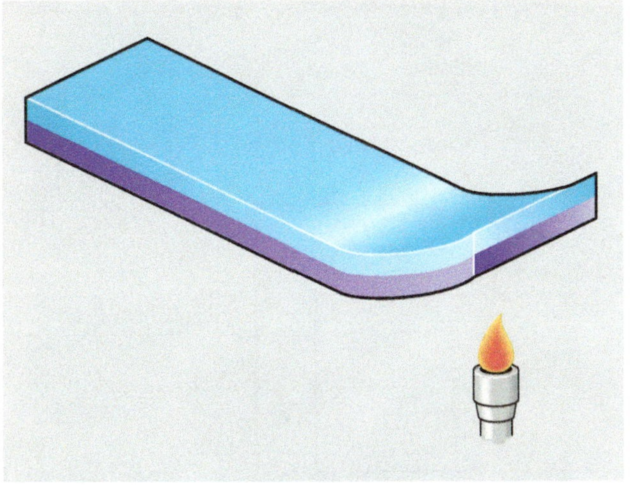

Figure 17–6 A bimetal strip warps with a change of temperature.

spiral shape as shown in Figure 17–7. The spiral permits a longer bimetal strip to be used in a small space. A long bimetal strip is desirable because it exhibits a greater amount of movement with a change of temperature.

If one end of the strip is mechanically held and a pointer is attached to the center of the spiral, a change in temperature will cause the pointer to rotate. If a calibrated scale is placed behind the pointer, it becomes a thermometer. If the center of the spiral is held in position and a contact is attached to the end of the bimetal strip, it becomes a thermostat. A small permanent magnet is used to provide a snap action for the contacts (Figure 17–8). When the moving contact reaches a point that is close to the stationary contact, the magnet attracts the metal strip and causes a sudden closing of the contacts. When the bimetal strip cools, it pulls away from the magnet. When the force of the bimetal strip becomes strong enough, it overcomes the force of the magnet and the contacts snap open.

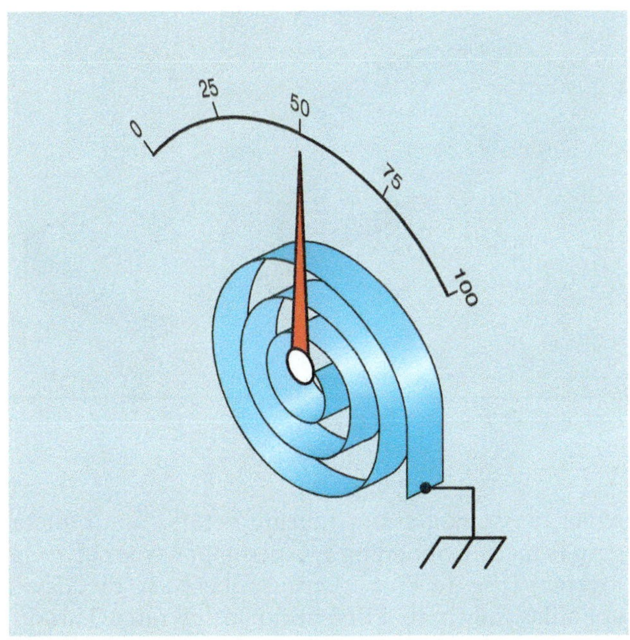

Figure 17–7 A bimetal strip used as a thermometer.

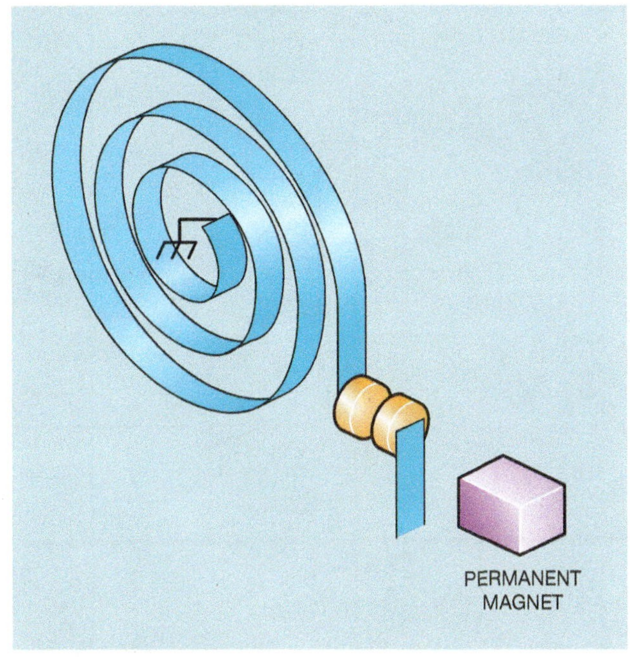

Figure 17–8 A bimetal strip used to operate a set of contacts.

Thermocouples

In 1822, a German scientist named Seebeck discovered that when two dissimilar metals are joined at one end, and that junction is heated, a voltage is produced (Figure 17–9). This is known as the *Seebeck effect*. The device produced by the joining of two dissimilar metals for the purpose of producing electricity with heat is called a *thermocouple*. The amount of voltage produced by a thermocouple is determined by:

1. The type of materials used to produce the thermocouple.
2. The temperature difference of the two junctions.

The chart in Figure 17–10 shows common types of thermocouples. The different metals used in the construction of thermocouples as well as their normal temperature ranges are shown.

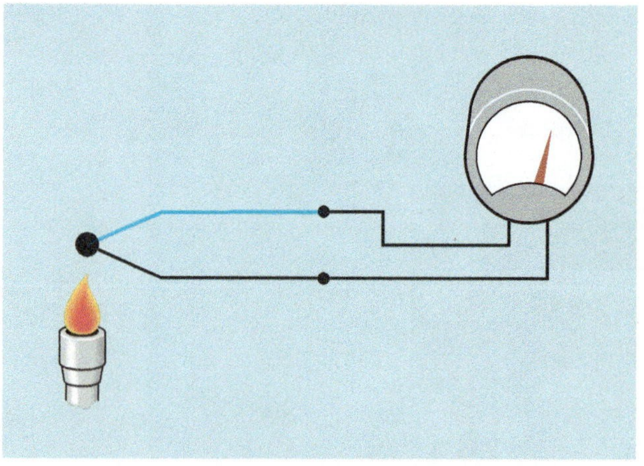

Figure 17–9 Thermocouple.

The amount of voltage produced by a thermocouple is small, generally in the order of millivolts (1 millivolt = 0.001 volt). The polarity of the voltage of some thermocouples is determined by the temperature. For example, a type "J" thermocouple produces 0 volts at about 32°F. At temperatures above 32°F, the iron wire is positive and the constantan wire is negative. At temperatures below 32°F, the iron wire becomes negative and the constantan wire becomes positive. At a temperature of +300°F, a type "J" thermocouple will produce a voltage of about +7.9 millivolts. At a temperature of −300°F, it will produce a voltage of about −7.9 millivolts.

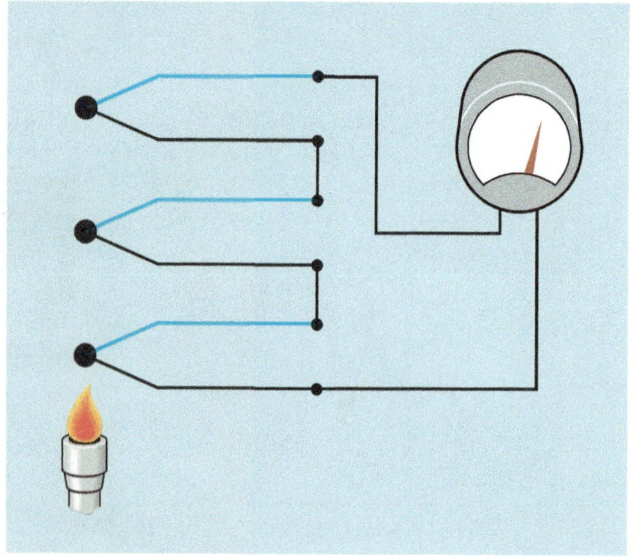

Figure 17–11 Thermopile.

Because thermocouples produce such low voltages, they are often connected in series as shown in Figure 17–11. This connection is referred to as a *thermopile*. Thermocouples and thermopiles are generally used

TYPE	MATERIAL		DEGREES F	DEGREES C
J	IRON	CONSTANTAN	-328 to +32 +32 to +1432	-200 to 0 0 to +778
K	CHROMEL	ALUMEL	-328 to +32 +32 to +2472	-200 to 0 0 to +1356
T	COPPER	CONSTANTAN	-328 to +32 +32 to +752	-200 to 0 0 to +400
E	CHROMEL	CONSTANTAN	-328 to +32 +32 to +1832	-200 to 0 0 to +1000
R	PLATINUM 13% RHODIUM	PLATINUM	+32 to +3232	0 to +1778
S	PLATINUM 10% RHODIUM	PLATINUM	+32 to +3232	0 to +1778
B	PLATINUM 30% RHODIUM	PLATINUM 6% RHODIUM	+992 to +3352	+533 to +1800

Figure 17–10 Thermocouple chart.

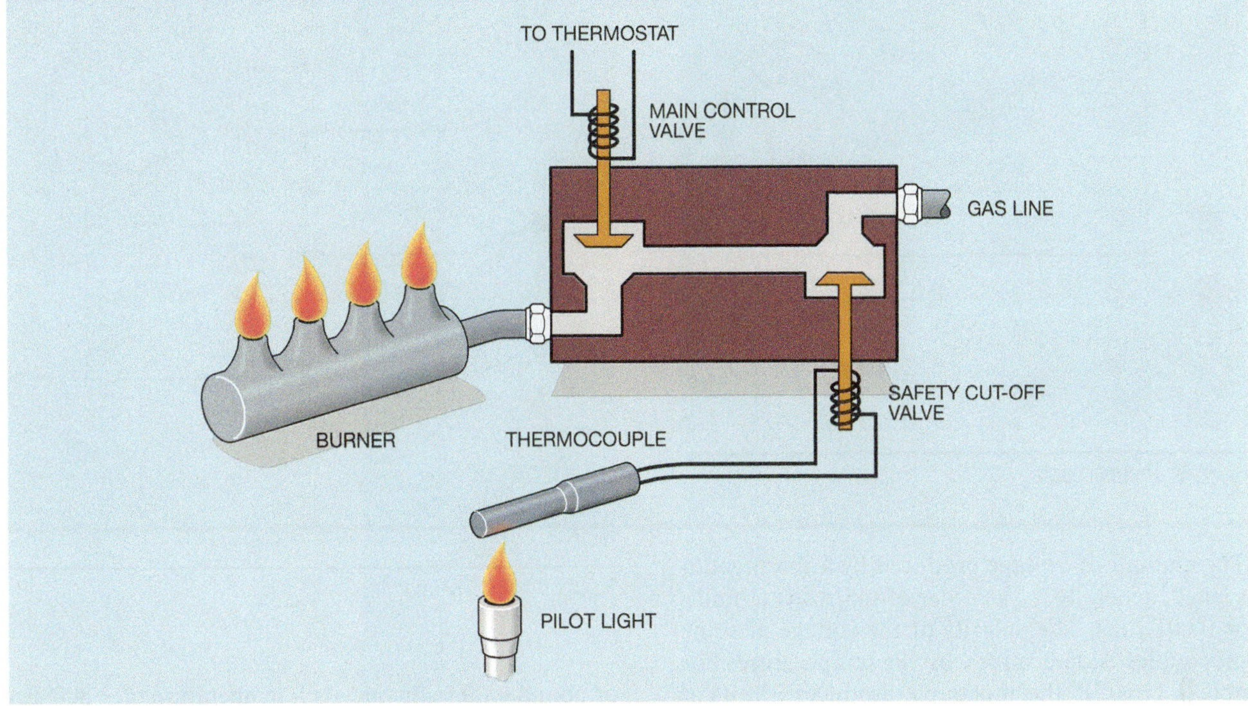

Figure 17–12 A thermocouple provides power to the safety cut-off valve.

for making temperature measurements and are sometimes used to detect the presence of a pilot light in appliances that operate with natural gas. The thermocouple is heated by the pilot light. The current produced by the thermocouple is used to produce a magnetic field that holds a gas valve open and permits gas to flow to the main burner. If the pilot light should go out, the thermocouple ceases to produce current and the valve closes (Figure 17–12).

Resistance Temperature Detectors

The *resistance temperature detector* (RTD) is made of platinum wire. The resistance of platinum changes greatly with temperature. When platinum is heated, its resistance increases at a very predictable rate; this makes the RTD an ideal device for measuring temperature very accurately. RTDs are used to measure temperatures that range from −328 to +1166 degrees Fahrenheit (−200° to +630 °C). RTDs are made in different styles to perform different functions. Figure 17–13 illustrates a typical RTD used as a probe. A very small coil of platinum wire is encased inside a copper tip. Copper is used to provide good thermal contact. This permits the probe to be very fast-acting. The chart in Figure 17–14 shows resistance versus temperature for a typical RTD probe. The temperature is given in degrees Celsius and the resistance is given in ohms. RTDs in several different case styles are shown in Figure 17–15.

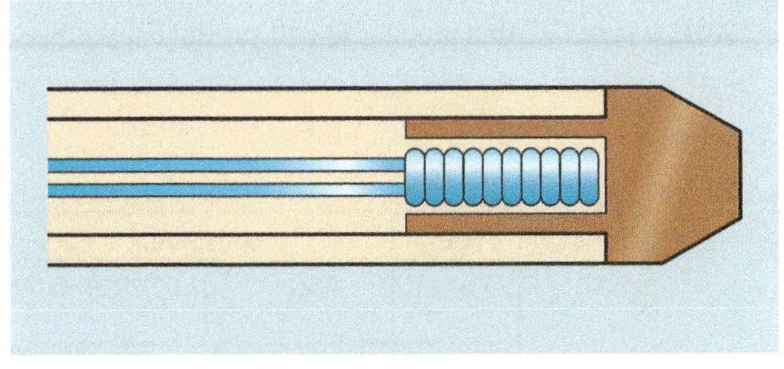

Figure 17–13 Resistance temperature detector.

Thermistors

The term *thermistor* is derived from the words "thermal resistor." **Thermistors** are actually thermally sensitive semiconductor devices. There are two basic types of thermistors: one type has a negative temperature coefficient (NTC) and the other has a positive temperature coefficient (PTC). A thermistor that has a negative

DEGREES C	RESISTANCE (Ω)
0	100
50	119.39
100	138.5
150	157.32
200	175.84
250	194.08
300	212.03
350	229.69
400	247.06
450	264.16
500	280.93
550	297.44
600	313.65

Figure 17–14 Temperature and resistance for a typical RTD.

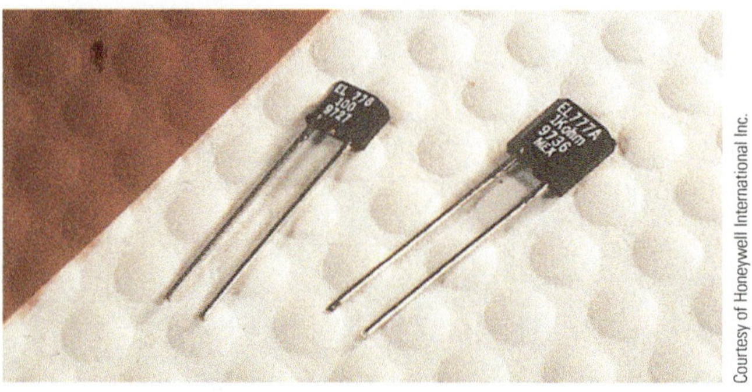

Courtesy of Honeywell International Inc.

Figure 17–15 RTDs in different case styles.

temperature coefficient will decrease its resistance as the temperature increases. A thermistor that has a positive temperature coefficient will increase its resistance as the temperature increases. The NTC thermistor is the most widely used.

Thermistors are highly nonlinear devices. For this reason they are difficult to use for measuring temperature. Devices that measure temperature with a thermistor must be calibrated for the particular type of thermistor being used. If the thermistor is ever replaced, it has to be an exact replacement or the circuit will no longer operate correctly. Because of their nonlinear characteristics, thermistors are often used as *set point detectors* as opposed to actual temperature measurement. A set point detector is a device that activates some process or circuit when the temperature reaches a certain level. For example, assume a thermistor has been placed inside the stator winding of a motor. If the motor should become overheated, the windings could become severely damaged or destroyed. The thermistor can be used to detect the temperature of the windings. When the temperature reaches a certain point, the resistance value of the thermistor changes enough to cause the starter coil to drop out and disconnect the motor from the line. Thermistors can be operated in temperatures that range from about −100° to +300°F.

One common use for thermistors is in the solid-state starting relays used with small refrigeration compressors (Figure 17–16). Starting relays are used with hermetically sealed motors to disconnect the start windings from the circuit when the motor reaches about 75 percent of its full speed. Thermistors can be used for this application because they exhibit an extremely rapid change of resistance with a change of temperature. A schematic diagram showing the connection for a solid-state relay is shown in Figure 17–17.

When power is first applied to the circuit, the thermistor is cool and has a relatively low resistance. This permits current to flow through both the start and run windings of the motor. The temperature of the thermistor increases because of the current flowing through it. The increase of temperature causes the resistance to change from a very low value of 3 or 4 ohms to several thousand ohms. This increase of resistance is very sudden and has the effect of opening a set of contacts connected in series with the start winding. Although the start winding is never completely disconnected from the power line, the amount of current flow through it is very small, typically 0.03 to 0.05 amps, and does not affect the operation of the motor. This small amount of *leakage current* maintains the temperature of the thermistor and prevents it from returning to a low resistance. After power has been disconnected from the motor, a cool-down period of about 2 minutes should be allowed before restarting the motor. This cool-down period is needed for the thermistor to return to a low value of resistance.

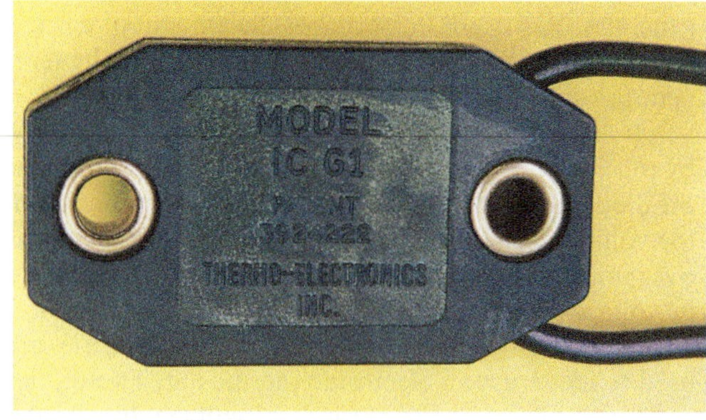

Figure 17–16 Solid-state starting relay.

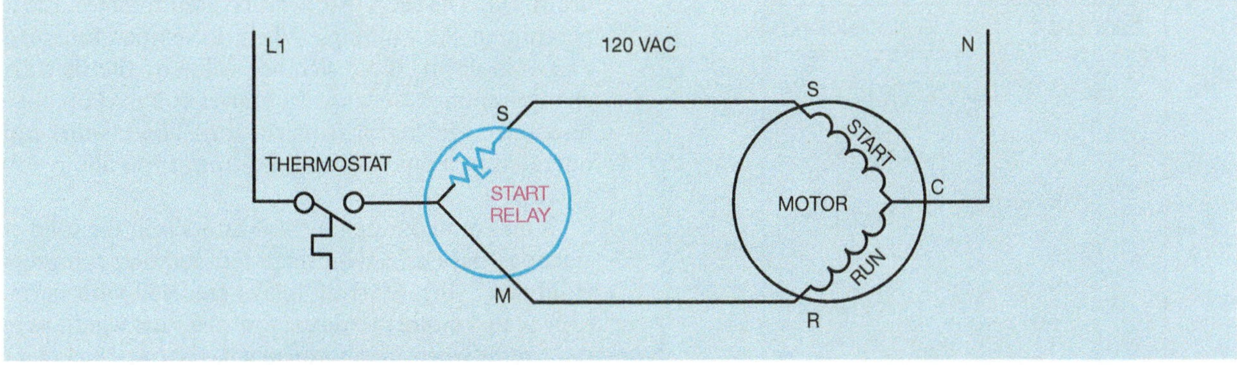

Figure 17–17 Connection of solid-state starting relay.

The PN Junction

Another device that has the ability to measure temperature is the PN junction or diode. The diode is becoming a very popular device for measuring temperature because it is accurate and linear.

When a silicon diode is used as a temperature sensor, a constant current is passed through the diode. Figure 17–18 illustrates this type of circuit. In this circuit, resistor R1 limits the current flow through the transistor and sensor diode. The value of R1 also determines the amount of current that flows through the diode. Diode D1 is a 5.1 volt zener used to produce a constant voltage drop between the base and emitter of the PNP transistor. Resistor R2 limits the amount of current flow through the zener diode and the base of the transistor. D1 is a common silicon diode. It is being used as the temperature sensor for the circuit. If a digital voltmeter is connected across the diode, a voltage drop between 0.8 and 0 volts can be seen. The amount of voltage drop is determined by the temperature of the diode.

Another circuit that can be used as a constant current generator is shown in Figure 17–19. In this circuit, a field effect transistor (FET) is used to produce a current generator. Resistor R1 determines the amount of current that will flow through the diode. Diode D1 is the temperature sensor.

If the diode is subjected to a lower temperature, say, by touching it with a piece of ice, the voltage drop across the diode increases. If the diode temperature is increased, the voltage drop decreases because the diode has a negative temperature coefficient. As its temperature increases, its voltage drop becomes less.

In Figure 17–20, two diodes connected in a series are used to construct an electronic thermostat. Two diodes are used to increase the amount of voltage drop as the temperature changes. A field effect transistor and resistor are used to provide a constant current to the two diodes used as the heat sensor. An operational amplifier is used

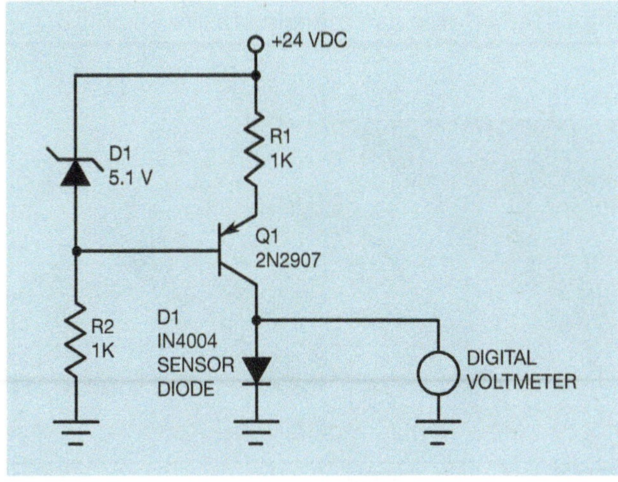

Figure 17–18 Constant current generator.

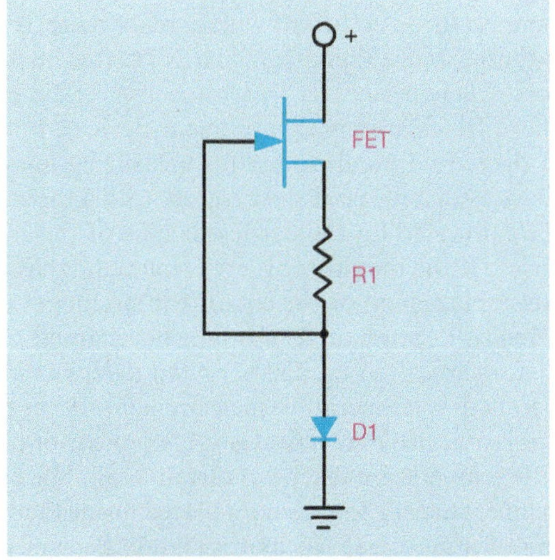

Figure 17–19 Field effect transistor used to produce a constant current generator.

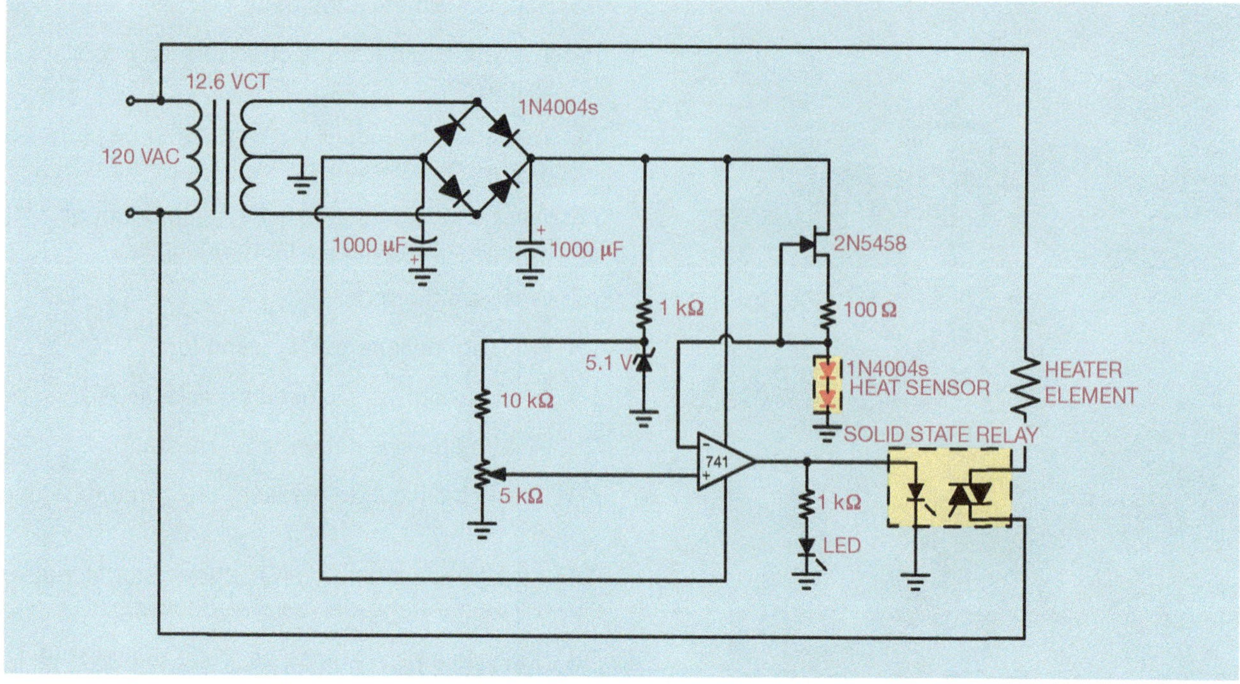

Figure 17–20 Solid-state thermostat using diodes as heat sensors.

to turn a solid-state relay on or off as the temperature changes. In the example shown, the circuit operates as a heating thermostat. The output of the amplifier turns on when the temperature decreases sufficiently. The circuit can be converted to a cooling thermostat by reversing the connections of the inverting and noninverting inputs of the amplifier.

Expansion Due to Pressure

Another common method of sensing a change of temperature is by the increase of pressure of some chemicals. Refrigerants confined in a sealed container, for example, will increase the pressure in the container with an increase of temperature. If a simple bellows is connected to a line containing refrigerant (Figure 17–21), the bellows will expand as the pressure inside the sealed system increases. When the surrounding air temperature decreases, the pressure inside the system decreases and the bellows contracts. When the air temperature increases, the pressure increases and the bellows expands. If the bellows controls a set of contacts, it becomes a bellows type thermostat. A bellows thermostat and the standard NEMA symbols used to represent a temperature operated switch are shown in Figure 17–22.

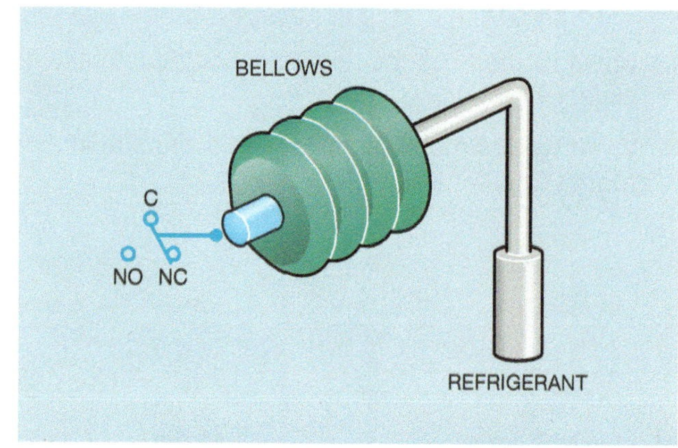

Figure 17–21 Bellows contracts and expands with a change of refrigerant pressure.

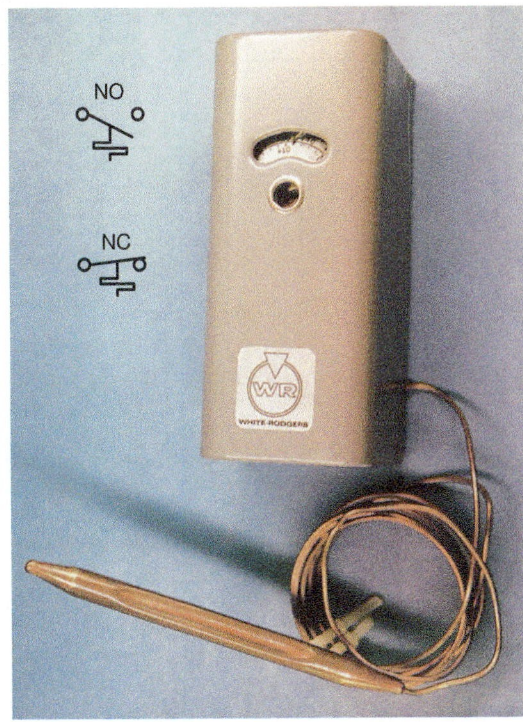

Figure 17–22 Industrial temperature switch.

Review Questions

1. Should a metal bar be heated or cooled to make it expand?
2. What type of metal remains in a liquid state at room temperature?
3. How is a bimetal strip made?
4. Why are bimetal strips often formed into a spiral shape?
5. Why should electrical contacts never be permitted to open or close slowly?
6. What two factors determine the amount of voltage produced by a thermocouple?
7. What is a thermopile?
8. What do the letters RTD stand for?
9. What type of wire is used to make an RTD?
10. What material is a thermistor made of?
11. Why is it difficult to measure temperature with a thermistor?
12. If the temperature of a NTC thermistor increases, will its resistance increase or decrease?
13. How can a silicon diode be made to measure temperature?
14. Assume that a silicon diode is being used as a temperature detector. If its temperature increases, will its voltage drop increase or decrease?
15. What type of chemical is used to cause a pressure change in a bellows type thermostat?

Chapter 18

HALL EFFECT SENSORS

Objectives

After studying this chapter the student will be able to:

- Describe the Hall effect.
- Discuss the principles of operation of a Hall generator.
- Discuss applications in which Hall generators can be used.

Principles of Operation

The Hall effect is a simple principle that is widely used in industry today. The Hall effect was discovered by Edward H. Hall at Johns Hopkins University in 1879. Hall originally used a piece of pure gold to produce the Hall effect, but today a piece of semiconductor material is used because semiconductor material works better and is less expensive to use. The device is often referred to as the Hall generator.

Figure 18–1 illustrates how the Hall effect is produced. A constant current power supply is connected to opposite sides of a piece of semiconductor material. A sensitive voltmeter is connected to the other two sides. If the current flows straight through the semiconductor material, no voltage is produced across the voltmeter connection.

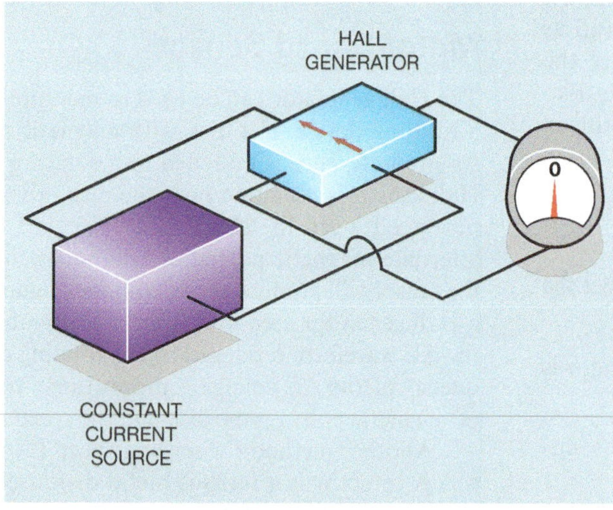

Figure 18–1 Constant current flows through a piece of semiconductor material.

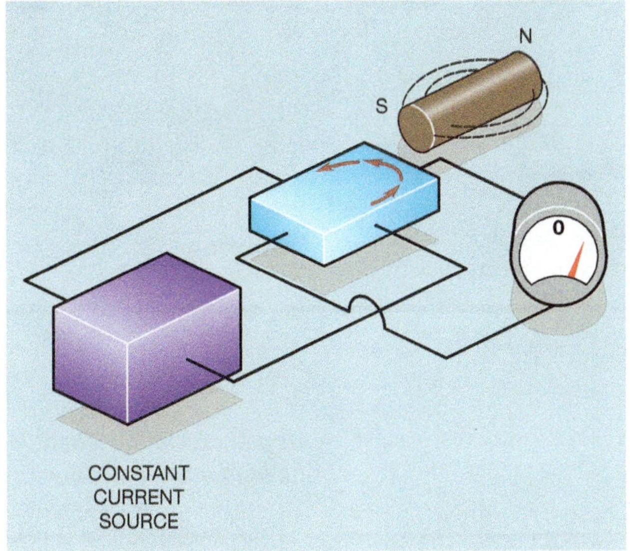

Figure 18–2 A magnetic field deflects the path of current flow through the semiconductor.

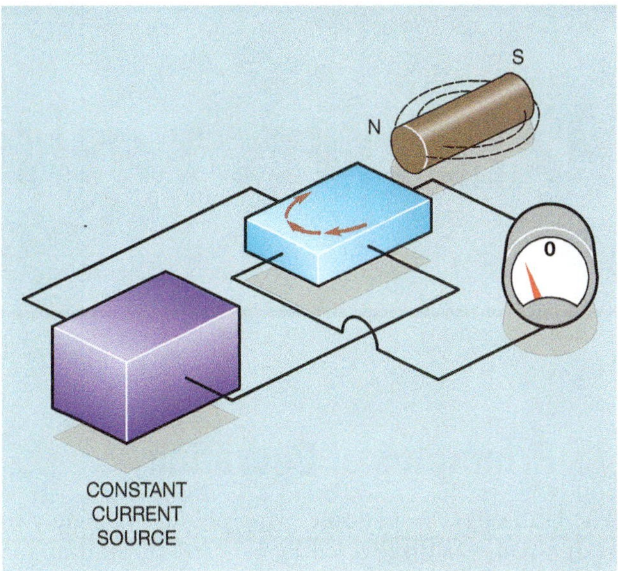

Figure 18–3 The current path is deflected in the opposite direction.

Figure 18–2 shows the effect of bringing a magnetic field near the semiconductor material. The magnetic field causes the current flow path to be deflected to one side of the material. This causes a potential or voltage to be produced across the opposite sides of the semiconductor material.

If the polarity of the magnetic field is reversed, the current path is deflected in the opposite direction as shown in Figure 18–3. This causes the polarity of the voltage produced by the Hall generator to change. Two factors determine the polarity of the voltage produced by the Hall generator:

1. The direction of current flow through the semiconductor material.
2. The polarity of the magnetic field used to deflect the current.

The amount of voltage produced by the Hall generator is determined by:

1. The amount of current flowing through the semiconductor material.
2. The strength of the magnetic field used to deflect the current path.

The Hall generator has many advantages over other types of sensors. Because it is a solid-state device, it has no moving parts or contacts to wear out. It is not affected by dirt, oil, or vibration. The Hall generator is an integrated circuit that is mounted in many different types and styles of cases.

Hall Generator Applications

Motor Speed Sensor

The Hall generator can be used to measure the speed of a rotating device. If a disk with magnetic poles around its circumference is attached to a rotating shaft, and a Hall sensor is mounted near the disk, a voltage will be produced when the shaft turns. Because the disk has alternate magnetic polarities around its circumference, the sensor will produce an AC voltage. Figure 18–4 shows a Hall generator used in this manner. Figure 18–5 shows the AC waveform produced by the rotating disk. The frequency of the AC voltage is proportional to the number of magnetic poles on the disk and the speed of rotation.

Another method for sensing speed is to use a *reluctor*. A reluctor is a ferrous metal disk used to shunt a magnetic field away from some other object. This type of sensor uses a notched metal disk attached to a rotating shaft. The disk separates a Hall sensor and a permanent magnet (Figure 18–6). When the notch is between the sensor and the magnet, a voltage is produced by the Hall generator. When the solid metal part of the disk

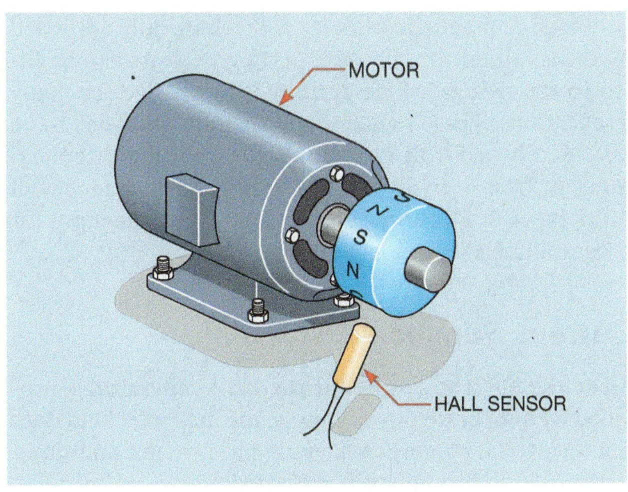

Figure 18–4 An AC voltage is produced by the rotating magnetic disk.

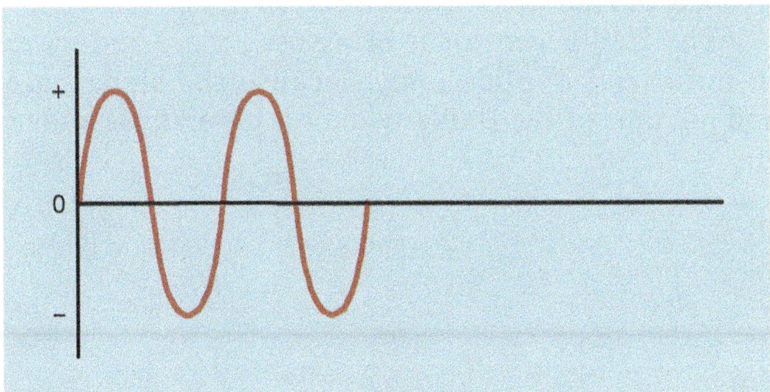

Figure 18–5 AC waveform produced by Hall effect device.

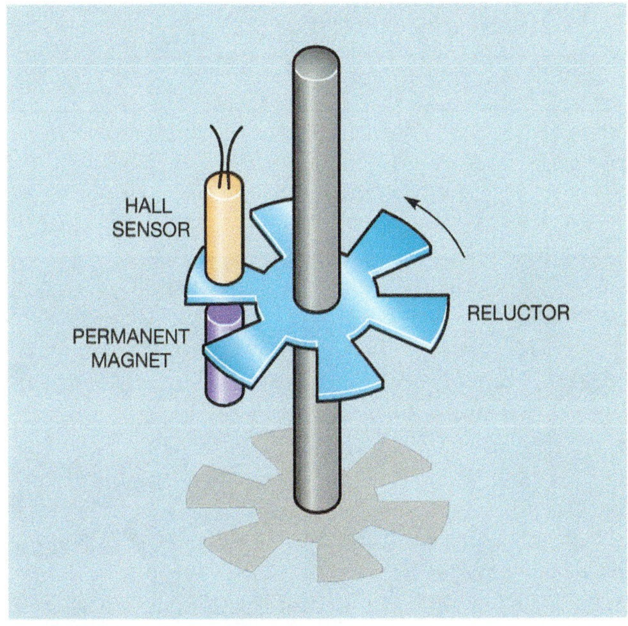

Figure 18–6 Reluctor shunts magnetic field away from sensor.

is between the sensor and magnet, the magnetic field is shunted away from the sensor. This causes a significant drop in the voltage produced by the Hall generator.

Because the polarity of the magnetic field does not change, the voltage produced by the Hall generator is pulsating direct current instead of alternating current. Figure 18–7 shows the DC pulses produced by the generator. The number of pulses produced per second is proportional to the number of notches on the reluctor and the speed of the rotating shaft.

Position Sensor

The Hall generator can be used in a manner similar to a limit switch. If the sensor is mounted beside a piece of moving equipment, and a permanent magnet is attached to the moving equipment, a voltage will be produced when the magnet moves near the sensor (Figure 18–8).

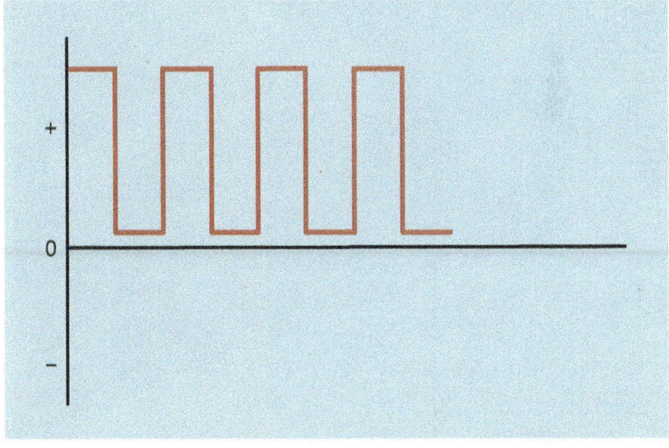

Figure 18–7 Square wave pulses produced by the Hall generator.

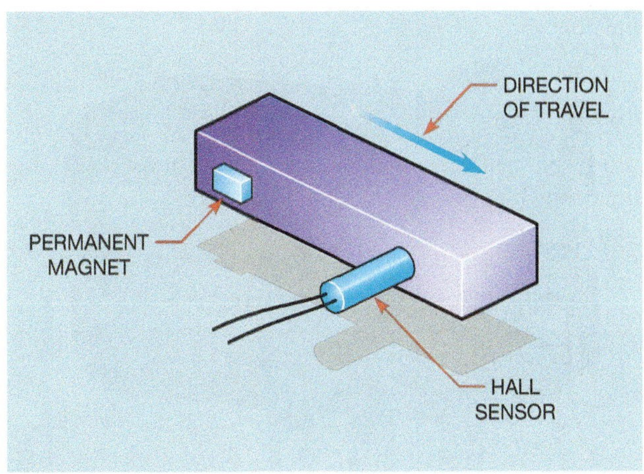

Figure 18–8 Hall generator used to sense position of moving device.

The advantages of the Hall sensor are that it has no lever arm or contacts to wear like a common limit switch, and it can operate through millions of operations of the machine.

A Hall effect position sensor is shown in Figure 18–9. Notice that these type sensors vary in size and style to fit almost any application. Position sensors operate as a digital device in that they sense the presence or absence of magnetic field. They do not have the ability to sense the intensity of the field.

Hall Effect Limit Switches

Another Hall effect device used in a very similar application is the Hall effect limit switch (Figure 18–10). This limit switch uses a Hall generator instead of a set of contacts. A magnetic plunger is mechanically activated by the small button. Different types of levers can be fitted to the switch, which permits it to be used for many applications. These switches are generally intended to be operated by a 5-volt DC supply for TTL logic applications, or by a 6- to 24-volt DC supply for interface with other types of electronic controls or to provide input for programmable controllers.

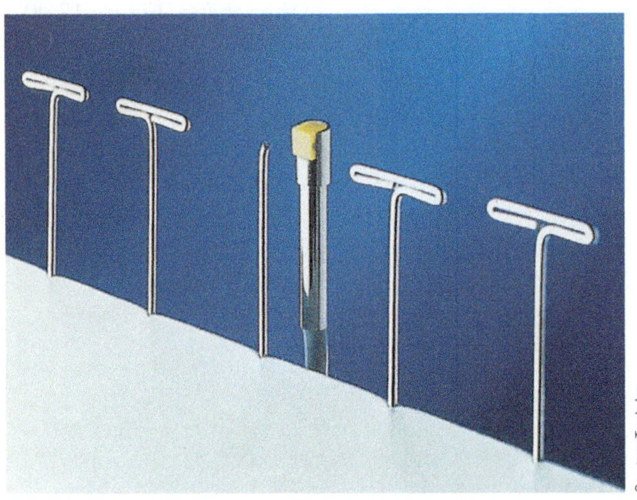

Courtesy Turck, Inc.

Figure 18–9 Hall effect position sensors.

Courtesy of Honeywell International Inc.

Figure 18–10 Hall effect limit switch.

Current Sensor

Since the current source for the Hall generator is provided by a separate power supply, the magnetic field does not have to be moving or changing to produce an output voltage. If a Hall sensor is mounted near a coil of wire, a voltage will be produced by the generator when current flows through the wire. Figure 18–11 shows a Hall sensor used to detect when a DC current flows through a circuit. A Hall effect sensor is shown in Figure 18–12.

The Hall generator is being used more and more in industrial applications. Because the signal rise and fall time of the Hall generator is generally less than

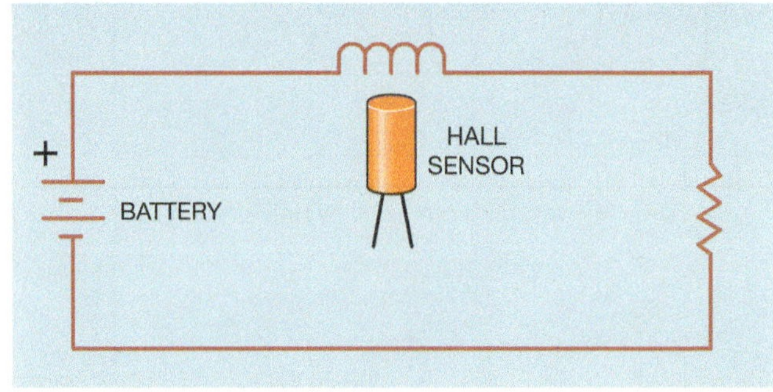

Figure 18–11 Hall sensor detects when DC current flows through the circuit.

Courtesy Turck, Inc.

Figure 18–12 Hall effect sensor.

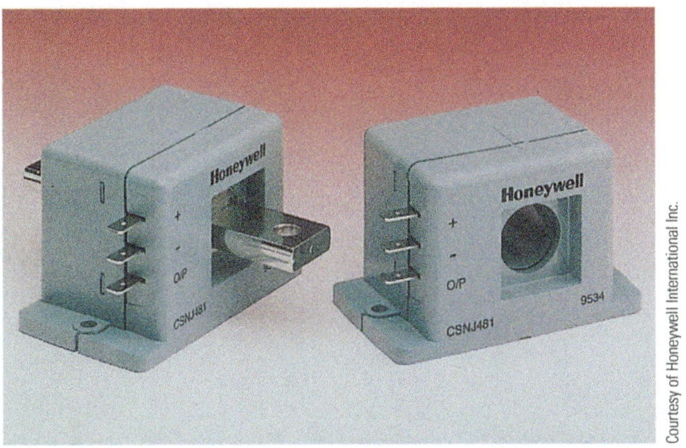

Figure 18–13 Hall effect linear transducer.

10 microseconds, it can operate at pulse rates as high as 100,000 pulses per second. This makes it especially useful in industry.

Linear Transducers

Linear transducers are designed to produce an output voltage that is proportional to the strength of a magnetic field. Input voltage is typically 8 to 16 volts, but the amount of output voltage is determined by the type of transducer used. Hall effect linear transducers can be obtained that have two types of outputs. One type has a *regulated* output and produces voltages of 1.5 to 4.5 volts. The other type has a *ratiometric* output and produces an output voltage that is 25 percent to 75 percent of the input voltage. A Hall effect linear transducer is shown in Figure 18–13.

Another common device that generally employs the use of a Hall Effect sensor is the clamp-type ammeter that can measure both AC and DC currents, Figure 18–14.

Most clamp-type ammeters depend on the continuous change of a magnetic field caused by alternating current to induce a voltage into a transformer. The meter determines the current by the amount of induced voltage. The amount of induced voltage is proportional to the strength of the magnetic field caused by the current flowing through the conductor. Because direct current does not cause a continuous change in the magnetic field, most clamp on type meters cannot measure DC current. The Hall effect sensor, however, does not depend on the movement of a magnetic field. The current value is determined by the strength of the magnetic field only.

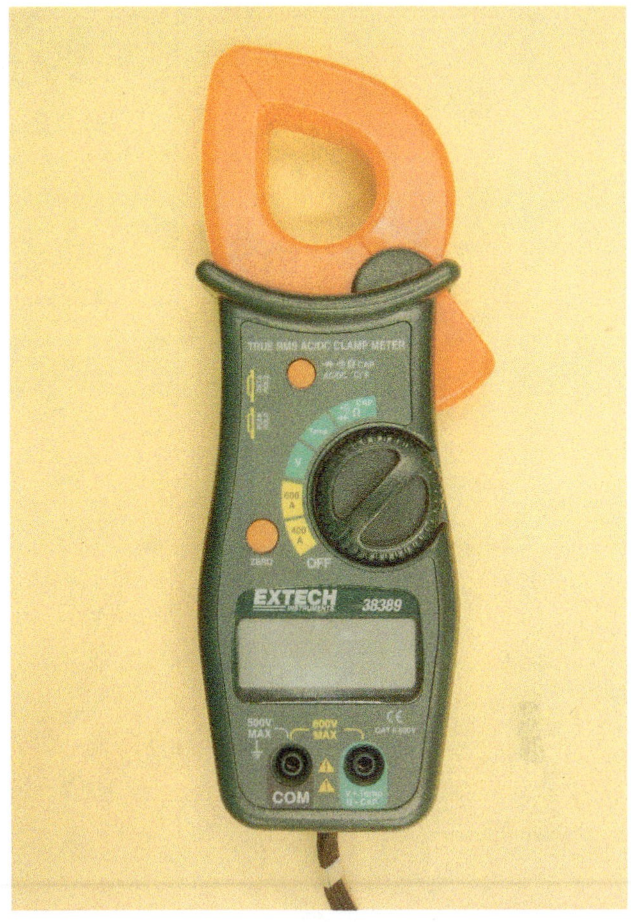

Figure 18–14 Clamp-type ammeters that can measure both AC and DC currents generally employ the use of a Hall effect sensor.

Review Questions

1. What material was used to make the first Hall generator?
2. What two factors determine the polarity of the output voltage produced by the Hall generator?
3. What two factors determine the amount of voltage produced by the Hall generator?
4. What is a reluctor?
5. Why does a magnetic field not have to be moving or changing to produce an output voltage in the Hall generator?

Chapter 19

PROXIMITY DETECTORS

Objectives

After studying this chapter the student will be able to:

» Describe the operation of proximity detectors.

» Discuss the difference between proximity detectors that can sense ferrous metal and detectors that can sense all types of metal.

» Discuss different types of proximity detectors.

Applications

Proximity detectors are used to sense the presence of an object without physically touching it. This prevents wear on the detector and gives it the ability to sense red hot metals. Proximity detectors operate on different principles. Some are metal detectors that sense ferrous metals only and others can sense all types of metals. Other proximity detectors sense objects by a change of capacitance. Some proximity detectors employ sound waves and can be used to detect the distance to an object or to determine the position of an object.

Metal Detectors

There are several methods of constructing a proximity detector that is intended to detect the presence of metal. One method is shown in Figure 19–1. This is a very simple circuit intended to illustrate the principle of operation for a metal detector. The sensor coil is connected through a series resistor to an oscillator. An oscillator is a device used to generate alternating current at a desired frequency. A voltage detector, in this illustration a voltmeter, is connected across the resistor. Because AC voltage is applied to this circuit, the amount of current flow is determined by the resistance of the resistor and the inductive reactance of the coil. The voltage drop across the resistor is proportional to its resistance and the amount of current flow.

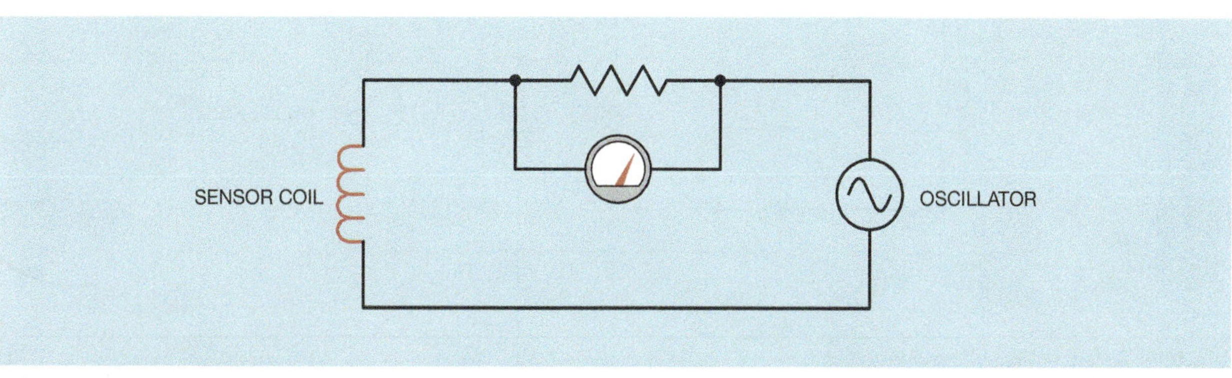

Figure 19–1 Simple proximity detector.

If ferrous metal is placed near the sensor coil, its inductance increases in value. This causes an increase in inductive reactance and a decrease in the amount of current flow through the circuit. When the current flow through the resistor is decreased, the voltage drop across the resistor decreases also (Figure 19–2). The drop in voltage can be used to turn relays or other devices on or off.

This method of detecting metal does not work well for all conditions. Another method that is more sensitive to small amounts of metal is shown in Figure 19–3. This detector uses a tank circuit tuned to the frequency of the oscillator. The sensor head contains two coils instead of one. This type of sensor is a small transformer. When the tank circuit is tuned to the frequency of the oscillator, current flow around the tank loop is high. This causes a high voltage to be induced into the secondary coil of the sensor head.

When ferrous metal is placed near the sensor as shown in Figure 19–4, the inductance of the coil increases. When the inductance of the coil changes, the tank circuit no longer resonates to the frequency of the oscillator. This causes the current flow around the loop to decrease significantly. The decrease of current flow through the sensor coil causes the secondary voltage to drop also.

Notice that both types of circuits depend on a ferrous metal to change the inductance of a coil. If a detector is to be used to detect nonferrous metals, some means other than changing the inductance of the coil must be used. An all-metal detector uses a tank circuit as shown in Figure 19–5. All-metal detectors operate at radio frequencies, and the balance of the tank circuit is used to keep the oscillator running. If the tank circuit becomes unbalanced, the oscillator stops operating. When a nonferrous metal, such as aluminum, copper, or brass, is placed near the sensor coil, eddy currents are induced into the surface of the metal. The induction of eddy currents into the metal causes the tank circuit to become unbalanced and the oscillator to stop operating. When the

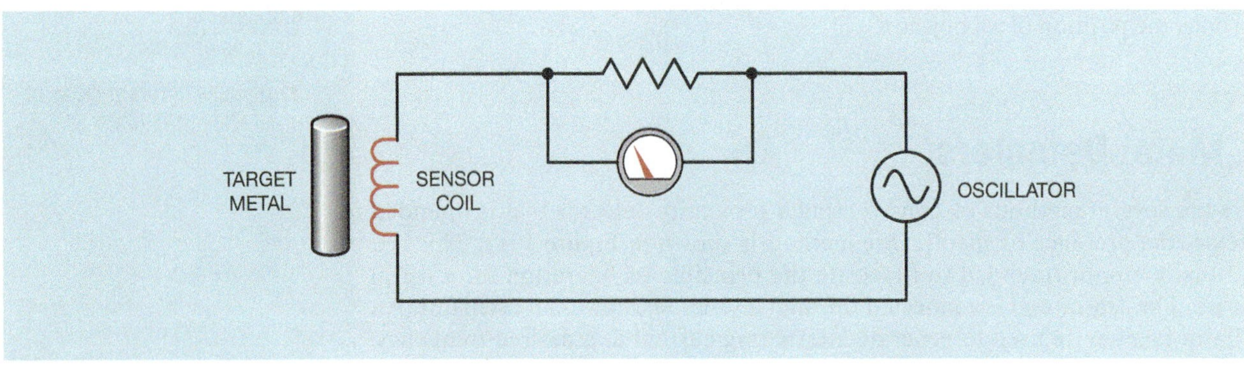

Figure 19–2 The presence of metal causes a decrease of voltage across the register.

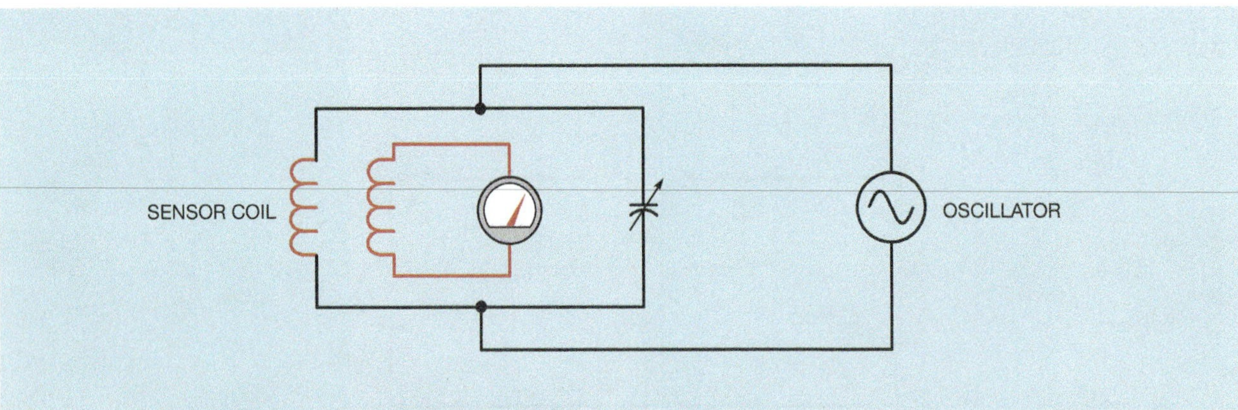

Figure 19–3 Tuned tank circuit used to detect metal.

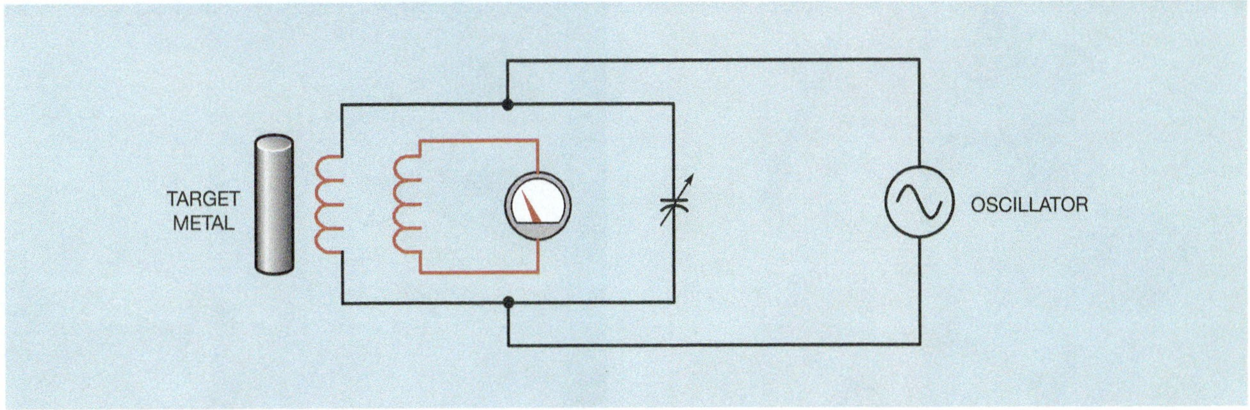

Figure 19–4 The presence of metal detunes the tank circuit.

Figure 19–5 Balance of the tank circuit permits the oscillator to operate.

oscillator stops operating, some other part of the circuit signals an output to turn on or off.

Proximity detectors used to sense all types of metals will sense ferrous metals better than nonferrous. A ferrous metal can be sensed at about three times the distance of a nonferrous metal.

Mounting

Some proximity detectors are made as a single unit. Other detectors use a control unit that can be installed in a relay cabinet and a sensor that is mounted at a remote location. Figure 19–6 shows different types of proximity detectors. Regardless of the type of detector used, care and forethought should be used when mounting the sensor. The sensor must be near enough to the target metal to provide a strong positive signal, but it should not be so near that there is a possibility of the sensor being hit by the metal object. One advantage of the proximity detector is that no physical contact is necessary between the detector and the metal object for the detector to sense the object.

Sensors should be mounted as far away from other metals as possible. This is especially true for sensors used with units designed to detect all types of metals. In some cases it may be necessary to mount the sensor unit on a nonmetal surface such as wood or plastic. If proximity detectors are to be used in areas that contain metal shavings or metal dust, an effort should be made to place the sensor in a position that will prevent the shavings or dust from collecting around it. In some installations it may be necessary to periodically clean the metal shavings or dust away from the sensor.

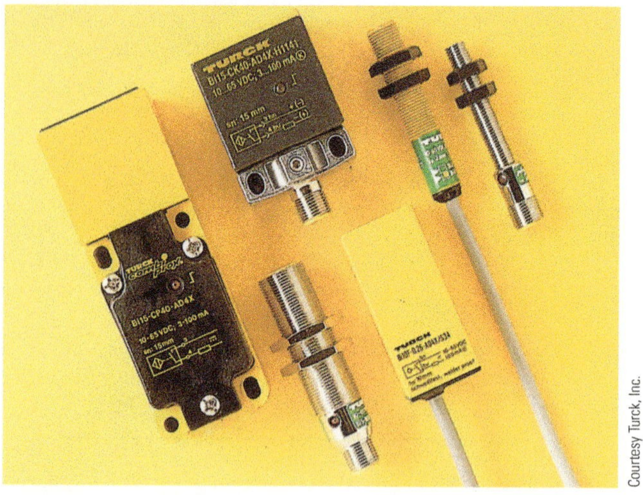

Courtesy Turck, Inc.

Figure 19–6 Proximity detectors.

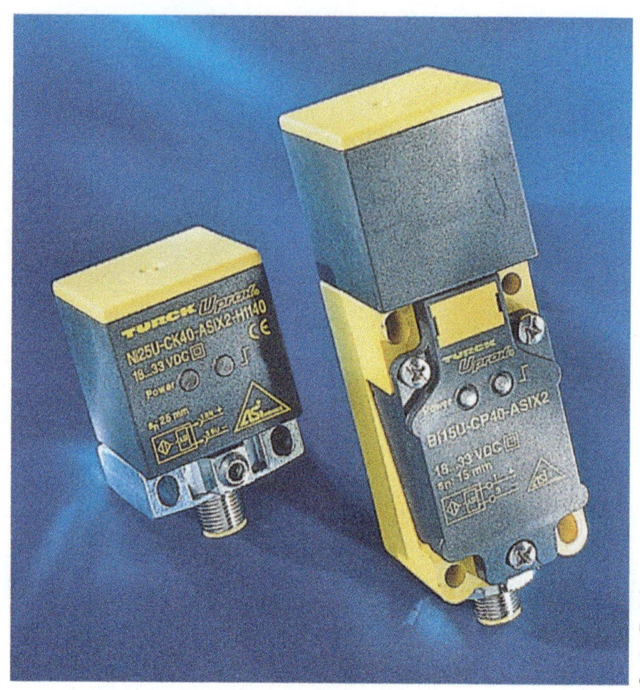

Courtesy Turck, Inc.

Figure 19–7 Capacitive proximity detectors.

Capacitive Proximity Detectors

Although proximity detectors are generally considered to be metal detectors, other types of detectors sense the presence of objects that do not contain metal of any kind. One type of these detectors operates on a change of capacitance. When an object is brought into the proximity of one of these detectors, a change of capacitance causes the detector to activate. Several different types of capacitive proximity detectors are shown in Figure 19–7.

Because capacitive proximity detectors do not depend on metal to operate, they will sense virtually any material such as wood, glass, concrete, plastic, and sheet rock. They can even be used to sense liquid levels through a sight glass. One disadvantage of capacitive proximity detectors is that they have a very limited range. Most cannot sense objects over approximately 1 inch or 25 millimeters away. Many capacitive proximity detectors are being used to replace mechanical limit switches because they do not have to make contact with an object to sense its position. Most can be operated with a wide range of voltages such as 2–250 VAC, or 20–320 VDC.

Ultrasonic Proximity Detectors

Another type of proximity detector that does not depend on the presence of metal for operation is the *ultrasonic detector*. Ultrasonic detectors operate by emitting a pulse of high frequency sound and then detecting the echo when it bounces off an object (Figure 19–8). These detectors can be used to determine the distance to the object by measuring the time interval between the emission of the pulse and the return of the echo. Many ultrasonic sensors have an analog output of voltage or current, the value of which is determined by the distance to the object. This feature permits them to be used in applications where it is necessary to sense the position of an object (Figure 19–9). An ultrasonic proximity detector is shown in Figure 19–10.

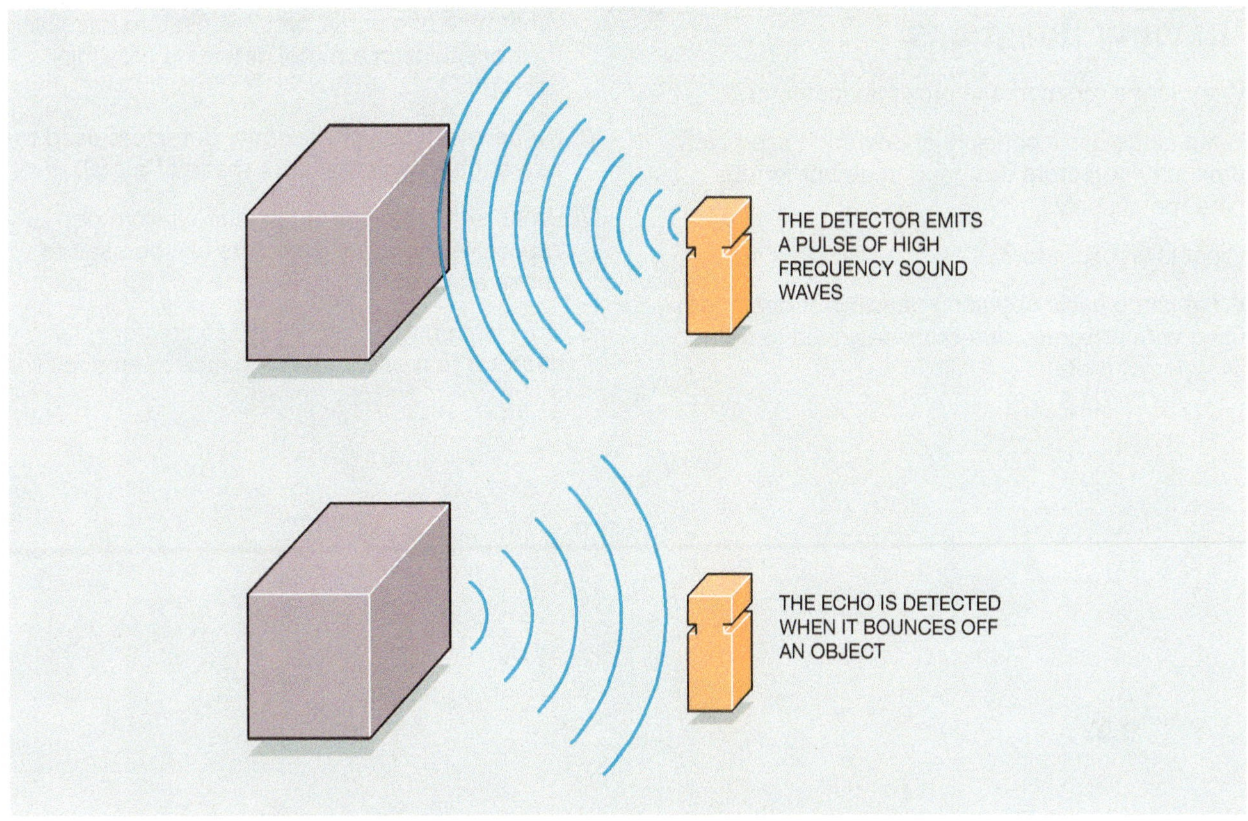

Figure 19–8 Ultrasonic proximity detectors operate by emitting high frequency sound waves.

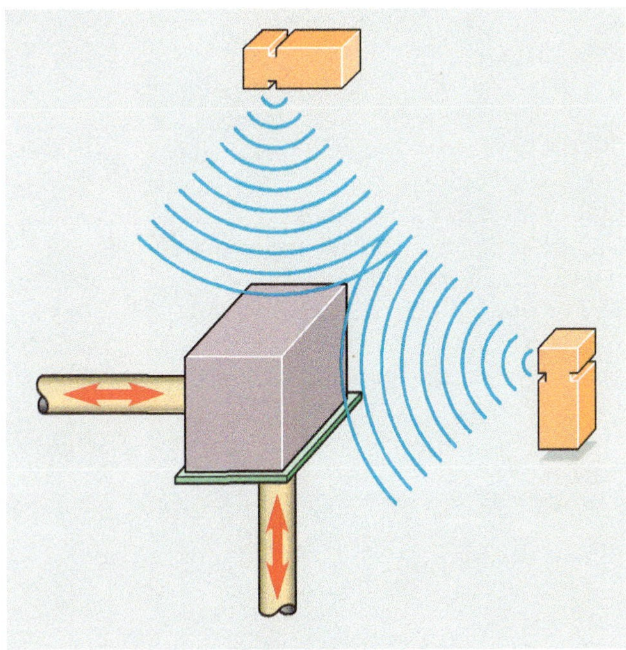

Figure 19–9 Ultrasonic proximity detectors used as position sensors.

Figure 19–10 Ultrasonic proximity detector.

Review Questions

1. What is the function of a proximity detector?
2. What is the basic principle of operation used with proximity detectors designed to detect ferrous type metals only?
3. What is an oscillator?
4. What is the basic operating principle of operation used with proximity detectors designed to detect all types of metal?
5. What type of electric circuit is used to increase the sensitivity of a metal detecting proximity detector?
6. Name two types of proximity detectors used to detect objects that are not made of metal.
7. What is the maximum range at which most capacitive proximity detectors can be used to sense an object?
8. How is it possible for an ultrasonic proximity detector to measure the distance to an object?

Chapter 20

PHOTODETECTORS

Objectives

After studying this chapter the student will be able to:

» List different devices used as light sensors.

» Discuss the advantages of photo-operated controls.

» Describe different methods of installing photodetectors.

Applications

Photodetectors are widely used in today's industry. They can be used to sense the presence or absence of almost any object. Photodetectors do not have to make physical contact with the object they are sensing, so there is no mechanical arm to wear out. Many photodetectors can operate at speeds that cannot be tolerated by mechanical contact switches. They are used in almost every type of industry, and their uses are increasing steadily.

Types of Detectors

Photo-operated devices fall into one of three categories: photovoltaic, photoemissive, and photoconductive.

Photovoltaic

Photovoltaic devices are more often called solar cells. They are made of silicon and have the ability to produce a voltage in the presence of light. The amount of voltage produced by a cell is determined by the material it is made of. When silicon is used, the solar cell produces 0.5 volts in the presence of direct sunlight. If there is a complete circuit connected to the cell, current will flow through the circuit. The amount of current produced by a solar cell is determined by the surface area of the cell. For instance, assume a solar cell has a surface area of 1 square inch, and another cell has a surface area of 4 square inches. If both cells are made of silicon, both will produce 0.5 volts when in direct sunlight. The larger cell, however, will produce four times as much current as the small one.

Figure 20–1 shows the schematic symbol for a photovoltaic cell. Notice that the symbol is the same as the symbol used to represent a single cell battery except for the arrow pointing toward it. The battery symbol means the device has the ability to produce a voltage, and the arrow means that it must receive light to do so.

Photovoltaic cells have the advantage of being able to operate electrical equipment without external power. Because silicon solar cells produce only 0.5 volts, it is often necessary to connect several of them together to obtain enough voltage and current to operate the desired device. For example, assume that solar cells are to be used to operate a DC relay coil that requires 3 volts at 250 milliamps. Now assume that the solar cells to be used have the ability to produce 0.5 volts at 150 milliamps.

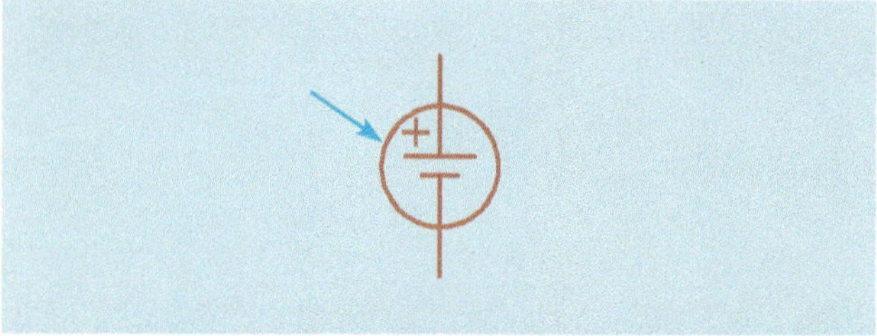

Figure 20–1 Schematic symbol for a photovoltaic cell.

If six solar cells are connected in series, they will produce 3 volts at 150 milliamps (Figure 20–2). The voltage produced by the connection is sufficient to operate the relay, but the current capacity is not. Therefore, six more solar cells must be connected in series. This connection is then connected parallel to the first connection producing a circuit that has a voltage rating of 3 volts and a current rating of 300 milliamps, which is sufficient to operate the relay coil.

Although solar cells are sometimes used in industrial control applications, they are more often used to provide electric power in remote locations where connection to power lines is not available. Banks of solar cells charge batteries during daylight hours (Figure 20–3). The DC voltage provided by the batteries is then converted into AC voltage to operate desired equipment.

Figure 20–3 Solar cells used to provide power at the Anvick River Lodge located in a remote region of Alaska.

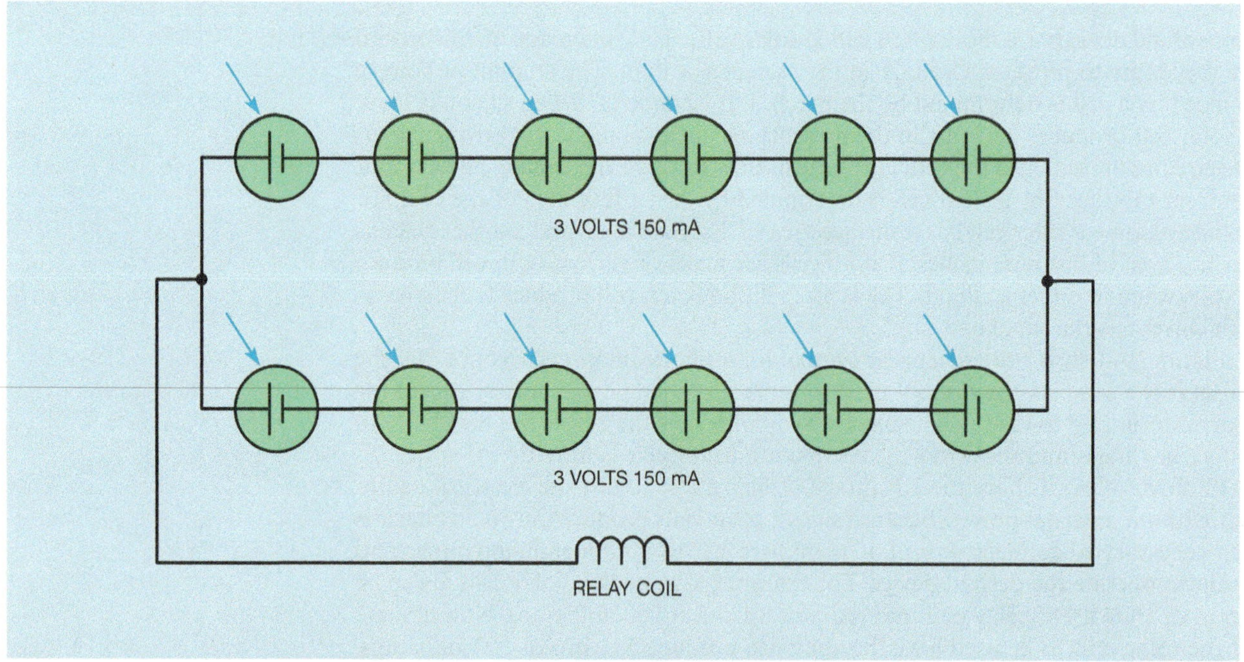

Figure 20–2 Series-parallel connection of solar cells produces 3 volts at 300 milliamps.

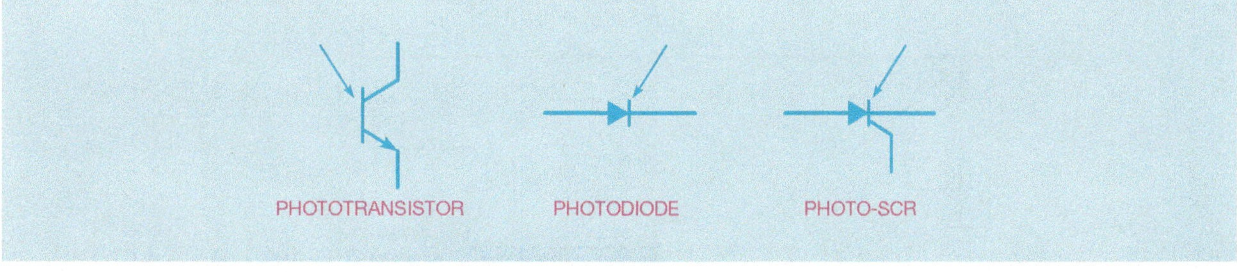

Figure 20–4 Schematic symbols for the phototransistor, the photodiode, and the photo-SCR.

Photoemissive Devices

Photoemissive devices emit electrons when in the presence of light. They include such devices as the phototransistor, the **photodiode**, and the photo-SCR. The schematic symbols for these devices are shown in Figure 20–4. The emission of electrons is used to turn these solid-state components on. The circuit in Figure 20–5 shows a phototransistor used to turn on a relay coil. When the phototransistor is in darkness, no electrons are emitted by the base junction, and the transistor is turned off. When the phototransistor is in the presence of light, it turns on and permits current to flow through the relay coil. The diode connected parallel to the relay coil is known as a kickback or freewheeling diode. Its function is to prevent induced voltage spikes from occurring when the current suddenly stops flowing through the coil and the magnetic field collapses.

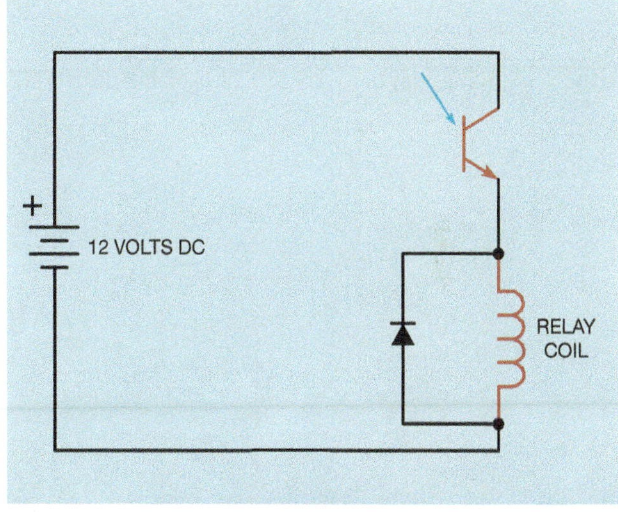

Figure 20–5 Phototransistor controls relay coil.

In the circuit shown in Figure 20–5, the relay coil will turn on when the phototransistor is in the presence of light, and turn off when the phototransistor is in darkness. Some circuits may require the reverse operation. This can be accomplished by adding a resistor and a junction transistor to the circuit (Figure 20–6). In this circuit a common junction transistor is used to control the current flow through the relay coil. Resistor R1 limits the current flow through the base of the junction transistor. When the phototransistor is in darkness, it has a very high resistance. This permits current to flow to the base of the junction transistor and turn it on. When the phototransistor is in the presence of light, it turns on and connects the base of the junction transistor to the negative side of the battery. This causes the junction transistor to turn off. The phototransistor in the circuit is used as a **stealer transistor**. A stealer transistor steals the base current away from some other transistor to keep it turned off.

Some circuits may require the phototransistor to have a higher gain than it has under normal conditions. This can be accomplished by using the phototransistor as the driver for a Darlington amplifier circuit, Figure 20–7. A Darlington amplifier circuit generally has a gain of over 10,000.

Photodiodes and photo-SCRs are used in circuits similar to those shown for the phototransistor. The photodiode will permit current to flow through it in the presence of light. The photo-SCR has the same operating characteristics as a common junction SCR. The only difference is that light is used to trigger the gate when using a photo-SCR.

Regardless of the type of photoemissive device used, or the type circuit it is used in, the greatest advantage of the photoemissive device is speed. A photoemissive device can turn on or off in a few microseconds. Photovoltaic or photoconductive devices generally require several milliseconds to turn on or off. This makes the use of photoemissive devices imperative in high speed switching circuits.

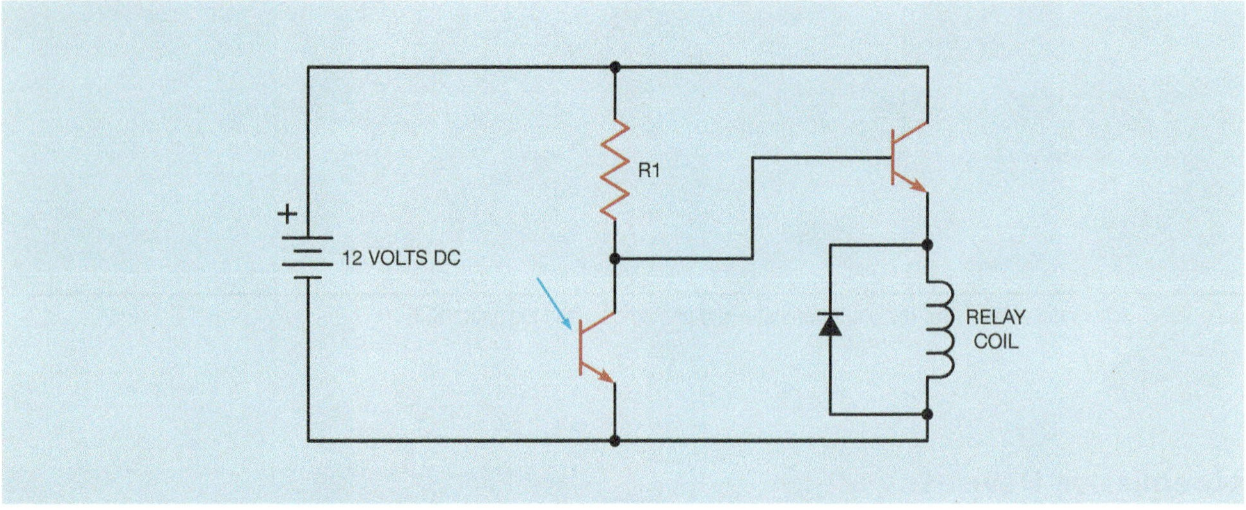

Figure 20–6 The relay turns on when the phototransistor is in darkness.

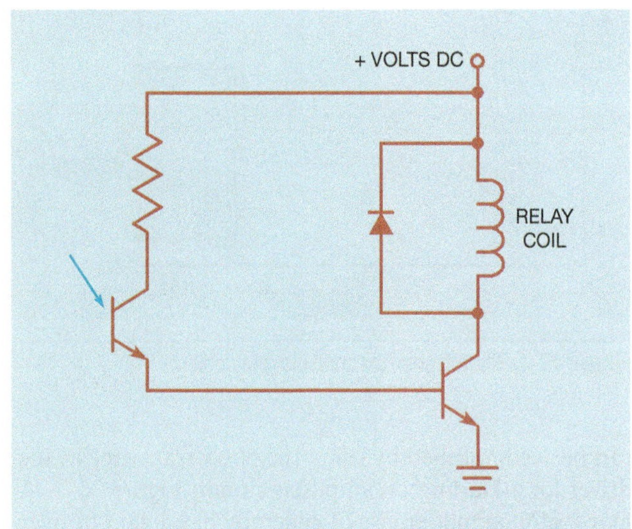

Figure 20–7 The phototransistor is used as the driver for a Darlington amplifier.

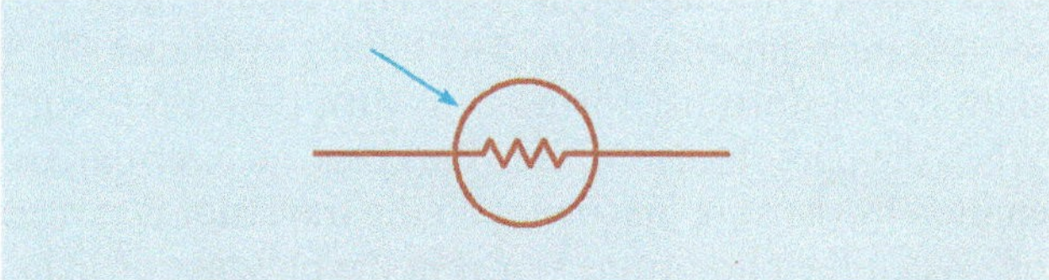

Figure 20–8 Schematic symbol for a cad cell.

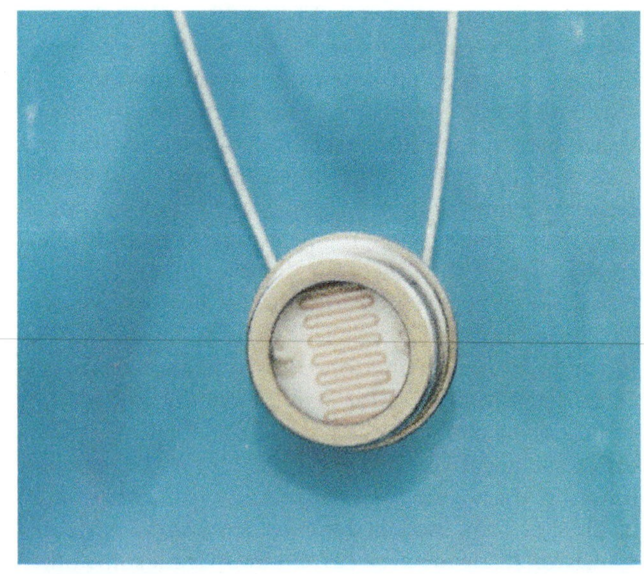

Figure 20–9 Cad cell.

Photoconductive Devices

Photoconductive devices exhibit a change of resistance due to the presence or absence of light. The most common photoconductive device is the cadmium sulfide cell or **cad cell.** The cad cell has a resistance of about 50 ohms in direct sunlight and several hundred thousand ohms in darkness. It is generally used as a light sensitive switch. The schematic symbol for a cad cell is shown in Figure 20–8. Figure 20–9 shows a typical cad cell.

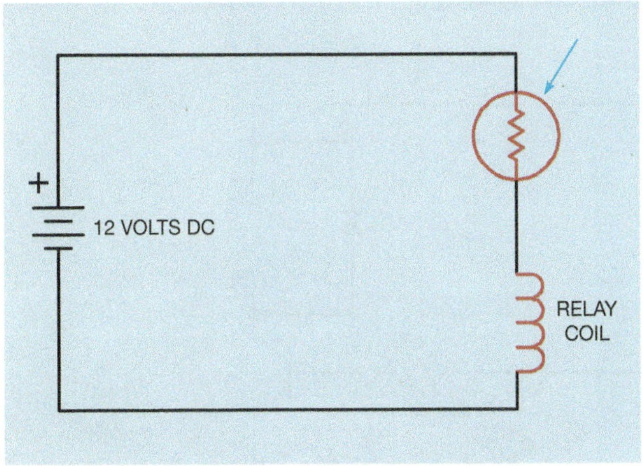

Figure 20–10 Cad cell controls relay coil.

Figure 20–10 shows a basic circuit of a cad cell being used to control a relay. When the cad cell is in darkness, its resistance is high. This prevents the amount of current needed to turn the relay on from flowing through the circuit. When the cad cell is in the presence of light, its resistance is low. The amount of current needed to operate the relay can now flow through the circuit.

Although this circuit will work if the cad cell is large enough to handle the current, it has a couple of problems.

1. There is no way to adjust the sensitivity of the circuit. Photo-operated switches are generally located in many different areas of a plant. The surrounding light intensity can vary from one area to another. It is, therefore, necessary to be able to adjust the sensor for the amount of light needed to operate it.
2. The sense of operation of the circuit cannot be changed. The circuit shown in Figure 20–10 permits the relay to turn on when the cad cell is in the presence of light. There may be conditions that would make it desirable to turn the relay on when the cad cell is in darkness.

Figure 20–11 shows a photodetector circuit that uses a cad cell as the sensor and an operational amplifier as the control circuit. The circuit operates as follows: Resistor R1 and the cad cell form a voltage divider circuit that is connected to the inverting input of the amplifier. Resistor R2 is used as a potentiometer to preset a positive voltage at the noninverting input. This control adjusts the sensitivity of the circuit. Resistor R3 limits the current to a light-emitting diode (LED). The LED is mounted on the outside of the case of the photodetector

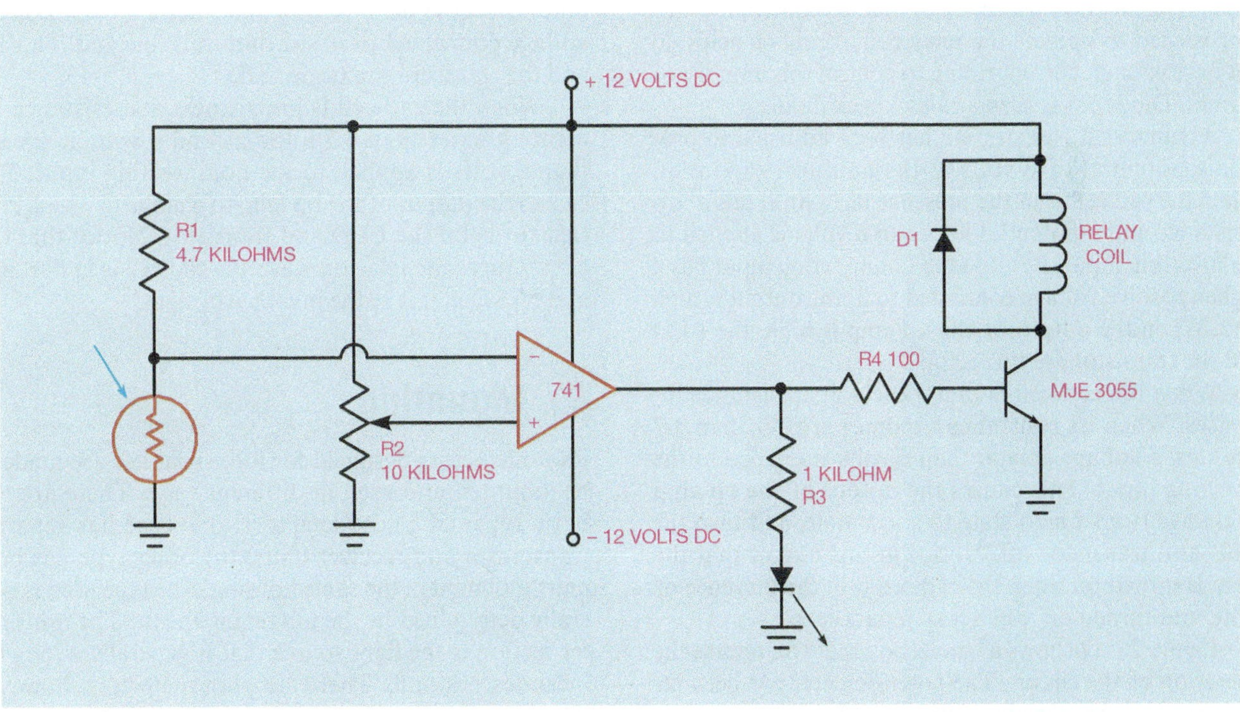

Figure 20–11 The relay coil is energized when the cad cell is in light.

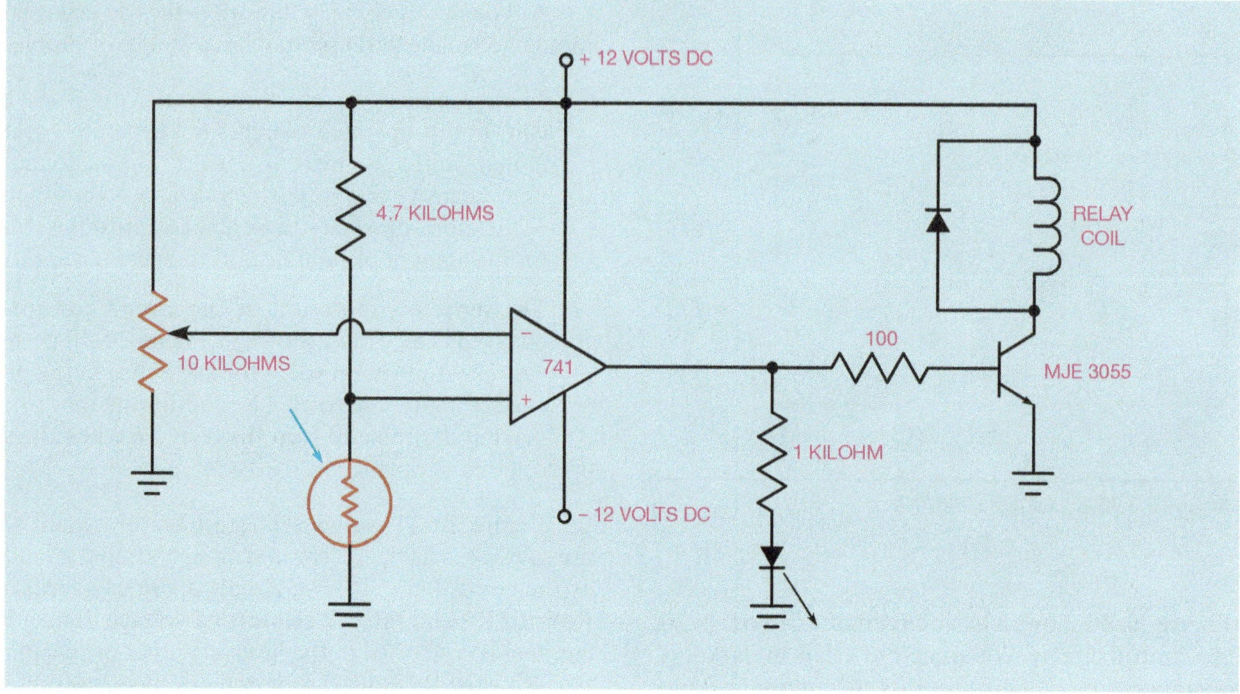

Figure 20–12 The relay is energized when the cad cell is in darkness.

and is used to indicate when the relay coil is energized. Resistor R4 limits the base current to the junction transistor. The junction transistor is used to control the current needed to operate the relay coil. Many **op amps** do not have enough current rating to control this amount of current. Diode D1 is used as a kickback diode.

Assume that Resistor R2 has been adjusted to provide a potential of 6 volts at the noninverting input. When the cad cell is in the presence of light, it has a low resistance and a potential less than 6 volts is applied to the inverting input. Because the noninverting input has a higher positive voltage connected to it, the output is high also. When the output of the op amp is high, the LED and the transistor are turned on.

When the cad cell is in darkness, its resistance increases. When its resistance becomes greater than 4.7 kilohms, a voltage greater than 6 volts is applied to the inverting input. This causes the output of the op amp to change from a high state to a low state, and turn the LED and transistor off. Notice in this circuit that the relay is turned on when the cad cell is in the presence of light, and turned off when it is in darkness.

Figure 20–12 shows a connection that will reverse the operation of the circuit. The potentiometer has been reconnected to the inverting input, and the voltage divider circuit has been connected to the noninverting input. To understand the operation of this circuit, assume that a potential of 6 volts has been preset at the inverting input.

When the cad cell is light, it has a low resistance and a voltage less than 6 volts is applied to the noninverting input. Because the inverting input has a greater positive voltage connected to it, the output is low and the LED and the transistor are turned off.

When the cad cell is in darkness, its resistance becomes greater than 4.7 kilohms and a voltage greater than 6 volts is applied to the noninverting input. This causes the output of the op amp to change to a high state that turns on the LED and transistor. Notice that this circuit turns the relay on when the cad cell is in darkness and off when it is in the presence of light.

Mounting

Photodetectors designed for industrial use are made to be mounted and used in different ways. There are two basic types of photodetectors: one type has separate transmitter and receiver units; the other type has both units mounted in the same housing. The type used is generally determined by the job requirements. The transmitter section is the light source that is generally a long life incandescent bulb. There are photodetectors, however, that use an infrared transmitter. These cannot be seen by the human eye and are often used in burglar alarm systems. The receiver unit houses the photodetector and, generally, the circuitry required to operate the system.

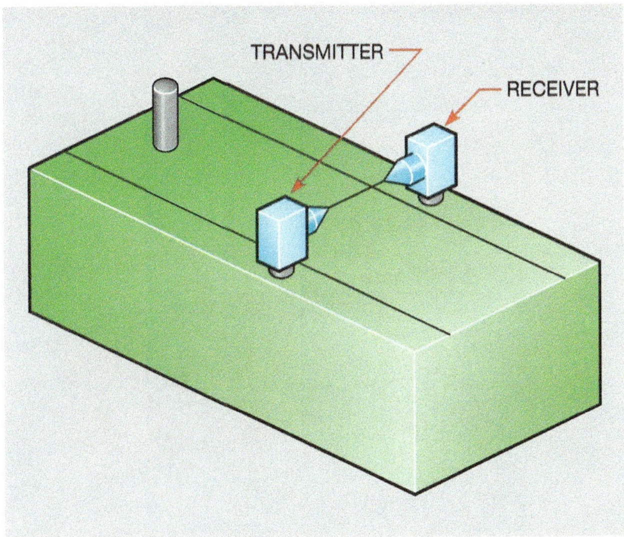

Figure 20–13 Photodetector senses presence of object on conveyor line.

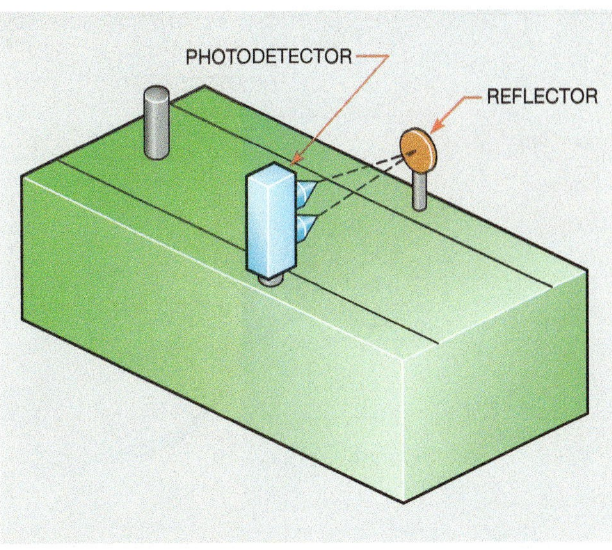

Figure 20–15 The object is sensed when it passes between the photodetector and the reflector.

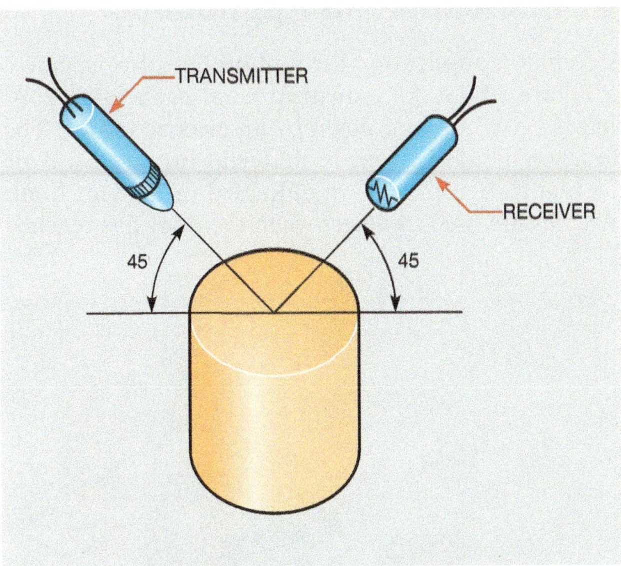

Figure 20–14 Object is sensed by reflecting light off a shiny surface.

Figure 20–13 shows a photodetector used to detect the presence of an object on the conveyor line. When the object passes between the transmitter and receiver units, the light beam is broken and the detector activates. Notice that no physical contact was necessary for the photodetector to sense the presence of the object.

Figure 20–14 illustrates another method of mounting the transmitter and receiver. In this example, an object is sensed by reflecting light off of a shiny surface. Notice that the transmitter and receiver must be mounted at the same angle with respect to the object to be sensed. This type of mounting will only work with objects that have the same height, such as cans on a conveyor line.

Photodetectors that have both the transmitter and the receiver units mounted in the same housing depend on a reflector for operation. Figure 20–15 shows this type of unit mounted on a conveyor line. The transmitter is aimed at the reflector. The light beam is reflected back to the receiver. When an object passes between the photodetector unit and the reflector, the light to the receiver is interrupted. This type of unit has the advantage of needing electrical connection at only one piece of equipment. This permits easy mounting of the photodetector unit, and mounting of the reflector in hard to reach positions that would make running control wiring difficult. Many of these units have a range of 20 feet and more.

Another type of unit that operates on the principle of reflected light uses an optical fiber cable. The fibers in the cable are divided in half. One-half of the fibers is connected to the transmitter, and the other half is connected to the receiver (Figure 20–16). This unit has the advantage of permitting the transmitter and the receiver to be mounted in a very small area. Figure 20–17 illustrates a common use for this type of unit. The unit is used to control a label cutting machine. The labels are printed on a large roll and must be cut for individual packages. The label roll contains a narrow strip on one side which is dark colored except for shiny sections spaced at regular intervals. The optical fiber cable is located above this narrow strip. When the dark surface of the strip is passing

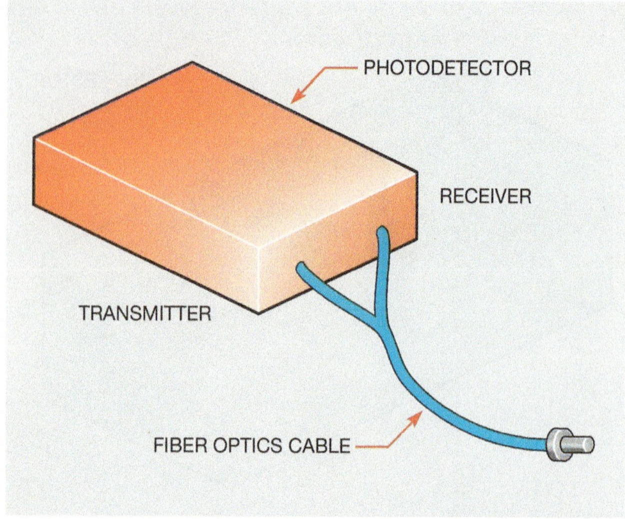

Figure 20–16 Optical cable is used to transmit and receive light.

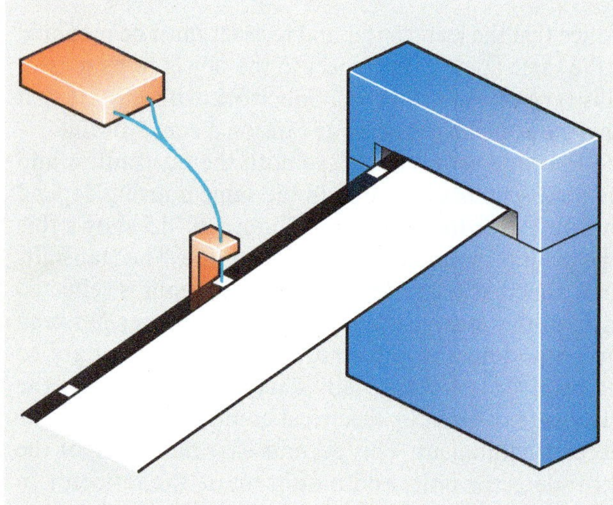

Figure 20–17 Optical cable detects shiny area on one side of label.

beneath the optical cable, no reflected light returns to the receiver unit. When the shiny section passes beneath the cable, light is reflected back to the receiver unit. The photodetector sends a signal to the control circuit and tells it to cut the label.

Photodetectors are very dependable and have an excellent maintenance and service record. They can be used to sense almost any object without making physical contact with it, and can operate millions of times without damage or wear. A photodetector is shown in Figure 20–18.

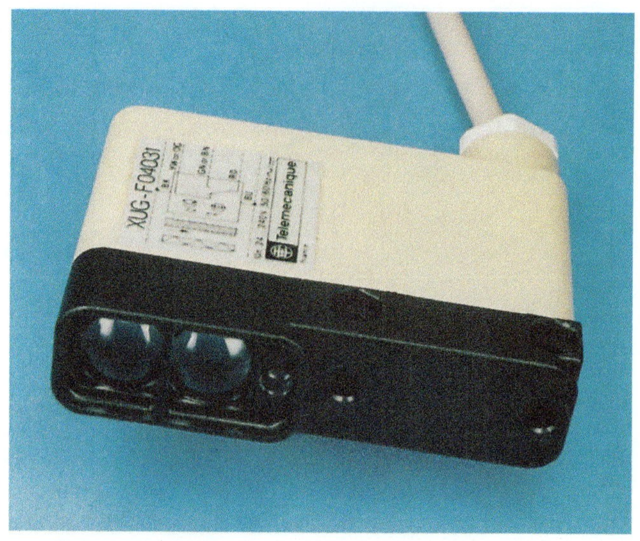

Figure 20–18 Photodetector unit with both transmitter and receiver units.

Photodetector Application

A common application for a photodetector is shown in Figure 20–19. An industrial plant has a large door that is raised up and down by an electric motor. The doorway is used to move material into and out of the plant when trucks are unloaded or loaded. Limit switches are used to detect when the door has reached

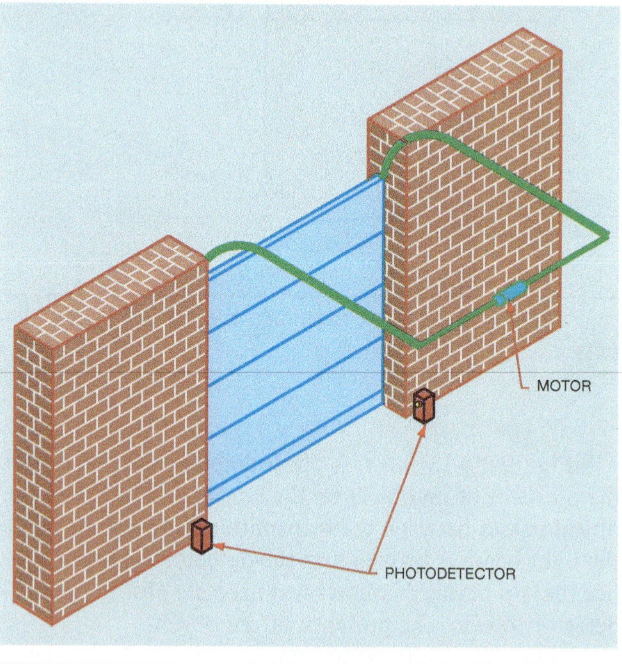

Figure 20–19 Photodetectors stop door if light beam is broken.

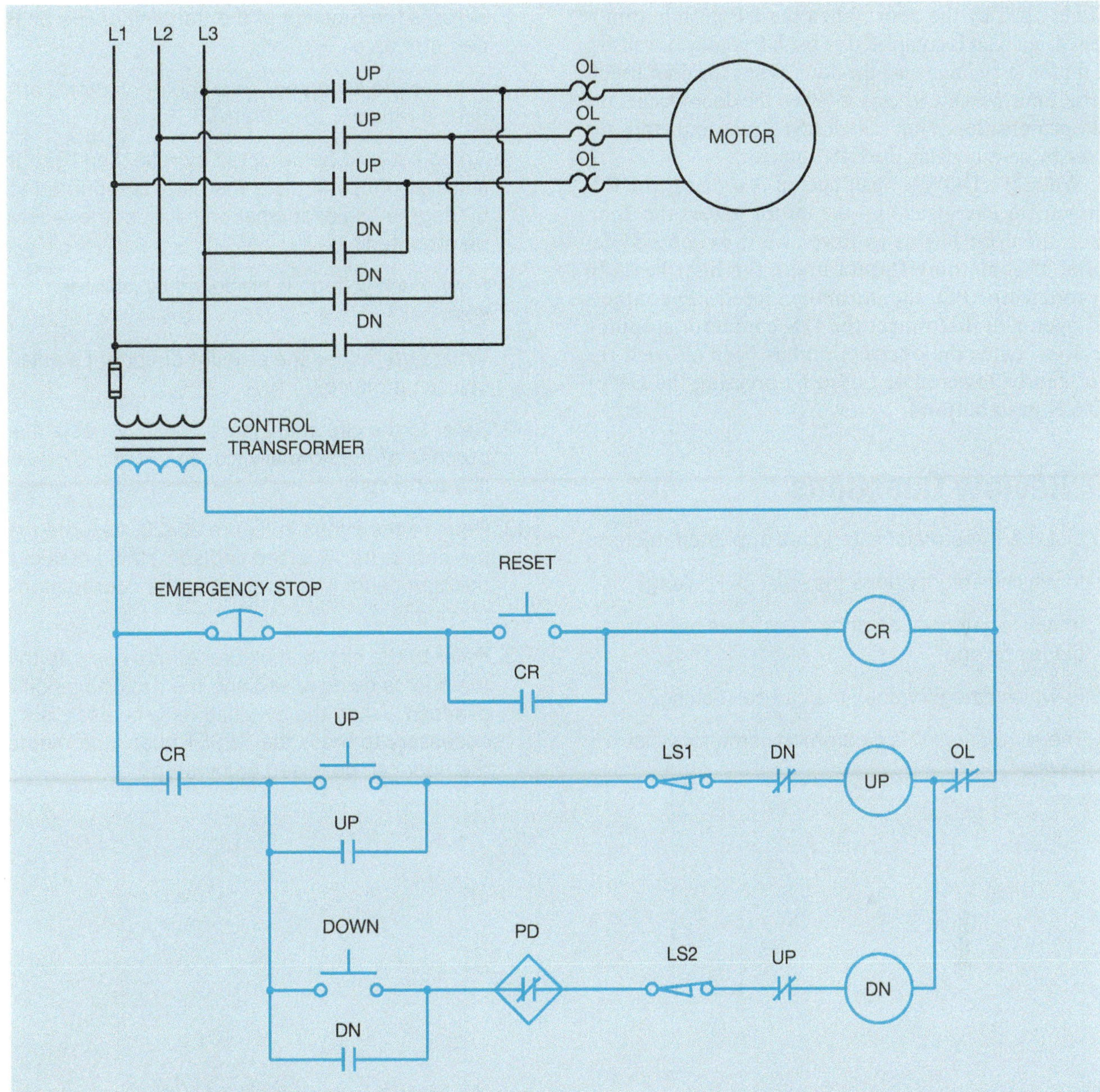

Figure 20–20 Door opener control circuit.

its maximum distance of travel in the up or open position, or in the down or closed position. Limit switch LS1 detects when the door is fully open or up and limit switch LS2 detects when the door is completely closed or down. Due to the danger of someone walking through or driving a fork truck through the door when it is being lowered, a photodetector is installed on each side of the door. If the light beam is broken when the door is being lowered, it will immediately stop the motor.

The circuit to control the operation of this door is shown in Figure 20–20. Because the motor must run in one direction to raise the door and run in the opposite the direction to lower the door, the circuit is basically a forward–reverse control. An emergency STOP button can be used to stop the motor at anytime. Once the emergency STOP button has been pressed, the circuit must be reset before the door can be raised or lowered. To understand the operation of the circuit, assume the door to be in the closed position. At this point in time, limit switch LS2 will be held open by the door. When the UP push button is pressed, a circuit is completed to the UP contactor causing the motor to begin raising the door. When the door begins to rise, limit switch LS2 closes. When the door reaches the full open position, limit switch LS1 opens and stops the motor by de-energizing the UP contactor.

When the DOWN push button is pressed, the DN contactor is energized and the motor lowers the door. When the door begins to lower, limit switch LS1 recloses. If something should break the light beam to photodetector PD, the normally closed PD contacts will open and disconnect the DN contactor stopping the door. Once the obstruction has been cleared, the door can be lowered or raised by pressing the UP or DOWN push buttons.

Review Questions

1. List the three major categories of photodetectors.
2. In which category does the solar cell belong?
3. In which category do phototransistors and photodiodes belong?
4. In which category does the cad cell belong?
5. The term *cad cell* is a common name for what device?
6. What is the function of the transmitter in a photodetector unit?
7. What is the advantage of a photodetector that uses a reflector to operate?
8. An object is to be detected by reflecting light off a shiny surface. If the transmitter is mounted at a 60 degree angle, at what angle must the receiver be mounted?
9. How much voltage is produced by a silicon solar cell?
10. What determines the amount of current a solar cell can produce?
11. Refer to the circuit in Figure 20–20. What is the purpose of the normally closed DN and UP auxiliary contacts?
12. Refer to the circuit in Figure 20–20. Assume that the door is being raised and something breaks the light beam. Will this cause the door to stop moving?
13. Refer to the circuit in Figure 20–20. Assume that the door is being raised and the motor trips on overload. When the overload relay is reset, is it necessary to press the RESET push button before the door can be raised or lowered?

Chapter 21

READING LARGE SCHEMATIC DIAGRAMS

Objectives

After studying this chapter the student will be able to:

» Discuss notations written on large schematics.

» Find contacts that are controlled by specific coils on different lines.

» Find contacts that are controlled by specific coils on different electrical prints.

The schematics presented in this book so far have been small and intended to teach circuit logic and how basic control systems operate. Schematics in industry, however, are often much more complicated and may contain several pages. Notation is generally used to help the electrician interpret the meaning of certain components and find contacts that are controlled by coils. The schematic shown in Figure 21–1 is part of a typical industrial control schematic. Refer to this schematic to locate the following information provided about the control system.

1. At the very top left-hand side find the notation: (2300V 3θ 60 Hz.). This indicates that the motor is connected to a 2300-volt three phase 60 hertz power line (Figure 21–2).
2. To the right of the first notation locate the notation: (200A 5000V DISCONNECT SWITCH). This indicates that there is a 200 ampere disconnect switch rated at 5000 volts that can be used to disconnect the motor from the power line. Also, notice that there are six contacts for this switch, two in each line. This is common for high voltage disconnect switches.
3. At the top of the schematic, locate the two current transformers, 1CT and 2CT. These two current transformers are used to detect the amount of motor current. Current transformers produce an output current of 5 amps under a short-circuit condition. The notation beside each CT indicates that it has a ratio of 150 to 5. The secondary of 1CT is connected to 1OL and 3OL. The secondary of 2CT is connected to 2OL and 4OL. Overload coils 1OL and 3OL are connected in series. That forces each to have the same current flow. Also note that coil symbols are used for the overloads, not heater symbols. This indicates that these overload relays are magnetic, not thermal.
4. Locate the two 10A fuses connected to the primary of the control transformer (Figure 21–3). The control transformer is rated at 2 kVA (2000 volt-amps). The high voltage winding is rated at 2300 volts and the secondary winding is rated at 230 volts. Also, note that the secondary winding contains a center tap (X3). The center tap can be used to provide 120 volts from either of the other X terminals. Terminals X1 and X2 are connected to 30A fuses.
5. To the left of the 30A fuse connected to terminal X1, locate the notation (EP12246-00) in Figure 21–4. This notation indicates that you are looking at Electrical Print number 12246 and line number 00. Most multipage schematics use a similar form of notation to indicate the page and line number you are viewing.

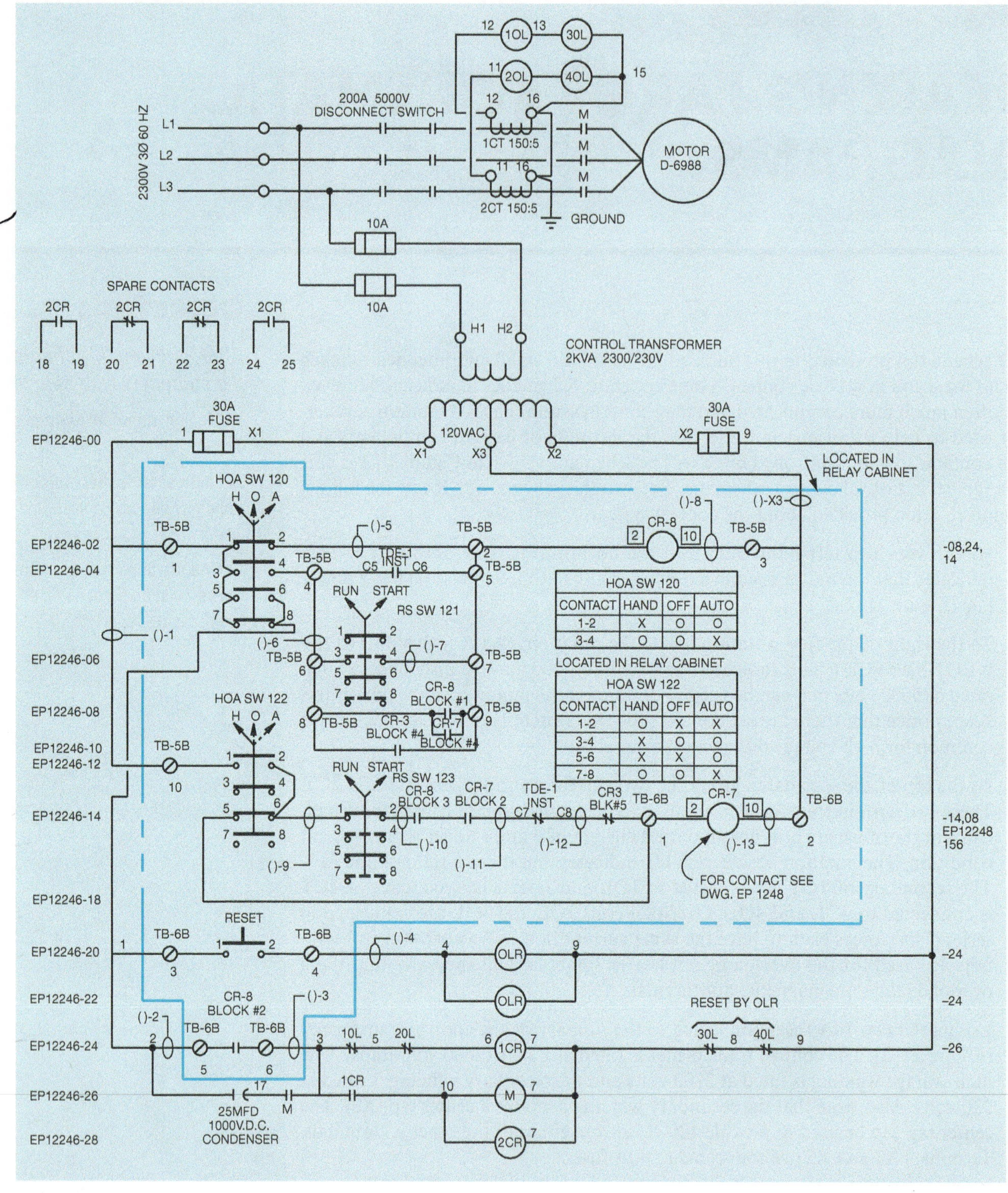

HOA SW 120

CONTACT	HAND	OFF	AUTO
1-2	X	O	O
3-4	O	O	X

LOCATED IN RELAY CABINET

HOA SW 122

CONTACT	HAND	OFF	AUTO
1-2	O	X	X
3-4	X	O	O
5-6	X	X	O
7-8	O	O	X

Figure 21–1 Typical industrial schematic diagram.

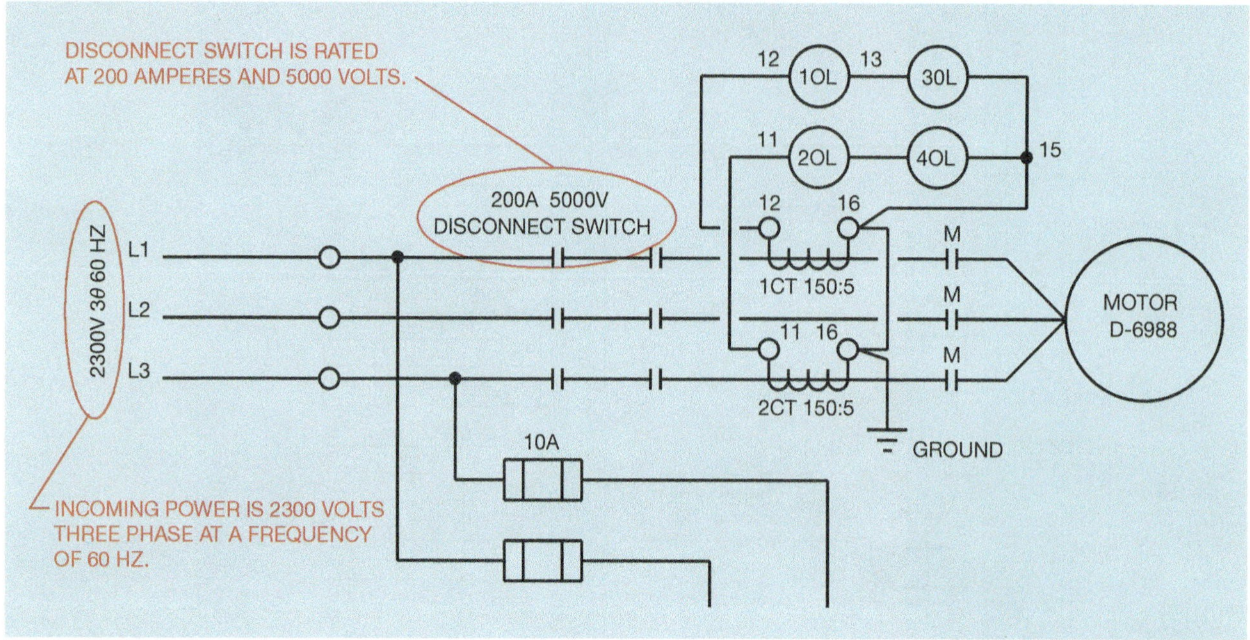

Figure 21–2 Full voltage section of the schematic.

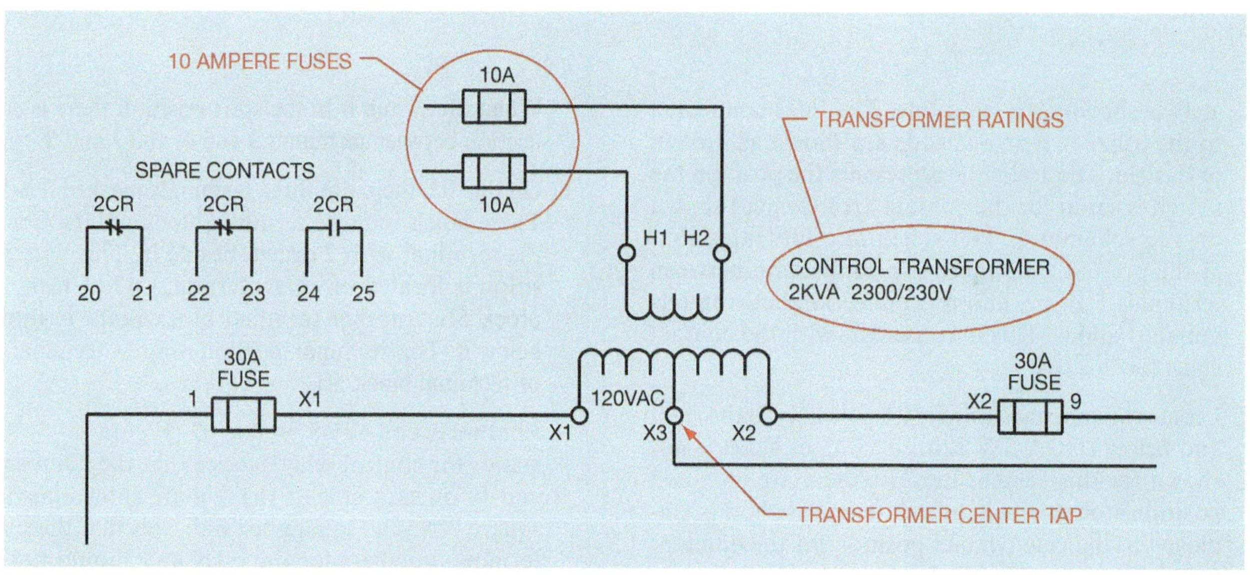

Figure 21–3 Full voltage is converted to control voltage.

6. On line number EP12246-02, locate the hand-off-automatic switch labeled (HOA SW 120). Also locate the contact chart for this switch just to the right of the center of the schematic. The chart indicates connection between specific terminals for different settings of the switch. An X indicates connection between terminals and 0 indicates no connection. Notice that in the hand position, there is connection between terminals 1 and 2. There is no connection between terminals 3 and 4. In the off position, there is no connection between any of the terminals. In the auto position there is connection between terminals 3 and 4 but no connection between terminals 1 and 2.

Referring back to the switch itself, notice that three arrows are drawn at the top of the switch. One arrow points to H, one points to O, and one points to A. The line connected to the arrowhead pointed

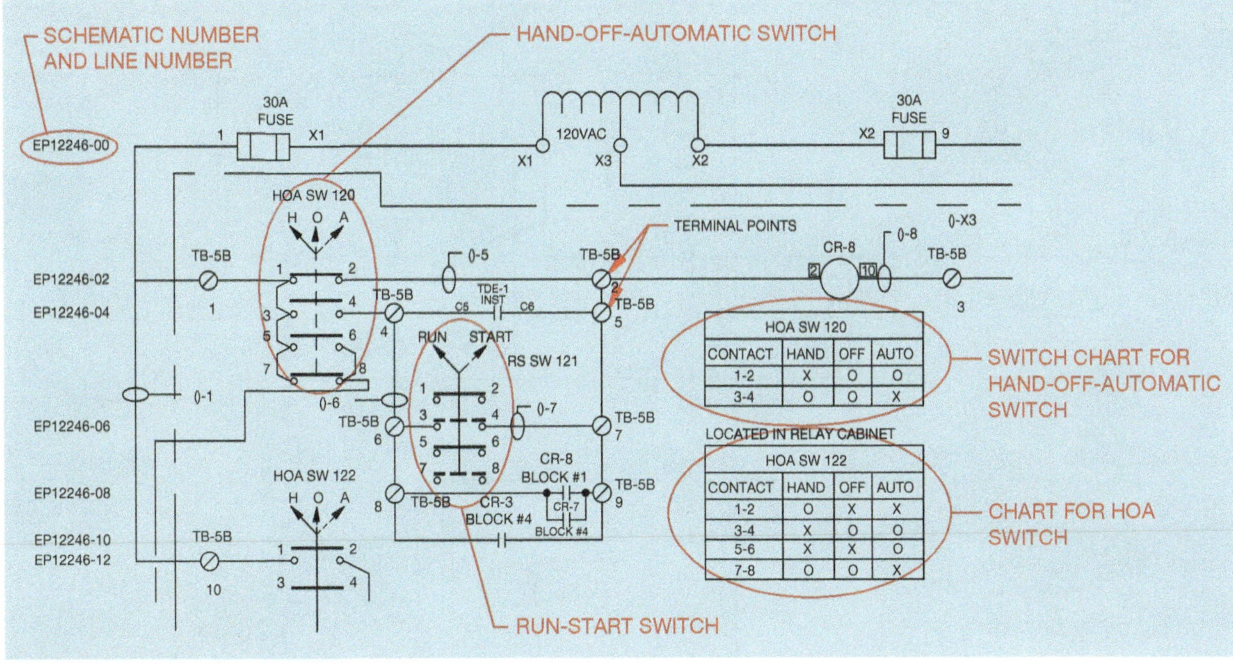

Figure 21–4 Charts show connection of switches in different positions.

at H is shown as a solid line. The lines connected to the other two arrowheads are shown as broken or dashed. The solid line represents the position the switch is set in for the contact arrangement shown on the schematic. The schematic indicates that at the present time there is a connection between terminals 1 and 2, and no connection between terminals 3 and 4. This is consistent with the contact chart for this switch.

7. Locate the run-start switch (RS SW 121) to the right and below (HOA SW 120). A contact chart is not shown for this switch. Because there are only two positions for this switch, a different method is employed to indicate contact position for the different switch positions. Notice that arrowheads at the top of the switch are pointing at the run and start positions. The line drawn to the run position is solid and the line drawn to the start position is shown as broken or dashed. The schematic shows a solid line between switch terminals 1 and 2, and 5 and 6. Dashed lines are shown between terminals 3 and 4, and 7 and 8. When the switch is set in the run position, there is a connection between terminals 1 and 2, and 5 and 6. When the switch is in the start position there is connection between terminals 3 and 4, and 7 and 8.

8. On line 02, there are three terminals marked TB-5B. These labels indicate terminal block points. Locate the terminal with 2 drawn beside it. This wire position is located on screw terminal #2 of terminal block 5B. Another terminal block point is shown below it. This terminal location is screw terminal #5 of terminal block 5B.

9. Find relay coil CR-8 on line 02 (Figure 21–5). CR stands for control relay. Notice that the numbers 2 and 10 on each side of the coil are shown inside a square box. The square box indicates that these are terminal numbers for the relay and should not be confused with wire numbers. Terminals 2 and 10 are standard coil connections for relays designed to fit into an 11 pin tube socket. If you were trying to physically locate this relay, the pin numbers would be a strong hint as to what you are trying to find.

10. Beside pin number 10 of relay coil CR-8 is a circle with a line connected to it. The line goes to a symbol that looks like ( )-8. This indicates a test point. Test

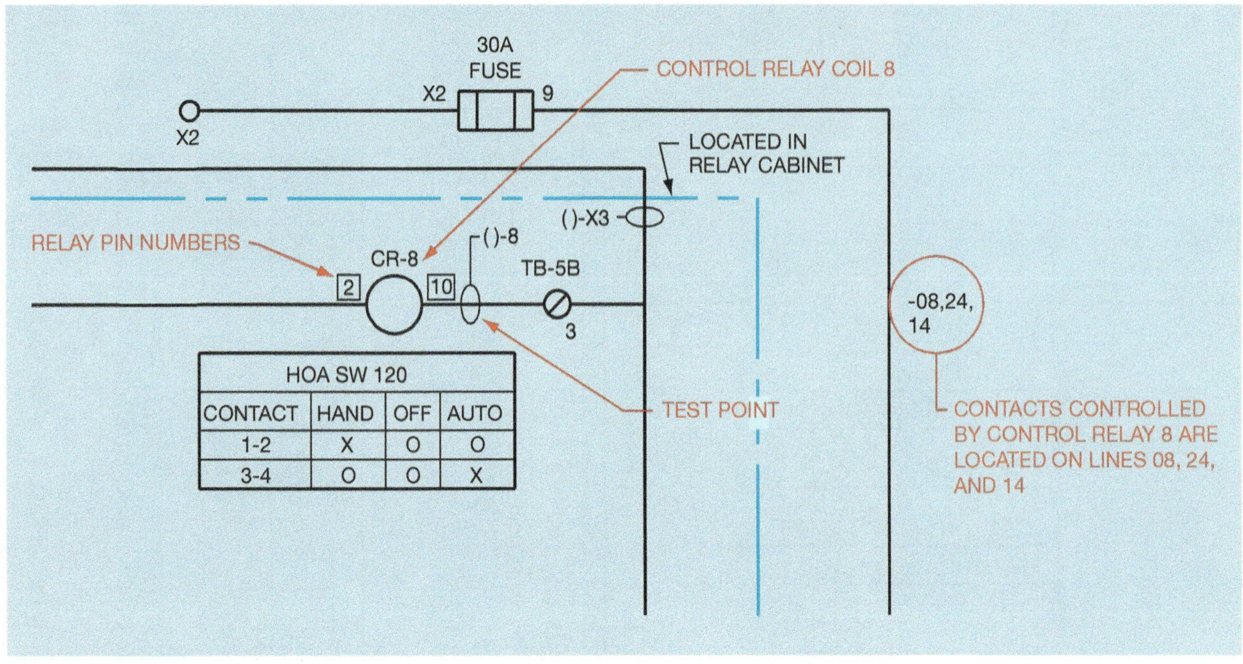

Figure 21–5 Notations on the schematic can be a great help.

points are often placed at strategic points to aid in troubleshooting when necessary.

11. At the far right-hand side of line 02 is the notation (−08, 24, 14). These numbers indicate the lines on the schematic where contacts controlled by relay coil CR-8 can be found. Find the contacts labeled CR-8 on these lines of the schematic.

12. Locate coil CR-7 on line 14 (Figure 21–6). At the far right-hand side find the notation (−14, 08, EP12248 156). This notation again indicates the places where contacts controlled by coil CR-7 can be found. CR-7 contacts are located on lines 14 and 08 in this schematic and on line 156 of Electrical Print #12248.

13. At the right side of the schematic between lines 00 and 02 is the notation (LOCATED IN RELAY CABINET). An arrow is pointing at a dashed line. This gives the physical location of such control components as starters, relays, and terminal blocks. Push buttons, HOA switches, pilot lights, and so on are generally located on a control terminal where an operator has access to them.

These are notations that are common to many industrial control schematics. Nothing is standard, however. Many manufacturers use their own numbering and notation system, which is specific to their company. Some use the NEMA symbols that are discussed in this textbook and others do not. With practice and an understanding of basic control logic and schematics, most electricians can determine what these different symbols mean by the way they are used in a circuit. The old saying "Practice makes perfect" certainly applies to reading schematic diagrams.

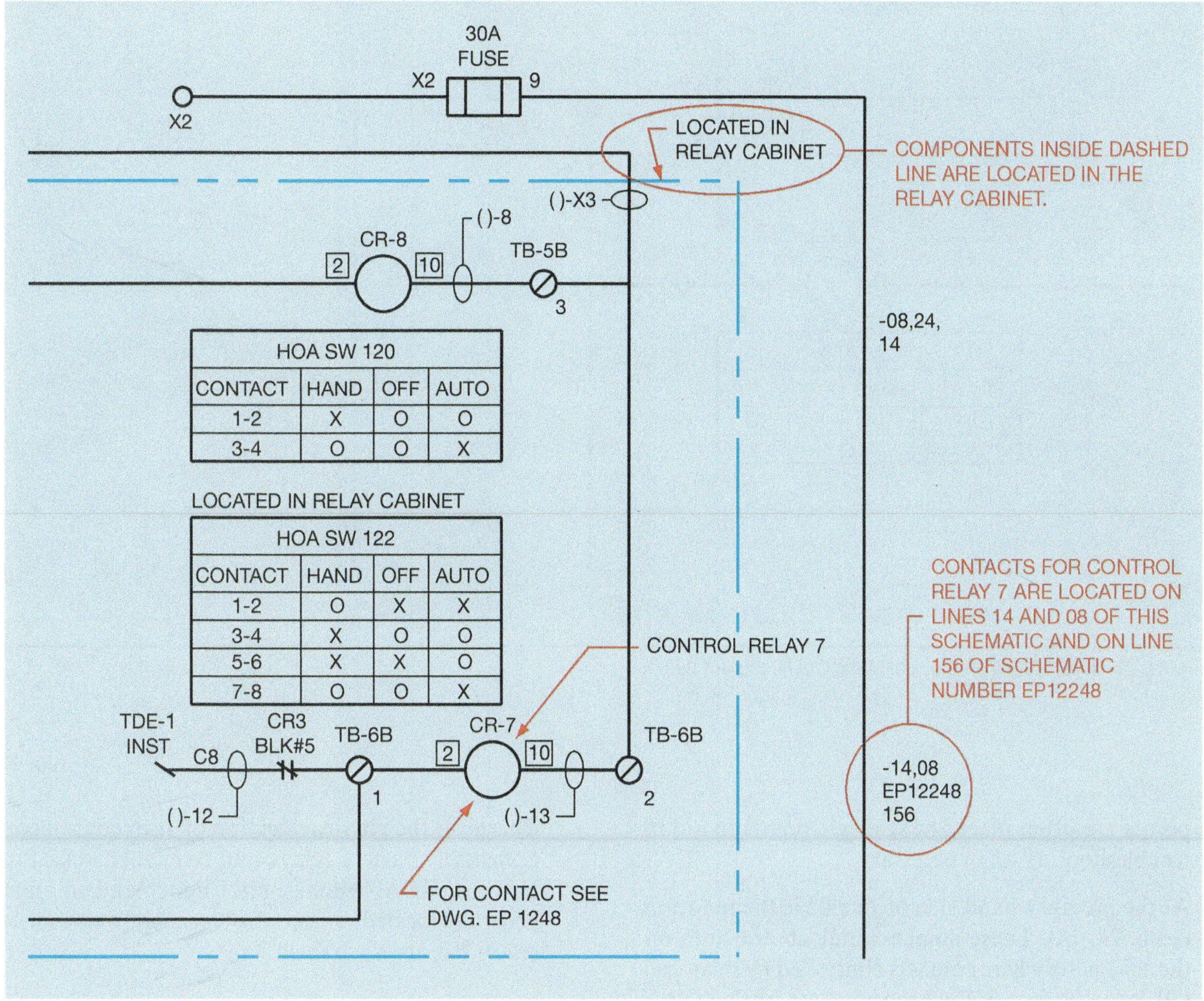

Figure 21–6 Multi-page schematics generally employ some type of notation to indicate where the contacts controlled by a relay are located.

Review Questions

Refer to Figure 21–1 to answer the following questions.

1. When switch HOA SW 122 is in the off position, which contacts have connection between them?
2. How much voltage will be applied to coil 1CR when it is energized?
3. Referring to switch (RS SW 123), in what position must the switch be set to make connection between terminals 3 and 4?
4. What are the terminal numbers for the two normally open spare contacts controlled by coil 2CR?
5. How much voltage is applied to coil CR-7 when it is energized?
6. What contact(s) are located between screw numbers 8 and 9 of terminal block 5B?
7. Relay coil CR-7 is located between what terminal block and screw numbers?
8. Assume that HOA SW 120 has been set in the auto position. List four ways by which coil CR-8 could be energized.
9. In what position must switch SW 123 be set to make connection between terminals 3 and 4?
10. If one of the magnetic overload relays should open its contact, how can it be reset?

Chapter 22

INSTALLING CONTROL SYSTEMS

Objectives

After studying this chapter the student will be able to:

- Discuss methods of installing a control system.
- Use wire numbers to troubleshoot a control circuit.
- Explain advantages of using the point-to-point system of connecting a circuit.
- Explain advantages of using terminal strips for connecting a circuit.

There are different ways in which control systems can be installed. Wiring diagrams can sometimes be misleading because they show all the components grouped together in a small area, when in reality they may be located some distance apart. Some control components, such as limit switches, photodetectors, or flow switches, may be located on the machine itself, while other components such as control relays and motor starters may be located inside a cabinet. Control components such as push buttons, pilot lights, and manual switches may be located on a control panel that is convenient for an operator. Components such as float switches and pressure switches will generally be located on tanks.

The control circuit shown in Figure 22–1 is used to fill a tank with water and then add air pressure if needed. A stop-run switch permits power to be supplied to the circuit or not. Float switch FS1 senses when the tank is full of water. A second float switch, FS2, mounted close to the bottom of the tank senses when the water level is low. Two float switches are necessary in this circuit because the tank is to be pressurized. This prevents the use of a rod or chain type float switch because the tank must be sealed. When the water level drops sufficiently, FS2 contacts close and energize the coil of motor starter M. This causes all M contacts to change position. The M load contacts connect the pump motor to the power line and start the pump. The normally open auxiliary contact connected in parallel with float switch FS2 closes to maintain the circuit when the water level begins to rise and the FS2 contact reopens. When the tank is full, the normally closed FS1 contacts open and de-energize coil M. In order to increase the water pressure, an air line is added to the tank (Figure 22–2). If there is not sufficient pressure in the tank, pressure switch PS1 closes and energizes the solenoid valve permitting air to pressurize the tank. When the pressure is sufficient, contact PS1 opens and de-energizes the solenoid valve.

The only time that air pressure should be permitted to enter the tank is when the tank is full of water. Float switch FS1 senses when the tank is full. The normally open contact of float switch FS1 is used to prevent the solenoid coil from energizing unless the tank is full of water.

Notice that two float switches are connected with a dashed line. One is normally closed and marked FS1. The other is shown normally open. Also notice that one end of each of these two float switches is connected to the same electrical point. The

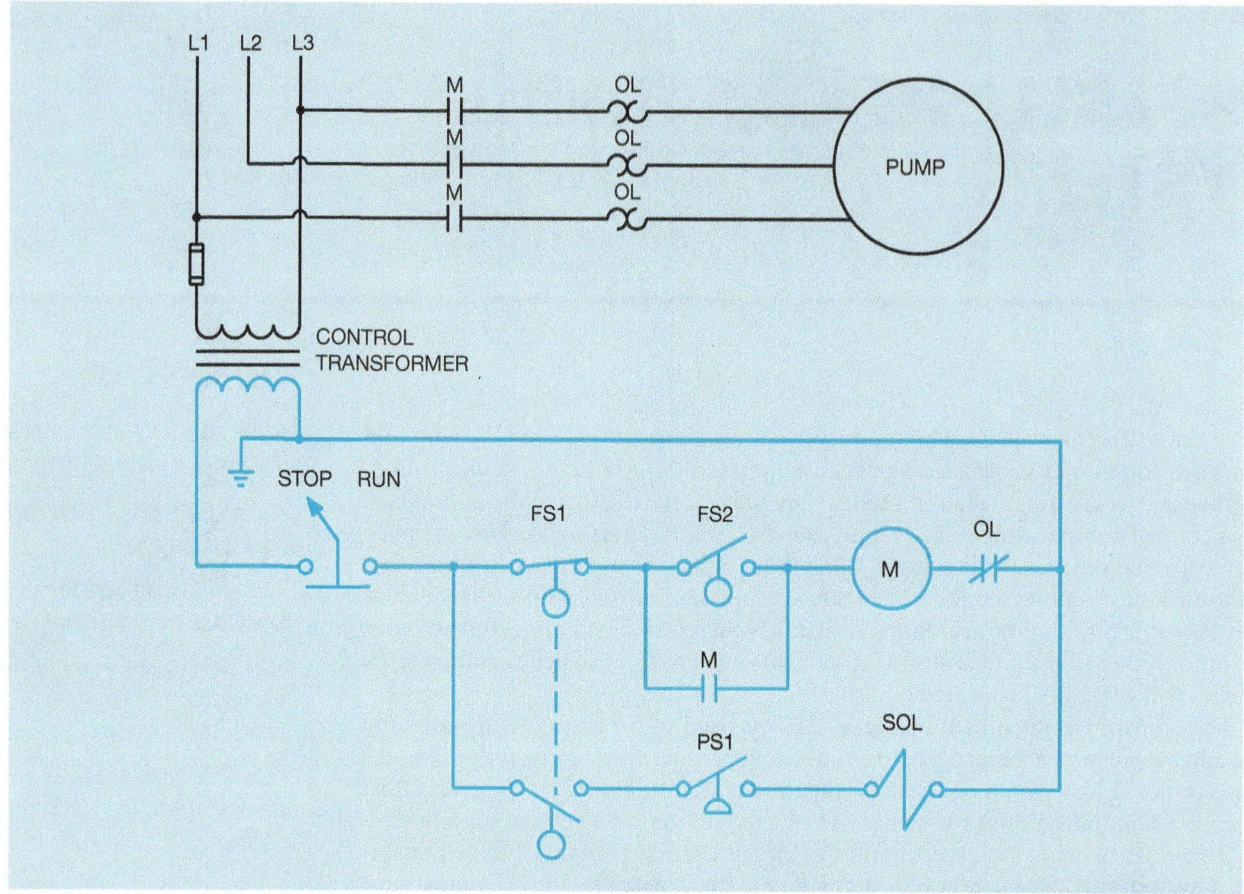

Figure 22–1 Pump control circuit.

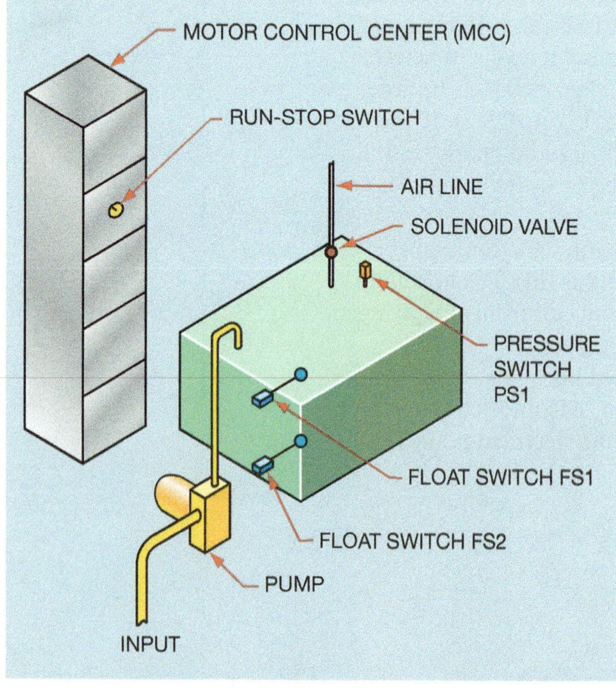

Figure 22–2 Component location.

common terminal of the float switch would be connected to this electrical point. The normally closed section of the switch should connect to one side of float switch FS2 and one side of M auxiliary contact. The normally open section of the FS1 should connect to one side of the pressure switch.

Component Location

As mentioned previously, wiring diagrams can sometimes be misleading because they show components located in close proximity to each other. In reality, the circuit components may be located some distance from each other. The location of the components in the circuit shown in Figure 22–1 is shown in Figure 22–2. The two float switches, the pressure switch, and the solenoid valve are located on the pressurized water tank. The pump is located beside the tank, and the manual run-stop switch, motor starter, and control transformer are located in the motor control center. Conductors are run in conduit from the motor control center to the tank to make actual connection.

Point-to-Point Connection

A couple of methods can be employed to connect the circuit. The first step will be to place wire numbers on the schematic. Wire numbers will be added to the control circuit only. The load side of the circuit will not be numbered. These wire numbers are shown in Figure 22–3. The circuit components are shown in Figure 22–4. The components located inside the dashed lines are located inside the motor control center. Note that wire numbers that correspond to the schematic in Figure 22–3 were added to the circuit components.

One method of connecting this circuit is the point-to-point method. Components are connected from one point to another depending on the proximity of one component to another. This method generally results in a saving of wire. Some of the connections are made inside the MCC and others must be made away from the MCC. The connections made inside the MCC are shown with dashed lines and the ones outside the MCC are made with solid lines. When making connections, wire numbers should be placed on the wire at any point of termination. This greatly aids in troubleshooting the circuit. Each component should also be labeled with a tag to identify the component. Point-to-point connection of this circuit is shown in Figure 22–5. Note that notation is used to identify the wire number.

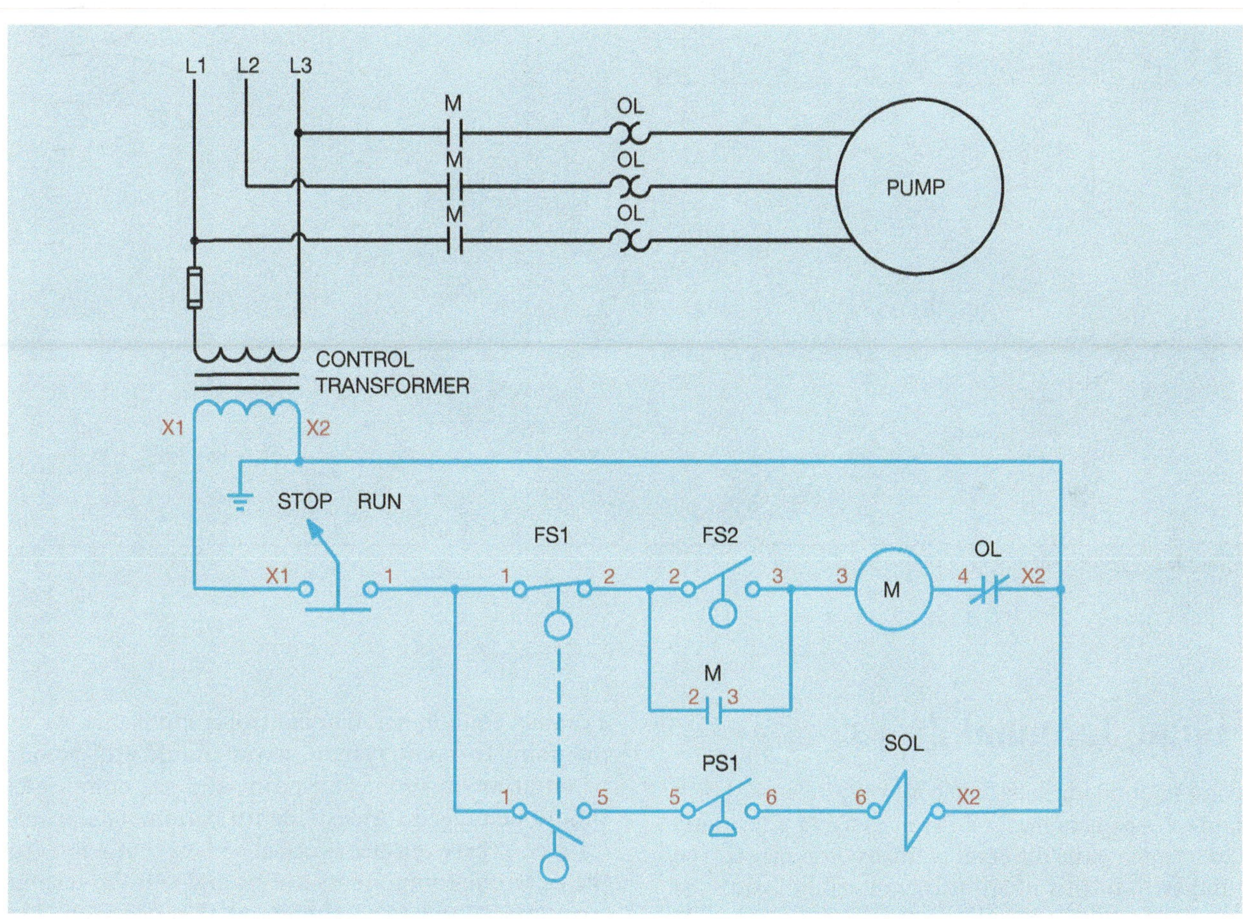

Figure 22–3 Adding wire numbers to the schematic.

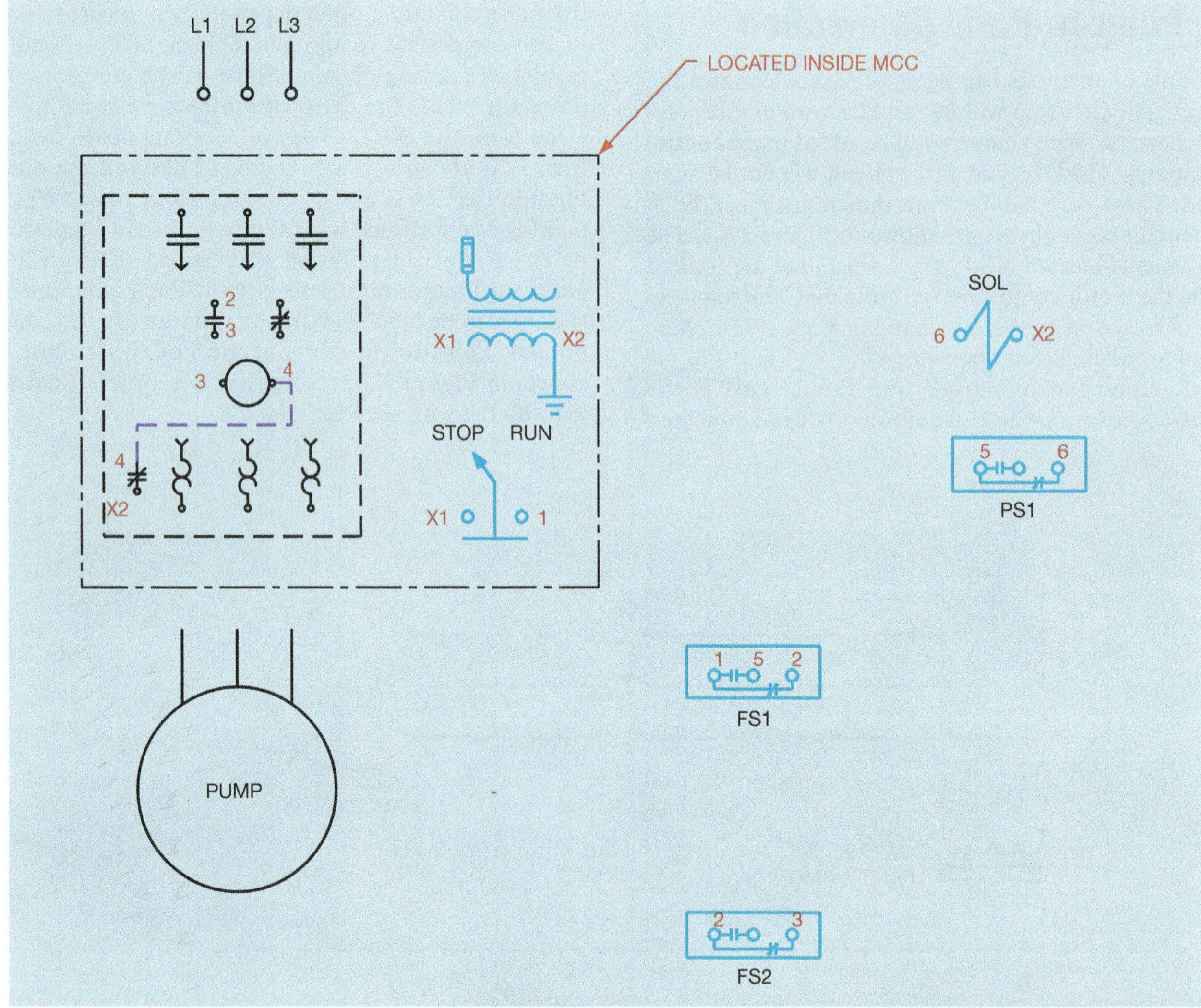

Figure 22–4 Circuit components.

Using Terminal Strips

A second method for installing control systems is shown in Figure 22–6. This method involves using a terminal strip to terminate the different control components. The terminal strip in this illustration is located inside the motor control center, but they are often located inside a cabinet containing other control components. As with the point-to-point system, wires should still be numbered at any termination point and the components should be tagged for easy identification. The primary difference between the two methods is that wires from the different components are brought to the terminal strip and connection is made at that location. Note

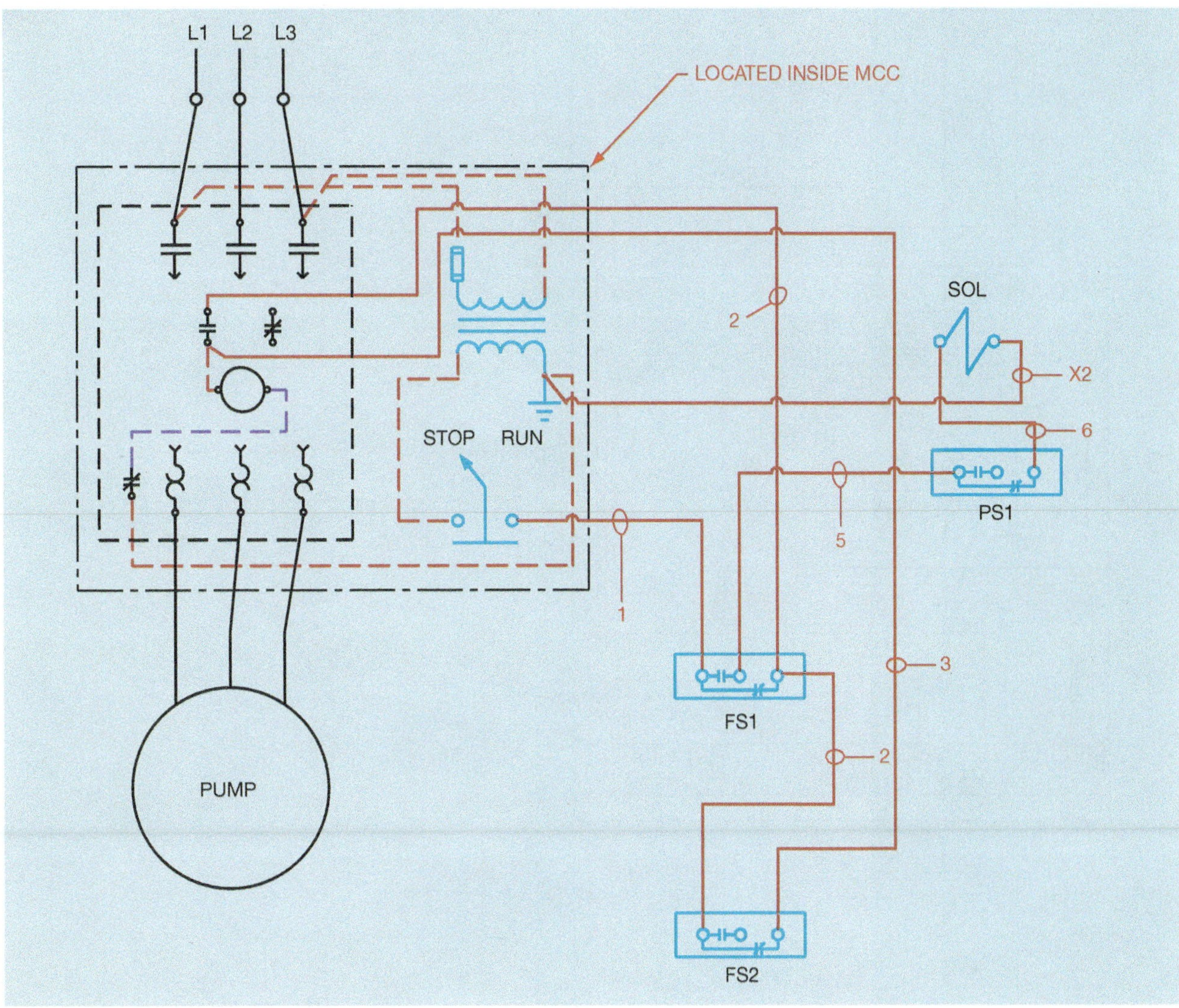

Figure 22–5 Point-to-point wiring method.

that some wiring is still made from point-to-point inside the motor control center. Although connecting components to a terminal strip generally involves using a greater amount of wire, it can be the less expensive method when it is necessary to troubleshoot the circuit. If the electrician needs to know if the normally open switch on float switch FS1 is open or closed, he or she can test it from the terminal strip without having to find FS1, remove the cover, and check the condition of the contacts.

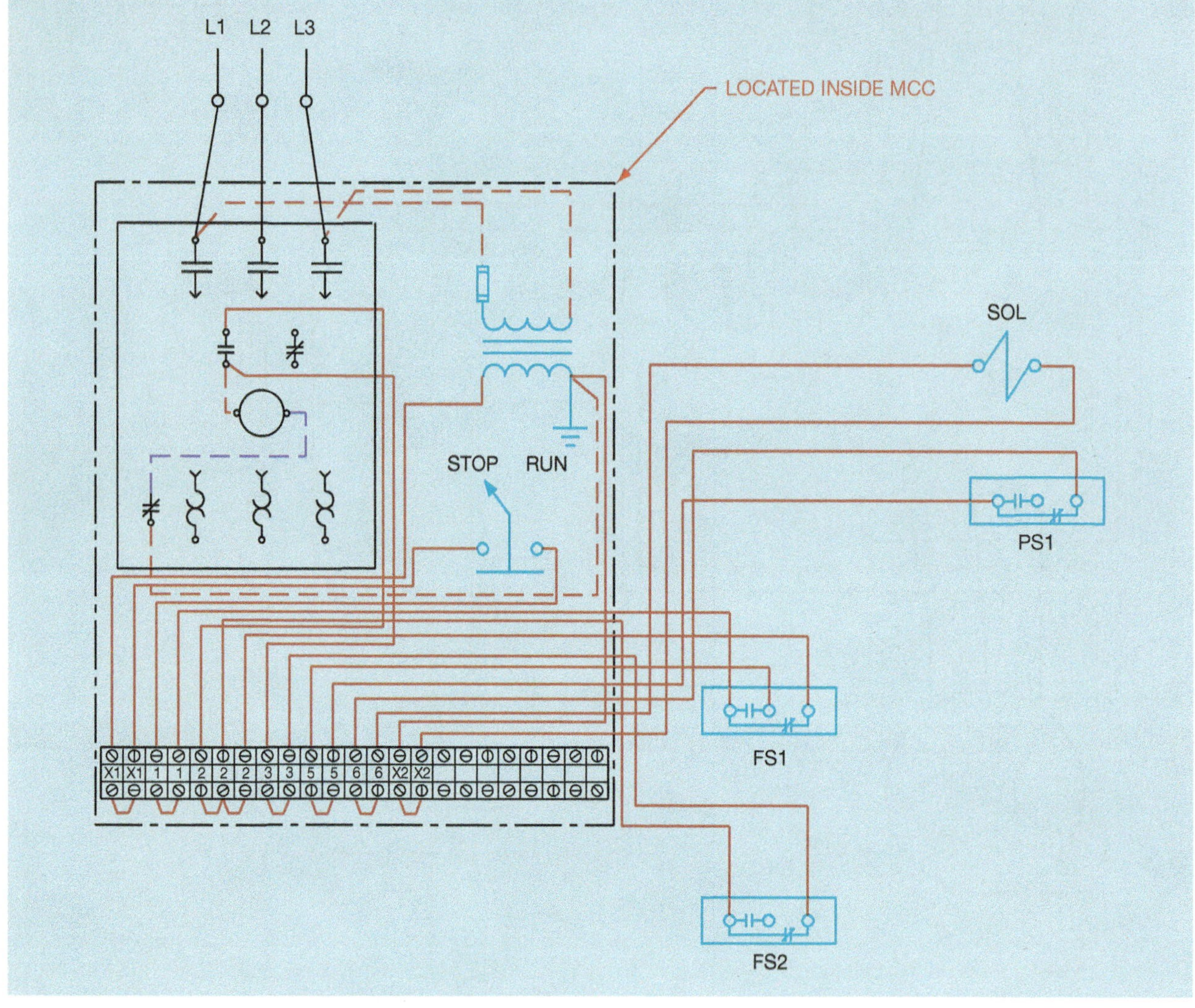

Figure 22–6 Connections are made to a terminal strip.

Review Questions

1. What is an advantage of the point-to-point method of connecting circuit components?
2. When connecting a control system, what should be done each time a wire termination is made to any component?
3. What should be done to each component to help identify it?
4. What is the disadvantage of wiring components to a terminal strip?
5. What is the main advantage of making connections at a terminal strip?
6. Refer to the circuit shown in Figure 22–1. Should float switch FS2 be wired as normally open or normally closed? Explain your answer.
7. What does the dashed line between the two float switch contacts labeled FS1 indicate?

Section 4

STARTING AND BRAKING METHODS

Chapter 23

ACROSS-THE-LINE STARTING

Objectives

After studying this chapter the student will be able to:

» Discuss across-the-line starting for alternating current motors.

» Discuss across-the-line starting for direct current motors.

Across-the-line starting is the simplest of all starting methods. It is accomplished by connecting the motor directly to the power line. Across-the-line starting is used for both alternating current and small direct current motors. The size motor that can be started across-the-line can vary from one area to another, depending on the power limitations of the electrical service. In heavily industrialized areas, motors of over 1000 horsepower are often started across-the-line. In other areas, motors of less than 100 horsepower may require some type of starter that limits the amount of starting current.

Three-Phase Alternating Current Motors

Three-phase motors are by far the most common type of motor found throughout industry. They can be divided into three main types: *squirrel cage*, *wound rotor*, and *synchronous*. Of the three types the squirrel cage is the most common. The components of a three-phase squirrel cage motor are shown in Figure 23–1. A typical three-phase squirrel cage motor is shown in Figure 23–2.

Three-phase motors have several advantages over other types of electric motors:

- They produce more horsepower per case size and weight than other motors.
- They do not need the separate set of start windings that is required by many single-phase motors.
- They do not require a commutator and brushes.
- They exhibit very long life spans when operated within their power rating.
- They are more energy efficient than single-phase motors.

Regardless of the type, all three-phase motors operate on the principle of a rotating magnetic field. The stator winding consists of three separate winding arranged around the inside of a cylindrical core, Figure 23–3. The example in Figure 23–3 shows six separate pole pieces, two for each phase. This drawing is intended to help illustrate how the three windings are arranged around the stator. In actual practice, the inside of the stator is relatively smooth as shown in Figure 23–4.

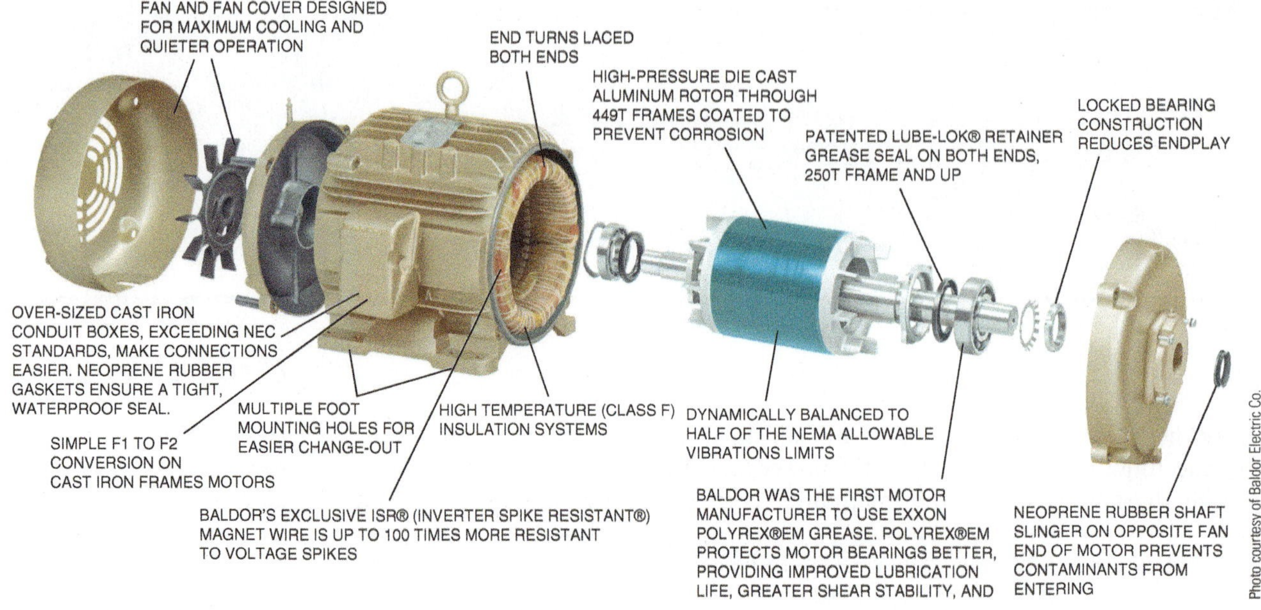

Figure 23–1 Components of a three-phase squirrel cage motor.

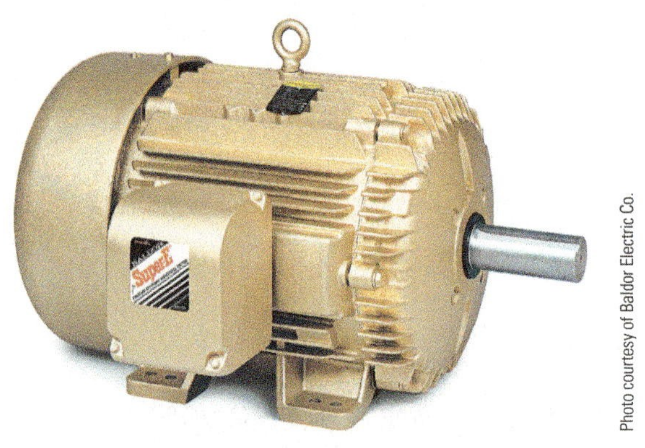

Figure 23–2 A three-phase squirrel cage motor.

When power is applied to the stator of the three-phase motor, a magnetic field will rotate around the inside of the stator core. There are three factors that cause the field to rotate:

1. The arrangement of the three windings around the inside of the stator core.
2. The fact that the three voltages are 120° out-of-phase with each other.
3. The fact that the voltages change polarity at regular intervals.

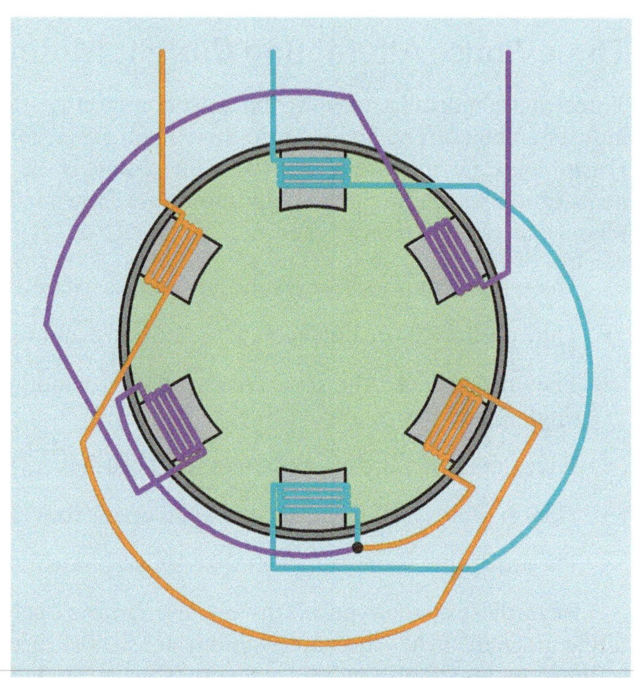

Figure 23–3 Stator winding of a two-pole, three-phase motor.

The speed of the rotating magnetic field is called *synchronous speed*, and is determined by the number of stator poles per phase and the frequency of the applied

Photo courtesy of Baldor Electric Co.

Figure 23–4 Stator windings form a smooth inner surface.

voltage. The formula for determining synchronous speed is:

$$s = \frac{120f}{P}$$

Where s = speed in RPM
f = frequency in Hz.
P = number of stator poles per phase.

EXAMPLE: What is the synchronous speed of a three-phase motor that contains six poles per phase and is connected to 60 Hz?

$$s = \frac{120 \times 60}{6}$$

$$s = \frac{7200}{6}$$

$$s = 1200\ RPM$$

All three types of three-phase motors use the same basic stator winding to produce a rotating magnetic field. The differences among the motors are the types of rotors or rotating members. The rotors of both wound rotor and synchronous motors contain windings. The windings of a wound-rotor motor are connected to three slip rings mounted on the motor shaft and brushes are used to allow different values of resistance to be connected to the rotor circuit. The rotor winding of a synchronous motor must have direct current applied to it in order to produce a rotor magnetic field.

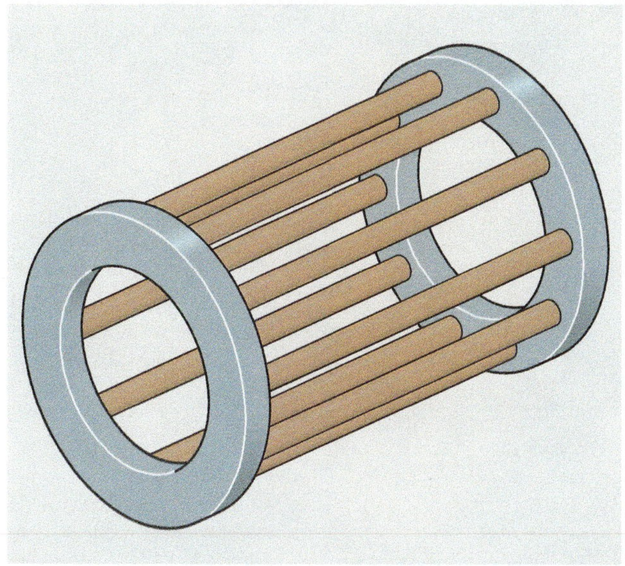

Figure 23–5 Basic squirrel cage rotor without laminations.

Photo courtesy of Baldor Electric Co.

Figure 23–6 When laminations are added to the squirrel cage winding, the rotor looks like a solid piece of metal.

The most common type of three-phase motor is the squirrel cage. They are so named because the actual rotor winding is made by joining metal bars together at each end, Figure 23–5. With the laminations removed, the winding resembles the circular wheel placed inside the cage of a small animal such as a squirrel or hamster. The animals can run inside the wheel for exercise. When laminations are added, the rotor looks like a solid metal object, Figure 23–6. If the rotor is cut in half, Figure 23–7, the squirrel cage winding can be seen.

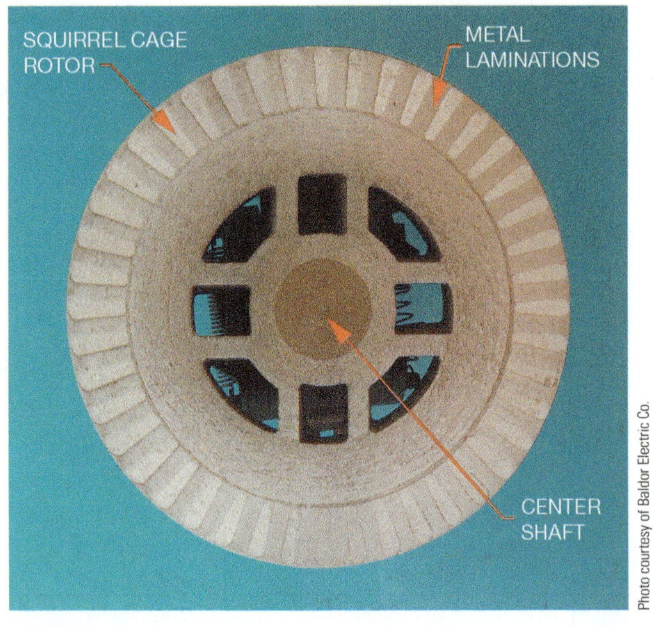

Figure 23–7 The squirrel cage winding is visible if the rotor is cut in half.

The type of bars used to construct the rotor has a great effect on the operating characteristics of the motor. The type of rotor is identified by a code letter on the nameplate of a motor. Code letters range from A through V. Table 430.7(B) of the *National Electrical Code®* lists these code letters (Figure 23–8).

It may be sometimes necessary to determine the amount of in-rush current when installing a motor. This is especially true in areas where the power company limits the amount of current it can supply. In-rush current is referred to a *locked rotor current*, because it is the amount of current that would flow if the rotor were to be locked so it could not turn and the power turned on. To determine in-rush current for a motor, find the code letter on the motor nameplate. Do not confuse the rotor code letter with the NEMA code letter found on many motors. The nameplate will generally state one as CODE and the other as NEMA CODE. Once the code letter has been determined, it is possible to calculate the starting current for the motor. A simple across-the-line starting circuit for an AC motor is shown in Figure 23–9.

Code letters	Kilovolt-Amperes per Horsepower with Locked Rotor
A	0–3.14
B	3.15–3.54
C	3.55–3.99
D	4.0–4.49
E	4.5–4.99
F	5.0–5.59
G	5.6–6.29
H	6.3–7.09
J	7.1–7.99
K	8.0–8.99
L	9.0–9.99
M	10.0–11.19
N	11.2–12.49
P	12.5–13.99
R	14.0–15.99
S	16.0–17.99
T	18.0–19.99
U	20.0–22.39
V	22.4 and up

Figure 23–8 Table 430.7(B) of the *NEC®*.

Direct Current Motors

Direct current motors do not depend on inductive reactance to limit current flow as alternating current motors do. Initial in-rush current is limited by the resistance of the windings. When the armature begins to turn, counter EMF is developed in the armature. Counter EMF is the main current limiting force in a direct current motor. Small DC motors can be started across-the-line because

EXAMPLE: Assume a 200 hp three-phase squirrel cage motor is connected to 480 volts and has a code letter J. Table 430.7(B) lists 7.1 – 7.99 kVA per horsepower for a motor with code letter J. To determine maximum starting current, multiply 7.99 by the horsepower.

$$7.99 \times 200 = 1598 \text{ kVA}$$

Because the motor is three phase, the formula shown can be used to calculate the starting current.

$$I = \frac{VA}{E \times \sqrt{3}}$$

$$I = \frac{1{,}598{,}000 \text{ VA}}{480 \times 1.732}$$

$$I = 1922.15 \text{ amperes}$$

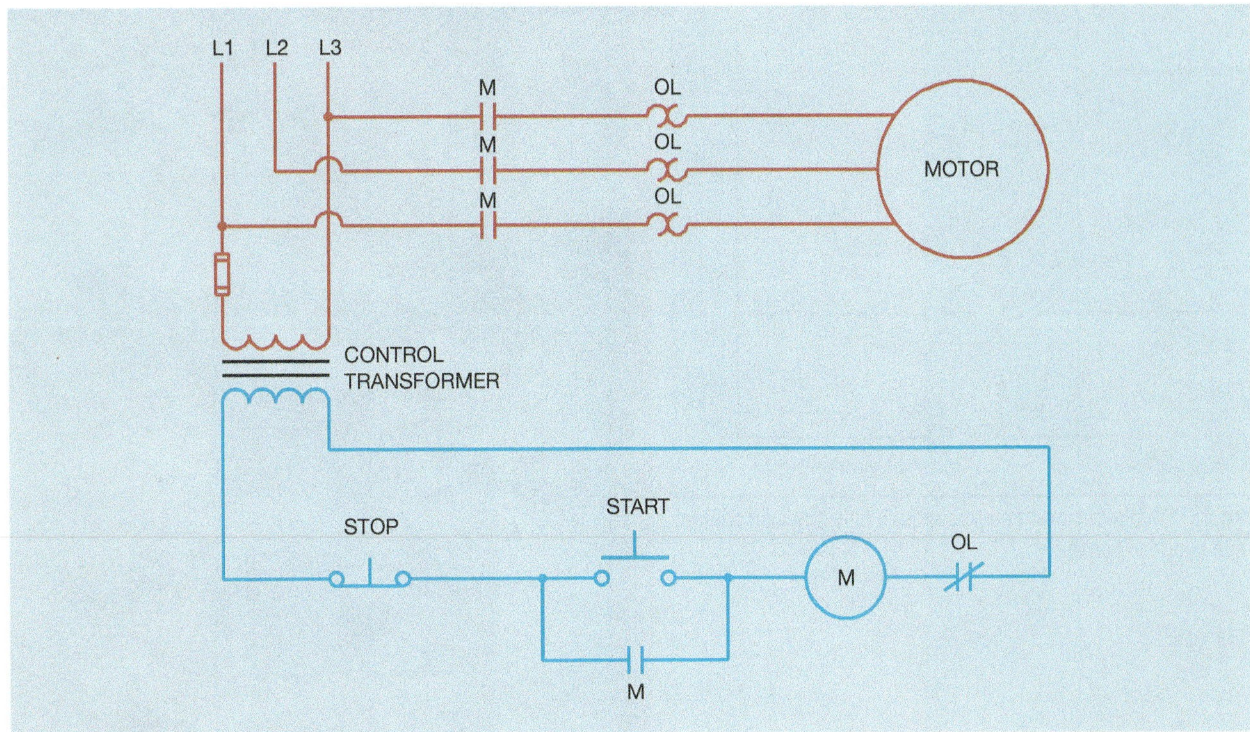

Figure 23–9 Basic circuit for across-the-line starting of an AC motor.

they have low inertia and the armature will rapidly gain enough speed to produce the counter EMF needed to limit current.

Contactors intended to connect direct current motors to the line often have arc suppression and magnetic blowout coils as discussed in Chapter 5. Some DC contactors contain two separate coils (Figure 23–10). One is the holding coil and remains energized as long as power is applied to the contactor. The other coil is called the pick-up coil and is used to increase the magnetic field strength when the contactor is first energized. When the armature makes connection with the electromagnet, a limit switch opens the circuit to the pick-up coil. The pick-up coil is intended for momentary duty and not continuous operation. If the pick-up is not disconnected when the contactor is energized, it will generally be damaged by excessive heat.

Direct current motors require only one load contact to connect or disconnect them to the power line and one overload sensing device (Figure 23–11). Table 430.37 of the *National Electrical Code®* states that single-phase AC and DC motors require protection in only one conductor in an ungrounded system. In a grounded system, the current sensing device must be in the ungrounded conductor. A typical direct current motor starter is shown in Figure 23–12. Note the arc shield around the contact and the blowout coil located at the top of the contactor.

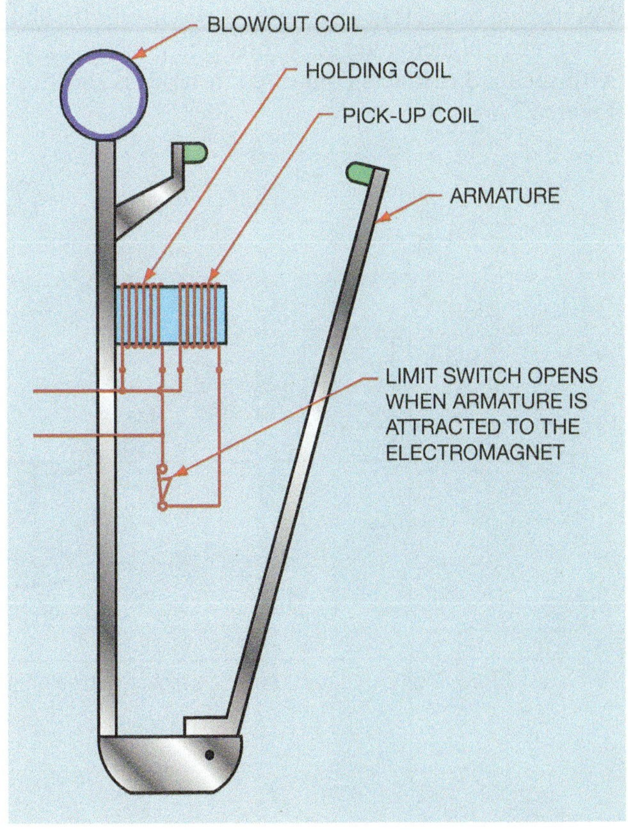

Figure 23–10 Some DC contactors contain a pick-up coil and holding coil.

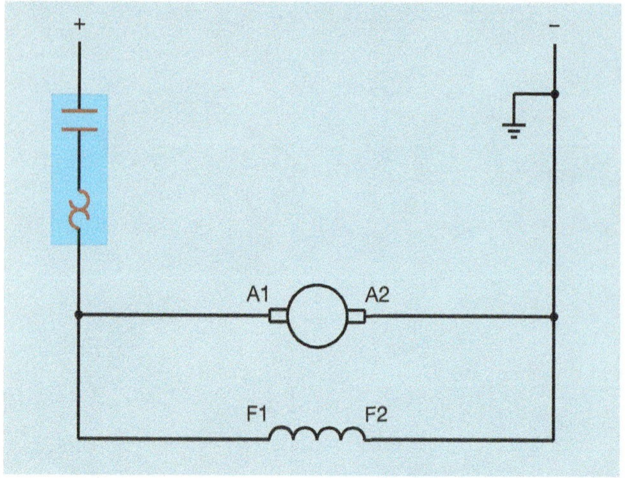

Figure 23–11 Direct current motors require only one load contact and one overload sensor.

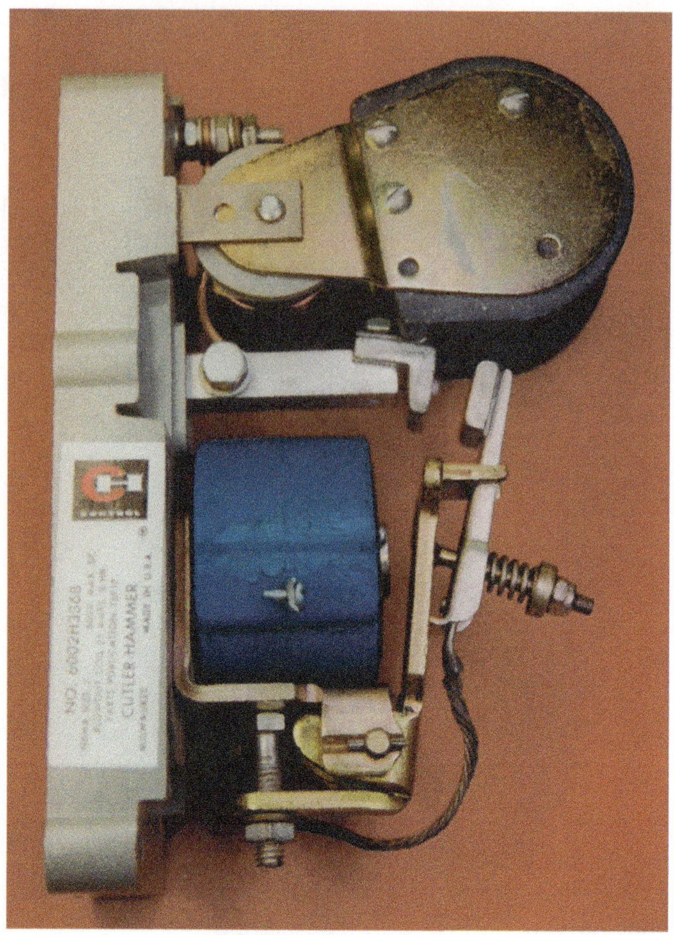

Figure 23–12 Typical direct current motor starter.

Direct current starters and contactors that do not utilize arc suppressors or magnetic blowout coils often use multiple contacts to break the circuit (Figure 23–13). Connecting contacts in series gives the contactor the ability to break higher voltages. A three-phase motor starter with the load contacts connected in series is shown in Figure 23–14.

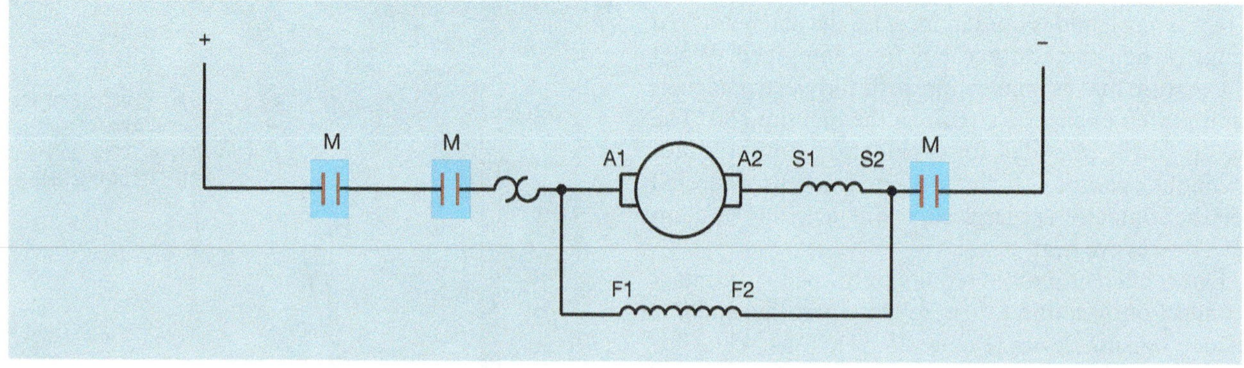

Figure 23–13 Multiple contacts provide a higher voltage rating.

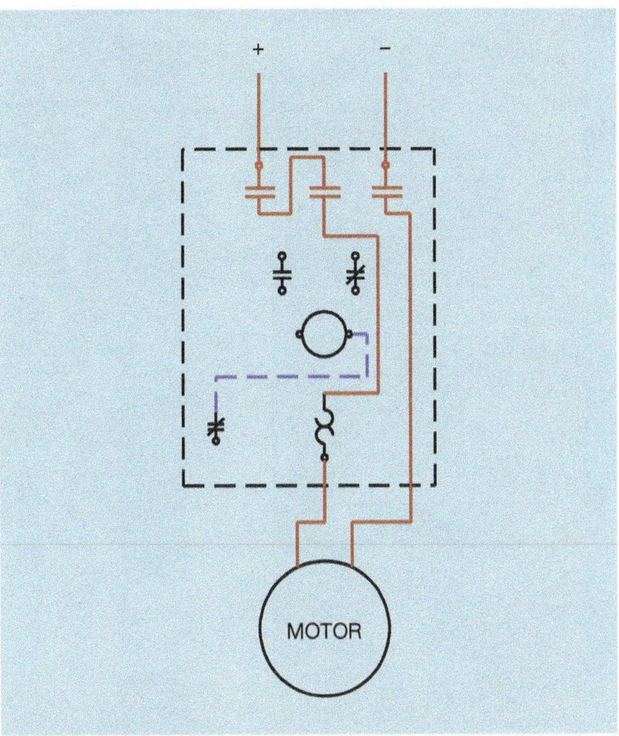

Figure 23–14 Load contacts are connected in a series.

Review Questions

1. How is across-the-line starting accomplished?
2. How many overload sensing devices are required for single-phase AC and DC motors?
3. If a direct current motor is connected to a grounded DC power system, should the over current sensing device be placed in the grounded or ungrounded conductor?
4. Some direct current contactors contain two coils, the hold coil and the pick-up coil. Explain the function of each coil.
5. What method is used to disconnect the pick-up coil of a DC contactor?

Chapter 24

RESISTOR AND REACTOR STARTING FOR AC MOTORS

Objectives

After studying this chapter the student will be able to:

- Discuss why resistor and reactor starting is used for starting alternating current motors.
- Discuss how resistor starting is used with direct current motors.
- Discuss different methods of accomplishing resistor and reactor starting.
- Explain the difference between resistor and reactor starting.

There are conditions where it is not possible to start a motor across-the-line due to excessive starting current or excessive torque. Some power systems are not capable of providing the high initial currents produced by starting large horsepower motors. When this is the case, some means must be employed to reduce the amount of starting current. Two of the most common methods are resistor and reactor starting. Although these two methods are very similar, they differ in the means used to reduce the amount of starting current.

Resistor Starting

Resistor starting is accomplished by connecting resistors in series with the motor during the starting period (Figure 24–1). When the START button is pressed, motor starter coil M energizes and closes all M contacts. The three M load contacts connect the motor and resistors to the line. Because the resistors are connected in series with the motor, they limit the amount of in-rush current. When the M auxiliary contact connected in series with coil TR closes, the timer begins its timing sequence. At the end of the time period, timed contact TR closes and energizes contactor R. This connection causes the three R contacts connected in parallel with the resistors to close. The R contacts shunt the resistors out of the line and the motor is now connected to full power.

The circuit in Figure 24–1 uses time delay to shunt the resistors out of the circuit after some period of time. Time delay is one of the most popular methods of determining when to connect the motor directly to the power line because it is simple and inexpensive, but it is not the only method. Some control circuits sense motor speed to determine when to shunt the resistors out of the power line (Figure 24–2). In the illustration, permanent magnets are attached to the motor shaft and a Hall effect sensor determines the motor speed. When the motor speed reaches a predetermined level, contactor R energizes and shunts the resistors out of the line (Figure 24–3).

Another way of determining when to shunt the resistors out of the circuit is by sensing motor current. Current transformers are used to sense the amount of motor current (Figure 24–4). In this circuit the current sensor contacts are normally closed. When the motor starts, high current causes the sensor contacts to open. An on-delay timer provides enough time delay to permit the motor to begin starting before contactor coil R can energize. This timer is generally set for a very short time delay. As the motor speed increases, the current drops. When the current drops to a low enough level, the current sensor contact recloses and permits contactor R to energize.

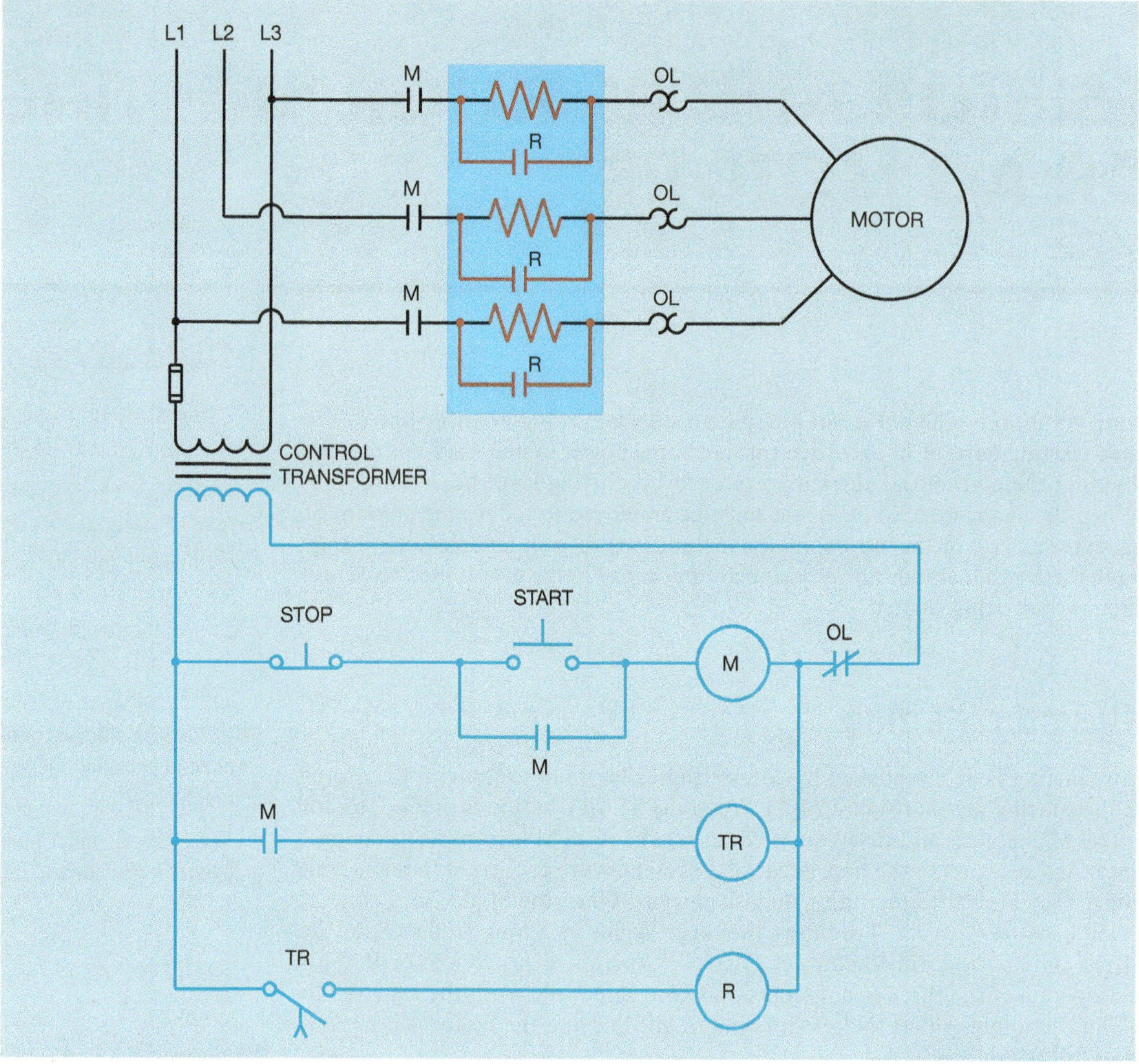

Figure 24–1 Resistor starting circuit.

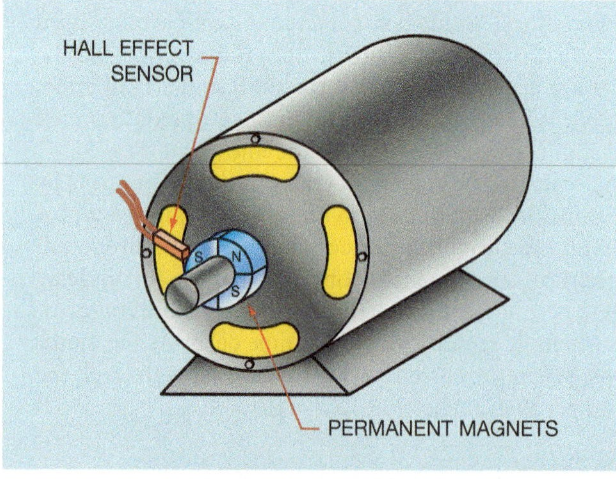

Figure 24–2 Hall effect sensor is used to determine motor speed.

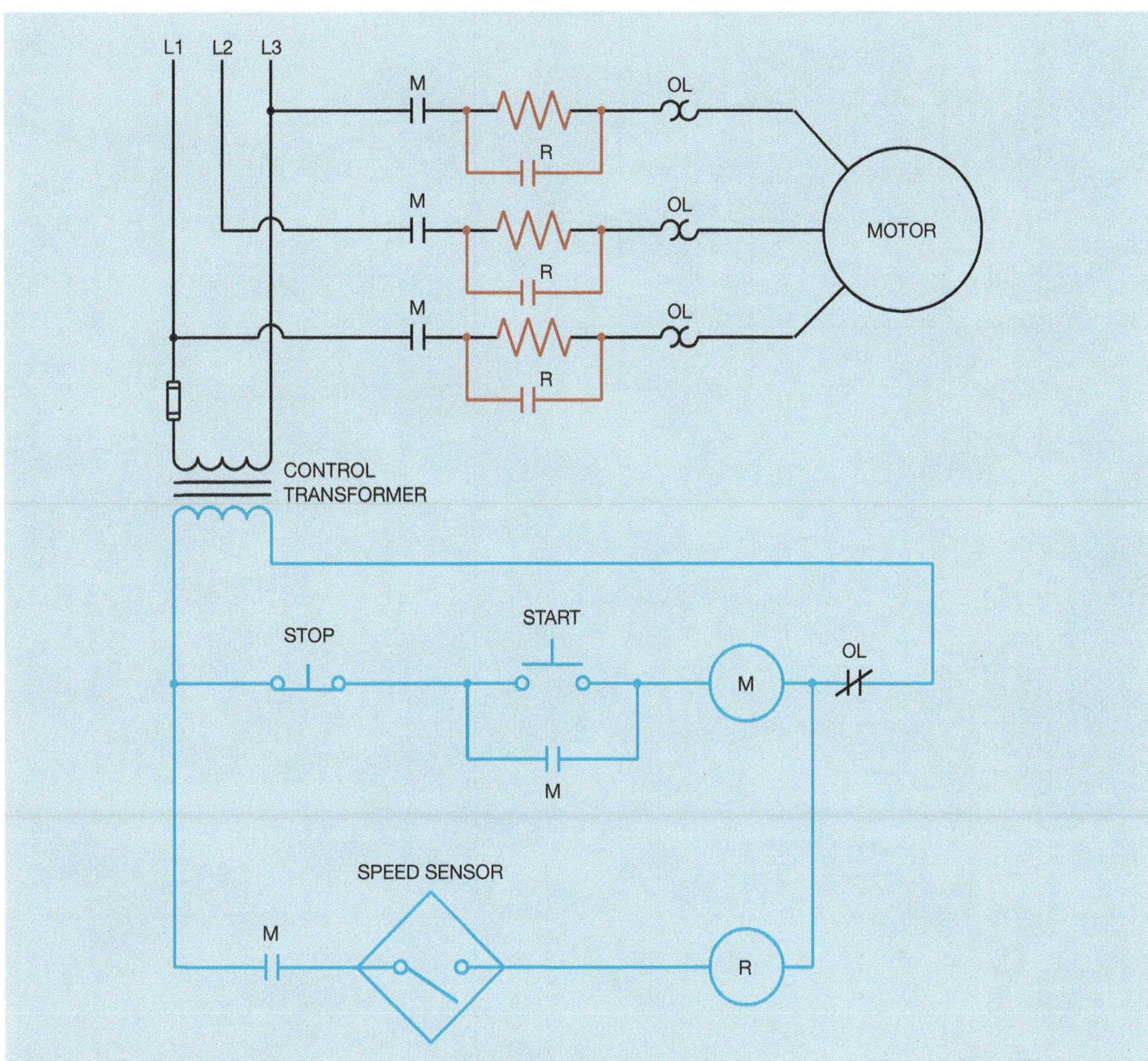

Figure 24–3 Speed sensor is used to shunt resistors out of the circuit.

Reactor Starting

A reactor starter is the same basic control except that reactors or choke coils are used to limit in-rush current instead of resistors (Figure 24–5). Reactors limit in-rush current with inductive reactance instead of resistance. Reactors have an advantage in current limiting circuits because of the rise time of current in an inductive circuit. In a resistive circuit, the current will reach its full Ohm's law value instantly. In an inductive circuit, the current must rise at an exponential rate (Figure 24–6). This exponential rise time of current further reduces the in-rush current.

Step Starting

Some resistor and reactor starters use multiple steps of starting. This is accomplished by tapping the resistor or reactor to provide different values, of resistance or inductive reactance (Figure 24–7). When the START button is pressed, M load contacts close and connect the motor and inductors to the line. The M auxiliary contact closes and starts timer TR1. After a time delay, TR1 contact closes and energizes S1 coil. This causes half of the series inductor to be shunted out, reducing the inductive reactance connected in series with the motor.

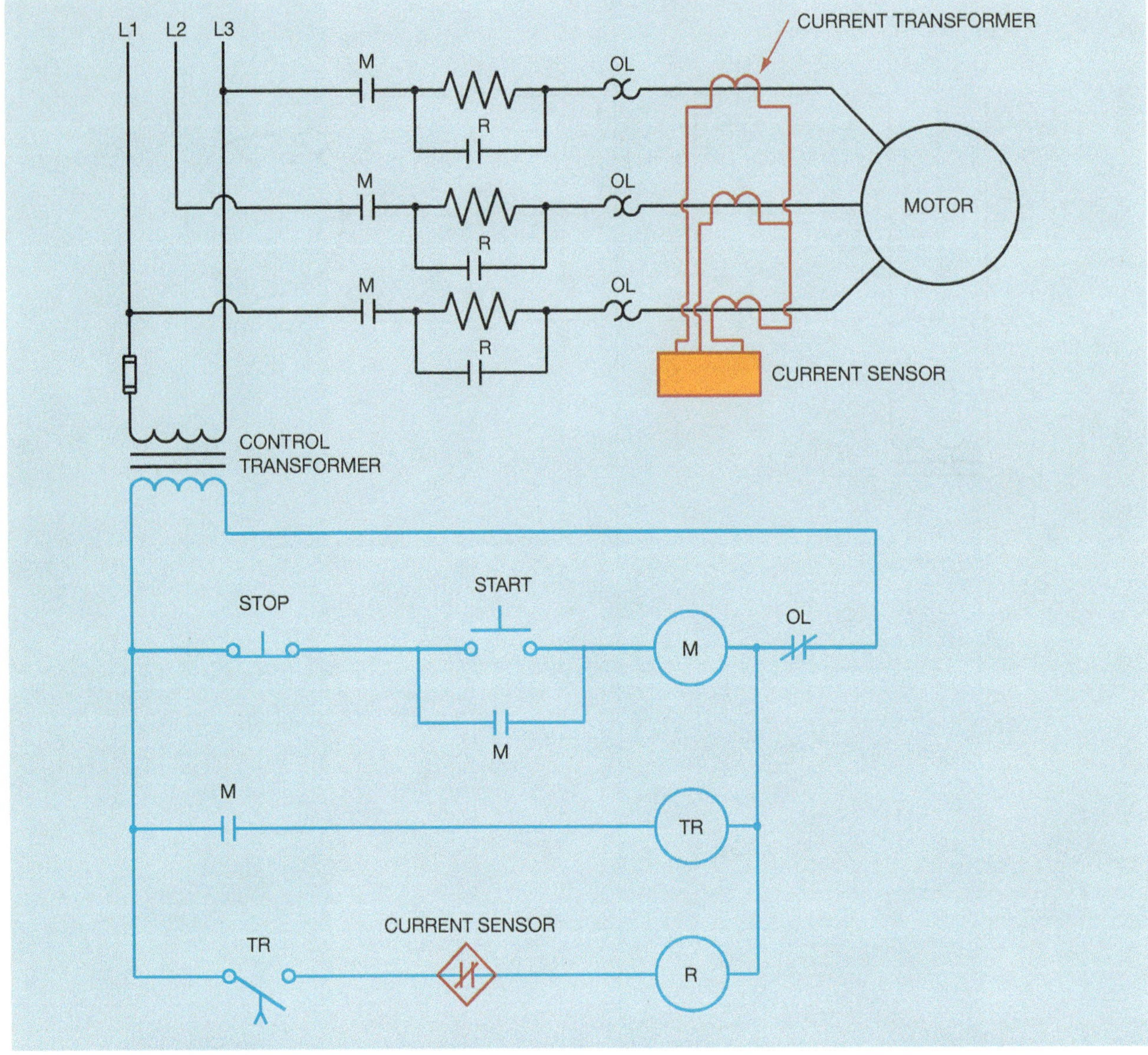

Figure 24–4 Current sensor determines when resistors are to be shunted out of the line.

Motor current increases, causing the motor speed to increase. The S1 auxiliary contact closed at the same time causing timer TR2 to start its timing sequence. When TR2 contact closes, contactor S2 energizes, causing all of the inductance to be shunted out. The motor is now connected directly to the power line. Some circuits may use several steps of starting depending on the circuit requirements.

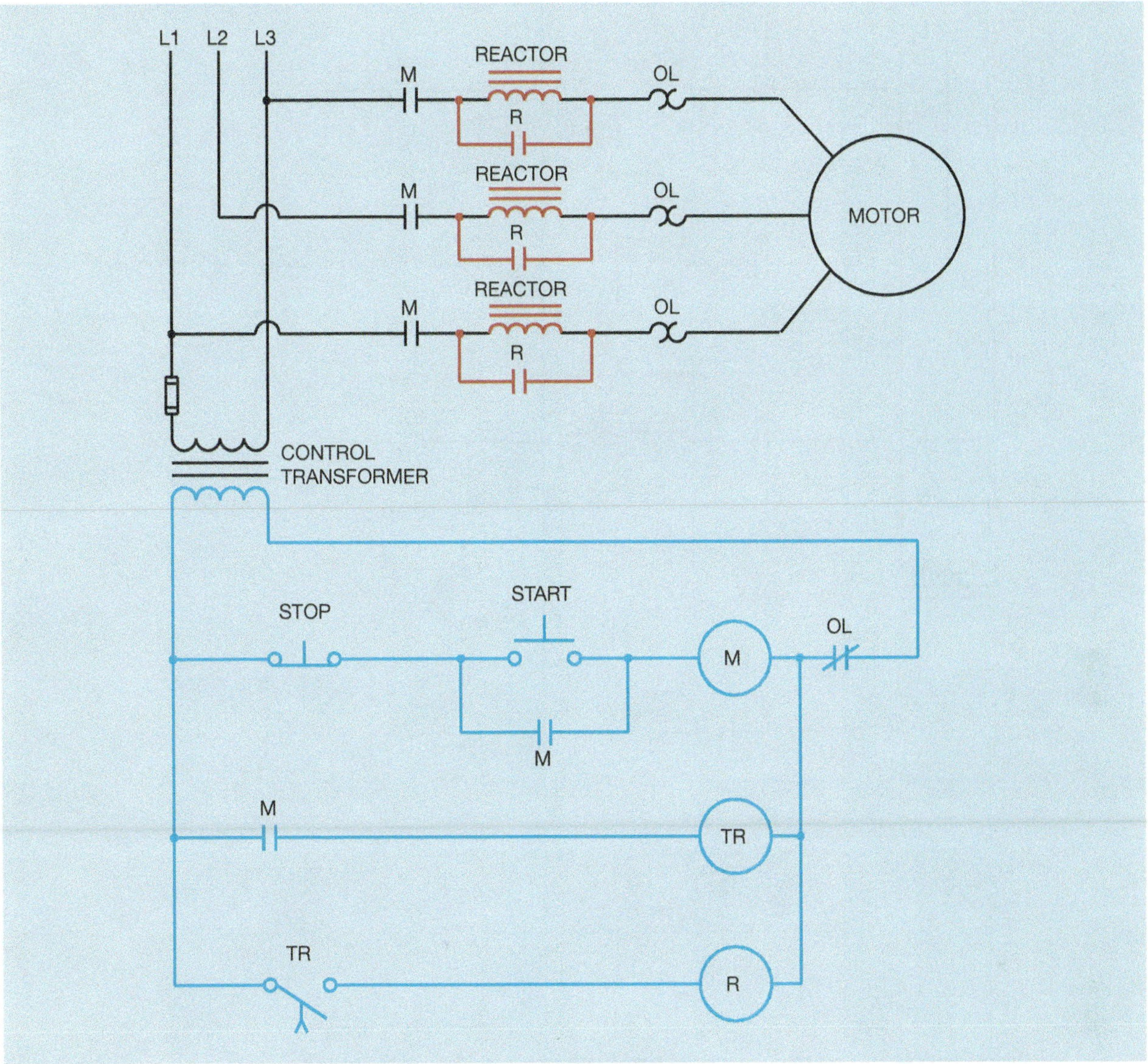

Figure 24–5 Reactor starting circuit.

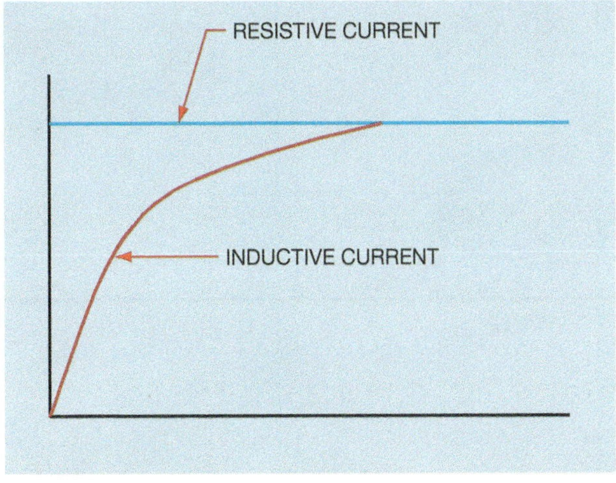

Figure 24–6 Current in an inductive circuit rises at an exponential rate.

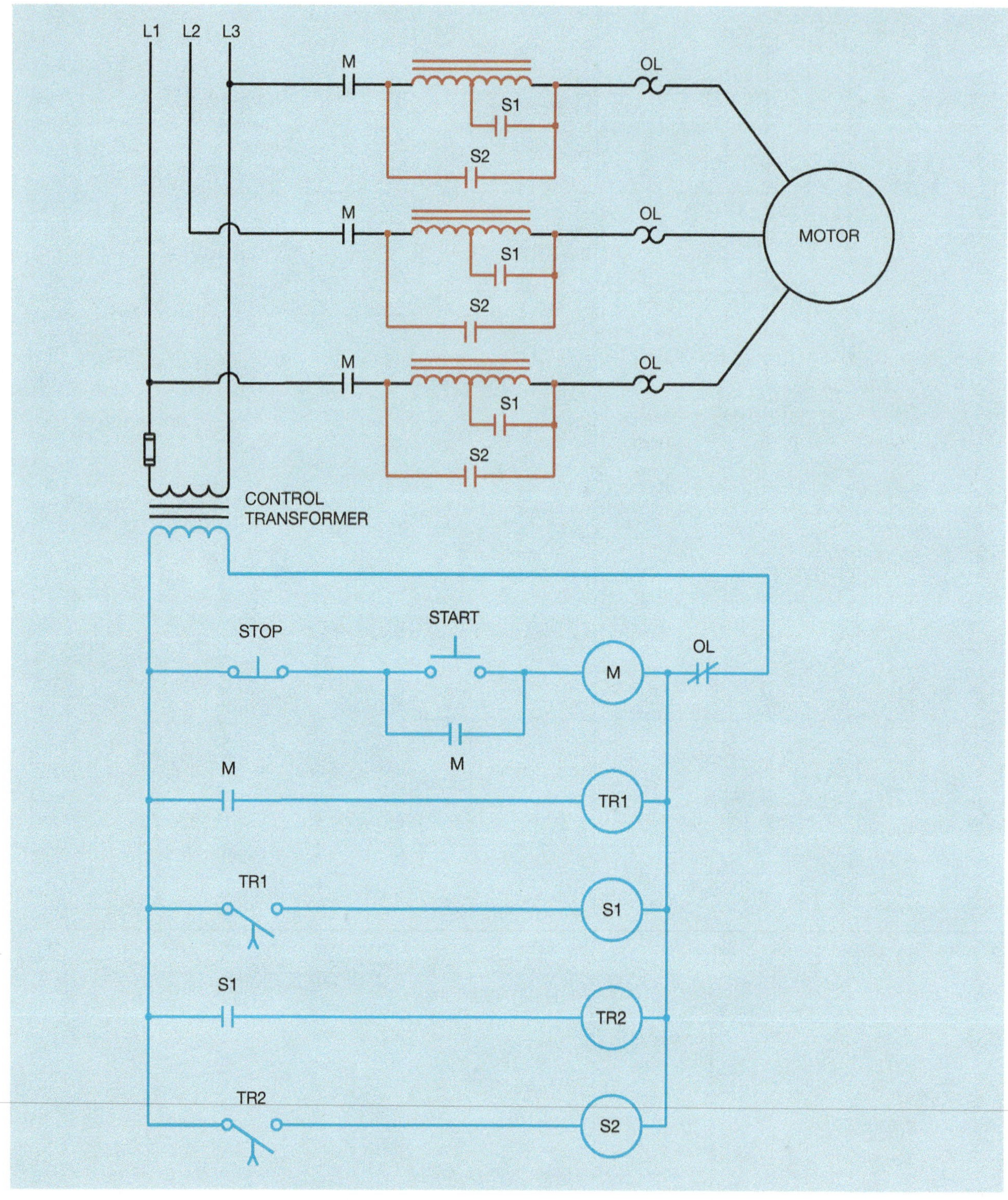

Figure 24–7 Three-step reactor starting circuit.

Review Questions

1. What two electrical components are commonly connected in series with a motor to limit starting current?
2. What advantage does a reactor have when limiting in-rush current that is not available with a resistor?
3. Refer to the circuit shown in Figure 24–1. Assume that timer TR is set for a delay of 10 seconds. When the START button is pressed the motor starts in low speed. After a delay of 30 seconds the motor is still in its lowest speed and has not accelerated to normal speed. Which of the following could *not* cause this condition?
   a. The START button is shorted.
   b. Timer coil TR is open.
   c. Contactor coil R is open.
   d. Timed contact TR did not close after a delay of 10 seconds.
4. Refer to the circuit shown in Figure 24–7. Assume that each timer is set for a delay of 5 seconds. When the START button is pressed, the motor starts in its lowest speed. After a delay of 5 seconds the motor accelerates to second speed. After another delay of 5 seconds, the motor stops running. During troubleshooting, you discover that the control transformer fuse is blown. Which of the following could cause this condition?
   a. TR1 coil is shorted.
   b. S1 coil is open.
   c. S2 coil is shorted.
   d. TR2 coil is open.
5. Refer to the circuit shown in Figure 24–7. Assume that each timer is set for a delay of 5 seconds. When the START button is pressed, the motor starts in its highest speed. Which of the following could cause this condition?
   a. The STOP button is shorted.
   b. TR1 timer coil is open.
   c. S1 auxiliary contact is shorted.
   d. TR2 timer coil is shorted.

Chapter 25

AUTOTRANSFORMER STARTING

Objectives

After studying this chapter the student will be able to:

- Discuss autotransformer starting.
- Discuss different types of autotransformer starters.
- Connect an autotransformer starter.
- Define closed and open transition starting.

Autotransformer starters reduce the amount of in-rush current by reducing the voltage applied to the motor during the starting period. Many autotransformer starters contain taps that can be set for 50, 65, or 80 percent of the line voltage. Reducing the voltage applied to the motor not only reduces the amount of in-rush current but also reduces the motor torque. If 50 percent of the normal voltage is connected to the motor, the in-rush current will drop to 50 percent also. This decrease produces a torque that is 25 percent of the value when full voltage is connected to the motor. If the motor torque is insufficient to start the load when the 50 percent tap is used, the 65 or 80 percent taps are available.

Autotransformer starters are generally employed to start squirrel cage type motors. Wound rotor type motors and synchronous motors do not generally use this type of starter. Autotransformer starters are inductive type loads and will affect the power factor during the starting period.

Most autotransformer starters use two transformers connected in open-delta to reduce the voltage applied to the motor during the period of acceleration (Figure 25–1). During the starting period, the motor is connected to the reduced voltage taps on the transformers. After the motor has accelerated to about 75 percent of normal speed, the motor is connected to full voltage. A time delay starter of this type is shown in Figure 25–2. To understand more clearly the operation of the autotransformer starter, refer to the schematic diagram shown in Figure 25–2. When the START button is pressed, a circuit is completed to the coil of control relay CR, causing all CR contacts to close. One contact is employed to hold CR coil in the circuit when the START button is released. Another completes a circuit to the coil of TR timer, which permits the timing sequence to begin. The CR contact connected in series with the normally closed TR contact supplies power to the coil of S (start) contactor. The fourth CR contact permits power to be connected to R (run) contactor when the normally open timed TR contact closes.

When the coil of S contactor energizes, all S contacts change position. The normally closed S contact connected in series with R coil opens to prevent both S and R contactors from ever being energized at the same time. This is the same interlocking used with reversing starters. When the S load contacts close, the motor is connected to the power line through the autotransformers. The autotransformers supply 65 percent of the line voltage to the motor. This reduced voltage produces less in-rush current during starting and also reduces the starting torque of the motor.

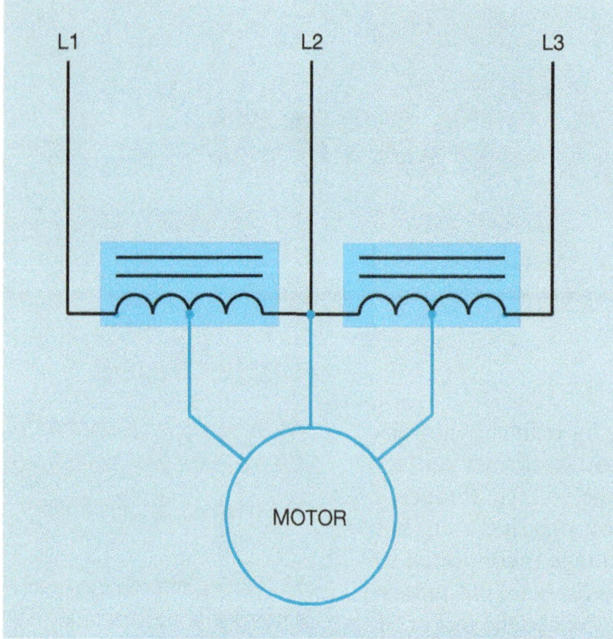

Figure 25–1 Transformers are connected in open-delta.

When the time sequence for TR timer is completed, both TR contacts change position. The normally closed TR contact opens and disconnects contactor S from the line causing all S contacts to return to their normal position. The normally open TR contact closes and supplies power through the now closed S contact to coil R. When contactor R energizes, all R contacts change position. The normally closed R contact connected in series with S coil opens to provide interlocking for the circuit. The R load contacts close and connect the motor to full voltage.

When the STOP button is pressed, control relay CR de-energizes and opens all CR contacts. This disconnects all other control components from the power line and the circuit returns to its normal position. A wiring diagram for this circuit is shown in Figure 25–3.

Open and Closed Transition Starting

Open transition starting is generally used on starters of size 5 and smaller. Open transition simply means that there is a brief period of time when the motor is disconnected from power when the start contactor opens and the run contactor closes. The circuits shown in Figures 25–2 and 25–3 are examples of open transition starters.

Closed transition starting is generally used on starters size 6 and larger. For closed transition starting, two separate start contactors are used (Figure 25–4). When the motor is started, both S1 and S2 contactors close their contacts. The S1 contacts open first and separately from the S2 contacts. At this point, part of the autotransformer windings are connected in series with the motor and act as series inductors. This permits the motor to accelerate to a greater speed before the R contacts close and the S2 contacts open. Although the R and S2 contacts are closed at the same time, the interval of time between the R contacts closing and the S2 contacts opening is so short that it does not damage the autotransformer winding.

Notice that the circuit in Figure 25–4 contains three current transformers (CTs). This is typical in circuits that control large horsepower motors. The CTs reduce the current to a level that common overload heaters can be used to protect the motor. A schematic diagram of a timed circuit for closed transition starting is shown in Figure 25–5. When the motor reaches the run stage, it is connected directly to the power line and the autotransformer is completely disconnected from the circuit. This is done to conserve energy and extend the life of the transformers. A typical autotransformer starter is shown in Figure 25–6.

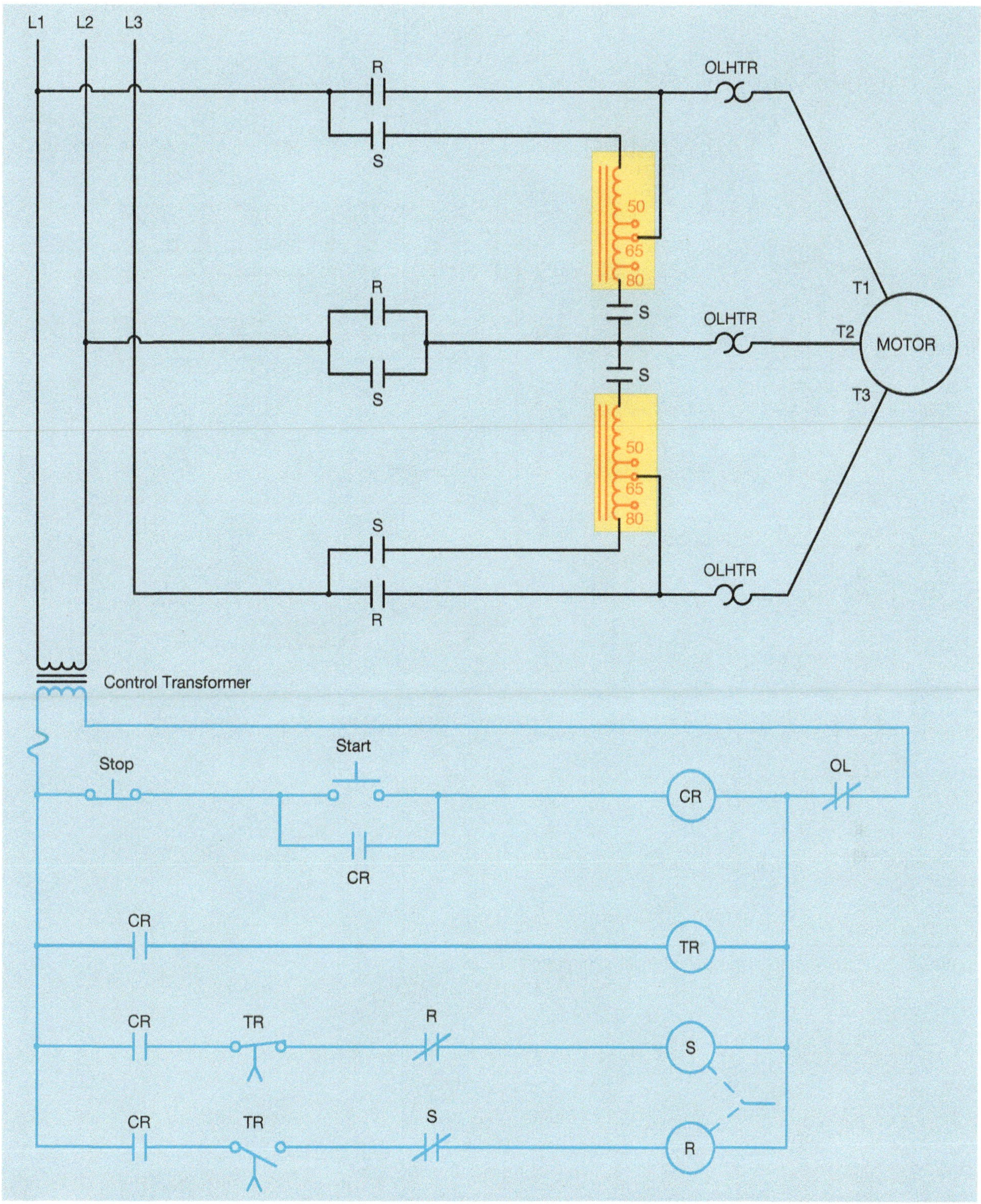

Figure 25–2 Autotransformer starters provide greater starting torque per ampere drawn from the line than any other type of reduced voltage starter. This is a schematic diagram for a time-controlled autotransformer starter.

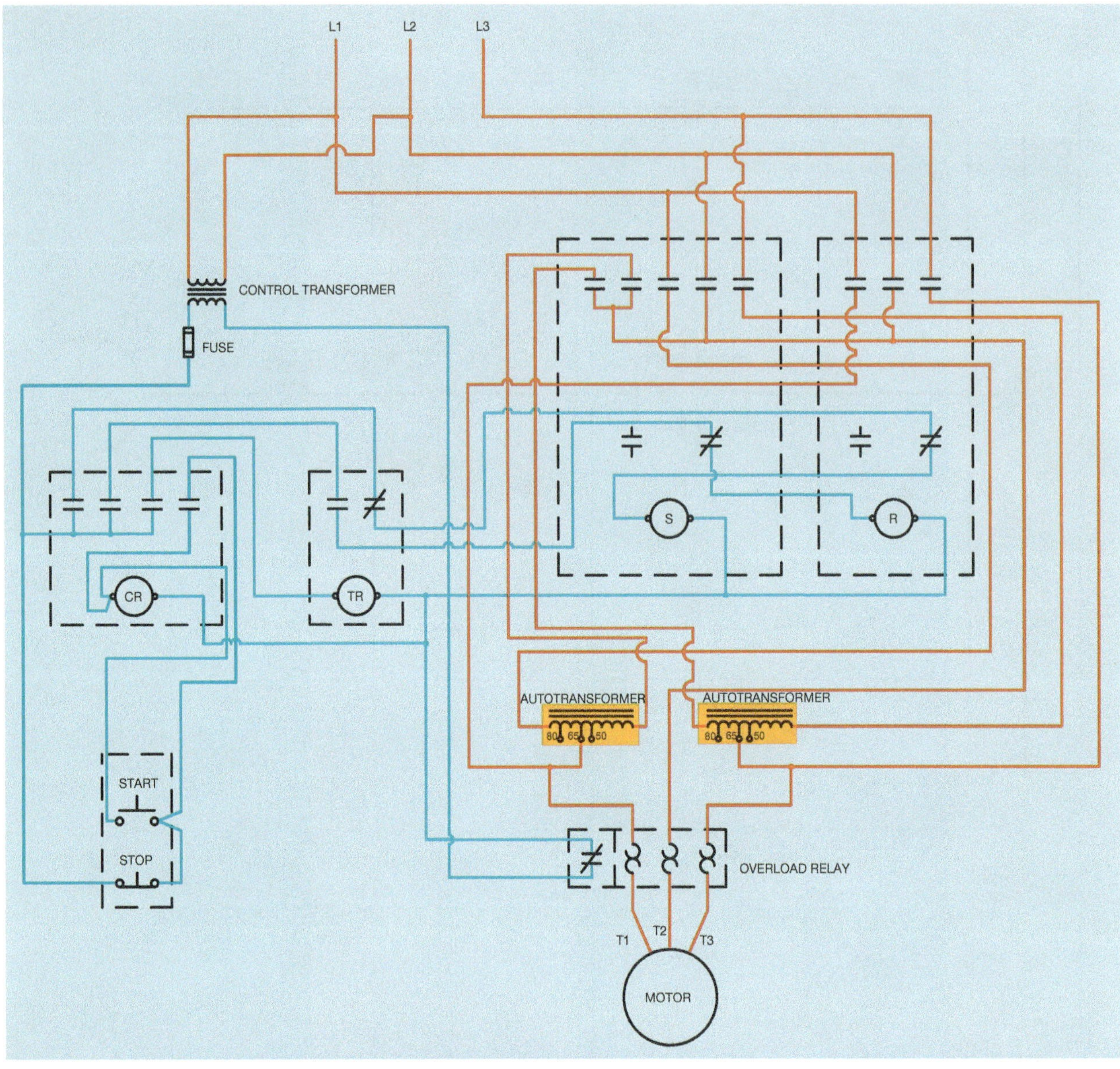

Figure 25–3 Typical wiring diagram for a typical autotransformer starter.

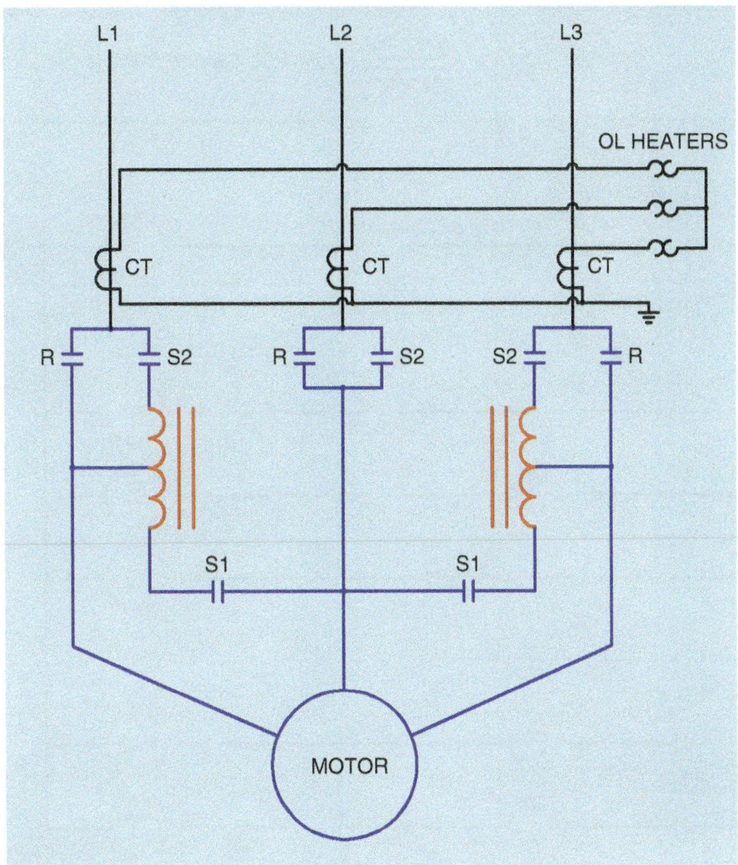

Figure 25–4 Closed transition starting uses two separate start contactors.

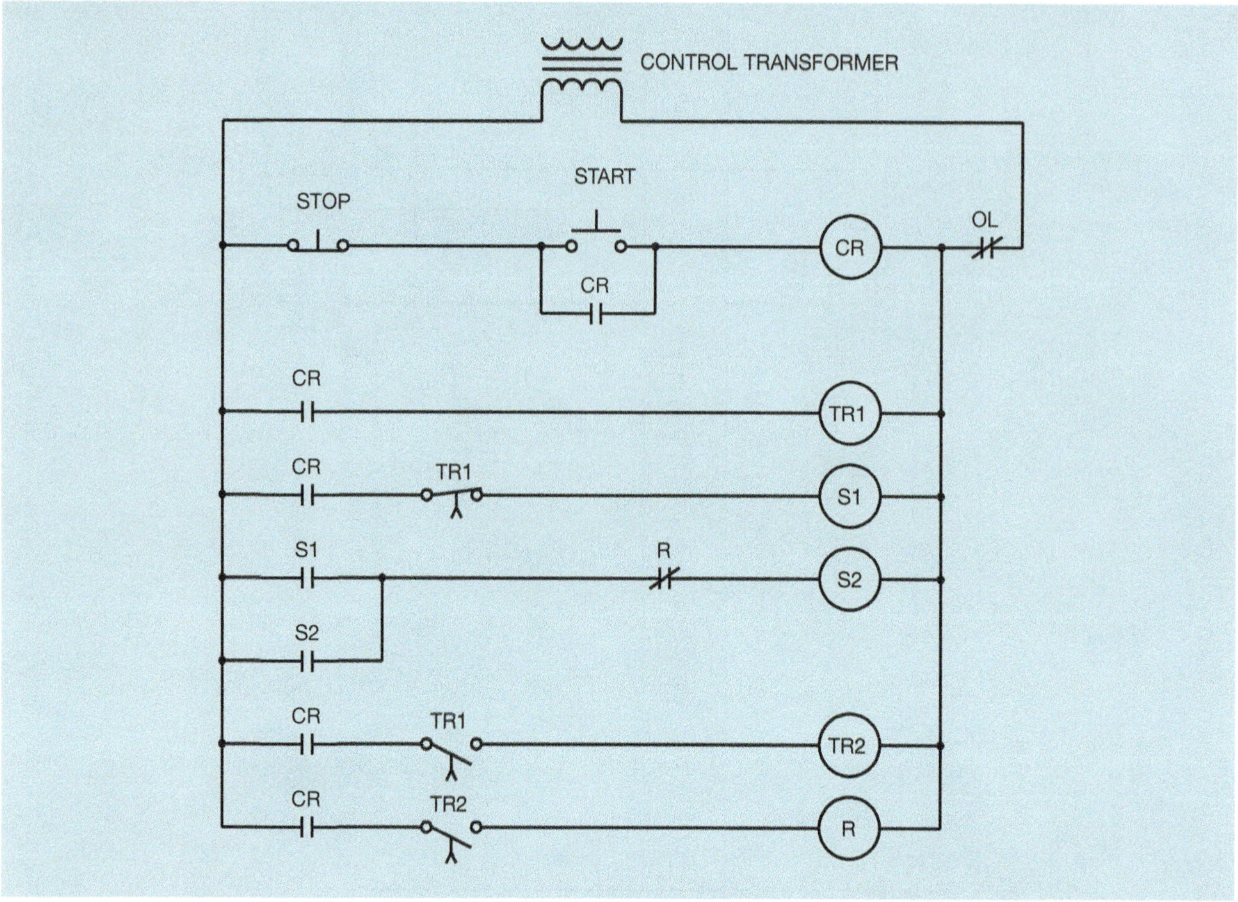

Figure 25–5 Closed transition starting circuit.

Courtesy Schneider Electric USA, Inc.

Figure 25–6 Typical autotransformer starter.

Review Questions

1. Why is it desirable to disconnect the autotransformer from the circuit when the motor reaches the run stage?
2. Explain the differences between open and closed transition starting.
3. Autotransformers often contain taps that permit different percentages of line voltages to be connected to the motor during starting. What are three common percentages?
4. Refer to the circuit shown in Figure 25–2. Assume that timer TR is set for a time delay of 10 seconds. When the START button is pressed, the motor does not start. After a period of 10 seconds, the motor starts with full line voltage applied to it. Which of the following could cause this condition?

a. Timer TR coil is open.
b. CR coil is open.
c. Contactor S coil is open.
d. Contactor R coil is open.

5. Refer to the circuit shown in Figure 25–2. Assume that timer TR is set for a delay of 10 seconds. Assume that contactor coil R is open. Explain the operation of the circuit if the START button is pressed.
6. Refer to the circuit shown in Figure 25–5. Assume that timer TR1 is set for a delay of 10 seconds and timer TR2 is set for a delay of 5 seconds. After the START button is pressed, how long is the time delay before the S1 contacts open?
7. Refer to the circuit shown in Figure 25–5. Assume that timer TR1 is set for a delay of 10 seconds and timer TR2 is set for a delay of 5 seconds. From the time the START button is pressed, how long will it take the motor to be connected to full line voltage?
8. Refer to the circuit shown in Figure 25–5. Explain the steps necessary for coil S2 to energize.
9. Refer to the circuit shown in Figure 25–5. What causes contactor coil S2 to de-energize after the motor reaches the full run stage?
10. Refer to the circuit shown in Figure 25–5. Assume that timer TR1 is set for a delay of 10 seconds and timer TR2 is set for a delay of 5 seconds. When the START button is pressed, the motor starts. After 10 seconds the S1 contacts open and the motor continues to accelerate, but never reaches full speed. After a delay of about 30 seconds, the motor trips out on overload. Which of the following could cause this problem?
    a. TR1 coil is open.
    b. S2 coil is open.
    c. S1 coil is open.
    d. R coil is open.

Chapter 26

WYE-DELTA STARTING

Objectives

After studying this chapter the student will be able to:

- Calculate starting current for a motor with its windings connected in delta.
- Calculate starting current for a motor with its windings connected in wye.
- List requirements for wye-delta starting.
- Connect a motor for wye-delta starting.
- Discuss open and closed transition starting.

Wye-delta starting is often used with large horsepower motors to reduce in-rush current during the starting period and reduce starting torque. Wye-delta starting is accomplished by connecting the motor stator windings in wye or star during the starting period and then reconnecting them in delta during the run period. This is sometimes called *soft starting*. If the stator windings of a motor are connected in delta during the starting period, the starting current will be three times the value if the windings were connected in wye. Assume that a motor is to be connected to a 480-volt three-phase power line. Also, assume that the motor windings have an impedance of 0.5 ohms when the motor is first started. If the stator windings are connected in delta (Figure 26–1), the voltage across each phase will be 480 volts because line voltage and phase voltage are the same in a **delta connection.** The amount of current flow in each phase winding (stator winding) can be determined with Ohm's law.

$$I_{PHASE} = \frac{E_{PHASE}}{Z_{PHASE}}$$

$$I_{PHASE} = \frac{480}{0.5}$$

$$I_{PHASE} = 960 \text{ amperes}$$

In a delta connection, the line current is greater than the phase current by a value of the square root of 3 ($\sqrt{3}$) or 1.732. Therefore, the amount of line current will be:

$$I_{LINE} = I_{PHASE} \times 1.732$$

$$I_{LINE} = 960 \times 1.732$$

$$I_{LINE} = 1662.72 \text{ amperes}$$

If the stator windings are connected in wye (Figure 26–2), the voltage across each phase windings will be 277 volts, because in a wye connected load, the phase voltage is less than the line voltage by a factor of the square root of 3 or 1.732. The amount of in-rush current can be determined using Ohm's law.

$$I_{PHASE} = \frac{E_{PHASE}}{Z_{PHASE}}$$

$$I_{PHASE} = \frac{277}{0.5}$$

$$I_{PHASE} = 554 \text{ amperes}$$

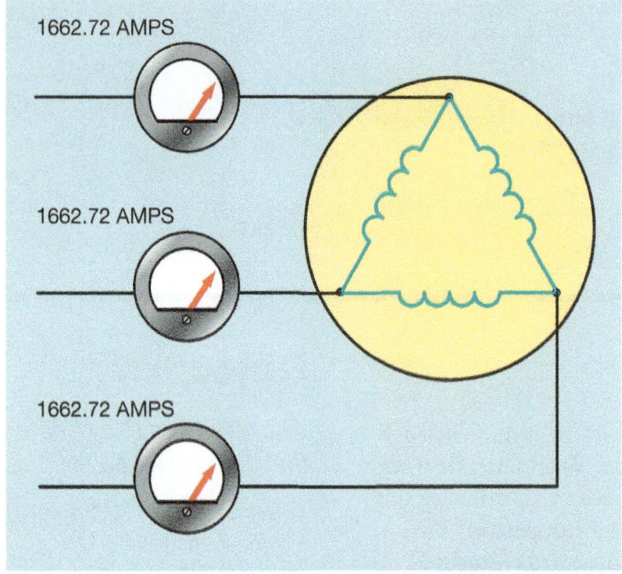

Figure 26–1 Stator windings are connected in delta during the starting period.

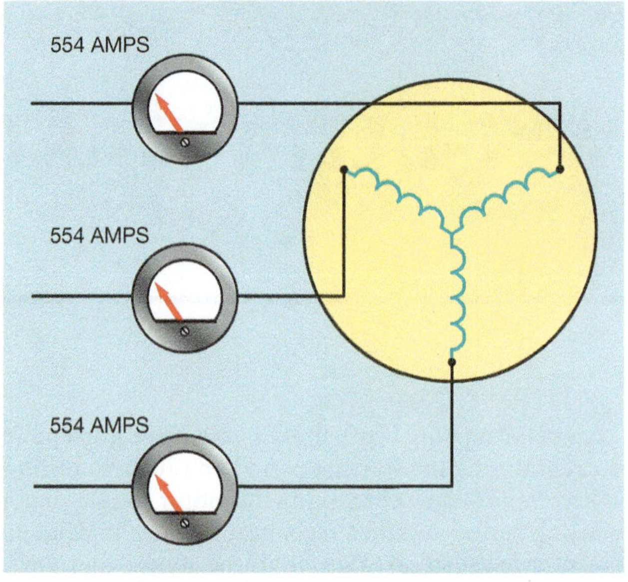

Figure 26–2 Stator windings are connected in wye or star during the starting period.

In a **wye connection,** the line current and phase current are the same. Therefore, the starting current has been reduced from 1662.72 amperes to 554 amperes by connecting the stator windings in wye instead of delta during the starting period.

Wye-Delta Starting Requirements

Two requirements must be met before wye-delta starting can be used.

1. **The motor must be designed for the stator windings to be connected in delta during the run period.** Motors can be designed to operate with their stator windings connected in either wye or delta. The actual power requirements are the same, depending on motor horsepower. The speed of a three-phase induction motor is determined by the number of stator poles per phase and the frequency of the applied voltage. Therefore, the motor operates at the same speed regardless which connection is used when the motor is designed.
2. **All stator windings leads must be accessible.** Motors designed to operate on a single voltage commonly supply three leads labeled T1, T2, and T3 at the terminal connection box located on the motor. Dual voltage motors commonly supply nine leads labeled T1 through T9 at the terminal connection box. If a motor is designed to operate on a single voltage, six terminal leads must be provided. The numbering for these six leads is shown in Figure 26–3. Notice that the lead numbers are standardized for each of the three phases. The opposite end of terminal lead T1 is T4, the opposite end of T2 is T5, and the opposite end of T3 is T6. If the stator windings are to be connected in delta, terminals T1 and T6 are connected together, T2 and T4 are connected, and T3 and T5 are connected. If the stator windings are to be connected in wye, T4, T5, and T6 are connected together. Motors not intended for wye-delta starting would have these connections made internally and only three leads would be supplied at the terminal connection box. A motor with a delta connected stator, for example, would have T1 and T6 connected internally and a single lead labeled T1 would be provided for connection to the power line. Wye connected motors have T4, T5, and T6 connected internally.

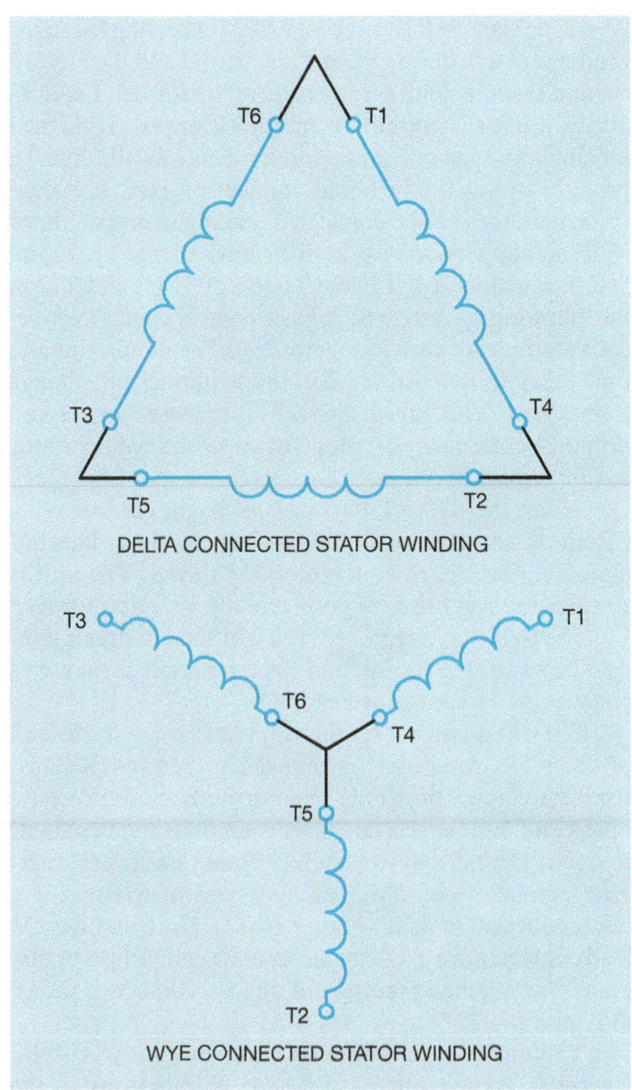

Figure 26–3 Standard lead numbers for single voltage motors.

Figure 26–4 Standard lead numbers for dual voltage motors.

Dual Voltage Connections

Motors that are intended to operate on two voltages such as 240 or 480 volts contain two separate windings for each phase (Figure 26–4). Notice that dual voltage motors actually contain 12 T leads. Dual voltage motors not intended for wye-delta connection will have certain terminals tied internally as shown in Figure 26–5. Although all three-phase dual voltage motors actually contain 12 T leads, only terminal leads T1 through T9 are brought out to the terminal connection box for motors not intended for wye-delta starting.

If the motor is to be operated on the higher voltage, the stator leads will be connected in series as shown in Figure 26–6. If the motor is to be connected for operation on the lower voltage the stator windings will be connected in parallel as shown in Figure 26–7.

Although dual voltage motors designed for wye-delta starting will supply all 12 T leads at the terminal connection box, it is necessary to make the proper connections for high or low voltage. The connection diagrams for dual voltage motors with 12 T leads are shown in Figure 26–8. Note that the diagrams do not show connection to power leads. These connections are made as part of the control circuit.

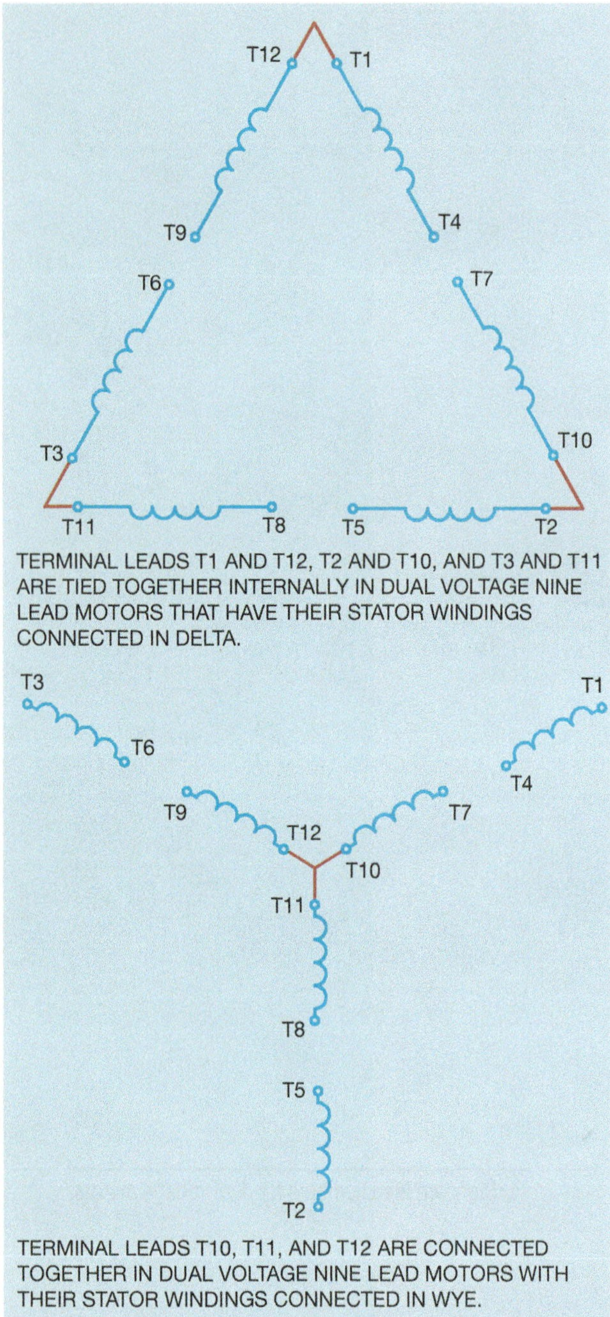

Figure 26–5 Nine lead dual voltage motors have some stator windings connected together internally.

Connecting the Stator Leads

Wye-delta starting is accomplished by connecting the stator windings in wye during the starting period and then reconnecting them in delta for normal run operation. For simplicity, it is assumed that the motor illustrated is designed for single voltage operation and has leads T1 through T6 brought out at the terminal connection box. If a dual voltage motor is to be connected, make the proper stator winding connections for high or low voltage operation and then change T4, T5, and T6 to T10, T11, and T12 in the following connections. A basic control circuit for wye-delta starting is shown in Figure 26–9. This circuit employs time delay to determine when the windings will change from wye to delta. Starting circuits that sense motor speed or motor current to determine when to change the stator windings from wye to delta are also common.

When the START button is pressed, control relay CR energizes, causing all CR contacts to close. This immediately energizes contactors 1M and S. The motor stator windings are now connected in wye as shown in Figure 26–10 (on page 274). The 1M load contacts connect power to the motor, and the S contacts form a wye connection for the stator windings.

The 1M auxiliary contact supplies power to the coil of timer TR. After a preset time delay, the two TR timed contacts change position. The normally closed contact opens and disconnects coil S, causing the S load contacts to open. The normally open TR contact closes and energizes contactor coil 2M. The motor stator windings are now connected in delta (Figure 26–11). Note that the 2M load contacts are used to make the delta connection. A diagram showing the connection of all load contacts is shown in Figure 26–12.

The most critical part of connecting a wye-delta starter is making the actual load connections to the motor. An improper connection generally results in the motor stopping and reversing direction when transition is made from wye to delta. It is recommended that the circuit and components be numbered to help avoid mistakes in connection (Figure 26–13 on page 275).

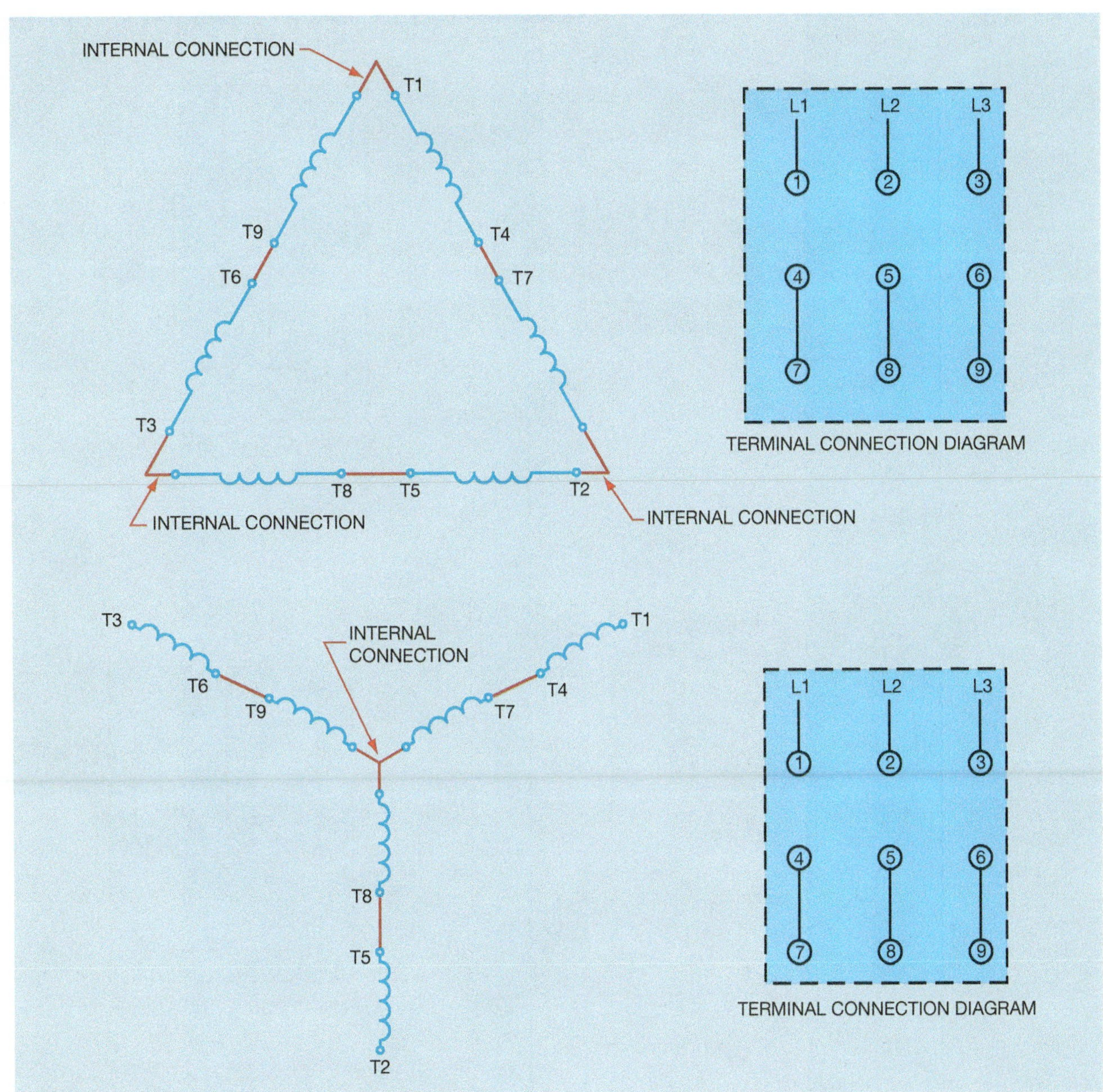

Figure 26–6 High voltage connection for nine lead motors.

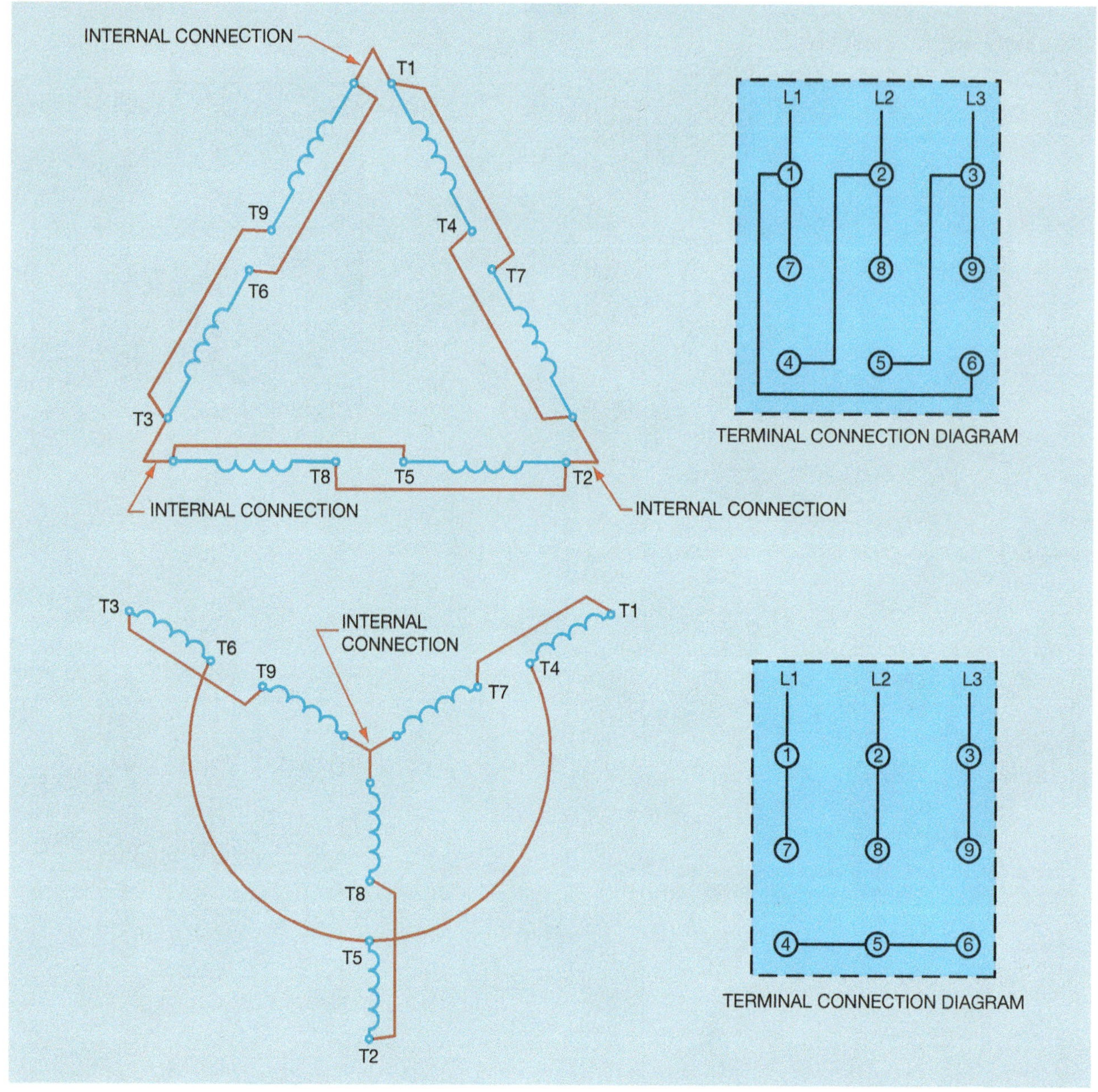

Figure 26–7 Low voltage connection for nine lead motors.

Closed Transition Starting

The control circuit discussed so far uses open transition starting. This means that the motor is disconnected from the power line during the transition from wye to delta. This may be objectionable in some applications if the transition causes spikes on the power line when the motor changes from wye to delta. Another method that does not disconnect the motor from the power line is called closed transition starting. Closed transition starting is accomplished by adding another three pole contactor and resistors to the circuit, Figure 26–14 (on page 276). The added contactor, designated as 1A, energizes momentarily to connect resistors between the power line and motor when the transition is made from wye to delta. Also note that an on-delay timer (TR2) with a delay of 1 second has been added to the control circuit. The purpose of this timer is to prevent a contact race between contactors

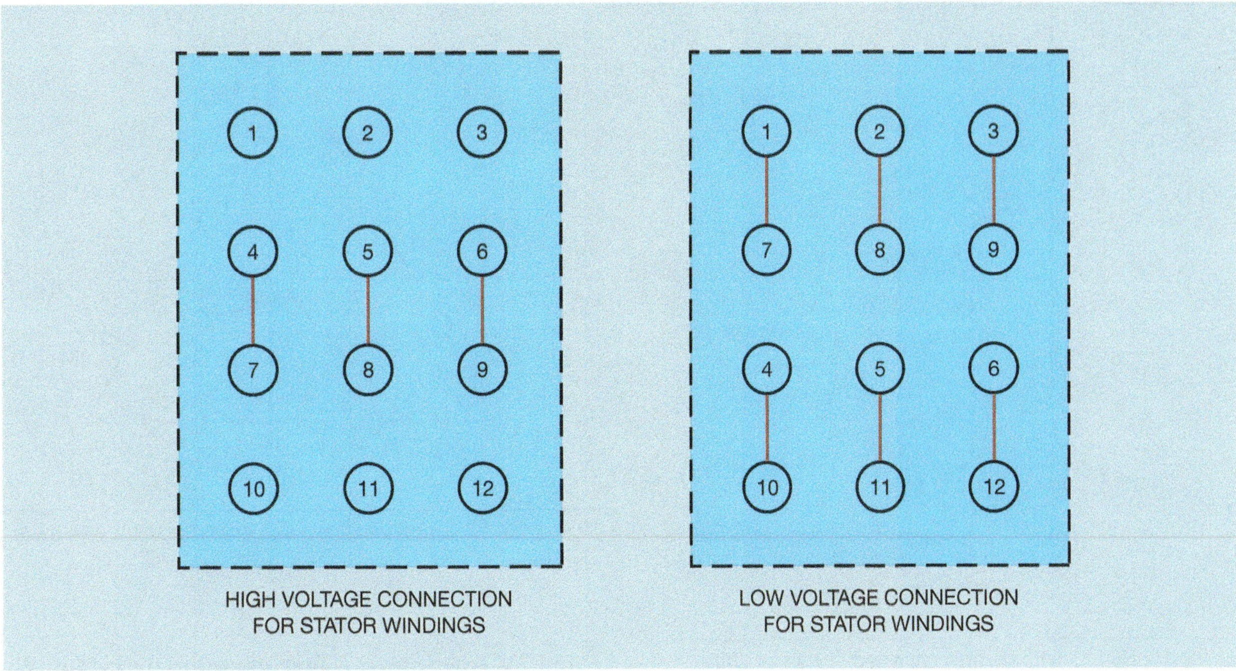

Figure 26–8 Stator winding connection for dual voltage 12 lead motors.

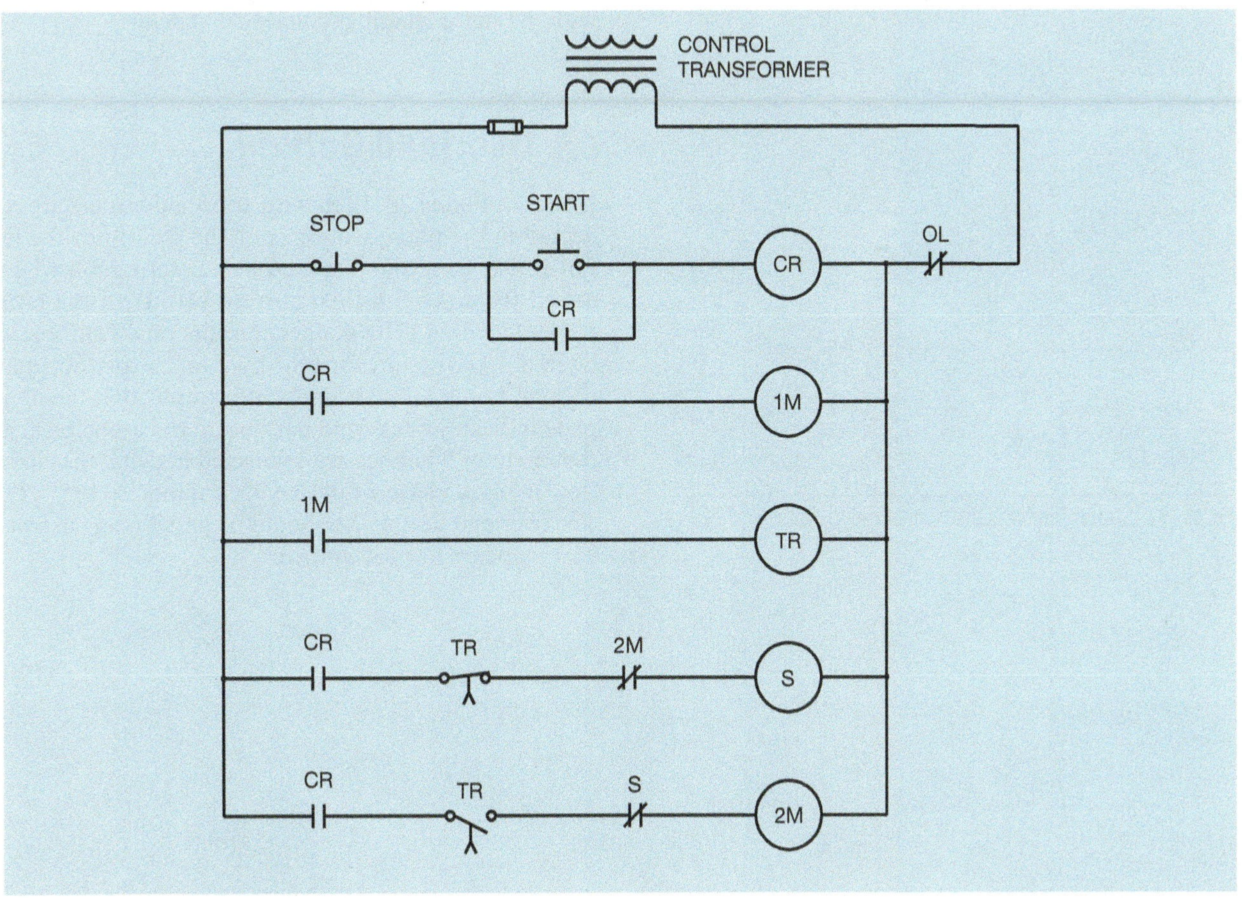

Figure 26–9 Basic control circuit for wye-delta starter using time delay.

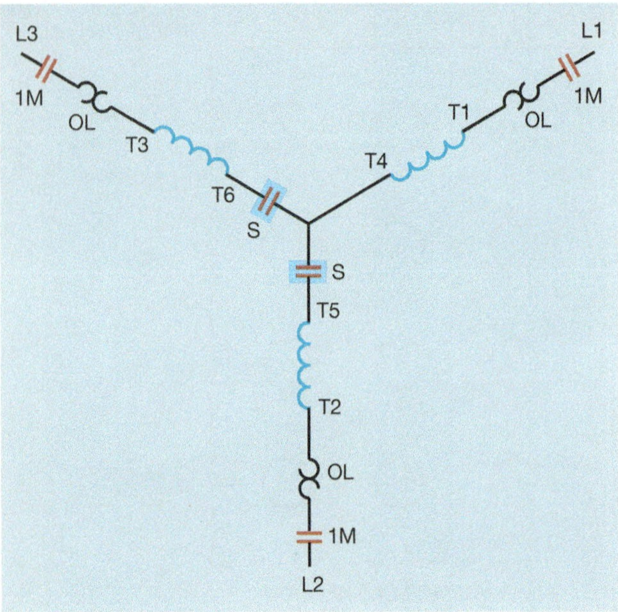

Figure 26–10 Stator windings are connected in wye for starting.

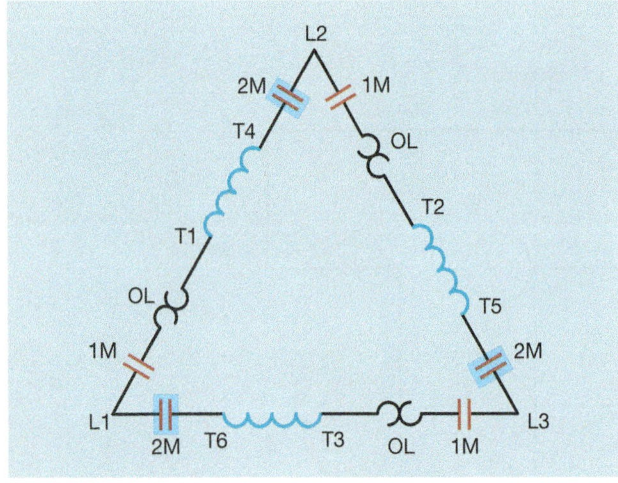

Figure 26–11 Stator windings are connected in delta.

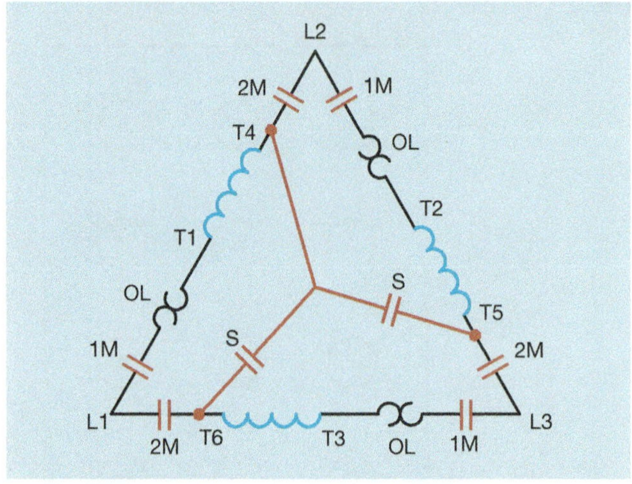

Figure 26–12 Stator windings with all contacts for wye-delta starting.

S and 2M when power is first applied to the circuit. Without timer TR2, it would be possible for contactor 2M to energize before contactor S. This would prevent the motor from being connected in wye. The motor would start with the stator windings connected in delta.

Overload Setting

Notice in Figure 26–12 that the overload heaters are connected in the phase windings of the delta, not the line. For this reason, the overload heater rating must be reduced from the full load current rating on the motor nameplate. In a delta connection, the phase current will be less than the line current by a factor of the square root of 3 or 1.732. Assume, for example, that the nameplate indicates a full load current of 165 amperes. If the motor stator windings are connected in delta, the current flow in each phase would be 95.3 amperes (165/1.732). The overload heater size should be based on a current of 95.3 amperes, not 165 amperes.

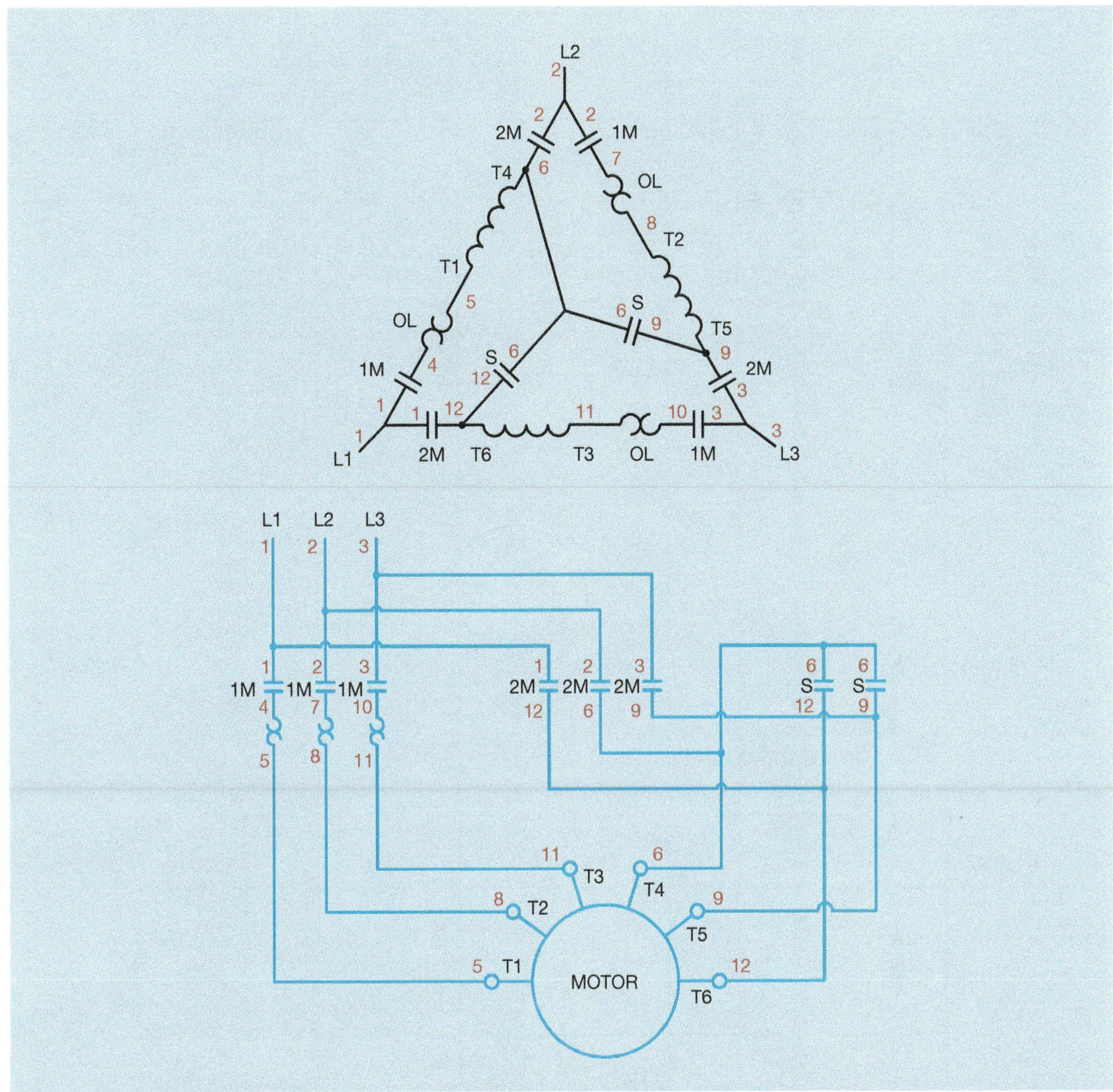

Figure 26–13 Load circuit connection for wye-delta starting.

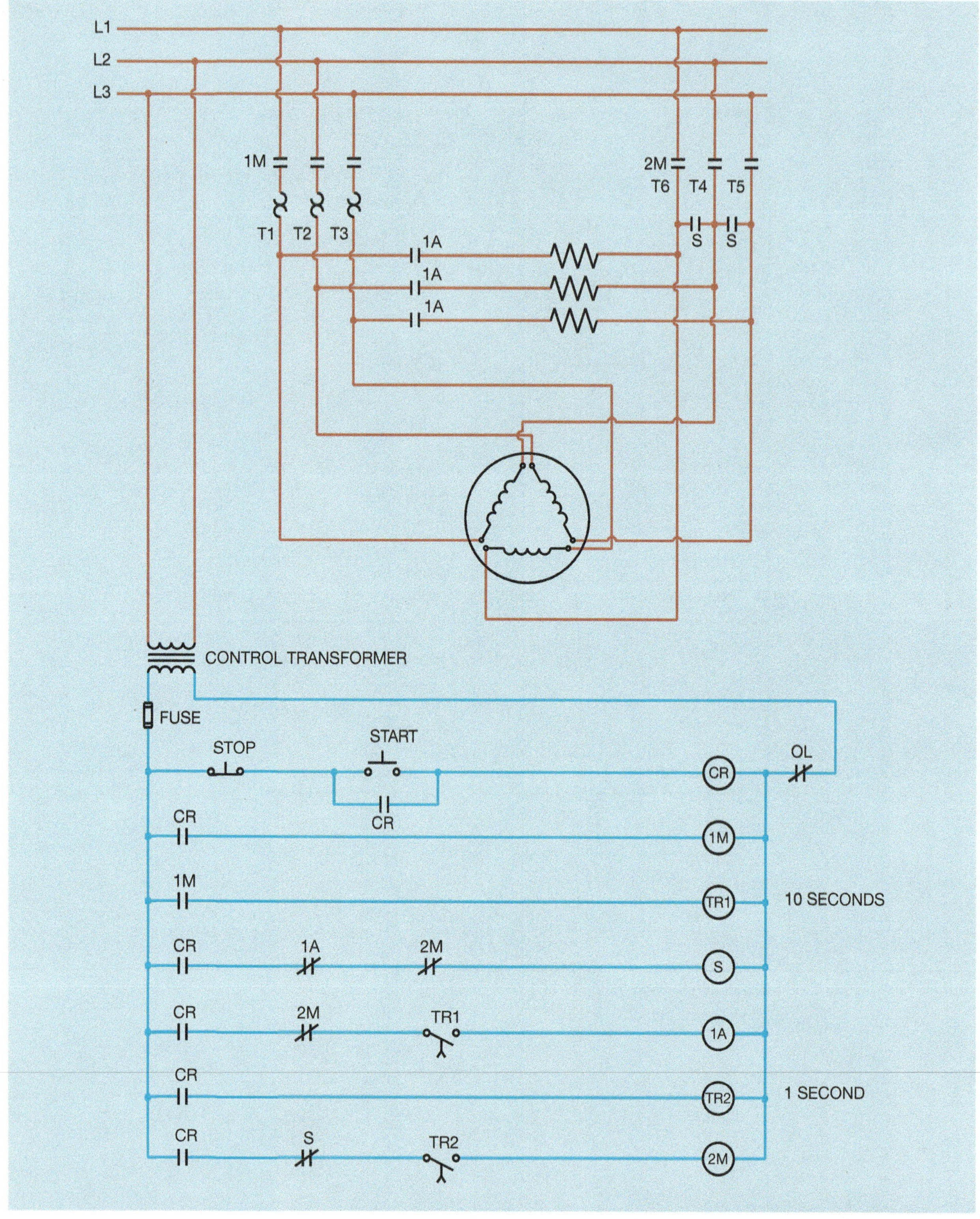

Figure 26–14 Basic schematic diagram Sizes 1, 2, 3, 4, and 5 wye-delta starters with closed transition starting.

Review Questions

1. Name two requirements that must be met before a motor can be used for wye-delta starting.

2. The stator windings of a 2300 volt motor have an impedance of 6 omega when the motor is first started. What would be the in-rush current if the stator windings were connected in delta?

3. What would be the amount of in-rush current if the motor described in question 2 had the stator windings connected in wye?

4. Refer to the circuit shown in Figure 26–9. Assume that timer TR is set for a delay of 10 seconds. When the START button is pressed, the motor starts with its windings connected in wye. After a period of 1 minute, the motor has not changed from wye to delta. Which of the following could cause this condition?
   a. TR timer coil is open.
   b. S contactor coil is open.
   c. 1M starter coil is open.
   d. The control transformer fuse is blown.

5. Refer to the circuit shown in Figure 26–9. Assume that timer TR is set for a delay of 10 seconds. When the START button is pressed, the motor does not start. After a delay of 10 seconds, the motor suddenly starts with its stator windings connected in delta. Which of the following could cause this problem?
   a. TR timer coil is open.
   b. 2M contactor coil is open.
   c. S contactor coil is open.
   d. 1M starter coil is open.

6. Refer to the circuit shown in Figure 26–9. What is the purpose of the normally closed 2M and S contacts in the schematic?

7. The motor nameplate of a wye-delta started motor has a full load current of 287 amperes. What current rating should be used to determine the proper overload heater size?

8. Refer to the circuit shown in Figure 26–14. When the motor changes from wye to delta, what causes contactor coil S to de-energize and open S contacts?

9. Refer to the circuit shown in Figure 26–14. What is the purpose of timer TR2?

10. Refer to the circuit shown in Figure 26–14. When the START button is pressed, the control transformer fuse blows immediately. Which of the following could *not* cause this problem?
    a. Control relay coil CR is shorted.
    b. Starter coil 1M is shorted.
    c. Contactor coil S is shorted.
    d. Contactor coil 2M is shorted.

Chapter 27

PART WINDING STARTERS

Objectives

After studying this chapter the student will be able to:

- Describe the construction of a motor designed to be used for part winding starting.
- Discuss three step starting for part winding motors.
- Draw a control circuit for a part winding starter.
- Connect a motor for part winding starting.

Part winding starting is another method of reducing the starting current of squirrel cage induction motors. Motors designed to be used for part winding starting contain two separate stator windings (Figure 27–1). The stator windings may be wye or delta connected depending on the manufacturer. These two windings are designed to be connected in parallel with each other. When the motor is started, only one of the windings is connected to the power line. Because only half the motor winding is used during starting, this method of starting is called *part winding* starting. Part winding starting reduces the normal locked rotor current to approximately 66 percent of the value if both windings are connected during starting and the torque is reduced to approximately

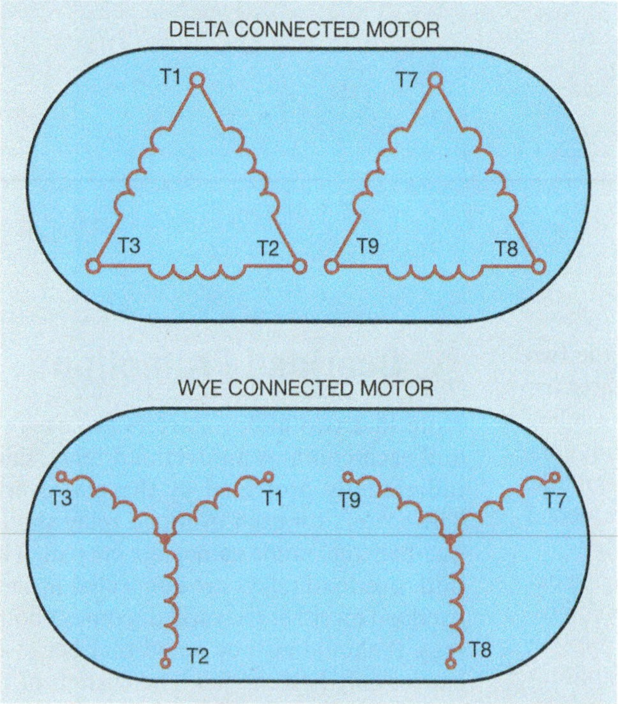

Figure 27–1 Motors designed for part winding starting contain two stator windings intended to be connected in parallel with each other.

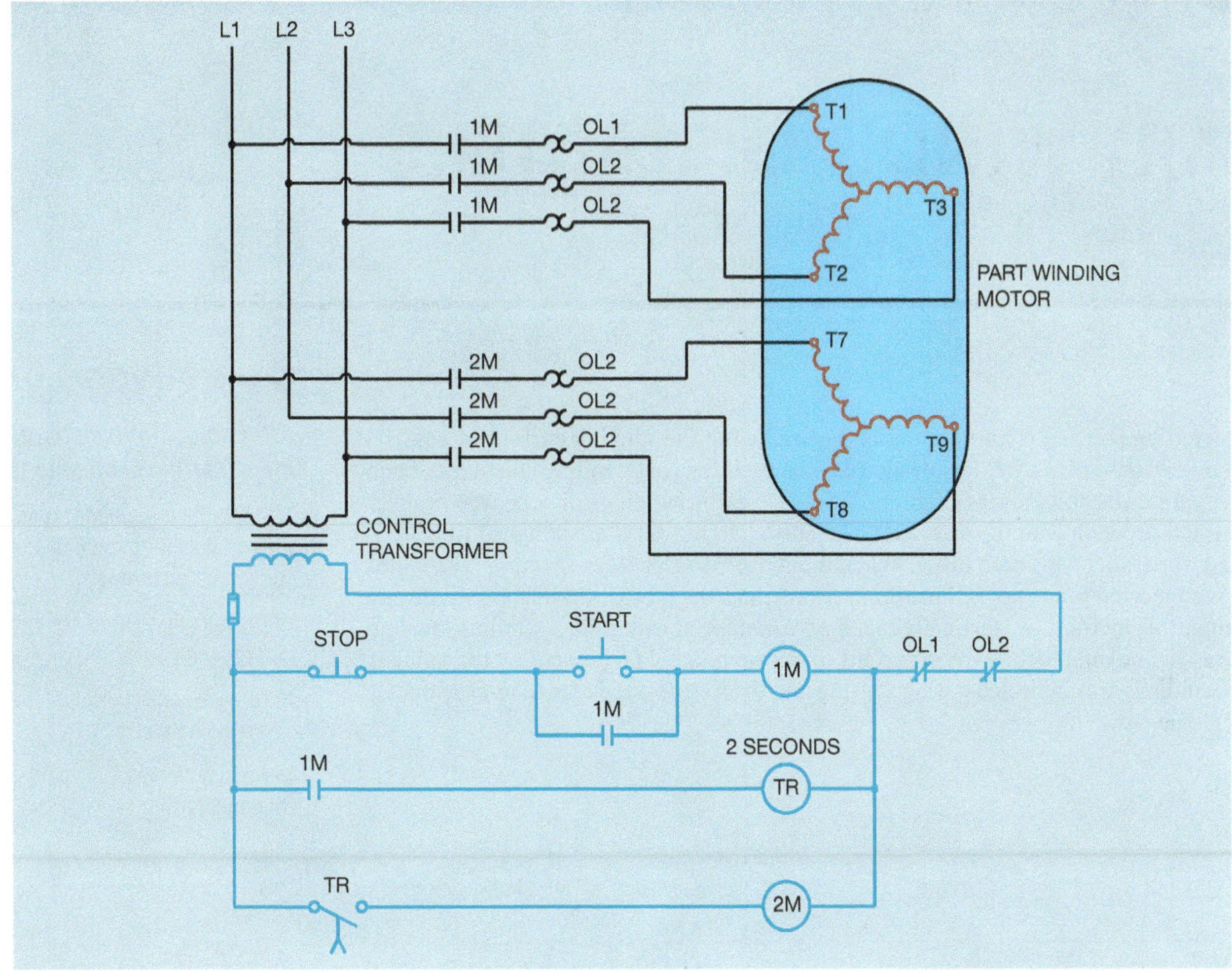

Figure 27–2 Part winding starter.

50 percent. It should be noted that neither of the two windings is individually capable of withstanding the starting current for more than a few seconds. The first winding will overheat rapidly if the second windings is not connected within a very short period of time. As a general rule, a time delay of 2 to 3 seconds is common before the second winding is connected in parallel with the first.

Part winding starting is accomplished by bringing out both sets of motor leads so that external connection is possible (Figure 27–2). When the START button is pressed, motor starter 1M energizes and connects the first motor windings to the line. The normally open 1M auxiliary contact closes and starts on-delay timer TR. After a 2-second time delay, timed contact TR closes and energizes motor starter 2M. This causes the 2M load contacts to close and connect the second stator winding to the power line.

Overload Protection

Note that two motor starters are used in the circuit, and each contains an overload relay. Each winding is individually protected by thermal overload heaters. The heaters for each overload relay should be sized at one-half the motor nameplate current. The contacts of both overload relays are connected in series so that an overload on either relay will disconnect both motor windings. It should also be noted that because each starter carries only half the full load current of the motor, the starter size can generally be reduced from what would be required for a single starter. Another advantage of part winding starters is that they provide closed transition starting because the motor is never disconnected from the power line during the starting time.

Dual Voltage Motors

Some, but not all, dual voltage motors may be used for part winding starting. The manufacturer should be contacted before an attempt is made to use a dual voltage motor in this application. Delta connected dual voltage motors are not acceptable for part winding starting. When dual voltage motors are used, the motor must be operated on the low voltage setting of the motor. A 240/480-volt motor, for example, could only be operated on 240 volts. A dual voltage motor connection is shown in Figure 27–3. Motor terminal leads T4, T5, and T6 are connected together forming a separate wye connection for the motor.

Motor Applications

Part winding starting is typically used for motors that supply the moving force for centrifugal pumps, fans, and blowers. They are often found in air conditioning and refrigeration applications. They are not generally employed to start heavy inertia loads that require an excessive amount of starting time.

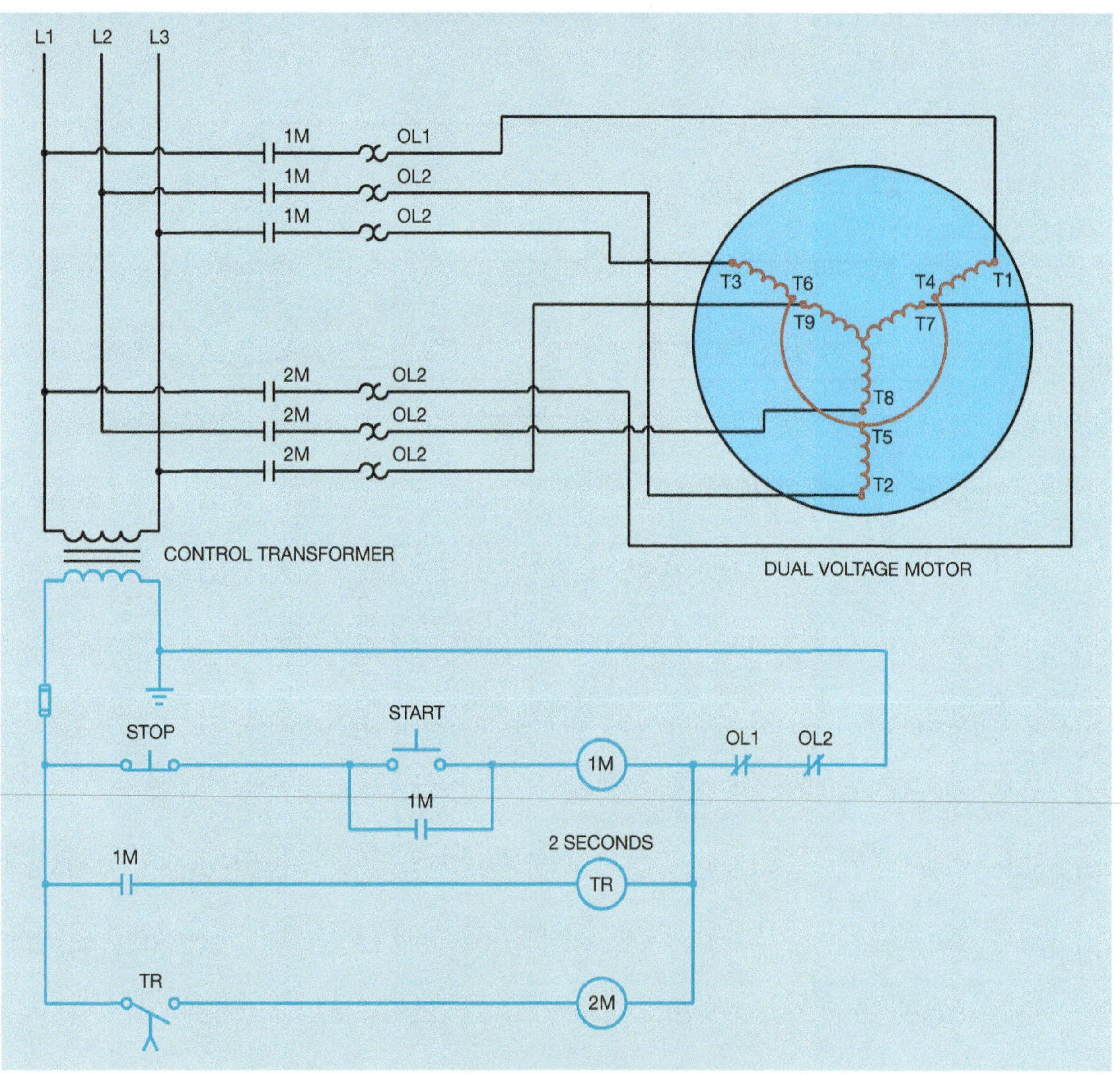

Figure 27–3 Dual voltage motor used for part winding starting.

Three Step Starting

The thermal capacity of the stator windings greatly limits the length of starting time for a part winding motor. To help overcome this problem, it is possible to provide a third step in the starting process and further limit starting current. This is accomplished by connecting resistance in series with the stator winding during the starting period (Figure 27–4). The resistors are generally sized to provide about 50 percent of the line voltage to the stator winding when the motor is first started. This provides approximately three equal increments of starting for the motor. In the circuit shown in Figure 27–4, when the START button is pressed, motor starter 1M energizes and connects one of the stator windings to the power line through the series resistors. After a delay of 2 seconds, TR1 timed contact closes and energizes contactor S. The S load contacts close and shunt the resistors out of the line. One stator winding is now connected to full line voltage.

After another 2-second delay, motor starter 2M energizes and connects the second stator winding to the power line. The motor now has both stator windings connected to full voltage.

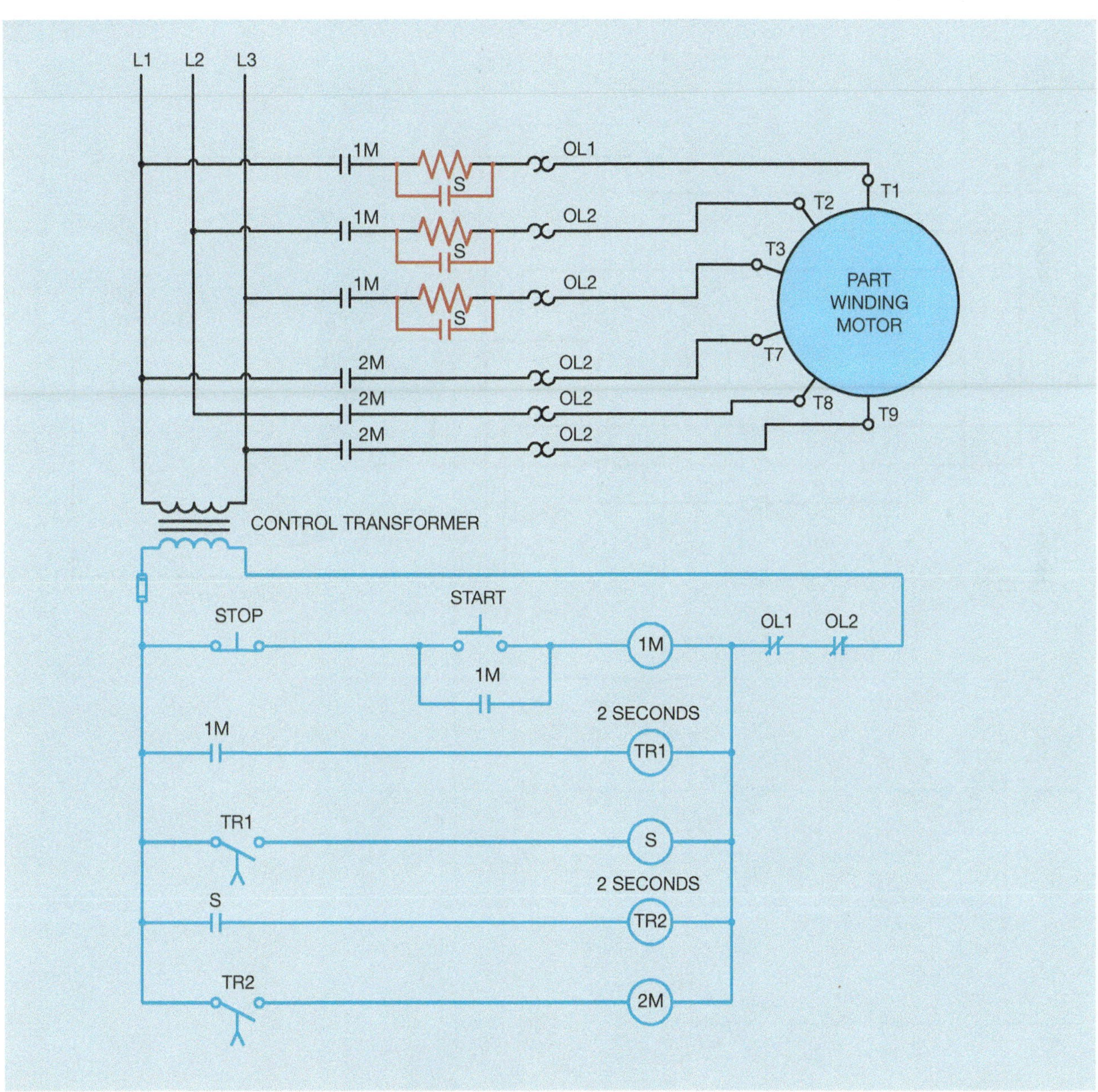

Figure 27–4 Three step starting for part winding motor.

Automatic Shutdown

Part winding motors are very sensitive to the length of time that one winding can be connected before thermal damage occurs. If the second winding is not connected to the power line within a short period of time, the first winding can be severely damaged. To help prevent damaging the first winding, some circuits contain a timer that disconnects power to the motor if the second winding is not energized within a predetermined time. This timer is often called a **watchdog timer** because its function is to watch for proper operation of the circuit each time the motor is started. A circuit with a watchdog timer is shown in Figure 27–5. Watchdog timers are often set for twice the amount of time necessary for the second winding to energize. When the START button is pushed, the watchdog timer begins its count. If the circuit operates properly, the normally closed 2M auxiliary contact will disconnect the timer before it times out and de-energizes control relay CR.

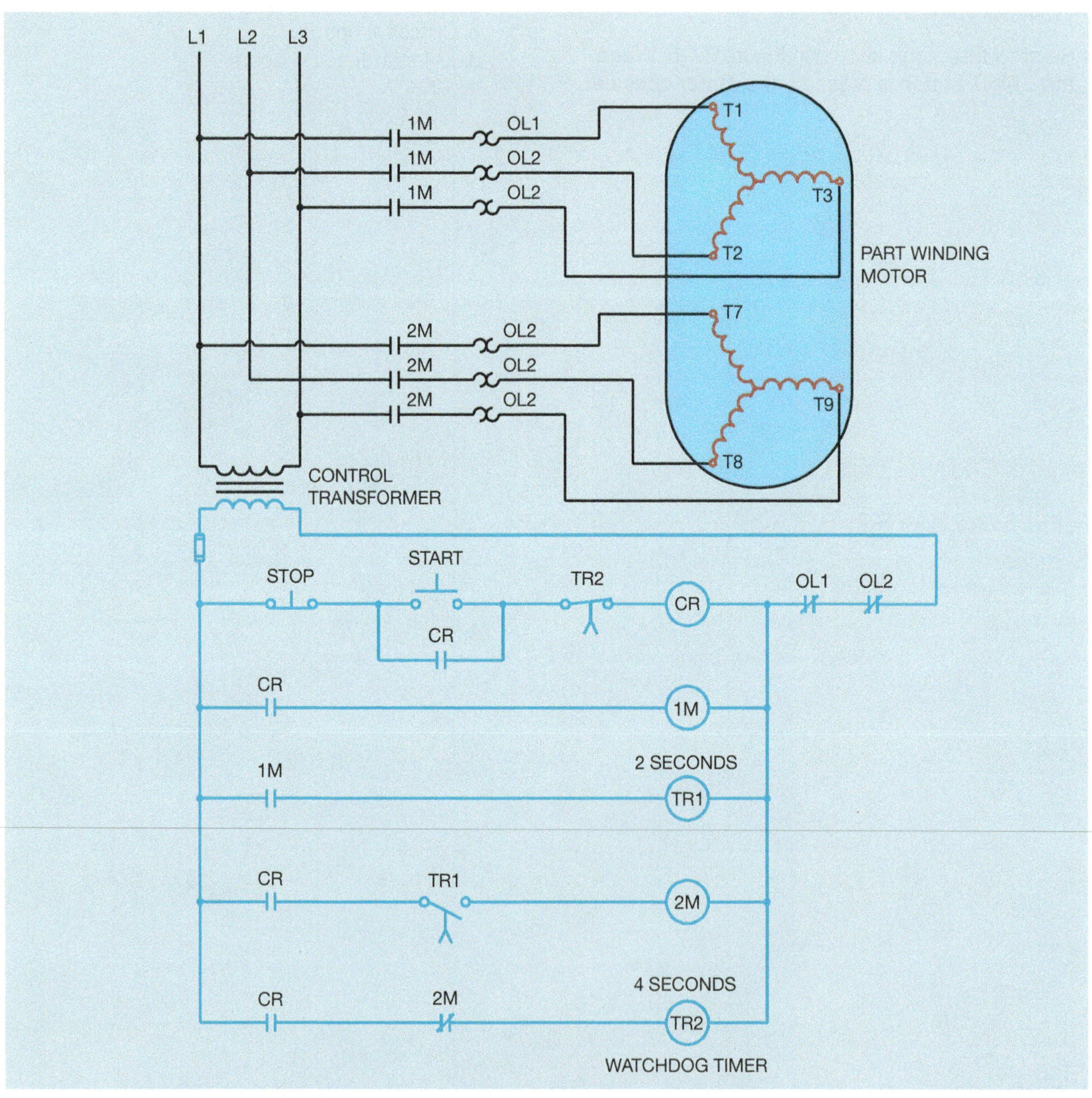

Figure 27–5 Watchdog timer disconnects the motor if the second winding does not energize.

Review Questions

1. A dual voltage 240/480-volt motor is to be used for part winding starting. Which voltage must be used and why?
2. Are the stator windings of a motor designed for part winding starting connected in parallel or series?
3. The nameplate of a part winding motor indicates a full load current rating of 72 amperes. What current rating should be used when sizing the overload heaters?
4. What is a watchdog timer?
5. Refer to the circuit shown in Figure 27–5. When the START button is pressed, the motor does not start. Which of the following could not cause this problem.
   a. The control transformer fuse is blown.
   b. Overload contact #2 is open.
   c. TR1 timer coil is open.
   d. Control relay coil CR is open.
6. Refer to the circuit shown in Figure 27–5. When the START button is pressed, the motor does not start. After a 4-second time delay control relay CR de-energizes. Which of the following could cause this problem?
   a. TR1 timer coil is open.
   b. 1M starter coil is open.
   c. CR coil is open.
   d. 2M starter coil is open.

Chapter 28

DIRECT CURRENT MOTORS

Objectives

After studying this chapter the student will be able to:

- List the major types of direct current motors.
- Explain the difference between series and shunt field windings.
- Discuss different types of armature windings.
- Describe operating characteristics of different types of direct current motors.
- Discuss methods of limiting in-rush current.
- Describe the function of a field loss relay.
- Discuss methods of manual speed control for direct current motors.
- Determine the direction of rotation of a direct current motor.

Direct current motors have been used in industry for many years. Their speed can be controlled from zero to maximum RPM. and they have the ability to develop very high torque at low speed. Direct current motors operate on the principle of attraction and repulsion of magnetism, not rotating **magnetic fields** as do alternating current poly-phase motors. Direct current motors can be divided into three basic types—series, shunt, and compound. The operating characteristics of each depend on the type of field and armature windings in the motor.

Field Windings

Two types of field windings are used in DC motors: series and shunt. Series field windings are constructed with a relatively few turns of large wire and have a very low resistance. They are connected in series with the armature. Series field windings are marked S1 and S2 in the terminal connection box.

Shunt field windings contain many turns of relatively small wire. They are connected in parallel with the armature and exhibit a much higher resistance than the series field winding. Shunt field windings are marked F1 and F2 in the terminal connection box.

Armature Windings

The three basic types of armature windings are lap, wave, and frogleg. Lap wound armatures are constructed with a relatively few turns of large wire. They are commonly used in machines that are intended to operate on low voltage and high current, such as starter motors in automobiles, streetcars, and trolleys. Lap wound armatures have their windings connected in parallel with each other (Figure 28–1).

Wave wound armatures are intended for used in high voltage, low current machines, such as high voltage generators. The armature windings are connected in series (Figure 28–2). In a generator, the voltage produced in each winding combines, to increase the total output voltage. In a motor, the voltage applied to the circuit is divided across each winding.

Frogleg wound armatures are probably most used in large industrial machines. They are intended for use in moderate voltage circuits. The windings in a frogleg wound armature are connected in series-parallel (Figure 28–3). Regardless of the type armature used, the brush or armature leads are identified as A1 and A2 in the terminal connection box.

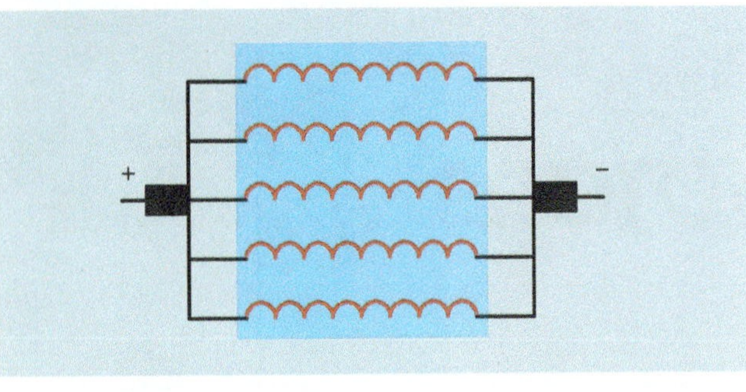

Figure 28–1 Lap wound armatures have their windings connected in parallel.

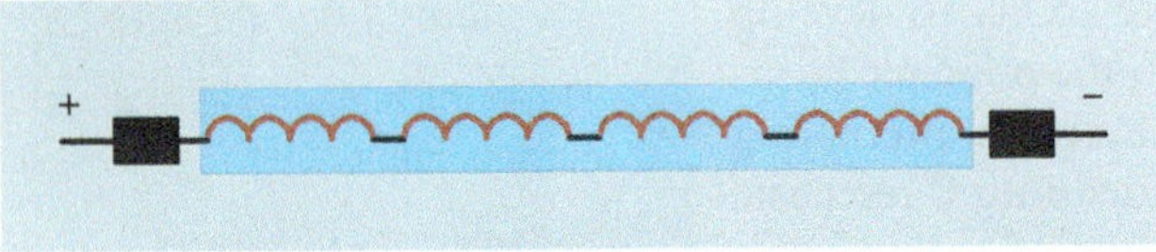

Figure 28–2 Wave wound armatures have windings connected in series.

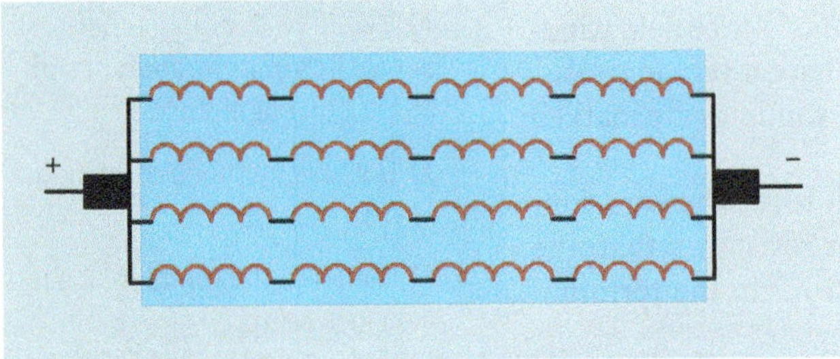

Figure 28–3 Frogleg wound armatures have windings connected in series-parallel.

Series Motors

Series motors contain only a series field connected in series with the armature (Figure 28–4). Because the armature and series fields are connected in series, they will have the same amount of current flow. An increase in armature current produces a corresponding increase in field current. Series motors can develop very high torque and are used as automobile starters and to power electric streetcars and trolleys. For many years they have been used to drive the wheels of diesel-electric locomotives.

Series motors operate on the principle of attraction and repulsion of magnetism. The amount of torque they develop depends on the magnetic field strength or flux density of the armature magnetic field and the magnetic field produced by the pole pieces (Figure 28–5). The motor's torque increases by the square of the motor current. Each time the motor's current doubles, the torque increases four times.

Series type motors do have one undesirable operating characteristic. They have almost no upper speed limit if operated without a load. Series motors should be connected either directly to the load or with gears. They should not be connected with belts or chains that can break. Large series motors can attain a great enough speed that centrifugal force will sling the windings out of the armature slots and the bars out of the commutator. In an unloaded condition, the only speed limiting factors are windage loss, bearing friction, and brush friction.

Speed control for a series motor is generally obtained by connecting resistance in series with the armature

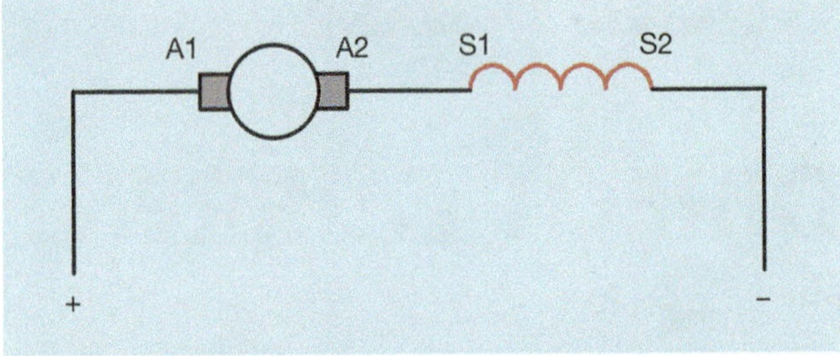

Figure 28–4 Series DC machines contain a series field connected in series with the armature.

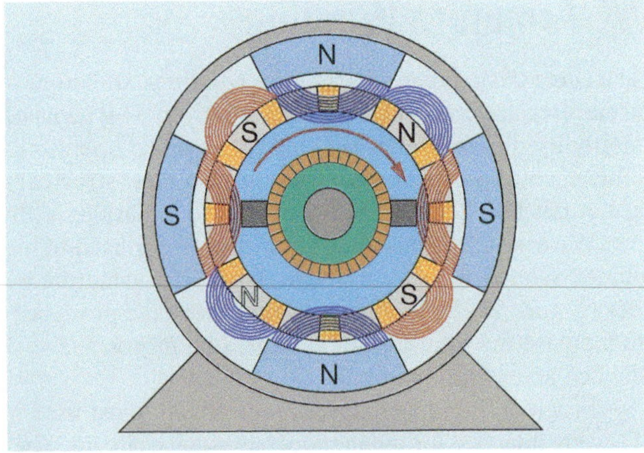

Figure 28–5 Torque of a direct current motor is determined by the flux density of the armature magnetic field and the pole pieces.

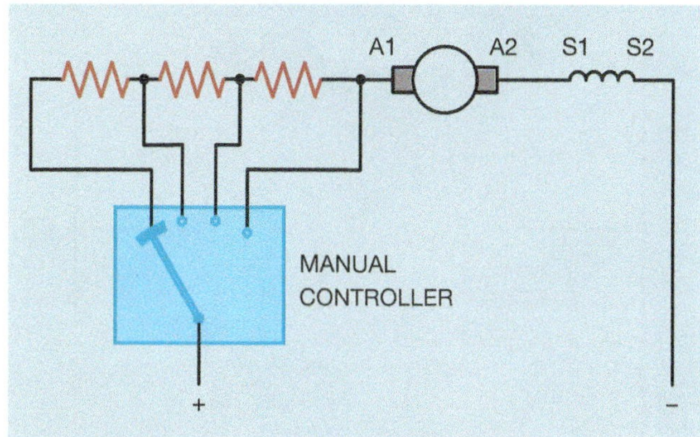

Figure 28–6 Series motors are controlled by connecting resistance in series with the armature.

(Figure 28–6). Manual type controllers are often used to determine the amount of resistance that is connected in series. The controller illustrated in Figure 28–6 is a make-before-break type. The controller is designed to make connection with another resistor before it breaks connection with the one to which it is connected. This provides closed transition starting for the motor.

Shunt Motors

Shunt motors have the shunt field connected in parallel with the armature (Figure 28–7). Because the shunt field is connected in parallel with the armature, the armature voltage supplies the current to the field winding. This connection produces a relatively constant current flow through the field, providing a stable magnetic field in the pole pieces. Because the magnetic field of the pole pieces is stable, the motor speed is relatively constant from no load to full load. Shunt motors are often called constant speed motors. Adding load to the motor results in a reduction of speed (Figure 28–8), but the shunt motor's speed is very constant compared to that of a series motor. The differential in speed from no load to full load is referred to as *speed regulation*. The speed regulation of a direct current motor is proportional to the armature resistance. Motors that have a very low armature resistance will have a better speed regulation.

If reducing the starting current of a shunt motor is necessary, it is accomplished by connecting resistance in series with the armature (Figure 28–9). In the circuit shown, the motor has three steps of starting. When the START button is pressed, both resistors are connected

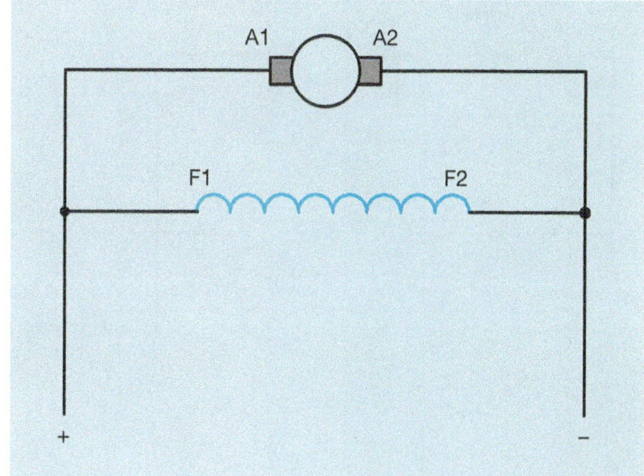

Figure 28–7 A shunt motor has the shunt field connected in parallel with the armature.

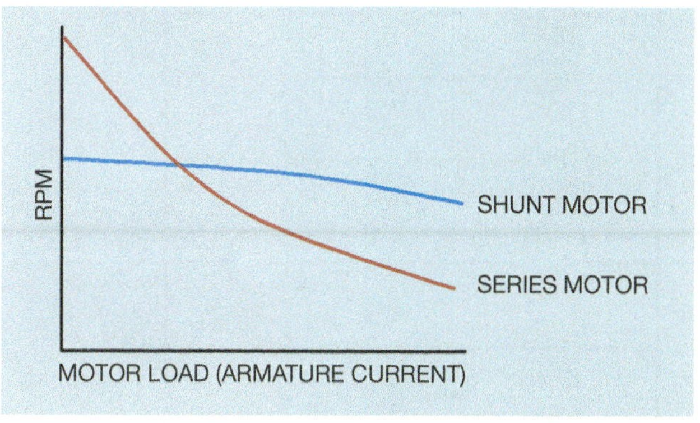

Figure 28–8 Speed characteristics of series and shunt motors.

in series with the armature. The resistors are shunted out of the circuit with a 3-second time delay between each step. In the example circuit, the motor is connected to a source of direct current, but the control system is controlled by alternating current. This is a common practice in industry.

When full line voltage is applied to both the armature and shunt field, the motor is said to operate at base or normal speed. If full voltage is applied to the shunt field and reduced voltage is applied to the armature, the motor operates at below normal speed or under speed (Figure 28–10). If full voltage is applied to the armature and reduced voltage is applied to the shunt field, the motor operates at above normal speed or over speed (Figure 28–11).

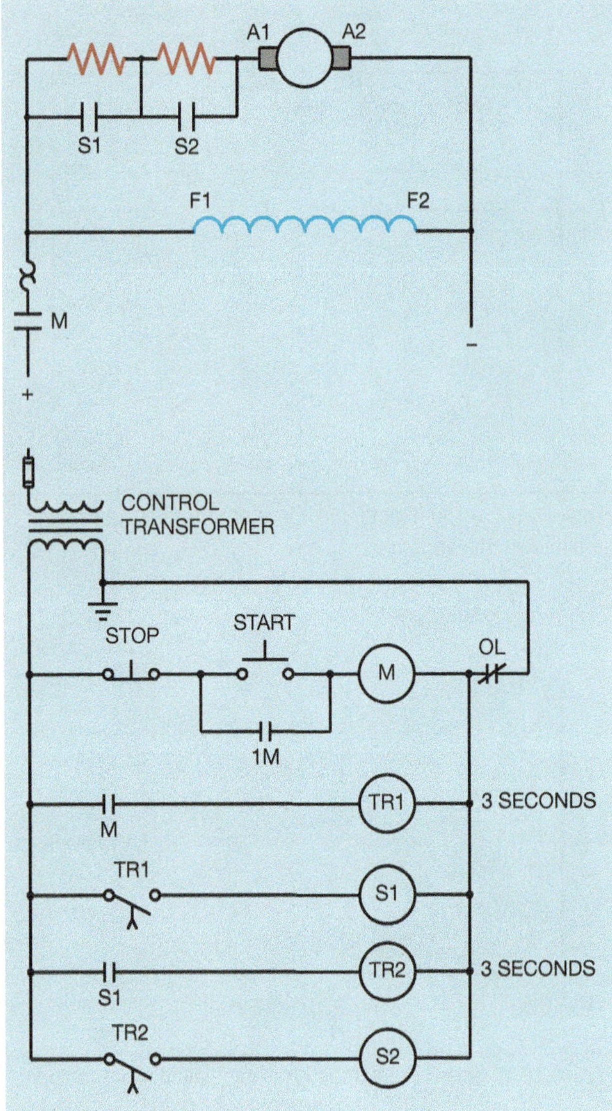

Figure 28–9 Three steps of starting for a DC shunt motor.

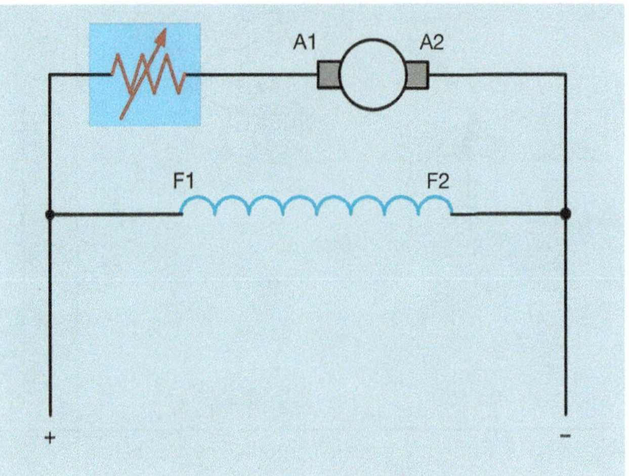

Figure 28–10 Under speed is accomplished by connecting full voltage to the shunt field and reduced voltage to the armature.

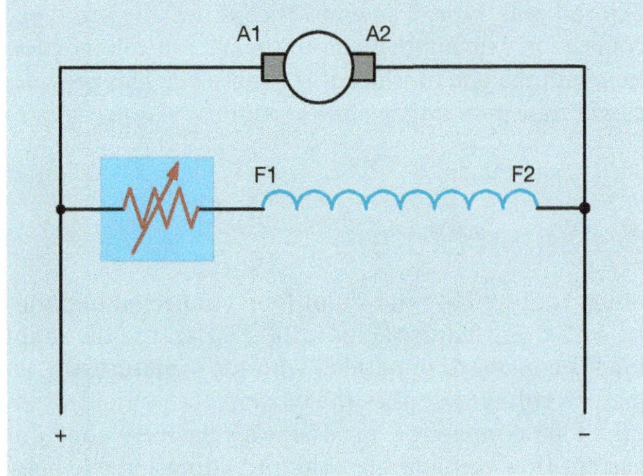

Figure 28–11 Over speed is accomplished by connecting full voltage to the armature and reduced voltage to the shunt field.

Connecting resistance in the shunt field causes the motor to operate at a higher speed because:

- Added resistance causes less current flow through the shunt field.
- Less current flow reduces the flux density in the pole pieces.
- Less flux density results in a reduction of counter EMF produced in the armature.
- Less counter EMF causes an increase of armature current.
- Increased armature current produces a stronger magnetic field around the armature.
- Although the field flux has been reduced, the armature flux increases by a much greater amount resulting in a net gain in torque.
- Increased torque causes the motor speed to increase.

Compound Motors

Compound motors are so named because they contain both a series and shunt fields (Figure 28–12). Compound motors are the most used DC motors in industry. They combine the high torque characteristics of series motors with the speed regulation of shunt motors. Although their speed regulation is not as good as a shunt motor, it is much better than the series motor (Figure 28–13). The constant voltage applied to the shunt field provides a stable amount of flux to the pole pieces to control speed. As load is added to the motor, armature current increases, causing the series field to become stronger and produce more torque.

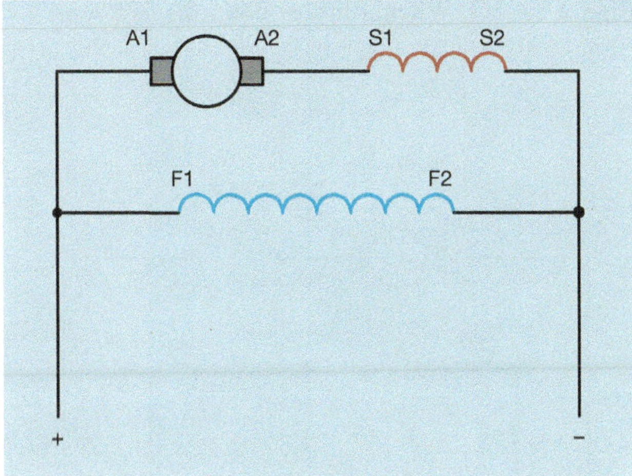

Figure 28–12 Compound motors contain both series and shunt field windings.

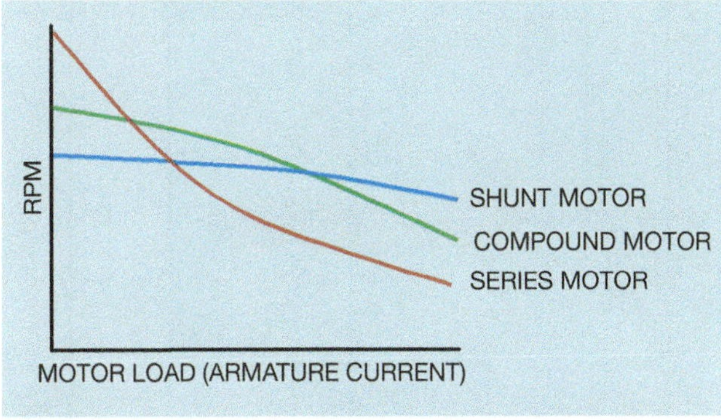

Figure 28–13 Speed characteristics of series, shunt, and compound motors.

Field Loss Relay

Compound motors generally employ a field loss relay (FLR) that is intended to disconnect the motor from the power line in the event that shunt field current should be interrupted or drop below an acceptable level (Figure 28–14). The field loss relay is a current relay. The coil of the FLR connects in series with the shunt field windings. When enough current flows through the FLR, a normally open contact closes. If the current should drop below a predetermined level, the contact will open. In the circuit shown in Figure 28–14, when the START button is pressed, a circuit is completed to M starter coil causing both the M load contact and M auxiliary contact to close. When the M load contact closes, it provides power to both the armature and shunt field. The current flow through the coil of the field loss relay causes the FLR contact to

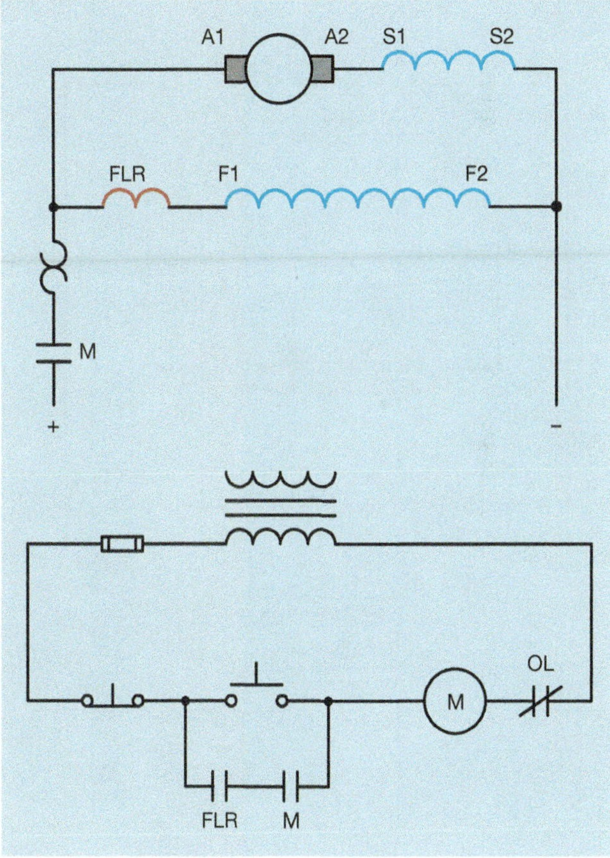

Figure 28–14 The field loss relay opens the circuit in the event that current flow through the shunt field drops below an acceptable level.

close. After the START button is released, both the FLR and M auxiliary contacts must remain closed to permit M starter to remain energized. If current flow through the shunt field should drop below an acceptable level, the FLR contact opens and de-energizes M starter.

Because compound motors contain both a series and shunt field winding, if current flow through the shunt field should be interrupted, the motor suddenly becomes a series motor. This change could cause the motor to over speed and damage the motor or driven machinery.

External Shunt Field Control

Many compound motors employ a separate DC power supply to provide the current for the shunt field and armature (Figure 28–15). Shunt field current and armature current can be controlled by the amount of voltage produced by the power supply. Most large direct current motors are controlled in this way. Solid-state power supplies provide the power necessary to operate the motor. If the shunt field is connected to a separate power source, it is common practice to leave the shunt field energized even when the motor is not operating. The heat produced by the shunt field prevents moisture from forming in the motor windings.

Some direct current compound motors actually contain two separate shunt fields. This is often the case if the motor is intended to be operated above the normal or base speed. The use of two shunt field windings ensures that one is in the circuit to provide some amount of constant magnetic flux to the pole pieces. The second winding is connected to a variable voltage power supply and is used to control the motor above normal speeds (Figure 28–16). When a motor contains two separate

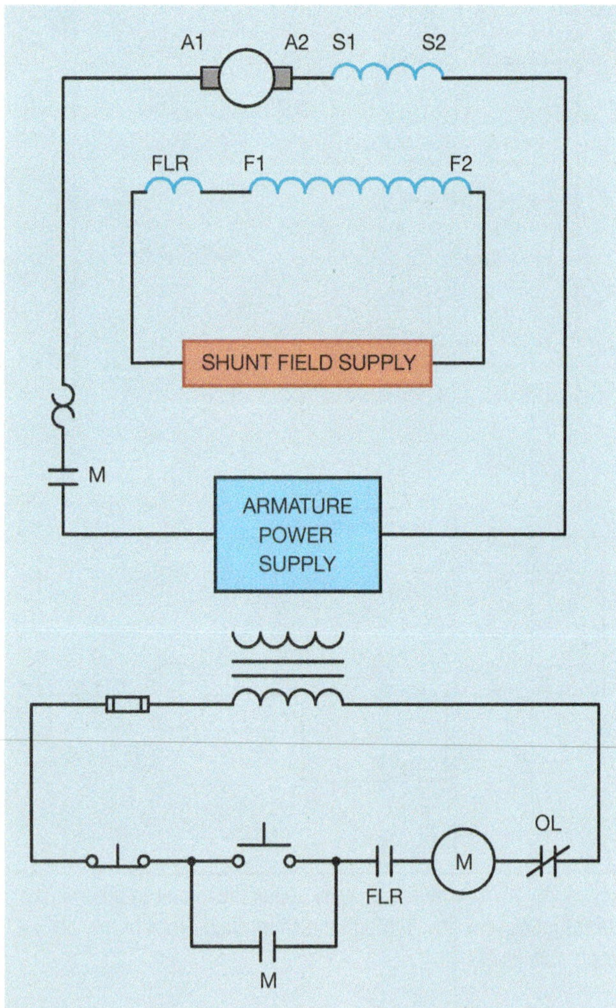

Figure 28–15 The shunt field current is provided by a separate power supply.

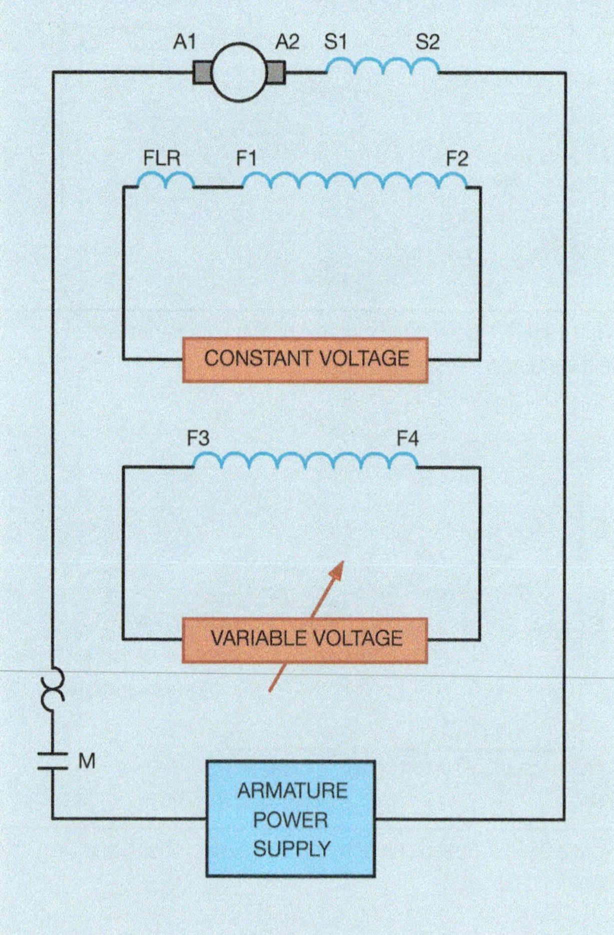

Figure 28–16 Two separate shunt fields are often used when the motor is to be operated above normal speed or base speed.

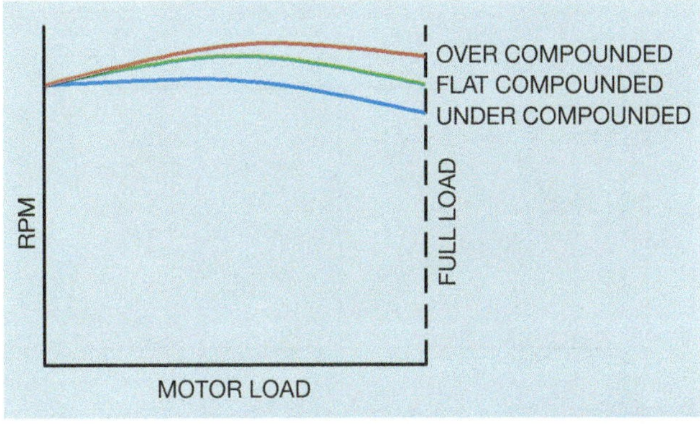

Figure 28–17 The compounding determines the speed regulation of a compound motor.

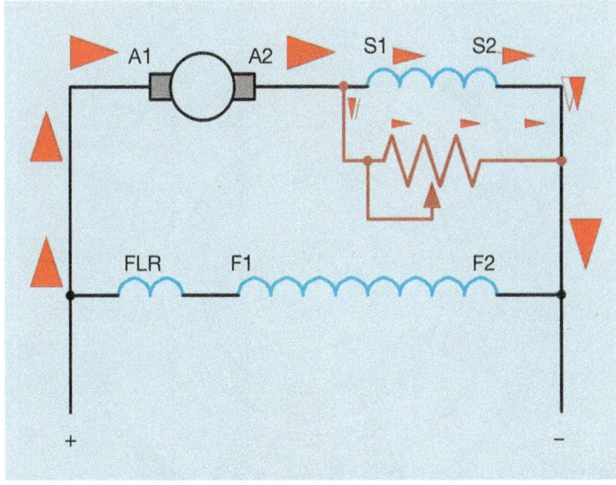

Figure 28–18 The series field shunt rheostat diverts part of the current away from the series field.

shunt fields, one is labeled F1 and F2, and the other is labeled F3 and F4.

Controlling Compounding

Compounding is the term used to indicate the relative strength of the series and shunt fields. If the series and shunt fields are balanced, the series field compensates for any motor losses and causes the motor to operate at almost constant speed from no load to full load. This condition is called *flat compounding* (Figure 28–17). If the series field is too strong and produces an excessive amount of magnetic flux as load is added to the motor, the speed will be greater at full load than it was at no load. This is called *over compounding.* If the series field does not produce enough magnetic flux as load is added to the motor, it will not compensate for motor losses and the full load speed will be less than the no load speed. This is called *under compounding*.

Because the amount of magnetic flux produced by the series field is the factor that determines compounding, controlling the current flow through the series field will control the amount of compounding. Controlling compounding is accomplished by connecting a resistor in parallel with the series field (Figure 28–18). This resistor is called the *series field shunt rheostat* or the *series field diverter*. Its function is to divert part of the current away from the series field making it produce less magnetic flux.

Cumulative and Differential Compounding

Compound direct current motors can also be connected as cumulative or differential compound. When a direct current compound motor is constructed, both the series and shunt field windings are wound on each pole piece (Figure 28–19). When both fields are connected in such a manner that current flows in the same direction through each, they aid each other in the production of magnetism because each field produces the same magnetic polarity (Figure 28–20). This condition is called *cumulative compounding*.

If the series and shunt fields are connected in such a manner that the current flows in opposite directions through the fields, they oppose each other in the production of magnetism because each field produces the opposite magnetic polarity (Figure 28–21). This condition is called *differential compounding*. Although there are some applications for differentially compounded motors, it is generally a connection to be avoided. If a differentially compounded motor is started at no load, the shunt field determines the direction or rotation. As load is added to the motor, the series field becomes stronger as armature current increases. The motor

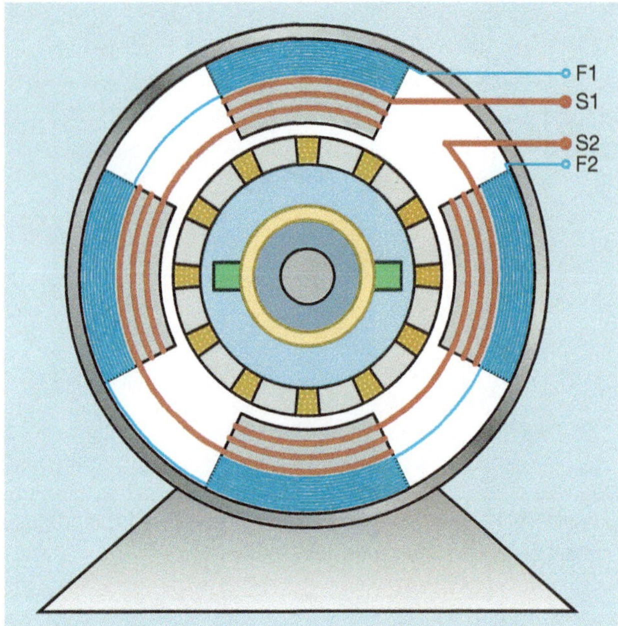

Figure 28–19 Both series and shunt field windings are wound on the same pole piece.

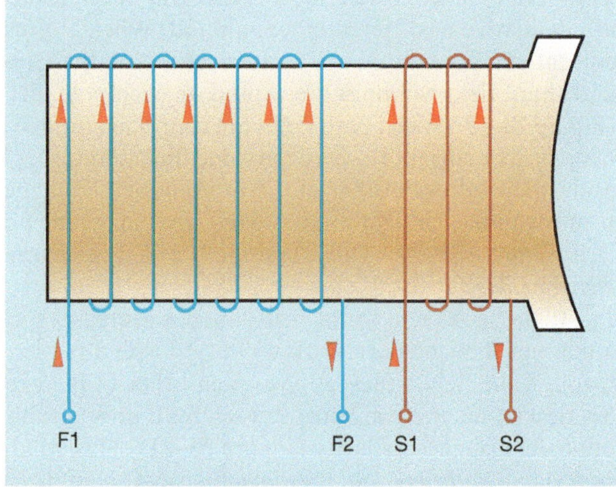

Figure 28–20 Motors are cumulative compounded when the current flow through both the series and shunt fields produce the same magnetic polarity.

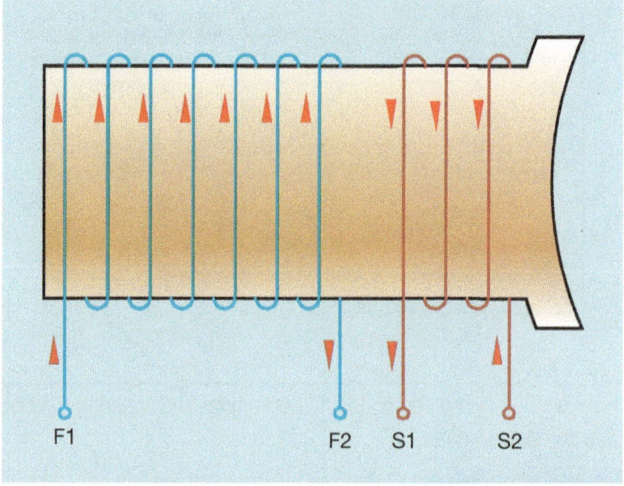

Figure 28–21 Motors are differential compounded when the current flow through the series and shunt fields produce the opposite magnetic polarity.

speed reduces rapidly as load is added to the motor. If the load becomes great enough, the series field becomes strong enough to change the polarity of the pole pieces. This causes the motor to reverse direction and operate as a series motor.

Testing the Motor for Cumulative or Differential Compounding

When compound motors are installed, they should be tested to determine whether they are connected for cumulative or differential compounding. Although the series and shunt field markings should indicate the proper connection, in some cases these markings have been found to be incorrect. As stated previously, when a compound motor is operated without a connected load, the shunt field determines the direction of rotation. To test a compound motor for cumulative or differential compounding:

- Disconnect the motor from the load.
- Disconnect the shunt field at the terminal connection box.
- The motor is now connected as a series motor. "Bump" the motor by momentarily connecting it to power just long enough to observe the direction of rotation.
- Reconnect the shunt field and connect the motor to power.
- If the motor turns in the same direction, the fields are connected cumulative compound. If the motor turns in the opposite direction, the fields are connected differential compound. The compounding can be changed by reversing connection to either the series or shunt field winding.

Direction of Rotation

The direction of rotation of a direct current motor can be reversed by changing the magnetic polarity of either the pole pieces or the armature, but not both. Small shunt motors often employ a switch that reverses the direction of current flow through the field winding (Figure 28–22). This is often done with shunt type motors because the amount of field current is generally less than the armature current, permitting a switch with a lower current rating to be used.

The most common way of reversing the direction of rotation for a direct current motor is to reverse the direction of current flow through the armature. This is especially true of compound motors because the motor contains both series and shunt fields. Reversing the armature leads removes any danger of changing a cumulative compound motor into a differential compound motor. A forward–reverse control for a direct current compound motor is shown in Figure 28–23. The control circuit is the same as that used for reversing the direction of rotation of a poly-phase motor discussed previously.

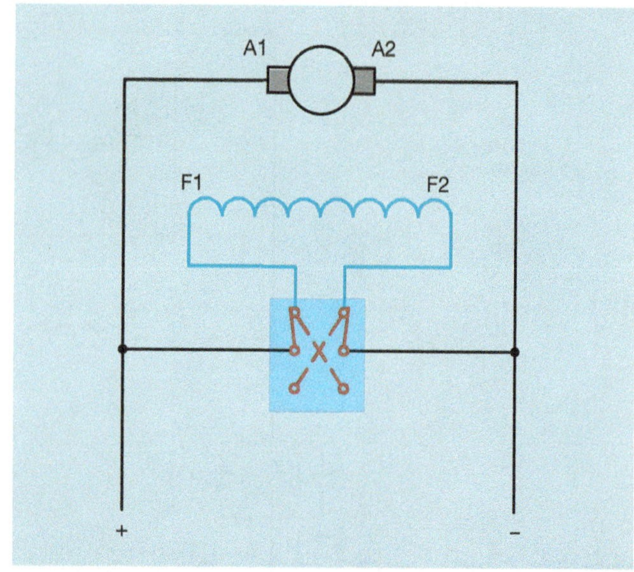

Figure 28–22 A double-pole double-throw (DPDT) switch can be used to reverse the current direction through the shunt field.

The load part of the circuit, however, causes the direction of current flow through the armature to change depending on which contactor is energized.

Determining the Direction of Rotation

If the leads of a direct current motor have been labeled in a standard manner, the direction of rotation can be determined when the motor is connected. Direction of rotation is determined by facing the commutator end of the motor, which is generally the rear of the motor. The standard connections for series, shunt, and compound motors are shown in Figure 28–24. A cut-away view of a direct current motor showing the commutator, brushes, armature windings, and field windings is shown in Figure 28–25.

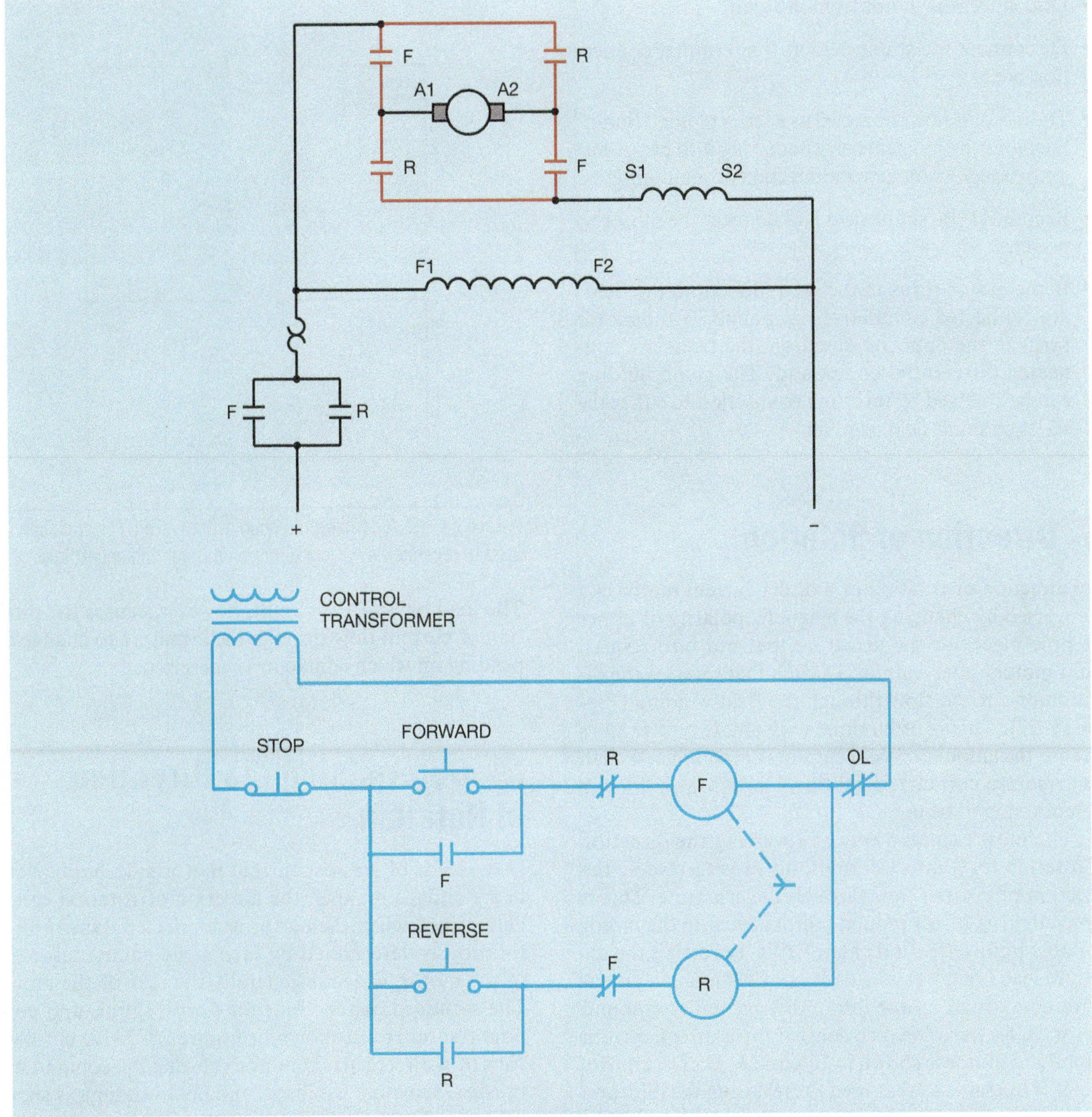

Figure 28–23 Forward–reverse control for a direct current compound motor.

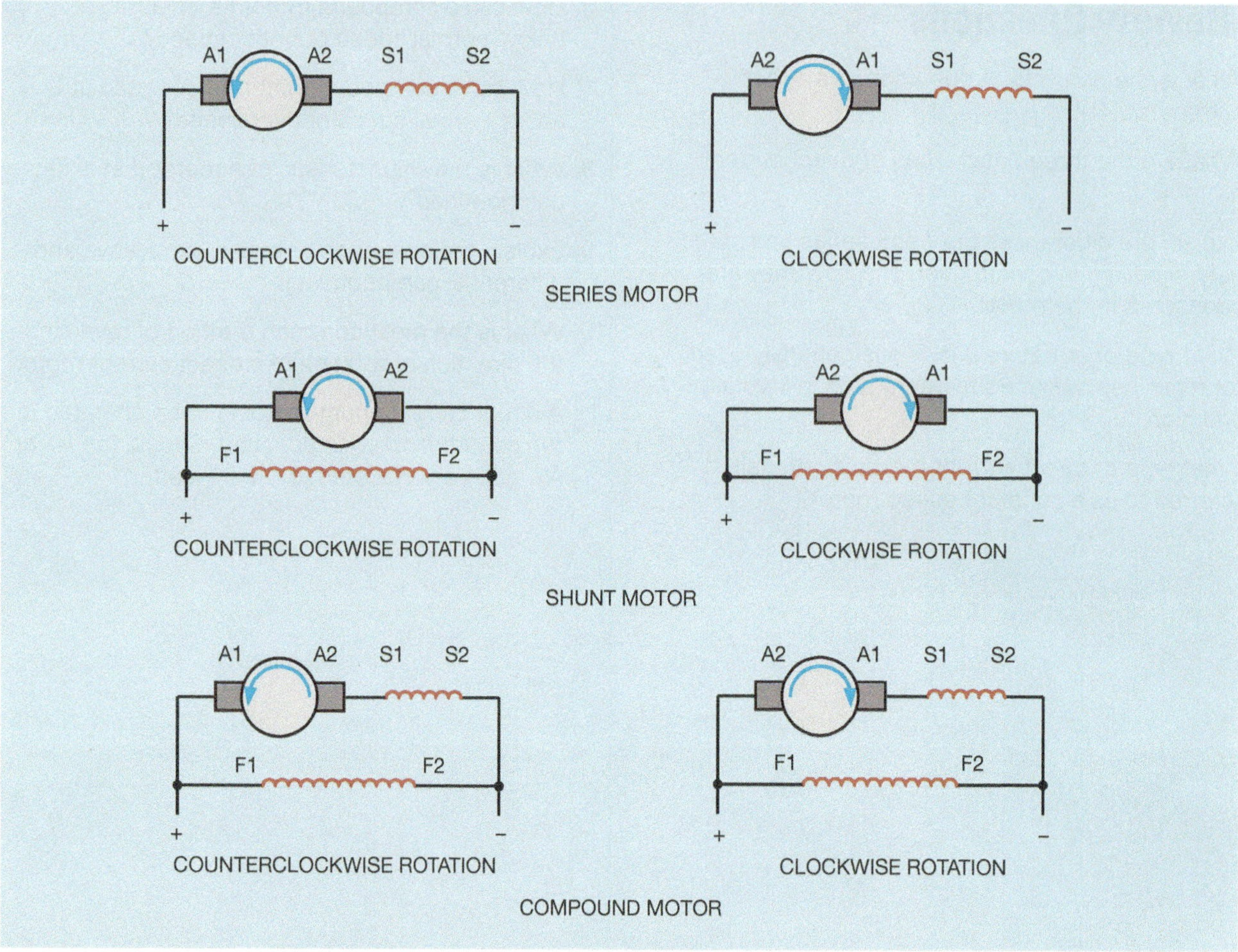

A1 A2 S1 S2
\+ −
COUNTERCLOCKWISE ROTATION

A2 A1 S1 S2
\+ −
CLOCKWISE ROTATION

SERIES MOTOR

A1 A2
F1 F2
\+ −
COUNTERCLOCKWISE ROTATION

A2 A1
F1 F2
\+ −
CLOCKWISE ROTATION

SHUNT MOTOR

A1 A2 S1 S2
F1 F2
\+ −
COUNTERCLOCKWISE ROTATION

A2 A1 S1 S2
F1 F2
\+ −
CLOCKWISE ROTATION

COMPOUND MOTOR

Figure 28–24 Determining direction of rotation of direct current motors.

Photo courtesy of Baldor Electric Co.

Figure 28–25 A DC motor with visible commutator, brushes, armature windings, and field windings.

Review Questions

1. What is the principle of operation of a direct current motor?
2. What are the three basic types of direct current motors?
3. Explain the differences between series and shunt field windings in construction and how they are connected in the motor.
4. What type of armature winding is generally used for machines designed to operate on low voltage and high current?
5. What type of direct current motor is generally referred to as a constant speed motor?
6. How can a compound motor be operated at below normal speed or under speed?
7. How can a compound motor be operated at above normal speed or over speed?
8. What is the characteristic of a motor that is flat compounded?
9. Explain the difference between cumulative and differential compounding.
10. What is the most common method of reversing the direction of rotation of a direct current motor?
11. A direct current shunt motor is to be operated in the counterclockwise direction. Should the A1 or A2 lead be connected to the F1 lead?

Chapter 29

SINGLE PHASE MOTORS

Objectives

After studying this chapter, the student will be able to:

- Discuss the operation of single phase motors.
- List different types of split phase motors.
- Discuss different starting methods for split phase motors.
- Reverse the direction of rotation for a split phase motor.
- Connect a dual voltage motor for operation on either 240 or 120 volts.

Although most motors used in industry are three phase, there are many applications for single phase motors. They operate on either 120 or 240 volts, are small horsepower, and almost all are intended to be started across-the-line. Some are designed for single voltage operation, and others are dual voltage. Unlike three phase motors that operate on the principle of a rotating magnetic field, different types of single phase motors operate on different principles. Some do operate on the principle of a rotating magnetic field, but others do not. The most prevalent type of single phase motor that may require different starting methods is the split phase motor. **Split phase** motors do operate on the principle of a rotating magnetic field. In order to produce a rotating magnetic field, there must be more than one phase. Split phase motors are so named because they split one phase into two phases to imitate a two phase power system. A two phase power system produces two separate voltages 90 degrees out of phase with each other (Figure 29–1).

In order to split one phase into two, split phase motors contain two separate windings: the run winding and the start winding (Figure 29–2). The run and start windings are connected in parallel with each other.

The three types of split phase motors are **resistance start induction run**, capacitor start induction run, and capacitor-start capacitor-run. The resistance start induction run and the capacitor start capacitor run motors must disconnect the start winding when the motor reaches about 75 percent of its rated speed. The capacitor start induction run motor has a capacitor connected in series with the start winding to increase the starting torque of the motor (Figure 29–3). The capacitor increases the starting torque by producing a greater phase shift between the current in the run winding and the current in the start winding. Maximum starting torque is developed when the run-winding current and start-winding current are 90 degrees out of phase with each other. The start windings of resistance start induction run and capacitor start induction run motors are designed to remain in the circuit for only a very short period of time and will be damaged if they are not disconnected. Capacitor-start capacitor-run motors are so designed that the start windings are not disconnected from the circuit. These motors operate very similar to true two phase motors. They are generally used as the compressor motor for central air conditioning systems in homes and to power ceiling fans and other appliances that do not require some method of disconnecting the start winding.

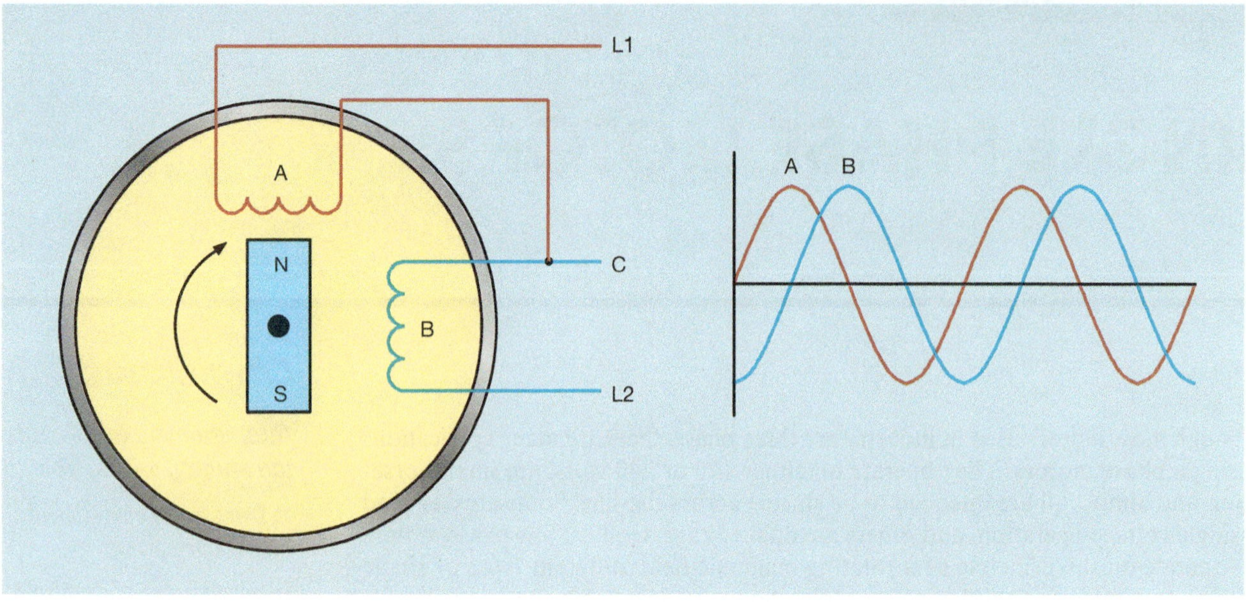

Figure 29–1 Two phase alternator.

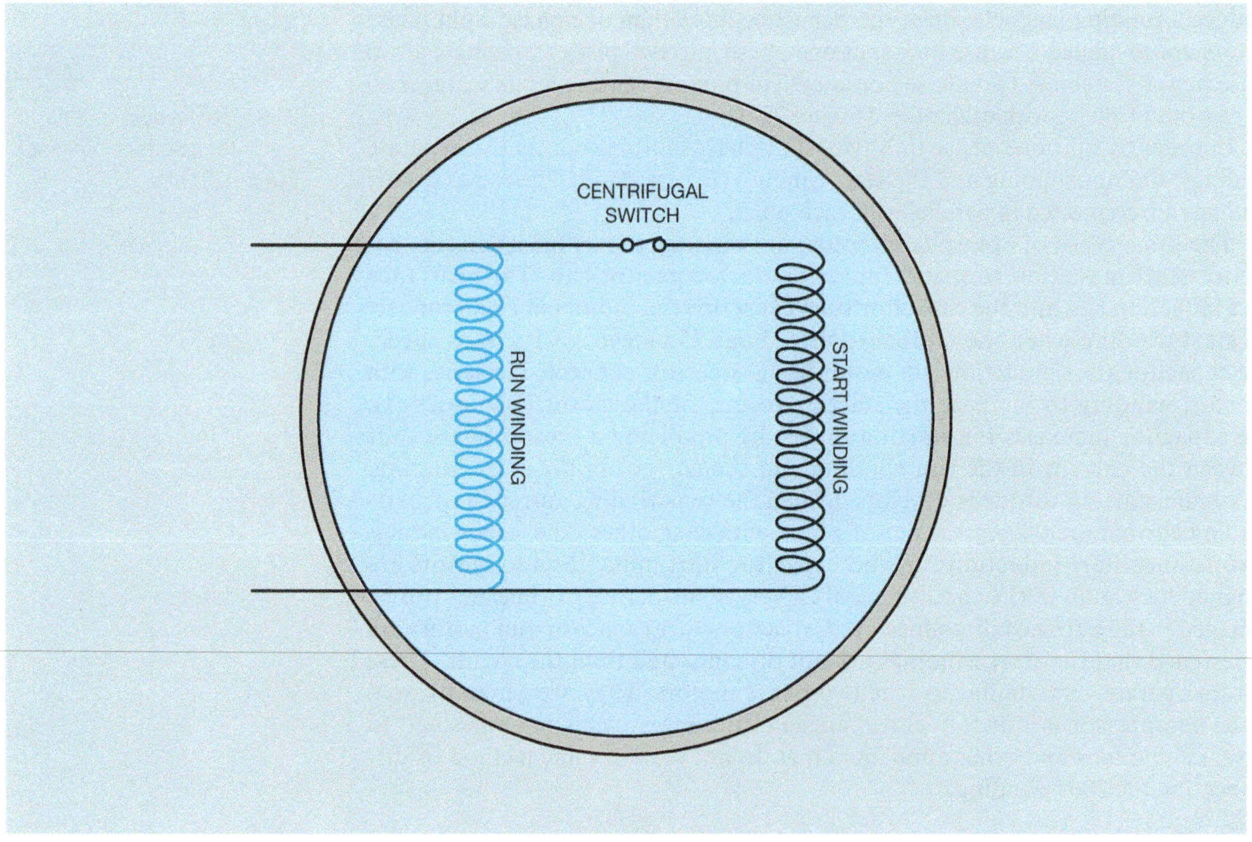

Figure 29–2 The run winding and start winding are connected in parallel with each other.

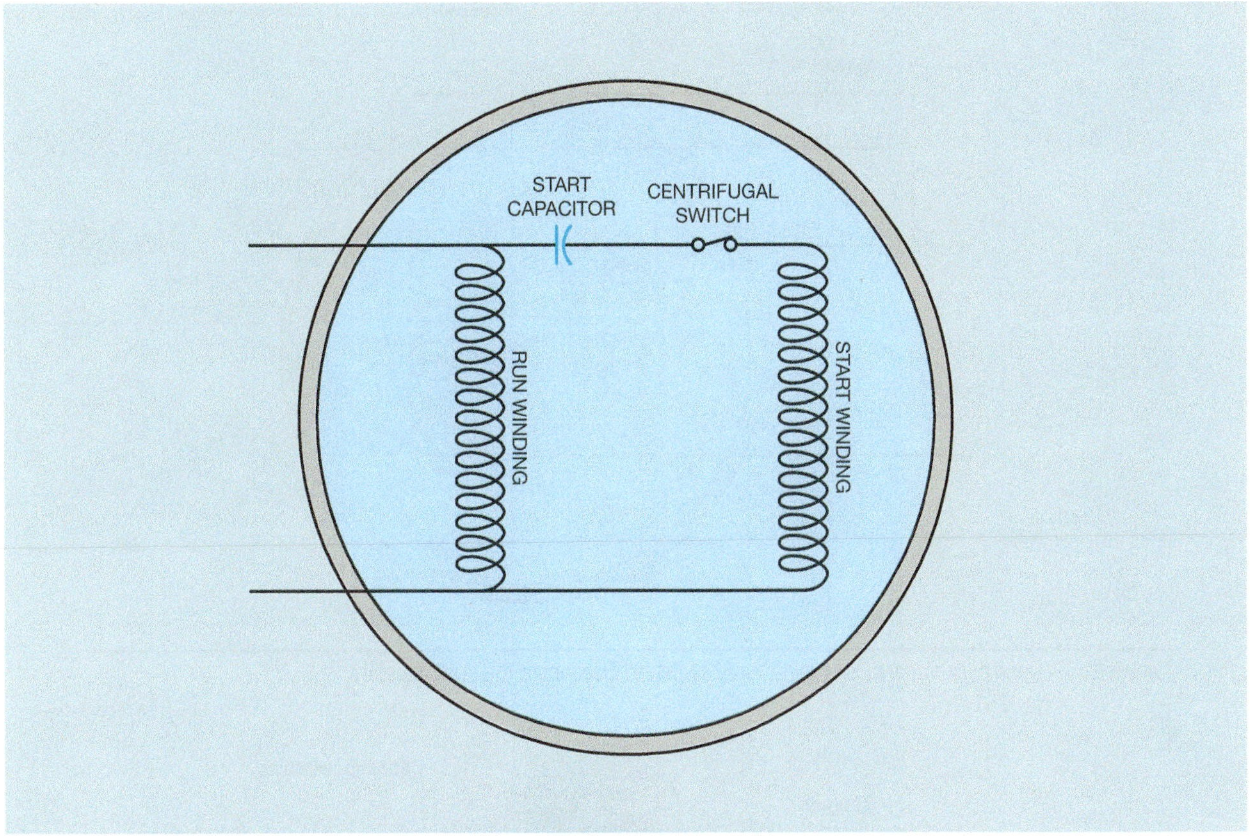

Figure 29–3 The starting capacitor produces a 90 degree phase shift between run-winding current and start-winding current.

Centrifugal Switch

Split phase motors intended to operate in the open using a centrifugal switch connected to the shaft of the rotor (Figure 29–4). The centrifugal switch is operated by springloaded counterweights. When the rotor reaches a certain speed, the counterweights overcome the springs and open the switch, disconnecting the start winding from the power line.

Some capacitor-start capacitor-run motors, sometimes referred to as permanent split capacitor motors, employ the use of a starting capacitor as well as a running capacitor. The starting capacitor is used to provide extra torque during the starting period and must be disconnected after the motor reaches about 75 percent of rated speed. Open case motors generally use a centrifugal switch to perform this function, Figure 29–5. An exploded view of a capacitor-start capacitor-run motor with extra starting capacitor is shown in Figure 29–6.

Figure 29–4 Centrifugal switch.

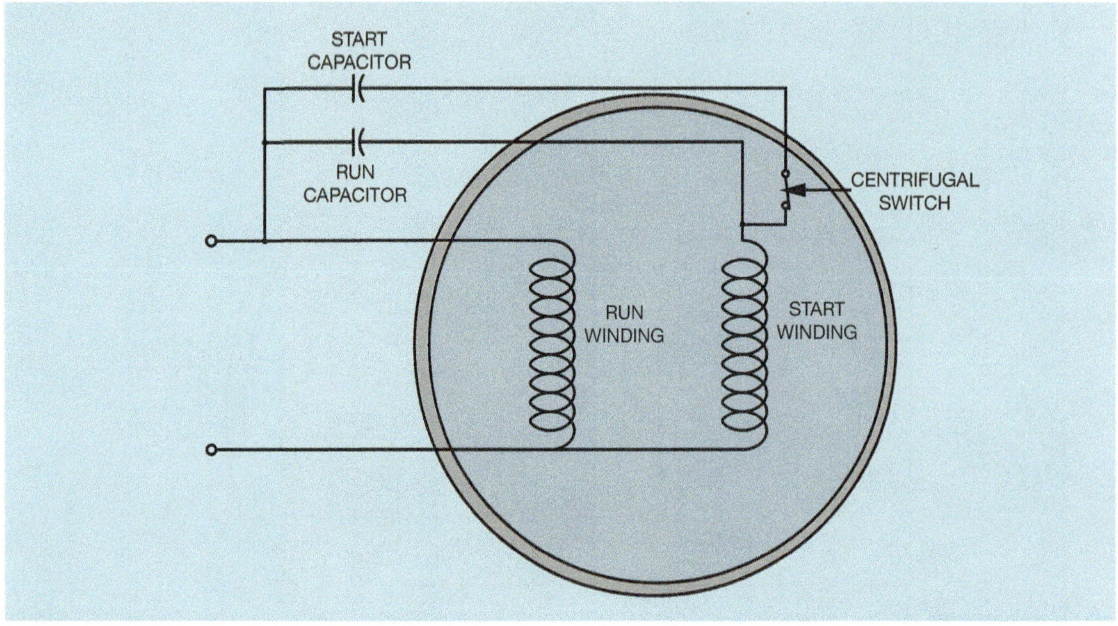

Figure 29–5 Open case motors use a centrifugal switch to disconnect the start capacitor.

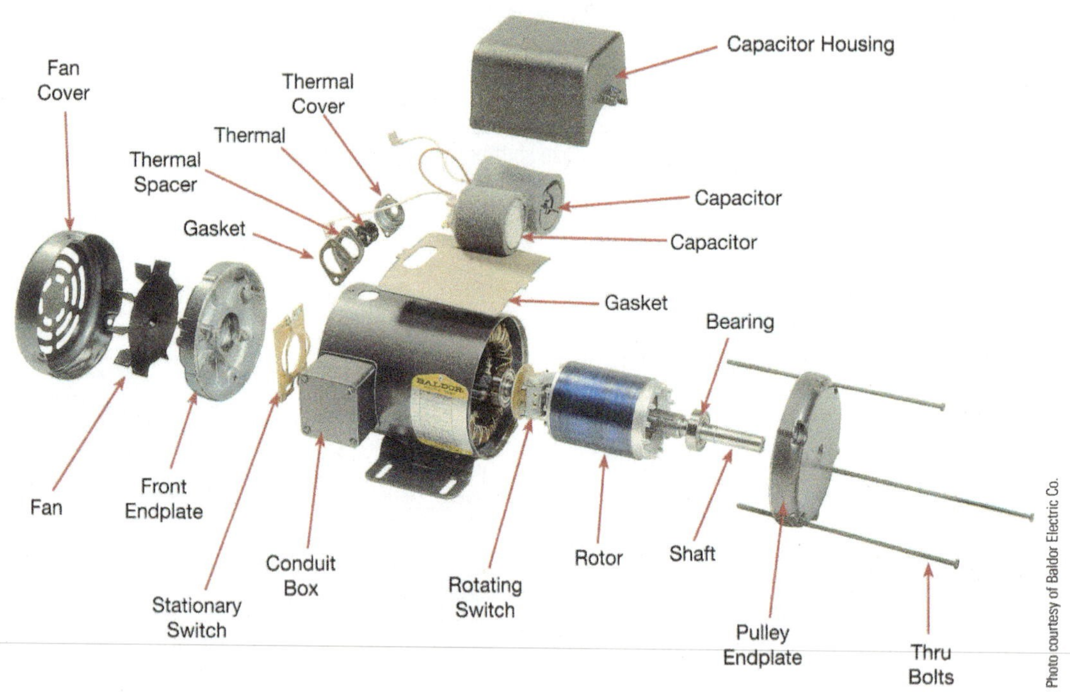

Figure 29–6 Exploded view of a capacitor-start capacitor-run motor with extra starting capacitor.

Hot-Wire Starting Relay

Centrifugal switches cannot be used on all types of split phase motors, however. Hermetically sealed motors used in refrigeration and air conditioning, or submerged pump motors, must use some other means to disconnect the start winding. Although the *hot-wire relay* is seldom used anymore, it is found on some older units that are still in service. The hot-wire relay functions as both a starting relay and an overload relay. In the circuit shown in Figure 29–7, it is assumed that a thermostat controls the operation of the motor. When the thermostat closes,

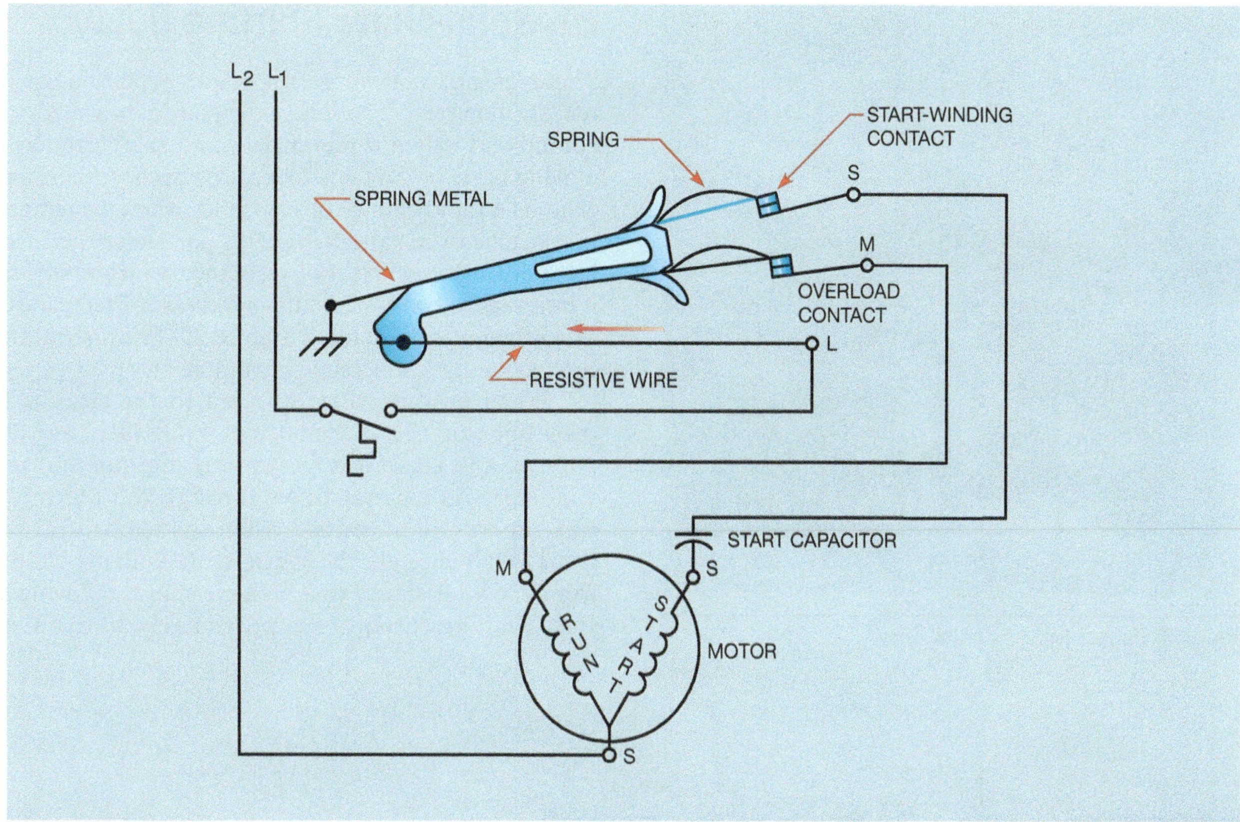

Figure 29–7 Hot-wire relay connection.

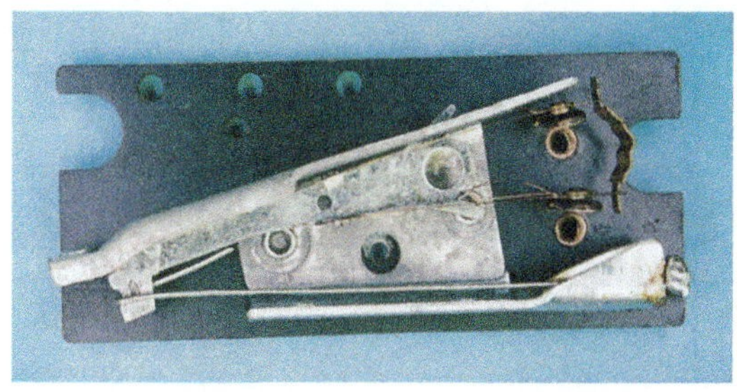

Figure 29–8 Hot-wire type of starting relay.

current flows through a resistive wire and two normally closed contacts connected to the start and run windings of the motor. The starting current of the motor is high, which rapidly heats the resistive wire causing it to expand. The expansion of the wire causes the springloaded start-winding contact to open and disconnect the start winding from the circuit, reducing motor current. If the motor is not overloaded, the resistive wire never becomes hot enough to cause the overload contact to open, and the motor continues to run. If the motor should become overloaded, however, the resistive wire expands enough to open the overload contact and disconnect the motor from the line (Figure 29–8).

Current Relay

The **current relay** operates by sensing the amount of current flow in the circuit. This type of relay operates on the principle of a magnetic field instead of expanding metal. The current relay contains a coil with a few turns of large wire and a set of normally open contacts (Figure 29–9). The coil of the relay is connected in series with the run winding of the motor, and the contacts are connected in series with the start winding as shown in Figure 29–10. When the thermostat contact closes, power is applied to the run winding of the motor. Because the start winding is open, the motor cannot start. This causes a high current to flow in the run-winding circuit. This high current flow produces a strong magnetic field in the coil of the relay, causing the normally open contacts to close and connect the start winding to the circuit. When the motor starts, the run-winding current is greatly reduced, permitting the start contacts to reopen and disconnect the start winding from the circuit.

Figure 29–9 Current type of starting relay.

Solid-State Starting Relay

The *solid-state starting relay* is rapidly replacing the current starting relay. The solid-state relay uses a solid-state component called a **thermistor**, and therefore has no moving parts or contacts to wear or burn. A thermistor exhibits a rapid change of resistance when the temperature reaches a certain point. This particular thermistor has a positive coefficient of resistance, which means that it increases its resistance with an increase of temperature. The schematic diagram in Figure 29–11 illustrates the connection for a solid-state starting relay.

When power is first applied to the circuit, the resistance of the thermistor is relatively low, 3 or 4 ohms, and current flows to both the run and start windings. As current flows through the thermistor its temperature increases. When the temperature becomes high enough, the thermistor suddenly changes from a low resistance to a high resistance reducing the start-winding current to approximately 30 to 50 mA.

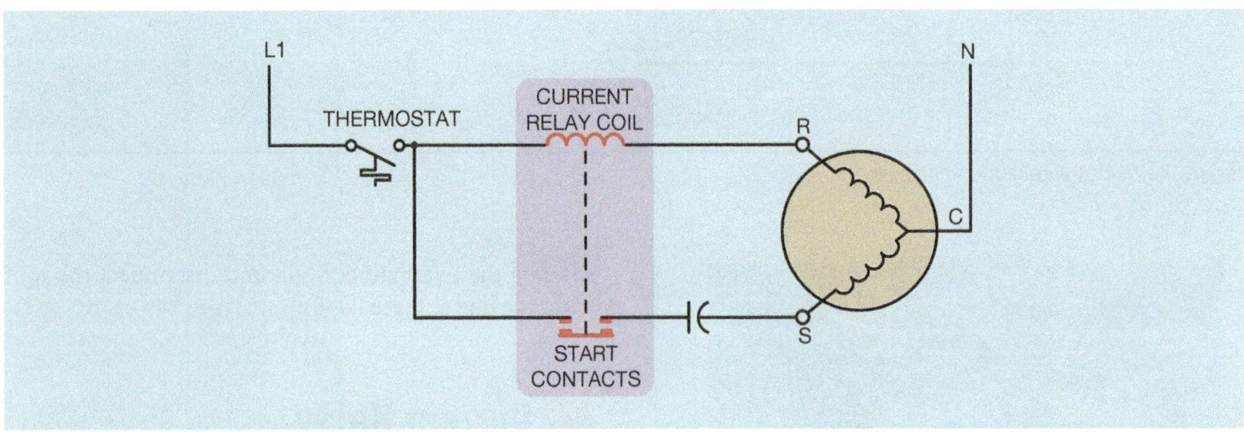

Figure 29–10 Current relay connection.

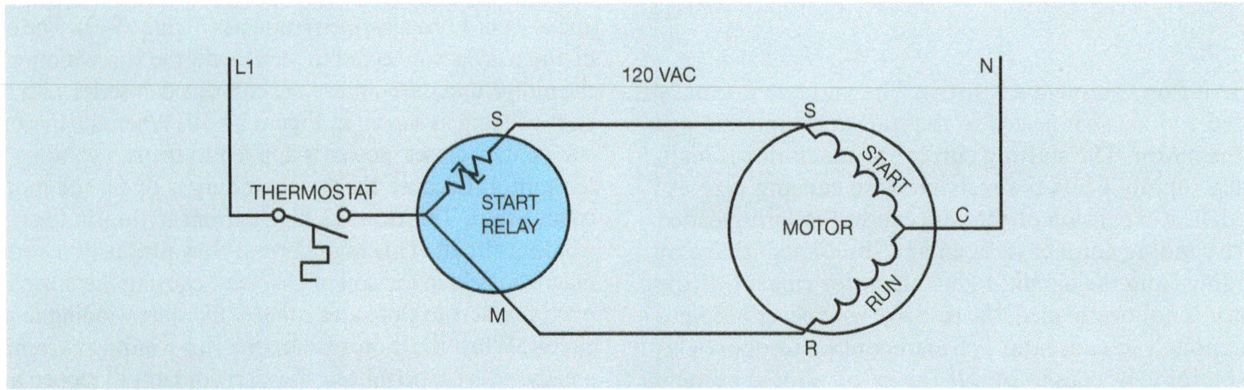

Figure 29–11 Solid-state starting relay circuit.

This has the effect of disconnecting the start winding from the circuit. Although a small amount of *leakage current* continues to flow, it has no effect on the operation of the motor. This leakage current maintains the temperature of the thermistor and prevents it from returning to a low resistance while the motor is in operation. When the motor is stopped, a cooldown period of 2 or 3 minutes should be allowed to permit the thermistor to return to a low resistance.

Potential Starting Relay

The *potential starting relay* is used with the capacitor-start capacitor-run or permanent split capacitor motor. This type of split phase motor does not disconnect the start windings from the circuit. Because the start winding remains energized, it operates very similarly to a true two phase motor. All of these motors contain a run capacitor that remains connected in the start winding circuit at all times. Many of these motors contain a second capacitor that is used during the starting period only. This capacitor must be disconnected from the circuit when the motor reaches about 75 percent of its rated speed. Open case motors generally use a centrifugal switch to perform this function, but hermetically sealed motors generally depend on a potential starting relay (Figure 29–12). The potential relay operates by sensing the increase of voltage induced in the start winding when the motor is in operation. The coil of the relay is connected in parallel with the start winding of the motor. A normally closed SR contact is connected in series with the starting capacitor. When power is connected to the motor, both the run and start windings are energized. At this time, both the run and start capacitors are connected in the start-winding circuit.

The rotating magnetic field of the stator induces a current in the rotor of the motor. As the rotor begins to turn, its magnetic field induces a voltage into the start winding, increasing the total voltage across the winding. Because the coil of the potential relay is connected in parallel with the start winding, this voltage increase is applied to it also, causing the normally closed contact connected in series with the starting capacitor to open and disconnect the starting capacitor from the circuit (Figure 29–13).

Dual Voltage Motors

Some split phase motors can be connected to operate on either 120 or 240 volts. This is accomplished by providing two sets of run windings. Some motors also provide two sets of start windings, and other motors use only one start winding. When two sets of start windings are used, the run windings are labeled T1 through T4, and the start windings are labeled T5 through T8 (Figure 29–14).

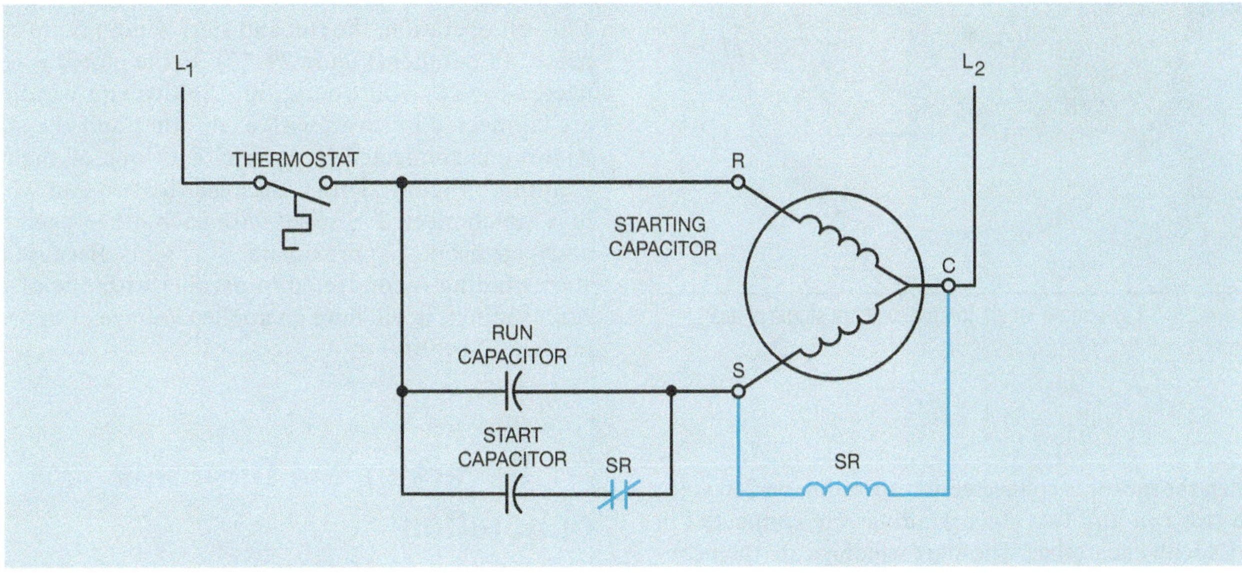

Figure 29–12 Potential relay connection.

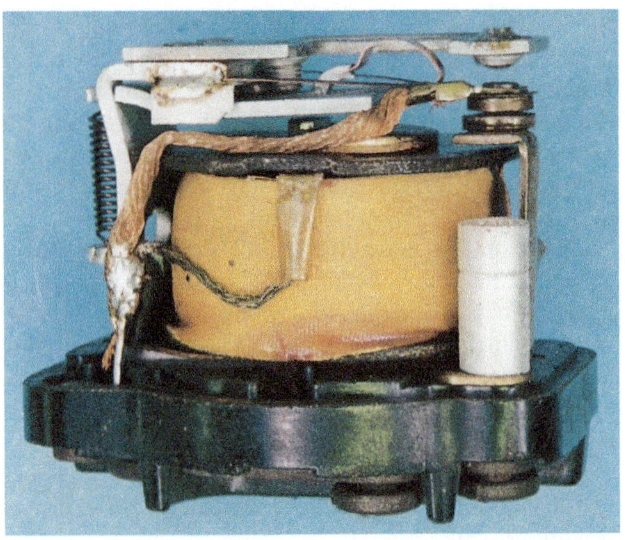

Figure 29–13 Potential starting relay.

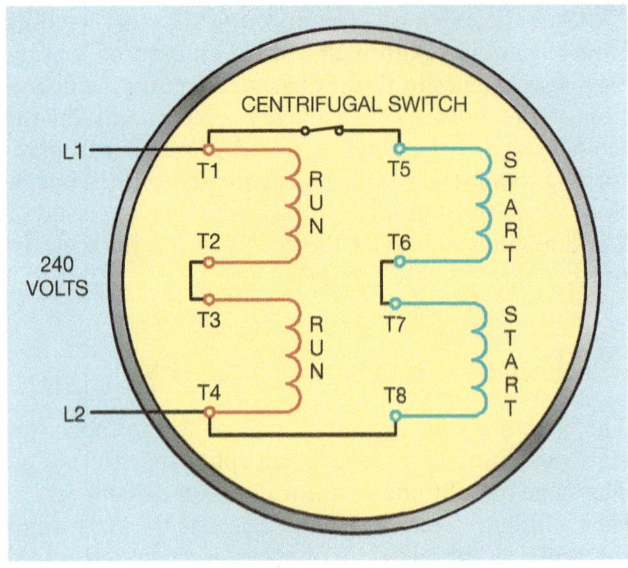

Figure 29–15 A 240-volt connection for split phase motor with two run and two start windings.

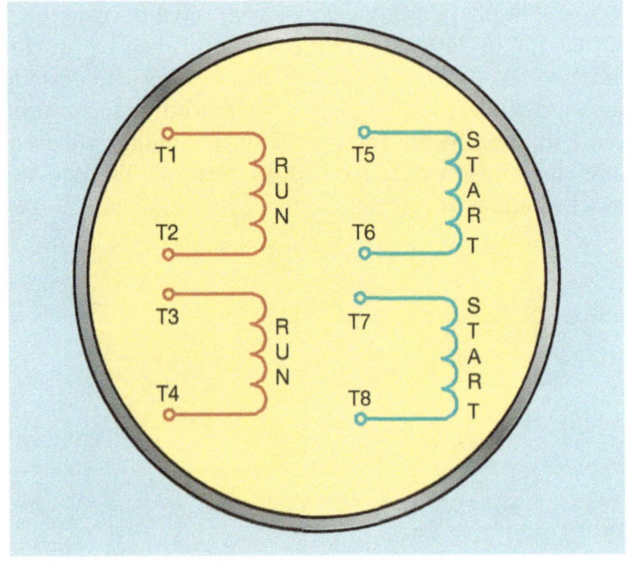

Figure 29–14 Standard labels for dual voltage single phase motors.

When the motor is connected for operation on 240 volts, the two run and two start windings are connected in series with each other. The start windings are then connected in parallel with the run windings (Figure 29–15). If the motor is to be connected for 120-volt operation, the run and start windings are connected in parallel (Figure 29–16).

Some dual voltage split phase type motors contain only one start winding instead of two. These motors will have six T leads in the terminal connection box instead of eight. The run windings are labeled T1 through T4, but the terminals of the start windings may be labeled as T5 and T6 or as T5 and T8 depending on the manufacturer. If the motor is connected for 120-volt operation, the run and start windings are connected in parallel (Figure 29–17). If the motor is connected for 240-volt operation, the two run windings are connected in series with each other and the start winding is connected in parallel with one of the run windings (Figure 29–18). Because the two run windings are connected in series with each other, each has a voltage drop of approximately 120 volts. Because the start winding is connected in parallel with one of the run windings, it will have an applied voltage of approximately 120 volts.

Reversing the Direction of Rotation

The direction of rotation of a split phase motor can be reversed by changing the run or start windings in relation to each other. Motors that permit reversal of direction bring the run-winding or start-winding leads out at the

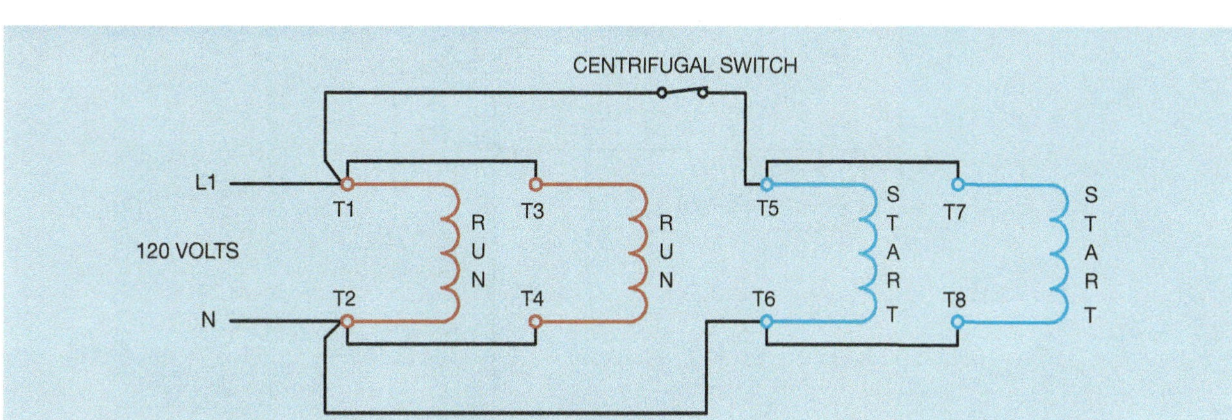

Figure 29–16 A 120-volt connection for split phase motor with two run and two start windings.

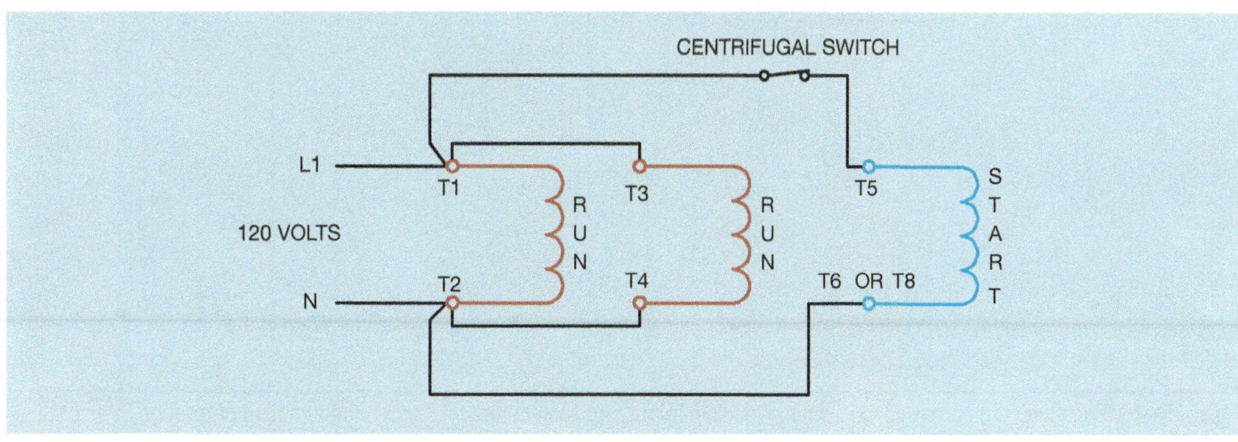

Figure 29–17 A 120-volt connection for split phase motor with two run windings and one start winding.

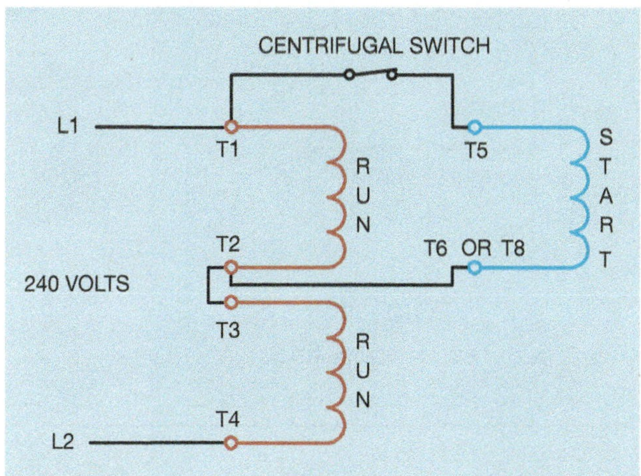

Figure 29–18 A 240-volt connection for split phase motor with two run windings and one start winding.

terminal connection box. The most common method is to reverse the start winding because this does not interfere with the run-winding connection of a dual voltage motor (Figure 29–19).

Direction of rotation is determined by facing the rear of the motor. If the windings are labeled in a standard manner, the direction of rotation can be determined when the motor is connected (Figure 29–20). Connecting the T5 lead to the T1 lead results in clockwise rotation. Connecting the T5 lead to the T4 lead will result in counterclockwise rotation.

Multispeed Motors

Some single phase motors are designed to be operated at more than one speed. There are two ways of making a motor that operates on more than one speed. One

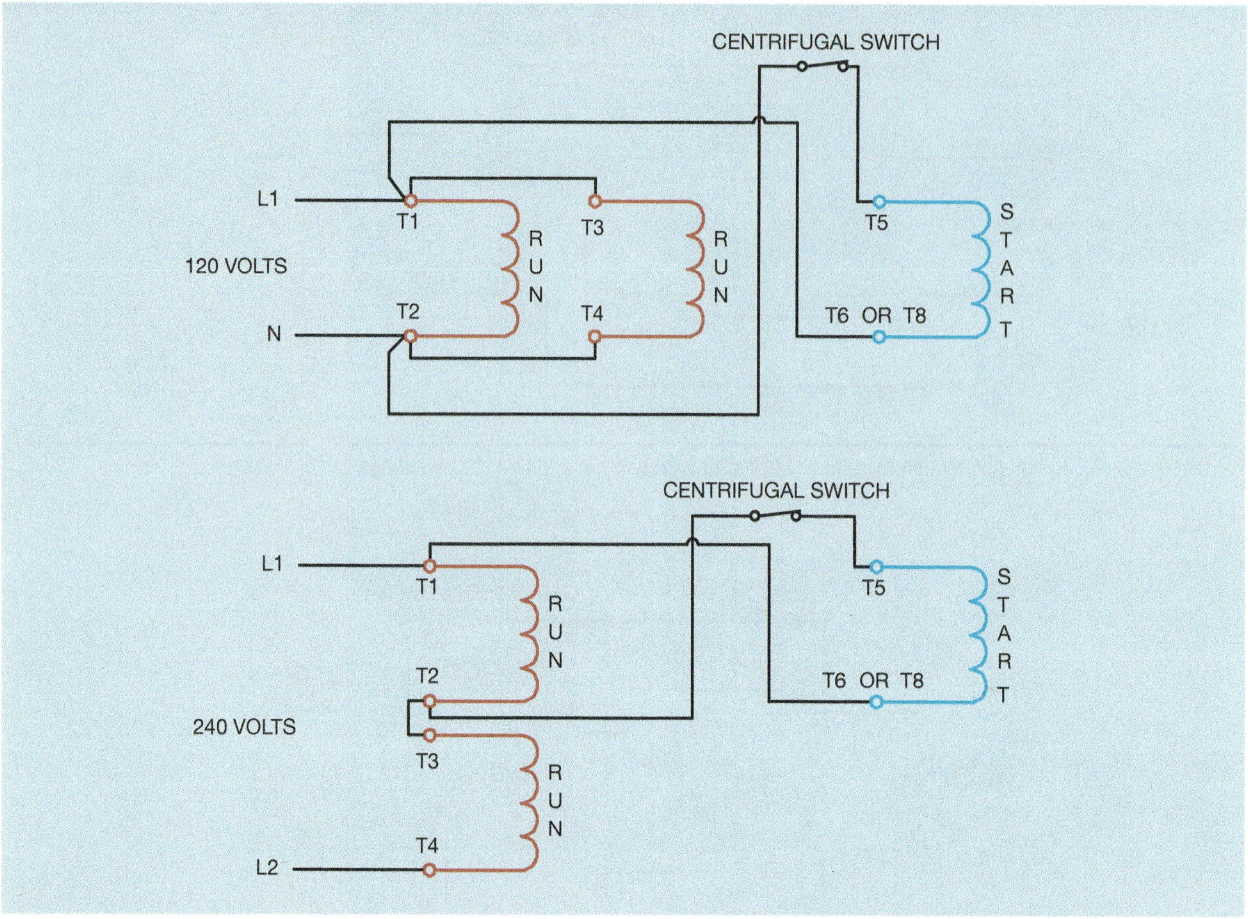

Figure 29–19 Reversing the start- and run-winding leads in relation to each other will reverse the direction of rotation of a split phase motor.

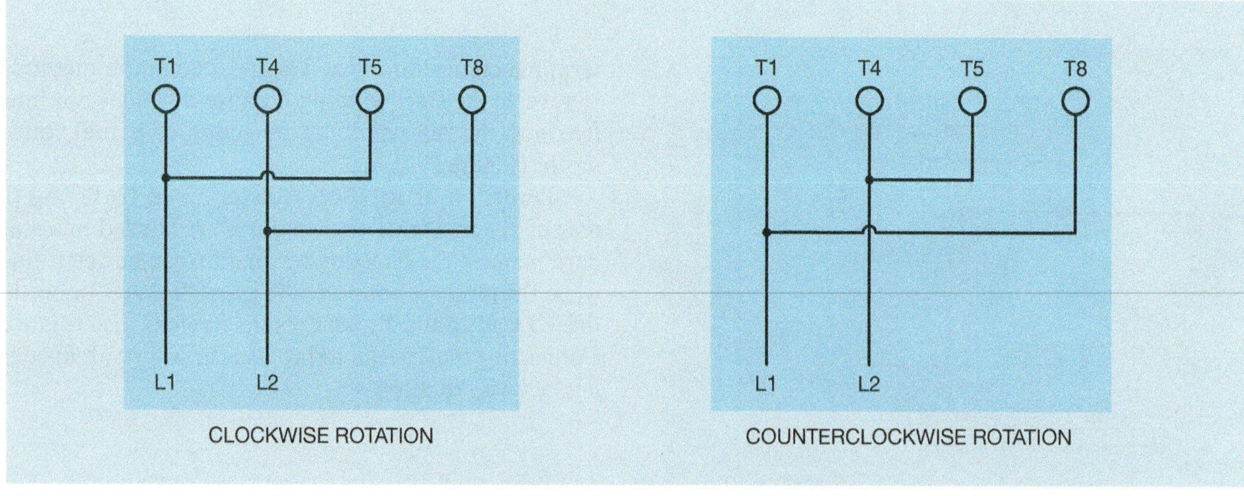

Figure 29–20 Determining direction of rotation for a split phase motor.

method is to change the number of stator poles in the motor. The speed of a rotating magnetic field is called the *synchronous* speed and is determined by two factors:

- The number of stator poles
- Frequency of the applied voltage

Motors that change speed by changing the number of stator poles are called *consequent pole* motors. They accomplish this by changing the direction of current flow through the run windings. The windings shown in Figure 29–21 have current flowing through them in the same direction. This produces the same polarity of magnetic field in each, and they basically form one pole. If the windings are reconnected in such a manner that current flows through each in a different direction, two magnetic poles are formed instead of one (Figure 29–22). The number of poles has now been doubled, causing the synchronous speed to decrease.

Consequent pole motors have the advantage of maintaining high torque when the speed is reduced. For this reason, they are used in two-speed air conditioning compressors for central cooling system.

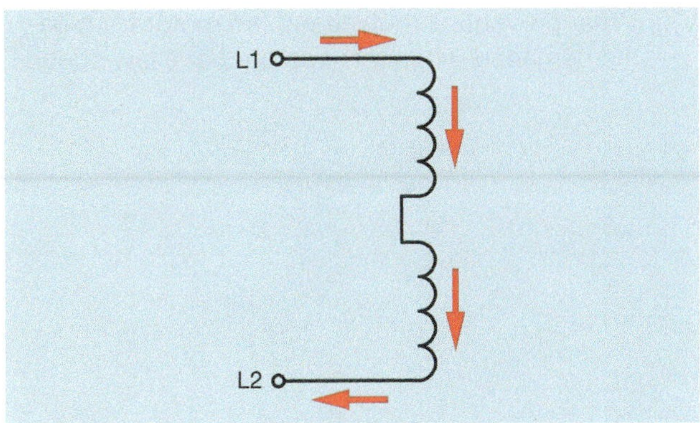

Figure 29–21 The direction of current flow through each winding is the same, forming one magnetic pole.

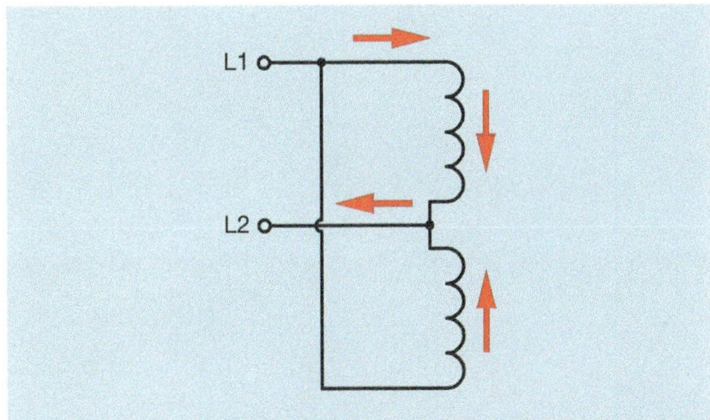

Figure 29–22 The direction of current flow through each winding is different, forming two different magnetic poles.

Multispeed Fan Motors

Another very common multispeed single phase motor is used to operate fans and blowers. These motors generally have three to five speeds and cannot be used for applications that require high torque. Multiple speeds are obtained by connecting inductance in series with the main run winding (Figure 29–23). In order for motors of this type to operate without damage to the stator, they are wound with high impedance windings. For this reason, they do not have the ability to develop high torque. A selector is often used to change speeds by inserting one or more windings in series with the main run winding.

Multispeed motors of this type cannot employ a centrifugal switch to disconnect the start winding. The two types of single phase motors used for this application are capacitor start capacitor run and shaded pole. **Shaded pole motors** are used in applications where it is not necessary to reverse the direction of rotation of the motor. Multispeed motors of this type are generally used to operate ceiling fans, small rotating fans, and as the blower motor for many central heating and air conditioning systems. The stator winding and rotor of a multispeed shaded pole motor are shown in Figure 29–24.

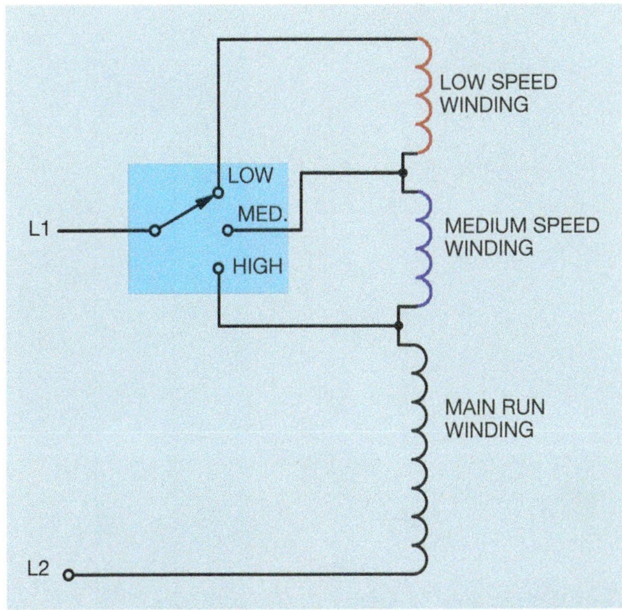

Figure 29–23 Speed is changed by connecting inductance in series with the main run winding.

Figure 29–24 Stator winding and rotor of a multispeed shaded pole motor.

Review Questions

1. What are the three major types of split phase motors?
2. What type of split phase motor does not have to disconnect the start winding when the motor reaches approximately 75 percent of its rated speed?
3. What device is generally used to disconnect the start windings of a split phase motor for open type motors that are not hermetically sealed?
4. What is the advantage of a capacitor start induction run motor over a resistance start induction run motor?
5. Maximum starting torque for a split phase motor is developed when the run-winding current and start-winding current are how many degrees out of phase with each other?
6. How is the direction of rotation of a split phase motor changed?
7. What is synchronous speed?
8. What two factors determine synchronous speed?
9. What type of motor changes speed by changing the number of stator poles?
10. What prevents a multispeed fan motor from being damaged when it is operated at a low speed?

Chapter 30

BRAKING

Objectives

After studying this chapter the student will be able to:

- Discuss mechanical type brakes.
- Connect a mechanical brake circuit.
- Discuss dynamic braking for DC and AC motors.
- Connect a plugging circuit.

Motors are generally permitted to slow to a stop when disconnected from the power line, but in some instances that is not an option or convenient. Several methods can be employed to provide braking for a motor. Some of these are:

- Mechanical brakes
- Dynamic braking
- Plugging

Mechanical Brakes

Mechanical brakes can be obtained in two basic types: drum and disk. Drum brakes use brake shoes to apply pressure against a drum (Figure 30–1). A metal cylinder, called the drum, is attached to the motor shaft. Brake shoes are placed around the drum. A spring is used to adjust the amount of pressure the brake shoes exert against the drum to control the amount of braking that takes place when the motor is stopped. When the motor is operating, a **solenoid** is energized to release the pressure of the brake shoes. When the motor is stopped, the brakes engage immediately. A circuit of this type is shown in Figure 30–2. Mechanical brakes work by converting the kinetic (moving) energy of the load into thermal (heat) energy when the motor is stopped. Mechanical type brakes have an advantage in that they can hold a suspended load. For this reason, mechanical brakes are often used on cranes.

Disc brakes work in a very similar manner to drum brakes. The only real difference is that brake pads are used to exert force against a spinning disc instead of a cylindrical drum. A combination disc brake and magnetic clutch is shown in Figure 30–3.

Dynamic Braking

Dynamic braking can be used to slow both direct and alternating current motors. Dynamic braking is sometimes referred to as **magnetic braking** because in both instances it employs the use of magnetic fields to slow the rotation of a motor. The advantage of dynamic braking is that there are no mechanical brake shoes to wear out. The disadvantage is that dynamic brakes cannot hold a suspended load. Although dynamic braking can be used for both direct and alternating current motors, the principles and methods used for each are very different.

Figure 30–1 Drum brake.

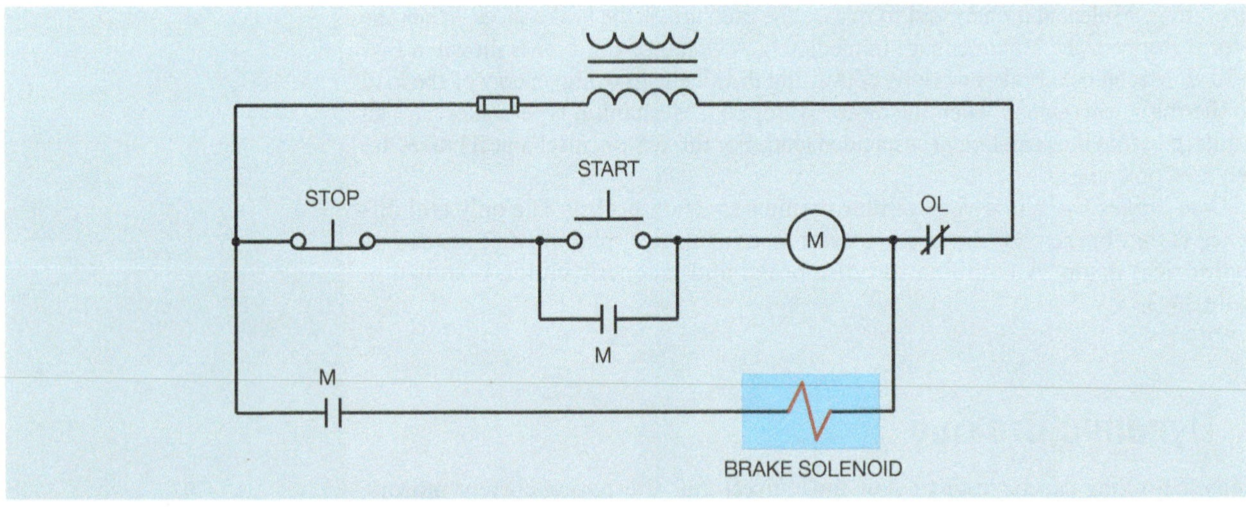

Figure 30–2 The brake is applied automatically when the motor is not operating.

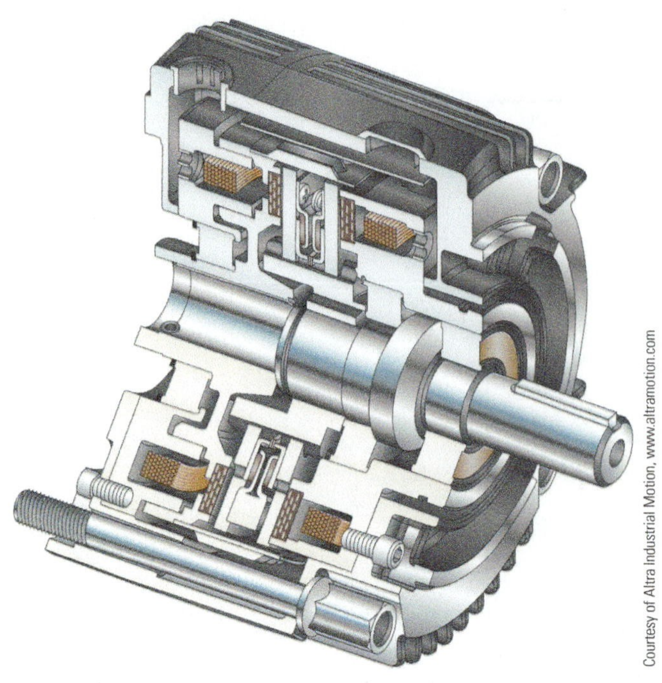

Courtesy of Altra Industrial Motion, www.altramotion.com

Figure 30–3 Cut-away view of a combination electromagnetic disk brake and clutch.

Dynamic Braking for Direct Current Motors

A direct current machine can be used as either a motor or generator. When used as a motor, electrical energy is converted into mechanical energy. When used as a generator, mechanical energy is converted into electrical energy. The principle of dynamic braking for a direct current motor is to change the motor into a generator. When a generator produces electrical power, it produces *counter torque*, making the armature hard to turn. The amount of counter torque produced by the generator is proportional to the armature current.

Dynamic braking for a DC motor is accomplished by permitting power to remain connected to the shunt field when the motor is stopped, and reconnecting the armature to a high wattage resistor (Figure 30–4). The resistor may actually be more than one resistor depending on motor size, length of braking time, and armature current. The braking time can be controlled by adjusting the resistance value. If current remains connected to the shunt field, the pole pieces retain their magnetism. Connecting a resistance across the armature terminals causes the motor to become a generator.

Dynamic braking for a DC motor is very effective, but the braking effect becomes weaker as the armature slows down. Counter torque in a generator is proportional to the magnetic field strength of the pole pieces and armature. Although the flux density of the pole pieces remains constant as long as shunt field current is constant, the armature magnetic field is proportional to armature current. Armature current is proportional to the amount of induced voltage and the resistance of the connected load. Three factors determine induced voltage:

- Strength of magnetic field. (In this instance, the flux density of the pole pieces.)
- Length of conductor. (Also stated as number of turns of wire. In this instance it is the number of turns on wire in the armature winding.)
- Speed of the cutting action. (Armature speed)

As the armature slows, less voltage is induced in the armature windings causing a decrease of armature current.

Dynamic Braking for Alternating Current Motors

Dynamic braking for alternating current motors is accomplished in a different way than described for direct current motors. Dynamic braking for an AC motor can be accomplished by connecting direct current to the stator winding. This connection causes the stator magnetic field to maintain a constant polarity instead of reversing polarity each time the current changes direction. As the rotor of a squirrel cage motor spins through the stationary magnetic field, a current is induced into the rotor bars. The current flow in the rotor causes a magnetic field to form around the rotor bars. The rotor magnetic field is attracted to the stator field causing the rotor to slow down. The amount of braking force is proportional to the magnetic field strength of the stator field and the rotor field. The braking force can be controlled by the amount of direct current supplied to the stator.

When direct current is applied to the stator winding, there is no inductive reactance to limit stator current. The only current limiting effect is the wire resistance of the stator winding. Dynamic braking circuits for alternating current motors generally include a **step-down transformer** to lower the voltage to the rectifier and often include a series resistor to control the current applied to the stator winding (Figure 30–5).

In the circuit shown, an off-delay timer is used to determine the length of braking time for the circuit. When the START button is pushed, motor starter M energizes and closes all M load contacts to connect the motor to the line. The M auxiliary contacts change position at

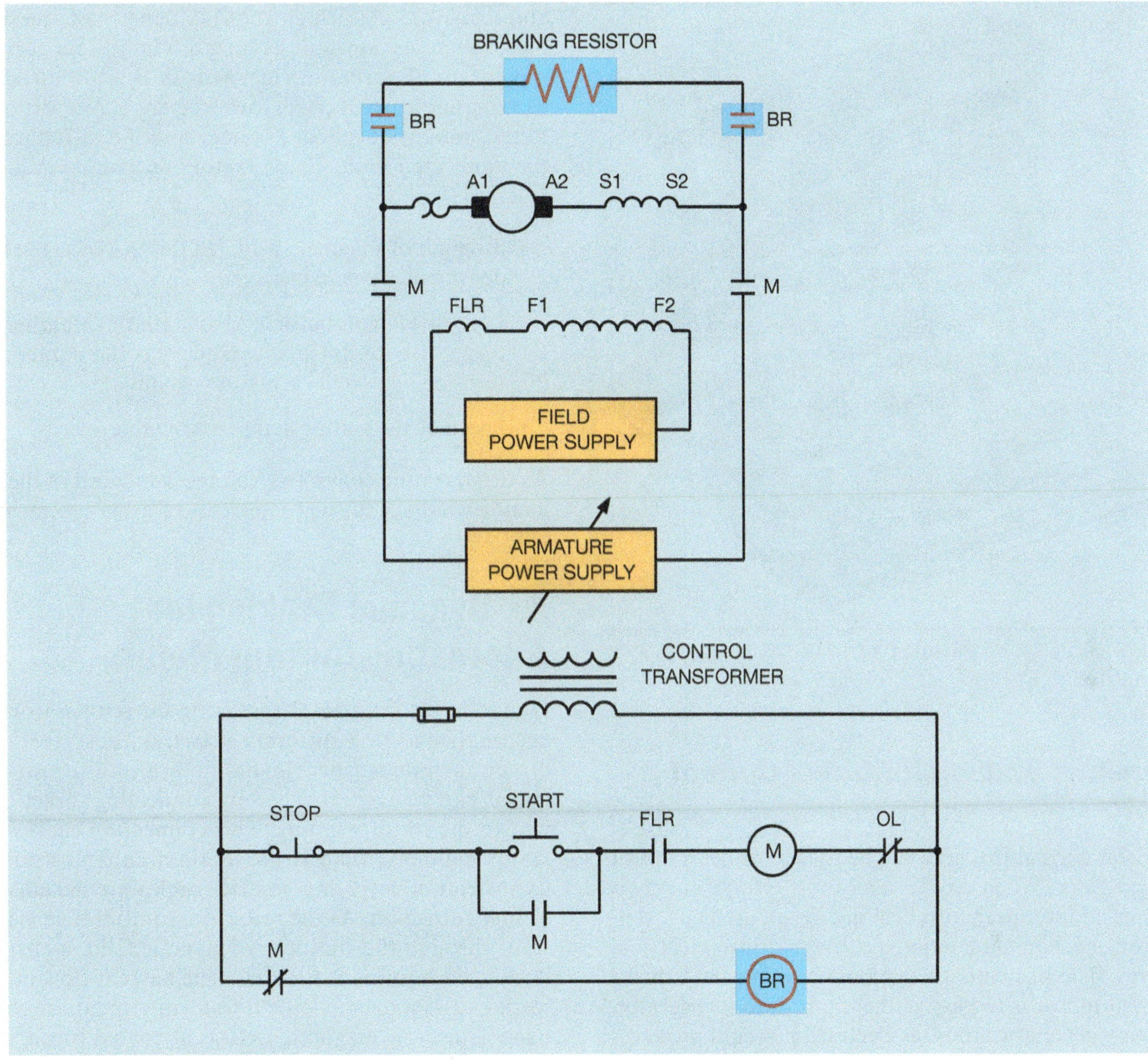

Figure 30–4 Dynamic braking circuit for a direct current motor.

the same time. The normally closed M contact opens to prevent power being applied to the dynamic brake relay (DBR). The two normally open M auxiliary contacts close, sealing the circuit and supplying power to the coil of off-delay timer TR. Because TR is an off-delay timer, the TR timed contacts close immediately. The circuit remains in this position until the STOP button is pressed. At that time, motor starter M de-energizes and disconnects the motor from the line. The normally open M auxiliary contact connected in series with timer coil TR opens, starting the timing sequence. The normally closed M auxiliary contact closes and provides a current path through the now closed TR timed contact to the coil of DBR. This causes the DBR contacts to close and connect the step-down transformer and **rectifier** to the power line. Direct current is now supplied to the stator winding. Direct current will be supplied to the stator winding until the timed TR contact opens and de-energizes coil DBR.

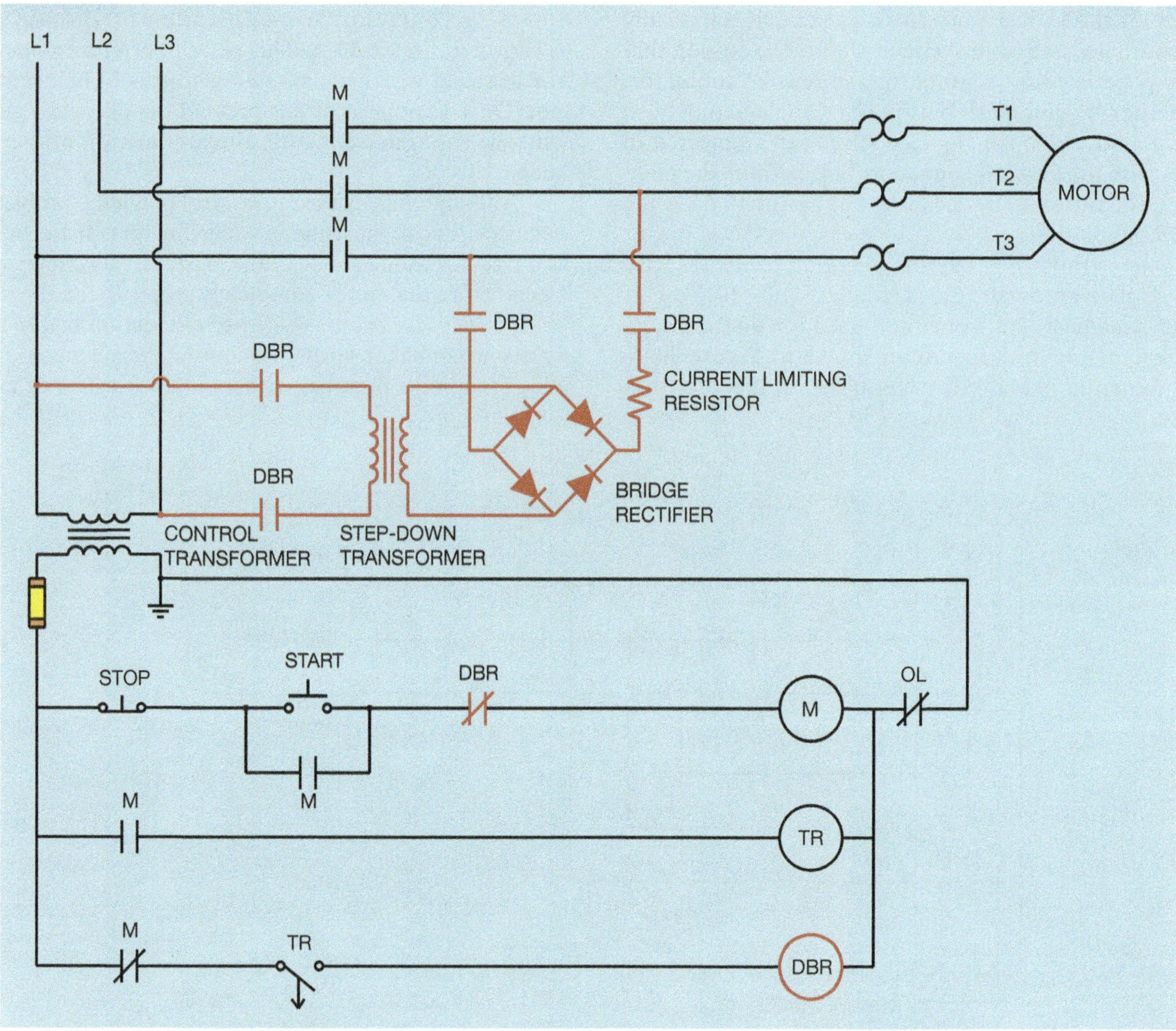

Figure 30–5 Dynamic braking for AC motors is accomplished by supplying direct current to the stator winding.

Plugging

Plugging is defined by NEMA as *a system of braking, in which the motor connections are reversed so that the motor develops a counter torque that acts as a retarding force.* Plugging can be used with direct current motors, but is more often used with three phase squirrel cage motors. Plugging is accomplished with three phase motors by disconnecting the motor from the power line and momentarily reversing the direction of rotation. As a general rule, the reversing contactor is of a larger size than the forward contactor because of the increased plugging current. Several methods can be employed when a plugging control is desired.

Manual Plugging

One type of plugging control depends on an operator to manually perform the operation. A manual plugging

control is shown in Figure 30–6. The circuit is basically a forward–reverse control circuit with the exception that there is no holding contact for the reverse contactor. Also, the plugging push button is a double-acting push button with the normally closed section connected in series with the forward contactor. This permits the plugging push button to be used without having to press the STOP button first.

One method of providing plugging control is with the use of an automatic timed circuit (Figure 30–7). This is the same basic control circuit used for time controlling a dynamic braking circuit shown in Figure 30–5. The dynamic brake relay has been replaced with a reversing contactor. A modification of this circuit is shown in Figure 30–8. This circuit permits an operator to select whether a plugging stop is to be used or not. Once the operator has pressed the plugging push button, the timer controls the amount of plugging time.

Although time is used to control plugging, problems can occur due to the length of plugging time. If the timer is not set for a long enough time, the reversing circuit will open before the motor completely stops. If the timer is set too long, the motor will reverse direction before the reversing contactor opens. The most accurate method of plug stopping a motor is with a *plugging switch* or *zero*

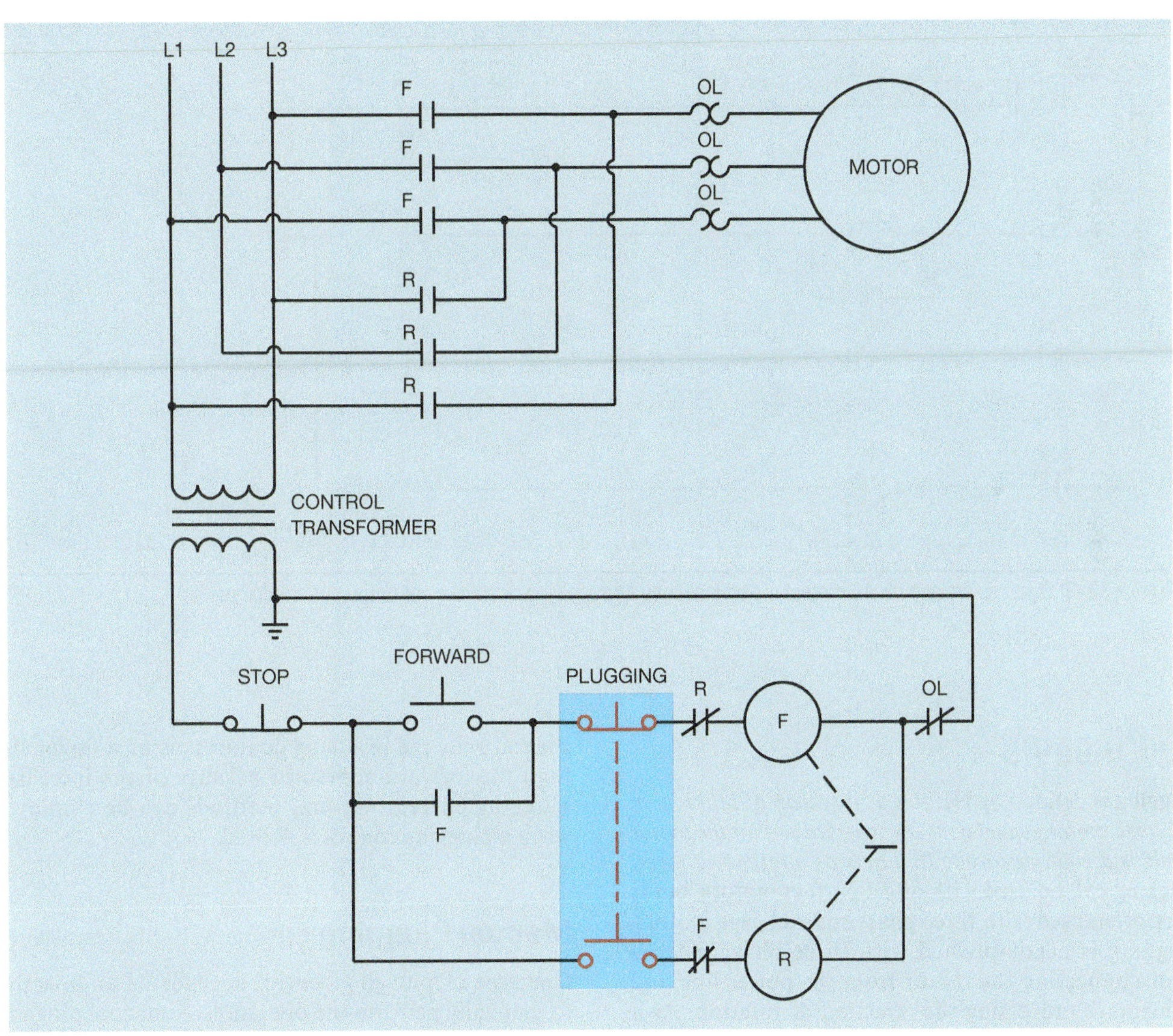

Figure 30–6 Manual plugging control.

speed switch (Figure 30–9). The plugging switch is connected to the motor shaft or the shaft of the drive machine. The motion of the rotating shaft is transmitted to the plugging switch either by a centrifugal mechanism or by an eddy current induction disc inside the switch. The plugging switch contact is connected to the coil of the reversing starter (Figure 30–10). When the motor is started, the forward motion of the motor causes the normally open plugging switch contact to close. When the STOP button is pressed, the normally closed F contact connected in series with the reversing contactor recloses and reverses the direction of rotation of the motor. When the shaft of the motor stops rotating, the plugging switch contact reopens and disconnects the reversing contactor.

Plugging switches with two normally open contacts can be obtained for use with forward–reverse controls. These switches permit a plugging stop in either direction when the STOP button is pressed (Figure 30–11). The direction of motor rotation determines which switch closes. The switch symbol indicates the direction of rotation necessary to cause the switch contacts to close.

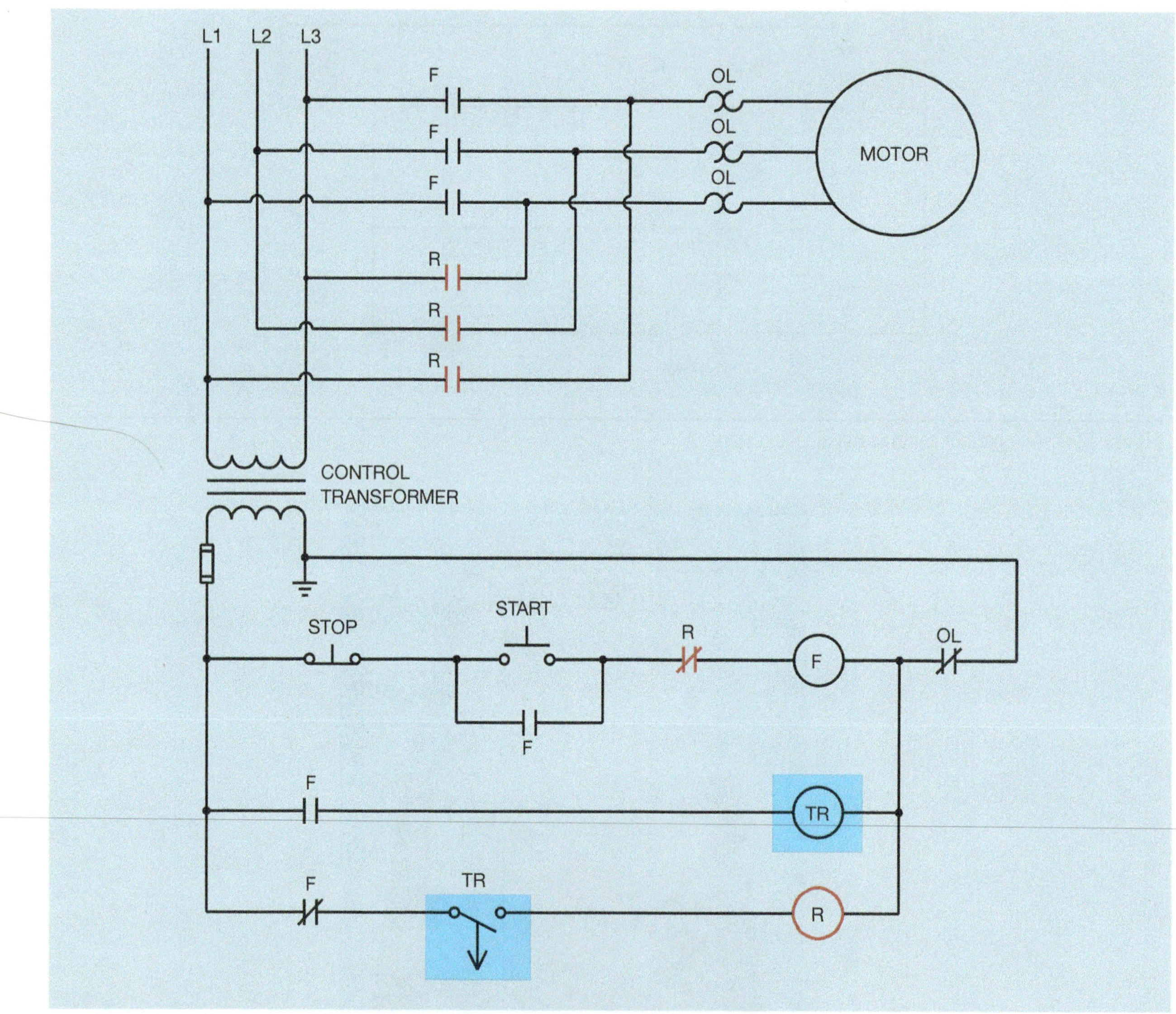

Figure 30–7 Timed controlled plugging circuit.

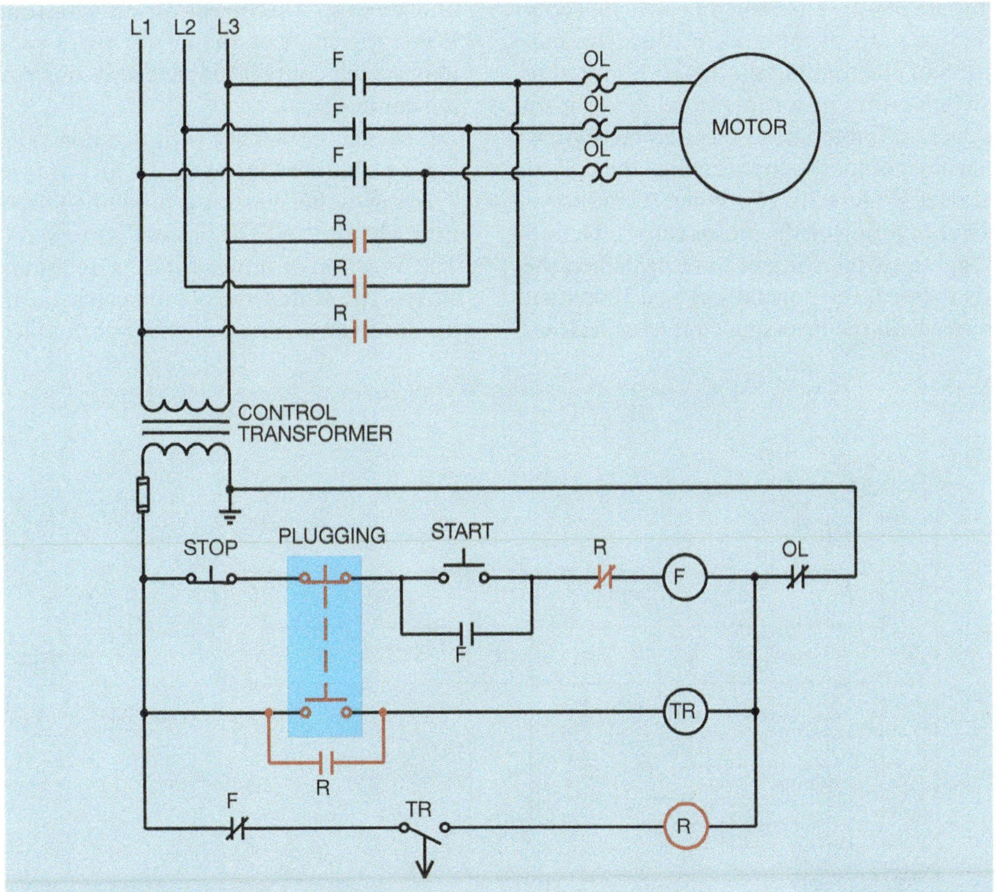

Figure 30–8 An operator controls the plugging stop.

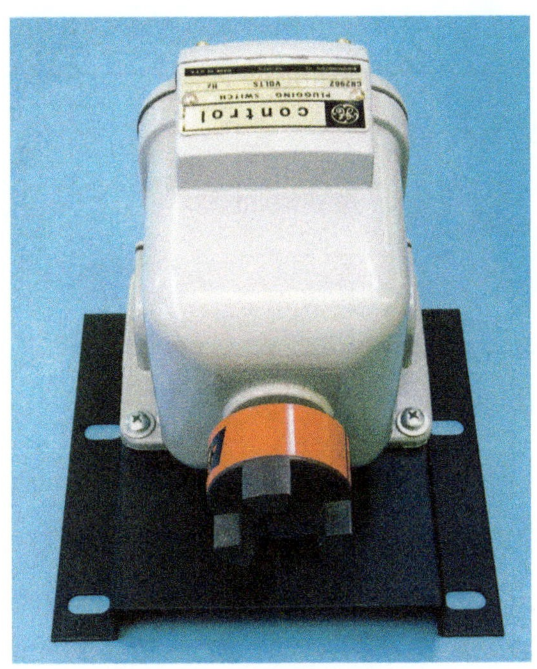

Figure 30–9 Plugging switch or zero speed switch.

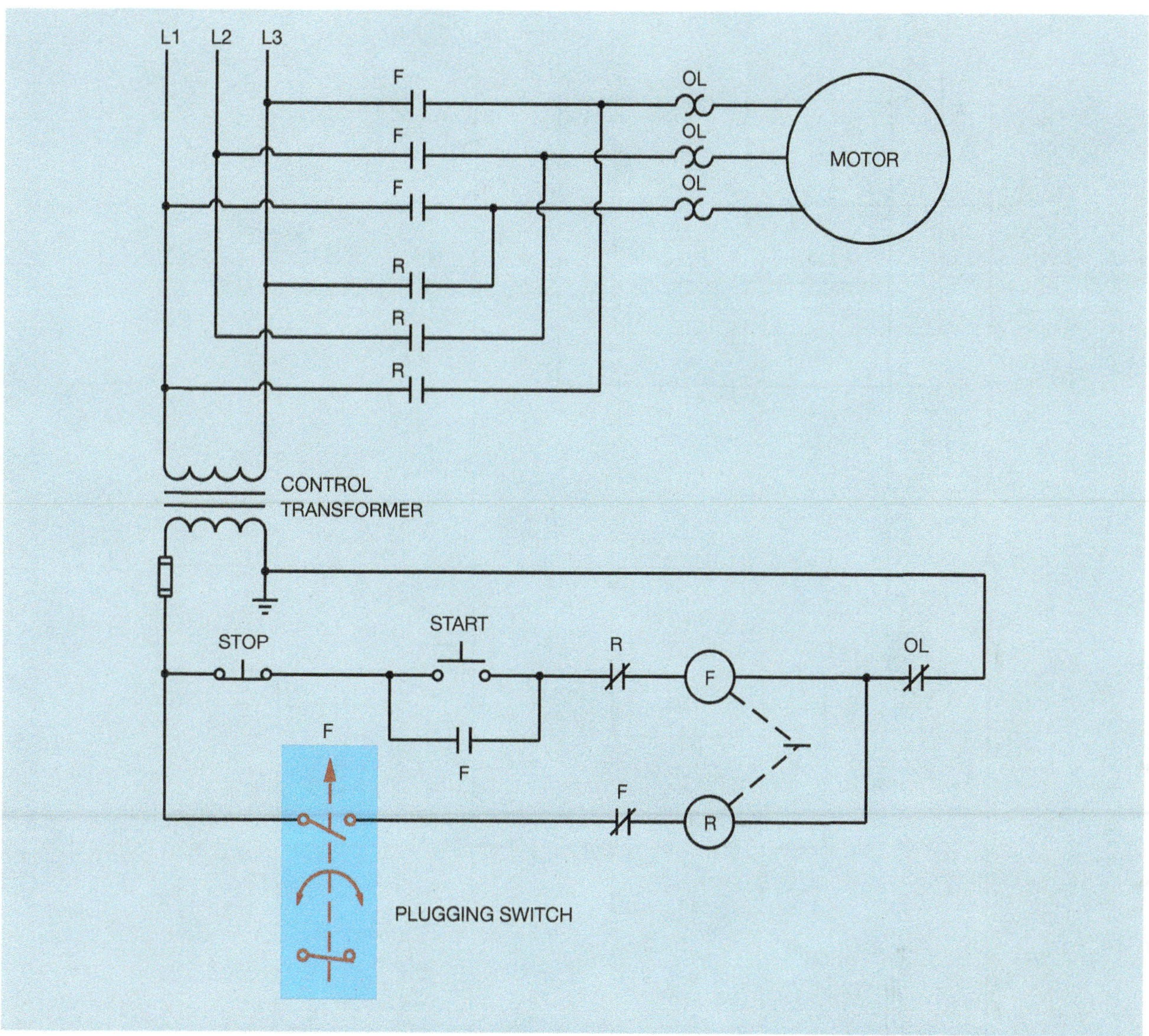

Figure 30–10 Plugging switch controls the operation of the reversing contactor.

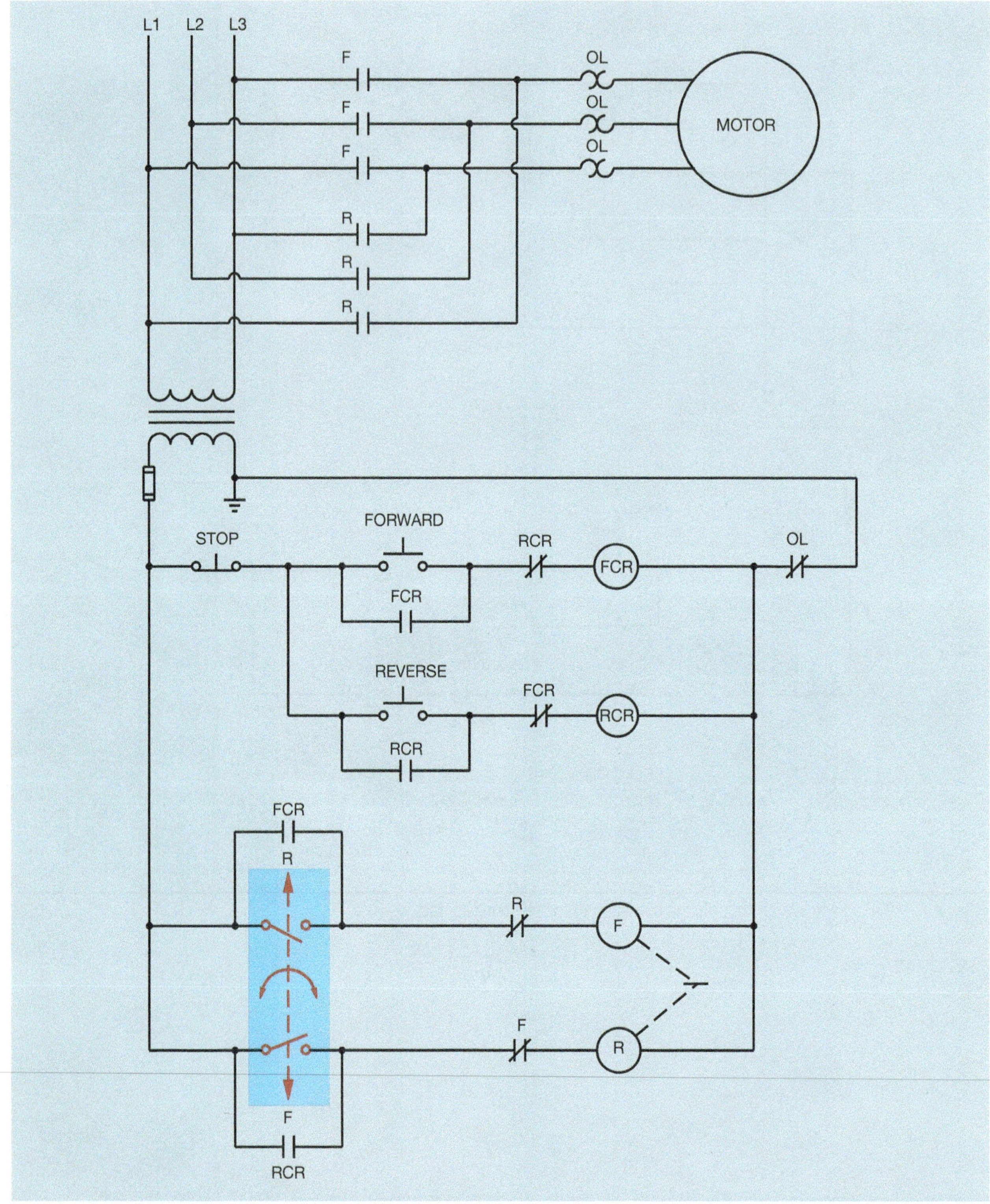

Figure 30–11 Plugging switch used with forward–reverse control.

Review Questions

1. Name three methods of braking a motor.
2. How is the braking force of drum type brakes controlled?
3. Why are mechanical brakes often used on cranes?
4. What is the advantage of dynamic brakes over mechanical brakes?
5. What is the disadvantage of dynamic brakes when compared to mechanical brakes?
6. The amount of counter torque developed by a direct current generator is proportional to what?
7. When using dynamic braking for a direct current motor, how is the braking time controlled?
8. Name three factors that determine the amount of induced voltage.
9. How is dynamic braking for direct current motors accomplished?
10. How is the dynamic braking force of an alternating current motor controlled?
11. How is a plugging stop accomplished?
12. What device is generally used to accurately stop a motor when a plugging stop is used?
13. Refer to the circuit shown in Figure 30–10. When the START button is pushed and the motor starts in the forward direction, the plugging switch closes. What prevents the reversing contactor from energizing when the plugging switch contact closes?

Section 5

WOUND ROTOR, SYNCHRONOUS, AND CONSEQUENT POLE MOTORS

Chapter 31

WOUND ROTOR MOTORS

Objectives

After studying this chapter the student will be able to:

- » Identify the terminal markings of a wound rotor induction motor.
- » Discuss the operating characteristics of wound rotor motors.
- » Connect a wound rotor motor for operation.
- » Discuss speed control of wound rotor motors.

The wound rotor induction motor is one of the three major types of three phase motors. It is often called the *slip ring* motor because of the three slip rings on the rotor shaft. The stator winding of a wound rotor motor is identical to the squirrel cage motor. The difference between the two motors lies in the construction of the rotor. The rotor of a squirrel cage motor is constructed of bars connected together at each end by shorting rings. The rotor of a wound rotor induction motor is constructed by winding three separate windings in the rotor (Figure 31–1).

The wound rotor motor was the first alternating current motor that permitted speed control. It has a higher starting torque per amp of starting current than any other type of three phase motor. It can be started in multiple steps to provide smooth acceleration from 0 RPM to maximum RPM. Wound rotor motors are typically employed to operate conveyers, cranes, mixers, pumps, variable speed fans, and a variety of other devices. They are often used to power gear-driven machines because they can be started without supplying a large amount of torque, which can damage and even strip the teeth off gears.

The three phase rotor winding contains the same number of poles as the stator winding. One end of each rotor winding is connected together inside the rotor to form a wye connection and the other end of each winding is connected to one of the slip rings mounted on the rotor shaft. The slip rings permit external resistance

Courtesy Wazee Companies, LLC Denver, CO

Figure 31–1 Rotors of wound rotor induction motors.

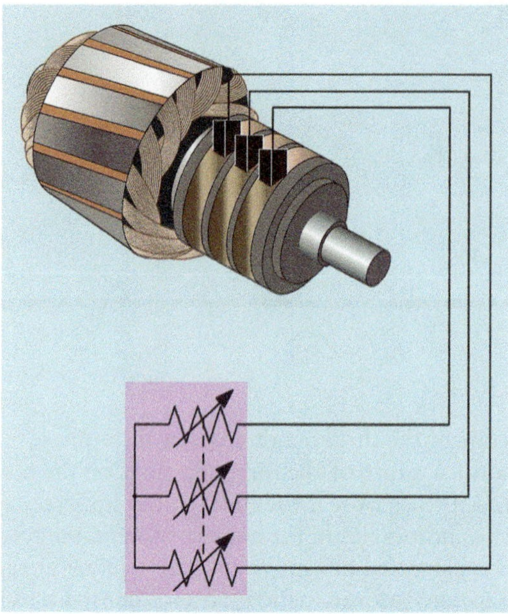

Figure 31–2 The rotor of a wound rotor induction motor is connected to external resistors.

to be connected to the rotor circuit (Figure 31–2). Placing external resistance in the rotor circuit allows control of the amount of current that can flow through the rotor windings during both the starting and running of the motor. Three factors determine the amount of torque developed by a three phase induction motor:

- Strength of the magnetic field of the stator.
- Strength of the magnetic field of the rotor.
- Phase angle difference between rotor and stator flux.

Because an induction motor is basically a transformer, controlling the amount of rotor current also controls the amount of stator current. This feature permits the wound rotor motor to control the in-rush current during the starting period. Limiting the in-rush current also limits the amount of starting torque the motor produces.

The third factor that determines the amount of torque developed is the phase angle difference between stator and rotor flux. Maximum torque is developed when the magnetic fields of the stator and rotor are in-phase with each other. Imagine two bar magnets with their north and south poles connected together. If the magnets are placed so there is no angular difference between them (Figure 31–3A), the attracting force is maximized. If the magnets are broken apart so there is an angular difference between them, there is still a force of attraction, but it is less than when they are connected together (Figure 31–3B). The greater the angle of separation, the less the force of attraction becomes (Figure 31–3C).

Adding resistance to the rotor circuit causes the induced current in the rotor to be more in-phase with the stator current. This produces a very small phase angle difference between the magnetic fields of the rotor and stator. This is the reason that the wound rotor induction motor produces the greatest amount of starting torque per amp of starting current of any three phase motor.

The stator windings of a wound rotor motor are marked in the same manner as any other three phase motor—T1, T2, and T3 for single voltage motors. Dual voltage motors have nine T leads like squirrel cage motors. The rotor leads are labeled M1, M2, and M3. The M2 lead is located on the center slip ring, and the M3 lead is connected to the slip ring closest to the rotor windings. The schematic symbol for a wound rotor induction motor is shown in Figure 31–4.

Manual Control of a Wound Rotor Motor

The starting current and speed of a wound rotor induction motor is controlled by adding or subtracting the amount of resistance connected in the rotor circuit. Small wound rotor motors are often controlled manually by a three-pole make-before-break rotary switch.

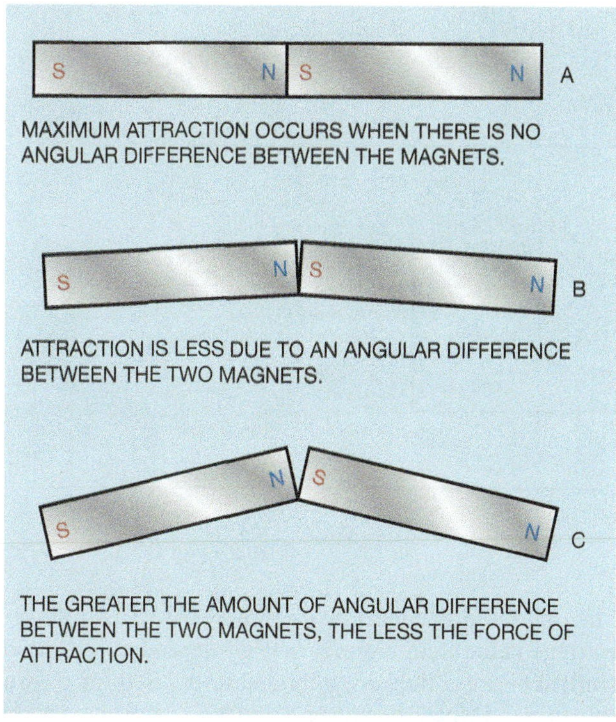

Figure 31–3 The force of attraction is proportional to the flux density of the two magnets and the angle between them.

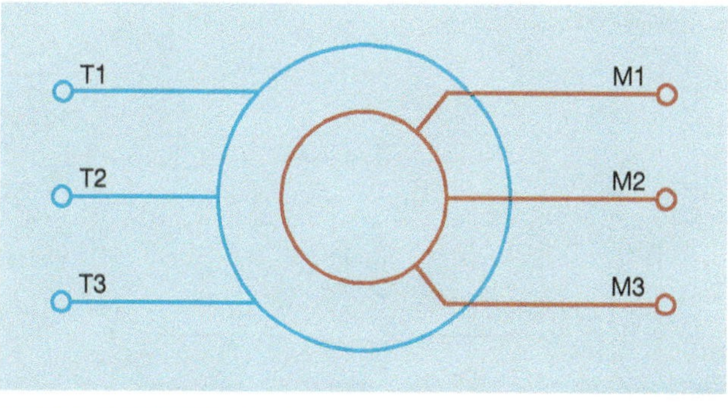

Figure 31–4 Schematic symbol for a wound rotor induction motor.

The switch contains as many contacts as there are steps of resistance (Figure 31–5). A micro limit switch senses when the controller is set for maximum resistance. Most controllers will not start unless all resistance is in the rotor circuit, forcing the motor to start in its lowest speed. Once the motor has been started, the resistance can then be adjusted to increase the motor speed. When all the resistance has been removed from the circuit and the M leads are shorted together, the motor will operate at full speed. The operating characteristics of a wound

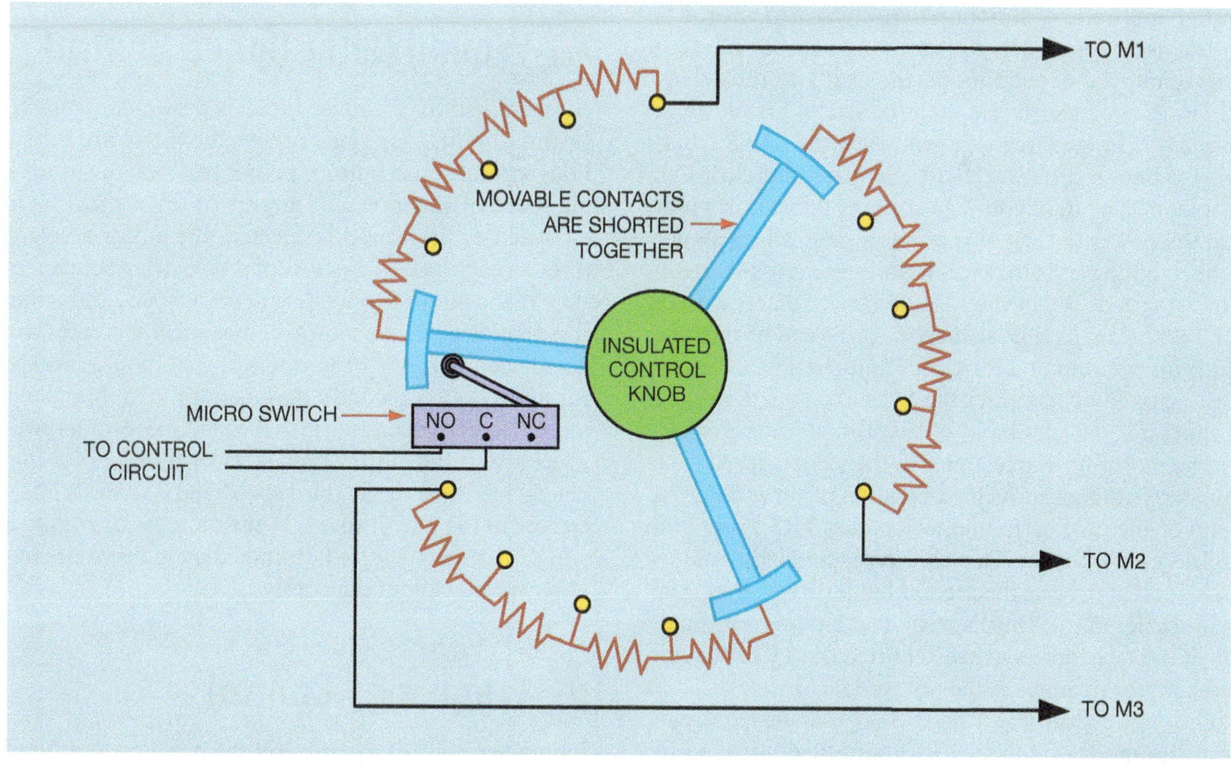

Figure 31–5 Manual controller for a wound rotor induction motor.

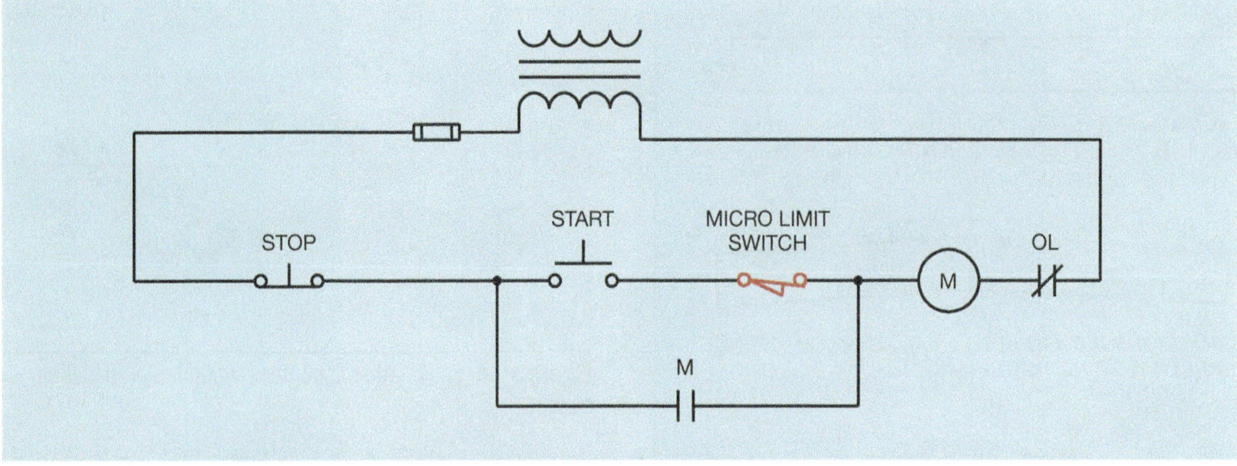

Figure 31–6 Control circuit for a manually controlled wound rotor motor.

rotor motor with the rotor leads shorted together are very similar to those of a squirrel cage motor. A circuit for use with a manual controller is shown in Figure 31–6.

Timed Controlled Starting

Another method of starting a wound rotor motor is with the use of time delay relays. Any number of steps can be employed depending on the needs of the driven machine. A circuit with four steps of starting is shown in Figure 31–7. In the circuit shown, when the START button is pressed, motor starter M energizes and closes all M contacts. The load contacts connect the stator winding to the power line. At this point in time, all resistance is connected in the rotor circuit and the motor starts in its lowest speed. When the M auxiliary contacts closed, timer TR1 began its time sequence. At the end of the time period, timed contact TR1 closes and energizes the coil of contactor S1. This causes the S1 load contacts to close and short out the first bank of resistors in the rotor circuit. The motor now accelerates to the second speed. The S1 auxiliary contact starts the operation of timer TR2. At the end of the time period, timed contact TR2 closes and energizes contactor S2. This causes the S2 load contacts to close and short out the second bank of resistors. The motor accelerates to third speed. The process continues until all the resistors have been shorted out of the circuit and the motor operates at the full speed.

The circuit shown in Figure 31–7 is a starter circuit in that the speed of the motor cannot be controlled by permitting resistance to remain in the circuit. Each time the START button is pressed, the motor accelerates through each step of speed until it reaches full speed. Starting circuits generally employ resistors of a lower wattage value than circuits that are intended for speed control because they are intended to be used for only a short period when the motor is started. Controllers must employ resistors that have a high enough wattage rating to remain in the circuit at all times.

Wound Rotor Speed Control

A time-operated controller circuit is shown in Figure 31–8. In this circuit, four steps of speed control are possible. Four separate push buttons permit selection of the operating speed of the motor. If any speed other than the lowest speed or first speed is selected, the motor accelerates through each step with a 3-second time delay between each step. If the motor is operating at a low speed and a higher speed is selected, the motor immediately increases to the next speed if it has been operating in its present speed for more than 3 seconds. Assume, for example, that the motor has been operating in the second speed for more than 3 seconds. If the fourth speed is selected, the motor immediately increases to the third speed and 3 seconds later increases to the fourth speed. If the motor is operating and a lower speed is selected, it immediately decreases to the lower speed without time delay.

Frequency Control

Frequency control operates on the principle that the frequency of the induced voltage in the motor secondary (rotor) decreases as the speed of the rotor increases. The rotor windings contain the same number of poles as the

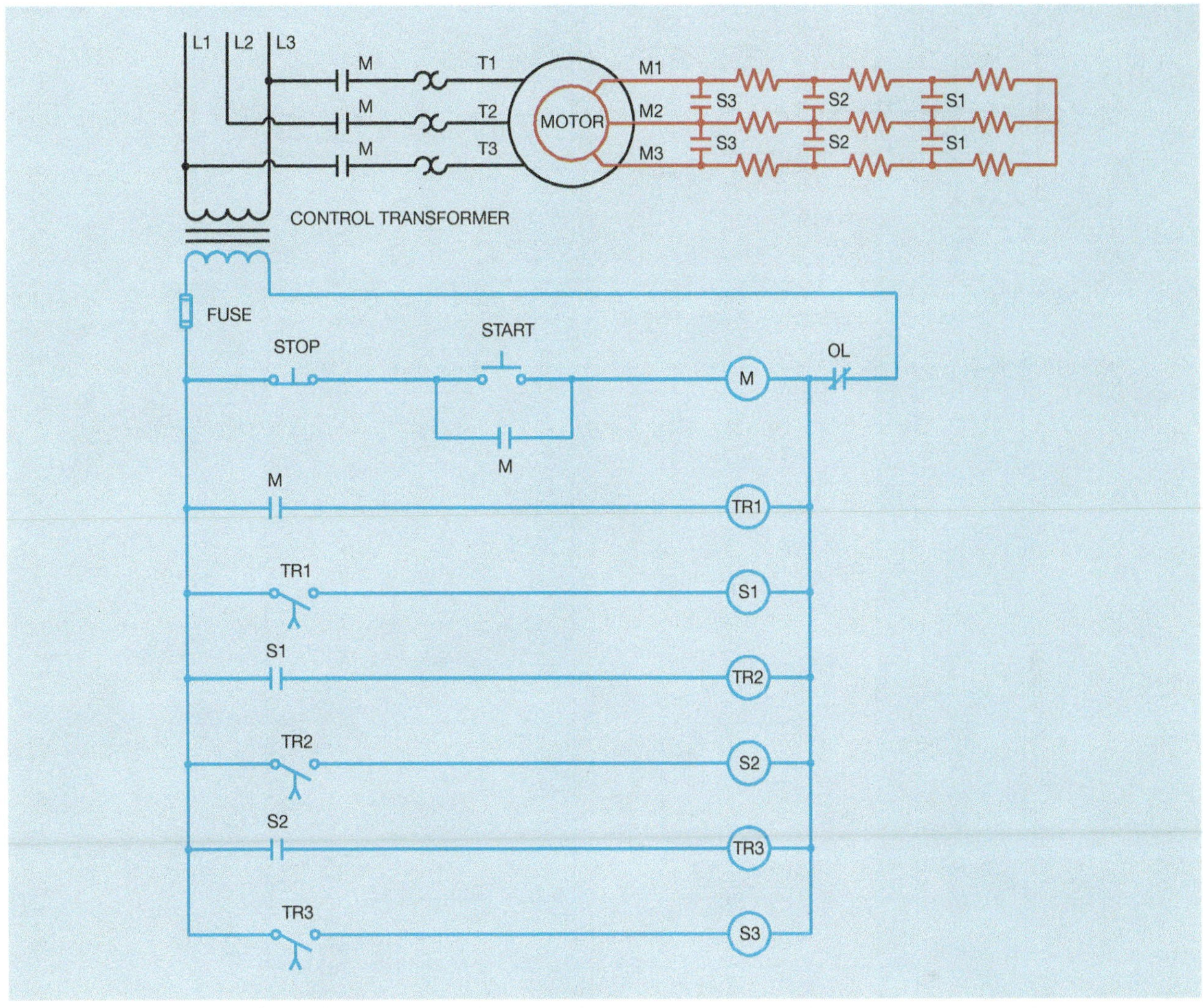

Figure 31–7 Timed starting for a wound rotor induction motor.

stator. When the motor is stopped and power is first applied to the stator windings, the voltage induced into the rotor will have the same frequency as the power line. This will be 60 Hz throughout the United States and Canada. When the rotor begins to turn, there is less cutting action between the rotating magnetic field of the stator and the windings in the rotor. This causes a decrease in both induced voltage and frequency. The greater the rotor speed becomes, the lower the frequency and amount of the induced voltage. The difference between rotor speed and synchronous speed (speed of the rotating magnetic field) is called **slip** and is measured as a percentage. Assume that the stator winding of a motor has four poles per phase. This would result in a synchronous speed of 1800 RPM when connected to 60 Hz. Now, assume that the rotor is turning at a speed of 1710 RPM. This is a difference of 90 RPM. This results in a 5 percent slip for the motor.

$$\text{Slip} = \frac{90}{1800}$$

$$\text{Slip} = 0.05$$

$$\text{Slip} = 5\%$$

A 5 percent slip would result in a rotor frequency of 3 Hz.

$$F = 60 \text{ Hz} \times 0.05$$

$$F = 3 \text{ Hz}$$

or

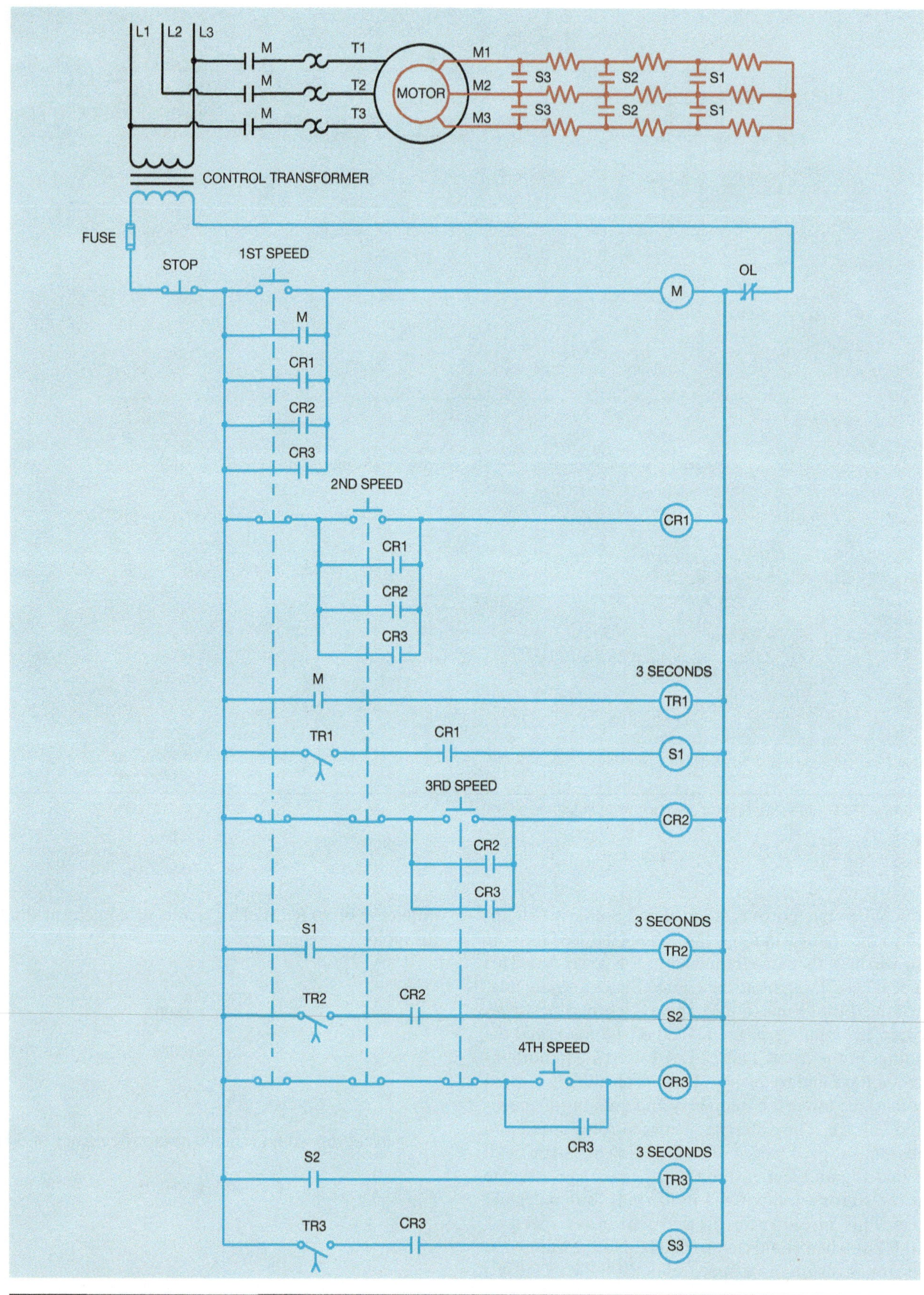

Figure 31–8 Time-operated speed control for a wound rotor induction motor.

$$F = \frac{PS}{120}$$

$$F = \frac{4 \times 90}{120}$$

$$F = \frac{360}{120}$$

$$F = 3\ \text{Hz}$$

Where F = frequency in hertz
P = number of poles per phase
S = speed in RPM
120 = constant

A diagram of a wound rotor motor starter using frequency relays is shown in Figure 31–9. Note that the frequency relays are connected to the secondary winding of the motor, and that the load contacts are connected normally closed instead of normally open. Also note that a capacitor is connected in series with one of the frequency relays. In an alternating current circuit, the current limiting effect of a capacitor is called *capacitive reactance*. Capacitive reactance is inversely proportional to the frequency. A decrease in frequency causes a corresponding increase in capacitive reactance.

When the START button is pressed, M contactor energizes and connects the stator winding to the line. This causes a voltage to be induced into the rotor circuit at a frequency of 60 Hz. The 60 Hz frequency causes both S1 and S2 contactors to energize and open their load contacts. The rotor is now connected to maximum resistance and starts in the lowest speed. As the frequency decreases, capacitive reactance increases, causing contactor S1 to de-energize first and reclose the S1 contacts. The motor now increases in speed, causing a further reduction of both induced voltage and frequency. When contactor S2 de-energizes, the S2 load contacts reclose and short out the second bank of resistors. The motor is now operating at its highest speed.

The main disadvantage of frequency control is that some amount of resistance must remain in the circuit at all times. The load contacts of the frequency relays are closed when power is first applied to the motor. If a set of closed contacts were connected directly across the M leads, no voltage would be generated to operate the coils of the frequency relays and they would never be able to open their normally closed contacts.

Frequency control does have an advantage over other types of control in that it is very responsive to changes in

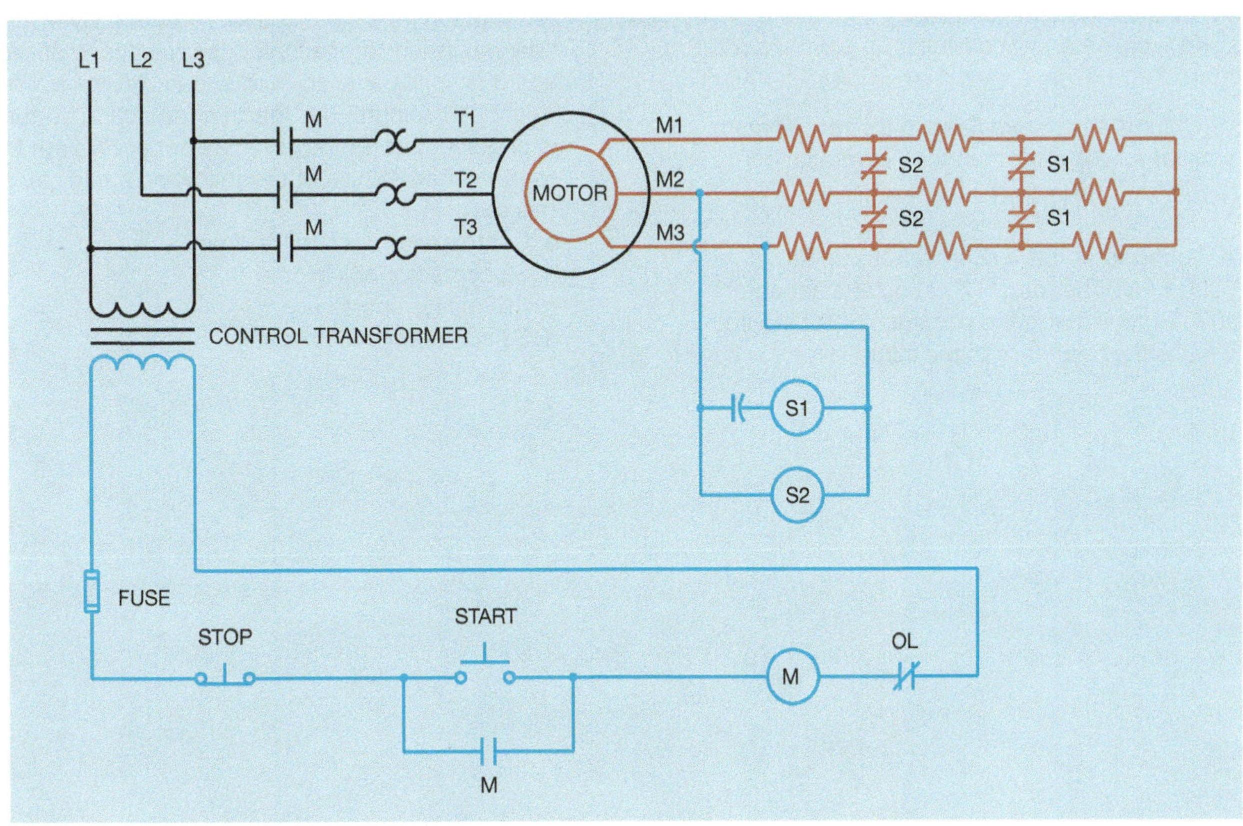

Figure 31–9 Frequency control for a wound rotor induction motor.

motor load. If the motor is connected to a light load, the rotor gains speed rapidly, causing the motor to accelerate rapidly. If the load is heavy, the rotor gains speed at a slower rate, causing a more gradual increase in speed to help the motor overcome the inertia of the load.

Review Questions

1. How many slip rings are on the rotor shaft of a wound rotor motor?
2. What is the purpose of the slip rings located on the rotor shaft of a wound rotor motor?
3. A wound rotor induction motor has a stator that contains six poles per phase. How many poles per phase are in the rotor circuit?
4. Name three factors that determine the amount of torque developed by a wound rotor induction motor.
5. Explain why the wound rotor motor produces the greatest amount of starting torque per amp of starting current of any three phase motor.
6. Explain why controlling the rotor current also controls the stator current.
7. What is the function of a micro limit switch when used with a manual controller for a wound rotor motor?
8. Why are the resistors used in the rotor circuit smaller for a starter than for a controller?
9. What is rotor slip?
10. A wound rotor has a synchronous speed of 1200 RPM. The rotor is rotating at a speed of 1075 RPM. What is the percent of rotor slip and what is the frequency of the induced rotor voltage?
11. Refer to the circuit shown in Figure 31–6. Assume that the motor is running at full speed and the STOP button is pressed. The motor stops running. When the manual control knob is returned to the highest resistance setting, the motor immediately starts running in its lowest speed. Which of the following could cause this problem?
    a. The STOP push button is shorted.
    b. The START push button is shorted.
    c. M auxiliary contact is shorted.
    d. The micro limit switch contact did not reclose when the control was returned to the highest resistance setting.
12. Refer to the circuit shown in Figure 31–7. Assume that the timers are set for a delay of 3 seconds each. When the START button is pressed, the motor starts in its lowest speed. After 3 seconds the motor accelerates to second speed, but never reaches third speed. Which of the following *cannot* cause this problem?
    a. TR1 timer coil is open.
    b. S1 contactor coil is open.
    c. TR2 timer coil is open.
    d. S2 contactor coil is open.
13. Refer to the schematic diagram in Figure 31–8. Assume that the motor is not running. When the third speed push button is pressed, the motor starts in its lowest speed. After a delay of 3 seconds, the motor accelerates to second speed and 3 seconds later to third speed. After a period of about 1 minute, the fourth speed push button is pressed, but the motor does not accelerate to fourth speed. Which of the following could cause this problem?
    a. Control relay CR2 coil is open.
    b. S2 contactor coil is open.
    c. CR3 coil is shorted.
    d. S3 contactor coil is open.

Chapter 32

SYNCHRONOUS MOTORS

Objectives

After studying this chapter the student will be able to:

- Discuss the operation of a synchronous motor.
- List differences between synchronous motors and squirrel cage motors.
- Explain the purpose of an amortisseur winding.
- Discuss how a synchronous motor can produce a leading power factor.
- Discuss the operation of a brushless exciter.

Synchronous motors are so named because of their ability to operate at synchronous speed. They are able to operate at the speed of the rotating magnetic field because they are *not* induction motors. They exhibit other characteristics that make them different than squirrel cage or wound rotor induction motors. Some of these characteristics are:

- They can operate at synchronous speed.
- They operate at a constant speed from no load to full load. Synchronous motors either operate at synchronous speed or they stall and stop running.
- They can produce a leading power factor.
- They are sometimes operated without load to help correct plant power factor. In this mode of operation they are called synchronous condensers.
- The rotor must be excited with an external source of direct current.
- They contain a special squirrel cage winding called the amortisseur winding, which is used to start the motor.

Starting a Synchronous Motor

A special squirrel cage winding, called the amortisseur winding, is used to start a synchronous motor. The rotor of a synchronous motor is shown in Figure 32–1. The amortisseur winding is very similar to a type A squirrel cage winding. It provides good starting torque and a relatively low starting current. Once the synchronous motor has accelerated to a speed close to that of the rotating magnetic field, the rotor is excited by connecting it to a source of direct current. Exciting the rotor causes pole pieces wound in the rotor to become electromagnets. These electromagnets lock with the rotating magnetic field of the stator and the motor runs at **synchronous speed**. *A synchronous motor should* never *be started with excitation applied to the rotor.* The magnetic field of the pole pieces will be alternately attracted and repelled by the rotating magnetic field resulting in no torque being produced in either direction. High induced voltage, however, may damage the rotor windings and other components connected in the rotor circuit. The excitation current should be connected to the rotor only after it has accelerated to a speed that is close to synchronous speed.

Figure 32–1 Rotor of a synchronous motor.

Excitation Current

There are several ways that excitation current can be supplied to the rotor of a synchronous motor such as slip rings, a brushless exciter, and DC generator. Small synchronous motors generally contain two slip rings on the rotor shaft. A set of brushes are used to supply direct current to the rotor (Figure 32–2).

If manual starting is employed, an operator will manually excite the rotor after it has accelerated close to synchronous speed. During this acceleration process, a high voltage can be induced into the windings of the rotor. A resistor, called the *field discharge resistor*, is connected in parallel with the rotor winding. Its function is to limit the amount of induced voltage when the motor is started and limit the amount of induced voltage caused by the collapsing magnetic field when the motor is stopped and the excitation current is disconnected. A switch called the *field discharge switch* is used to connect the excitation current to the rotor. The switch is designed so that when it is closed it will make connection to the direct current power supply before it breaks connection with the field discharge resistor. When the switch is opened, it will make connection to the field discharge resistor before it breaks connection with the direct current power supply. This permits the field discharge resistor to always be connected to the rotor when DC excitation is not being applied to the rotor.

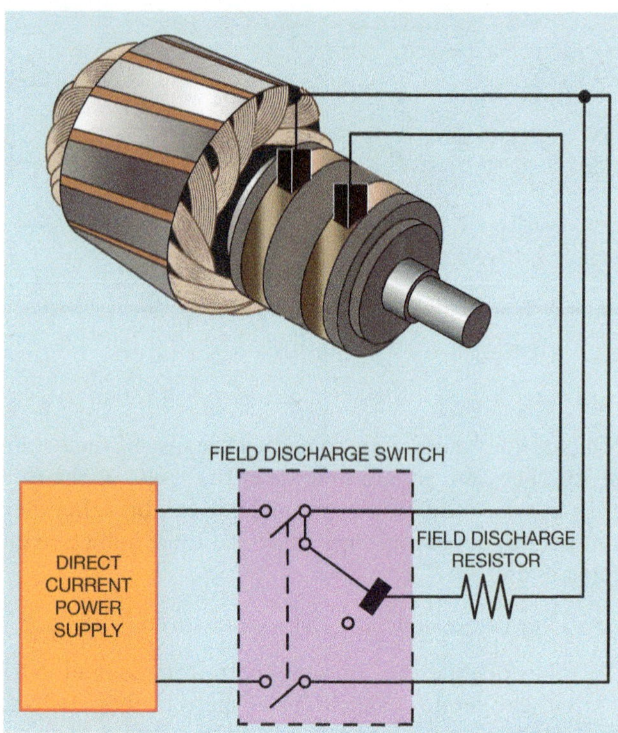

Figure 32–2 A field discharge resistor is connected in parallel with the rotor winding during starting.

The Brushless Exciter

A second method of supplying excitation current to the rotor is with a brushless exciter. The brushless exciter has an advantage in that there are no brushes or slip rings to wear. The brushless exciter is basically a small three phase alternator winding and three phase rectifier located on the shaft of the rotor. Refer to the photograph in Figure 32–1. At the back of the rotor a small winding can be seen. This is the winding of the brushless exciter. Electromagnets are placed on either side of the winding (Figure 32–3). A three phase rectifier and fuses are also located on the rotor shaft. The rectifier converts the three phase alternating current produced in the alternator winding into direct current before it is supplied to the rotor winding (Figure 32–4). The amount of excitation current supplied to the rotor winding is controlled by the amount of direct current supplied to the electromagnets. The output voltage of the alternator winding is controlled by the flux density of the pole pieces.

Direct Current Generator

Another method of supplying excitation current is with the use of a self-excited direct current generator mounted on the rotor shaft. The amount of excitation current is

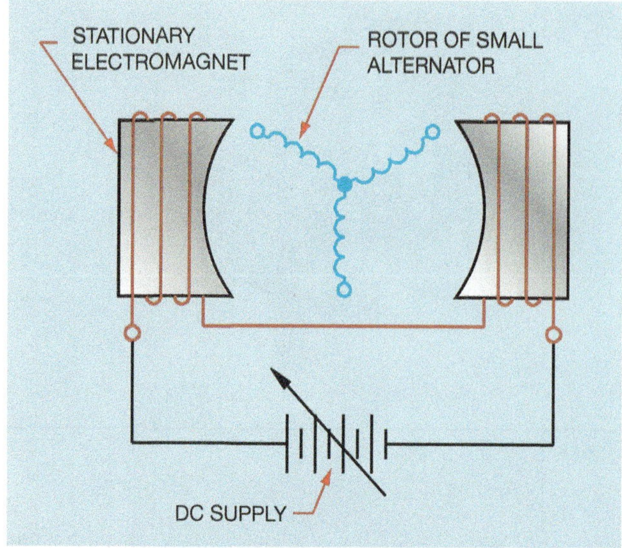

Figure 32–3 The brushless exciter contains stationary electromagnets.

adjusted by controlling the field current of the generator. The output of the armature supplies the excitation current for the rotor. Because the generator is self-excited, it does not require an external source of direct current. Although that is an advantage over supplying the excitation current through slip rings or with a brushless exciter, the generator does contain a commutator and brushes. The generator generally requires more maintenance than the other methods.

Automatic Starting for Synchronous Motors

Synchronous motors can be automatically started as well as manually started. One of the advantages of a synchronous motor is that it provides good starting torque with a relatively low starting current. Many large motors are capable of being started directly across-the-line because of this feature. If the power company will not permit across-the-line starting, synchronous motors can also employ autotransformer starting, reactor starting, or wye-delta starting. Regardless of the method employed to connect the stator winding to the power line, the main part of automatic control for a synchronous motor lies in connecting excitation current to the rotor at the proper time. The method employed is determined by the manner in which excitation is applied to the rotor. In the case of manual excitation, the field discharge switch is used. Brushless exciter circuits often employ electronic devices for sensing the rotor speed to connect DC excitation to the rotor at the proper time. If a direct current generator is employed to provide excitation current, a special field contactor, out-of-step relay, and polarized field frequency relay are generally used.

The Field Contactor

The field contactor looks very similar to a common three-pole contactor (Figure 32–5). This is not a standard contactor, however. The field contactor contains

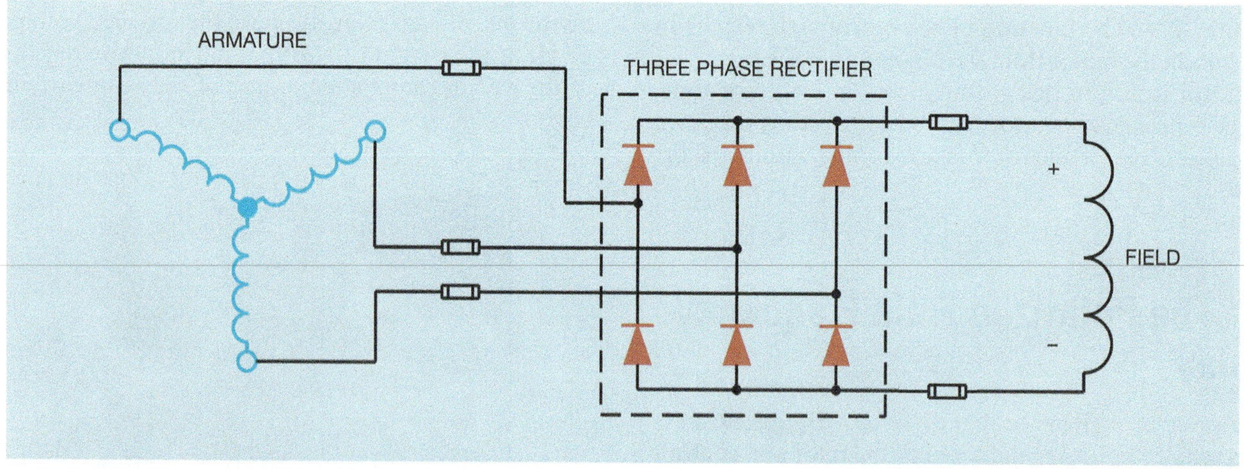

Figure 32–4 Basic brushless exciter circuit.

Courtesy of Rockwell Automation, Inc.

Figure 32–5 Field contactor used in the starting of a synchronous motor.

Courtesy of Rockwell Automation, Inc.

Figure 32–6 Out-of-step relay.

a DC coil and is energized by the excitation current of the rotor. The field contactor serves the same function as the field discharge switch discussed previously. The two outside contacts connect and disconnect the excitation current to the rotor circuit. The middle contact connects and disconnects the field discharge resistor at the proper time.

Out-of-Step Relay

The out-of-step relay is actually a timer that contains a current operated coil instead of a voltage operated coil. The coil is connected in series with the field discharge resistor. The timer can be pneumatic, dashpot, or electronic. A dashpot type out-of-step relay is shown in Figure 32–6. The function of the out-of-step relay is to disconnect the motor from the power line in the event that the rotor is not excited within a certain length of time. Large synchronous motors can be damaged by excessive starting current if the rotor is not excited within a short time.

The Polarized Field Frequency Relay

The polarized field frequency relay (Figure 32–7) is responsible for sensing the speed of the rotor and controlling the operation of the field contactor. The polarized field frequency relay (PFR) is used in conjunction with a reactor. The reactor is connected in the rotor circuit of the synchronous motor. The polarized field frequency relay contains two separate coils, one DC and one AC (Figure 32–8). Coil A is the DC coil and is connected to the source of direct current excitation. Its function is to polarize the magnetic core material of the relay. Coil B is the AC coil. This coil is connected in parallel with the reactor (Figure 32–9). To understand the operation of the circuit, first consider the path magnetic flux takes if only the DC coil of the PFR is energized (Figure 32–10). Note that the flux path is through the crossbar, not the ends of the relay. Because the flux does not reach the ends of the pole piece, the armature is not attracted and the contact remains closed.

When the synchronous motor is started, however, the rotating magnetic field of the stator induces an AC voltage into the rotor winding. A current path exists through the reactor, field discharge resistor, and coil of the out-of-step relay. Because the induced voltage is 60 Hz at the instant of starting, the inductive reactance of the reactor causes a major part of the rotor current to

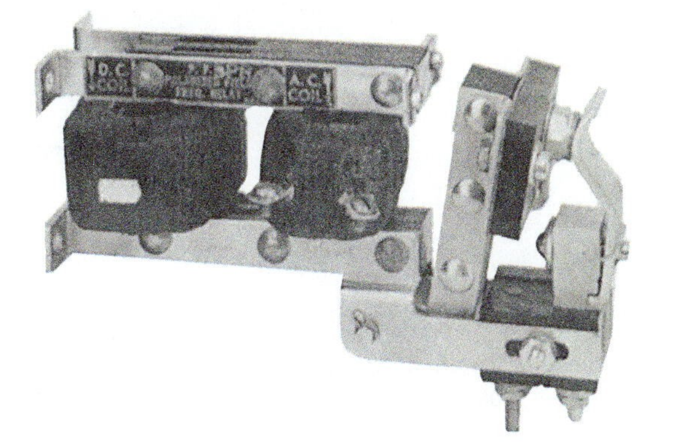

Courtesy of Rockwell Automation, Inc.

Figure 32–7 Polarized field frequency relay.

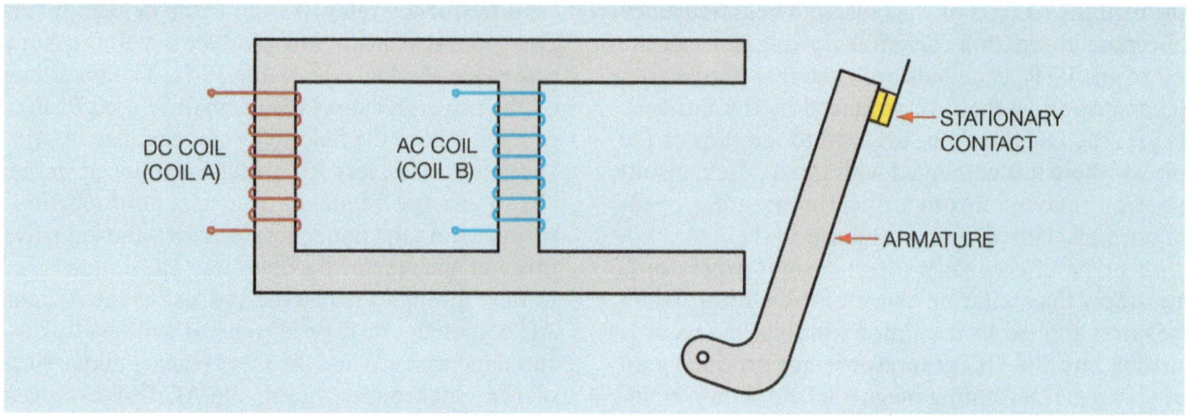

Figure 32–8 The polarized field frequency relay contains both a DC and AC coil.

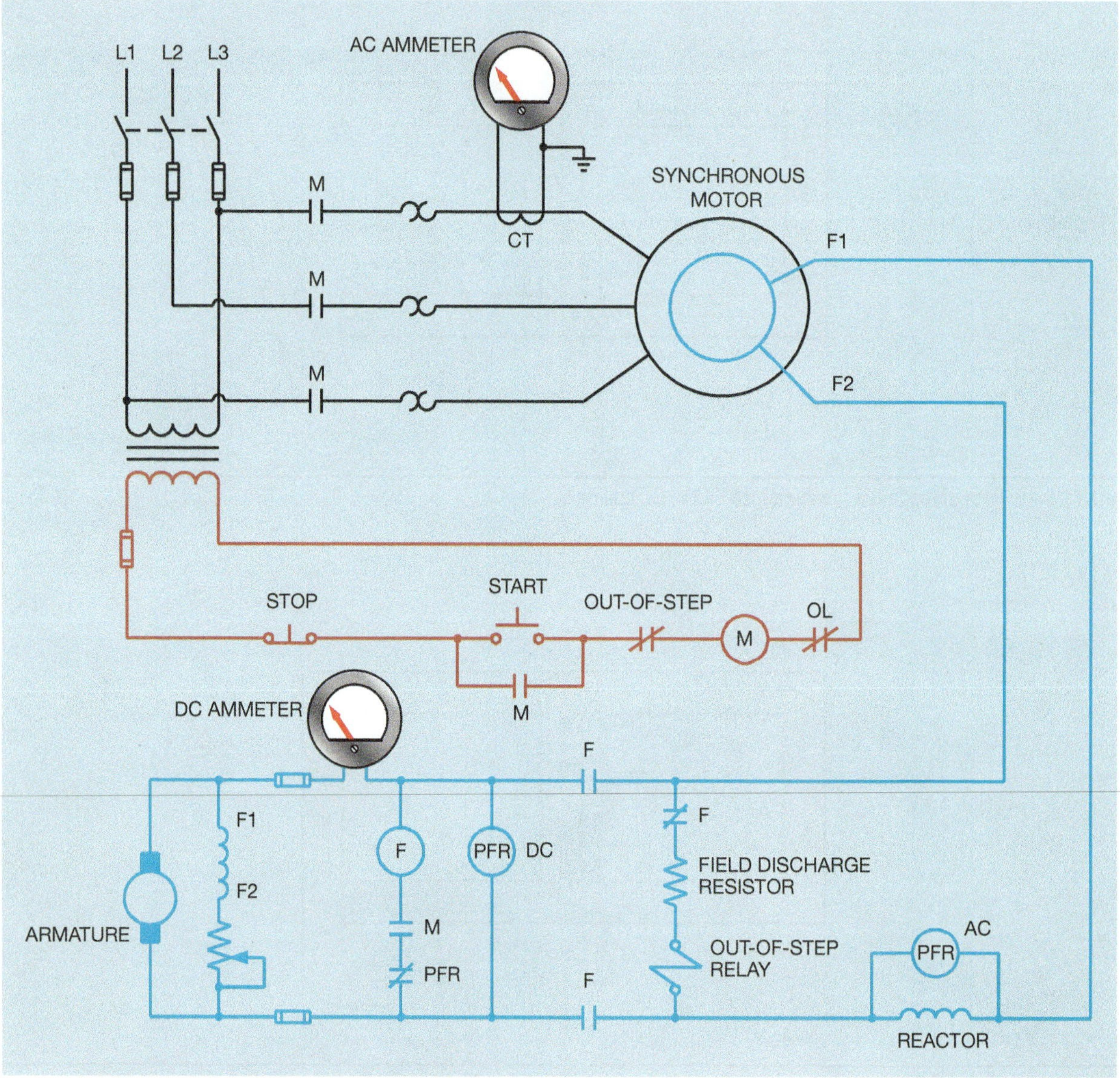

Figure 32–9 Control circuit for a synchronous motor.

flow through the AC coil of the polarized field frequency relay. Because alternating current is flowing through the AC coil of the PFR, each half cycle the flux produces in the AC coil opposes the flux produced by the DC coil. This causes the DC flux to be diverted to the ends of the pole pieces where it is combined with the AC flux, resulting in a strong enough flux to attract the armature, opening the normally closed contact (Figure 32–11).

In this type of control, a direct current generator is used to supply the excitation current for the rotor. When power is first applied to the stator winding, the rotor is not turning and the DC generator is not producing an output voltage. The rotating magnetic field, however, induces a high voltage into the rotor windings, supplying a large amount of current for the AC coil of the polarized field frequency relay. As the rotor begins to turn, the DC generator begins to produce a voltage, supplying power for the DC coil of the PFR. The combined flux of the two coils causes the normally closed PFR contact to open before the field relay can energize. As the rotor speed increases, less AC voltage is induced in the rotor circuit and the frequency decreases in proportion to rotor speed. As the frequency decreases, the inductive reactance of the reactor becomes less, causing more current to flow through the reactor and less to the AC coil. The AC coil of the PFR produces less and less flux as rotor speed increases. When the rotor reaches about 90 percent of the synchronous speed, the AC flux can no longer maintain the current path through the PFR armature and the DC flux returns to the path as shown in Figure 32–10.

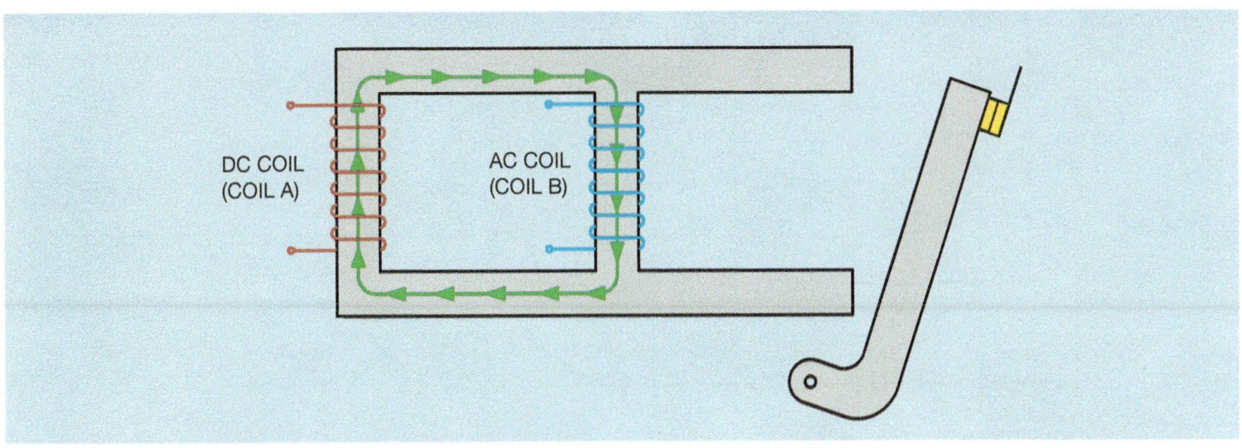

Figure 32–10 Path of magnetic flux produced by the DC coil only.

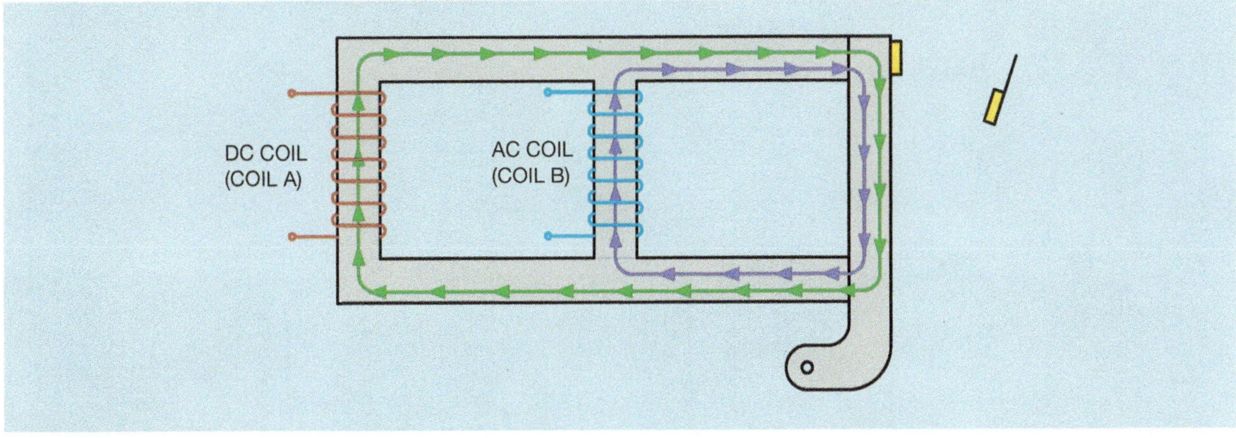

Figure 32–11 Flux of AC and DC coils combines to attract the armature.

When the armature drops away, it recloses the PFR contact and connects the coil of the field relay to the line. When the field relay energizes, direct current is connected to the rotor circuit and the field discharge resistor and out-of-step relay are disconnected from the line.

Power Factor Correction

As stated previously, synchronous motors can be made to produce a leading power factor. A synchronous motor can be made to produce a leading power factor by overexciting the rotor. If the rotor is underexcited, the motor will have a lagging power factor similar to a squirrel cage or wound rotor induction motor. The reason is when the DC excitation current is too low, part of the AC current supplied to the stator winding is used to magnetize the iron in the motor.

Normal excitation is achieved when the amount of excitation current is sufficient to magnetize the iron core of the motor and no alternating current is required. Two conditions indicate when normal excitation has been achieved:

1. The current supplied to the motor drops to its lowest level.
2. The power factor will be 100 percent or unity.

If more than normal excitation current is supplied, overexcitation occurs. In this condition, the DC excitation current overmagnetizes the iron of the motor and part of the AC line current is used to de-magnetize the iron. The de-magnetizing process causes the AC line current to lead the voltage in the same manner as a capacitor.

Applications

Due to their starting characteristics and ability to correct power factor, synchronous motors are generally employed where large horsepower motors are needed. They often provide the power for pumps, compressors, centrifuges, and large grinders. A 1500 HP synchronous motor is shown in Figure 32–12.

Photo courtesy of WEG Electric Corp.

Figure 32–12 A 1500 HP synchronous motor.

Review Questions

1. What is a synchronous motor called when it is operated without load and used for power factor correction?
2. What is an amortisseur winding and what function does it serve?
3. Should the excitation current be applied to the rotor of a synchronous motor before it is started?
4. What is the function of a field discharge resistor?
5. What controls the output voltage of the alternator when a brushless exciter is used to supply the excitation current of the rotor?
6. What is the purpose of the DC coil on a polarized field frequency relay?
7. What is the purpose of an out-of-step relay?
8. Why is it possible for a synchronous motor to operate at the speed of the rotating magnetic field?
9. Name two factors that indicate when normal excitation current is being applied to the motor.
10. How can a synchronous motor be made to produce a leading power factor?

Chapter 33

CONSEQUENT POLE MOTORS

Objectives

After studying this chapter the student will be able to:

» Identify terminal markings for two-speed one-winding consequent pole motors.

» Discuss how speed of a consequent pole motor is changed.

» Connect a two-speed, one-winding consequent pole motor.

» Discuss the construction of three-speed consequent pole motors.

» Discuss different types of four-speed consequent pole motors.

Consequent pole motors have the ability to change speed by changing the number of stator poles. Two factors determine the synchronous speed of an AC motor:

1. Frequency of the applied voltage
2. Number of stator poles per phase

A chart showing synchronous speed of 60 and 50 Hz motors with different numbers of poles is shown in Figure 33–1. A three phase, two-pole motor contains six actual poles. The magnetic field makes one revolution of a two-pole motor each complete cycle. If the stator of a motor were to be cut and laid out flat, the magnetic field would traverse the entire length in one cycle (Figure 33–2A). If the number of stator poles is doubled to four per phase (Figure 33–2B), the magnetic field traverses the same number of stator poles during one cycle. Because the number of poles has been doubled, the magnetic field will travel only half as far during one complete cycle. Consequent pole motors have an advantage over some others types of variable speed alternating current motors in that they maintain a high torque when speed is reduced.

The number of stator poles is changed by redirecting the current through pairs of poles (Figure 33–3). If the current travels in the same direction through two pole pieces, both produce the same magnetic polarity and are essentially one pole piece. If the current direction is opposite through each pole piece, they produce opposite magnetic polarities and are essentially two poles.

Two-speed consequent pole motors contain one reconnectable stator winding. A two-speed motor contains six T leads in the terminal connection box. The motor

STATOR POLES PER PHASE	SPEED IN RPM	
	60 HZ	50 HZ
2	3600	3000
4	1800	1500
6	1200	1000
8	900	750

Figure 33–1 Synchronous speed is determined by the frequency and number of stator poles per phase.

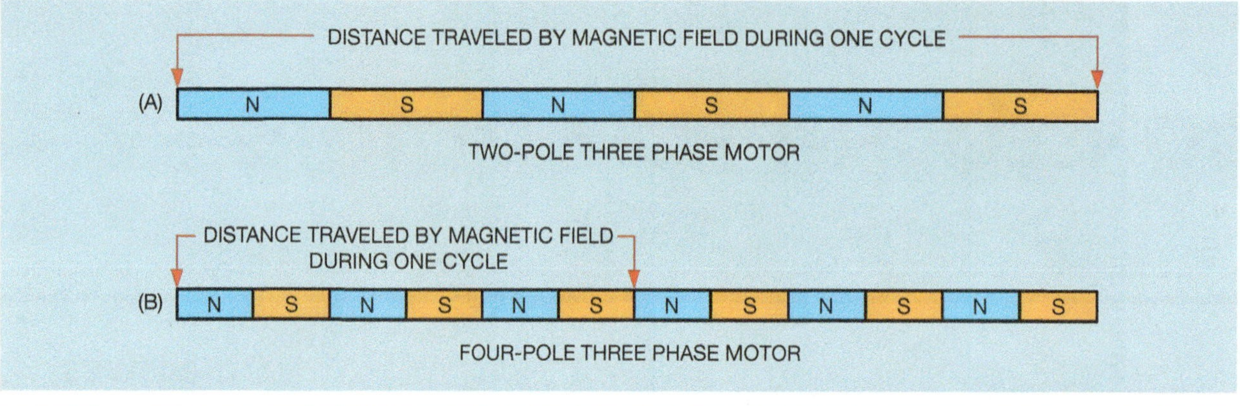

Figure 33–2 The magnetic field travels through the same number of poles during each complete cycle.

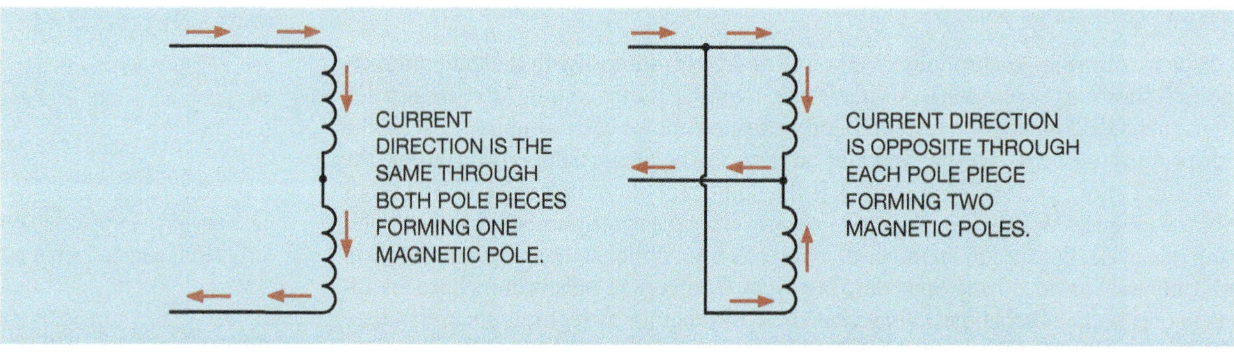

Figure 33–3 The direction of current flow determines the number of poles.

can be connected to form a series delta or parallel wye (Figure 33–4). If the motor is wound in such a way that the series delta connection gives the high speed and the parallel wye gives the low speed, the horsepower will be the same for either connection. If the winding is such that the series delta gives the low speed and the parallel wye gives the high speed, the torque will be the same for both speeds.

Two-speed consequent pole motors provide a speed ratio of 2:1. For example, a two-speed consequent pole motor could provide synchronous speeds of 3600 and 1800 RPM, or 1800 and 900 RPM, or 1200 and

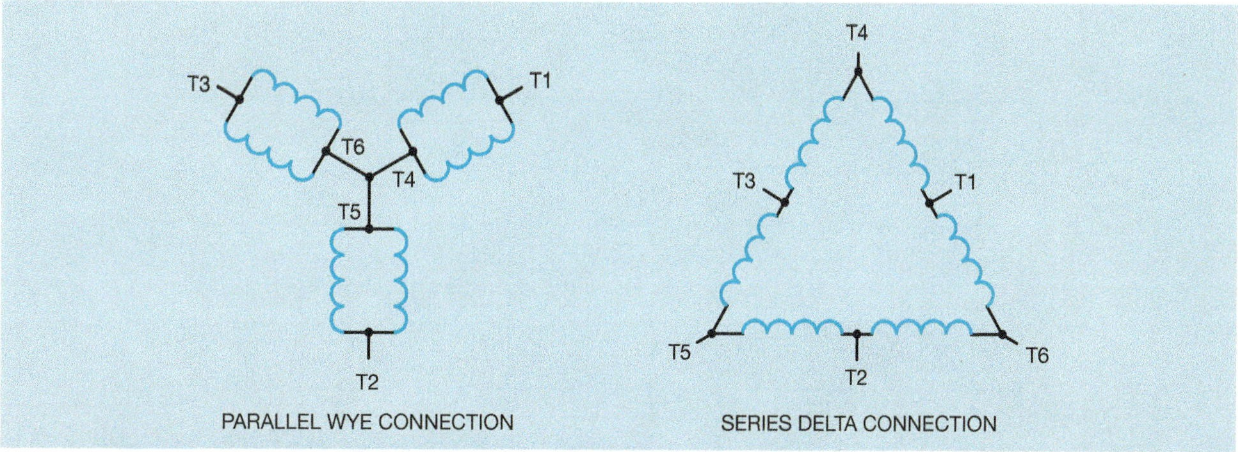

Figure 33–4 Stator windings can be connected as either parallel wye or series delta.

600 RPM. The connection diagram for a two-speed consequent pole motor is shown in Figure 33–5. A typical controller for a two-speed motor is shown in Figure 33–6. Note that the low speed connection requires six load contacts, three to connect the L1, L2, and L3 to T1, T2, and T3, and three to short leads T4, T5, and T6 together. Although contactors with six load contacts can be obtained, it is common practice to employ a separate three-pole contactor to short T4, T5, and T6 together.

In the circuit shown in Figure 33–6, the STOP button must be pressed before a change of speed can be made. Another control circuit is shown in Figure 33–7 that forces the motor to start in low speed before it can be accelerated to high speed. The STOP button does not have to be pressed before the motor can be accelerated to the second speed. A permissive relay (PR) is used to accomplish this logic. The motor can be returned to the low speed by pressing the LOW push button after the motor has been accelerated to high speed. The load connections are the same as shown in Figure 33–6.

SPEED	L1	L2	L3	OPEN	TOGETHER
LOW	T1	T2	T3	———	T4, T5, T6
HIGH	T4	T5	T6	ALL OTHERS	

Figure 33–5 Connection diagram for a two-speed consequent pole motor.

The control circuit shown in Figure 33–8 permits the motor to be started in either high or low speed. The speed can be changed by pressing either the high or low speed push buttons. It is not necessary to press the stop button first.

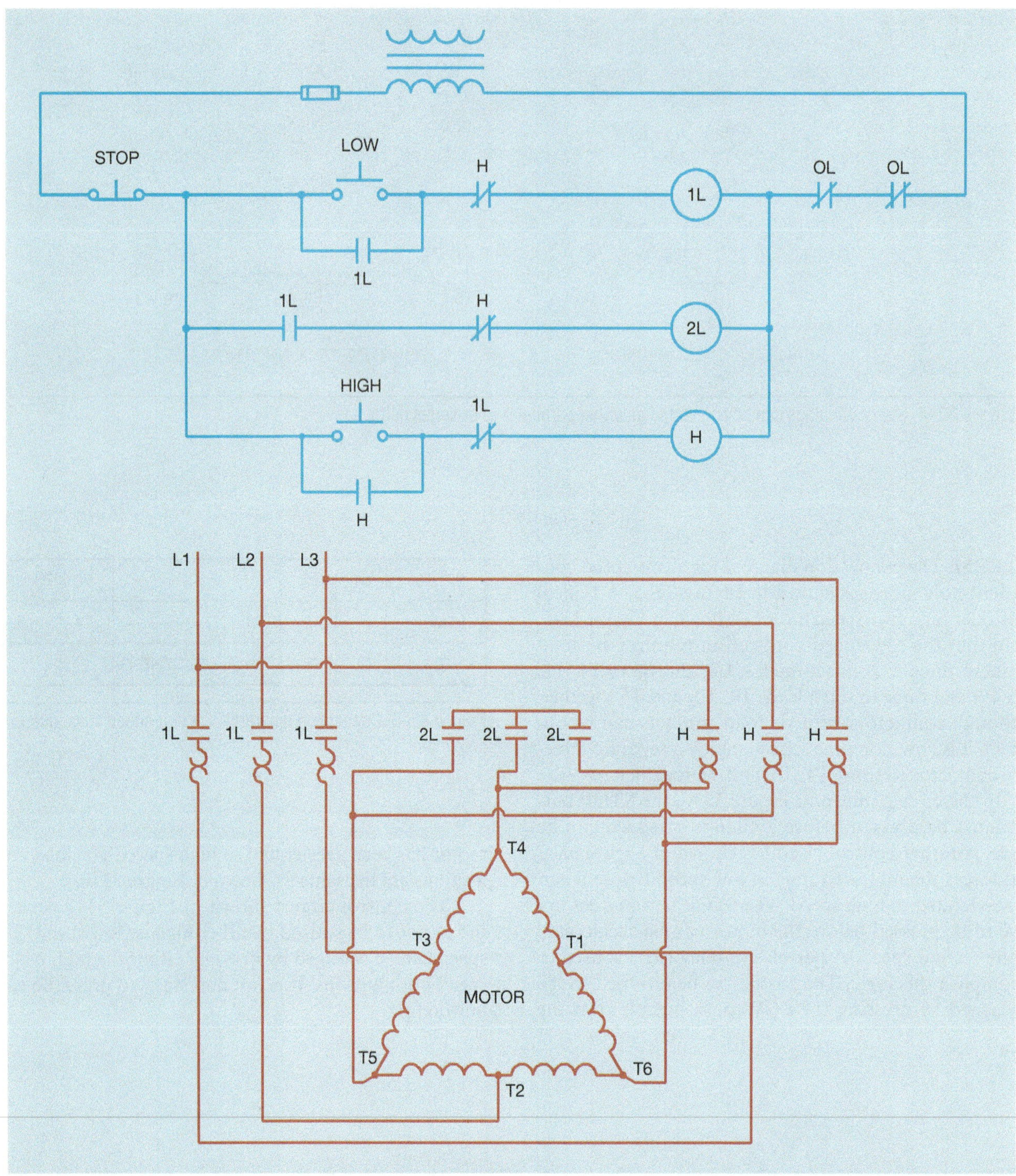

Figure 33–6 Two-speed control for a consequent pole motor.

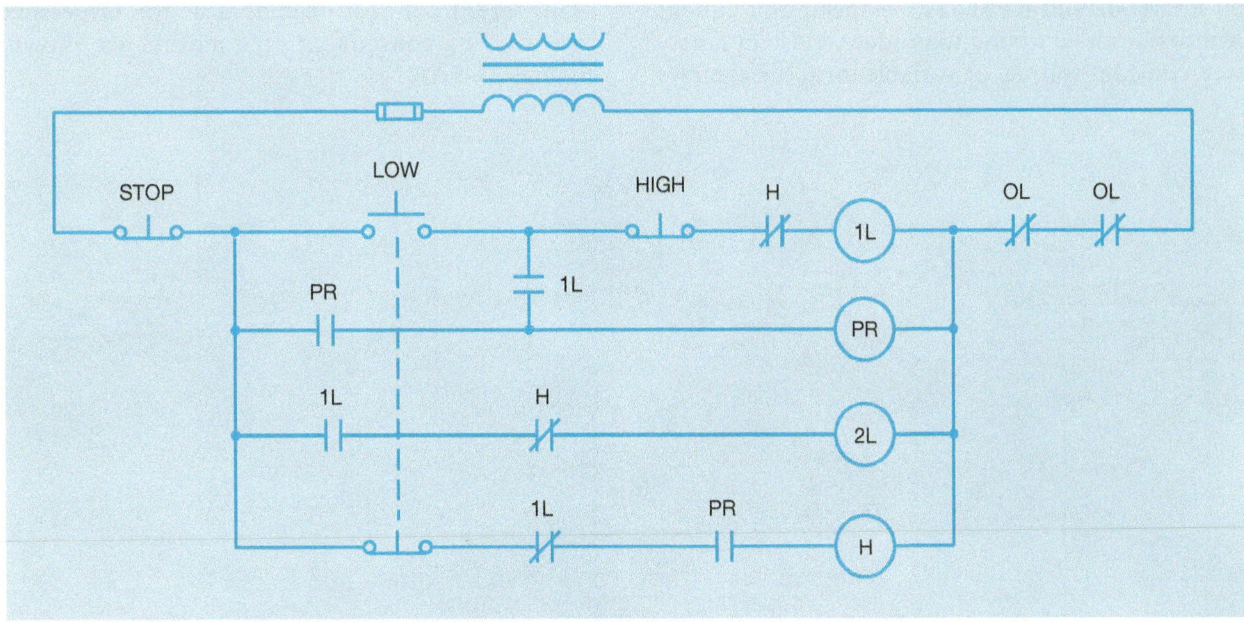

Figure 33–7 The motor must be started in low speed before it can be accelerated to high speed.

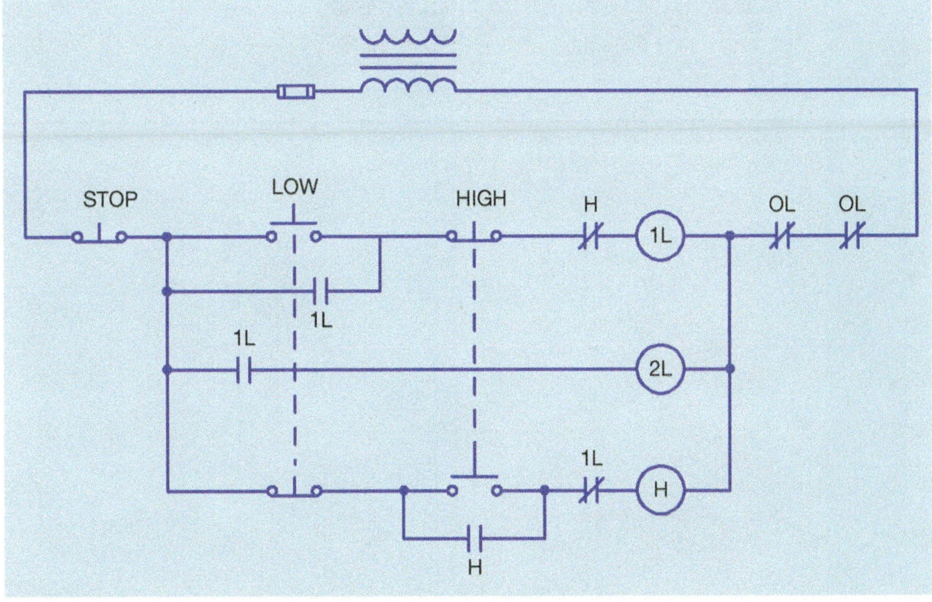

Figure 33–8 Motor speed can be changed without pressing the stop button.

Three-Speed Consequent Pole Motors

Consequent pole motors that are intended to operate with three speeds contain two separate stator windings. One winding is reconnectable like the winding in a two-speed motor. The second winding is wound for a certain number of poles and is not reconnectable. If one stator winding were wound with six poles and the second were reconnectable for two or four poles, the motor would develop synchronous speeds of 3600 RPM, 1800 RPM, or 1200 RPM when connected to a 60 Hz line. If the reconnectable winding were to be wound for four- or eight-pole connection, the motor would develop synchronous speeds of 1800 RPM,

1200 RPM, or 900 RPM. Three-speed consequent pole motors can be wound to produce constant horsepower, constant torque, or variable torque. Examples of different connection diagrams for three-speed, two-winding consequent pole motors are shown in Figures 33–9A–I.

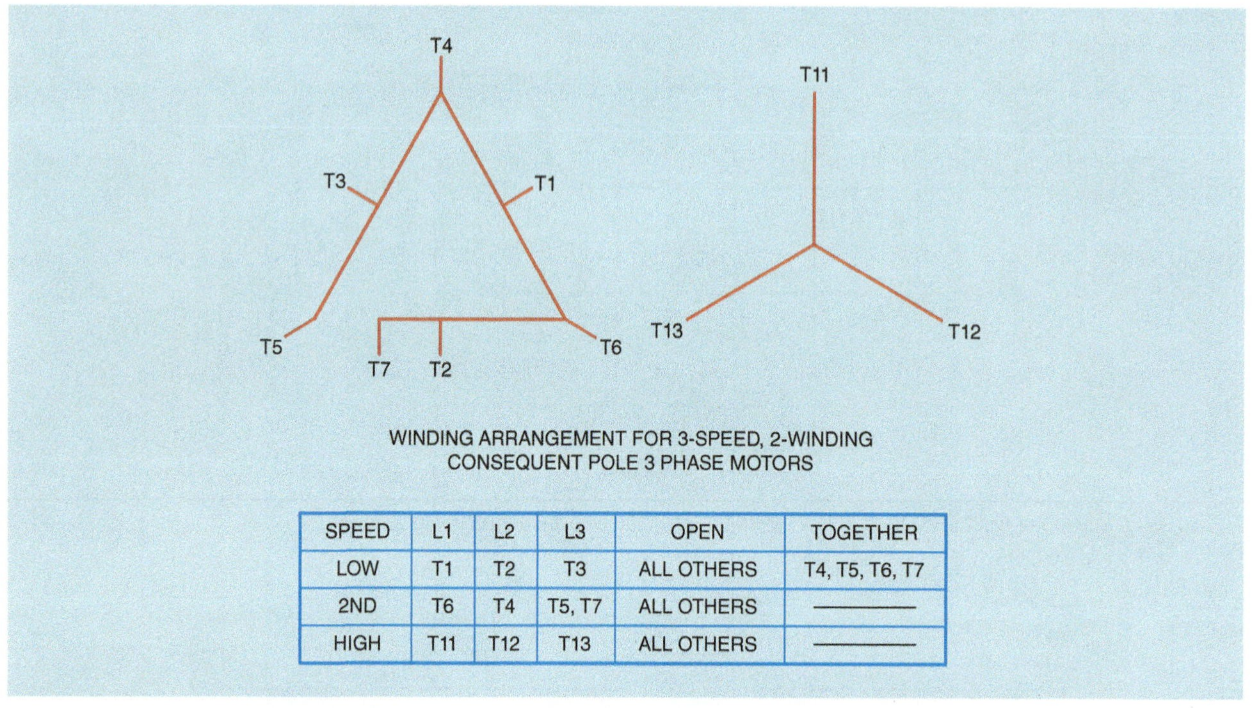

SPEED	L1	L2	L3	OPEN	TOGETHER
LOW	T1	T2	T3	ALL OTHERS	T4, T5, T6, T7
2ND	T6	T4	T5, T7	ALL OTHERS	———
HIGH	T11	T12	T13	ALL OTHERS	———

Figure 33–9A Constant horsepower.

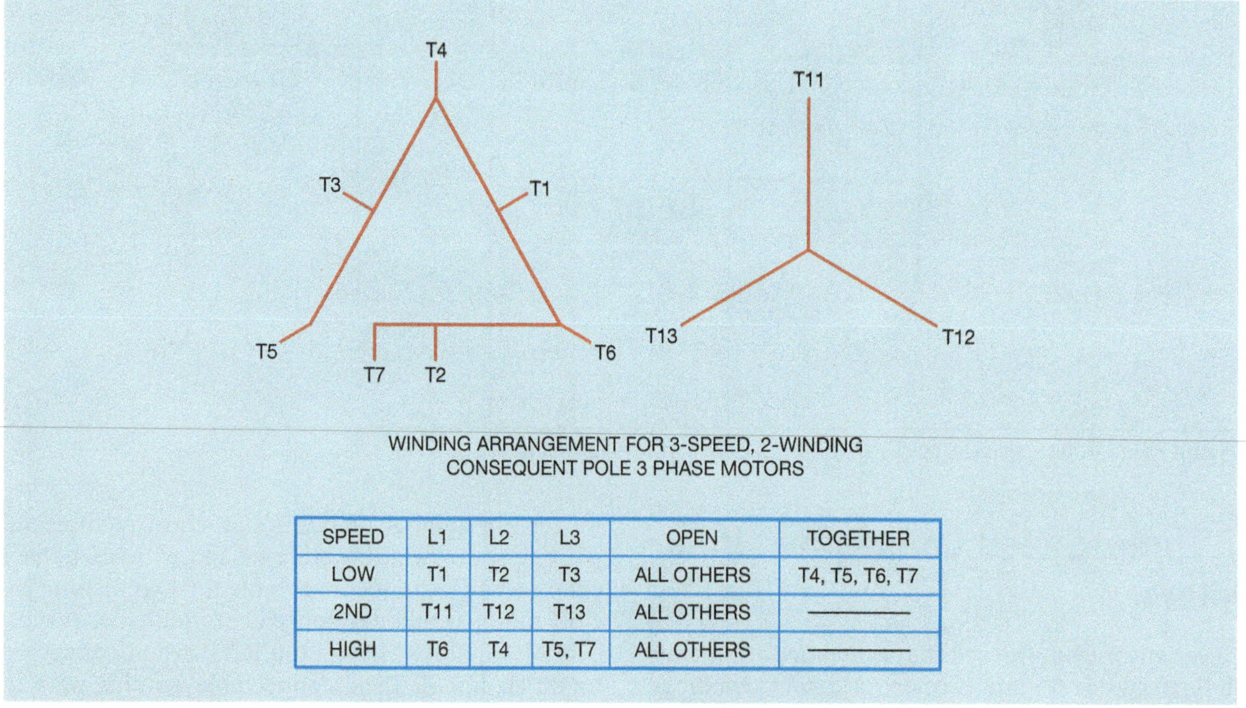

SPEED	L1	L2	L3	OPEN	TOGETHER
LOW	T1	T2	T3	ALL OTHERS	T4, T5, T6, T7
2ND	T11	T12	T13	ALL OTHERS	———
HIGH	T6	T4	T5, T7	ALL OTHERS	———

Figure 33–9B Constant horsepower.

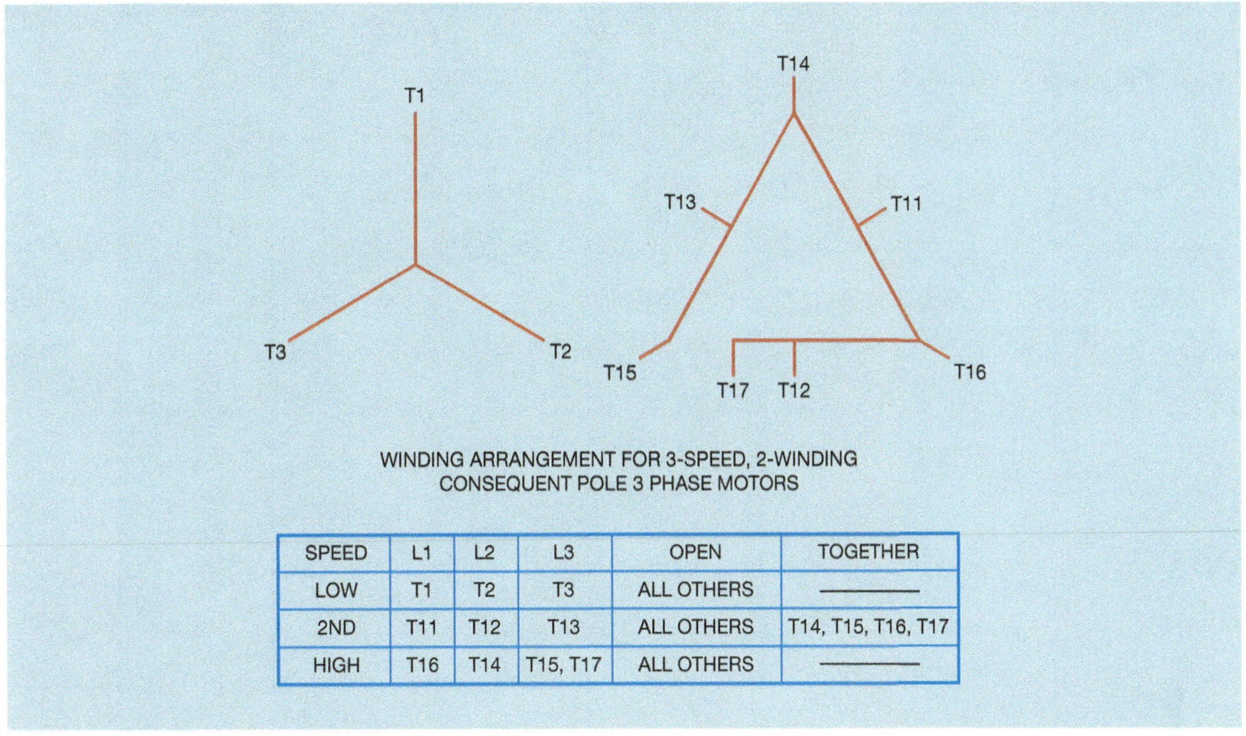

SPEED	L1	L2	L3	OPEN	TOGETHER
LOW	T1	T2	T3	ALL OTHERS	————
2ND	T11	T12	T13	ALL OTHERS	T14, T15, T16, T17
HIGH	T16	T14	T15, T17	ALL OTHERS	————

Figure 33–9C Constant horsepower.

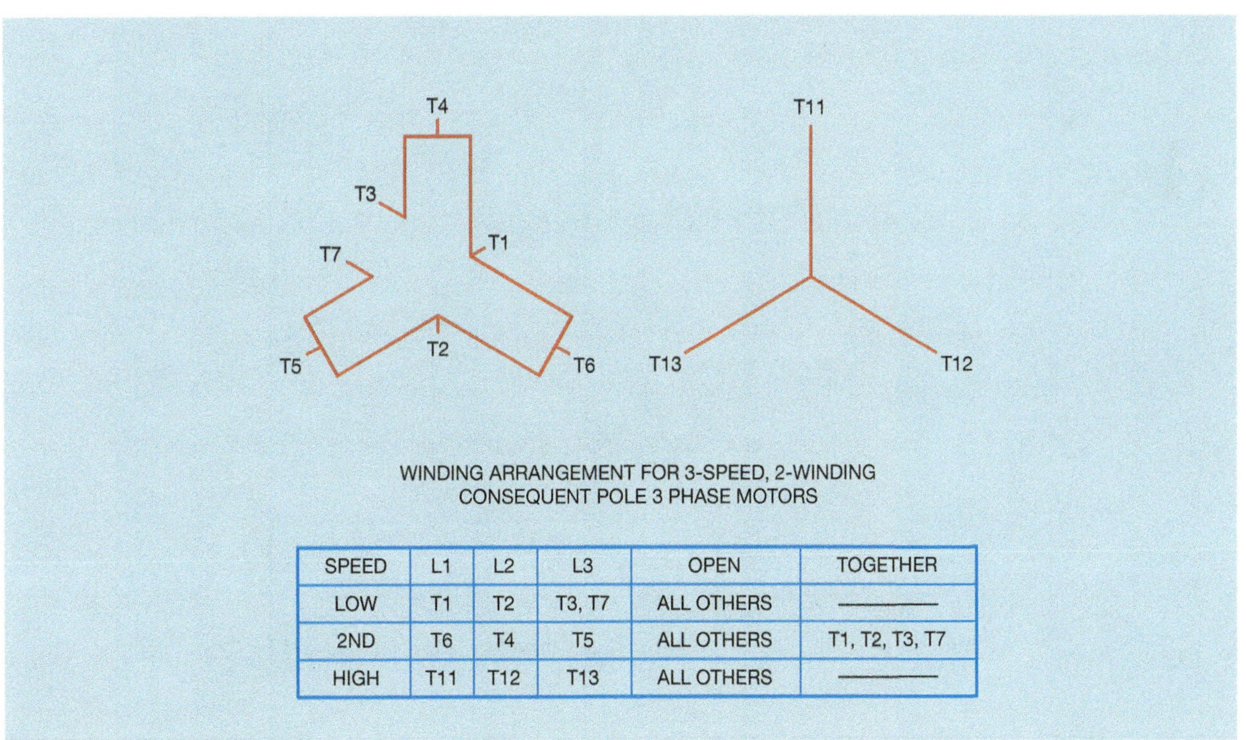

SPEED	L1	L2	L3	OPEN	TOGETHER
LOW	T1	T2	T3, T7	ALL OTHERS	————
2ND	T6	T4	T5	ALL OTHERS	T1, T2, T3, T7
HIGH	T11	T12	T13	ALL OTHERS	————

Figure 33–9D Constant torque.

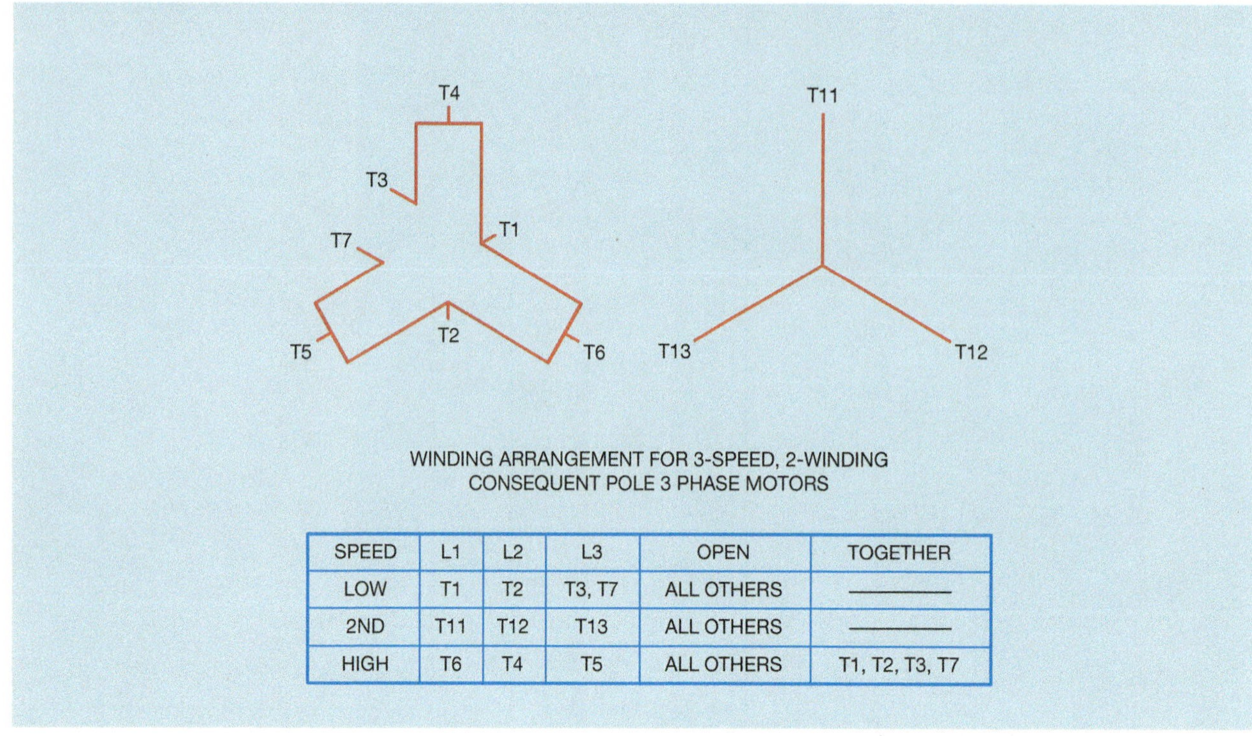

SPEED	L1	L2	L3	OPEN	TOGETHER
LOW	T1	T2	T3, T7	ALL OTHERS	————
2ND	T11	T12	T13	ALL OTHERS	————
HIGH	T6	T4	T5	ALL OTHERS	T1, T2, T3, T7

Figure 33–9E Constant torque.

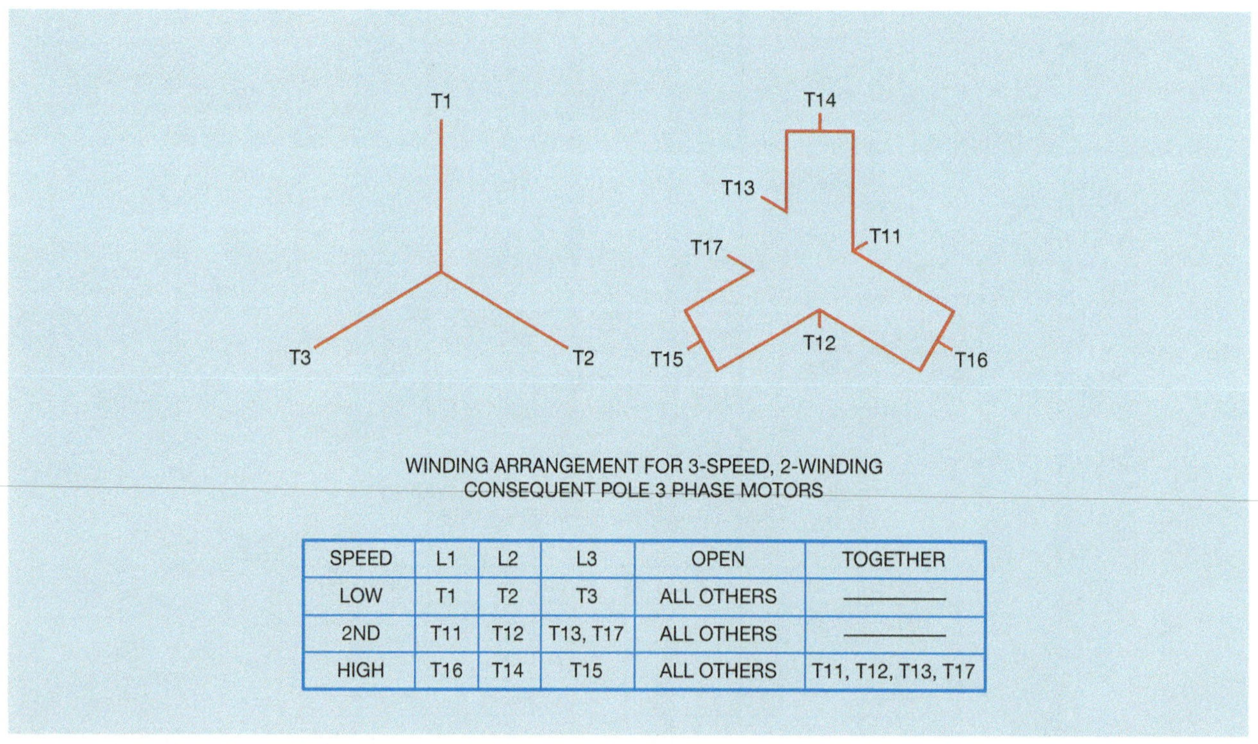

SPEED	L1	L2	L3	OPEN	TOGETHER
LOW	T1	T2	T3	ALL OTHERS	————
2ND	T11	T12	T13, T17	ALL OTHERS	————
HIGH	T16	T14	T15	ALL OTHERS	T11, T12, T13, T17

Figure 33–9F Constant torque.

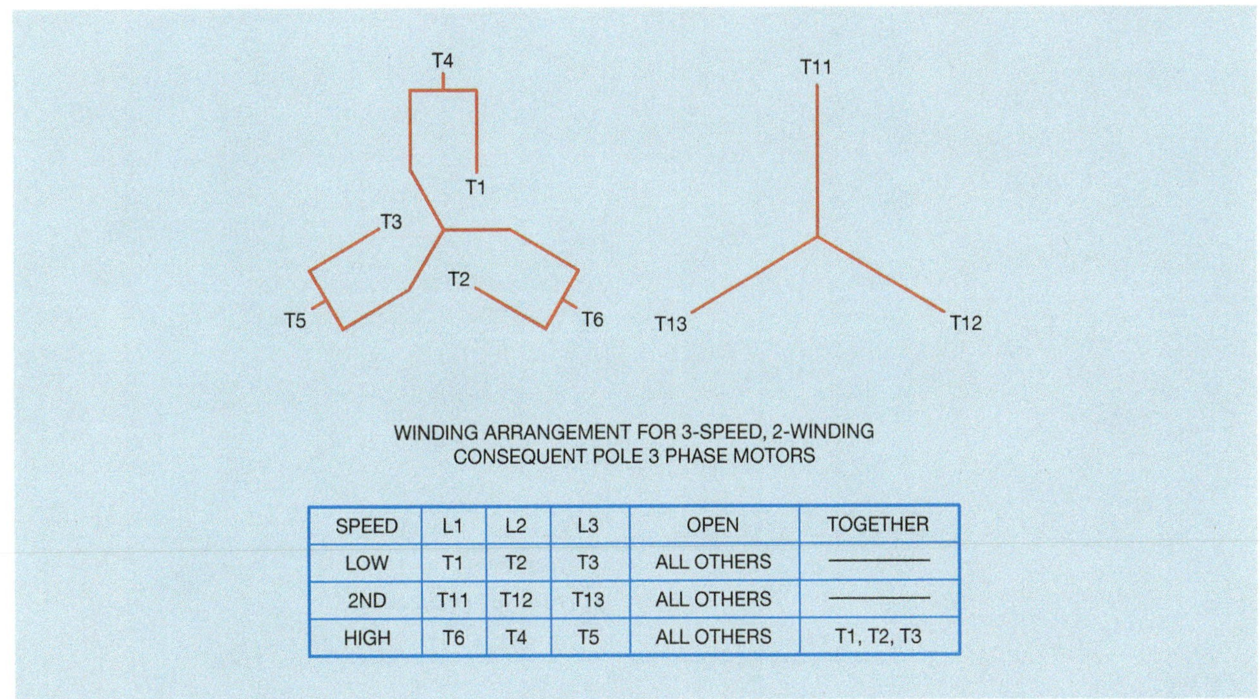

SPEED	L1	L2	L3	OPEN	TOGETHER
LOW	T1	T2	T3	ALL OTHERS	———
2ND	T11	T12	T13	ALL OTHERS	———
HIGH	T6	T4	T5	ALL OTHERS	T1, T2, T3

Figure 33–9G Variable torque.

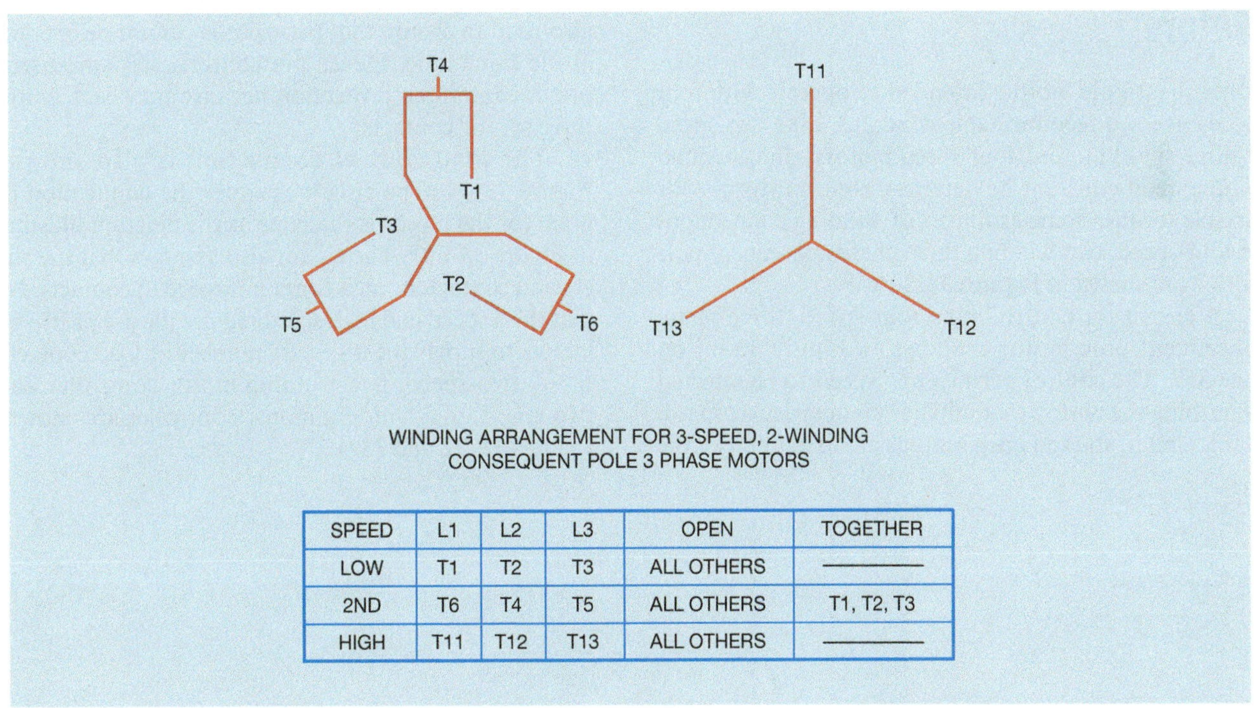

SPEED	L1	L2	L3	OPEN	TOGETHER
LOW	T1	T2	T3	ALL OTHERS	———
2ND	T6	T4	T5	ALL OTHERS	T1, T2, T3
HIGH	T11	T12	T13	ALL OTHERS	———

Figure 33–9H Variable torque.

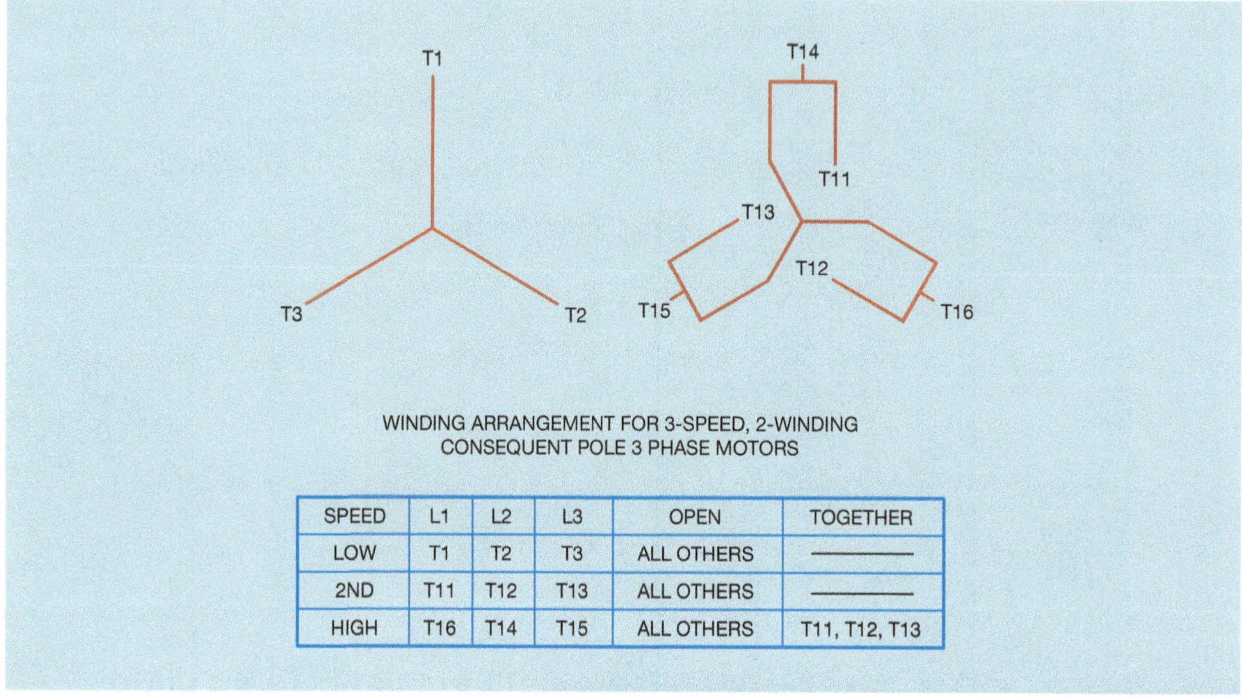

SPEED	L1	L2	L3	OPEN	TOGETHER
LOW	T1	T2	T3	ALL OTHERS	————
2ND	T11	T12	T13	ALL OTHERS	————
HIGH	T16	T14	T15	ALL OTHERS	T11, T12, T13

Figure 33–9I Variable torque.

Four-Speed Consequent Pole Motors

Consequent pole motors intended to operate with four speeds use two reconnectable windings. Like two-speed or three-speed motors, four-speed motors can be wound to operate at constant horsepower, constant torque, or variable torque. Some examples of winding connections for four-speed, two-winding three phase consequent pole motors are shown in Figures 33–10A–F.

A circuit for controlling a four-speed, three phase consequent pole motor is shown in Figure 33–11 on page 352. The control permits any speed to be selected by pushing the button that initiates that particular speed. In this circuit, stacked push buttons are used to break the circuit to any other speed before the starter that controls the selected speed is energized. Electrical interlocks are also used to ensure that two-speeds cannot be energized at the same time. Eleven pin control relays are used to provide interlock protection because they each contain three sets of contacts.

The load contact connection is also shown in Figure 33–11. The circuit assumes the connection diagram for the motor is the same as the diagram illustrated in Figure 33–10F. The circuit also assumes that the starters and contactors each contain three load contacts. Note that third speed and high speed require the use of two contactors to supply the necessary number of load contacts.

A two-speed, two-winding motor controller and a two-speed, one-winding motor controller are shown in Figures 33–12 and 33–13.

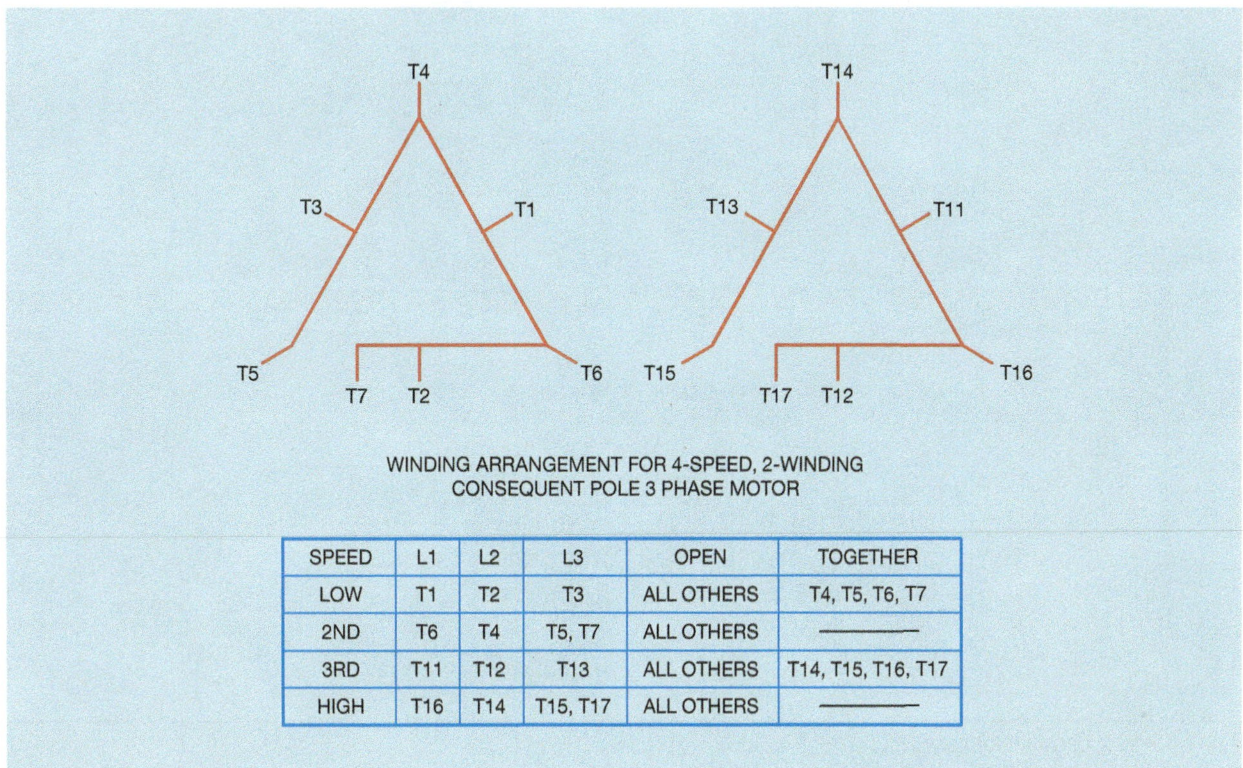

SPEED	L1	L2	L3	OPEN	TOGETHER
LOW	T1	T2	T3	ALL OTHERS	T4, T5, T6, T7
2ND	T6	T4	T5, T7	ALL OTHERS	————
3RD	T11	T12	T13	ALL OTHERS	T14, T15, T16, T17
HIGH	T16	T14	T15, T17	ALL OTHERS	————

Figure 33–10A Constant horsepower.

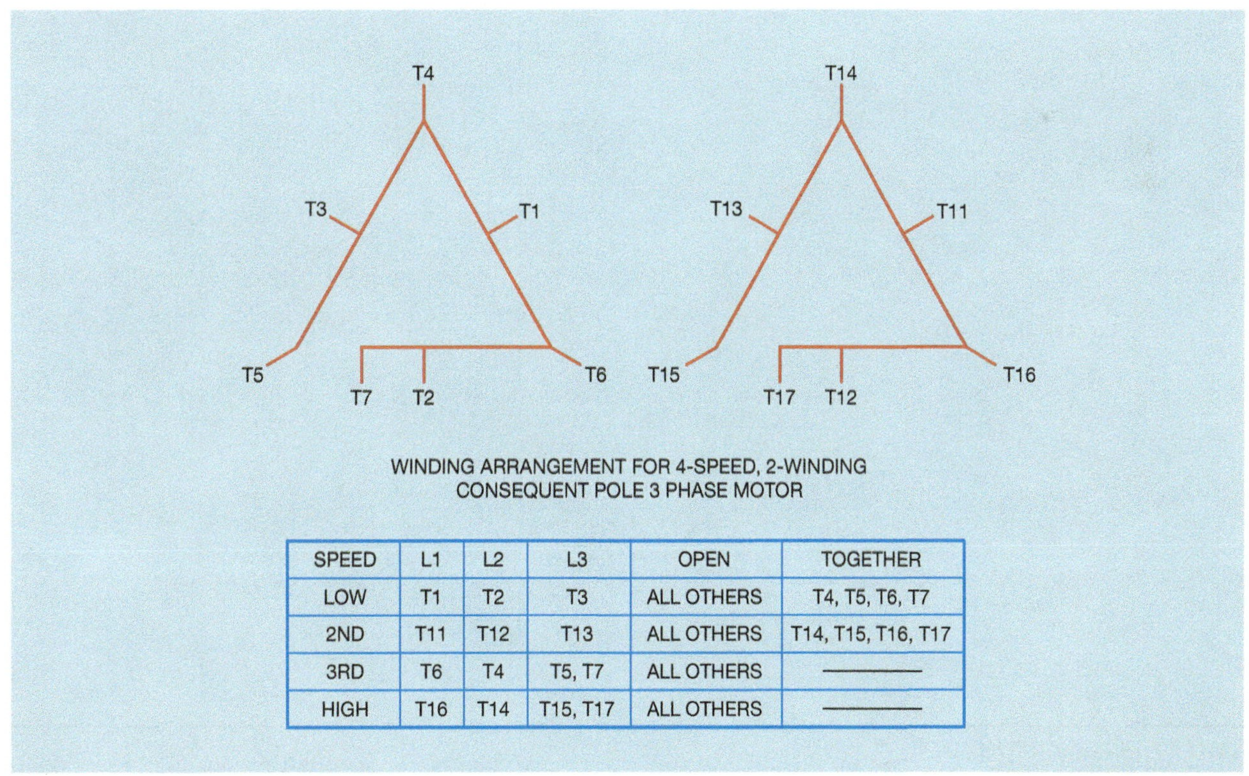

SPEED	L1	L2	L3	OPEN	TOGETHER
LOW	T1	T2	T3	ALL OTHERS	T4, T5, T6, T7
2ND	T11	T12	T13	ALL OTHERS	T14, T15, T16, T17
3RD	T6	T4	T5, T7	ALL OTHERS	————
HIGH	T16	T14	T15, T17	ALL OTHERS	————

Figure 33–10B Constant horsepower.

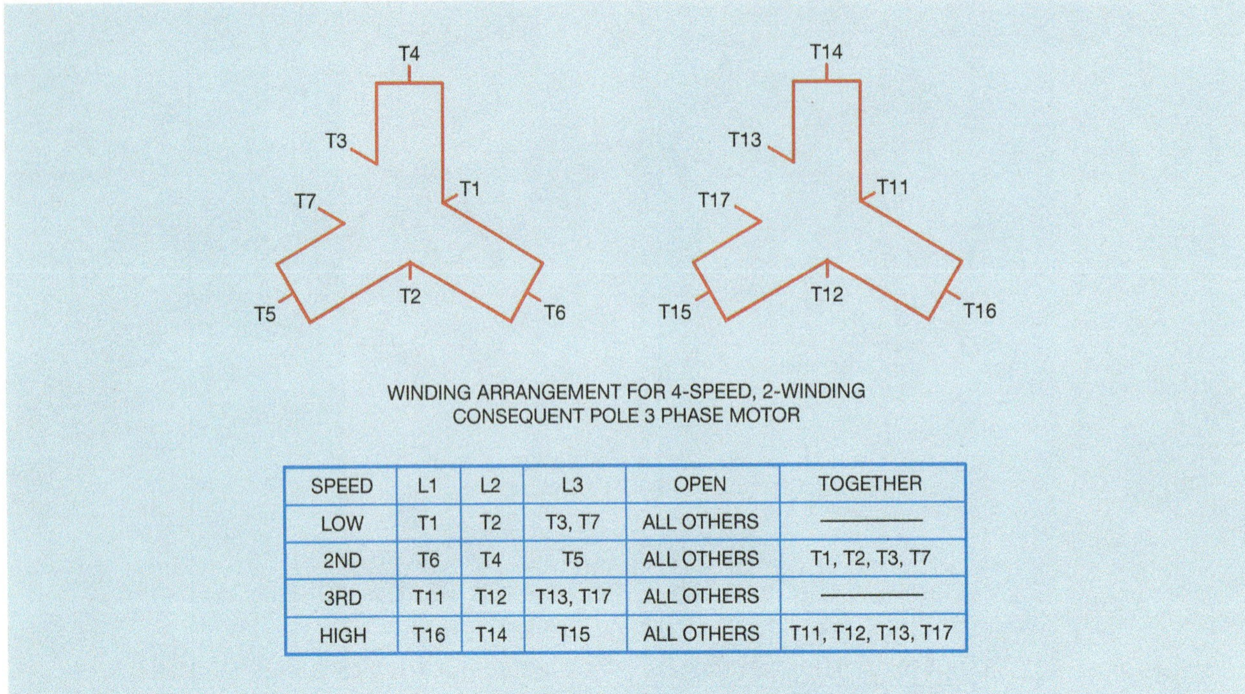

SPEED	L1	L2	L3	OPEN	TOGETHER
LOW	T1	T2	T3, T7	ALL OTHERS	————
2ND	T6	T4	T5	ALL OTHERS	T1, T2, T3, T7
3RD	T11	T12	T13, T17	ALL OTHERS	————
HIGH	T16	T14	T15	ALL OTHERS	T11, T12, T13, T17

Figure 33–10C Constant torque.

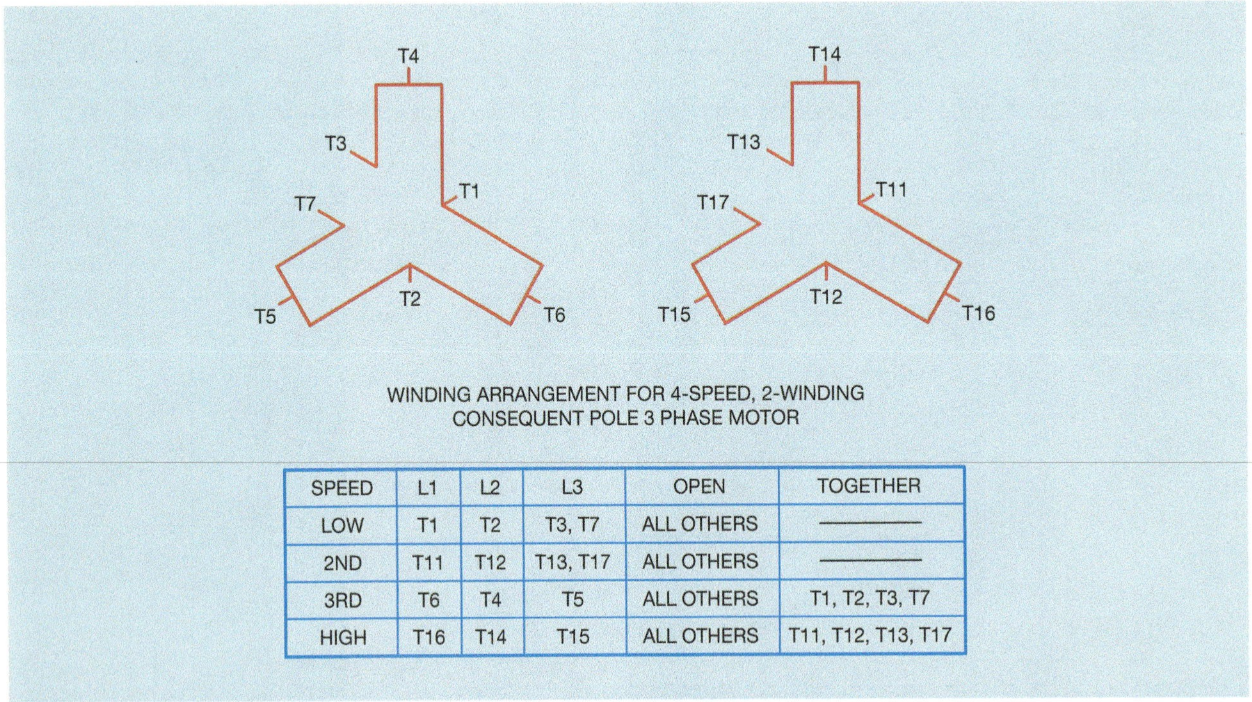

SPEED	L1	L2	L3	OPEN	TOGETHER
LOW	T1	T2	T3, T7	ALL OTHERS	————
2ND	T11	T12	T13, T17	ALL OTHERS	————
3RD	T6	T4	T5	ALL OTHERS	T1, T2, T3, T7
HIGH	T16	T14	T15	ALL OTHERS	T11, T12, T13, T17

Figure 33–10D Constant torque.

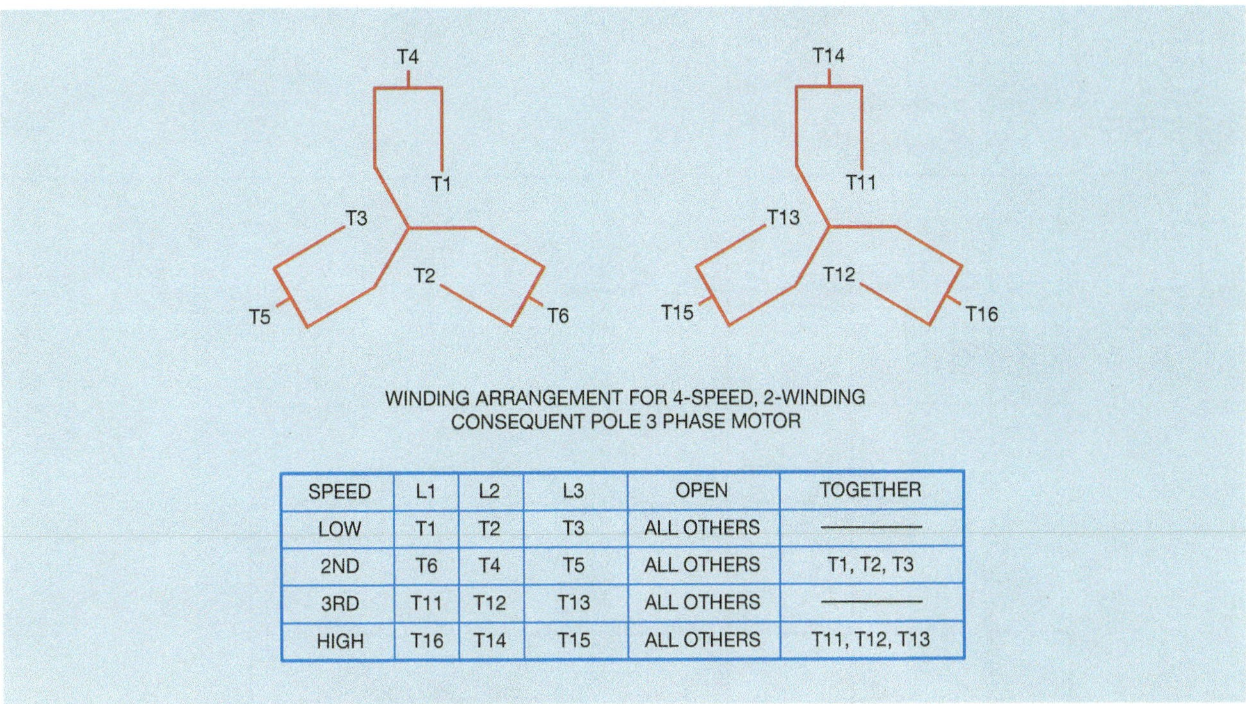

SPEED	L1	L2	L3	OPEN	TOGETHER
LOW	T1	T2	T3	ALL OTHERS	————
2ND	T6	T4	T5	ALL OTHERS	T1, T2, T3
3RD	T11	T12	T13	ALL OTHERS	————
HIGH	T16	T14	T15	ALL OTHERS	T11, T12, T13

Figure 33–10E Variable torque.

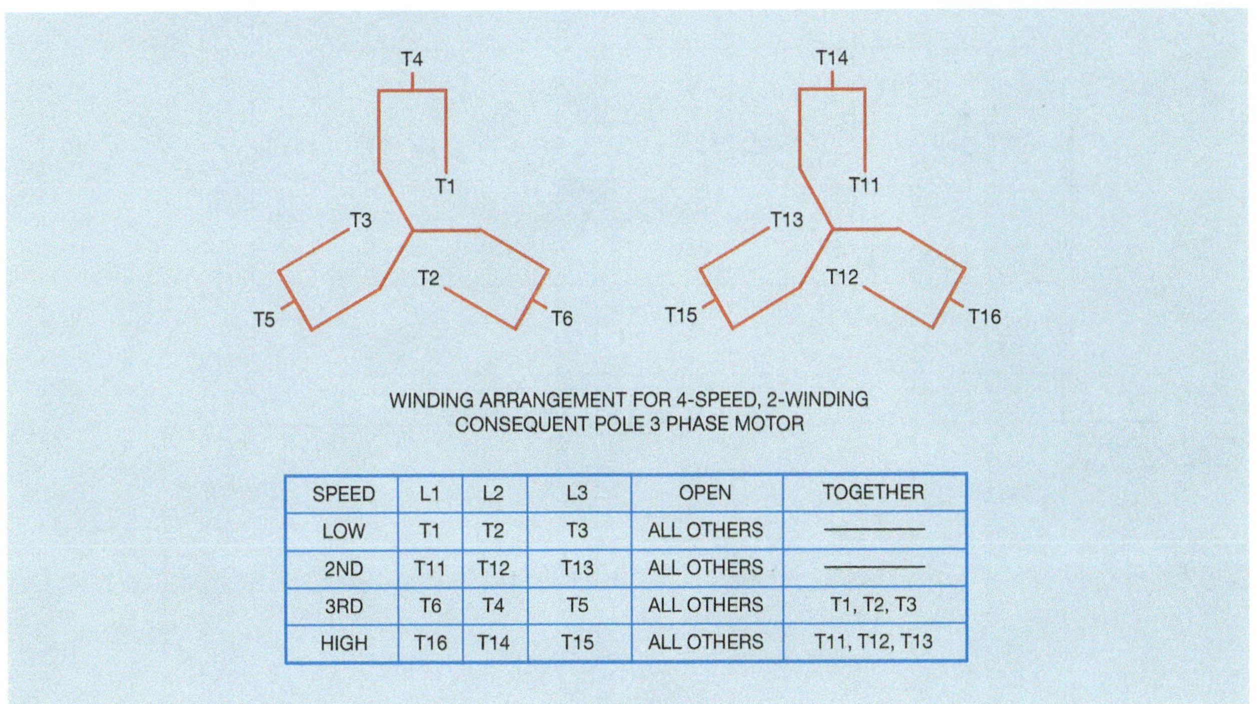

SPEED	L1	L2	L3	OPEN	TOGETHER
LOW	T1	T2	T3	ALL OTHERS	————
2ND	T11	T12	T13	ALL OTHERS	————
3RD	T6	T4	T5	ALL OTHERS	T1, T2, T3
HIGH	T16	T14	T15	ALL OTHERS	T11, T12, T13

Figure 33–10F Variable torque.

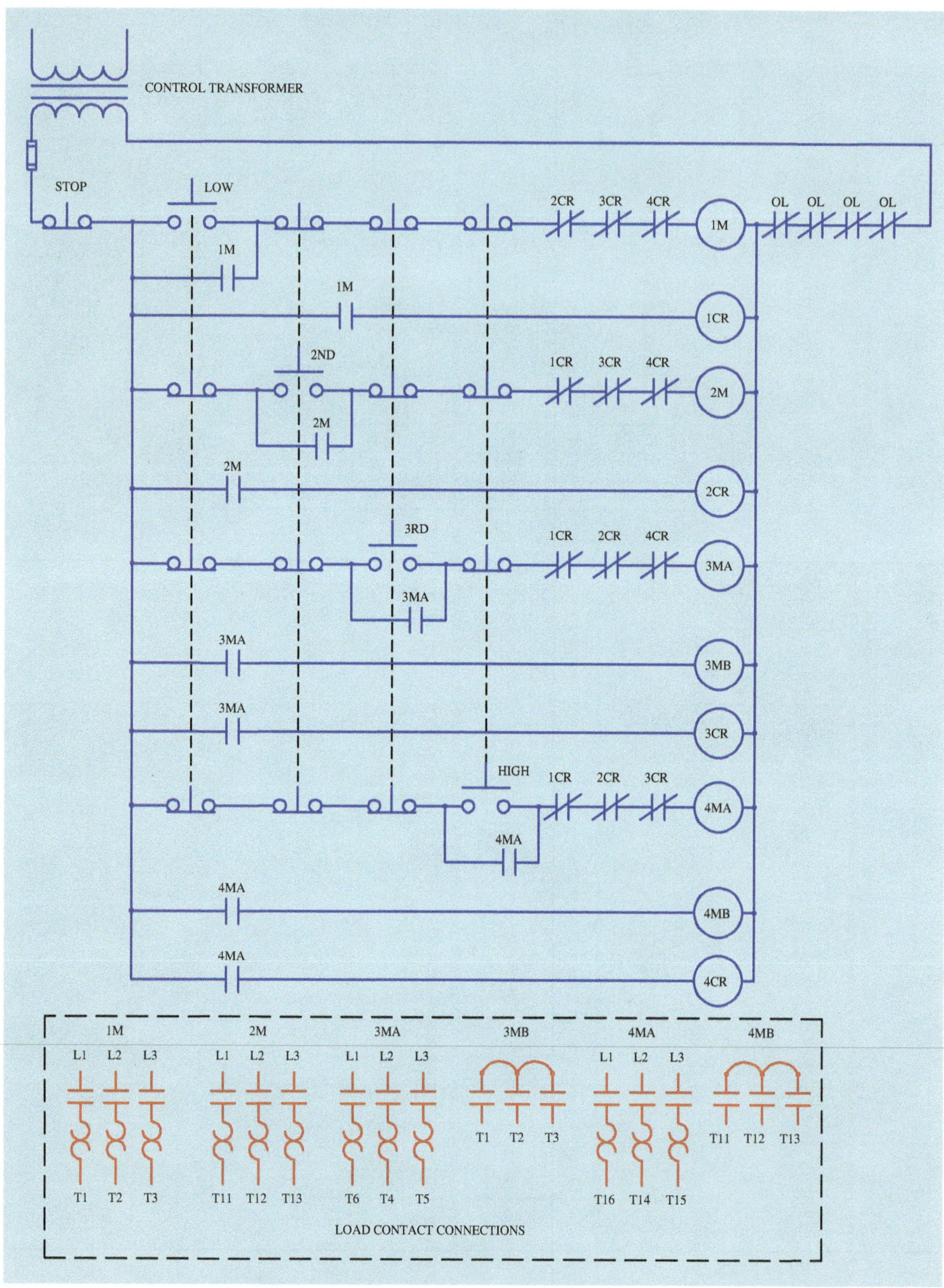

Figure 33–11 Push button control for a four-speed consequent pole three phase motor.

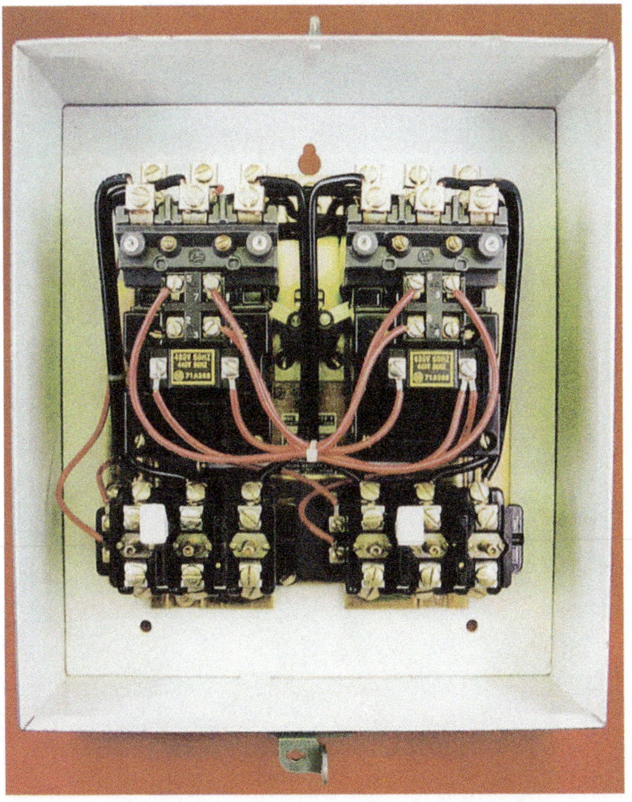

Figure 33–12 Two-speed, two-winding motor controller.

Figure 33–13 Two-speed, one-winding motor controller.

Review Questions

1. Name two factors that determine the synchronous speed of a motor.
2. How many speeds can be obtained from a consequent pole motor that contains only one stator winding?
3. What is the advantage of consequent pole motors over some other types of variable speed motors?
4. A consequent pole motor has synchronous speeds of 1800, 1200, and 900 RPM. How many stator windings does this motor have?
5. Refer to the circuit shown in Figure 33–6. You are to install this control system. How many auxiliary contacts should starter 1L contain and how many are normally open and how many are normally closed?
6. Refer to the circuit shown in Figure 33–6. What is the function of contactor 2L?
7. Refer to the circuit shown in Figure 33–7. When the low speed push button is pressed, the motor begins to run in low speed. When the high speed push button is pressed, the motor stops running. Which of the following could cause this problem?
   a. The 1L contactor coil is open
   b. H contactor coil is open
   c. PR relay coil is open
   d. The 2L contactor coil is open
8. Refer to the circuit shown in Figure 33–11. Assume that coil 2CR is shorted. Would it be possible to run the motor in the third speed?
9. Refer to the circuit shown in Figure 33–11. Explain the action of the circuit if coil 2CR is shorted and the second speed push button is pressed.
10. Refer to the circuit shown in Figure 33–11. You are to construct this circuit on the job. Would it be possible to used an 11 pin control relay for 4CR?

Section 6

VARIABLE SPEED DRIVES

Chapter 34

VARIABLE VOLTAGE AND MAGNETIC CLUTCHES

Objectives

After studying this chapter the student will be able to:

» Discuss the types of motors that can be controlled with variable voltage.

» Discuss requirements for motors that are controlled with variable voltage.

» Discuss the operation of a magnetic clutch.

Chapter 33 discussed the operation of consequent pole motors that change speed by changing the number of stator poles per phase. Although this is one method of controlling the speed of a motor, it is not the only method. Many small single phase motors change speed by varying the amount of voltage applied to the motor. This method does not change the speed of the rotating magnetic field of the motor, but it does cause the field to become weaker. As a result, the rotor slip becomes greater, causing a decrease of motor speed.

Variable voltage control is used with small fractional horsepower motors that operate light loads, such as fans and blowers. Motors that are intended to operate with variable voltage are designed with high impedance stators. The high impedance of the stator prevents the current flow from becoming excessive as the rotor slows down. The disadvantage of motors that contain high impedance stator windings is that they are very limited in the amount of torque they can produce. When load is added to a motor of this type, its speed decreases rapidly.

Motors that used a centrifugal switch to disconnect the start windings cannot be used with variable voltage control. This limits the type of induction motors to capacitor start capacitor run and shaded pole motors. Capacitor start capacitor run motors are employed in applications where it is desirable to reverse the direction of rotation of the motor.

Another type of alternating current motor that can use variable voltage for speed control is the universal or AC series motor. These motors are commonly used in devices such as power drills, skill saws, vacuum cleaners, household mixers, and many other appliances. They can generally be recognized by the fact that they contain a commutator and brushes similar to a direct current motor. Universal motors are so named because they can operate on AC or DC voltage. These motors are used with solid-state speed control devices to operate electric drill, routers, reciprocating saws, and other variable speed tools.

Voltage Control Methods

There are different methods of obtaining a variable AC voltage. One method is with the use of an autotransformer with a sliding tap (Figure 34–1). The sliding tap causes a change in the turns ratio of the **transformer** (Figure 34–2). An autotransformer is probably the most efficient and reliable method of supplying variable AC voltage, but it is expensive and requires a large amount of space for mounting.

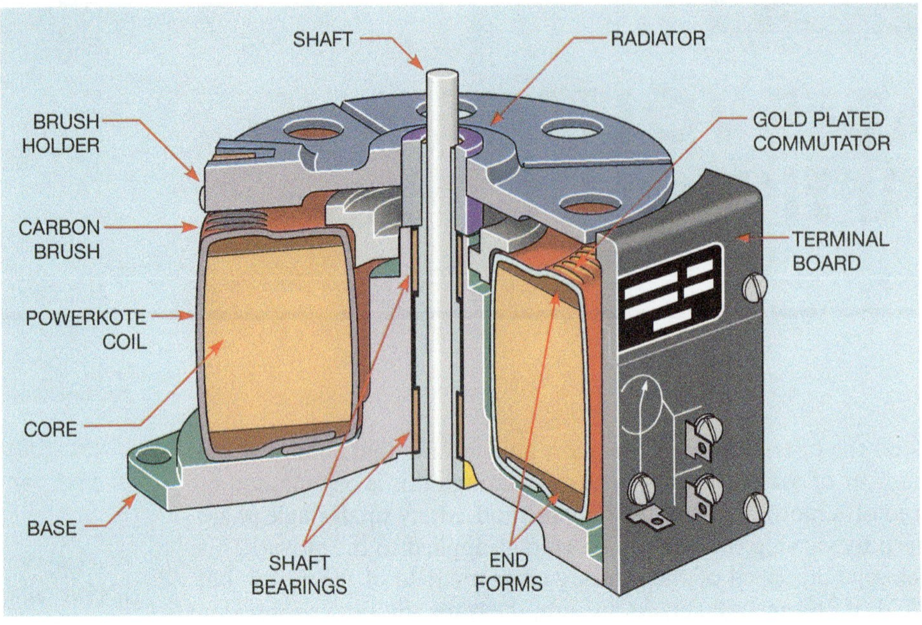

Figure 34–1 Cutaway view of a variable autotransformer.

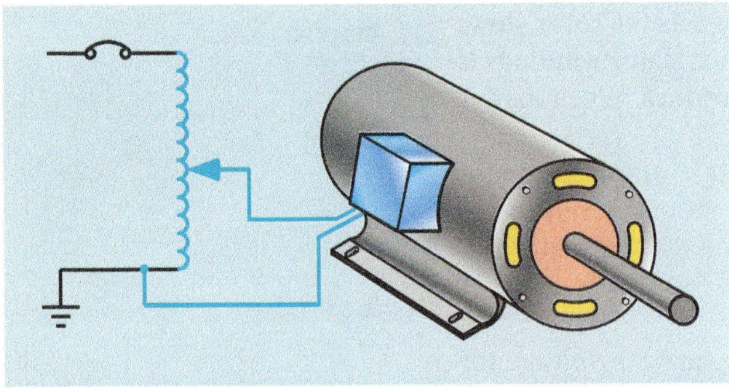

Figure 34–2 An autotransformer supplies variable voltage to a motor.

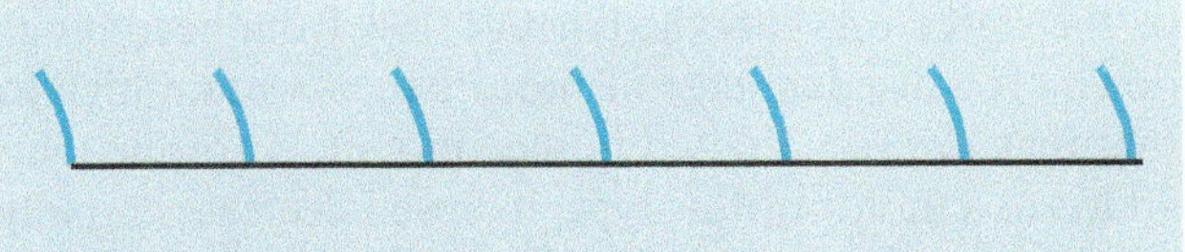

Figure 34–3 Conducting part of a waveform produces pulsating direct current.

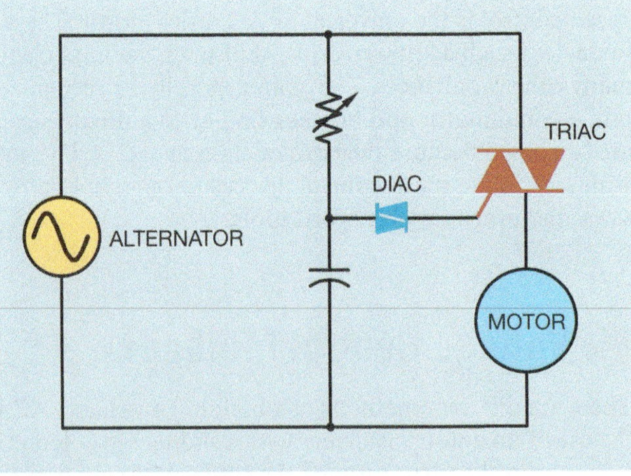

Figure 34–4 Basic triac control circuit.

Another method involves the use of a solid-state device called a **triac.** A triac is a solid-state device similar to a silicon controlled rectifier (SCR), except that it will conduct both the positive and negative portions of a waveform. Triacs are commonly used in dimmers employed to control incandescent lighting. Triac light dimmers have a characteristic of conducting one-half to the waveform before the other half begins conducting. Because only one-half of the waveform is conducting, the output voltage is DC, not AC (Figure 34–3). Resistive loads such as incandescent lamps are not harmed when direct current is applied to them, but a great deal of harm can occur when DC voltage is applied to an inductive device such as a motor. Only triac controls that are designed for use with inductive loads should be used to control a motor. A basic triac control circuit is shown in Figure 34–4. A triac variable speed control for small AC motors is shown in Figure 34–5.

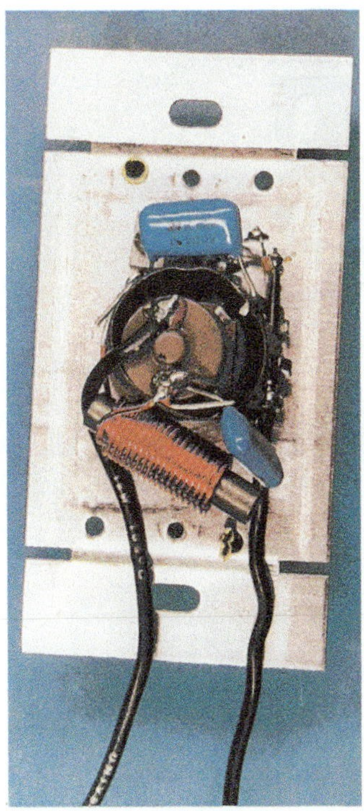

Figure 34–5 Variable speed control using a triac to control the voltage applied to a motor.

Magnetic Clutches

Magnetic clutches are used in applications where it is desirable to permit a motor to reach full speed before load is applied. Clutches can provide a smooth start for loads that can be damaged by sudden starting, or for high inertia loads such as centrifuges or flywheels. Electromagnetic clutches are divided into three components: the field, which contains the coil winding; the rotor, which contains the friction material; and the armature, which is the second friction surface. When direct current is applied to the field coil, magnetism is created; this causes the friction face of the rotor to become a magnet. The magnetic field of the rotor attracts the armature disc, which causes the armature and rotor to clamp together, connecting the motor to the load. The force of the clutch can be controlled by adjusting the voltage supplied to the field winding. The amount of slip determines how rapidly the clutch can accelerate the driven load and the amount of torque applied to the load. When power is removed, a spring forces the armature and rotor apart. Some magnetic clutches employ the use of slip rings and brushes to supply power to the field coil, but most contain a stationary electromagnet as shown in Figure 34–6. A magnetic clutch is shown in Figure 34–7.

The clutch illustrated in Figure 34–6 is a single-face clutch, which means that it contains only one clutch disc. Clutches intended to connect large horsepower motors to heavy loads often contain multiple clutch faces. Double-face clutches have both the armature and field discs mounted on the same hub. A double-face friction lining is sandwiched between them. When the field winding is energized, the field disc and armature disc are drawn together with the double-face friction lining between them. Double-face clutches can be obtained in sizes up to 78 inches in diameter.

Some clutches are intended to provide tension control and are operated with a large amount of slippage between the driving and driven members. These clutches produce an excessive amount of heat because of the friction between clutch discs. Many of these clutches are water cooled to help remove the heat.

Eddy Current Clutches

Eddy current clutches are so named because they induce eddy currents into a metal cylinder or drum. One part of the clutch contains slip rings and a winding (Figure 34–8A). The armature or rotor is so constructed that when the winding is excited with direct current, magnetic pole pieces are formed. The rotor is mounted inside the metal drum that forms the output shaft of the clutch (Figure 34–8B). The rotor is the input of the clutch and is connected to an AC induction motor. The motor provides the turning force for the clutch (Figure 34–9). When direct current is applied to the rotor, the spinning electromagnets induce eddy currents into the metal drum. The induced eddy currents form magnetic poles inside the drum. The magnetic fields of the rotor and drum are attracted to each other and the clutch turns in the same direction as the motor.

The main advantage of an eddy current clutch is that there is no mechanical connection between the rotor and drum. Because there is no mechanical connection, there is no friction to produce excessive heat and there is no wear as is the case with mechanical clutches. The speed of the clutch can be controlled by varying the amount of direct current applied to the armature or rotor. Because the output speed is determined by the amount of slip between the rotor and drum, when load is added, the slip becomes greater causing a decrease in speed. This decrease can be compensated for by increasing the amount of direct current applied to the rotor. Many eddy current clutch circuits contain a speed sensing device that automatically increases or decreases the DC excitation current when load is added or removed.

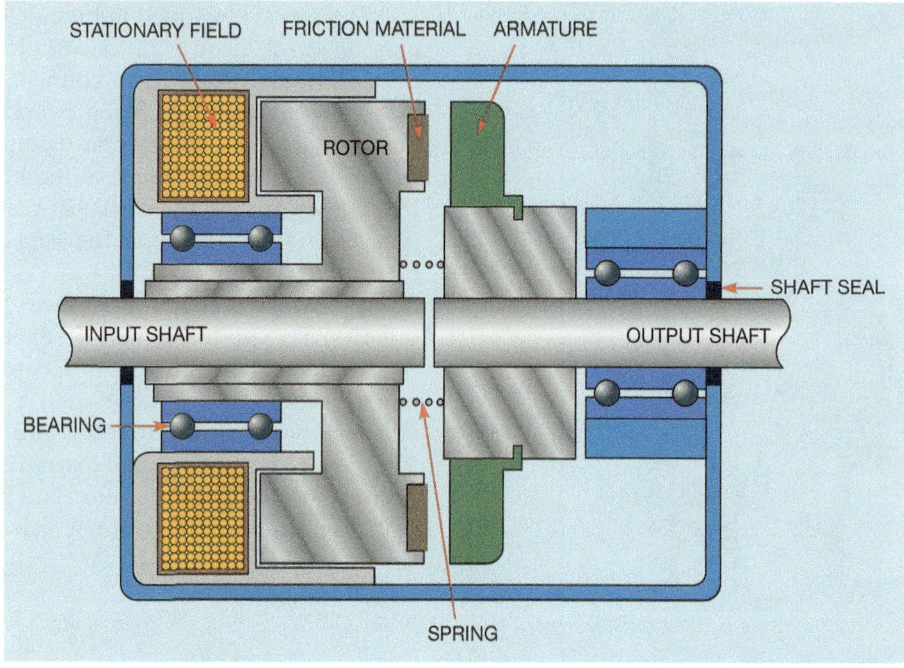

Figure 34–6 Magnetic clutch with stationary filed winding.

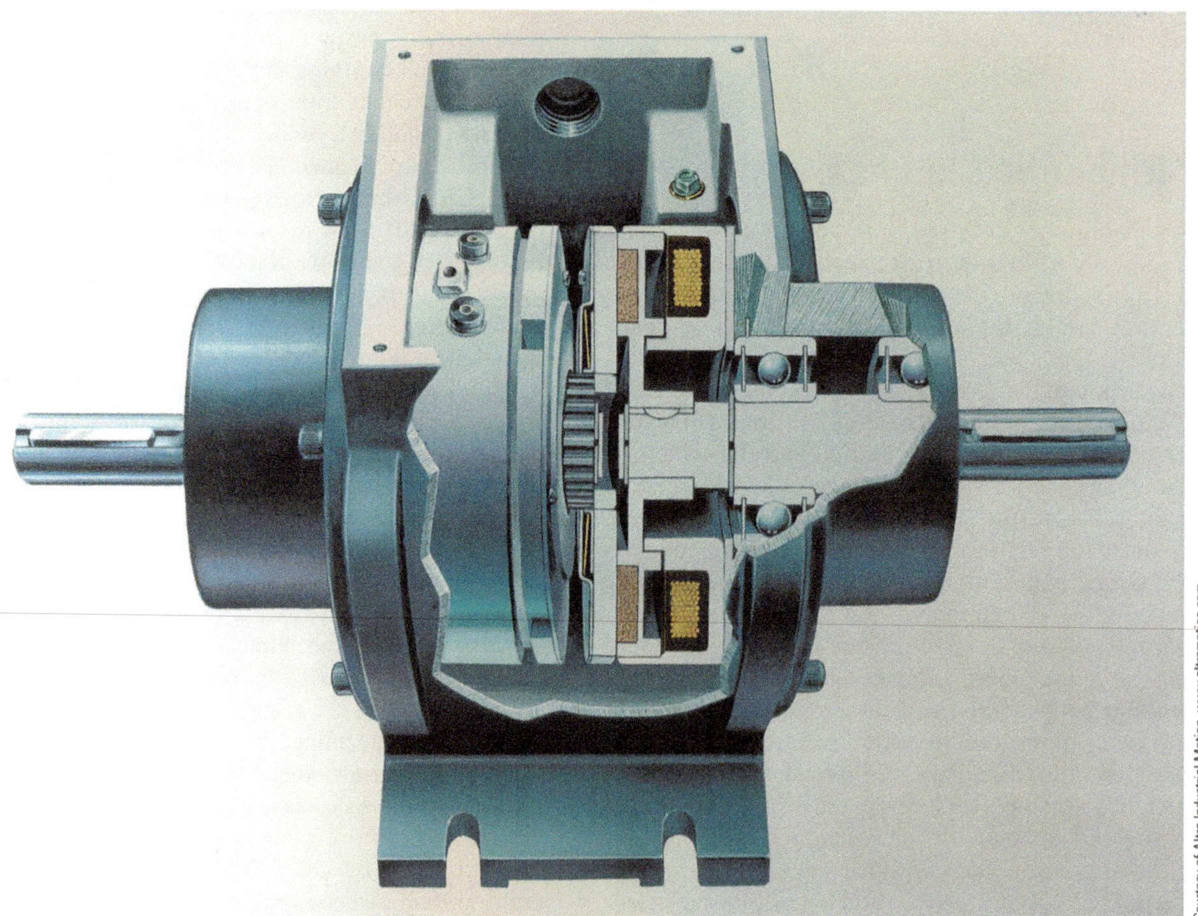

Courtesy of Altra Industrial Motion, www.altramotion.com

Figure 34–7 Cutaway view of a magnetic clutch.

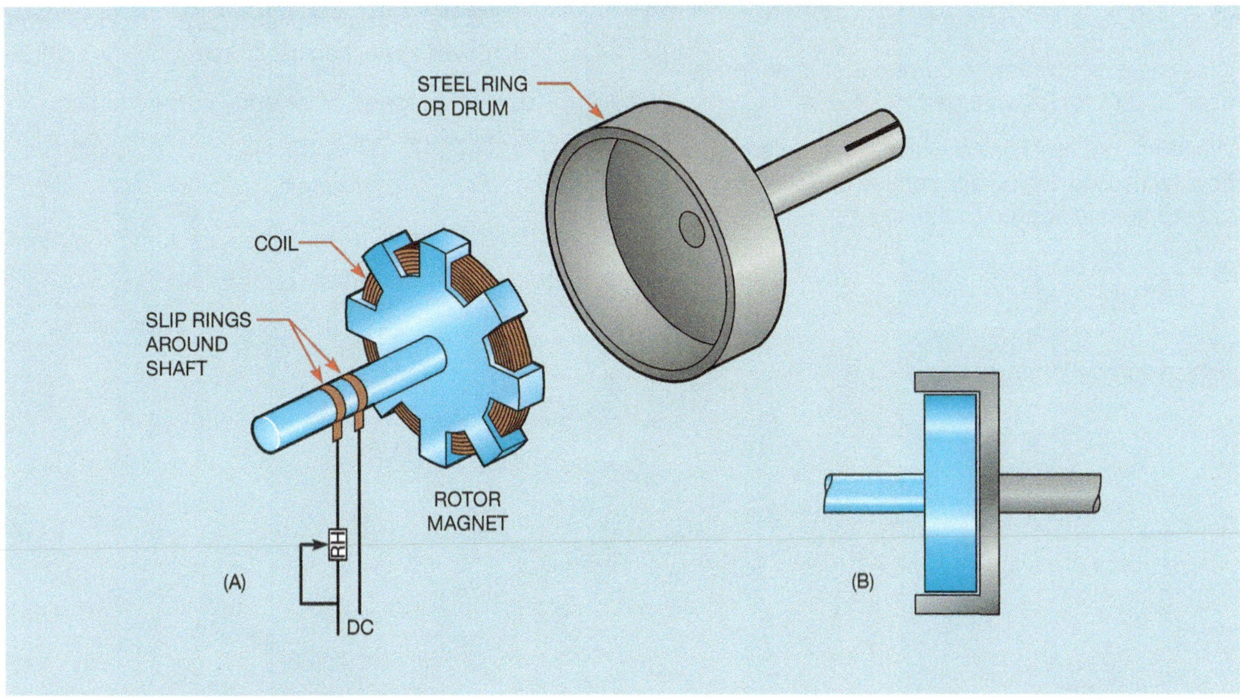

Figure 34–8 Diagram A shows magnetic armature or rotor and drum. Diagram B shows rotor mounted inside the drum. The rotor is the input shaft of the clutch and the drum is the output shaft.

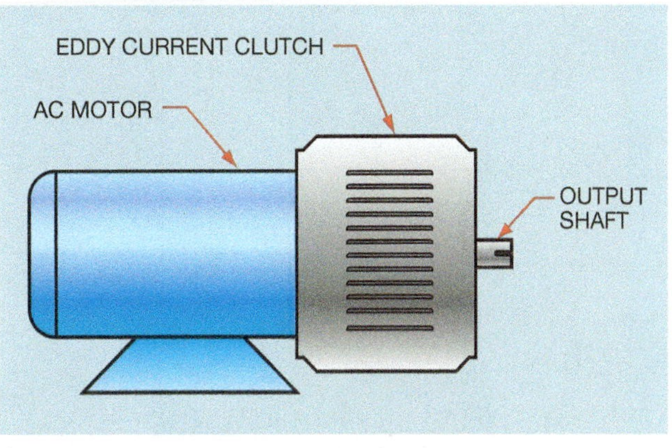

Figure 34–9 An AC motor is coupled to an eddy current clutch.

Review Questions

1. Does varying the voltage to an AC induction motor cause a change in synchronous speed?
2. Why do induction motors that are intended to be controlled by variable voltage contain high impedance stator windings?
3. What is the disadvantage of a motor that contains a high impedance stator winding?
4. What type of AC induction motor is used with variable voltage control when it is desirable for the motor to reverse direction?
5. What type of motor, which can be controlled with variable voltage, is used to operate power drills, vacuum cleaners, or routers?

6. Why are universal motors so named?
7. What type of solid-state component is generally used to control AC voltage?
8. When using a mechanical clutch, what determines how fast a load can be accelerated and the amount of initial torque applied to the load?
9. What is the primary advantage of an eddy current clutch over a mechanical clutch?
10. How is the speed of an eddy current clutch controlled?

Chapter 35

SOLID-STATE DC MOTOR CONTROLS

Objectives

After studying this chapter the student will be able to:

- Describe armature control.
- Discuss DC voltage control with a three phase bridge rectifier.
- Describe methods of current limit control.
- Discuss feedback for constant speed control.

Direct current motors are used throughout much of industry because of their ability to produce high torque at low speed, and because of their variable speed characteristics. A DC motor is generally operated at or below *normal speed.* Normal speed for a DC motor is obtained by operating the motor with full rated voltage applied to the field and armature. The motor can be operated at below normal speed by applying rated voltage to the field and reduced voltage to the armature.

In Chapter 28, resistance was connected in series with the armature to limit current and, therefore, speed. Although this method does work and was used in industry for many years, it is seldom used today. When resistance is used for speed control, much of the power applied to the circuit is wasted in heating the resistors, and the speed control of the motor is not smooth because resistance is taken out of the circuit in steps.

Speed control of a DC motor is much smoother if two separate *power supplies,* which convert the AC voltage to DC voltage, are used to control the motor instead of resistors connected in series with the armature (Figure 35–1). Notice that one power supply is used to supply a constant voltage to the shunt field of the motor, and the other power supply is variable and supplies voltage to the armature only.

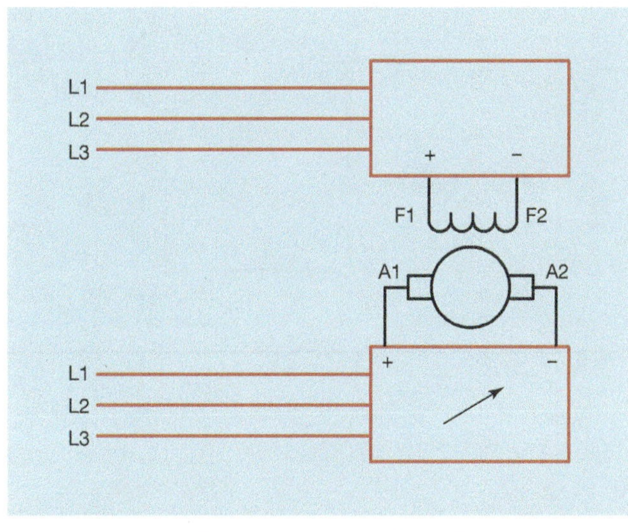

Figure 35–1 Separate power supplies used to control armature and field.

The Shunt Field Power Supply

Most solid-state DC motor controllers provide a separate DC power supply which is used to furnish excitation current to the shunt field. The shunt field of most industrial motors requires a current of only a few amps to excite the field magnets; therefore, a small power supply can be used to fulfill this need. The shunt field power supply is generally designed to remain turned on even when the main (armature) power supply is turned off. If power is connected to the shunt field even when the motor is not operating, the shunt field acts as a small resistance heater for the motor. This heat helps prevent moisture from forming in the motor due to condensation.

The Armature Power Supply

The armature power supply is used to provide variable DC voltage to the armature of the motor. This power supply is the heart of the solid-state motor controller. Depending on the size and power rating of the controller, armature power supplies can be designed to produce from a few amps to hundreds of amps. Most of the solid-state motor controllers intended to provide the DC power needed to operate large DC motors convert three phase AC voltage directly into DC voltage with a three phase bridge rectifier.

The diodes of the rectifier, however, are replaced with SCRs to provide control of the output voltage (Figure 35–2). Figure 35–3 shows SCRs used for this type of DC motor controller. A large diode is often connected across the output of the bridge. This diode is known as a *freewheeling* or **kickback diode** and is used to kill inductive spike voltages produced in the armature. If armature power is suddenly interrupted, the collapsing magnetic field induces a high voltage into the armature windings. The diode is reverse biased when the power supply is operating under normal conditions, but an induced voltage is opposite in polarity to the applied voltage. This means the kickback diode will be forward biased to any voltage induced into the armature. Because a silicon diode has a voltage drop of 0.6 to 0.7 volts in the forward direction, a high voltage spike cannot be produced in the armature.

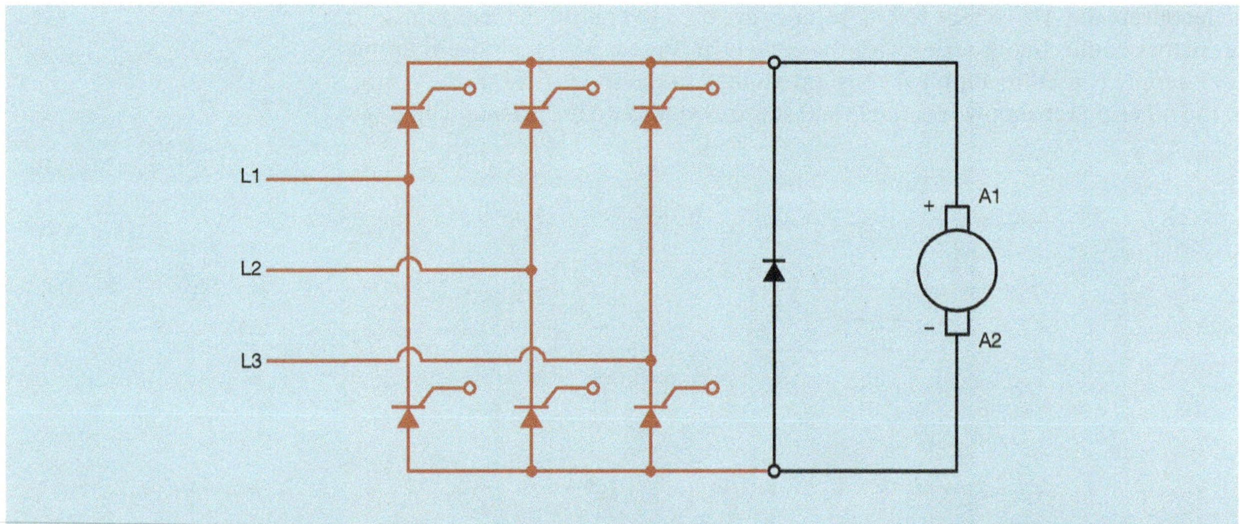

Figure 35–2 Three phase bridge rectifier.

Figure 35–3 This unit is designed to control a 150-HP DC motor. The fuses shown protect the three phase input lines. The large SCRs rectify the AC voltage into DC voltage.

Voltage Control

Output voltage control is achieved by phase shifting the SCRs. The phase shift control unit determines the output voltage of the rectifier (Figure 35–4). Because the phase shift unit is the real controller of the circuit, other sections of the circuit provide information to the phase shift control unit. Figure 35–5 shows a typical phase shift control unit.

Field Failure Control

As stated previously, if current flow through the shunt field is interrupted, a compound wound, DC motor will become a series motor and race to high speeds. Some method must be provided to disconnect the armature from the circuit in case current flow through the shunt field stops. Several methods can be used to sense current flow through the shunt field. In Chapter 28, a current relay was connected in series with the shunt field. A contact of the current relay was connected in series with the coil of a motor starter used to connect the armature to

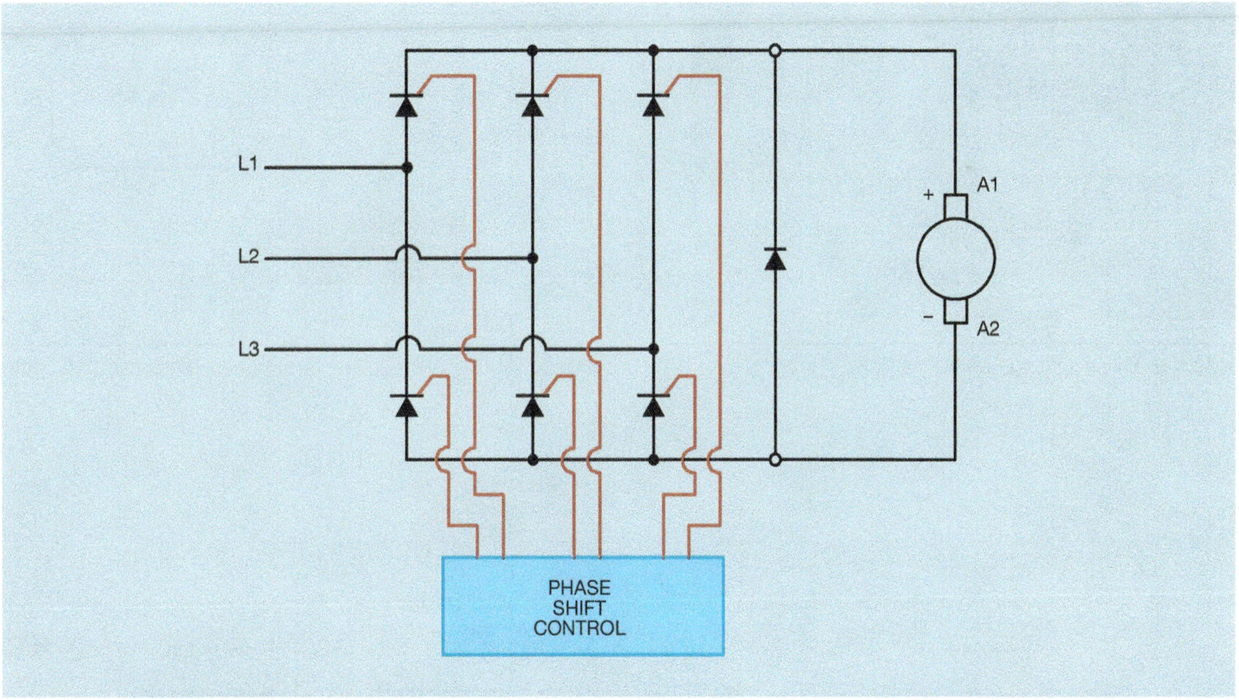

Figure 35–4 Phase shift controls output voltage.

the power line. If current flow were stopped, the contact of the current relay would open causing the circuit of the motor starter coil to open.

Another method used to sense current flow is to connect a low value of resistance in series with the shunt field (Figure 35–6). The voltage drop across the sense resistor is proportional to the current flowing through the resistor (E = I × R). Because the sense resistor is connected in series with the shunt field, the current flow through the sense resistor must be the same as the current flow through the shunt field. A circuit can be designed to measure the voltage drop across the sense resistor. If this voltage falls below a certain level, a signal is sent to the phase shift control unit and the SCRs are turned off (Figure 35–7).

Current Limit Control

The armature of a large DC motor has a very low resistance, typically less than 1 ohm. If the controller is turned on with full voltage applied to the armature, or if the motor stalls while full voltage is applied to the armature, a very large current will flow. This current can damage the armature of the motor or the electronic components of the controller. For this reason, most solid-state, DC motor controls use some method to limit the current to a safe value.

One method of sensing the current is to insert a low value of resistance in series with the armature circuit. The amount of voltage dropped across the sense resistor is proportional to the current flow through the resistor.

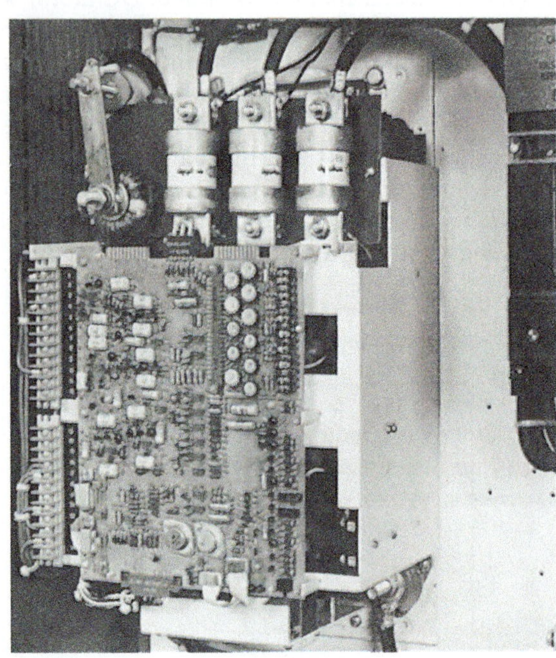

Courtesy of Eaton Corporation

Figure 35–5 Phase shift control for the SCRs.

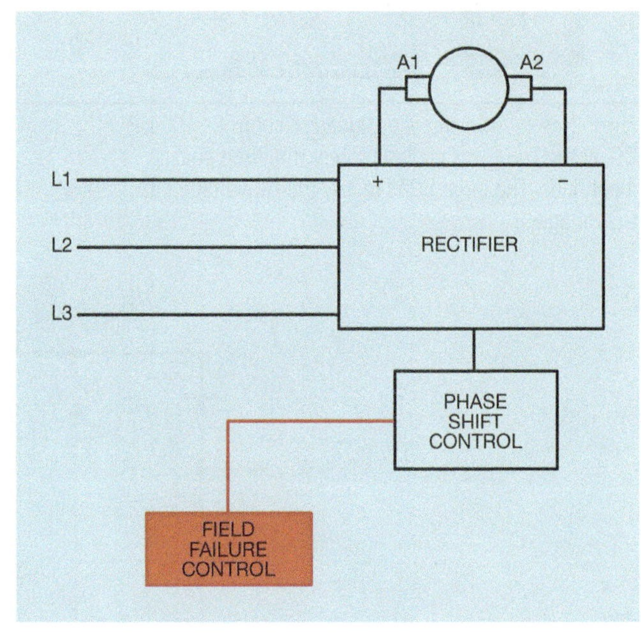

Figure 35–7 Field failure control signals the phase shift control.

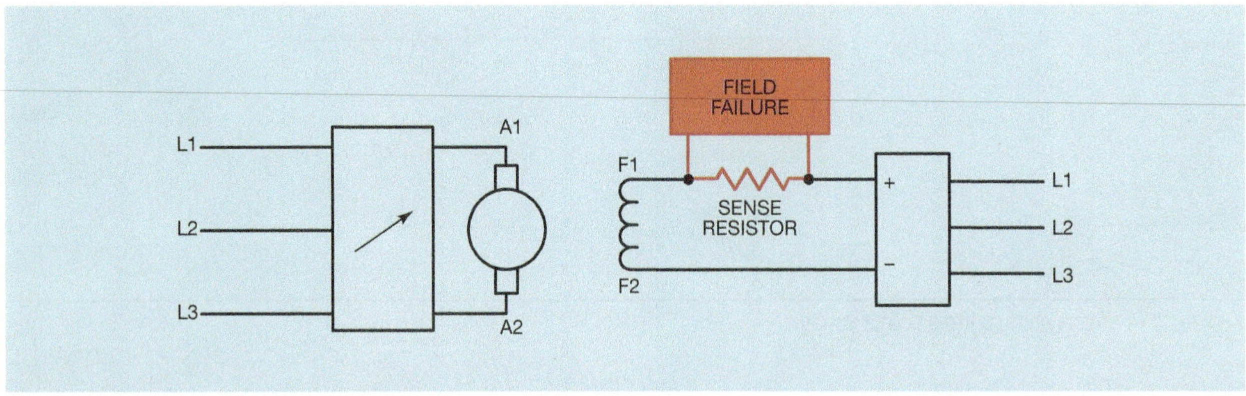

Figure 35–6 Resistor used to sense current flow through field.

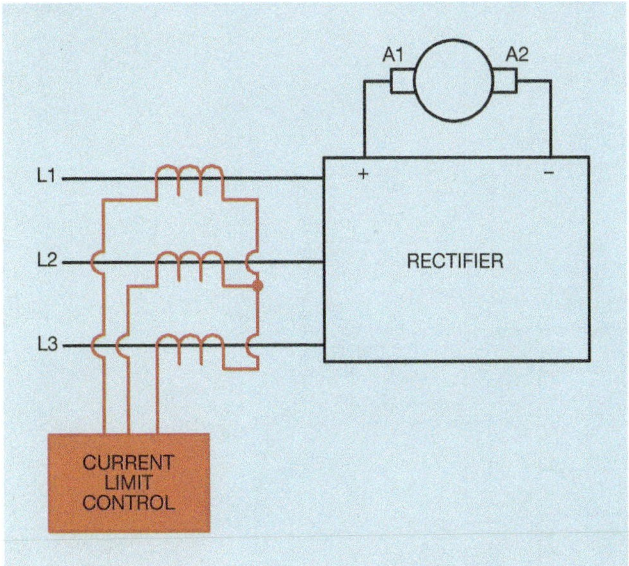

Figure 35–8 Current transformers measure AC line current.

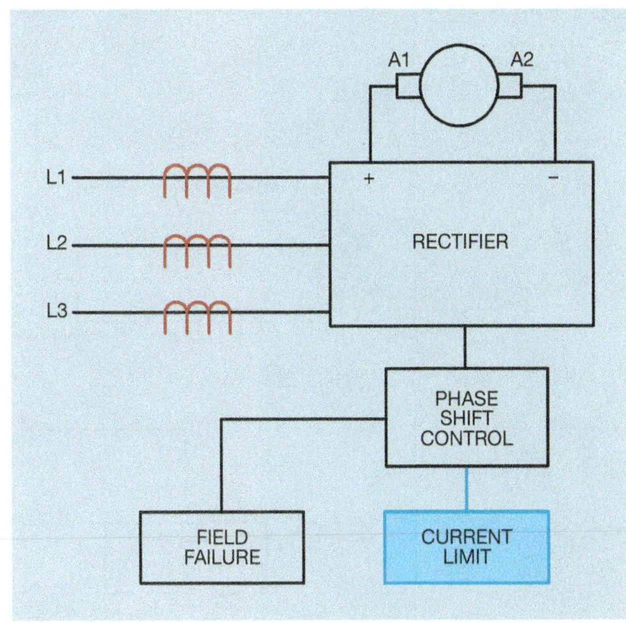

Figure 35–9 Current flow to armature is limited.

When the voltage drop reaches a certain level, a signal is sent to the phase shift control telling it not to permit any more voltage to be applied to the armature.

When DC motors of about 25 HP or larger are to be controlled, resistance connected in series with the armature can cause problems. Therefore, another method of sensing armature current can be used (Figure 35–8). In this circuit, current transformers are connected to the AC input lines. The current supplied to the rectifier will be proportional to the current supplied to the armature. When a predetermined amount of current is detected by the current transformers, a signal is sent to the phase shift control telling it not to permit the voltage applied to the armature to increase (Figure 35–9). This method of sensing the armature current has the advantage of not adding resistance to the armature circuit. Regardless of the method used, the current limit control signals the phase shift control, and the phase shift control limits the voltage applied to the armature.

Speed Control

The greatest advantage of using direct current motors is their variable speed characteristic. Although the ability to change motor speed is often desirable, it is generally necessary that the motor maintain a constant speed once it has been set. For example, assume that a DC motor can be adjusted to operate at any speed from 0 to 1800 RPM. Now assume that the operator has adjusted the motor to operate at 1200 RPM. The operator controls are connected to the phase shift control unit (Figure 35–10). If the operator desires to change speed, a signal is sent to the phase shift control unit and the phase shift control permits the voltage applied to the armature to increase or decrease.

A DC motor, like many other motors, will change speed if the load is changed. If the voltage connected to the armature remains constant, an increase in load causes the motor speed to decrease, or a decrease in load causes the motor speed to increase. Because the phase shift unit controls the voltage applied to the armature, it can be used to control motor speed. If the motor speed is to be held constant, some means must be used to detect the speed of the motor. A very common method of detecting motor speed is with the use of an *electrotachometer* (Figure 35–11). An electrotachometer is a small, permanent, magnet generator connected to the motor shaft. The output voltage of the generator is proportional to its speed. The output voltage of the generator is connected to the phase shift control unit (Figure 35–12). If load is added to the motor, the motor speed decreases. When the motor speed decreases, the output voltage of the electrotachometer drops. The phase shift unit detects the voltage drop of the tachometer and increases the armature voltage until the tachometer voltage returns to the proper value.

If the load is removed, the motor speed increases. An increase in motor speed causes an increase in the output voltage of the tachometer. The phase shift unit

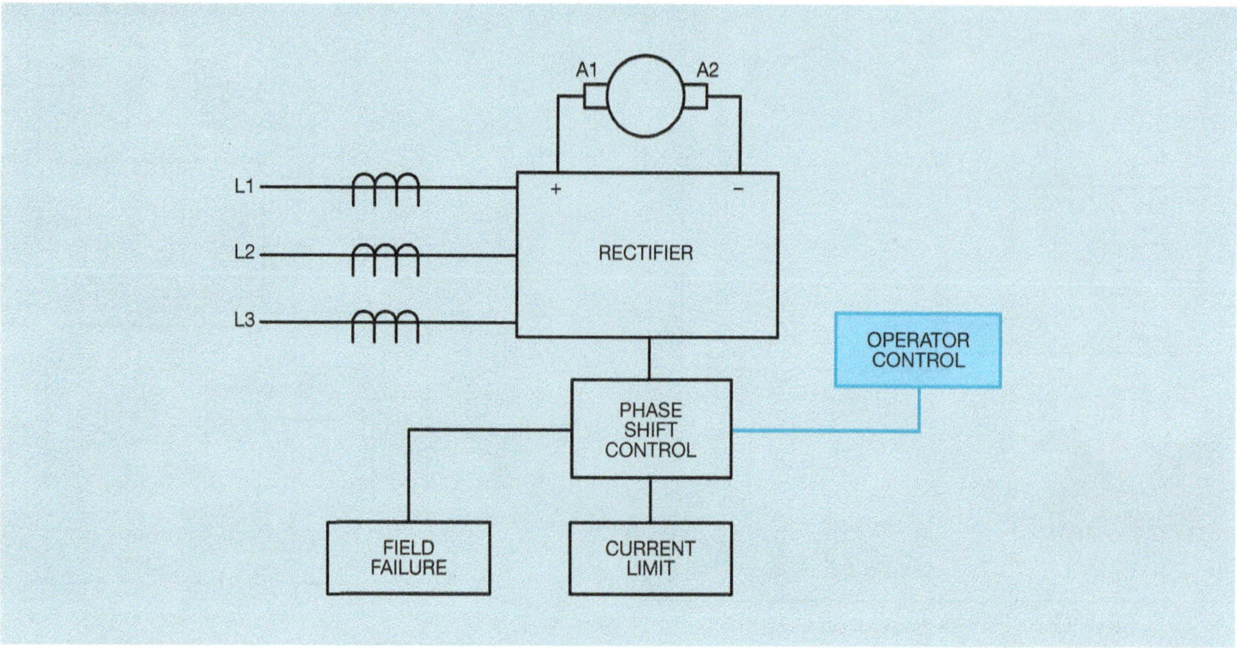

A1 A2

L1

L2

L3

\+ −

RECTIFIER

OPERATOR CONTROL

PHASE SHIFT CONTROL

FIELD FAILURE

CURRENT LIMIT

Figure 35–10 Operator control is connected to the phase shift control unit.

Courtesy of Rockwell Automation, Inc.

Figure 35–11 DC motor with tachometer attached.

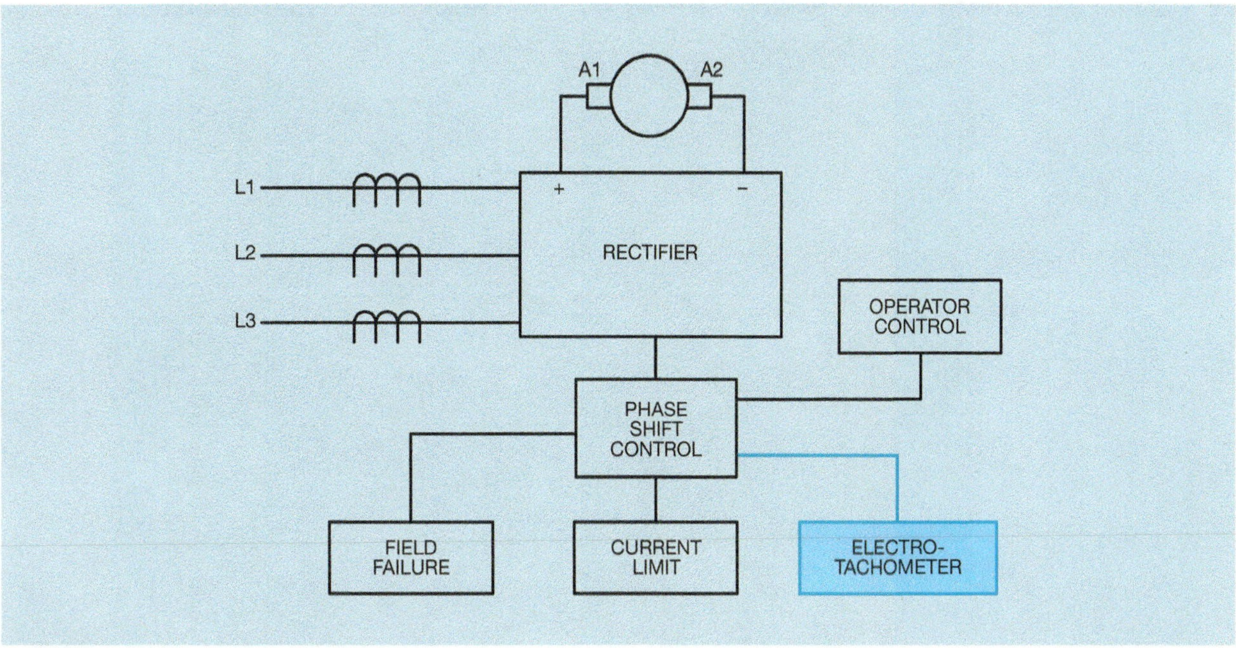

Figure 35–12 Electrotachometer measures motor speed.

Courtesy of Eaton Corporation

Figure 35–13 An SCR motor control unit mounted in a cabinet.

detects the increase of tachometer voltage and causes a decrease in the voltage applied to the armature. Electronic components respond so fast that there is almost no noticeable change in motor speed when load is added or removed. An SCR motor control unit is shown in Figure 35–13.

Review Questions

1. What electronic component is generally used to change the AC voltage into DC voltage in large DC motor controllers?
2. Why is this component used instead of a diode?
3. What is a freewheeling or kickback diode?
4. Name two methods of sensing the current flow through the shunt field.
5. Name two methods of sensing armature current.
6. What unit controls the voltage applied to the armature?
7. What device is often used to sense motor speed?
8. If the motor speed decreases, does the output voltage of the electrotachometer increase or decrease?

Chapter 36

VARIABLE FREQUENCY CONTROL

Objectives

After studying this chapter the student will be able to:

- Explain how the speed of an induction motor can be changed with a change of frequency.
- Discuss different methods of controlling frequency.
- Discuss precautions that must be made when the frequency is lowered.
- Define the terms *ramping* and *volts* per hertz.

The speed of a three phase induction motor can be controlled by either changing the number of stator poles per phase, as is the case with consequent pole motors, or by changing the frequency of the applied voltage. Both methods produce a change in the synchronous speed of the rotating magnetic field. The chart shown in Figure 36–1 indicates that when the frequency is changed, a corresponding change in synchronous speed results.

Changing frequency, however, causes a corresponding change in the inductive reactance of the windings ($X_L = 2\pi fL$). Because a decrease in frequency produces a decrease in inductive reactance, the amount of voltage applied to the motor must be reduced in proportion to the decrease of frequency to prevent overheating the windings due to excessive current. Any type of variable frequency control must also adjust the output voltage with a change in frequency. There are two basic methods of achieving variable frequency control—alternator and solid state.

Alternator Control

Alternators are often used to control the speed of several induction motors that require the same change in speed, such as motors on a conveyer line (Figure 36–2). The alternator is turned by a DC motor or an AC motor coupled to an eddy current clutch. The output frequency of the alternator is determined by the speed of the rotor. The output voltage of the alternator is determined by the amount of DC excitation current applied to the rotor. Because the output voltage must change with a change of frequency, a variable voltage DC supply is used to provide excitation current. Most controls of this type employ some method of sensing alternator speed and make automatic adjustments to the excitation current.

POLES PER PHASE	SYNCHRONOUS SPEED IN RPM					
	60 HZ	50 HZ	40 HZ	30 HZ	20 HZ	10 HZ
2	3600	3000	2400	1800	1200	600
4	1800	1500	1200	900	600	300
6	1200	1000	800	600	400	200
8	900	750	600	450	300	150

Figure 36–1 Synchronous speed is determined by the number of stator poles and frequency.

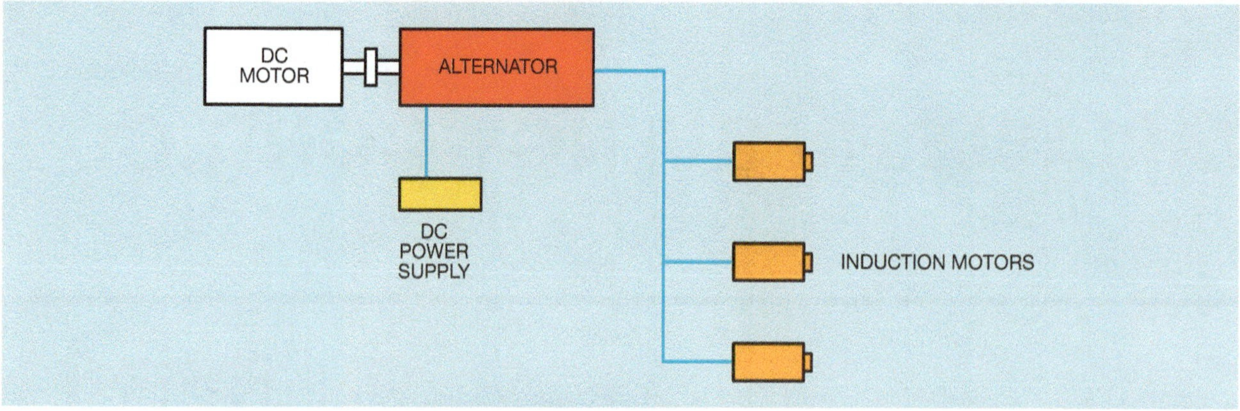

Figure 36–2 An alternator controls the speed of several induction motors.

Solid-State Control

Most variable frequency drives operate by first changing the AC voltage into DC and then changing it back to AC at the desired frequency. A variable frequency drive designed to control motors up to 500 HP depending on model is shown in Figure 36–3. Several methods are used to change the DC voltage back into AC. The manufacturer, age of the equipment, and the motor horsepower the drive must control generally determine the method employed. Variable frequency drives intended to control the speed of motors up to 500 HP generally use transistors. In the circuit shown in Figure 36–4, a three phase bridge rectifier changes the alternating current into direct current. The bridge rectifier uses six SCRs (Silicon Controlled Rectifiers). The SCRs permit the output voltage of the rectifier to be controlled. As the frequency decreases, the SCRs fire later in the cycle and lower the output voltage to the transistors. A choke coil and capacitor bank are used to filter the output voltage before transistors Q1 through Q6 change the DC voltage back into AC. An electronic control unit is connected to the bases of transistors Q1 through Q6. The control unit converts the DC voltage back into three phase alternating current by turning transistors on or off at the proper time and in the proper sequence. Assume, for example, that transistors Q1 and Q4 are switched on at the same time. This permits stator winding T1 to be connected to a positive voltage and T2 to be connected to a negative voltage. Current can flow through Q4 to T2, through the motor stator winding and through T1 to Q1.

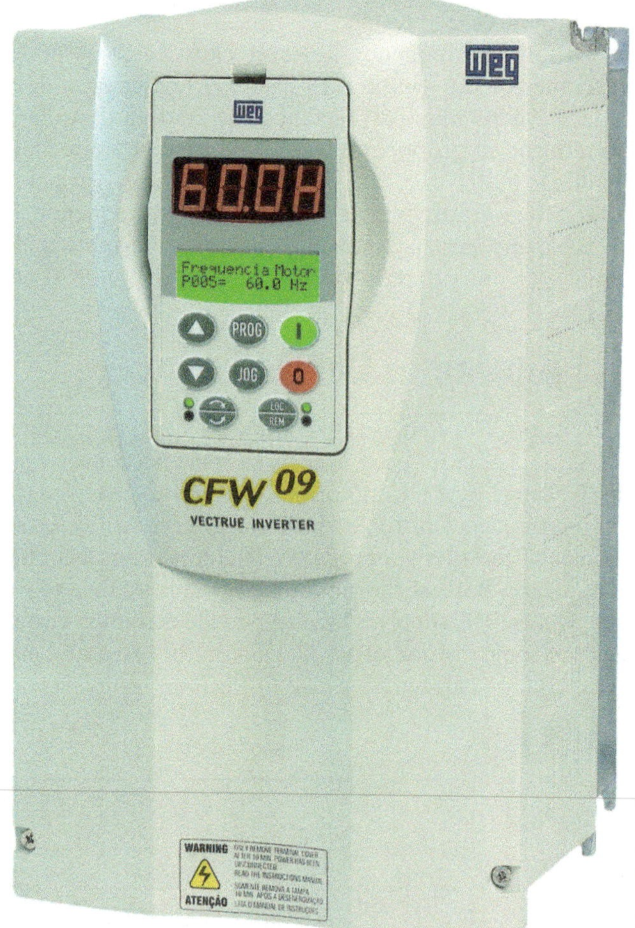

Photo courtesy WEG Electric Corp.

Figure 36–3 A CFW09 variable frequency drive can control motors up to 500 HP depending on the model of the drive.

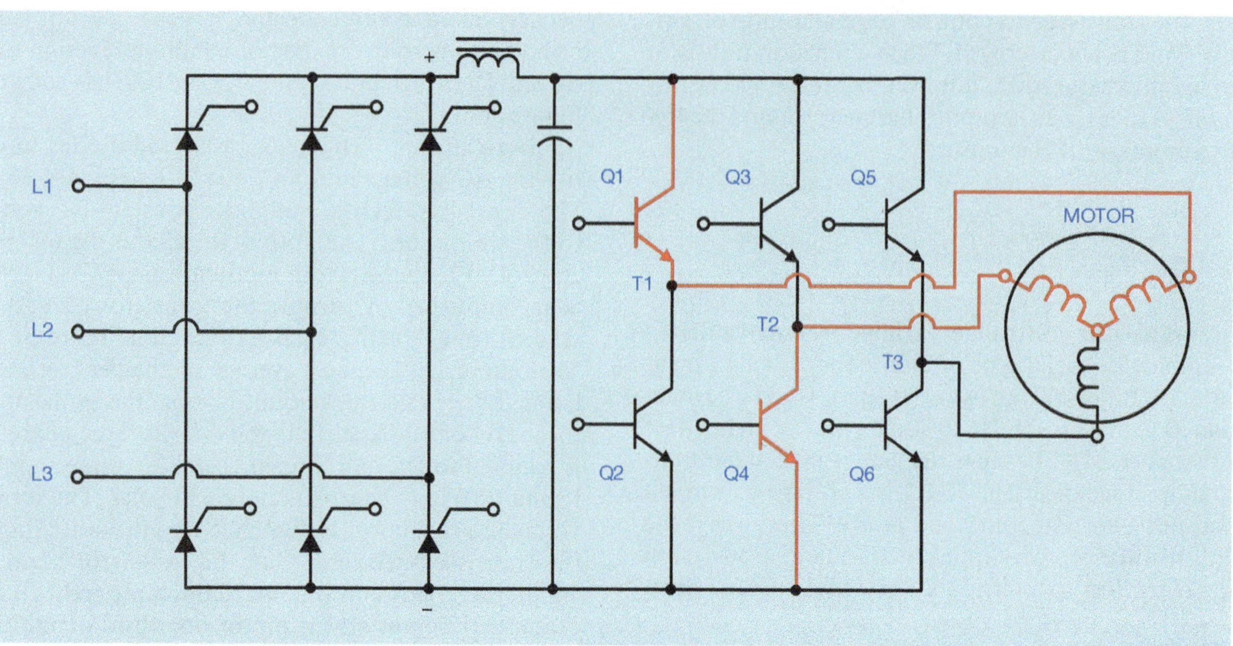

Figure 36–4 Solid-state variable frequency control using junction transistors.

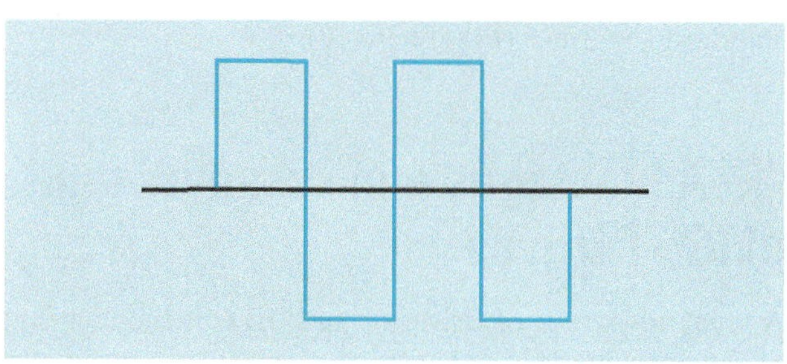

Figure 36–5 Square wave.

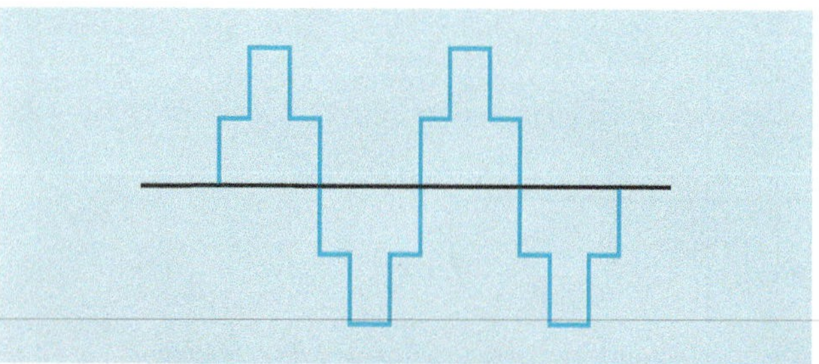

Figure 36–6 Stepped wave.

Now assume that transistors Q1 and Q4 are switched off and transistors Q3 and Q6 are switched on. Current now flows through Q6 to stator winding T3, through the motor to T2, and through Q3 to the positive of the power supply.

Because the transistors are turned completely on or off, the waveform produced is a square wave (Figure 36–5) instead of a sine wave. Induction motors operate on a square wave without a great deal of problem. Some manufacturers design units that produce a stepped waveform as shown in Figure 36–6. The stepped waveform is used because it more closely approximates a sine wave.

Some Related Problems

The circuit illustrated in Figure 36–4 employs the use of SCRs in the power supply and junction transistors in the output stage. SCR power supplies control the output voltage by chopping the incoming waveform. This can cause harmonics on the line that cause overheating of transformers and motors, and can cause fuses to blow and circuit breakers to trip. When bipolar junction transistors are employed as switches, they are generally driven into saturation by supplying them with an excessive amount of base-emitter current. Saturating the transistor causes the collector-emitter voltage to drop to between 0.04 and 0.03 volts. This small voltage drop

allows the transistor to control large amounts of current without being destroyed. When a junction transistor is driven into saturation, however, it cannot recover or turn off as quickly as normal. This greatly limits the frequency response of the transistor.

IGBTs

Many transistor controlled variable frequency drives now employ a special type of transistor called an Insulated Gate Bipolar Transistor (IGBT). IGBTs have an insulated gate very similar to some types of field effect transistors (FETs). Because the gate is insulated, it has very high impedance. The IGBT is a voltage controlled device, not a current controlled device. This gives it the ability to turn off very quickly. IGBTs can be driven into saturation to provide a very low voltage drop between **emitter** and **collector**, but they do not suffer from the slow recovery time of common junction transistors. The schematic symbol for an IGBT is shown in Figure 36–7.

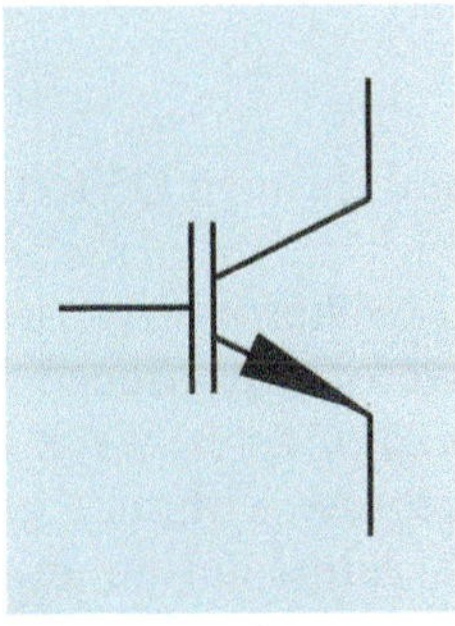

Figure 36–7 Schematic symbol for an insulated gate bipolar transistor.

Drives using IGBTs generally used diodes to rectify the AC voltage into DC, not SCRs (Figure 36–8). The three phase rectifier supplies a constant DC voltage to the transistors. The output voltage to the motor is controlled by pulse width modulation (PWM). PWM is accomplished by turning the transistor on and off several times during each half cycle (Figure 36–9). The output voltage is an average of the peak or maximum voltage and the amount of time the transistor is turned on or off. Assume that 480 volts three phase AC is rectified to DC and filtered. The DC voltage applied to the IGBTs is approximately 630 volts. The output voltage to the motor is controlled by the switching of the transistors. Assume that the transistor is on for 10 microseconds and off for 20 microseconds. In this example the transistor is on for one-third of the time and off for two-thirds of the time. The voltage applied to the motor would be 210 volts (630/3). The speed at which IGBTs can operate permits pulse width modulation to produce a stepped wave that is very similar to a standard sine wave (Figure 36–10).

Advantages and Disadvantages of IGBT Drives

A great advantage of drives using IGBTs is the fact that SCRs are generally not used in the power supply and this greatly reduces problems with line harmonics. The

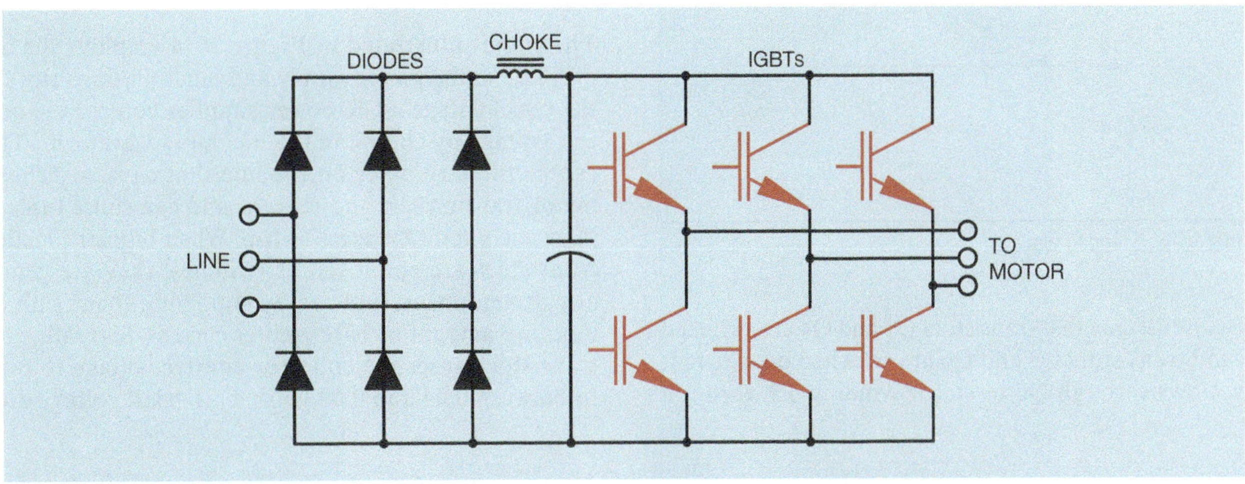

Figure 36–8 Variable frequency drives using IGBTs generally use diodes in the rectifier instead of SCRs.

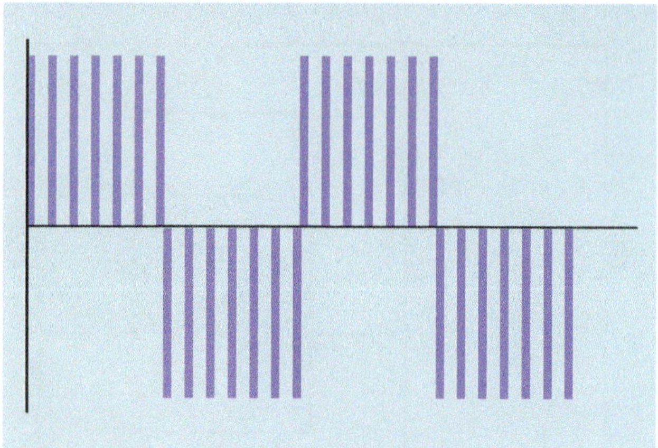

Figure 36–9 Pulse width modulation is accomplished by turning the voltage on and off several times during each half cycle.

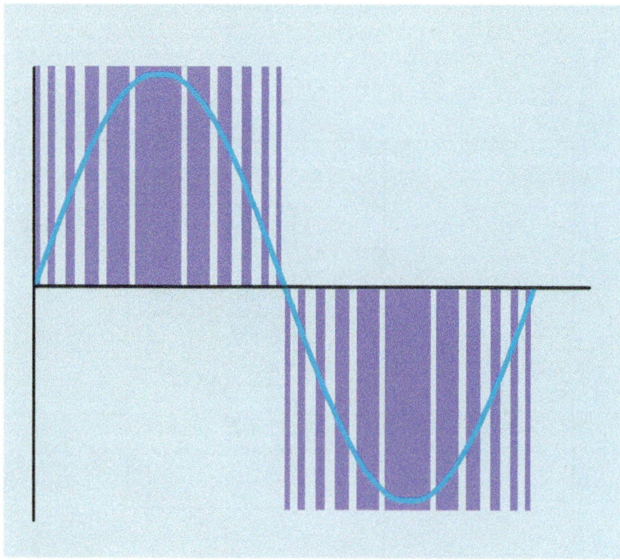

Figure 36–10 The speed of the IGBTs can produce a stepped wave that is similar to a sine wave.

greatest disadvantage is that the fast switching rate of the transistors can cause voltage spikes in the range of 1600 volts to be applied to the motor. These voltage spikes can destroy some motors. Line length from the drive to the motor is of great concern with drives using IGBTs. Short line lengths are preferred.

Inverter Rated Motors

Due to the problem of excessive voltage spikes caused by IGBT drives, some manufacturers produce a motor that is **inverter rated.** These motors are specifically designed to be operated by variable frequency drives. They differ from standard motors in several ways:

- Many inverter rated motors contain a separate blower to provide continuous cooling for the motor regardless of the speed. Many motors use a fan connected to the motor shaft to help draw air though the motor. When the motor speed is reduced, the fan cannot maintain sufficient air flow to cool the motor.
- Inverter rated motors generally have insulating paper between the windings and the stator core (Figure 36–11). The high voltage spikes produce high currents that produce a strong magnetic field. This increased magnetic field causes the motor windings to move because like magnetic fields repel each other. This movement can eventually cause the insulation to wear off the wire and produce a grounded motor winding.
- Inverter rated motors generally have phase paper added to the terminal leads. Phase paper is insulating paper added to the terminal leads that exit the motor. The high voltage spikes affect the beginning lead of a coil much more than the wire inside the coil. The coil is an inductor that naturally opposes a change of current. Most of the insulation stress caused by high voltage spikes occurs at the beginning of a winding.
- The magnet wire used in the construction of the motor windings has a higher rated insulation than other motors.
- The case size is larger than most three phase motors. The case size is larger because of the added insulating paper between the windings and the stator core. Also, a larger case size helps cool the motor by providing a larger surface area for the dissipation of heat.

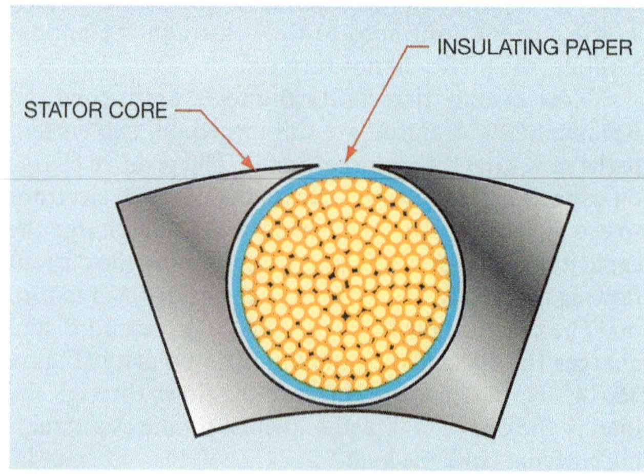

Figure 36–11 Insulating paper is between the windings and the stator frame.

Variable Frequency Drives Using SCRs and GTOs

Variable frequency drives intended to control motors over 500 HP generally used SCRs or GTO (Gate Turn Off) devices. GTOs are similar to SCRs except that conduction through the GTO can be stopped by applying a negative voltage, negative with respect to the cathode, to the gate. SCRs and GTOs are thyristors and have the ability to handle a greater amount of current than transistors. **Thyristors** are solid-state devices that exhibit only two states of operation, completely turned on or completely turned off. An example of a single phase circuit used to convert DC voltage to AC voltage with SCRs is shown in Figure 36–12. In this circuit, the SCRs are connected to a phase shift unit that controls the sequence and rate at which the SCRs are gated on. The circuit is constructed so that SCRs A and A' are gated on at the same time and SCRs B and B' are gated on at the same time. Inductors L1 and L2 are used for filtering and wave shaping. Diodes D1 through D4 are clamping diodes and are used to prevent the output voltage from becoming excessive. Capacitor C1 is used to turn one set of SCRs off when the other set is gated on. This capacitor must be a true AC capacitor because it will be charged to the alternate polarity each half cycle. In a converter intended to handle large amounts of power, capacitor C1 will be a bank of capacitors. To understand the operation of the circuit, assume that SCRs A and A' are gated on at the same time. Current flows through the circuit as shown in Figure 36–13. Notice the direction of current flow through the load. And that capacitor C1 has been charged to the polarity shown. When an SCR is gated on, it can only be turned off by permitting the current flow through the anode-cathode section to drop below a certain level called the holding current level. As long as the current continues to flow through the anode-cathode, the SCR will not turn off.

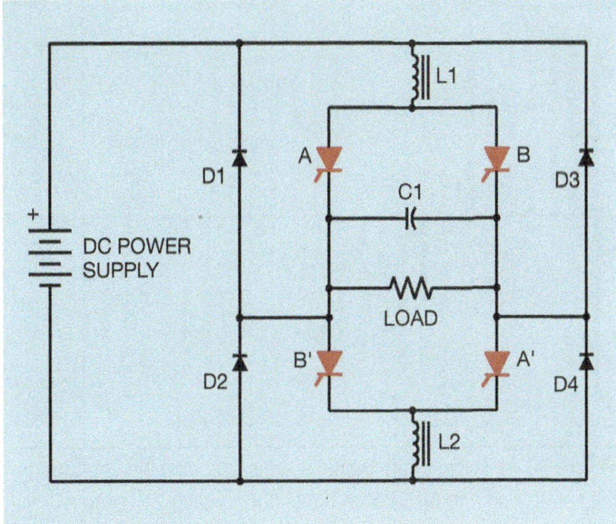

Figure 36–12 Changing DC into AC using SCRs.

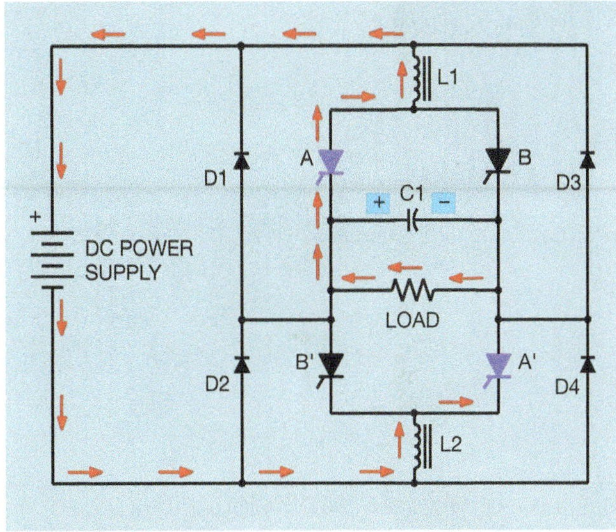

Figure 36–13 Current flows through SCRs A and A'.

Now assume that SCRs B and B' are turned on. Because SCRs A and A' are still turned on, two current paths now exist through the circuit. The positive charge on capacitor C1, however, causes the negative electrons to see an easier path. The current rushes to charge the capacitor to the opposite polarity, stopping the current flowing through SCRs A and A', permitting them to turn off. The current now flows through SCRs B and B' and charges the capacitor to the opposite polarity (Figure 36–14). Notice that the current now flows through the load in the opposite direction, which produces alternating current across the load.

To produce the next half cycle of AC current, SCRs A and A' are gated on again. The positively charged side of the capacitor now causes the current to stop flowing through SCRs B and B' permitting them to turn off. The current again flows through the load in the direction indicated in Figure 36–13. The frequency of the circuit is determined by the rate at which the SCRs are gated on. A variable frequency drive designed to control motors rated from 450 HP to 2500 HP is shown in Figure 36–15.

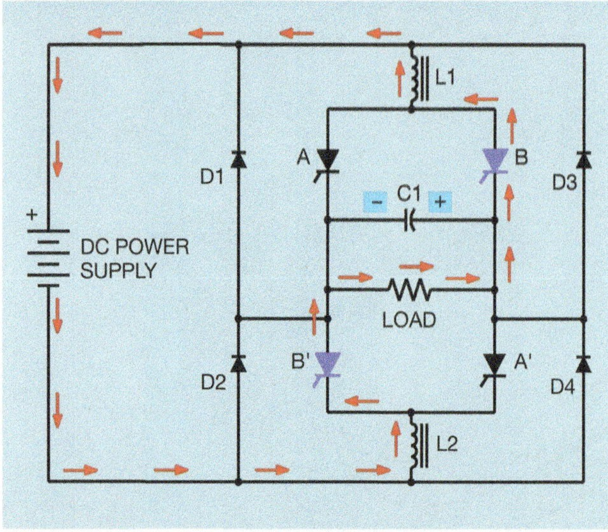

Figure 36–14 Current flows through SCRs B and B′.

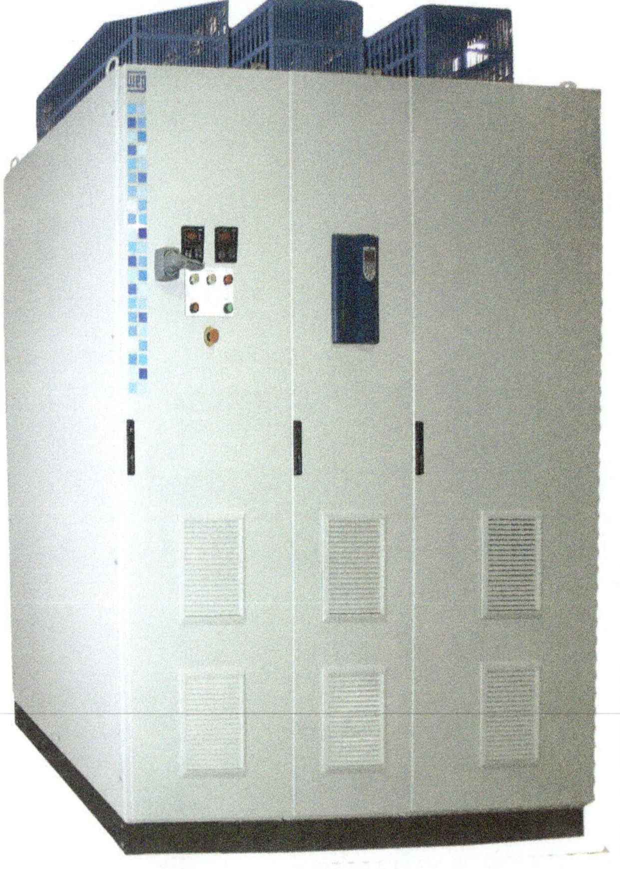

Figure 36–15 Variable frequency drive rated to control motors from 450 HP to 2500 HP with line voltages that range from 380 volts to 690 volts.

Features of Variable Frequency Control

Although the primary purpose of a variable frequency drive is to provide speed control for an AC motor, most drives provide functions that other types of controls do not. Many variable frequency drives can provide the low speed torque characteristic that is so desirable in DC motors. It is this feature that permits AC squirrel cage motors to replace DC motors for many applications.

Many variable frequency drives also provide current limit and automatic speed regulation for the motor. Current limit is generally accomplished by connecting current transformers to the input of the drive and sensing the increase in current as load is added. Speed regulation is accomplished by sensing the speed of the motor and feeding this information back to the drive (Figure 36–16).

Another feature of variable frequency drives is acceleration and deceleration control, sometimes called ramping. Ramping is used to accelerate or decelerate a motor over some period of time. Ramping permits the motor to bring the load up to speed slowly as opposed to simply connecting the motor directly to the line. Even if the speed control is set in the maximum position when the START button is pressed, ramping forces the motor to accelerate the load from 0 to its maximum RPM over several seconds. This feature can be a real advantage for some types of loads, especially gear drive loads. In some units, the amount of acceleration and deceleration time can be adjusted by setting **potentiometers** on the main control board (Figure 36–17). Other units are completely digitally controlled and the acceleration and deceleration times are programmed into the computer memory.

Some other adjustments that can usually be set by changing potentiometers or programming the unit are as follows:

Current Limit: This control sets the maximum amount of current the drive is permitted to deliver to the motor.

Volts per Hertz: This sets the ratio by which the voltage increases as frequency increases or decreases as frequency decreases.

Maximum Hertz: This control sets the maximum speed of the motor. Most motors are intended to operate between 0 and 60 Hz, but some drives permit the output frequency to be set above 60 Hz, which would permit the motor to operate at higher than normal speed. The maximum hertz control can also be set to limit the output frequency to a value

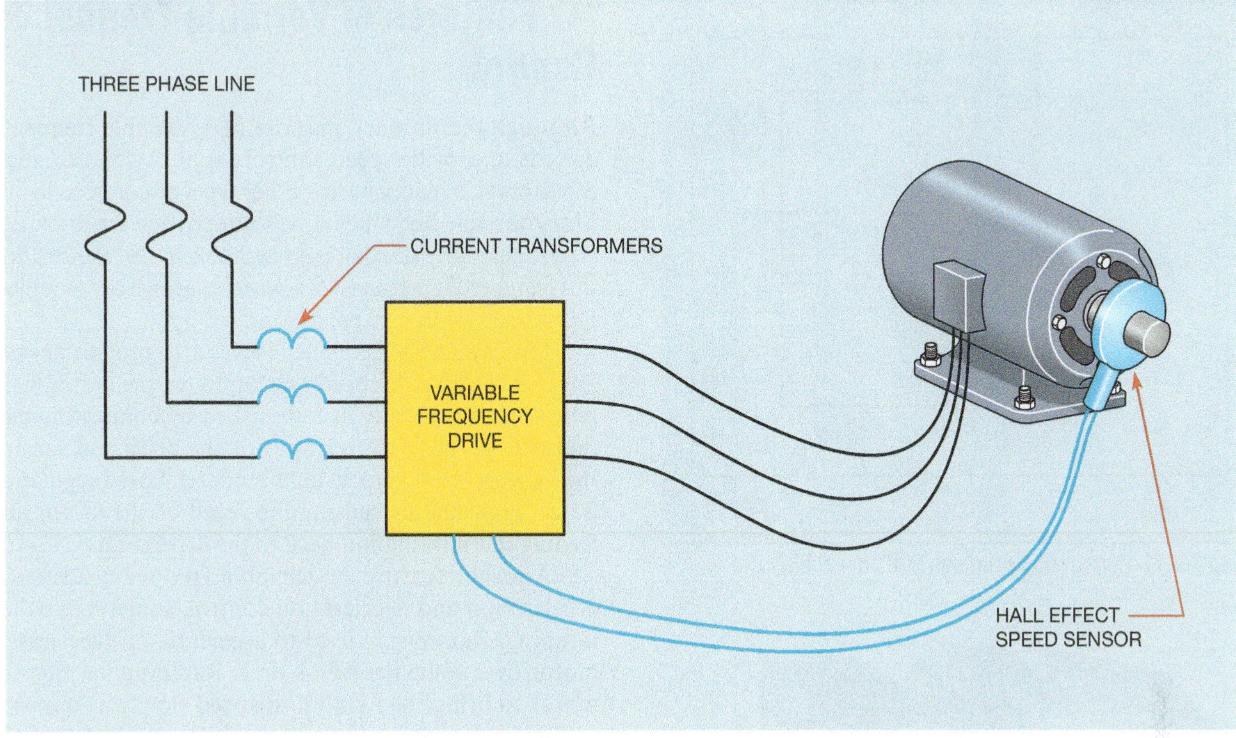

Figure 36–16 Most variable frequency drives provide current limit and speed regulation.

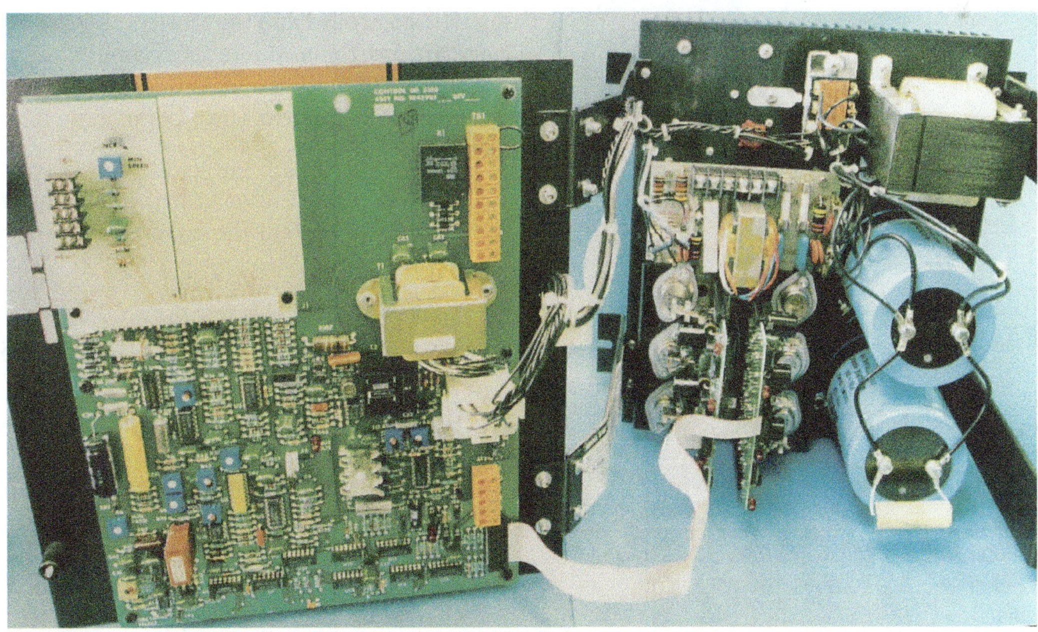

Figure 36–17 Some variable frequency drives permit setting to be made by making adjustments on a main control board.

less than 60 Hz, which would limit the motor speed to a value less than normal.

Minimum Hertz: This sets the minimum speed the motor is permitted to run.

Some variable frequency drives permit adjustment of current limit, maximum and minimum speed, or ramping time by adjustment of trim resistors located on the main control board. Other drives employ a microprocessor as the controller. The values of current limit, speed, or ramping time for these drives are programmed into the unit and are much easier to make and are generally more accurate than adjusting trim resistors. A programmable variable frequency drive is shown in Figure 36–18.

Other features of some variable frequency drive can include dynamic braking and in some cases regenerative braking. Regenerative braking permits the energy to be fed back into the power system. Most of the hybrid electric automobiles employ regenerative braking to permit power to be delivered back to the batteries when the brakes are applied.

Variable frequency drives are generally intended to control three phase motors. Some drives are designed to operate on single phase power and others are designed

Courtesy Toshiba International Corporation

Figure 36–18 A G9® low voltage variable frequency drive for severe duty applications. This frame size is 230 V/3 to 5 HP, 460 V/5 HP. Programmable variable frequency drives permit settings such as current limit, volts per hertz, maximum and minimum Hz, acceleration, and deceleration to be programmed into the unit.

to operate on three phase power. A drive designed to operate on three phase power can generally be operated on single phase provided certain limitations are observed.

- The output horsepower of the drive must be derated.
- The three phase motor connected to the drive should be designed to operate on the voltage supplying power to the drive. If the drive is connected to 240 volts single phase, the motor should be designed to operate on 240 volts three phase.

Review Questions

1. What is the synchronous speed of a six-pole motor operated with an applied voltage of 20 Hz?
2. Why is it necessary to reduce the voltage to a motor when the frequency is reduced?
3. If an alternator is used to provide variable frequency, how is the output voltage of the alternator controlled?
4. What solid-state device is generally used to produce variable frequency in drives designed to control motors up to 500 HP?
5. Why are SCRs used to construct a bridge rectifier in many solid-state variable frequency drives?
6. What is the main disadvantage of using SCRs in a variable frequency drive?
7. How are junction transistors driven into saturation and what is the advantage of driving a transistor into saturation?
8. What is the disadvantage of driving a junction transistor into saturation?
9. What is the advantage of an IGBT over a junction transistor?
10. In variable frequency drives that employ IGBTs, how is the output voltage to the motor controlled?
11. What type of motor is generally used with IGBT drives?
12. What is the primary difference between a GTO and an SCR?
13. What is a thyristor?
14. After an SCR has been turned on, what must be done to permit it to turn off again?
15. What is meant by ramping and why is it used?

Section 7

MOTOR INSTALLATION

Chapter 37

MOTOR INSTALLATION

Objectives

After studying this chapter the student will be able to:

» Determine the full load current rating of different types of motors using the *National Electrical Code (NEC)®*.

» Determine the conductor size for installing motors.

» Determine the overload size for different types of motors.

» Determine the size of the short-circuit protective device for individual motors and multimotor connections.

» Select the proper size starter for a particular motor.

Motor Nameplate Data

When it is necessary to install a motor in industry, one of the major sources of information concerning the motor is the nameplate. The National Electrical Manufacturers Association (NEMA) specifies that every motor nameplate must list specific items such as:

- Manufacturer's name
- Rated voltage
- Full load current
- Frequency
- Number of phases
- Full load speed
- Temperature rise or insulation system class
- Duty or time rating
- Horsepower
- Locked rotor indicating code letter
- Service factor
- Frame size
- Efficiency
- NEMA design code

It should be noted that not all motor manufacturers comply with NEMA specifications and their nameplates may or may not contain all the information specified by NEMA. In some instances, information not specified by NEMA may also be listed on a nameplate. A typical motor nameplate is shown in Figure 37–1. Each item on the nameplate will be discussed.

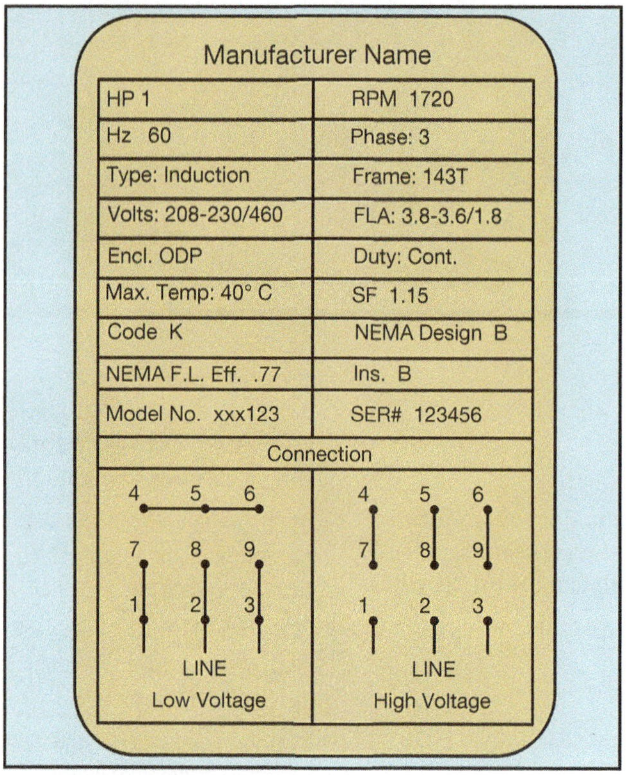

Manufacturer Name

HP 1	RPM 1720
Hz 60	Phase: 3
Type: Induction	Frame: 143T
Volts: 208-230/460	FLA: 3.8-3.6/1.8
Encl. ODP	Duty: Cont.
Max. Temp: 40° C	SF 1.15
Code K	NEMA Design B
NEMA F.L. Eff. .77	Ins. B
Model No. xxx123	SER# 123456
Connection	
4 5 6 / 7 8 9 / 1 2 3 / LINE / Low Voltage	4 5 6 / 7 8 9 / 1 2 3 / LINE / High Voltage

Figure 37–1 Motor nameplate.

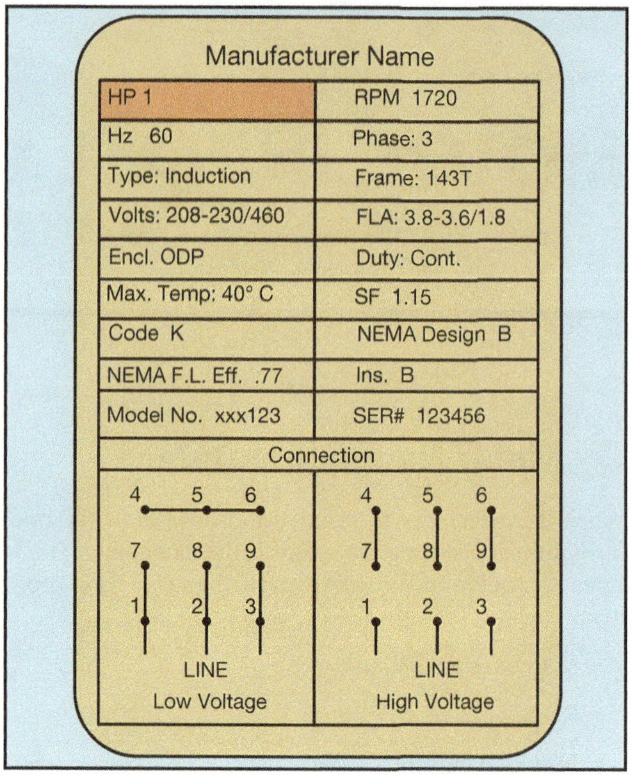

Manufacturer Name

HP 1	RPM 1720
Hz 60	Phase: 3
Type: Induction	Frame: 143T
Volts: 208-230/460	FLA: 3.8-3.6/1.8
Encl. ODP	Duty: Cont.
Max. Temp: 40° C	SF 1.15
Code K	NEMA Design B
NEMA F.L. Eff. .77	Ins. B
Model No. xxx123	SER# 123456
Connection	
4 5 6 / 7 8 9 / 1 2 3 / LINE / Low Voltage	4 5 6 / 7 8 9 / 1 2 3 / LINE / High Voltage

Figure 37–2 The motor nameplate indicates that the motor is 1 horsepower.

Manufacturer's Name

The very top of the nameplate shown in Figure 37–1 is the manufacturer's name. This lists the manufacturer of the motor.

Horsepower

Motors have a rate horsepower that is determined by the amount of torque they can produce at a specific speed under full load. The horsepower listed on the nameplate in this example is 1 horsepower, Figure 37–2.

When James Watt invented the steam engine he needed to rate its power in a way that the average person could understand. Through experimentation, he determined that the average horse could lift 550 pounds one foot in one second, or 1000 pounds 33 feet in one minute. Therefore, the definition of horsepower is:

$$\text{HP} = \text{ft. lb. per minute}/33{,}000$$

or

$$\text{HP} = \text{ft. lb. per second}/550$$

Torque is the twisting or turning force produced by the motor. It is rated in either foot-pounds or inch-pounds depending on the motor. Horsepower and torque are related as shown by the formula:

$$\text{HP} = (\text{Torque} \times \text{Speed})/\text{Constant}$$

If the torque is given in foot-pounds, the constant is 5252 ($33{,}000/2\pi$). If the torque is given in inch-pounds the constant is 63,025 (5252×12). Standard NEMA horsepower ratings are shown in Figure 37–3.

STANDARD NEMA HORSEPOWER RATINGS			
1	30	300	1250
1½	40	350	1500
2	50	400	1750
3	60	450	2000
5	75	500	2250
7½	100	600	2500
10	125	700	3000
15	150	800	3500
20	200	900	4000
25	250	1000	

Figure 37–3 The chart lists the standard NEMA horsepower ratings.

RPM

The RPM indicates the speed in Revolutions Per Minute that the motor will run at rate full load. The motor will run faster at light load or no load. The nameplate shown in Figure 37–4 indicates that the motor will have a speed of 1720 RPM when the motor is under full load.

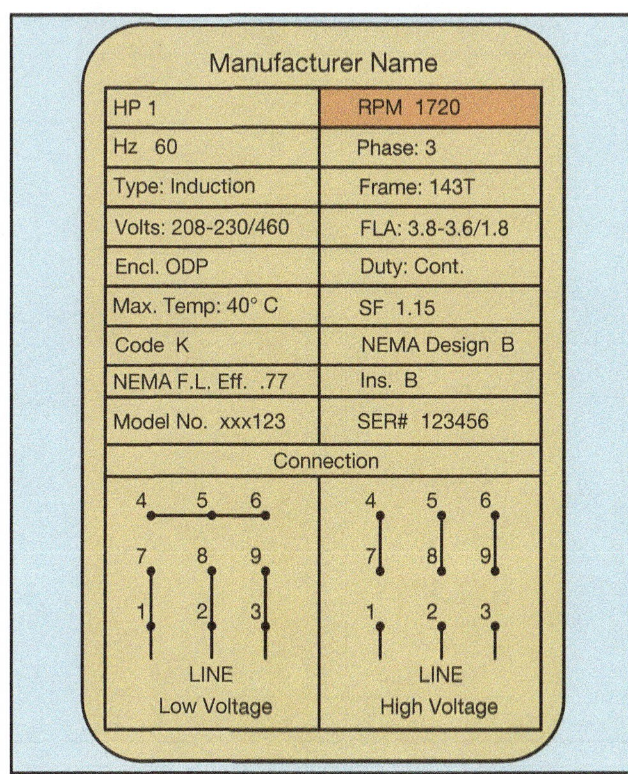

Figure 37–4 The motor has a full load speed of 1720 RPM.

Frequency

The frequency is measured in Hertz. The standard frequency used throughout the United States and Canada is 60 Hz. Some manufacturers, however, design motors for use in both the United States and Europe. The standard frequency in Europe is 50 Hz. Motors with a frequency rating of 50/60 Hz are not uncommon. The nameplate in Figure 37–5 indicates that the motor is designed to operate on a frequency of 60 Hz.

Phases

Phase indicates the number of phases on which the motor is designed to operate. Most industrial motors are three phase, which means that the power connected to them is three separate lines with the voltages 120° out-of-phase with each other. Other alternating current motors are generally single phase. Although there are some single-phase motors used in industrial applications, most are found in residential applications. Although two phase motors do exist, they are extremely rare in the United

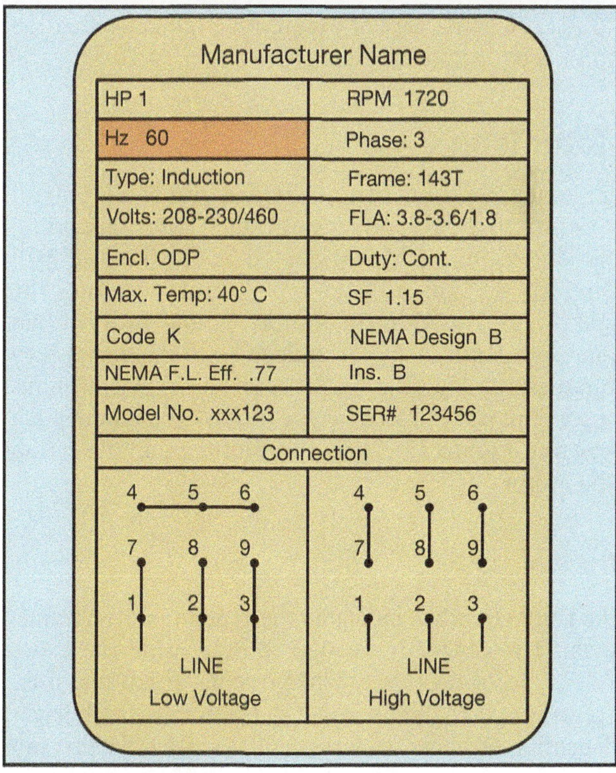

Figure 37–5 The nameplate indicates an operating frequency of 60 Hz.

Manufacturer Name

HP 1	RPM 1720
Hz 60	Phase: 3
Type: Induction	Frame: 143T
Volts: 208-230/460	FLA: 3.8-3.6/1.8
Encl. ODP	Duty: Cont.
Max. Temp: 40° C	SF 1.15
Code K	NEMA Design B
NEMA F.L. Eff. .77	Ins. B
Model No. xxx123	SER# 123456
Connection	
4 5 6 / 7 8 9 / 1 2 3 / LINE / Low Voltage	4 5 6 / 7 8 9 / 1 2 3 / LINE / High Voltage

Figure 37–6 The motor operates on three phase power.

States. The nameplate in Figure 37–6 indicates that the motor is to operate on three phase power.

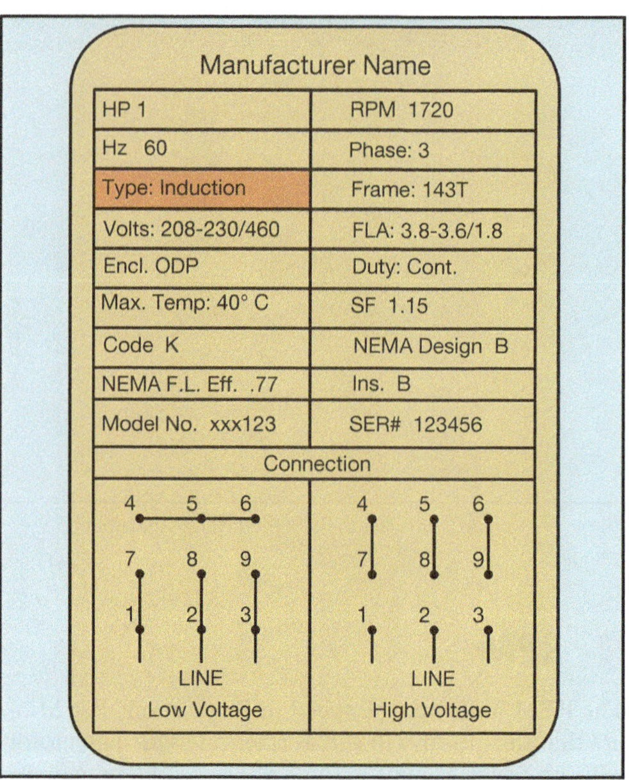

Figure 37–7 The motor is an induction type.

Motor Type

Three phase motors can be divided into three general types: squirrel cage induction, wound rotor induction, and synchronous. Motors listed as "induction" will generally be squirrel cage type, which describes the type of rotor used in the motor. Wound rotor induction motors are easily recognized by the fact that they contain three slip rings on the rotor shaft. Synchronous motors are not induction-type motors. The nameplate shown in Figure 37–7 lists this motor as an induction type motor.

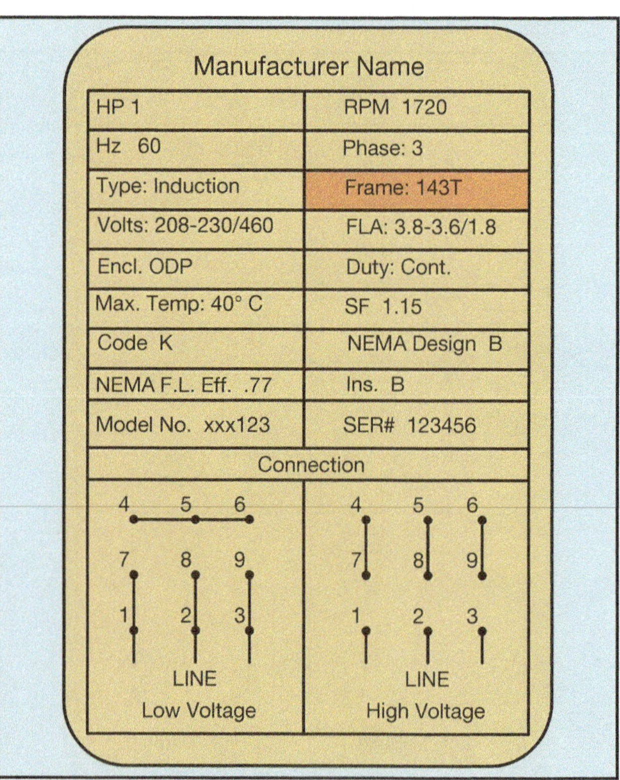

Figure 37–8 The motor has a 143T frame.

Frame

The frame number indicates the type of motor frame. A chart is generally needed to determine the exact dimensions of the frame. When dealing with frame sizes, a general rule of thumb is that the centerline shaft height (dimension D) above the bottom of the base is the first two digits of the frame number divided by 4. The nameplate in Figure 37–8 indicates that the motor has a 143T frame.

A frame 143T, for example, would have a shaft height of 3.5 inches (14/4) above the base of the motor. The chart in Figure 37–9 lists frame sizes for common U and T type motors.

In addition to frame numbers, letters also appear at the end of the numbers. These letters represent different frame styles.

C – The letter C designates a flange-mounted motor. C style is the most popular flange-mounted motor and have a specific bolt pattern on the shaft end of the motor that permits mounting to the driven device. C flange motors always have threaded bolt holes in the face of the motor.

D – Another type of flange mount motor is the D. The flange diameter of these motors is larger than the body of the motor and the bolt holes are not threaded. They are designed for bolts to pass through the holes.

H – These frames are used on some 56 frame motors. The H indicates that the base can be mounted in either 56, 143T, or 145T mounting positions.

J – J indicates that the motor is especially designed to mount to jet pumps. It has a threaded stainless steel shaft and a standard 56C face.

JM – The letters JM indicate that the pump shaft is designed for a mechanical seal. This motor also has a C face.

JP – Similar to the JM motor. The seal is designed for a packing type seal.

S – S indicates that the motor has a short shaft. They are generally intended to be directly coupled to a load. They are not intended to be used with belt drives.

T – T frame motors were standardized after 1964. Any motor with a T at the end of the frame size was made after 1964.

U – NEMA first standardized motor frames in 1952. Motors with a U in the frame number were manufactured between 1952 and 1964.

Y – Y indicates that the motor has a special mounting configuration. It does not indicate what the configuration is; only that it is non-standard.

Z – Z indicates that the motor has a special shaft. It could be longer, larger in diameter, threaded, or contain holes. Z indicates that the shaft is special in some undefined way.

Voltage

Volts indicate the operating voltage of the motor. The nameplate in Figure 37–10 indicates that the motor in this example is designed to operate on different voltages depending on the connection of the stator windings. These motors are generally referred to as "dual voltage motors." If the motor is connected for low voltage operation it will operate on 208 or 230 volts. If it is connected for high voltage operation it will operate on 460 volts. The 230-volt rating applies for voltage ranges of 220 to 240 volts. The 460-volt rating applies for voltage ranges of 440 to 480 volts.

Full Load Current

The ampere rating indicates the amount of current the motor should draw at full load. It will draw less current at light load or no load. Note that there are three currents listed. The first two currents (3.8 – 3.6) amperes indicate the amount of current the motor should draw when connected to 208 or 230 volts, respectively. The last current rating of 1.8 amperes indicates the amount of full load current the motor should draw when connected to 460 volts. The nameplate shown in Figure 37–11 lists the current as FLA (Full Load Amps). Some nameplates simply list the current as AMPS.

Enclosure

The nameplate in Figure 37–12 indicates that the motor has an ODP type enclosure. Motors have different types of enclosures depending on the application. Some of the common enclosures are:

ODP – Open Drip Proof – these are very common. The case has openings to permit ventilation through the motor windings.

TEFC – Totally Enclosed Fan Cooled – The motor case is sealed to prevent the entrance of moisture or dirt. A fan is used to help cool the motor.

TENV – Totally Enclosed Non-Vented – These motors are generally used in harsh environments such as chemical plants. They are designed to be hosed down.

EXP – Explosion Proof – Totally enclosed and non-vented. Designed to be used in areas that have hazardous atmospheres.

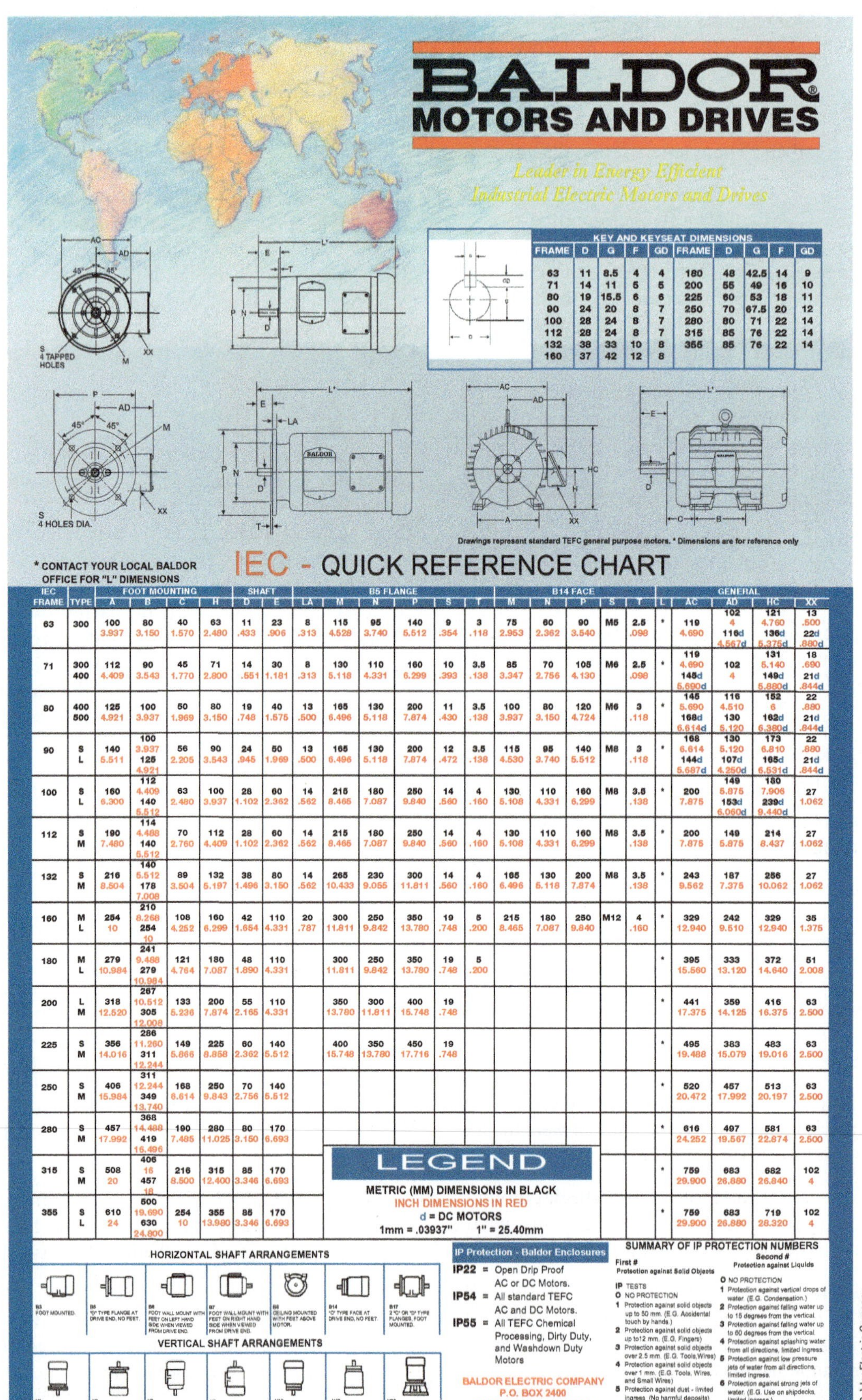

KEY AND KEYSEAT DIMENSIONS

FRAME	D	G	F	GD	FRAME	D	G	F	GD
63	11	8.5	4	4	180	48	42.5	14	9
71	14	11	5	5	200	55	49	16	10
80	19	15.5	6	6	225	60	53	18	11
90	24	20	8	7	250	70	67.5	20	12
100	28	24	8	7	280	80	71	22	14
112	28	24	8	7	315	85	76	22	14
132	38	33	10	8	355	85	76	22	14
160	37	42	12	8					

* CONTACT YOUR LOCAL BALDOR OFFICE FOR "L" DIMENSIONS

IEC - QUICK REFERENCE CHART

IEC		FOOT MOUNTING				SHAFT		B5 FLANGE						B14 FACE					GENERAL				
FRAME	TYPE	A	B	C	H	D	E	LA	M	N	P	S	T	M	N	P	S	T	L	AC	AD	HC	XX
63	300	100 3.937	80 3.150	40 1.570	63 2.480	11 .433	23 .906	8 .313	115 4.528	95 3.740	140 5.512	9 .354	3 .118	75 2.953	60 2.362	90 3.540	M5	2.5 .098	*	119 4.690	102 4 116d 4.567d	121 4.760 136d 5.375d	13 .500 22d .880d
71	300 400	112 4.409	90 3.543	45 1.770	71 2.800	14 .551	30 1.181	8 .313	130 5.118	110 4.331	160 6.299	10 .393	3.5 .138	85 3.347	70 2.756	105 4.130	M6	2.5 .098	*	119 4.690 145d 5.690d	102 4	131 5.140 149d 5.880d	18 .690 21d .844d
80	400 500	125 4.921	100 3.937	50 1.969	80 3.150	19 .748	40 1.575	13 .500	165 6.496	130 5.118	200 7.874	11 .430	3.5 .138	100 3.937	80 3.150	120 4.724	M6	3 .118	*	145 5.690 168d 6.614d	116 4.510 130 5.120	152 6 162d 6.380d	22 .880 21d .844d
90	S L	140 5.511	100 3.937 125 4.921	56 2.205	90 3.543	24 .945	50 1.969	13 .500	165 6.496	130 5.118	200 7.874	12 .472	3.5 .138	115 4.530	95 3.740	140 5.512	M8	3 .118	*	168 6.614 144d 5.687d	130 5.120 107d 4.250d	173 6.810 165d 6.531d	22 .880 21d .844d
100	S L	160 6.300	112 4.409 140 5.512	63 2.480	100 3.937	28 1.102	60 2.362	14 .562	215 8.465	180 7.087	250 9.840	14 .560	4 .160	130 5.108	110 4.331	160 6.299	M8	3.5 .138	*	200 7.875	149 5.875 153d 6.060d	180 7.906 239d 9.440d	27 1.062
112	S M	190 7.480	114 4.488 140 5.512	70 2.760	112 4.409	28 1.102	60 2.362	14 .562	215 8.465	180 7.087	250 9.840	14 .560	4 .160	130 5.108	110 4.331	160 6.299	M8	3.5 .138	*	200 7.875	149 5.875	214 8.437	27 1.062
132	S M	216 8.504	140 5.512 178 7.008	89 3.504	132 5.197	38 1.496	80 3.150	14 .562	265 10.433	230 9.055	300 11.811	14 .560	4 .160	165 6.496	130 5.118	200 7.874	M8	3.5 .138	*	243 9.562	187 7.375	256 10.062	27 1.062
160	M L	254 10	210 8.268 254 10	108 4.252	160 6.299	42 1.654	110 4.331	20 .787	300 11.811	250 9.842	350 13.780	19 .748	5 .200	215 8.465	180 7.087	250 9.840	M12	4 .160	*	329 12.940	242 9.510	329 12.940	35 1.375
180	M L	279 10.984	241 9.488 279 10.984	121 4.764	180 7.087	48 1.890	110 4.331		300 11.811	250 9.842	350 13.780	19 .748	5 .200						*	395 15.560	333 13.120	372 14.640	51 2.008
200	L M	318 12.520	267 10.512 305 12.008	133 5.236	200 7.874	55 2.165	110 4.331		350 13.780	300 11.811	400 15.748	19 .748							*	441 17.375	359 14.125	416 16.375	63 2.500
225	S M	356 14.016	286 11.260 311 12.244	149 5.866	225 8.858	60 2.362	140 5.512		400 15.748	350 13.780	450 17.716	19 .748							*	495 19.488	383 15.079	483 19.016	63 2.500
250	S M	406 15.984	311 12.244 349 13.740	168 6.614	250 9.843	70 2.756	140 5.512												*	520 20.472	457 17.992	513 20.197	63 2.500
280	S M	457 17.992	368 14.488 419 16.496	190 7.485	280 11.025	80 3.150	170 6.693												*	616 24.252	497 19.567	581 22.874	63 2.500
315	S M	508 20	406 16 457 18	216 8.500	315 12.400	85 3.346	170 6.693												*	759 29.900	683 26.880	682 26.840	102 4
355	S L	610 24	500 19.690 630 24.800	254 10	355 13.980	85 3.346	170 6.693												*	759 29.900	683 26.880	719 28.320	102 4

LEGEND

METRIC (MM) DIMENSIONS IN BLACK
INCH DIMENSIONS IN RED
d = DC MOTORS
1mm = .03937" 1" = 25.40mm

HORIZONTAL SHAFT ARRANGEMENTS

B3 FOOT MOUNTED. — B5 "D" TYPE FLANGE AT DRIVE END, NO FEET. — B6 FOOT WALL MOUNT WITH FEET ON LEFT HAND SIDE WHEN VIEWED FROM DRIVE END. — B7 FOOT WALL MOUNT WITH FEET ON RIGHT HAND SIDE WHEN VIEWED FROM DRIVE END. — B8 CEILING MOUNTED WITH FEET ABOVE MOTOR. — B14 "C" TYPE FACE AT DRIVE END, NO FEET. — B17 2 "C" OR "D" TYPE FLANGES, FOOT MOUNTED.

VERTICAL SHAFT ARRANGEMENTS

V1 "D" TYPE FLANGE AT DRIVE END, SHAFT DOWN, NO FEET. — V3 "D" TYPE FLANGE AT DRIVE END, SHAFT UP, NO FEET. — V5 VERTICAL FOOT, WALL MOUNTED, SHAFT DOWN. — V6 VERTICAL FOOT, WALL MOUNTED, SHAFT UP. — V18 "C" TYPE FACE AT DRIVE END, SHAFT DOWN, NO FEET. — V19 "C" TYPE FACE AT DRIVE END, SHAFT UP, NO FEET. — V22 SKIRT MOUNTING, SHAFT DOWN, NO FEET.

IP Protection - Baldor Enclosures

IP22 = Open Drip Proof AC or DC Motors.
IP54 = All standard TEFC AC and DC Motors.
IP55 = All TEFC Chemical Processing, Dirty Duty, and Washdown Duty Motors

BALDOR ELECTRIC COMPANY
P.O. BOX 2400
FORT SMITH, ARKANSAS
72902-2400 U.S.A.

SUMMARY OF IP PROTECTION NUMBERS

First # Protection against Solid Objects

IP TESTS
0 NO PROTECTION
1 Protection against solid objects up to 50 mm. (E.G. Accidental touch by hands.)
2 Protection against solid objects up to 12 mm. (E.G. Fingers)
3 Protection against solid objects over 2.5 mm. (E.G. Tools, Wires)
4 Protection against solid objects over 1 mm. (E.G. Tools, Wires, and Small Wires)
5 Protection against dust - limited ingress (No harmful deposits)
6 Totally protected against all dust.

Second # Protection against Liquids

0 NO PROTECTION
1 Protection against vertical drops of water. (E.G. Condensation.)
2 Protection against falling water up to 15 degrees from the vertical.
3 Protection against falling water up to 60 degrees from the vertical.
4 Protection against splashing water from all directions, limited ingress.
5 Protection against low pressure jets of water from all directions, limited ingress.
6 Protection against strong jets of water. (E.G. Use on shipdecks, limited ingress.)
7 Protection against immersion.
8 Protection against submersion.

Courtesy Baldor Electric Company

Figure 37–9 NEMA frame chart.

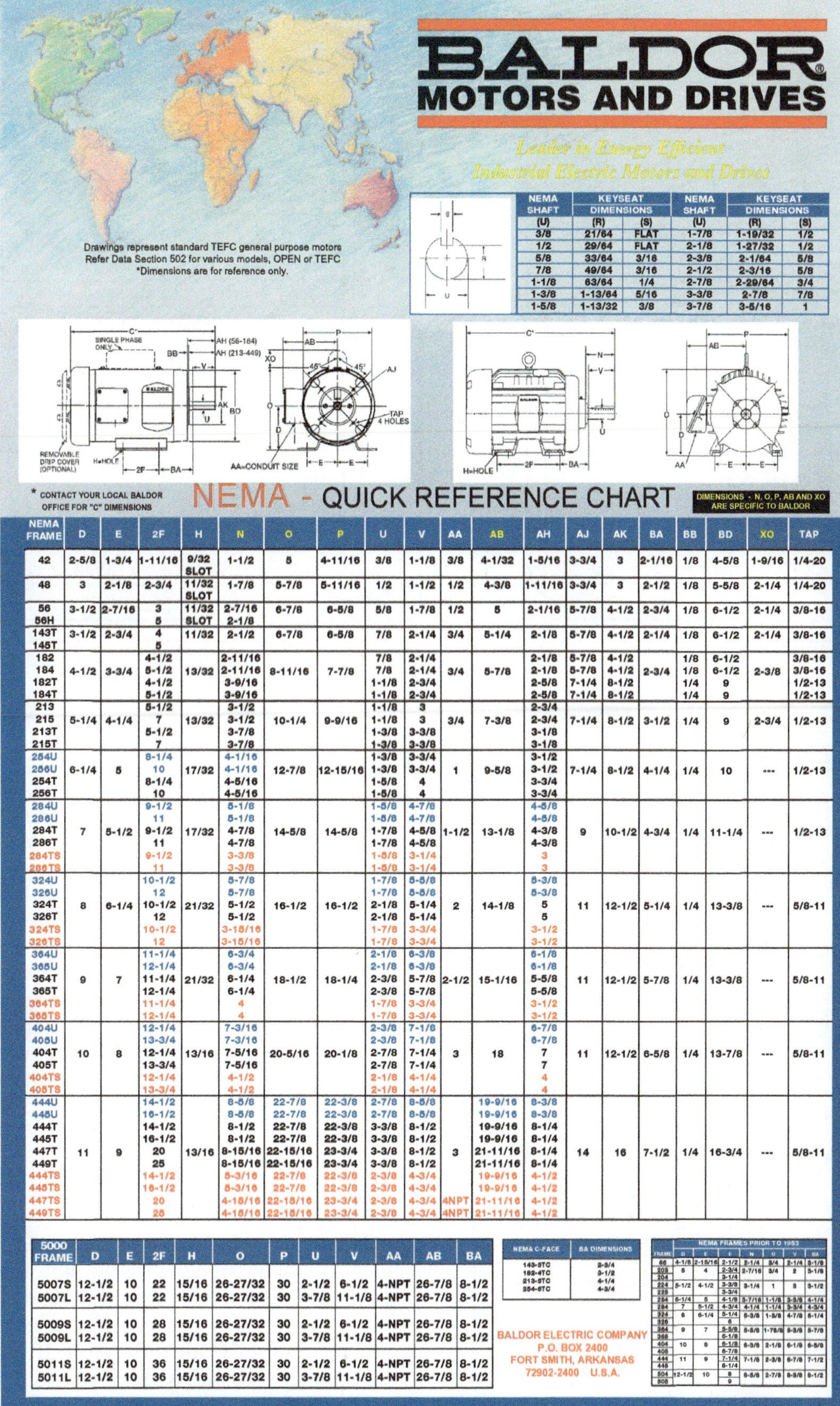

NEMA SHAFT (U)	KEYSEAT DIMENSIONS (R)	(S)	NEMA SHAFT (U)	KEYSEAT DIMENSIONS (R)	(S)
3/8	21/64	FLAT	1-7/8	1-19/32	1/2
1/2	29/64	FLAT	2-1/8	1-27/32	1/2
5/8	33/64	3/16	2-3/8	2-1/64	5/8
7/8	49/64	3/16	2-1/2	2-3/16	5/8
1-1/8	63/64	1/4	2-7/8	2-29/64	3/4
1-3/8	1-13/64	5/16	3-3/8	2-7/8	7/8
1-5/8	1-13/32	3/8	3-7/8	3-5/16	1

NEMA FRAME	D	E	2F	H	N	O	P	U	V	AA	AB	AH	AJ	AK	BA	BB	BD	XO	TAP
42	2-5/8	1-3/4	1-11/16	9/32 SLOT	1-1/2	5	4-11/16	3/8	1-1/8	3/8	4-1/32	1-5/16	3-3/4	3	2-1/16	1/8	4-5/8	1-9/16	1/4-20
48	3	2-1/8	2-3/4	11/32 SLOT	1-7/8	5-7/8	5-11/16	1/2	1-1/2	1/2	4-3/8	1-11/16	3-3/4	3	2-1/2	1/8	5-5/8	2-1/4	1/4-20
56	3-1/2	2-7/16	3	11/32 SLOT	2-7/16	6-7/8	6-5/8	5/8	1-7/8	1/2	5	2-1/16	5-7/8	4-1/2	2-3/4	1/8	6-1/2	2-1/4	3/8-16
56H			5		2-1/8														
143T	3-1/2	2-3/4	4	11/32	2-1/2	6-7/8	6-5/8	7/8	2-1/4	3/4	5-1/4	2-1/8	5-7/8	4-1/2	2-1/4	1/8	6-1/2	2-1/4	3/8-16
145T			5																
182	4-1/2	3-3/4	4-1/2	13/32	2-11/16	8-11/16	7-7/8	7/8	2-1/4	3/4	5-7/8	2-1/8	5-7/8	4-1/2	2-3/4	1/8	6-1/2	2-3/8	3/8-16
184			5-1/2		2-11/16			7/8	2-1/4			2-1/8	5-7/8	4-1/2		1/8	6-1/2		3/8-16
182T			4-1/2		3-9/16			1-1/8	2-3/4			2-5/8	7-1/4	8-1/2		1/4	9		1/2-13
184T			5-1/2		3-9/16			1-1/8	2-3/4			2-5/8	7-1/4	8-1/2		1/4	9		1/2-13
213	5-1/4	4-1/4	5-1/2	13/32	3-1/2	10-1/4	9-9/16	1-1/8	3	3/4	7-3/8	2-3/4	7-1/4	8-1/2	3-1/2	1/4	9	2-3/4	1/2-13
215			7		3-1/2			1-1/8	3			2-3/4							
213T			5-1/2		3-7/8			1-3/8	3-3/8			3-1/8							
215T			7		3-7/8			1-3/8	3-3/8			3-1/8							
254U	6-1/4	5	8-1/4	17/32	4-1/16	12-7/8	12-15/16	1-3/8	3-3/4	1	9-5/8	3-1/2	7-1/4	8-1/2	4-1/4	1/4	10	---	1/2-13
256U			10		4-1/16			1-3/8	3-3/4			3-1/2							
254T			8-1/4		4-5/16			1-5/8	4			3-3/4							
256T			10		4-5/16			1-5/8	4			3-3/4							
284U	7	5-1/2	9-1/2	17/32	5-1/8	14-5/8	14-5/8	1-5/8	4-7/8	1-1/2	13-1/8	4-5/8	9	10-1/2	4-3/4	1/4	11-1/4	---	1/2-13
286U			11		5-1/8			1-5/8	4-7/8			4-5/8							
284T			9-1/2		4-7/8			1-7/8	4-5/8			4-3/8							
286T			11		4-7/8			1-7/8	4-5/8			4-3/8							
284TS			9-1/2		3-3/8			1-5/8	3-1/4			3							
286TS			11		3-3/8			1-5/8	3-1/4			3							
324U	8	6-1/4	10-1/2	21/32	5-7/8	16-1/2	16-1/2	1-7/8	5-5/8	2	14-1/8	5-3/8	11	12-1/2	5-1/4	1/4	13-3/8	---	5/8-11
326U			12		5-7/8			1-7/8	5-5/8			5-3/8							
324T			10-1/2		5-1/2			2-1/8	5-1/4			5							
326T			12		5-1/2			2-1/8	5-1/4			5							
324TS			10-1/2		3-15/16			1-7/8	3-3/4			3-1/2							
326TS			12		3-15/16			1-7/8	3-3/4			3-1/2							
364U	9	7	11-1/4	21/32	6-3/4	18-1/2	18-1/4	2-1/8	6-3/8	2-1/2	15-1/16	6-1/8	11	12-1/2	5-7/8	1/4	13-3/8	---	5/8-11
365U			12-1/4		6-3/4			2-1/8	6-3/8			6-1/8							
364T			11-1/4		6-1/4			2-3/8	5-7/8			5-5/8							
365T			12-1/4		6-1/4			2-3/8	5-7/8			5-5/8							
364TS			11-1/4		4			1-7/8	3-3/4			3-1/2							
365TS			12-1/4		4			1-7/8	3-3/4			3-1/2							
404U	10	8	12-1/4	13/16	7-3/16	20-5/16	20-1/8	2-3/8	7-1/8	3	18	6-7/8	11	12-1/2	6-5/8	1/4	13-7/8	---	5/8-11
405U			13-3/4		7-3/16			2-3/8	7-1/8			6-7/8							
404T			12-1/4		7-5/16			2-7/8	7-1/4			7							
405T			13-3/4		7-5/16			2-7/8	7-1/4			7							
404TS			12-1/4		4-1/2			2-1/8	4-1/4			4							
405TS			13-3/4		4-1/2			2-1/8	4-1/4			4							
444U	11	9	14-1/2	13/16	8-5/8	22-7/8	22-3/8	2-7/8	8-5/8	3	19-9/16	8-3/8	14	16	7-1/2	1/4	16-3/4	---	5/8-11
445U			16-1/2		8-5/8	22-7/8	22-3/8	2-7/8	8-5/8		19-9/16	8-3/8							
444T			14-1/2		8-1/2	22-7/8	22-3/8	3-3/8	8-1/2		19-9/16	8-1/4							
445T			16-1/2		8-1/2	22-7/8	22-3/8	3-3/8	8-1/2		19-9/16	8-1/4							
447T			20		8-15/16	22-15/16	23-3/4	3-3/8	8-1/2		21-11/16	8-1/4							
449T			25		8-15/16	22-15/16	23-3/4	3-3/8	8-1/2		21-11/16	8-1/4							
444TS			14-1/2		5-3/16	22-7/8	22-3/8	2-3/8	4-3/4		19-9/16	4-1/2							
445TS			16-1/2		5-3/16	22-7/8	22-3/8	2-3/8	4-3/4		19-9/16	4-1/2							
447TS			20		4-15/16	22-15/16	23-3/4	2-3/8	4-3/4	4NPT	21-11/16	4-1/2							
449TS			25		4-15/16	22-15/16	23-3/4	2-3/8	4-3/4	4NPT	21-11/16	4-1/2							

5000 FRAME	D	E	2F	H	O	P	U	V	AA	AB	BA
5007S	12-1/2	10	22	15/16	26-27/32	30	2-1/2	6-1/2	4-NPT	26-7/8	8-1/2
5007L	12-1/2	10	22	15/16	26-27/32	30	3-7/8	11-1/8	4-NPT	26-7/8	8-1/2
5009S	12-1/2	10	28	15/16	26-27/32	30	2-1/2	6-1/2	4-NPT	26-7/8	8-1/2
5009L	12-1/2	10	28	15/16	26-27/32	30	3-7/8	11-1/8	4-NPT	26-7/8	8-1/2
5011S	12-1/2	10	36	15/16	26-27/32	30	2-1/2	6-1/2	4-NPT	26-7/8	8-1/2
5011L	12-1/2	10	36	15/16	26-27/32	30	3-7/8	11-1/8	4-NPT	26-7/8	8-1/2

NEMA C-FACE	BA DIMENSIONS
143-5TC	2-3/4
182-4TC	3-1/2
213-5TC	4-1/4
254-6TC	4-3/4

FRAME	NEMA FRAMES PRIOR TO 1953 D	E	F	N	U	V	BA
66	4-1/8	2-15/16	2-1/2	2-1/4	3/4	2-1/4	3-1/8
203	5	4	2-3/4	2-7/16	3/4	2	3-1/8
204			3-1/4				
224	5-1/2	4-1/2	3-3/8	3-1/4	1	3	3-1/2
225			3-3/4				
254	6-1/4	5	4-1/8	3-7/16	1-1/8	3-3/8	4-1/4
284	7	5-1/2	4-3/4	4-1/4	1-1/4	3-3/4	4-3/4
324	8	6-1/4	5-1/4	5-3/8	1-5/8	4-7/8	5-1/4
326			6				
364	9	7	5-5/8	5-5/8	1-7/8	5-3/8	5-7/8
365			6-1/8				
404	10	8	6-1/8	6-3/8	2-1/8	6-1/8	6-5/8
405			6-7/8				
444	11	9	7-1/4	7-1/8	2-3/8	6-7/8	7-1/2
445			8-1/4				
504	12-1/2	10	8	8-5/8	2-7/8	8-3/8	8-1/2
505			9				

Courtesy Baldor Electric Company

Figure 37–9 (Continued)

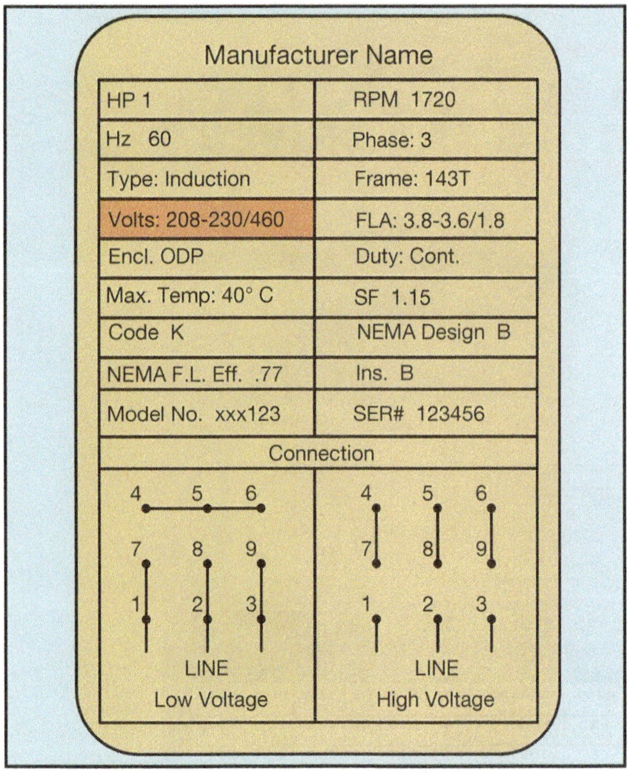

Manufacturer Name

HP 1	RPM 1720
Hz 60	Phase: 3
Type: Induction	Frame: 143T
Volts: 208-230/460	FLA: 3.8-3.6/1.8
Encl. ODP	Duty: Cont.
Max. Temp: 40° C	SF 1.15
Code K	NEMA Design B
NEMA F.L. Eff. .77	Ins. B
Model No. xxx123	SER# 123456
Connection	
4 5 6 7 8 9 1 2 3 LINE Low Voltage	4 5 6 7 8 9 1 2 3 LINE High Voltage

Figure 37–10 This motor can operate on different voltages.

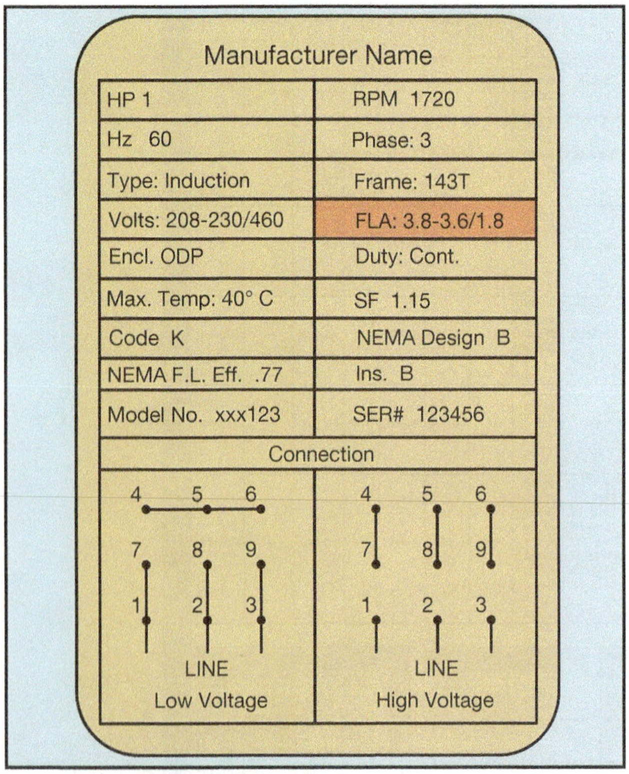

Manufacturer Name

HP 1	RPM 1720
Hz 60	Phase: 3
Type: Induction	Frame: 143T
Volts: 208-230/460	FLA: 3.8-3.6/1.8
Encl. ODP	Duty: Cont.
Max. Temp: 40° C	SF 1.15
Code K	NEMA Design B
NEMA F.L. Eff. .77	Ins. B
Model No. xxx123	SER# 123456
Connection	
4 5 6 7 8 9 1 2 3 LINE Low Voltage	4 5 6 7 8 9 1 2 3 LINE High Voltage

Figure 37–11 The nameplate lists the full load current when the motor is connected to different voltages.

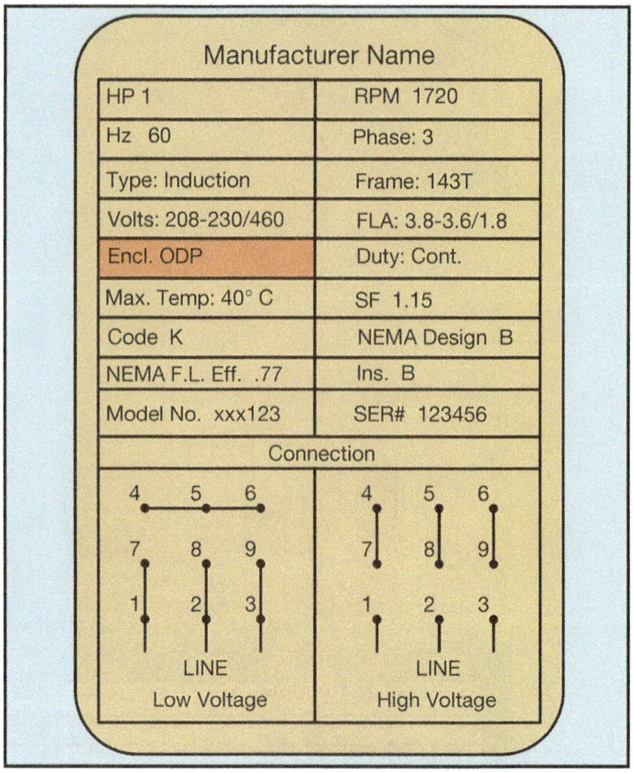

Manufacturer Name

HP 1	RPM 1720
Hz 60	Phase: 3
Type: Induction	Frame: 143T
Volts: 208-230/460	FLA: 3.8-3.6/1.8
Encl. ODP	Duty: Cont.
Max. Temp: 40° C	SF 1.15
Code K	NEMA Design B
NEMA F.L. Eff. .77	Ins. B
Model No. xxx123	SER# 123456
Connection	
4 5 6 7 8 9 1 2 3 LINE Low Voltage	4 5 6 7 8 9 1 2 3 LINE High Voltage

Figure 37–12 The nameplate indicates that the motor has an open-drip-proof enclosure.

Duty Cycle

The nameplate in Figure 37–13 lists the duty cycle as continuous. The duty cycle indicates the amount of time the motor is expected to operate. A motor with a continuous duty cycle is rated to run continuously at full load for 3 hours or more. Most motors are rated for continuous duty.

Intermittent duty motors are intended to operate for short periods of time. An example of an intermittent duty motor is the starter motor on an automobile. These motors develop a large amount of horsepower in a small case size. If these motors were to be operated continuously for a long period of time, they would be damaged by overheating.

Temperature Rise

The maximum temperature indicates the maximum amount of rise in temperature the motor will exhibit when operating continuously at full load. The nameplate in Figure 37–14 indicates a maximum temperature rise of 40°C for this motor. If the motor is operated in an area with a high ambient temperature, it could cause the motor to overheat. If the motor is

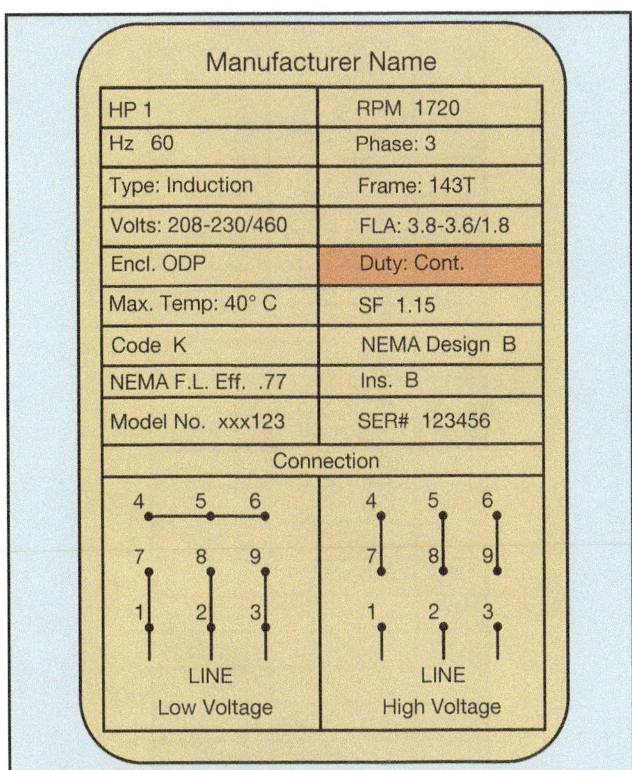

Figure 37–13 The motor is rated for continuous duty.

operated in an area with a low ambient temperature, it may operate at a lower temperature than that marked on the nameplate.

Service Factor

Motor Service Factor **(SF)** gives the allowable horsepower loading, which may be carried out under the conditions specified for the service factor at rated voltage and frequency. It is determined by multiplying the horsepower rating on the nameplate by the service factor. It gives some parameters in estimating horsepower needs and actual running horsepower requirements. The nameplate shown in Figure 37–15 indicates that the motor has a nameplate horsepower of 1 HP. The service factor, however, indicates that the motor is capable of producing 1.15 HP. (1 × 1.15). Selecting motors with a service factor greater than 1 allows for cooler winding temperatures at rated load, protects against intermittent heat rises, and helps to offset low or unbalanced line voltages.

If the motor is operated in the service factor range, however, it will cause a reduction in motor speed and efficiency, and increase motor temperature. This in turn

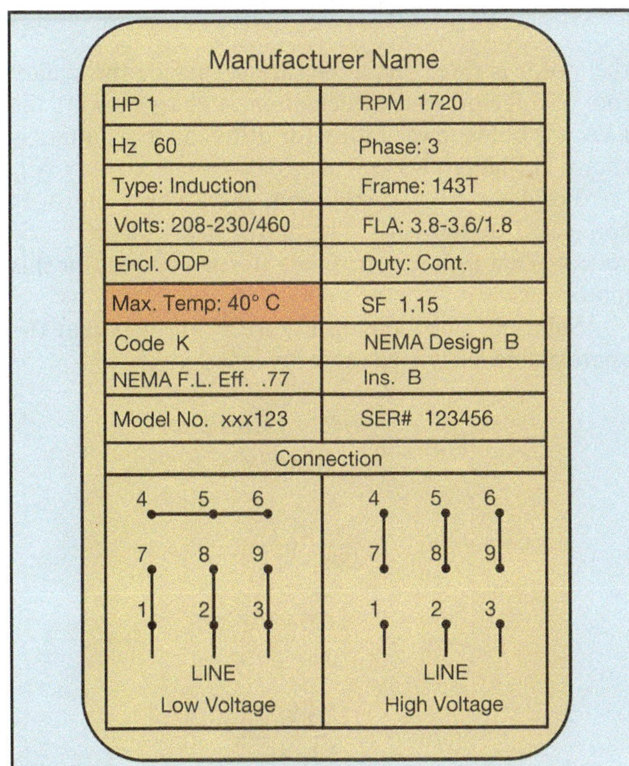

Figure 37–14 The nameplate lists a max temperature rise of 40°C.

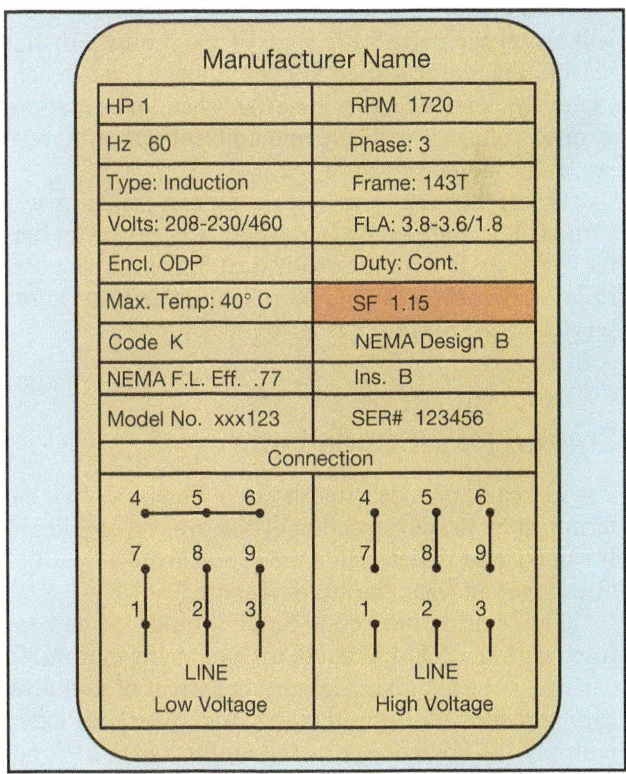

Figure 37–15 The motor has a service factor of 115 percent.

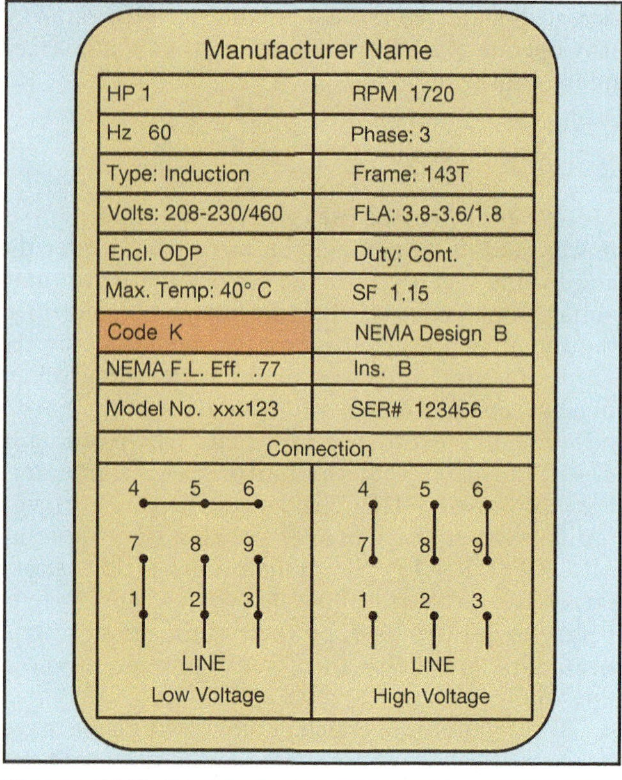

Figure 37–16 The motor has a locked rotor code letter K.

Code	kVA per Horsepower	Approx. Mid-Range Value
A	0.00–3.14	1.6
B	3.15–3.54	3.3
C	3.55–3.99	3.8
D	4.00–4.49	4.3
E	4.50–4.99	4.7
F	5.00–5.59	5.3
G	5.60–6.29	5.9
H	6.30–7.09	6.7
J	7.10–7.99	7.5
K	8.00–8.99	8.5
L	9.00–9.99	9.5
M	10.00–11.19	10.6
N	11.20–12.49	11.8
P	12.50–13.99	13.2
R	14.00–15.99	15
S	16.0–17.99	17
T	18.00–19.99	19
U	20.00–22.39	21
V	22.40–and up	?

Figure 37–17 Locked rotor code letter chart.

will lessen the overall life span of the motor. For this reason, the motor should not run in the SF range continuously. Service factors are established for operations at rated voltage, frequency, and ambient temperature at sea level conditions.

If the horsepower requirements fall between standard size horsepower ratings, it is generally better to purchase a motor of the next higher horsepower rating rather than depend on operating a motor in the service factor range.

Locked Rotor Code Letter

The locked rotor code letter shown in Figure 37–16 is determined by the construction of the squirrel cage rotor. It can be used to determine the approximate amount of inrush current when the motor is started.

The locked rotor code letter should not be confused with the NEMA design code on many motors. To determine the approximate starting current of a squirrel cage induction motor using the locked rotor code letter, multiply the horsepower of the motor by the kVA per horsepower factor and divide by the applied voltage. If the motor is three phase, be sure to include the square root of 3 factors in the calculation. A chart showing the kVA per horsepower rating for different code letters is shown in Figure 37–17.

Problem: A 15-horsepower, three phase motor is connected to 480 volts. The motor has a locked rotor code J. What is the approximate starting current for this motor?

Solution: The chart in Figure 37–17 lists that the approximate mid-range value for code J is 7.5.

$$Amps = \frac{HP \times kVA\ Factor}{Volts \times \sqrt{3}}$$

$$Amps = \frac{15 \times 7.5}{480 \times 1.732}$$

$$Amps = \frac{1125\ kVA}{480 \times 1.732}$$

$$Amps = \frac{112{,}500\ VA}{831.36}$$

$$Amps = 135.5$$

NEMA Design Code

Induction motors have different operating characteristics determined by their design. Such factors as the amount of iron used in the stator, the wire size, the number of turns of wire, and the rotor design all play a part in the operating characteristics of the motor. To obtain some uniformity in motor operating characteristics, NEMA assigns code letters to general purpose motors based on factors such as locked rotor torque, breakdown torque, slip, starting current, and other values. The NEMA code letters are: A, B, C, and D.

A – Motors with the code letter A exhibit normal starting torque and high to medium starting current. These motors are considered to exhibit normal starting torque and normal breakdown torque. They have a maximum slip of 5 percent and are suited for a variety of applications such as fans and pumps.

Percent slip is a measure of the decrease in speed that will occur when load is added to a motor. To determine the full load speed of an induction motor when the percent slip is known, multiply the synchronous speed by the percent slip and subtract that number from the synchronous speed. Synchronous speed is the speed of the rotating magnetic field.

EXAMPLE: A four-pole induction motor connected to 60 Hz. will have a synchronous speed of 1800 RPM. If the motor has a maximum percent slip of 5 percent, multiply 1800 by 5 percent.

$$1800 \text{ RPM} \times 0.05 = 90 \text{ RPM}$$

Subtract that number from the synchronous speed.

$$1800 \text{ RPM} - 90 \text{ RPM} = 1710 \text{ RPM}$$

Many motors will exhibit much less than 5 percent slip, but 5 percent is a maximum for motors with NEMA code A.

B – Design B motors are the most common. They exhibit high starting torque and low starting current. They have sufficient lock rotor starting torque to start most industrial loads. They exhibit normal breakdown torque and a maximum of 5 percent slip. They are commonly used in HVAC applications with fans, blowers, and pumps. The nameplate in Figure 37–18 indicates that this motor has a NEMA design code B.

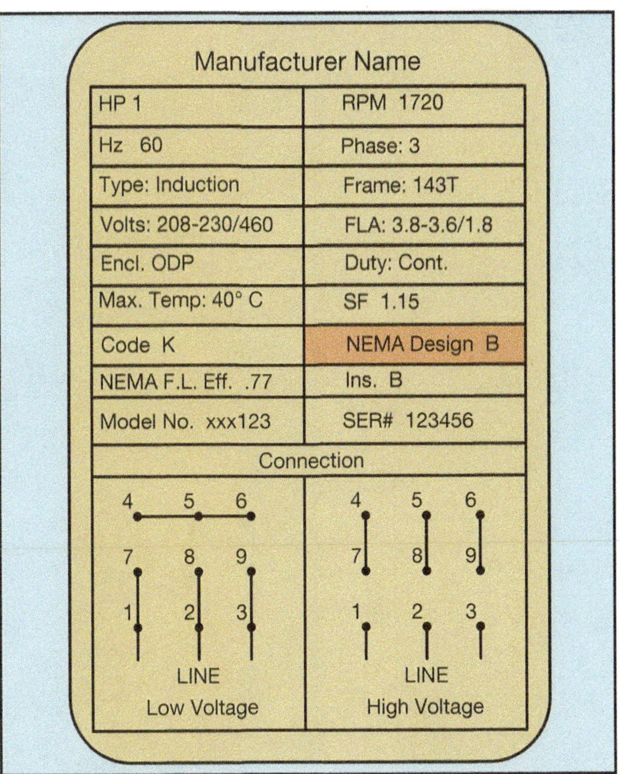

Figure 37–18 The nameplate indicates that this motor has a NEMA design code B.

C – Motors with the code letter C have high starting torques and low starting currents. They are used to start high inertia loads such as positive displacement pumps, centrifuges, and loads that require the use of flywheels. They have a maximum of 5 percent slip.

D – Design D motors exhibit very high starting torque and low starting current. However, these motors exhibit a large amount of rotor slip, in the range of 5 percent to 13 percent, when load is added. Design D motors are generally used for equipment that require high inertia starts such as cranes and hoists.

Motor Efficiency

The full load efficiency indicates the overall efficiency of the motor. The efficiency basically describes the amount of electrical energy supplied to the motor that is converted into kinetic energy. The remaining power is a loss and is mostly converted into heat. The nameplate in Figure 37–19 indicates that the motor in this example has an efficiency of 77 percent.

Manufacturer Name

HP 1	RPM 1720
Hz 60	Phase: 3
Type: Induction	Frame: 143T
Volts: 208-230/460	FLA: 3.8-3.6/1.8
Encl. ODP	Duty: Cont.
Max. Temp: 40° C	SF 1.15
Code K	NEMA Design B
NEMA F.L. Eff. .77	Ins. B
Model No. xxx123	SER# 123456
Connection	
4 5 6 / 7 8 9 / 1 2 3 / LINE / Low Voltage	4 5 6 / 7 8 9 / 1 2 3 / LINE / High Voltage

Figure 37–19 The motor in this example has an efficiency of 77 percent.

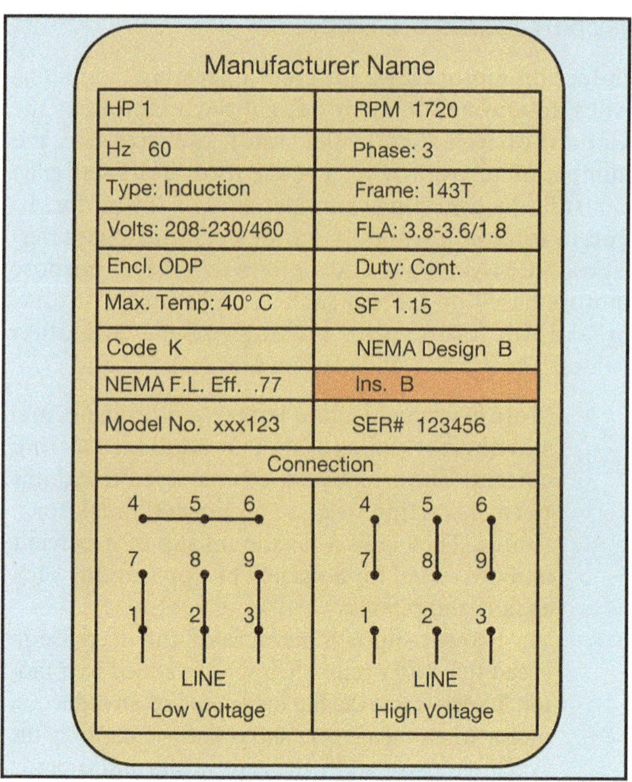

Figure 37–20 The motor has an insulation rating B.

Insulation Classification

The nameplate in Figure 37–20 indicates that this motor has a temperature classification B. The classification of insulation greatly affects the life span of the motor. Motor temperature is based on the hottest point in the motor under full load operation and is determined by the temperature rise of the motor and the surrounding ambient air temperature. Motors that operate in hotter climates should have a higher insulation temperature rating. The thermal capacity of different insulations is rated as A, B, F, and H. The chart shown in Figure 37–21 lists the amount of temperature each is designed to handle over a 20,000-hour period.

Class	20,000 Hour Life Temperature
A	105°C
B	130°C
F	155°C
H	180°C

Figure 37–21 The chart lists the temperature rating of different insulation types.

Model and Serial Numbers

The model number is assigned by the manufacturer. It can be used to purchase a motor with identical characteristics. Many motors with the same model number can generally be obtained. The serial number is also assigned by the manufacturer. The serial number, however, is used to identify a particular motor. No other motor should have the same serial number.

Connection Diagrams

The connection diagrams for both low and high voltage connections are given on the nameplate shown in Figure 37–22. Most dual voltage motors contain nine leads in the terminal connection box. These leads are numbered T1 through T9. The connection diagram is used to make high or low voltage connection for the motor. In the diagram shown, if the motor is to be operated on low voltage, T4, T5, and T6 should be connected together. T1 and T7 should be connected together, T2 and T8 should be connected together, and

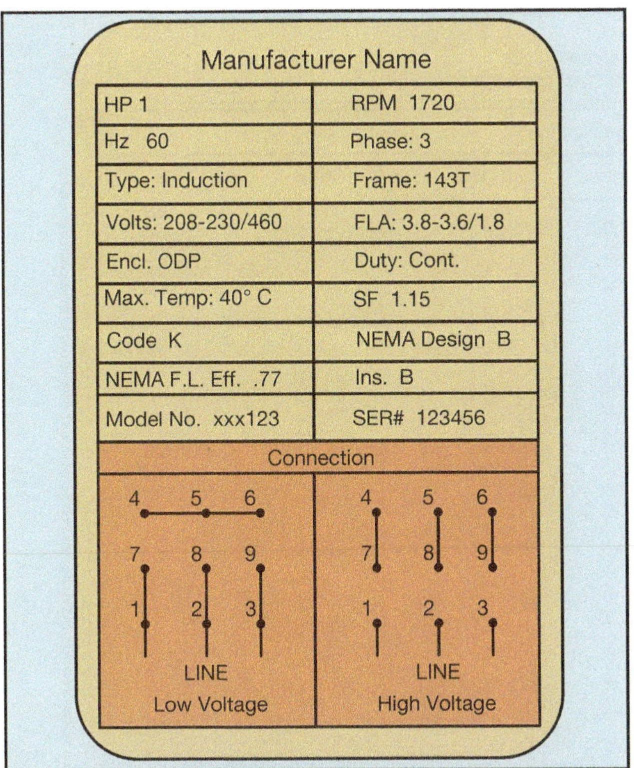

Figure 37–22 The connection diagram is used to connect the motor for high or low voltage operation.

T3 and T9 should be connected together. Power is connected to T1, T2, and T3.

If the motor is connected for high voltage operation, T4 and T7 are connected together, T5 and T8 are connected together, and T6 and T9 are connected together. Power is connected to T1, T2, and T3.

Determining Motor Current

There are different types of motors, such as direct current, single phase AC, two phase AC, and three phase AC. Different tables from the *National Electrical Code (NEC)®* are used to determine the running current for these different types of motors. *Table 430.247* in Figure 37–23 is used to determine the full load running current for direct current motors. *Table 430.248* in Figure 37–24 is used to determine the full load running current for single phase motors; *Table 430.249* in Figure 37–25 is used to determine the running current for two phase motors; and *Table 430.250* in Figure 37–26 is used to determine the full load running current for three phase motors. Note that the tables list the amount of current that the motor is expected to draw under a full load condition. The motor will exhibit less current draw if it is not under full load. These tables list the ampere rating of the motors according to horsepower (HP) and connected voltage. It should also be noted that *NEC® Section 430.6(A)* states these tables are to be used to determine *conductor size, short-circuit protection size,* and *ground fault protection size* instead of the nameplate rating of the motor. The motor overload size, however, is to be determined by the nameplate rating of the motor.

Direct Current Motors

Table 430.247 lists the full load running currents for direct current motors. The horsepower rating of the motor is given in the far left column. Rated voltages are listed across the top of the table. The table shows that a one (1) HP motor will have a full load current of 12.2 amperes when connected to 90 volts DC. If a 1 HP motor is designed to be connected to 240 volts, it will have a current draw of 4.7 amperes.

Single Phase AC Motors

The current ratings for single phase AC motors are given in *Table 430.248.* Particular attention should be paid to the statement preceding the table. The statement asserts that the values listed in this table are for motors that operate under normal speeds and torques. Motors especially designed for low speed and high torque, or multispeed motors, shall have their running current determined from the nameplate rating of the motor.

The voltages listed in the table are 115, 200, 208, and 230. The last sentence of the preceding statement says that the currents listed shall be permitted for voltages of 110 to 120 volts and 220 to 240 volts. This means that if the motor is connected to a 120-volt line, it is permissible to use the currents listed in the 115-volt column. If the motor is connected to a 220-volt line, the 230-volt column can be used.

EXAMPLE: A 3 HP single phase AC motor is connected to a 208-volt line. What will be the full load running current of this motor?

Locate 3 HP in the far left column of *Table 430.248.* Follow across to the 208-volt column. The full load current will be 18.7 amperes.

Table 430.247 Full-Load Current in Amperes, Direct-Current Motors
The following values of full-load currents* are for motors running at base speed.

	Armature Voltage Ratings*					
Horsepower	**90 Volts**	**120 Volts**	**180 Volts**	**240 Volts**	**500 Volts**	**550 Volts**
¼	4.0	3.1	2.0	1.6	—	—
⅓	5.2	4.1	2.6	2.0	—	—
½	6.8	5.4	3.4	2.7	—	—
¾	9.6	7.6	4.8	3.8	—	—
1	12.2	9.5	6.1	4.7	—	—
1½	—	13.2	8.3	6.6	—	—
2	—	17.0	10.8	8.5	—	—
3	—	25.0	16.0	12.2	—	—
5	—	40.0	27.0	20.0	—	—
7½	—	58.0	—	29.0	13.6	12.2
10	—	76.0	—	38.0	18.0	16.0
15	—	—	—	55.0	27.0	24.0
20	—	—	—	72.0	34.0	31.0
25	—	—	—	89.0	43.0	38.0
30	—	—	—	106	51.0	46.0
40	—	—	—	140	67.0	61.0
50	—	—	—	173	83.0	75.0
60	—	—	—	206	99.0	90.0
75	—	—	—	255	123	111
100	—	—	—	341	164	148
125	—	—	—	425	205	185
150	—	—	—	506	246	222
200	—	—	—	675	330	294

*These are average dc quantities.

Figure 37–23 Table 430.247 is used to determine the full load current for direct current motors.

Table 430.248 Full-Load Currents in Amperes, Single-Phase Alternating-Current Motors
The following values of full-load currents are for motors running at usual speeds and motors with normal torque characteristics. The voltages listed are rated motor voltages. The currents listed shall be permitted for system voltage ranges of 110 to 120 and 220 to 240 volts.

Horsepower	**115 Volts**	**200 Volts**	**208 Volts**	**230 Volts**
⅙	4.4	2.5	2.4	2.2
¼	5.8	3.3	3.2	2.9
⅓	7.2	4.1	4.0	3.6
½	9.8	5.6	5.4	4.9
¾	13.8	7.9	7.6	6.9
1	16	9.2	8.8	8.0
1½	20	11.5	11.0	10
2	24	13.8	13.2	12
3	34	19.6	18.7	17
5	56	32.2	30.8	28
7½	80	46.0	44.0	40
10	100	57.5	55.0	50

Figure 37–24 Table 430.248 is used to determine the full load current for single phase motors.

Table 430.249 Full-Load Current, Two-Phase Alternating-Current Motors (4-Wire)

The following values of full-load current are for motors with normal torque characteristics. Current in the common conductor of a 2-phase, 3 wire system will be 1.41 times the value given. The voltages listed are rated motor voltages. The currents listed shall be permitted for system voltage ranges of 110 to 120, 220 to 240, 440 to 480, and 550 to 1000 volts.

	Induction-Type Squirrel Cage and Wound Rotor (Amperes)				
Horsepower	**115 Volts**	**230 Volts**	**460 Volts**	**575 Volts**	**2300 Volts**
½	4.0	2.0	1.0	0.8	—
¾	4.8	2.4	1.2	1.0	—
1	6.4	3.2	1.6	1.3	—
1½	9.0	4.5	2.3	1.8	—
2	11.8	5.9	3.0	2.4	—
3	—	8.3	4.2	3.3	—
5	—	13.2	6.6	5.3	—
7½	—	19	9.0	8.0	—
10	—	24	12	10	—
15	—	36	18	14	—
20	—	47	23	19	—
25	—	59	29	24	—
30	—	69	35	28	—
40	—	90	45	36	—
50	—	113	56	45	—
60	—	133	67	53	14
75	—	166	83	66	18
100	—	218	109	87	23
125	—	270	135	108	28
150	—	312	156	125	32
200	—	416	208	167	43

Figure 37–25 Table 430.249 is used to determine the full load current for two phase motors.

Two Phase Motors

Although two phase motors are seldom used, *Table 430.249* lists the full load running currents for these motors. Like single phase motors, two phase motors that are especially designed for low speed, high torque applications and multispeed motors, use the nameplate rating instead of the values shown in the table. When using a two phase, three-wire system, the size of the neutral conductor must be increased by the square root of 2, or 1.41. The reason is that the voltages of a two phase system are 90 degrees out-of-phase with each other as shown in Figure 37–27. The principle of two phase power generation is shown in Figure 37–28. In a two phase alternator, the phase windings are arranged 90 degrees apart. The magnet is the rotor of the alternator. When the rotor turns, it induces voltage into the phase windings, which are 90 degrees apart. When one end of each phase winding is joined to form a common terminal, or neutral, the current in the neutral conductor will be greater than the current in either of the two phase conductors. An example is shown in Figure 37–29. In this example, a two phase alternator is connected to a two phase motor. The current draw on each of the phase windings is 10 amperes. The current flow in the neutral, however, is 1.41 times greater than the current flow in the phase windings or 14.1 amperes.

Table 430.250 Full-Load Current, Three-Phase Alternating-Current Motors
The following values of full-load currents are typical for motors running at speeds usual for belted motors and motors with normal torque characteristics.
The voltages listed are rated motor voltages. The currents listed shall be permitted for system voltage ranges of 110 to 120, 220 to 240, 440 to 480, and 550 to 1000 volts.

	Induction-Type Squirrel Cage and Wound Rotor (Amperes)							Synchronous-Type Unity Power Factor* (Amperes)			
Horsepower	115 Volts	200 Volts	208 Volts	230 Volts	460 Volts	575 Volts	2300 Volts	230 Volts	460 Volts	575 Volts	2300 Volts
½	4.4	2.5	2.4	2.2	1.1	0.9	—	—	—	—	—
¾	6.4	3.7	3.5	3.2	1.6	1.3	—	—	—	—	—
1	8.4	4.8	4.6	4.2	2.1	1.7	—	—	—	—	—
1½	12.0	6.9	6.6	6.0	3.0	2.4	—	—	—	—	—
2	13.6	7.8	7.5	6.8	3.4	2.7	—	—	—	—	—
3	—	11.0	10.6	9.6	4.8	3.9	—	—	—	—	—
5	—	17.5	16.7	15.2	7.6	6.1		—	—	—	—
7½	—	25.3	24.2	22	11	9	—	—	—	—	—
10	—	32.2	30.8	28	14	11	—	—	—	—	—
15	—	48.3	46.2	42	21	17	—	—	—	—	—
20	—	62.1	59.4	54	27	22	—	—	—	—	—
25	—	78.2	74.8	68	34	27	—	53	26	21	—
30	—	92	88	80	40	32	—	63	32	26	—
40	—	120	114	104	52	41	—	83	41	33	—
50	—	150	143	130	65	52	—	104	52	42	—
60	—	177	169	154	77	62	16	123	61	49	12
75	—	221	211	192	96	77	20	155	78	62	15
100	—	285	273	248	124	99	26	202	101	81	20
125	—	359	343	312	156	125	31	253	126	101	25
150	—	414	396	360	180	144	37	302	151	121	30
200	—	552	528	480	240	192	49	400	201	161	40
250	—	—	—	—	302	242	60	—	—	—	—
300	—	—	—	—	361	289	72	—	—	—	—
350	—	—	—	—	414	336	83	—	—	—	—
400	—	—	—	—	477	382	95	—	—	—	—
450	—	—	—	—	515	412	103	—	—	—	—
500	—	—	—	—	590	472	118	—	—	—	—

*For 90 and 80 percent power factor, the figures shall be multiplied by 1.1 and 1.25 respectively.

Figure 37–26 Table 430.250 is used to determine the full load current for three phase motors.

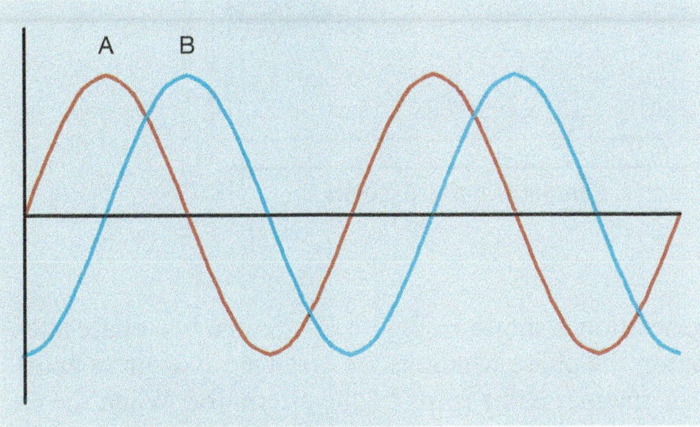

Figure 37–27 The voltages of a two phase system are 90 degrees out-of-phase with each other.

EXAMPLE: Compute the phase current and neutral current for a 60 HP, 460-volt two phase motor.

The phase current can be taken from *Table 430.249.*

Phase current = 67 amperes

The neutral current will be 1.41 times higher than the phase current.

Neutral current = 67 × 1.41

Neutral current = 94.5 amperes

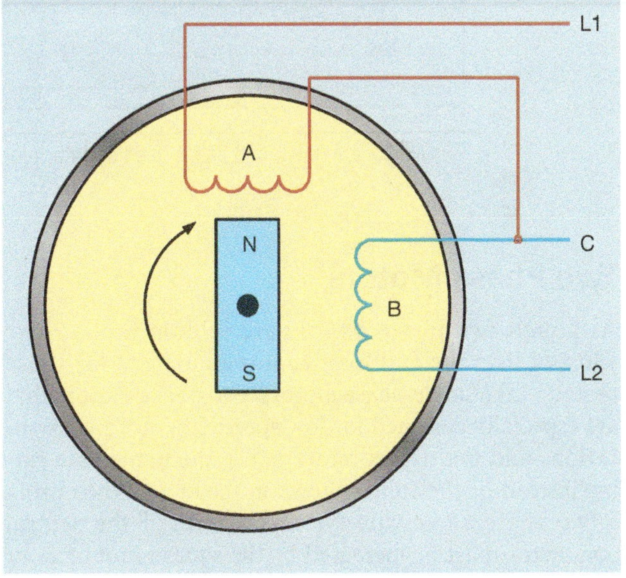

Figure 37–28 A two phase alternator produces voltages that are 90 degrees out-of-phase with each other.

Three Phase Motors

Table 430.250 is used to determine the full load current of three phase motors. The notes at the top of the table are very similar to the notes of *Tables 430.248* and *430.249*. The full load current of low

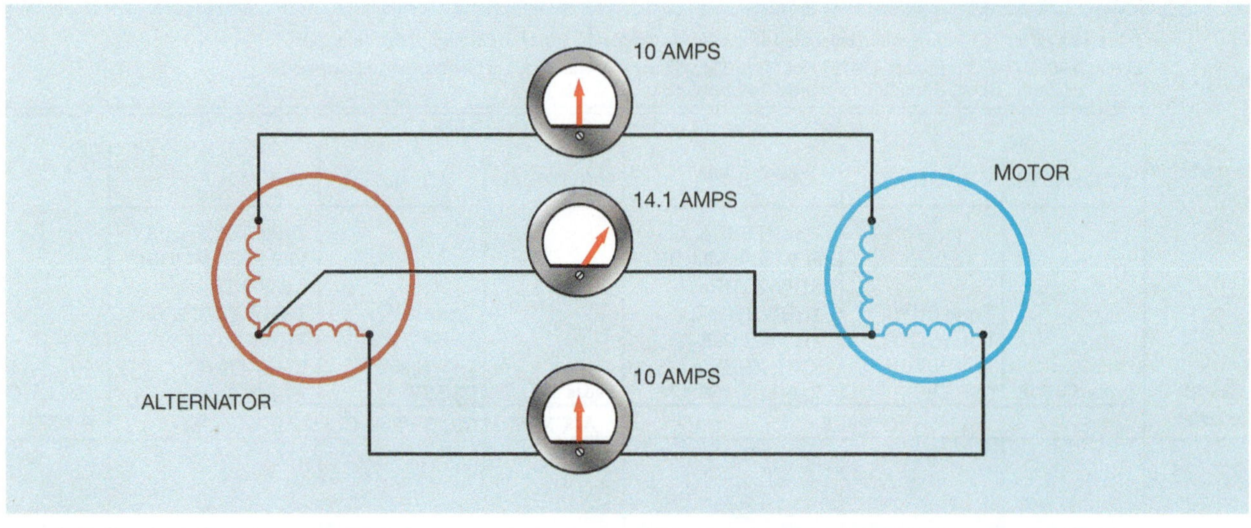

Figure 37–29 The neutral conductor of a two phase system has a greater current than the other two conductors.

speed, high torque and multispeed motors is to be determined from the nameplate rating instead of from the values listed in the table. *Table 430.250* has an extra note that deals with synchronous motors. Notice that the right side of *Table 430.250* is devoted to the full load currents of synchronous type motors. The currents listed are for synchronous type motors that are to be operated at unity or 100 percent power factor. Because synchronous motors are often made to have a leading power factor by overexcitation of the rotor current, the full load current rating must be increased when this is done. If the motor is operated at 90 percent power factor, the rated full load current in the table must be increased by 10 percent. If the motor is to be operated at 80 percent power factor, the full load current is to be increased by 25 percent.

EXAMPLE: A 150 HP, 460-volt synchronous motor is to be operated at 80 percent power factor. What will be the full load current rating of the motor?

The table indicates a current value of 151 amperes for this motor. To determine the running current at 80 percent power factor multiply this current by 125 percent or 1.25. (Multiplying by 1.25 results in the same answer that would be obtained by dividing by 0.80.)

$151 \times 1.25 = 188.75$ or 189 amperes

EXAMPLE: A 200 HP, 2300-volt synchronous motor is to be operated at 90 percent power factor. What will be the full load current rating of this motor?

Locate 200 horsepower in the far left column. Follow across to the 2,300-volt column listed under synchronous type motors. Increase this value by 10 percent.

$40 \times 1.10 = 44$ amperes

Determining Conductor Size for a Single Motor

NEC® Section 430.6(A) states that the conductor for a motor connection shall be based on the values from *Tables 430.247, 430.248, 430.249,* and *430.250* instead of the motor nameplate current. *Section 430.22(A)* states that conductors supplying a single motor shall have an ampacity of not less than 125 percent of the motor full load current. *NEC® Section 310* is used to select the conductor size after the ampacity has been determined. The exact table employed will be determined by the wiring conditions. Probably the most frequently used table is *Table 310.15(B)(16)*, Figure 37–30.

Termination Temperature

Another factor that must be taken into consideration when determining the conductor size is the temperature

Table 310.15(B)(16) Allowable Ampacities of Insulated Conductors Rated 0 Through 2000 Volts, 60°C Through 90°C (140°F Through 194°F) Not More Than Three Current-Carrying Conductors in Raceway, Cable, or Earth (Direct Buried), on Ambient Temperature of 30°C (86°F)

	Temperature Rating of Conductors [See Table 310.104(A)]						
	60°C (140°F)	75°C (167°F)	90°C (194°F)	60°C (140°F)	75°C (167°F)	90°C (194°F)	
Size AWG or kcmil	Types TW,UF	Types RHW, THHW,THW, THWN,XHHW, USE,ZW	Types TBS,SA,SIS, FEP,FEPB,MI,RHH, RHW-2,THHN, THHW,THW-2, THWN-2,USE-2, XHH, XHHW, XHHW-2, ZW-2	Types TW,UF	Types RHW, THHW,THW, THWN,XHHW, USE,ZW	Types TBS,SA,SIS, FEP,FEPB,MI,RHH, RHW-2,THHN, THHW,THW-2, THWN-2,USE-2, XHH, XHHW, XHHW-2, ZW-2	Size AWG or kcmil
	COPPER			ALUMINUM OR COPPER-CLAD ALUMINUM			
18	-	-	14	-	-	-	-
16	-	-	18	-	-	-	-
14*	20	20	25	-	-	-	-
12*	25	25	30	20	20	25	12*
10*	30	35	40	25	30	35	10*
8	40	50	55	30	40	45	8
6	55	65	75	40	50	60	6
4	70	85	95	55	65	75	4
3	85	100	110	65	75	85	3
2	95	115	130	75	90	100	2
1	110	130	150	85	100	115	1
1/0	125	150	170	100	120	135	1/0
2/0	145	175	195	115	135	150	2/0
3/0	165	200	225	130	155	175	3/0
4/0	195	230	260	150	180	205	4/0
250	215	255	290	170	205	230	250
300	240	285	320	190	230	255	300
350	260	310	350	210	250	280	350
400	280	335	380	225	270	305	400
500	320	380	430	260	310	350	500
600	355	420	475	285	340	385	600
700	385	460	520	310	375	420	700
750	400	475	535	320	385	435	750
800	410	490	555	330	395	450	800
900	435	520	585	355	425	480	900
1000	455	545	615	375	445	500	1000
1250	495	590	665	405	485	545	1250
1500	520	625	705	435	520	585	1500
1750	545	650	735	455	545	615	1750
2000	560	665	750	470	560	630	2000

Figure 37–30 Table 310.15(B)(16) is used to determine the ampacity of conductors.

rating of the devices and terminals as specified in *NEC®* *Section 110.14(C)*. This section states that the conductor is to be selected and coordinated as to not exceed the lowest temperature rating of any connected termination, any connected conductor, or any connected device. This means that regardless of the temperature rating of the conductor, the ampacity must be selected from a column that does not exceed the temperature rating of the termination. The conductors listed in the first column of *Table 310.15(B)(16)* have a temperature rating of 60°C, the conductors in the second column have a rating of 75°C, and the conductors in the third column have a rating of 90°C. The temperature ratings of devices such as circuit breakers, fuses, and terminals are found in the UL (Underwriters Laboratories) product directories. Occasionally, the temperature rating may be found on the piece of equipment, but this is the exception and not the rule. As a general rule the temperature rating of most devices will not exceed 75°C.

When the termination temperature rating is not listed or known, *NEC® Section 110.14(C)(1)(a)* states that for circuits rated at 100 amperes or less, or for #14 AWG through #1 AWG conductors, the ampacity of the wire, regardless of the temperature rating, will be selected from the 60°C column. This does not mean that only those types of insulation listed in the 60°C column can be used, but that the *ampacities* listed in the 60°C column must be used to select the conductor size. For example, assume that a copper conductor with type XHHW insulation is to be connected to a 50-ampere circuit breaker that does not have a listed temperature rating. According to *NEC® Table 310.15(B)(16)*, a #8 AWG copper conductor with XHHW insulation is rated to carry 55 amperes of current. Type XHHW insulation is located in the 90°C column, but the temperature rating of the circuit breaker is not known. Therefore, the wire size must be selected from the ampacity ratings in the 60°C column. A #6 AWG copper conductor with type XHHW insulation would be used.

NEC® Section 110.14(C)(1)(a)(4) has a special provision for motors with marked NEMA design codes B, C, D, or E. This section states that conductors rated at 75°C or higher may be selected from the 75°C column, even if the ampacity is 100 amperes or less. This code will not apply to motors that do not have a NEMA design code marked on their nameplate. Most motors manufactured before 1996 will not have a NEMA design code. The NEMA design code letter should not be confused with the code letter that indicates the type squirrel cage rotor used in the motor.

For circuits rated over 100 amperes, or for conductor sizes larger than #1 AWG, *Section 110.14(C)(1)(b)* states that the ampacity ratings listed in the 75°C column may be used to select wire sizes unless conductors with a 60°C temperature rating have been selected for use. For example, types TW and UF insulation are listed in the 60°C column. If one of these two insulation types has been specified, the wire size must be chosen from the 60°C column regardless of the ampere rating of the circuit.

EXAMPLE: A 30 HP, three phase squirrel cage induction motor is connected to a 480-volt line. The conductors are run in conduit to the motor. The motor does not have a NEMA design code listed on the nameplate. The termination temperature rating of the devices is not known. Copper conductors with THWN insulation are to be used for this motor connection. What size conductors should be used?

The first step is to determine the full load current of the motor. This is determined from *Table 430.250*. The table indicates a current of 40 amperes for this motor. The current must be increased by 25 percent according to *Section 430.22(A)*.

$$40 \times 1.25 = 50 \text{ amperes}$$

Table 310.15(B)(16) is used to determine the conductor size. Locate the column that contains THWN insulation in the copper section of the table. THWN is located in the 75°C column. Because this circuit is less than 100 amperes and the termination temperature is not known, and the motor does not contain a NEMA design code letter, the conductor size must be selected from the ampacities listed in the 60°C column. A #6 AWG copper conductor with type THWN insulation will be used.

Overload Size

When determining the overload size for a motor, the *nameplate* current rating of the motor is used instead of the current values listed in the tables [*NEC® Section 430.6(A)*]. Other factors such as the service factor (SF) or temperature rise (°C) of the motor are also to be considered when determining the overload size for a motor. The temperature rise of the motor is an indication of the type of insulation used on the motor windings and should not be confused with termination temperature discussed in *Section 110.14(C)*. *NEC® Section 430.32* (Figure 37–31), is used to determine the overload size for motors of one horsepower or more. The overload size is based on a percentage of the full load current of the motor listed on the motor nameplate.

EXAMPLE: A 25 HP, three phase induction motor has a nameplate rating of 32 amperes. The nameplate also shows a temperature rise of 30°C. Determine the percent rating of the overload for this motor.

NEC® Section 430.32(A)(1) indicates the overload size is 125 percent of the full load current rating of the motor.

If for some reason this overload size does not permit the motor to start without tripping out, *Section 430.32(C)* permits the overload size to be increased to a maximum of 140 percent for this motor. If this increase in overload size does not solve the starting problem, the overload may be shunted out of the circuit during the starting period in accordance with *Section 430.35(A)&(B)*.

430.32 Continuous-Duty Motors

(A) More than 1 Horsepower. Each motor used in a continuous duty application and rated more than 1 HP shall be protected against overload by one of the following means in 430.32(A)(1) through (A)(4).

(1) Separate Overload Device. A separate overload device that is responsive to motor current. This device shall be selected to trip or shall be rated at no more than the following percent of the motor nameplate full-load current rating:

Motors with a marked service factor 1.15 or greater	125%
Motors with a marked temperature rise 40°C or less	125%
All other motors	115%

Figure 37–31 Section 430-32 is used to determine overload size for motors.

Determining Locked-Rotor Current

There are two basic methods for determining the locked-rotor current (starting current) of a squirrel cage induction motor depending on the information available. If the motor nameplate lists code letters that range from A to V, they indicate the type of rotor bars used when the rotor was made. Different types of bars are used to make motors with different operating characteristics. The type of rotor bars largely determines the maximum starting current of the motor. *NEC® Table 430.7(B)* Figure 37–32 lists the different code letters and gives the locked-rotor kilovolt-amperes per horsepower. The starting current can be determined by multiplying the kVA rating by the horsepower rating and then dividing by the applied voltage.

Table 430.7(B) Locked-Rotor Indicating Code Letters

Code Letter	Kilovolt-Amperes per Horsepower with Locked Rotor
A	0–3.14
B	3.15–3.54
C	3.55–3.99
D	4.0–4.49
E	4.5–4.99
F	5.0–5.59
G	5.6–6.29
H	6.3–7.09
J	7.1–7.99
K	8.0–8.99
L	9.0–9.99
M	10.0–11.19
N	11.2–12.49
P	12.5–13.99
R	14.0–15.99
S	16.0–17.99
T	18.0–19.99
U	20.0–22.39
V	22.4 and up

Figure 37–32 Table 430.7(B) is used to determine locked-rotor current for motors that do not contain a NEMA code letter.

EXAMPLE: A 15 HP, three phase squirrel cage motor with a code letter of K is connected to a 240-volt line. Determine the locked-rotor current.

The table lists 8.0 to 8.99 kVA per horsepower for a motor with a code letter of K. An average value of 8.5 will be used.

$$8.5 \times 15 = 127.5 \text{ kVA or } 127{,}500 \text{ VA}$$

$$\frac{127{,}500}{240 \times \sqrt{3}} = 306.7 \text{ amperes}$$

The second method of determining locked-rotor current is to use *Tables 430.251(A)&(B)*, Figure 37–33 if the motor nameplate lists NEMA design codes. *Table 430.251(A)* lists the locked-rotor currents for single phase motors and *Table 430.251(B)* lists the locked-rotor currents for poly-phase motors.

Short-Circuit Protection

The rating of the short-circuit protective device is determined by *NEC® Table 430.52* Figure 37–34. The far left column lists the type of motor that is to be protected. To the right are four columns that list different types of

EXAMPLE: A 100 HP, three phase squirrel cage induction motor is connected to a 240-volt line. The motor does not contain a NEMA design code. A dual-element time delay fuse is to be used as the short-circuit protective device. Determine the size needed.

Table 430.250 lists a full load current of 248 amperes for this motor. *Table 430.52* indicates that a dual-element time delay fuse is to be calculated at 175 percent of the full load current rating for an AC poly-phase (more than one phase) squirrel cage motor, other than Design B. Because the motor does not list a NEMA design code on the nameplate, it will be assumed that the motor is Design B.

$$248 \times 1.75 = 434 \text{ amperes}$$

The nearest standard fuse size above the computed value listed in *Section 240.6* is 450 amperes, so 450 ampere fuses will be used to protect this motor.

If for some reason this fuse will not permit the motor to start without blowing, *NEC® Section 430.52 (C)(1) Exception 2(b)* states that the rating of a dual-element time delay fuse may be increased to a maximum of 225 percent of the full load motor current.

Table 430.251(A) Conversion Table of Single-Phase Locked-Rotor Currents for Selection of Disconnecting Means and Controllers as Determined from Horsepower and Voltage rating

For use only with 430.110, 440.12, 440.41, and 455.8(C)

Rated Horsepower	Maximum Locked-Rotor Current in Amperes, Single Phase		
	115 Volts	208 Volts	230 Volts
½	58.8	32.5	29.4
¾	82.8	45.8	41.4
1	96	53	48
1½	120	66	60
2	144	80	72
3	204	113	102
5	336	186	168
7½	480	265	240
10	1000	332	300

Figure 37–33 Tables 430.251(A) & (B) are used to determine locked-rotor current for motors that do contain NEMA code letters.

Table 430.251(B) Conversion Table of Polyphase Design B, C, and D Maximum Locked-rotor Currents for Selection of Disconnecting Means and Controllers as Determined from Horsepower and Voltage Rating and Design Letter
For use only with 430.110, 440.12, 440.41, and 455.8(C)

Rated Horsepower	Maximum Motor Locked-Rotor Current in Amperes, Two and Three-Phase Design B, C, and D*					
	115 Volts B,C,D	200 Volts B,C,D	208 Volts B,C,D	230 Volts B,C,D	460 Volts B,C,D	575 Volts B,C,D
½	40	23	22.1	20	10	8
¾	50	28.8	27.6	25	12.5	10
1	60	34.5	33	30	15	12
1½	80	46	44	40	20	16
2	100	57.5	55	50	25	20
3	—	73.6	71	64	32	25.6
5	—	105.8	102	92	46	36.8
7½	—	146	140	127	63.5	50.8
10	—	186.3	179	162	81	64.8
15	—	267	257	232	116	93
20	—	334	321	290	145	116
25	—	420	404	365	183	146
30	—	500	481	435	218	174
40	—	667	641	580	290	232
50	—	834	802	725	363	290
60	—	1001	962	870	435	348
75	—	1248	1200	1085	543	434
100	—	1668	1603	1450	725	580
125	—	2087	2007	1815	908	726
150	—	2496	2400	2170	1085	868
200	—	3335	3207	2900	1450	1160
250	—	—	—	—	1825	1460
300	—	—	—	—	2200	1760
350	—	—	—	—	2550	2040
400	—	—	—	—	2900	2320
450	—	—	—	—	3250	2600
500	—	—	—	—	3625	2900

*Design A motors are not limited to a maximum starting current or locked rotor current.

Figure 37–33 Tables 430.251(A) & (B) are used to determine locked-rotor current for motors that do contain NEMA code letters.

short-circuit protective devices; non–time delay fuses, dual-element time delay fuses, instantaneous trip circuit breakers, and inverse time circuit breakers. Although it is permissible to used non–time delay fuses and instantaneous trip circuit breakers, most motor circuits are protected by dual-element time delay fuses or inverse time circuit breakers.

Each of these columns lists the percentage of motor current that is to be used in determining the ampere rating of the short-circuit protective device. The current listed in the appropriate motor table is to be used instead of the nameplate current. *NEC® Section 430.52(C)(1)* states that the protective device is to have a rating or setting not exceeding the value calculated in accord with *Table 430.52. Exception No. 1* of this section, however, states that if the calculated value does not correspond to a standard size or rating of a fuse or circuit breaker then it shall be permissible to use the next higher standard size. The standard sizes of fuses and circuit breakers are listed in *NEC® Section 240.6* Figure 37–35.

Starting in 1996, *Table 430.52* lists squirrel cage motor types by NEMA design letters instead of code letters. *Section 430.7(A)(9)* requires that motor nameplates be marked with design letters B, C, D, or E. Motors manufactured before this requirement, however, do not list design letters on the nameplate. Most common squirrel cage motors used in industry actually fall in the Design B classification and for purposes of selecting the short-circuit protective device are considered to

Table 430.52 Maximum Rating or Setting of Motor Branch-Circuit Short-Circuit and Ground-Fault Protective Devices

	Percentage of Full-Load Current			
Type of Motor	**Nontime Delay Fuse**	**Dual Element (Time-Delay) Fuse**	**Instantaneous Trip Breaker**	**Inverse Time Breaker**
Single-phase motors	300	175	800	250
AC ployphase motors other than wound-rotor	300	175	800	250
Squirrel cage other than Design B energy-efficient	300	175	800	250
Design B energy-efficient	300	175	1100	250
Synchronous	300	175	800	250
Wound-rotor	150	150	800	150
DC (constant voltage)	150	150	250	150

Note: For certain exceptions to the values specified, see 430.54.
The values in the nontime Delay Fuse column apply to time delay Class CC fuses.
The values given in the last column also cover the ratings of nonadjustable inverse time types of circuit breakers that may be modified as in 430.42(C)(1). Exceptions No. 1 and No. 2.
Synchronous motors of the low-torque, low-speed type (usually 450 RPM or lower) such as used to drive reciprocating compressors, pumps, and so fourth that start unloaded do not require a fuse rating or circuit breaker setting in excess of 200 percent of full-load current.

Figure 37–34 Table 430.52 is used to determine the size of the short-circuit protective device for a motor.

240.6 Standard Ampere Ratings

(A) Fuses and Fixed-Trip Circuit Breakers. The standard ampere ratings for fuses and inverse time circuit breakers shall be considered 15, 20, 25, 30, 35, 40, 45, 50, 60, 70, 80, 90, 100, 110, 125, 150, 175, 200, 225, 250, 300, 350, 400, 450, 500, 600, 700, 800, 1000, 1200, 1600, 2000, 2500, 3000, 4000, 5000, and 6000 amperes. Additionally standard ampere ratings for fuses shall be 1, 3, 6, 10, and 601. The use of fuses and inverse time circuit breakers with nonstandard ampere ratings shall be permitted.

Figure 37–35 Section 240.6 lists standard fuse and circuit breaker sizes.

be Design B unless otherwise listed. Design B motors exhibit a little higher efficiency than other types of motors, but they also have a higher starting current.

Starter Size

Another factor that must be considered when installing a motor is the size starter used to connect the motor to the line. Starter sizes are rated by motor type, horsepower, and connected voltage. The two most common ratings are NEMA and IEC. A chart showing common NEMA size starters for alternating current motors is shown in Figure 37–36. A chart showing IEC starters for alternating current motors is shown in Figure 37–37. Each of these

Motor Starter Sizes and Ratings

NEMA Size	Load Volts	Maximum Horsepower Rating–Nonplugging and Nonjogging Duty		NEMA Size	Load Volts	Maximum Horsepower Rating–Nonplugging and Nonjogging Duty	
		Single Phase	Poly Phase			Single Phase	Poly Phase
00	115	½	. . .	3	115	7½	. . .
	200	. . .	1½		200	. . .	25
	230	1	1½		230	15	30
	380	. . .	1½		380	. . .	50
	460	. . .	2		460	. . .	50
	575	. . .	2		575	. . .	50
0	115	1	. . .	4	200	. . .	40
	200	. . .	3		230	. . .	50
	230	2	3		380	. . .	75
	380	. . .	5		460	. . .	100
	460	. . .	5		575	. . .	100
	575	. . .	5				
1	115	2	. . .	5	200	. . .	75
	200	. . .	7½		230	. . .	100
	230	3	7½		380	. . .	150
	380	. . .	10		460	. . .	200
	460	. . .	10		575	. . .	200
	575	. . .	10				
*1P				6	200	. . .	150
	115	3	. . .		230	. . .	200
	230	5	. . .		380	. . .	300
					460	. . .	400
					575	. . .	400
2	115	3	. . .	7	230	. . .	300
	200	. . .	10		460	. . .	600
	230	7½	15		575	. . .	600
	380	. . .	25	8	230	. . .	450
	460	. . .	25		460	. . .	900
	575	. . .	25		575	. . .	900

Tables are taken from NEMA Standards.
*1¾, 10 HP is available.

Figure 37–36 NEMA table of standard starter sizes.

I.E.C. MOTOR STARTERS (60 HZ)

SIZE	MAX AMPS	MOTOR VOLTAGE	MAX. HORSEPOWER 1 Ø	MAX. HORSEPOWER 3 Ø
A	7	115	1/4	
		200		1 1/2
		230	1/2	1 1/2
		460		3
		575		5
B	10	115	1/2	
		200		2
		230	1	2
		460		5
		575		7 1/2
C	12	115	1/2	
		200		3
		230	2	3
		460		7 1/2
		575		10
D	18	115	1	
		200		5
		230	3	5
		460		10
		575		15
E	25	115	2	
		200		5
		230	3	7 1/2
		460		15
		575		20
F	32	115	2	
		200		7 1/2
		230	5	10
		460		20
		575		25
G	37	115	3	
		200		7 1/2
		230	5	10
		460		25
		575		30
H	44	115	3	
		200		10
		230	7 1/2	15
		460		30
		575		40
J	60	115	5	
		200		15
		230	10	20
		460		40
		575		40
K	73	115	5	
		200		20
		230	10	25
		460		50
		575		50
L	85	115	7 1/2	
		200		25
		230	10	30
		460		60
		575		75

SIZE	MAX AMPS	MOTOR VOLTAGE	MAX. HORSEPOWER 1 Ø	MAX. HORSEPOWER 3 Ø
M	105	115	10	
		200		30
		230	10	40
		460		75
		575		100
N	140	115	10	
		200		40
		230	10	50
		460		100
		575		125
P	170	115		
		200		50
		230		60
		460		125
		575		125
R	200	115		
		200		60
		230		75
		460		150
		575		150
S	300	115		
		200		75
		230		100
		460		200
		575		200
T	420	115		
		200		125
		230		125
		460		250
		575		250
U	520	115		
		200		150
		230		150
		460		350
		575		250
V	550	115		
		200		150
		230		200
		460		400
		575		400
W	700	115		
		200		200
		230		250
		460		500
		575		500
X	810	115		
		200		250
		230		300
		460		600
		575		600
Z	1215	115		
		200		450
		230		450
		460		900
		575		900

Figure 37–37 I.E.C. motor starters rated by size, horsepower, and voltage for 60 Hz circuits.

charts lists the minimum size starter designed to connect the listed motors to the line. It is common to employ larger size starters than those listed. This is especially true when using IEC type starters because of their smaller load contact size.

Example Problems

In the following examples the conductor size, over current protection size, starter size, and overload heater size will be determined for different motors. When installing motors, a manufacturer's chart is used to select the proper overload heater in accord with the requirements of the circuit. A typical chart for selecting overload heaters is shown in Figure 37–38. These charts can differ from one manufacturer to another. The chart shown in Figure 37–38 differs from the chart shown in Figure 4–7 in several ways. The chart shown in Figure 4–7 lists heater sizes for starter sizes ranging from 00 to 1. The chart shown in Figure 37–38 lists starter sizes from 0 through 4. The heater sizes listed in the chart shown in Figure 4–7 list a range of full load currents for that heater. Heater XX26, for example, lists a range of full load currents from 2.84 amperes to 3.11 amperes. The heater sizes listed in the chart shown in Figure 37–38 lists a single current value instead of a range of currents. When using a chart of this type, the heater selected should never have a rating less than the full load current of the motor. Heater X55 lists a current of 12.8 amperes in the size 1 starter column. Heater X56 lists a current of 14.1 amperes in the same column. If a motor has a full load current of 13 amperes, heater size X56 would be selected because heater X55 has a current rating less than the full load current of the motor.

Example 1

A 40 HP, 240-volt DC motor has a nameplate current rating of 132 amperes. The conductors are to be copper with type TW insulation. The short-circuit protective device is to be an instantaneous trip circuit breaker. The termination temperature rating of the connected devices is not known. Determine the conductor size, overload heater size, and circuit breaker size for this installation. Refer to Figure 37–39.

The conductor size must be determined from the current listed in *Table 430.247.* This value is to be increased by 25 percent. Note: multiplying by 1.25 has the same effect as multiplying by 0.25 and then adding the product back to the original number (140 × 0.25 = 35) (35 + 140 = 175 amperes)

$$140 \times 1.25 = 175 \text{ amperes}$$

Overload heater selection for NEMA starter sizes 0 (00) -4. Heaters are calibrated for 115% of motor full load current. For heaters that correspond to 125% of motor full load current use the next size larger heater.

Heater type number	Size 0	Size 1	Size 2	Size 3	Size 4
X10	0.18	0.18	....	....	....
X11	0.20	0.20	....	....	....
X12	0.22	0.22	....	....	....
X13	0.24	0.24	....	....	....
X14	0.26	0.26	....	....	....
X15	0.29	0.29	....	....	....
X16	0.32	0.32	....	....	....
X17	0.35	0.35	....	....	....
X18	0.38	0.38	....	....	....
X19	0.42	0.42	....	....	....
X20	0.46	0.46	....	....	....
X21	0.51	0.51	....	....	....
X22	0.56	0.56	....	....	....
X23	0.62	0.62	....	....	....
X24	0.68	0.68	....	....	....
X25	0.75	0.75	....	....	....
X26	0.82	0.82	....	....	....
X27	0.90	0.90	....	....	....
X28	0.99	0.99	....	....	....
X29	1.09	1.09	....	....	....
X30	1.20	1.20	....	....	....
X31	1.32	1.32	....	....	....
X32	1.45	1.45	....	....	....
X33	1.59	1.59	....	....	....
X34	1.75	1.75	....	....	....
X35	1.93	1.93	....	....	....
X36	2.12	2.12	....	....	....
X37	2.33	2.33	....	....	....
X38	2.56	2.56	....	....	....
X39	2.81	2.81	....	....	....
X40	3.09	3.09	....	....	....
X41	3.40	3.40	....	....	....
X42	3.74	3.74	....	....	....
X43	4.11	4.11	....	....	....
X44	4.52	4.52	....	....	....
X45	4.97	4.97	....	....	....
X46	5.46	5.46	5.60	....	....
X47	6.01	6.01	6.15	....	....
X48	6.60	6.60	6.76	....	....
X49	7.26	7.26	7.43	....	....
X50	7.98	7.98	8.17	....	....
X51	8.78	8.78	8.98	....	....
X52	9.65	9.65	9.87	....	....
X53	10.6	10.6	10.8	....	....
X54	11.7	11.7	11.9	....	....
X55	12.8	12.8	13.1	....	....
X56	14.1	14.1	14.4	....	....
X57	15.4	15.4	15.7	....	....
X58	16.8	16.8	17.1	....	....
X59	18.3	18.3	18.6	....	....
X60	....	19.8	20.1	....	....
X61	....	21.3	21.7	25.5	....
X62	....	22.7	23.1	28.1	....
X63	....	24.4	24.8	31.0	32.0
X64	....	26.2	28.6	34.0	35.0
X65	....	28.2	30.5	37.0	38.5
X66	....	....	33.0	40.0	42.5
X67	....	....	35.5	43.5	46.5
X68	....	....	38.0	47.0	51.0
X69	....	....	40.5	51.0	55.0
X70	....	....	43.5	55.0	59.0
X71	....	....	47.0	59.0	64.0
X72	....	....	....	63.0	69.0
X73	....	....	....	67.0	74.0
X74	....	....	....	71.0	79.0
X75	....	....	....	76.0	84.0
X76	....	....	....	80.0	90.0
X77	....	....	....	85.0	96.0
X78	....	....	....	90.0	102
X79	....	....	....	....	107
X80	....	....	....	....	113
X81	....	....	....	....	118
X82	....	....	....	....	124
X83	....	....	....	....	130
X84	....	....	....	....	135

Figure 37–38 Overload Heater Chart.

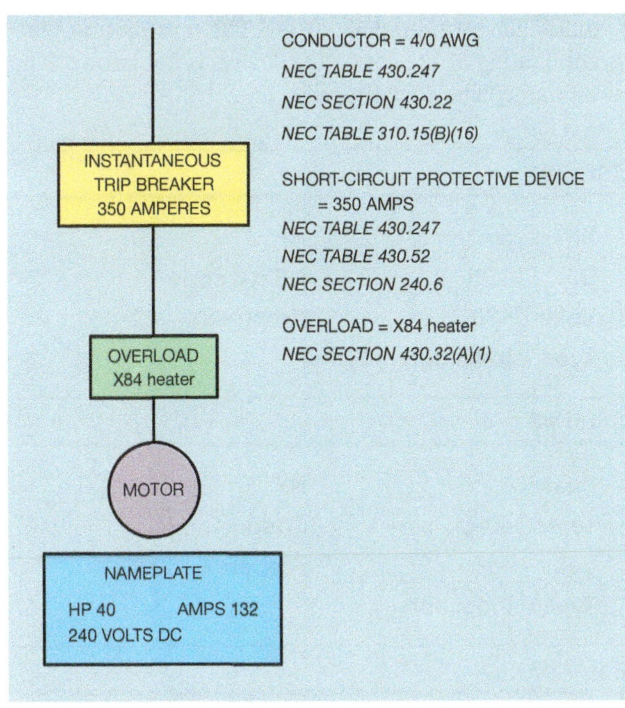

Figure 37–39 Example problem #1.

Table 310.15(B)(16) is used to find the conductor size. Although section 110.14(C) states that for currents of 100 amperes or greater that the ampacity rating of the conductor is to be determined from the 75°C column, in this instance, the insulation type is located in the 60°C column. Therefore, the conductor size must be determined using the 60°C column instead of the 75°C column. A 4/0 AWG copper conductor with type TW insulation will be used.

The overload heater size is determined from *NEC® Section 430.32(A)(1)*. Because there is no service factor or temperature rise listed on the motor nameplate, the heading *ALL OTHER MOTORS* will be used. The heater size will be selected on the basis of 115 percent of motor full load current. A size X84 heater is selected from the chart shown in Figure 37–38.

EXAMPLE: A 40 HP, three phase squirrel cage motor is connected to a 208-volt line. What are the minimum size NEMA and IEC starters that should be used to connect this motor to the line?

NEMA: The 200-volt listing is used for motors rated at 208 volts. Locate the NEMA size starter that corresponds to 200 volts and 40 horsepower. Because the motor is three phase, 40 HP will be in the poly-phase column. A NEMA size 4 starter is the minimum size for this motor.

IEC: As with the NEMA chart, the IEC chart lists 200 volts instead of 208 volts. A size N starter lists 200 volts and 40 HP in the three phase column.

The circuit breaker size is determined from *Table 430.52*. The current value listed in *Table 430.247* is used instead of the nameplate current. Under DC motors (constant voltage), the instantaneous trip circuit breaker rating is given at 250 percent.

$$140 \times 2.50 = 350 \text{ amperes}$$

Because 350 amperes is one of the standard sizes of circuit breakers listed in *NEC® Section 240.6*, that size breaker will be employed as the short-circuit protective device.

Example 2

A 150 HP, three phase squirrel cage induction motor is connected to a 440-volt line. The motor nameplate lists the following information:

Amps 175   SF 1.25   Code D   NEMA code B

The conductors are to be copper with type THHN insulation. The short-circuit protective device is to be an inverse time circuit breaker. The termination temperature rating is not known. Determine the conductor size, overload heater size, circuit breaker size, and minimum NEMA and IEC starter sizes. Refer to Figure 37–40.

The conductor size is determined from the current listed in *Table 430.250* and increased by 25 percent.

$$180 \times 1.25 = 225 \text{ amperes}$$

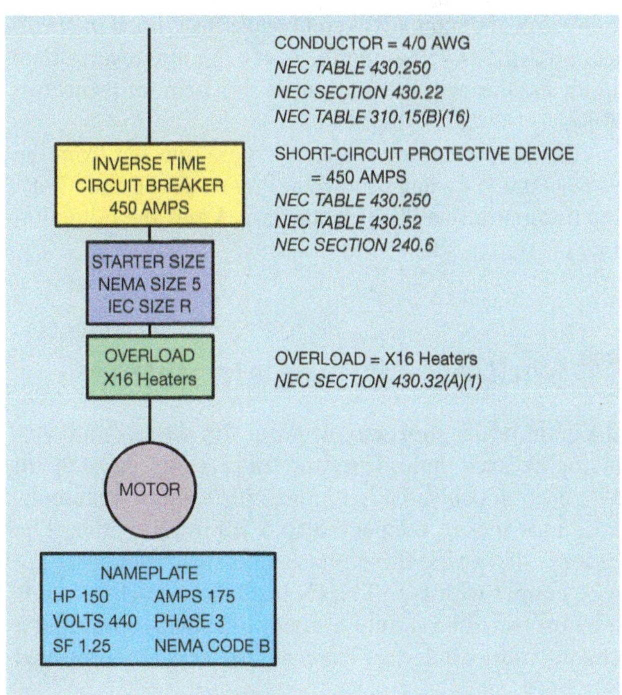

Figure 37–40 Example circuit #2.

Table 310.15(B)(16) is used to determine the conductor size. Type THHN insulation is located in the 90°C column. Because the motor nameplate lists NEMA code B, and the amperage is over 100 amperes, the conductor will be selected from the 75°C column. The conductor size will be 4/0 AWG.

The overload heater size is determined from the nameplate current and *NEC® Section 430.32(A)(1)*. The motor has a marked service factor of 1.25.

The overload heater size will be selected on the basis of 125 percent of motor full load current. Because the full load current of the motor is so high, current transformers will be employed to reduce the overload current rating. It is assumed that current transformers with a ratio of 300:5 will be used.

$$\frac{300}{5} = \frac{175}{X} \quad 300X = 875 \quad X = \frac{875}{300} \quad X = 2.9$$

The overload heater size will be selected on the basis of a motor with a full load current of 2.9 amperes. The chart in Figure 37–38 indicates that heater size X15 has a rating of 2.9 amperes. However, because the heater size is selected on the basis of 125 percent of the motor full load current, the next higher heater size will be selected. The motor will be protected by heaters X16.

The circuit breaker size is determined by *Tables 430.250 and 430.52. Table 430.52* indicates a factor of 250 percent for squirrel cage motors with NEMA design B. The value listed in *Table 430.250* will be increased by 250 percent.

$$180 \times 2.50 = 450 \text{ amperes}$$

One of the standard circuit breaker sizes listed in *NEC® Section 240.6* is 450 amperes. A 450 ampere inverse time circuit breaker will be used as the short-circuit protective device.

The proper motor starter sizes are selected from the NEMA and IEC charts shown in Figures 37–36 and 37–37. The minimum size NEMA starter is 5 and the minimum size IEC starter is R.

Multiple Motor Calculations

The main feeder short-circuit protective devices and conductor sizes for multiple motor connections are set forth in *NEC® Section 430.62(A)* and *430.24*. In this example, three motors are connected to a common feeder. The feeder is 480 volts, three phase, and the conductors are to be copper with type THHN insulation. Each motor is to be protected with dual-element time delay fuses and a separate overload device. The main feeder is also protected by dual-element time delay fuses. The termination temperature rating of the connected devices is not known. The motor nameplates state the following:

Motor #1

Phase	3	HP	20
SF	1.25	NEMA code	C
Volts	480	Amperes	23
Type	Induction		

Motor #2

Phase	3	HP	60
Temp.	40°C	Code	J
Volts	480	Amperes	72
Type	Induction		

Motor #3

Phase	3	HP	100
Code	A	Volts	480
Amperes	96	PF	90 percent
Type	Synchronous		

Motor #1 Calculation

The first step is to calculate the values for motor amperage, conductor size, overload heater size, short-circuit protection size, and starter size for each motor. Both NEMA and IEC starter sizes will be determined. The values for motor #1 are shown in Figure 37–41.

The ampere rating from *Table 430.250* is used to determine the conductor and fuse size. The amperage rating must be increased by 25 percent for the conductor size.

$$27 \times 1.25 = 33.75 \text{ amps}$$

The conductor size is chosen from *Table 310.15(B)(16)*. Although type THHN insulation is located in the 90°C column, the conductor size will be chosen from the 75°C column. Although the current is less than 100 amperes, *NEC® Section 110.14(C)(1)(d)*, permits the conductors to be chosen from the 75°C column if the motor has a NEMA design code.

$$33.75 \text{ amps} = \#10 \text{ AWG}$$

The fuse size is determined by using the motor current listed in *Table 430.250* and the demand factor from

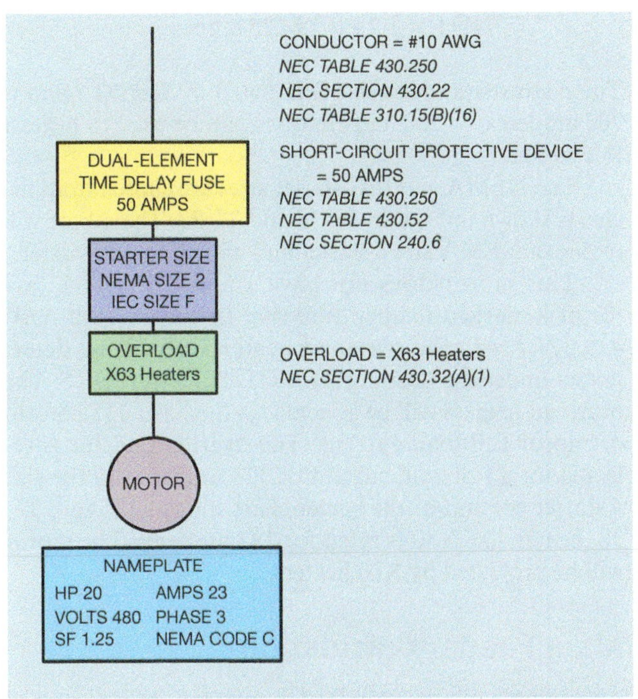

Figure 37–41 Motor #1 calculation.

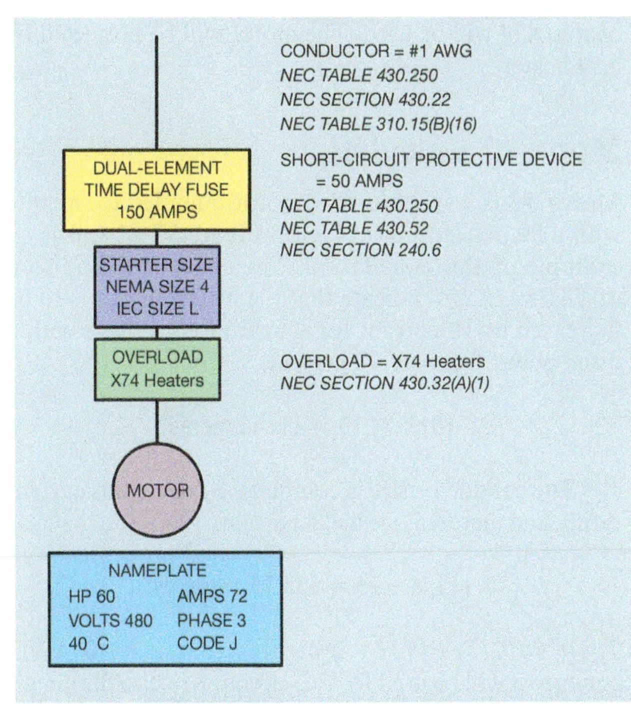

Figure 37–42 Motor #2 calculation.

Table 430.52. The percent of full load current for a dual-element time delay fuse protecting a squirrel cage motor listed as Design C is 175 percent. The current listed in *Table 430.250* will be increased by 175 percent.

$$27 \times 1.75 = 47.25 \text{ amperes}$$

The nearest standard fuse size listed in *Section 240.6* is 50 amperes, so 50 ampere fuses will be used.

The starter sizes are determined from the NEMA and IEC charts shown in Figures 37–36 and 37–37. A 20 HP motor connected to 480 volts would require a NEMA size 2 starter and an IEC size F starter.

The overload size is computed from the nameplate current. The demand factors in *Section 430.32(A)(1)* are used for the overload calculation. The heater is to be sized at 125 percent of the motor full load current. In the column for size 2 starters in the overload heater chart shown in Figure 37–38, overload heater size X62 lists a full load current of 23.1 amperes. Because the overload must be chosen on the basis of 125 percent of motor full load current, heater sizes X63 will be used.

Motor #2 Calculation

Figure 37–42 shows an example for the calculation for motor #2. *Table 430.250* lists a full load current of 77 amperes for this motor. This value of current is increased by 25 percent for the calculation of the conductor current.

$$77 \times 1.25 = 96.25 \text{ amperes}$$

Table 310.15(B)(16) indicates a #1 AWG conductor should be used for this motor connection. The conductor size is chosen from the 60°C column because the circuit current is less than 100 amperes in accord with *Section 110.14(C)*, and the motor nameplate does not indicate a NEMA design code. (The code J indicates the type of bars used in the construction of the rotor.)

The fuse size is determined from *Table 430.52*. The table current is increased by 175 percent for squirrel cage motors other than design B.

$$77 \times 1.75 = 134.25 \text{ amperes}$$

The nearest standard fuse size listed in *Section 240.6* is 150 amperes, so 150 ampere fuses will be used to protect this circuit.

The starter sizes are chosen from the NEMA and IEC starter charts. This motor would require a NEMA size 4 starter or a size L IEC starter.

The overload size is determined from *Section 430.32(A)(1)*. The motor nameplate lists a temperature rise of 40°C for this motor. The overload heater size will selected on the basis of 125 percent of motor full load current. In the size 4 starter column of the overload heater chart shown in Figure 37–38, overload heater X73 is rated for 74 amperes. Because the heater size must be rated at 125 percent of full load current the next size heater X74 will be used. The motor will be protected by X74 heaters.

Motor #3 Calculation

Motor #3 is a synchronous motor intended to operate with a 90 percent power factor. Figure 37–43 shows an example of this calculation. The notes at the bottom of *Table 430.250* indicate that the listed current is to be increased by 10 percent for synchronous motors with a listed power factor of 90 percent.

$$101 \times 1.10 = 111 \text{ amperes}$$

The conductor size is computed by using this current rating and increasing it by 25 percent.

$$111 \times 1.25 = 138.75 \text{ amperes}$$

Table 310.15(B)(16) indicates that a #1/0 AWG conductor will be used for this circuit. Because the circuit current is over 100 amperes, the conductor size is chosen from the 75°C column.

The fuse size is determined from *Table 430.52*. The percent of full load current for a synchronous motor is 175 percent.

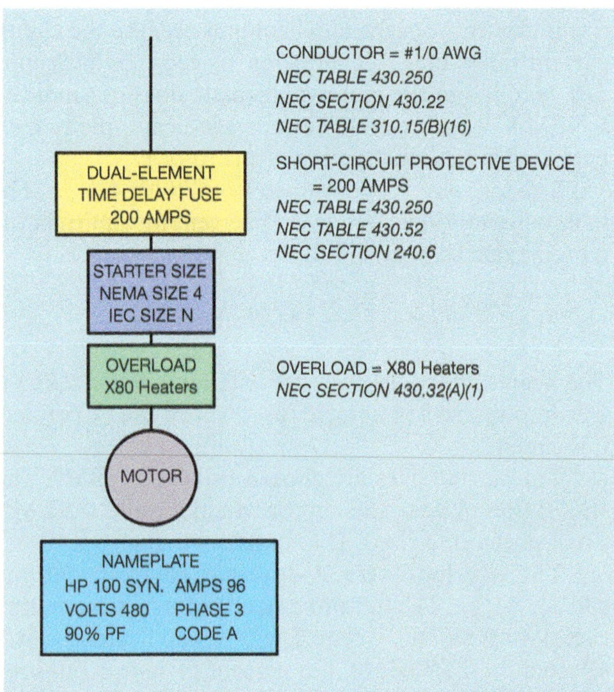

Figure 37–43 Motor #3 calculation.

$$111 \times 1.75 = 194.25 \text{ amperes}$$

The nearest standard size fuse listed in *Section 240.6* is 200 amperes, so 200 ampere fuses will be used to protect this circuit.

The NEMA and IEC starter sizes are chosen from the charts shown in Figures 37–36 and 37–37. The motor will require a NEMA size 4 starter and an IEC size N starter.

This motor does not have a marked service factor or a marked temperature rise. In accord with *NEC 430.32(A)(1)* the overload heater size will be determined under the heading ALL OTHER MOTORS. The overload heaters will be selected on the basis 115 percent of motor full load current. The overload heater is selected for a full load current of 111 amperes. In the size 4 starter column of the heater chart shown in Figure 37–38, heater size X80 is rated for 113 amperes. The motor will be protected by X80 heaters.

Main Feeder Calculation

An example of the main feeder connections is shown in Figure 37–44. The conductor size is computed in accord with *NEC® Section 430.24* by increasing the largest amperage rating of the motors connected to the feeder by 25 percent and then adding the ampere rating of the other motors to this amount. In this example, the 100 HP synchronous motor has the largest running current. This current will be increased by 25 percent and then the running currents of the other motors as determined from *Table 430.250* will be added.

$$111 \times 1.25 = 138.75 \text{ amperes}$$
$$138.75 + 77 + 27 = 242.75 \text{ amperes}$$

Table 310.15(B)(16) lists that 250 kcmil copper conductors are to be used as the main feeder conductors. The conductors were chosen from the 75°C column.

The size of the short-circuit protective device is determined by *Section 430.62(A)*. The code states that the rating or setting of the short-circuit protective device *shall not be greater than* the largest rating or setting of the largest branch circuit short-circuit and ground fault protective device for any motor supplied by the feeder plus the sum of the full load running currents of the other motors connected to the feeder. The largest fuse size in this example is the 100 HP synchronous motor. The fuse calculation for this motor is 200 amperes. The running currents of the other two motors will be added to this value to determine the fuse size for the main feeder.

$$200 + 77 + 27 = 304 \text{ amperes}$$

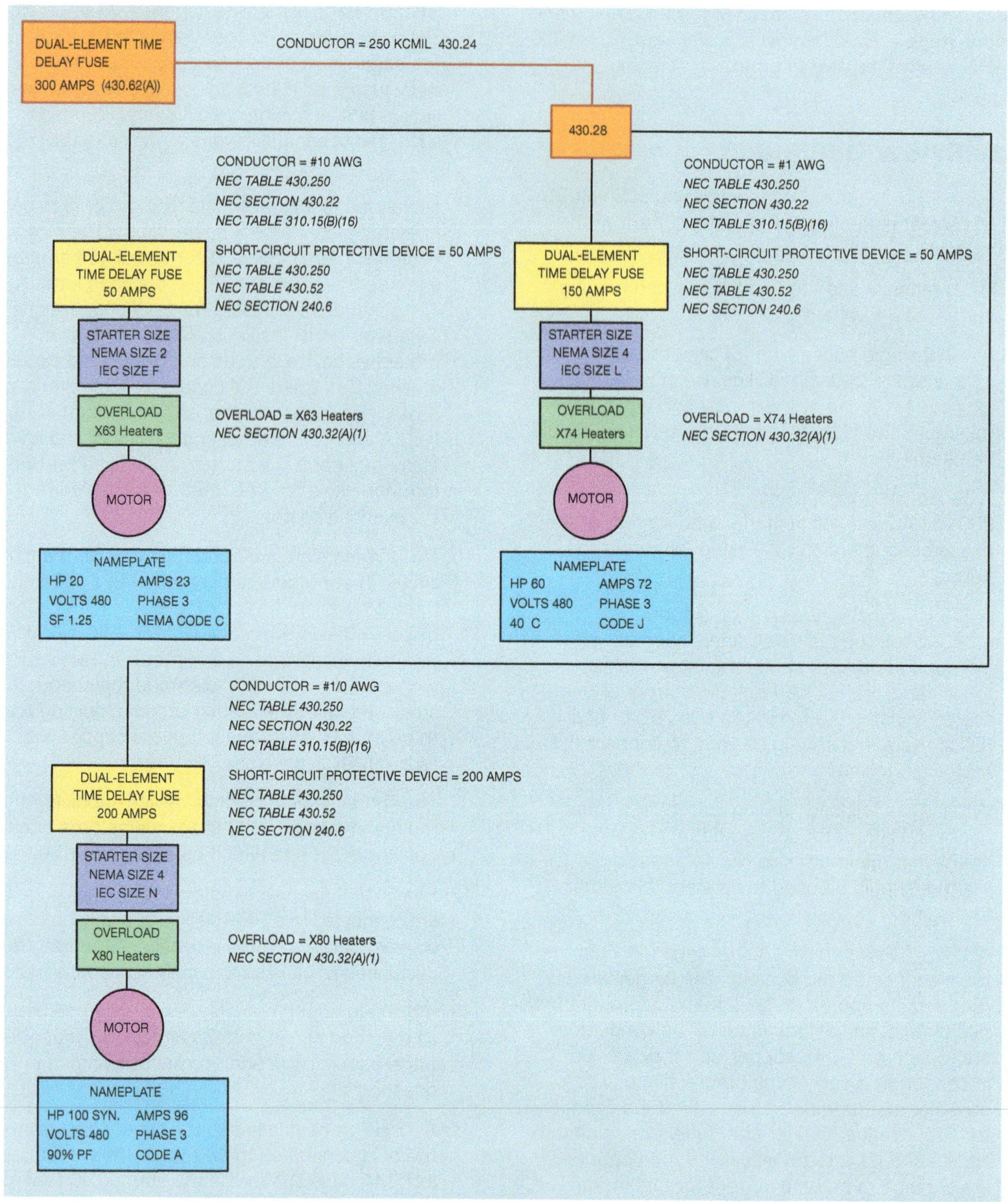

Figure 37–44 Main feeder calculation.

The closest standard fuse size listed in *Section 240.6* without going over 304 amperes is 300 amperes, so 300 ampere fuses will be used to protect this circuit.

Review Questions

1. A 20 HP, DC motor is connected to a 500-volt DC line. What is the full load running current of this motor?
2. What rating is used to find the full load running current of a torque motor?
3. A ¾ HP, single phase squirrel cage motor is connected to 240-volt AC line. What is the full load current rating of this motor and what is the minimum size NEMA and IEC starters that should be used?
4. A 30 HP, two phase motor is connected to a 230-volt AC line. What is the rated current of the phase conductors and the rated current of the neutral?
5. A 125 HP, synchronous motor is connected to a 230-volt, three phase AC line. The motor is intended to operate at 80 percent power factor. What is the full load running current of this motor? What is the minimum size NEMA and IEC starters that should be used to connect this motor to the line?
6. What is the full load running current of a three phase, 50 HP motor connected to a 560-volt line? What minimum size NEMA and IEC starters should be used to connect this motor to the line?
7. A 125 HP, three phase squirrel cage induction motor is connected to 560 volts. The nameplate current is 115 amperes. It has a marked temperature rise of 40°C and a code letter J. The conductors are to be type THHN copper, and they are run in conduit. The short-circuit protective device is dual-element time delay fuses. Find the conductor size, overload heater size, fuse size, minimum NEMA and IEC starter sizes, and the upper and lower range of starting current for this motor. The motor starter employs the use of current transformers with a ratio of 200:5 to reduce the current to the overload heaters.
8. A 7.5 HP, single phase squirrel cage induction motor is connected to 120 VAC. The motor has a code letter of H. The nameplate current is 76 amperes. The conductors are copper with type TW insulation. The short-circuit protection device is a non–time delay fuse. Find the conductor size, overload heater size, fuse size, minimum NEMA and IEC starter sizes, and upper and lower starting currents.
9. A 75 HP, three phase, synchronous motor is connected to a 230-volt line. The motor is to be operated at 80 percent power factor. The motor nameplate lists a full load current of 185 amperes, a temperature rise of 40°C, and a code letter A. The conductors are to be made of copper and have type THHN insulation. The short-circuit protective device is to be an inverse time circuit breaker. Determine the conductor size, overload heater size, circuit breaker size, minimum size NEMA and IEC starters, and the upper and lower starting current. The starter contains current transformers with a ratio of 300:5 to reduce current to the overload heaters.
10. Three motors are connected to a single branch circuit. The motors are connected to a 480-volt, three phase line. Motor #1 is a 50 HP induction motor with a NEMA code B. Motor #2 is 40 HP with a code letter of H, and motor #3 is 50 HP with a NEMA code C. Determine the conductor size needed for the branch circuit supplying these three motors. The conductors are copper with type THWN-2 insulation.
11. The short-circuit protective device supplying the motors in question #10 is an inverse time circuit breaker. What size circuit breaker should be used?
12. Five 5 HP, three phase motors with NEMA code B are connected to a 240-volt line. The conductors are copper with type THWN insulation. What size conductor should be used to supply all of these motors?
13. If dual-element time delay fuses are to be used as the short circuit protective device, what size fuses should be used to protect the circuit in question #12?
14. A 75 HP, three phase squirrel cage induction motor is connected to 480 volts. The motor has a NEMA code D. What is the starting current for this motor?
15. A 20 HP, three phase squirrel cage induction motor has a NEMA code E. The motor is connected to 208 volts. What is the starting current for this motor?

Section 8

PROGRAMMABLE LOGIC CONTROLLERS

Chapter 38

PROGRAMMABLE LOGIC CONTROLLERS

Objectives

After studying this chapter the student will be able to:

» List the principle parts of a programmable logic controller.

» Describe differences between programmable logic controllers and other types of computers.

» Discuss differences between the I/O rack, CPU, and program loader.

» Draw a diagram of how the input and output modules work.

Programmable logic controllers (PLCs) were first used by the automotive industry in the late 1960s. Each time a change was made in the design of an automobile, it was necessary to change the control system operating the machinery. This change consisted of physically rewiring the control system to make it perform the new operation. Rewiring the system was, of course, very time consuming and expensive. What the industry needed was a control system that could be changed without the extensive rewiring required to change relay control systems.

Differences between PLCs and PCs

One of the first questions generally asked is "Is a programmable logic controller a computer?" The answer to that question is yes. The PLC is a special type of computer designed to perform a special function. Although the programmable logic controller (PLC) and the personal computer (PC) are both computers, there are some significant differences. Both generally employ the same basic type of computer and memory chips to perform the tasks for which they are intended, but the PLC must operate in an industrial environment. Any computer that is intended for industrial use must be able to withstand extremes of temperature; ignore voltage spikes and drops on the power line; survive in an atmosphere that often contains corrosive vapors, oil, and dirt; and withstand shock and vibration.

Programmable logic controllers are designed to be programmed with schematic or ladder diagrams instead of common computer languages. An electrician that is familiar with ladder logic diagrams can generally learn to program a PLC in a few hours as opposed to the time required to train a person how to write programs for a standard computer.

Basic Components

Programmable logic controllers can be divided into four primary parts:

1. The power supply
2. The central processing unit (CPU)
3. The programming terminal or program loader
4. The I/O (pronounced eye-oh) rack

The Power Supply

The function of the power supply is to lower the incoming AC voltage to the desired level, rectify it to direct current, and then filter and regulate it. The internal logic of a PLC generally operates on 5 to 24 volts DC depending on the type of controller. This voltage must be free of voltage spikes and other electrical noise and be regulated to within 5 percent of the required voltage value. Some manufacturers of PLCs build a separate power supply and others build the power supply into the central processing unit.

Courtesy of Rockwell Automation, Inc.

Figure 38–1 A central processing unit.

The CPU

The CPU, or central processing unit, is the "brains" of the programmable logic controller. It contains the microprocessor chip and related integrated circuits to perform all the logic functions. The microprocessor chip used in most PLCs is the same as that found in most home and business personal computers.

The central processing unit often has a key located on the front panel (Figure 38–1). This switch must be turned on before the CPU can be programmed. Turning on the key prevents the circuit from being changed or deleted accidentally. Other manufacturers use a *software switch* to protect the circuit. A software switch is not a physical switch. It is a command that must be entered before the program can be changed or deleted. Whether a physical switch or a software switch is used, they both perform the same function. They prevent a program from being accidentally changed or deleted.

Plug connections on the central processing unit provide connection for the programming terminal and I/O racks (Figure 38–2). CPUs are designed so that once a program has been developed and tested, it can be stored on some type of medium such as tape, disc, CD, or other storage device. In this way, if a central processor unit fails and has to be replaced, the program can be downloaded from the storage medium. This eliminates the time-consuming process of having to reprogram the unit by hand.

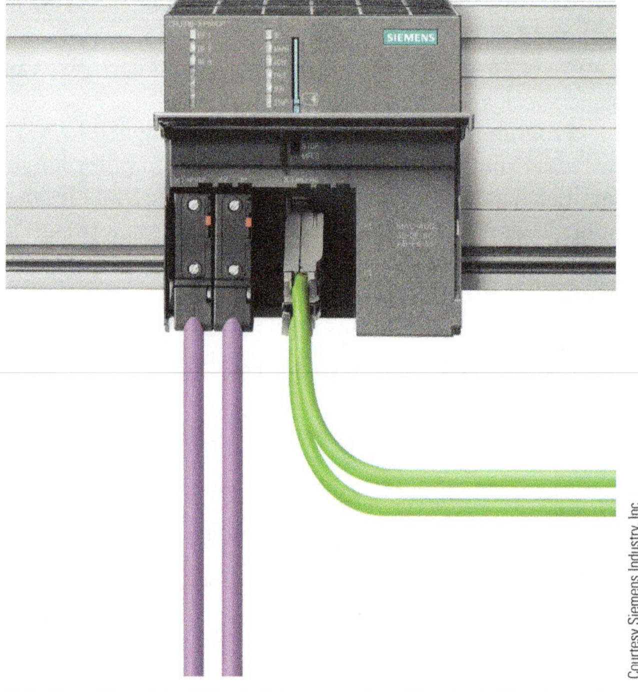

Courtesy Siemens Industry, Inc.

Figure 38–2 Plug connections located on the CPU.

The Programming Terminal

The programming terminal, or loading terminal, is used to program the CPU. The type of terminal used depends on the manufacturer and often the preference of the consumer. Some are small handheld devices that use a liquid crystal display or light emitting diodes to show the program (Figure 38–3). Some of these small units display one line of the program at a time and others require the program to be entered in a language called Boolean.

Another type of programming terminal contains a display and keyboard (Figure 38–4). This type of

Courtesy of Eaton Corporation

Figure 38–3 Handheld programming terminal and small programmable logic controller.

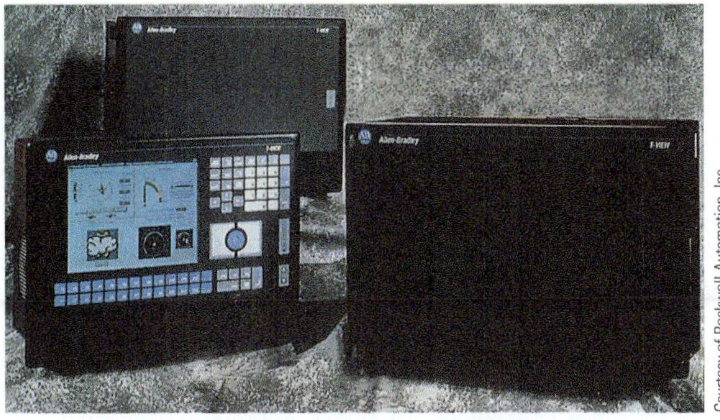
Courtesy of Rockwell Automation, Inc.

Figure 38–4 Programming terminal.

terminal generally displays several lines of the program at a time and can be used to observe the operation of the circuit as it is operating.

Many industries prefer to use a notebook or laptop computer for programming (Figure 38–5). An interface that permits the computer to be connected to the input of the PLC and software program is generally available from the manufacturer of the programmable logic controller.

The terminal is not only used to program the PLC, but is also used to troubleshoot the circuit. When the terminal is connected to the CPU, the circuit can be examined while it is in operation. Figure 38–6 illustrates a circuit typical of those seen on the display. Notice that this schematic diagram is different from the typical ladder diagram. All of the line components are shown as normally open or normally closed contacts. There are no NEMA symbols for push button, float switch, or limit switches. The programmable logic controller recognizes

Figure 38–5 A notebook computer is often used as the programming terminal for a PLC.

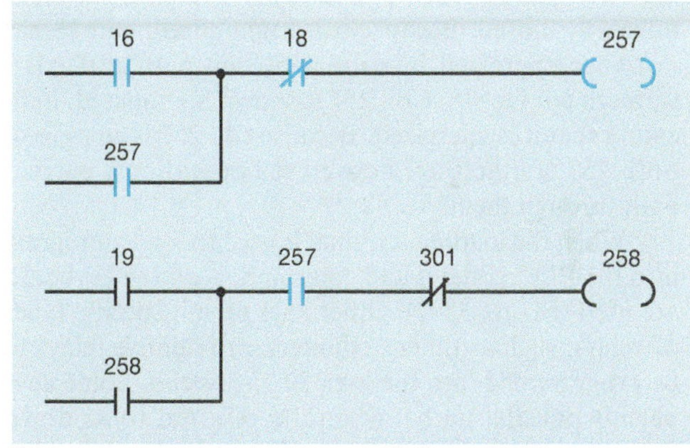

Figure 38–6 Analyzing circuit operation with a terminal.

only open or closed contacts. It does not know if a contact is connected to a push button, a limit switch, or a float switch. Each contact, however, does have a number. The number is used to distinguish one contact from another.

In this example, coil symbols look like a set of parenthesis instead of a circle as shown on most ladder diagrams. Each line ends with a coil, and each coil has a number. When a contact symbol has the same number as a coil, it means that the contact is controlled by that coil. The schematic in Figure 38–6 shows a coil numbered 257

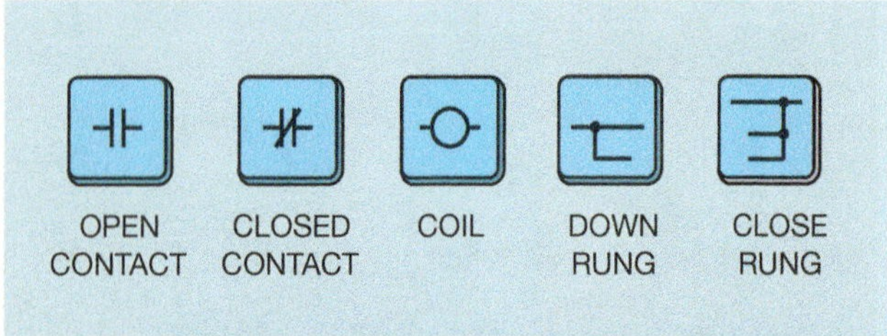

Figure 38–7 Symbols are used to program the PLC.

and two contacts numbered 257. When coil 257 is energized, the programmable logic controller interprets both contacts 257 to be closed.

A characteristic of interpreting a diagram when viewing it on the screen of most loading terminals is that when a current path exists through a contact or if a coil is energized, they will be highlighted on the display. In the example shown in Figure 38–6, coil 257, both 257 contacts, contact 16, and contact 18 are drawn with blue lines, illustrating that they are highlighted or illuminated on the display. Highlighting a contact does not mean that it has changed from its original state. It means that there is a complete circuit through that contact. Contact 16 is highlighted, indicating that coil 16 has energized and contact 16 is closed and providing a complete circuit. Contact 18, however, is shown as normally closed. Because it is highlighted, coil 18 has not been energized because a current path still exists through contact 18. Coil 257 is shown highlighted, indicating that it is energized. Because coil 257 is energized, both 257 contacts are now closed providing a current path through them.

When the loading terminal is used to load a program into the PLC, contact and coil symbols on the keyboard are used (Figure 38–7). Other keys permit specific types of relays, such as timers, counters, or retentive relays to be programmed into the logic of the circuit. Some keys permit parallel paths, generally referred to as down rungs, to be started and ended. The method employed to program a PLC is specific to the make and model of the controller. It is generally necessary to consult the manufacturer's literature if you are not familiar with the specific programmable logic controller.

The I/O Rack

The I/O rack is used to connect the CPU to the outside world. It contains input modules that carry information from control sensor devices to the CPU and output modules that carry instructions from the CPU to output devices in the field. I/O racks are shown in Figures 38–8A

Courtesy Schneider Electric USA, Inc.

Figure 38–8A I/O rack with input and output modules.

Courtesy of Rockwell Automation, Inc.

Figure 38–8B I/O rack with input and output modules.

and B. Input and output modules contain more than one input or output. Any number from 4 to 32 is common depending on the manufacturer and model of PLC. The modules shown in Figure 38–8A can each handle 16 connections. This means that each input module can handle 16 different input devices, such as push button, limit switches, proximity switches, or float switches. The output modules can each handle 16 external devices such as pilot lights, solenoid coils, or relay coils. The operating voltage can be either alternating or direct current depending on the manufacturer and model of controller, and is generally either 120 or 24 volts. The I/O rack shown in Figure 38–8A can handle 10 modules. Because each module can handle 16 input or output devices, the I/O rack is capable of handling 160 input and output devices. Many programmable logic controllers are capable of handling multiple I/O racks.

I/O Capacity

One factor that determines the size and cost of a programmable logic controller is its I/O capacity. Many small units may be intended to handle as few as 16 input and output devices. Large PLCs can generally handle several hundred. The number of input and output devices the controller must handle also affects the processor speed and amount of memory the CPU must have. A central processing unit with I/O racks is shown in Figure 38–9.

Figure 38–9 Central processor with I/O racks.

The Input Module

The central processing unit of a programmable logic controller is extremely sensitive to voltage spikes and electrical noise. For this reason, the input I/O uses opto-isolation to electrically separate the incoming signal from the CPU. Figure 38–10 shows a typical circuit used for the input. A metal oxide varistor (MOV) is connected across the AC input to help eliminate any voltage spikes that may occur on the line. The MOV is a voltage sensitive resistor. As long as the voltage across its terminals remains below a certain level, it exhibits a very high resistance. If the voltage should become too high, the resistance almost instantly changes to a very low value. A bridge rectifier changes the AC voltage into DC. A resistor is used to limit current to a light emitting diode. When power is applied to the circuit, the LED turns on. The light is detected by a phototransistor that signals the CPU that there is a voltage present at the input terminal.

When the module has more than one input, the bridge rectifiers are connected together on one side to form a common terminal. On the other side, the rectifiers are labeled 1, 2, 3, and 4. Figure 38–11 shows four bridge rectifiers connected together to form a common terminal. Figure 38–12 shows a limit switch connected to input 1, a temperature switch connected to input 2, a float switch connected to input 3, and a normally open push button connected to input 4. Notice that the pilot devices complete a circuit to the bridge rectifiers. In any switch closes, 120 volts AC will be connected to a bridge rectifier, causing the corresponding light emitting diode to turn on and signal the CPU that the input has voltage applied to it. When voltage is applied to an input, the CPU considers that input to be at a high level.

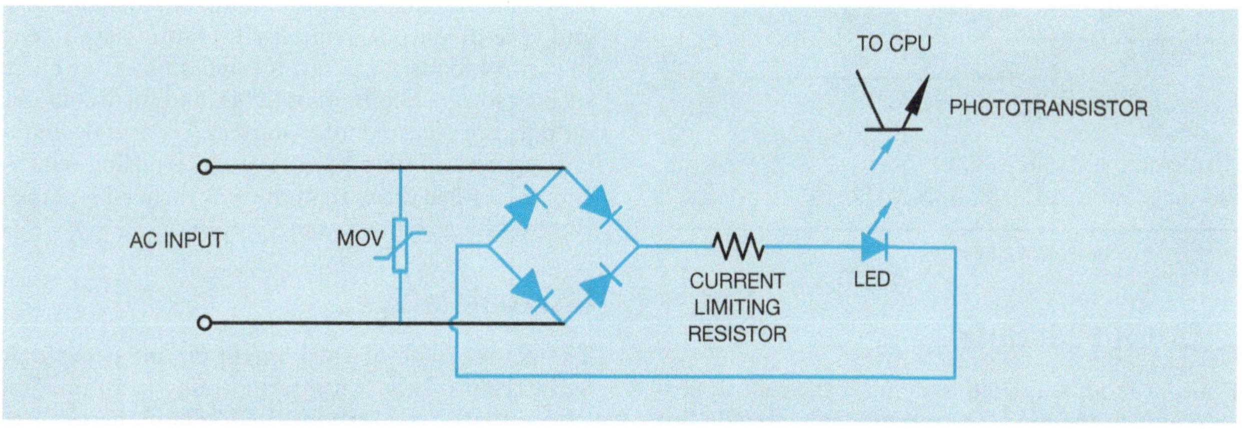

Figure 38–10 Input circuit.

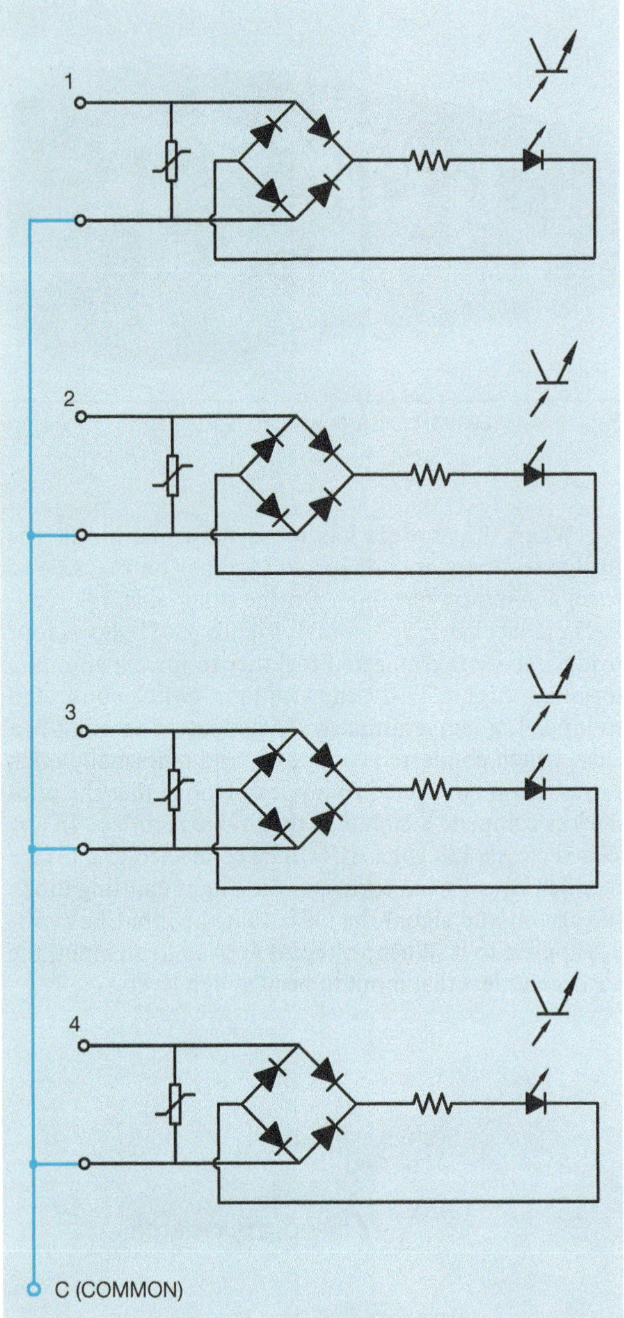

Figure 38–11 Four-input module.

The Output Module

The output module is used to connect the central processing unit to the load. Output modules provide line isolation between the CPU and the external circuit. Isolation is generally provided in one of two ways. The most popular is with optical isolation very similar to the input modules. In this case, the CPU controls a light emitting diode. The LED is used to signal a solid-state device to connect the load to the line. If the load is operated by direct current, a power phototransistor is used to connect the load to the line (Figure 38–13). If the load is an alternating current device, a triac is used to connect the load to the line (Figure 38–14). Notice that the CPU is separated from the external circuit by a light beam. No voltage spikes or electrical noise can be transmitted to the CPU.

The second method of controlling the output is with small relays (Figure 38–15). The CPU controls the relay coil. The contacts connect the load to the line. The advantage of this type of output module is that it is not sensitive to whether the voltage is AC or DC and can control 120 or 24 volt circuits. The disadvantage is that it does contain moving parts that can wear. In this instance, the CPU is isolated from the external circuit by a magnetic field instead of a light beam.

If the module contains more than one output, one terminal of each output device is connected together to form a common terminal similar to a module with multiple inputs (Figure 38–16). Notice that one side of each triac has been connected together to form a common point. The other side of each triac is labeled 1, 2, 3, or 4. If power transistors are used as output devices, the collectors or emitters of each transistor would be connected to form a common terminal. Figure 38–14 shows a relay coil connected to the output of a triac. Notice that the triac is used as a switch to connect the load to the line. The power to operate the load must be provided by an external source. *Output modules do not provide power to operate external loads.*

The amount of current an output can control is limited. The current rating of most outputs can range from 0.5 to about 3 amperes depending on the manufacturer and type of output being used. Outputs are intended to control loads that draw a small amount of current, such as solenoid coils, pilot lights, and relay coils. Some outputs can control motor starter coils directly and others require an interposing relay. Interposing relays are employed when the current draw of the load is above the current rating of the output.

Internal Relays

The actual logic of the control circuit is performed by **internal relays**. An internal relay is an imaginary device that exists only in the logic of the computer. It can have any number of contacts from one to several hundred, and the contacts can be programmed normally open or normally closed. Internal relays

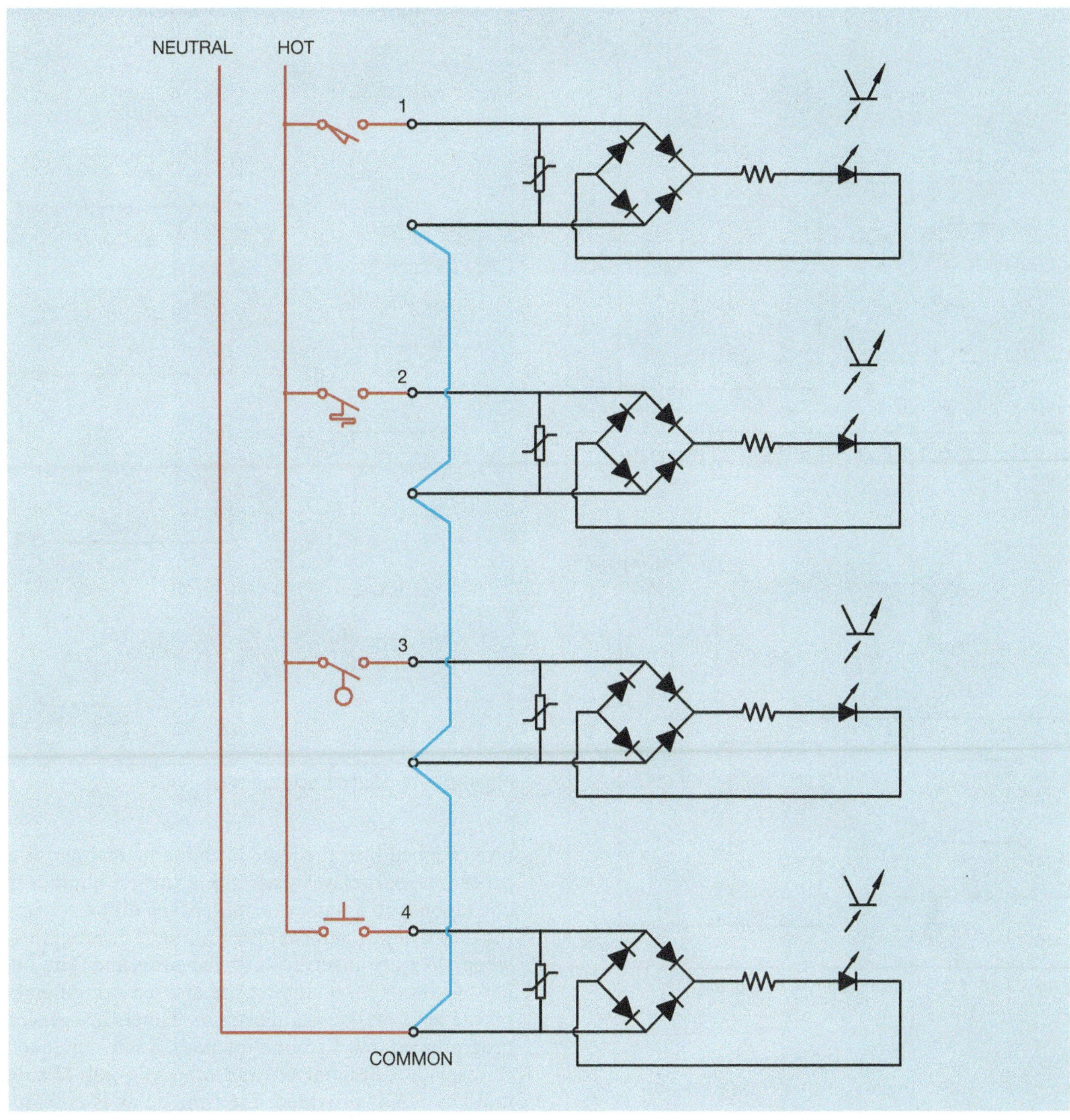

Figure 38–12 Pilot devices connected to input modules.

are programmed into the logic of the PLC by assigning them a certain number. Manufacturers provide a chart that lists which numbers can be used to program inputs and outputs, internal relay coils, timers, counters, and so on. When a coil is entered at the end of a line of logic and is given a number that corresponds to an internal relay, it will act like a physical relay. Any contacts given the same number as that relay will be controlled by that relay.

Timers and Counters

Timers and counters are internal relays also. There is no physical timer or counter in the PLC. They are

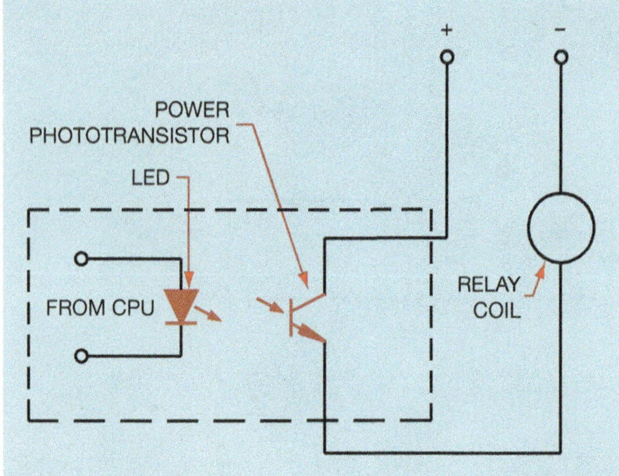

Figure 38–13 A power phototransistor connects a DC load to the line.

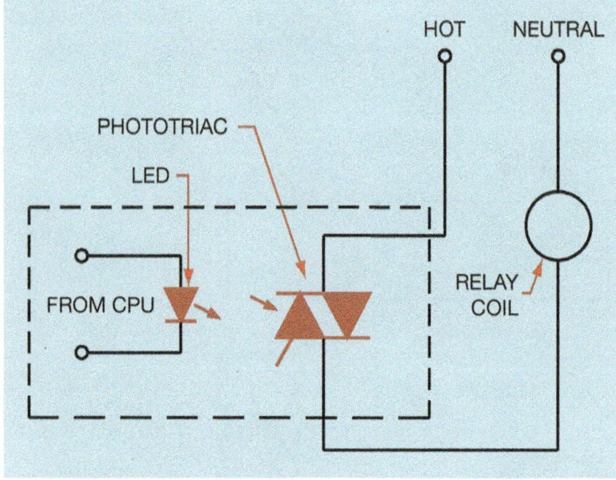

Figure 38–14 A triac connects an AC load to the line.

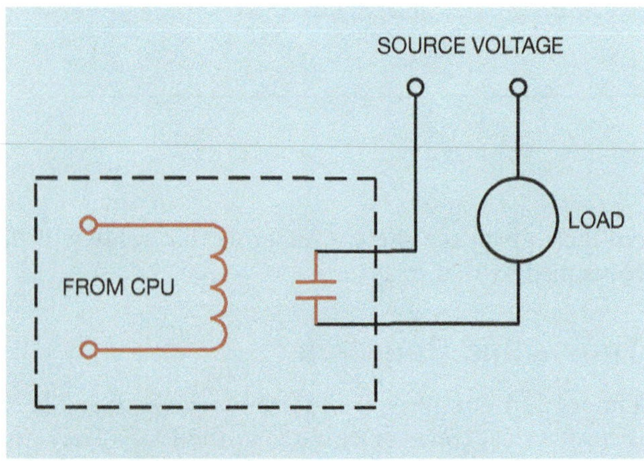

Figure 38–15 A relay connects the load to the line.

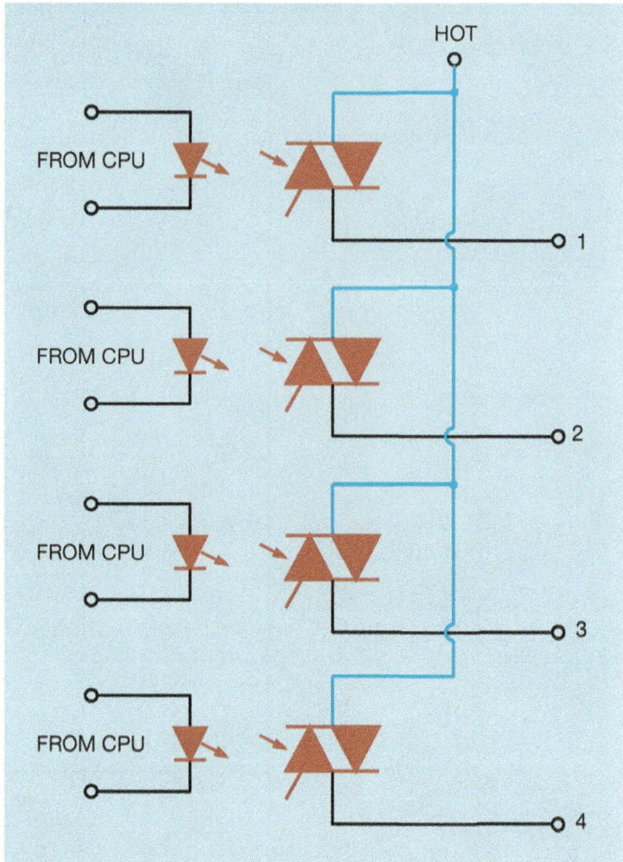

Figure 38–16 Multiple output module.

programmed into the logic in the same manner as any other internal relay, by assigning them a number that corresponds to a timer or counter. The difference is that the time delay or number of counts must be programmed when they are inserted into the program. The number of counts for a counter are entered using numbers on the keys on the load terminal. Timers are generally programmed in 0.1 second intervals. Some manufacturers provide a decimal key and others do not. If a decimal key is not provided, the time delay is entered as 0.1 second intervals. If a delay of 10 seconds is desired, for example, the number 100 would be entered because 100 tenths of a second equals 10 seconds.

Off-Delay Circuit

Some programmable logic controllers permit a timer to be programmed as on- or off-delay, but others permit only on-delay timers to be programmed. When a PLC permits only on-delay timers to be programmed, a simple circuit can be used to permit an on-delay timer to perform the function of an off-delay timer (Figure 38–17). To understand the action of the circuit, recall the

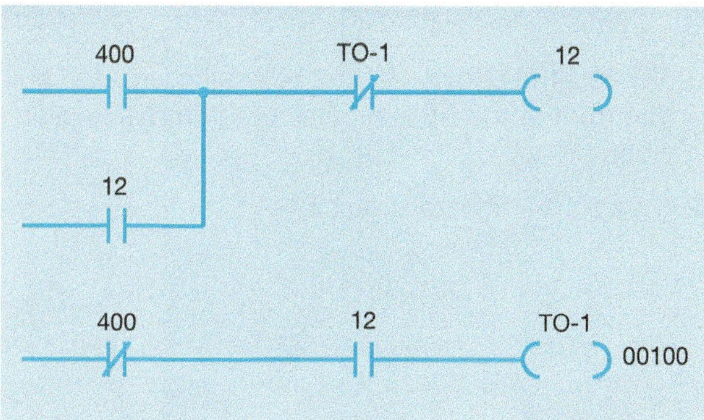

Figure 38–17 Off-delay timer circuit.

operation of an off-delay timer. When the timer coil is energized, the timed contacts change position immediately. When the coil is de-energized, the contacts remain in their energized state for some period of time before returning to their normal state. In the circuit shown in Figure 38–17, it is assumed that contact 400 controls the action of the timer. Coil 400 is an internal relay coil located somewhere in the circuit. Coil 12 is an output and controls some external device. Coil TO-1 is an on-delay timer set for 100 tenths of a second. When coil 400 is energized, both 400 contacts change position. The normally open 400 contact closes and provides a current path to coil 12. The normally closed 400 contact opens and prevents a circuit from being completed to coil TO-1 when coil 12 energizes. Note that coil 12 turned on immediately when contact 400 closed. When coil 400 is de-energized, both 400 contacts return to their normal position. A current path is maintained to coil 12 by the now-closed 12 contact in parallel with the normally open 400 contact. When the normally closed 400 contact returns to its normal position a current path is established to coil TO-1 through the now-closed 12 contact. This starts the time sequence of timer TO-1. After a delay of 10 seconds, the normally closed TO-1 contact opens and de-energizes coil 12, returning the two 12 contacts to their normal position. The circuit is now back in the state shown in Figure 38–17. Note the action of the circuit. When coil 400 was energized, output coil 12 turned on immediately. When coil 400 was de-energized, output 12 remained on for 10 seconds before turning off.

The number of internal relays and timers contained in a programmable logic controller is determined by the memory capacity of the computer. As a general rule, PLCs that have a large I/O capacity will have a large amount of memory. The use of programmable logic controllers has steadily increased since their invention in the late 1960s. A PLC can replace hundreds of relays and occupy only a fraction of the space. The circuit logic can be changed easily and quickly without requiring extensive hand rewiring. They have no moving parts or contacts to wear out, and their downtime is less than an equivalent relay circuit. When replacement is necessary they can be reprogrammed from a media storage device.

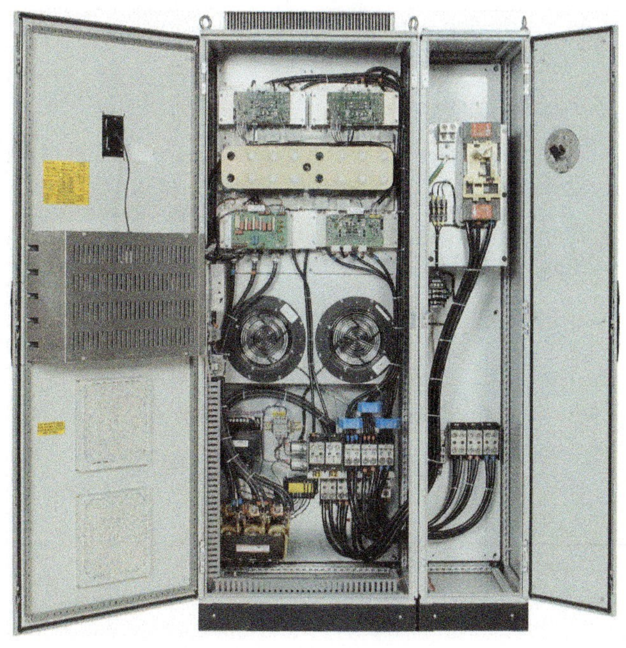
Courtesy of Baldor Electric Company

Figure 38–18 A DC drive unit controlled by a programmable logic controller.

The programming methods presented in this textbook are general because it is impossible to include examples of each specific manufacturer. The concepts presented in this chapter, however, are common to all programmable controllers. A programmable logic controller used to control a DC drive is shown in Figure 38–18.

Review Questions

1. What industry first started using programmable logic controllers?
2. Name two differences between PLCs and common home or business computers.
3. Name the four basic sections of a programmable logic controller.
4. In what section of the PLC is the actual logic performed?
5. What device is used to program a PLC?

6. What device separates the central processing unit from the outside world?
7. What is opto-isolation?
8. If an output I/O controls a DC voltage, what solid-state device is used to connect the load to the line?
9. If an output I/O controls an AC voltage, what solid-state device is used to connect the load to the line?
10. What is an internal relay?
11. What is the purpose of the key switch located on the front of the CPU in many programmable logic controllers?
12. What is a software switch?

Chapter 39

PROGRAMMING A PLC

Objectives

After studying this chapter the student will be able to:

» Convert a relay schematic to a schematic used for programming a PLC.

» Enter a program into a programmable logic controller.

In this chapter, a relay schematic will be converted into a diagram used to program a programmable logic controller. The process to be controlled is shown in Figure 39–1. A tank is used to mix two liquids. The control circuit operates as follows:

1. When the START button is pressed, solenoids A and B energize. This permits the two liquids to begin filling the tank.
2. When the tank is filled, the float switch trips. This de-energizes solenoids A and B and starts the motor used to mix the liquids together.
3. The motor is permitted to run for one minute. After one minute has elapsed, the motor turns off and solenoid C energizes to drain the tank.
4. When the tank is empty, the float switch de-energizes solenoid C.
5. A STOP button can be used to stop the process at any point.
6. If the motor becomes overloaded, the action of the entire circuit will stop.
7. Once the circuit has been energized it will continue to operate until it is manually stopped.

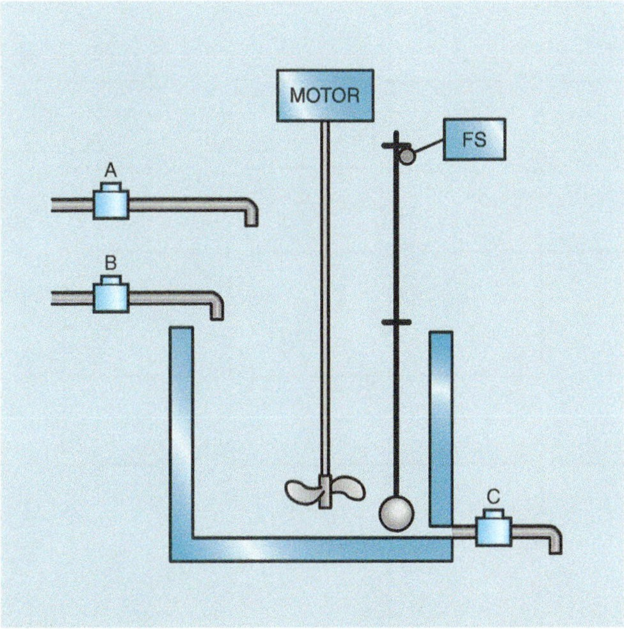

Figure 39–1 Tank used to mix two liquids.

Circuit Operation

A relay schematic that will perform the logic of this circuit is shown in Figure 39–2. The logic of this circuit is as follows:

1. When the START button is pushed, relay coil CR is energized. This causes all CR contacts to close. Contact CR-1 is a holding contact used to maintain the circuit to coil CR when the START button is released.
2. When contact CR-2 closes, a circuit is completed to solenoid coils A and B. This permits the two liquids that are to be mixed together to begin filling the tank.
3. As the tank fills, the float rises until the float switch is tripped. This causes the normally closed float switch contact to open and the normally open contact to close.
4. When the normally closed float switch opens, solenoid coils A and B de-energize and stop the flow of the two liquids into the tank.
5. When the normally open contact closes, a circuit is completed to the coil of a motor starter and the coil of an on-delay timer. The motor is used to mix the two liquids together.
6. At the end of the one-minute time period, all of the TR contacts change position. The normally closed TR-2 contact connected in series with the motor starter coil opens and stops the operation of the motor. The normally open TR-3 contact closes and energizes solenoid coil C, which permits liquid to begin draining from the tank. The normally closed TR-1 contact is used to assure that valves A and B cannot be re-energized until solenoid C de-energizes.
7. As liquid drains from the tank, the float drops. When the float drops far enough, the float switch trips and its contacts return to their normal positions. When the normally open float switch contact reopens and de-energizes coil TR, all TR contacts return to their normal positions.

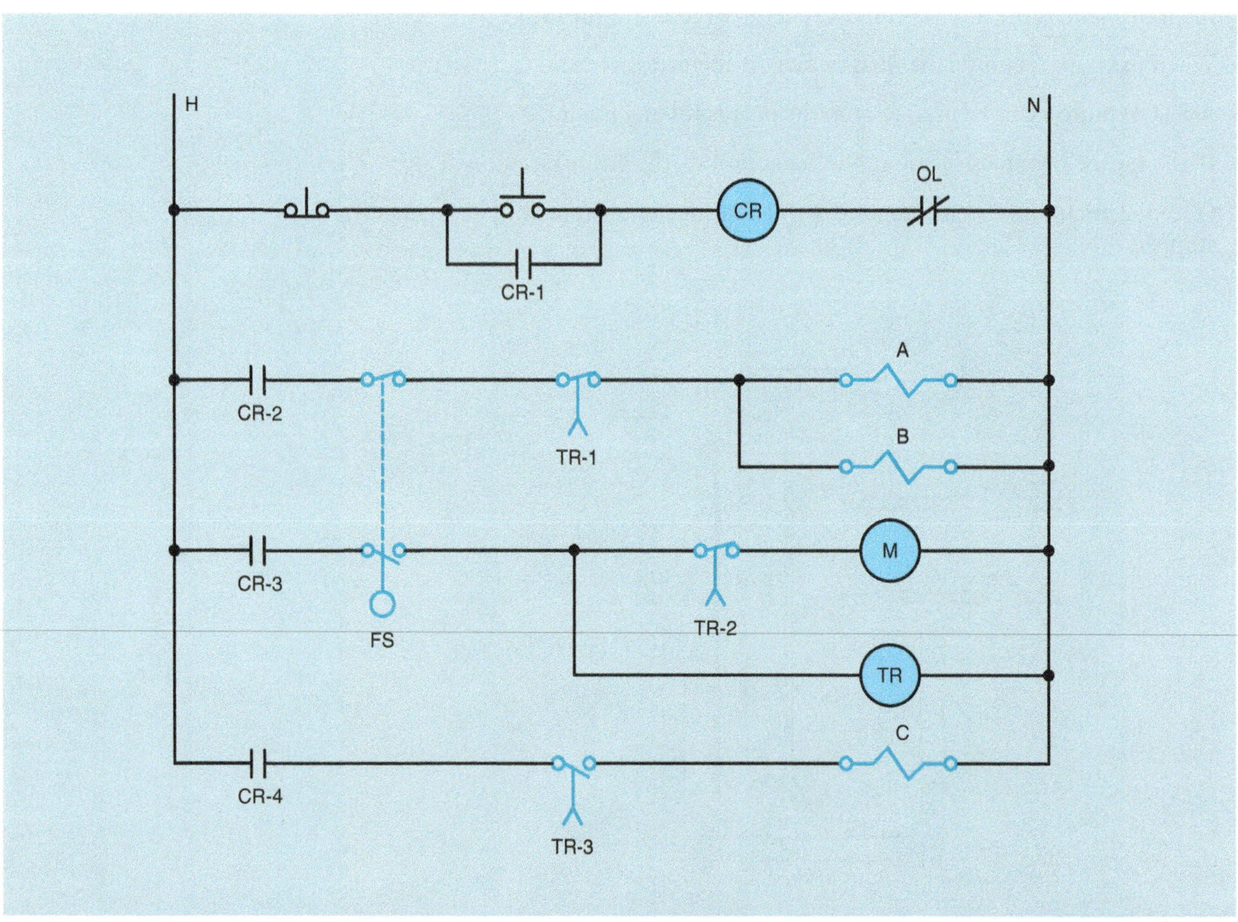

Figure 39–2 Relay schematic.

8. When the normally open TR-3 contact reopens, solenoid C de-energizes and closes the drain valve. Contact TR-2 recloses, but the motor cannot restart because of the normally open float switch contact. When contact TR-1 recloses, a circuit is completed to solenoids A and B. This permits the tank to begin refilling, and the process starts over again.
9. If the STOP button or overload contact opens, coil CR de-energizes and all CR contacts open. This de-energizes the entire circuit.

Developing a Program

This circuit will now be developed into a program that can be loaded into the programmable controller. Figure 39–3 shows a program being developed on a computer. Assume that the controller has an I/O capacity of 32, that I/O terminals 1 through 16 are used as inputs, and that terminals 17 through 32 are used as outputs.

Before a program can be developed for input into a programmable logic controller, it is necessary to assign which devices connect to the input and output terminals. This circuit contains four input devices and four output devices. It is also assumed that the motor starter for this circuit contains an overload relay that contains two contacts instead of one. One contact is normally closed and will be connected in series with the coil of the motor starter. The other contact is normally open and is used to supply an input to a programmable logic controller. If the motor should become overloaded, the normally closed contacts will open and disconnect the motor from the line. The normally open contacts will close and provide a signal to the programmable logic controller that the motor has tripped on overload. The input devices are as follows:

1. Normally closed STOP push button
2. Normally open START push button
3. Normally open overload contact
4. A float switch that contains both a normally open and normally closed contact

The four output devices are:

1. Solenoid valve A
2. Solenoid valve B
3. Motor starter coil M
4. Solenoid valve C.

The connection of devices to the inputs and outputs is shown in Figure 39–4. The normally closed STOP button is connected to input 1, the normally open START button is connected in input 2, the normally open overload contact is connected to input 3, and the float switch is connected to input 4.

The outputs for this PLC are 17 through 32. Output 17 is connected to solenoid A, output 18 is connected to solenoid B, output 19 is connected to the coil of the motor starter, and output 20 is connected to solenoid C. Note that the outputs *do not* supply the power to operate the output devices. The outputs simply complete a circuit. One side of each output device is connected to the grounded or neutral side of a 120 VAC power line. The ungrounded or hot conductor is connected to the common terminal of the four outputs. A good way to understand this connection is to imagine a set of contacts controlled by each output as shown in Figure 39–5. When programming the PLC, if a coil is given the same number as one of the outputs, it causes that contact to close and connect the load to the line.

Unfortunately, programmable logic controllers are not all programmed the same way. Almost every manufacturer employs a different set of coil numbers to perform different functions. It is necessary to consult the manual before programming a PLC with which you are unfamiliar. In order to program the PLC in this example, refer to the information in Figure 39–6. This chart indicates that numbers 1 through 16 are inputs. Any contact assigned a number between 1 and 16 will be examined each time the programmable logic controller scans the program. If an input has a low (0 volt) state, the contact assigned that number remains in the state it was programmed. If the input has a high (120 volt) state, the program interprets that contact as having changed

Figure 39–3 A program being developed on a programming terminal.

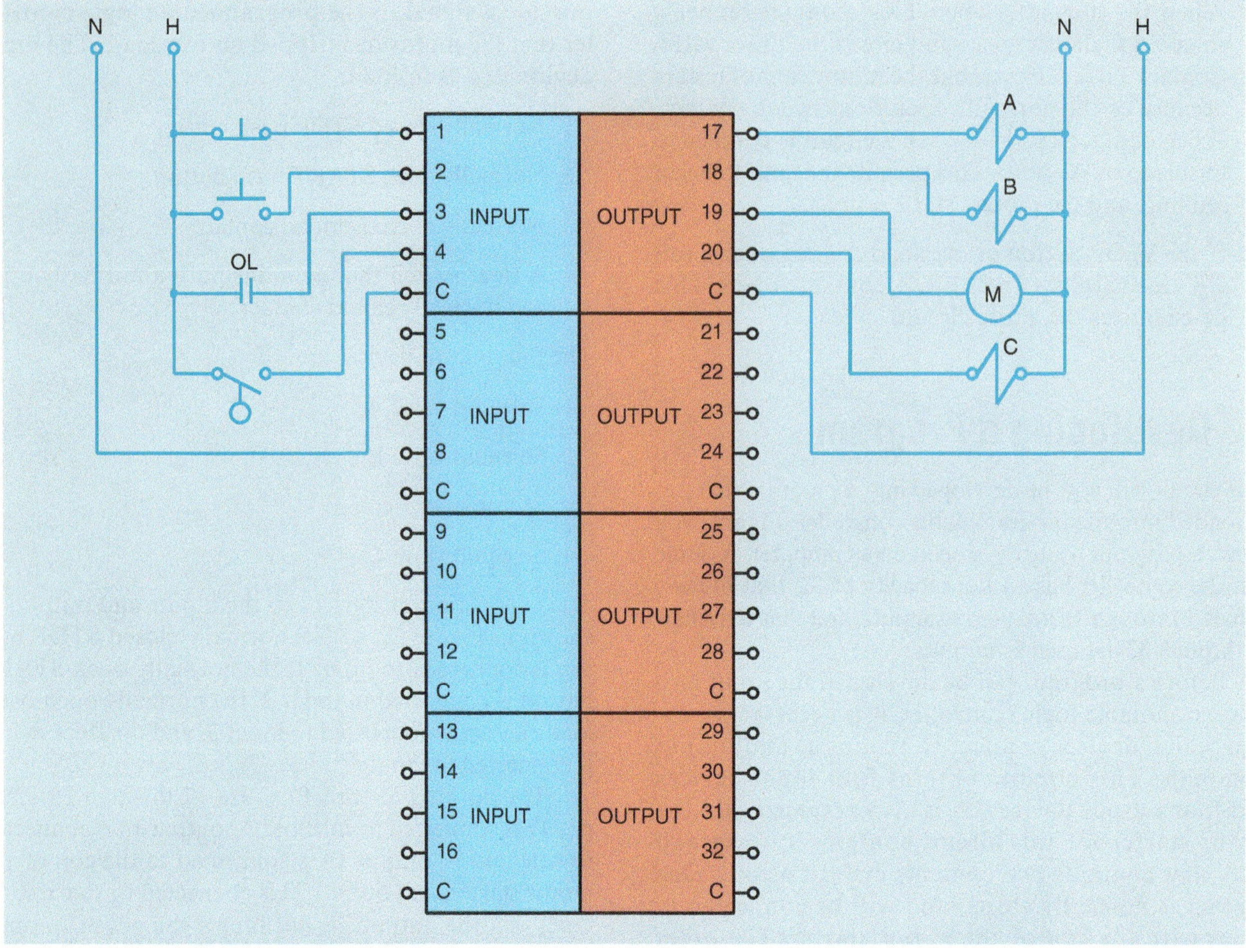

Figure 39–4 Component connection to I/O rack.

position. If it was programmed as open, the PLC now considers it closed.

Outputs are 17 through 32. Outputs are treated as coils by the PLC. If a coil is given the same number as an output, that output turns on (closes the contact) when the coil is energized. Coils that control outputs can be assigned internal contacts as well. Internal contacts are contacts that exist in the logic of the program only. They do not physically exist. Because they do not physically exist, a coil can be assigned as many internal contacts as desired and they can be normally open or normally closed.

The chart in Figure 39–6 also indicates that internal relays number from 33 to 103. Internal relays are like internal contacts. They do not physically exist. They exist as part of the program only. They are programmed into the circuit logic by inserting a coil symbol in the program and assigning it a number between 33 and 103.

Timers and counters are assigned coil numbers 200 through 264, and retentive relays are numbered 104 through 134.

Converting the Program

Developing a program for a programmable logic controller is a little different than designing a circuit with relay logic. Several rules must be followed with almost all programmable logic controllers.

1. Each line of logic must end with a coil. Some manufacturers permit coils to be connected in parallel and some do not.

2. As a general rule, coils cannot be connected in series.

3. The program will scan in the order that it is entered.

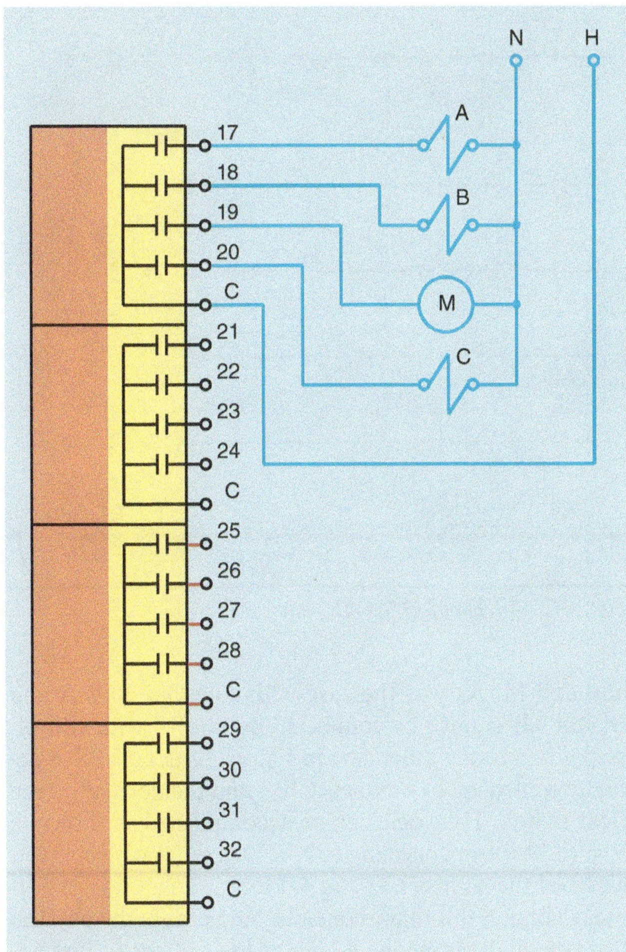

Figure 39–5 Output modules complete a circuit to connect the load to the line.

INPUTS	1–16
OUTPUTS	17–32
INTERNAL RELAYS	33–103
TIMERS AND COUNTERS	200–264
RETENTIVE RELAYS	104–134

Figure 39–6 Numbers that correspond to specific PLC functions.

4. Generally, coils cannot be assigned the same number. (Some programmable logic controllers require reset coils to reset counters and timers. These reset coils can be assigned the same number as the counter or timer they reset.)

The first two lines of logic for the circuit shown in Figure 39–2 can be seen in Figure 39–7. Notice that contact symbols are used to represent inputs instead of logic

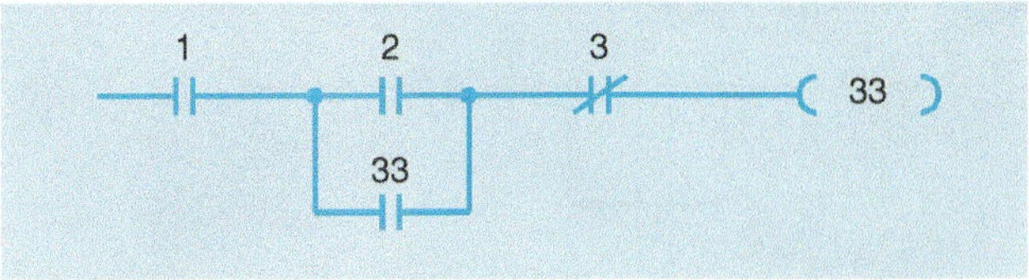

Figure 39–7 Lines 1 and 2 of the program.

symbols, such as push buttons or float switches. The programmable logic controller recognizes all inputs as open or closed contacts. It doesn't know what device is connected to which input. For this reason, you must first determine which device connects to which input before a program can be developed. Also, notice that input 1 is shown as a normally open contact. Referring to Figure 39–4, it can be seen that input 1 is connected to a normally closed push button. The input is programmed as normally open because the normally closed push button will supply a high voltage to input 1 in normal operation. Because input 1 is in a high state, the PLC changes the state of the open contact and considers it closed. When the STOP push button is pressed, the input voltage changes to low and the PLC changes the contact back to its original open state and causes coil 33 to de-energize.

Referring to the schematic in Figure 39–2, a control relay is used as part of the circuit logic. Because the control relay does not directly cause any output device to turn on or off, an internal relay will be used. The chart in Figure 39–6 indicates that internal relays number between 33 and 103. Coil 33 is an internal relay and does not physically exist. Any number of contacts can be assigned to this relay, and they can be open or closed. The 33 contact connected in parallel with input 2 is the holding contact labeled CR-1 in Figure 39–2.

The next two lines of logic are shown in Figure 39–8. The third line of logic in the schematic in Figure 39–2 contains a normally open CR-2 contact, a normally closed float switch contact, a normally closed on-delay timed contact, and solenoid coil A. The fourth line of logic contains solenoid coil B connected in parallel with solenoid coil A. Line three in Figure 39–8 uses a normally open contact assigned the #33 for contact CR-2. A normally closed contact symbol is assigned the number 4. Because the float switch is connected to input 4, it will control the action of this contact. As long as input 4 remains in a low state, the contact remains closed. If the float switch should close, input 4 becomes high and the 4 contact opens.

The next contact is timed contact TR-1. The chart in Figure 39–6 indicates that timers and counters are assigned numbers 200 through 264. In this circuit, timer TR is assigned 200. Line three ends with coil 17. When coil 17 becomes energized, it turns on output 17 and connects solenoid coil A to the line.

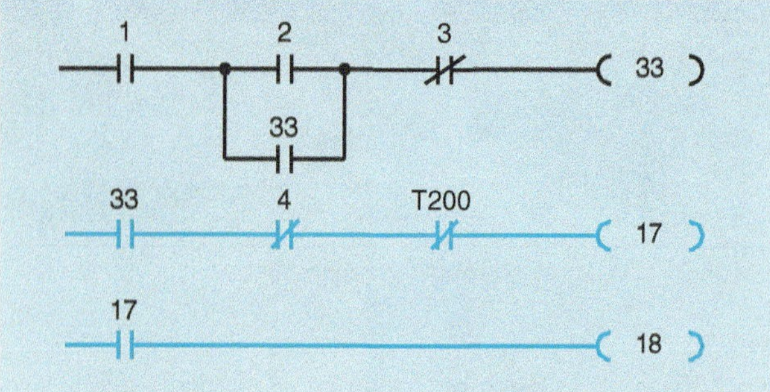

Figure 39–8 Lines 3 and 4 of the circuit are added.

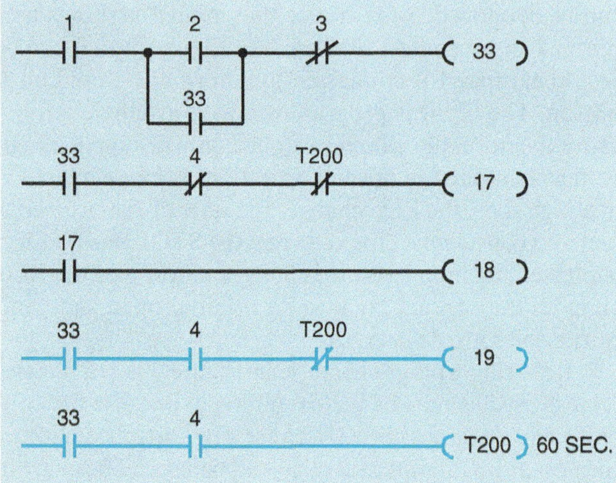

Figure 39–9 Lines 5 and 6 are added to the program.

The schematic in Figure 39–2 shows that solenoid coil B is connected in parallel with solenoid coil A. Programmable logic controllers do not permit coils to be connected in parallel. Each line of logic must end with its own coil. Because solenoid coil B is connected in parallel with solenoid A, they both operate at the same time. This logic can be accomplished by assigning an internal contact the same number as the coil controlling output 17. Notice in Figure 39–8 that when coil 17 energizes it causes contact 17 to close and energize output 18 at the same time.

In Figure 39–9, lines 5 and 6 of the schematic are added to the program. A normally open contact assigned number 33 is used as for contact CR-3. A normally open contact assigned the number 4 is controlled by the float switch, and a second normally closed timed contact controlled by timer 200 is programmed in line 5. The output coil is assigned the number 19. When this coil energizes it turns on output 19 and connects motor starter coil M to the line.

Line 6 contains timer coil TR. Notice in Figure 39–2 that coil TR is connected in parallel with contact TR-2 and coil M. As was the case with solenoid coils A and B, coil TR cannot be connected in parallel with coil M. According to the schematic in Figure 39–2, coil TR is actually controlled by contacts CR-3 and the normally open float switch. This logic can be accomplished as shown in Figure 39–9 by connecting coil T200 in series with contacts assigned the numbers 33 and 4. Float switches do not normally contain this many contacts, but because the physical float switch is supplying a high or low voltage to input 4, any number of contacts assigned the number 4 can be used.

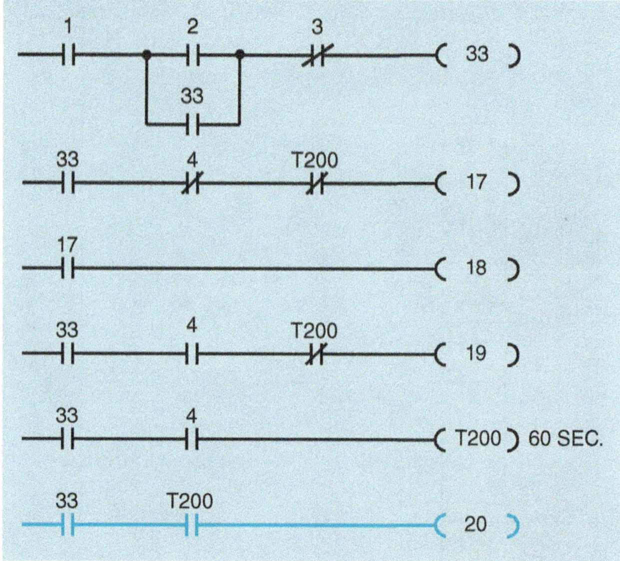

Figure 39–10 Line 7 of the program.

The last line of the program is shown in Figure 39–10. A normally open contact assigned the number 33 is used for contact CR-4 and a normally open contact controlled by timer T200 is used for the normally open timed contact labeled TR-3. Coil 20 controls the operation of solenoid coil C.

The circuit shown in Figure 39–2 has not been converted to a program that can be loaded into a programmable logic controller. The process is relatively simple if the rules concerning PLCs are followed.

Entering a Program

The manner in which a program is entered into the memory of the PLC is specific to the manufacturer and type of programming terminal used. Some programming terminals employ keys that contain contact, coil, and rung symbols to basically draw the program as it is entered. Small programming terminals may require that the program be entered in a language called Boolean. Boolean uses statements such as AND, OR,

NOT, OUT, etc. to enter programs. Contacts connected in series, for example, would be joined by AND statements and contacts that are connected in parallel with each other would be programmed with OR statements. In order to program a contact normally closed instead of normally open, the NOT statement is used. Different PLCs also require the use of different numbers to identify particular types of coils. One manufacturer may use any number between 600 and 699 to identify coils used as timer and counters. Another manufacturer may use any number between 900 and 999 to identify coils that can be used as timers and counters. When programming a PLC, it is always necessary to first become familiar with the programming requirements of the model and manufacturer of programmable logic controller being programmed.

Programming Considerations

When developing a program for a programmable logic controller, certain characteristics of a PLC should be considered. One of these characteristics is the manner in which a programmable logic controller performs its functions. Programmable logic controllers operate by scanning the program that has been entered into memory. This process is very similar to reading a book. It scans from top to bottom and from left to right. The computer scans the program one line at a time until it reaches the end of the program. It then resets any output conditions that have changed since the previous scan. The next step is to check all inputs to determine whether they are high (power applied to that input point) or low (no power applied to that input point). This information is available for the next scan. The next step is to update the display of the programming terminal if one is connected. The last step is to reset the "watchdog" timer. Most PLCs contain a timer that runs continually when the PLC is in the RUN mode. The function of this timer is to prevent the computer from becoming hung in some type of loop. If the timer is not reset at the end of each scan, the watchdog timer will reach zero and all outputs will be turned off. Although this process sounds long, it actually takes place in a few milliseconds. Depending on the program length, it may be scanned several hundred times each second. The watchdog timer duration is generally set for about twice the amount of time necessary to complete one scan.

Scanning can eliminate some of the problems with contact races that occur with relay logic. The circuit shown in Figure 39–11 contains two control relay coils. A normally closed contact, controlled by the opposite relay, is connected in series with each coil. When the switch is closed, which relay will turn on and which will be locked out of the circuit? This called a contact race. The relay that is turned on depends on which one managed to open its normally closed contact first and break the circuit to the other coil. There is no way to really know which relay will turn on and which will remain off. There is not even a guarantee that the same relay will turn on each time the switch is closed.

Programmable logic controllers eliminate the problem of contact races. Because the PLC scans the program in a similar manner to reading a book, if it is imperative that a certain relay turn on before another one, simply program the one that must turn on first ahead of the other one. A similar circuit is shown in Figure 39–12. When contact 1 closes, coil 100 will always be the internal relay that turns on because it is scanned before coil 101.

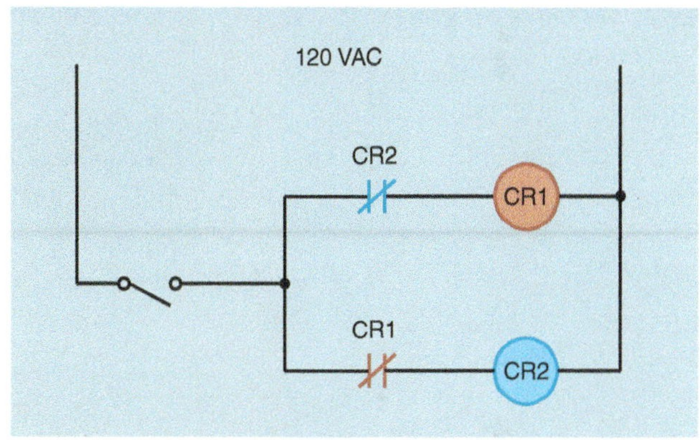

Figure 39–11 A contact race can exist in relay control circuits.

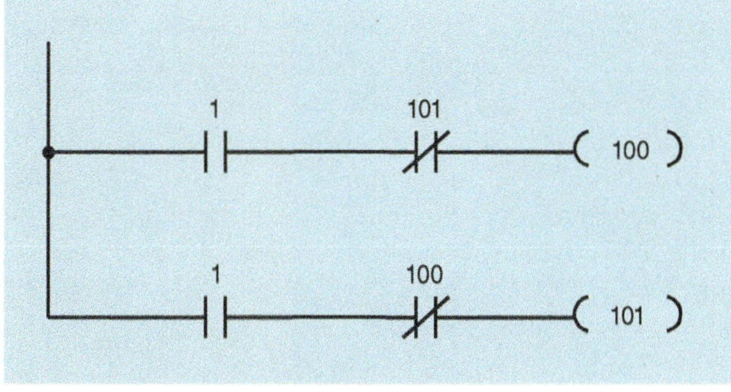

Figure 39–12 Scanning eliminates contact races in PLC logic.

Review Questions

1. Why are NEMA symbols such as push buttons, float switches, or limit switches not used in programmable logic controller schematics?
2. How are such components as coils and contacts identified and distinguished from others in a PLC schematic?
3. Why are normally closed components such as STOP push buttons programmed normally open instead of normally closed when entering a program into the memory of a PLC?
4. What is an internal relay?
5. Why is the output of a PLC used to energize the coil of a motor starter instead of energizing the motor directly?
6. List four basic rules for developing a program for a PLC.
7. A programmable logic controller requires that times be programmed in 0.1-second intervals. What number should be entered to produce a time delay of 3 minutes?
8. When programming in Boolean, what statement should be used to connect components in series?
9. When programming in Boolean, what statement should be used to connect components in parallel?
10. In a control circuit, it is imperative that a coil energize before another one. How can this be done when entering a program into the memory of the PLC?
11. What is the function of a watchdog timer?

Chapter 40

ANALOG SENSING FOR PROGRAMMABLE LOGIC CONTROLLERS

Objectives

After studying this chapter the student will be able to:

» Describe the differences between analog and digital inputs.

» Discuss precautions that should be taken when using analog inputs.

» Describe the operation of a differential amplifier.

Many of the programmable logic controllers found in industry are designed to accept analog as well as digital inputs. Analog means continuously varying. These inputs are designed to sense voltage, current, speed, pressure, temperature, or humidity. When an analog input is used, a special module mounts on the I/O rack of the PC. An analog sensor may be designed to operate between a range of settings, such as 50 to 300°C, or 0 to 100 psi. These sensors are used to indicate between a range of values instead of merely operating in an ON or OFF mode. An analog pressure sensor designed to indicate pressures between 0 and 100 psi would have to indicate when the pressure was 30 psi, 50 psi, or 80 psi. It would not just indicate whether the pressure had or had not reached 100 psi. A pressure sensor of this type can be constructed in several ways. One of the most common methods is to let the pressure sensor operate a current generator, which produces currents between 4 and 20 milliamperes. It is desirable for the sensor to produce a certain amount of current instead of a certain amount of voltage because it eliminates the problem of voltage drop on lines. For example, assume a pressure sensor is designed to sense pressures between 0 and 100 psi. Also assume that the sensor produces a voltage output of 1 volt when the pressure is 0 psi and a voltage of 5 volts when the pressure is 100 psi. Because this is an analog sensor, when the pressure is 50 psi, the sensor should produce a voltage of 3 volts. This sensor is connected to the analog input of a programmable controller Figure 40–1. The analog input has a sense resistance of 250Ω. If the wires between the sensor and the input of the programmable controller are short enough (so that there is almost no wire resistance), the circuit will operate without a problem. Because the sense resistor in the input of the programmable controller is the only resistance in the circuit, all of the output voltage of the pressure sensor will appear across it. If the pressure sensor produces a 3-volt output, 3 volts will appear across the sense resistor.

If the pressure sensor is located some distance away from the programmable controller, however, the resistance of the two wires running between the pressure sensor and the sense resistor can cause inaccurate readings. Assume that the pressure sensor is located far enough from the programmable controller so that the two conductors have a total resistance of 50Ω (Figure 40–2). This means that the total resistance of the circuit is now 300Ω (250 + 50 = 300). If the pressure sensor produces an output voltage of 3 volts when the pressure reaches 50 psi, the current flow in the circuit will be 0.010 amp (3/300 = 0.010). Because there is a current flow of 0.010 through the 250Ω sense resistor, a voltage of 2.5 volts will appear across it. This is substantially less than the 3 volts being produced by the pressure sensor.

If the pressure sensor is designed to operate a current generator with an output of 4 to 20 mA, the resistance of the wires will not cause an inaccurate reading at the sense resistor. Because the sense resistor and the resistance of the wire between

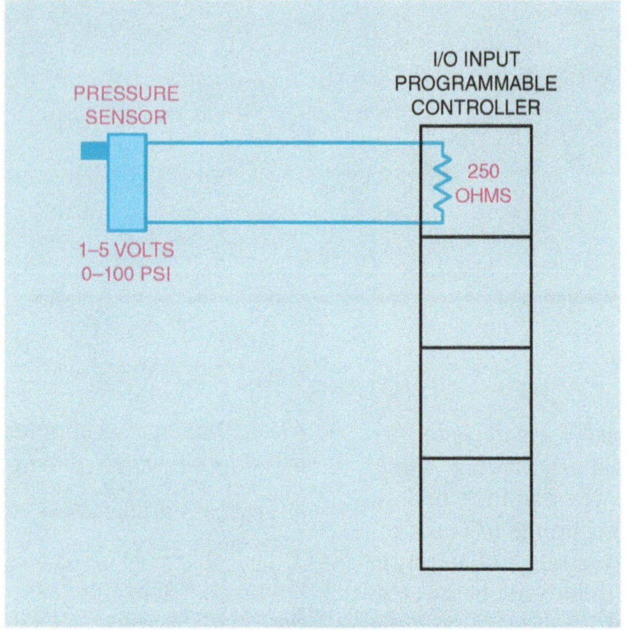

Figure 40–1 The pressure sensor produces 1 to 5 volts.

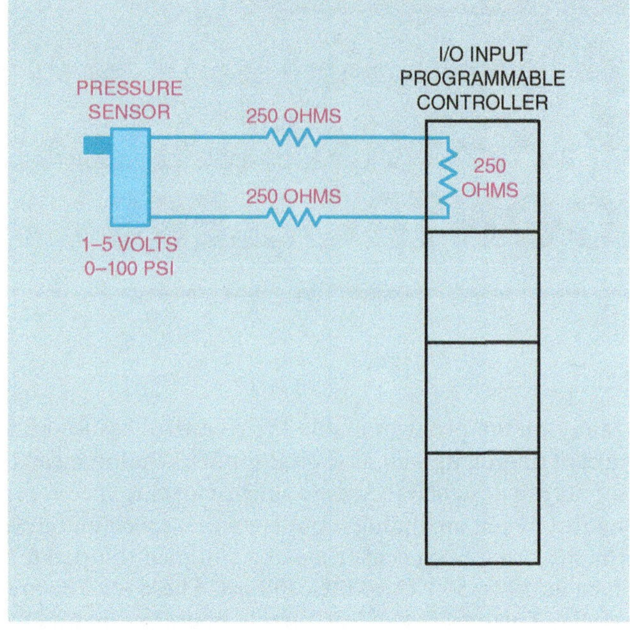

Figure 40–2 Resistance in the lines can cause problems.

the pressure sensor and the programmable controller form a series circuit, the current must be the same at the point in the circuit. If the pressure sensor produces an output current of 4 mA when the pressure is 0 psi and a current of 20 mA when the pressure is 100 psi, at 50 psi it will produce a current of 12 mA. When a current of 12 mA flows through the 250Ω sense resistor, a voltage of 3 volts will be dropped across it (Figure 40–3). Because the pressure sensor produces a certain amount of current instead of a certain amount of voltage with a change in pressure, the amount of wire resistance between the pressure sensor and programmable controller is of no concern.

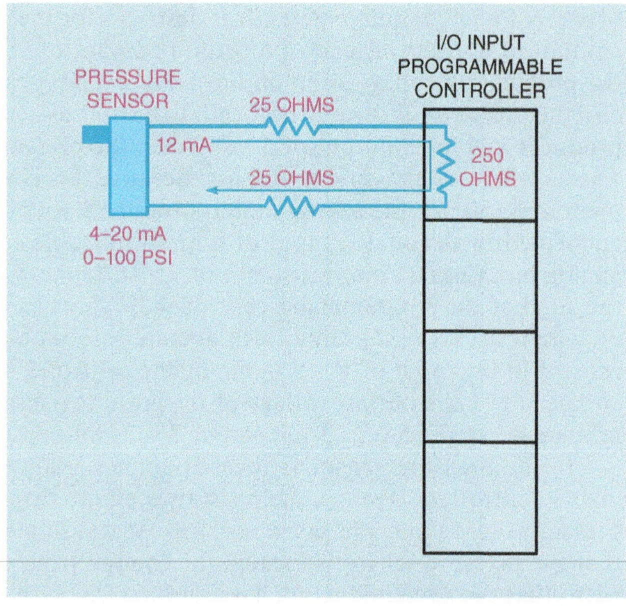

Figure 40–3 The current must be the same in a series circuit.

Installation

Most analog sensors can produce only very weak signals—0 to 10 volts or 4 to 20 milliamps are common. In an industrial environment where intense magnetic fields and large voltage spikes abound, it is easy to lose the input signal in the electrical noise. For this reason, special precautions should be taken when installing the signal wiring between the sensor and the input module. These precautions are particularly important when using analog inputs, but they should also be followed when using digital inputs.

Keep Wire Runs Short

Try to keep wire runs as short as possible. A long wire run has more surface area of wire to pick up stray electrical noise.

Plan the Route of the Signal Cable

Before starting, plan how the signal cable should be installed. *Never run signal wire in the same conduit with power wiring.* Try to run signal wiring as far away from power wiring as possible. When it is necessary to cross power wiring, install the signal cable so that it crosses at a right angle as shown in Figure 40–4.

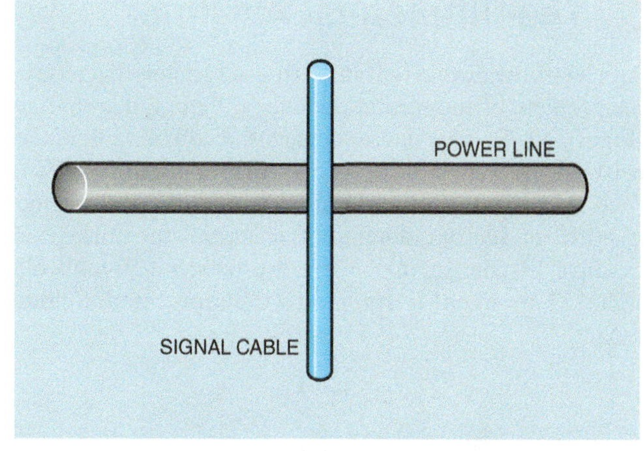

Figure 40–4 Signal cable crosses power line at right angle.

Use Shielded Cable

Shielded cable is used for the installation of signal wiring. One of the most common types, shown in Figure 40–5, uses twisted wires with a Mylar foil shield. The ground wire must be grounded if the shielding is to operate properly. This type of shielded cable can provide a noise reduction ratio of about 30,000 : 1.

Another type of signal cable uses a twisted pair of signal wires surrounded by a braided shield. This type of cable provides a noise reduction of about 300 : 1.

Common coaxial cable should be avoided. This cable consists of a single conductor surrounded by a braided shield. This type of cable offers very poor noise reduction.

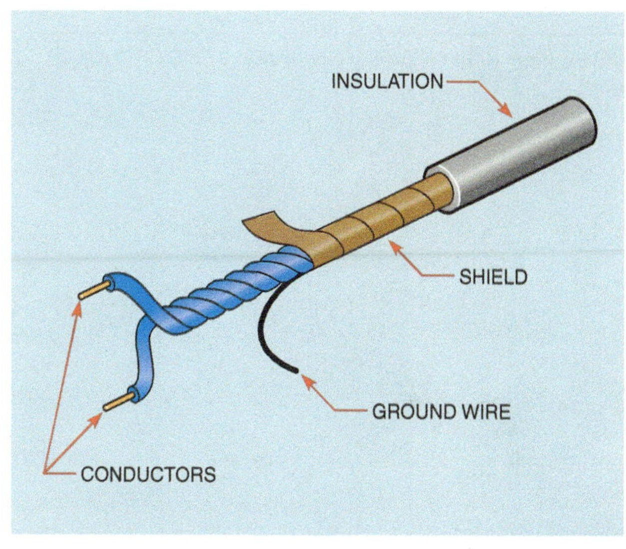

Figure 40–5 Shielded cable.

Grounding

Ground is generally thought of as being electrically neutral, or zero at all points. However, this may not be the case in practical application. It is common to find that different pieces of equipment have ground levels that are several volts apart (Figure 40–6).

To overcome this problem, large cable is sometimes used to tie the two pieces of equipment together. This forces them to exist at the same potential. This method is sometimes referred to as the brute-force method.

Where the brute-force method is not practical, the shield of the signal cable is grounded at only one end. The preferred method is generally to ground the shield at the sensor.

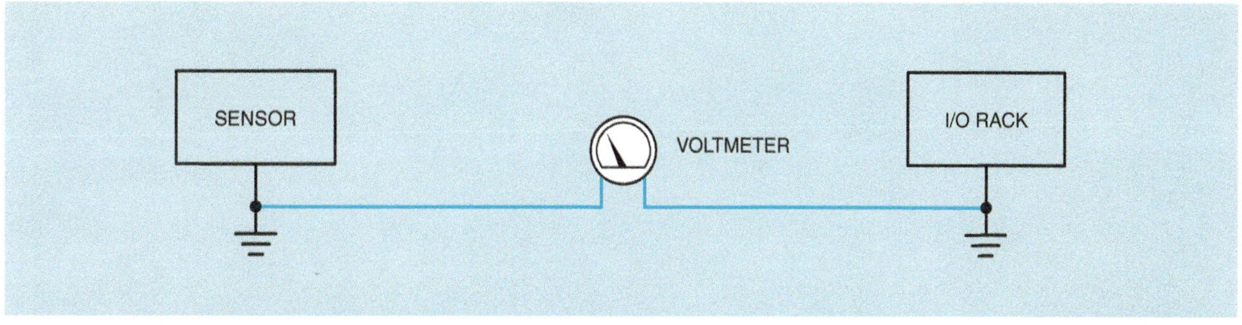

Figure 40–6 All grounds are not equal.

The Differential Amplifier

An electronic device that is often used to help overcome the problem of induced noise is the differential amplifier (Figure 40–7). This device detects the voltage difference between the pair of signal wires and amplifies this difference. Because the induced noise level should be the same in both conductors, the amplifier ignores the noise. For example, assume an analog sensor produces a 50-millivolt signal. This signal is applied to the input module, but induced noise is at a level of 5 volts. In this case the noise level is 100 times greater than the signal level. The induced noise level, however, is the same for both of the input conductors. Therefore, the differential amplifier ignores the 5-volt noise and amplifies only the voltage difference, which is 50 millivolts.

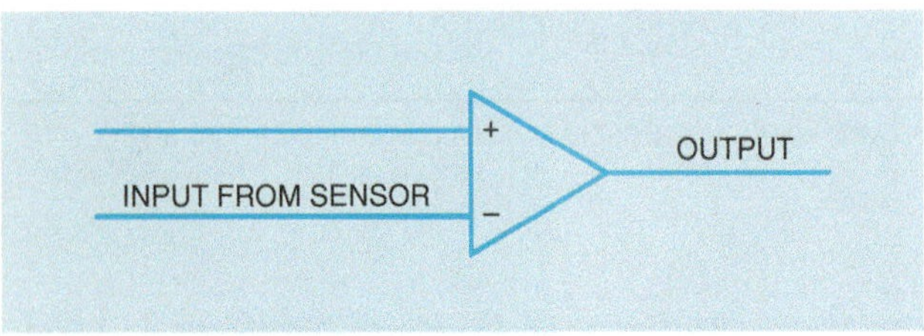

Figure 40–7 Differential amplifier detects difference in signal level.

Review Questions

1. Explain the difference between digital inputs and analog inputs.
2. Why should signal wire runs be kept as short as possible?
3. When signal wiring must cross power wiring, how should the wires be crossed?
4. Why is shielded wire used for signal runs?
5. What is the brute-force method of grounding?
6. Explain the operation of the differential amplifier.

Section 9

DEVELOPING CONTROL CIRCUITS AND TROUBLESHOOTING

Chapter 41

DEVELOPING CONTROL CIRCUITS

Objectives

After studying this chapter the student will be able to:

» Discuss steps for developing a motor control circuit from a list of requirements.

» Draw a control circuit using a list of requirements.

There are times when it becomes necessary to develop a motor control circuit to fulfill a particular need. The idea of designing a motor control circuit may seem almost impossible, but with practice and by following a logical procedure it is generally not as difficult as it would first appear. The best method of designing a motor control circuit is to solve one requirement at a time. When one part is operating, move to the next requirement. A man was once asked, "How do you eat an elephant?" His reply was, "One bite at a time." The same is true for designing a circuit. Don't try to fulfill all the requirements at once.

The following circuits illustrate a step-by-step procedure for designing a control circuit. Each illustration begins with a statement of the problem and the requirements for the circuit.

Developing Control Circuits

Circuit #1: Two Pump Motors

The water for a housing development is supplied by a central tank. The tank is pressurized by the water as it is filled. Two separate wells supply water to the tank, and each well has a separate pump. It is desirable that water be taken from each well equally, but it is undesirable that both pumps operate at the same time. A circuit is to be constructed that will let the pumps work alternately. Also, a separate switch must be installed that overrides the automatic control and lets either pump operate independently of the other in the event one pump fails. The requirements of the circuit are as follows:

1. The pump motors are operated by a 480-volt three phase system, but the control circuit must operate on a 120-volt supply.
2. Each pump motor contains a separate overload protector. If one pump overloads, it will not prevent operation of the second pump.
3. A manual ON-OFF switch can be used to control power to the circuit.
4. A pressure switch mounted on the tank controls the operation of the pump motors. When the pressure of the tank drops to a certain level, one of the pumps will be started. When the tank has been filled with water, the pressure switch will turn off the pump. When the pressure of the tank drops low enough again,

the other pump will start and run until the pressure switch is satisfied. Each time the pressure drops to a low enough level, the alternate pump motor will be used.

5. An override switch can be used to select the operation of a particular pump, or to permit the circuit to operate automatically.

When developing a control circuit, the logic of the circuit is developed one stage at a time until the circuit operates as desired. The first stage of the circuit is shown in Figure 41–1. In this stage, a control transformer has been used to step down the 480-volt supply line voltage to 120 volts for use by the control circuit. A fuse is used as short-circuit protection for the control wiring. A manually operated ON-OFF switch permits the control circuit to be disconnected from the power source. The pressure switch must close when the pressure drops. For this reason, it will be connected as normally closed. This is a normally closed held-open switch. A set of normally closed overload contacts are connected in series with coil 1M, which will operate the motor starter of pump motor #1.

To understand the operation of this part of the circuit, assume that the manual power switch has been set to the ON position. When the tank pressure drops sufficiently, pressure switch PS closes and energizes coil 1M, starting pump #1. As water fills the tank, the pressure increases. When the pressure has increased sufficiently, the pressure switch opens and disconnects coil 1M, stopping the operation of pump #1.

If pump #1 is to operate alternately with pump #2, some method must be devised to remember which pump operated last. This function will be performed by control relay CR. Because relay CR is to be used as a memory device, it must be permitted to remain energized when either or both of the motor starters are not energized. For this reason, this section of the circuit is connected to the input side of pressure switch PS. This addition to the circuit is shown in Figure 41–2.

The next stage of circuit development can be seen in Figure 41–3. In this stage of the circuit, motor starter 2M has been added. When pressure switch PS closes and energizes motor starter coil 1M, all 1M contacts change position. Contacts $1M_1$ and $1M_2$ close at the same time. When $1M_1$ contact closes, coil CR is energized, changing the position of all CR contacts. Contact CR_1 opens, but the current path to coil 1M is maintained by contact $1M_1$. Contact CR_2 is used as a holding contact around contact $1M_2$. Notice that each motor starter coil is protected by a separate overload contact. This fulfills the requirement that an overload on either motor will not prevent the operations of the other motor. Also notice that this section of the circuit has been connected to the output side of pressure switch PS. This permits the pressure switch to control the operation of both pumps.

To understand the operation of the circuit, assume pressure switch PS closes. This provides a current path to motor starter coil 1M. When coil 1M energizes, all 1M contacts change position and pump #1 starts. Contact $1M_1$ closes and energizes coil CR. Contact $1M_2$ closes to maintain a current path to coil 1M. Contact $1M_3$ opens to provide interlock with coil 2M, which prevents it from energizing whenever coil 1M is energized.

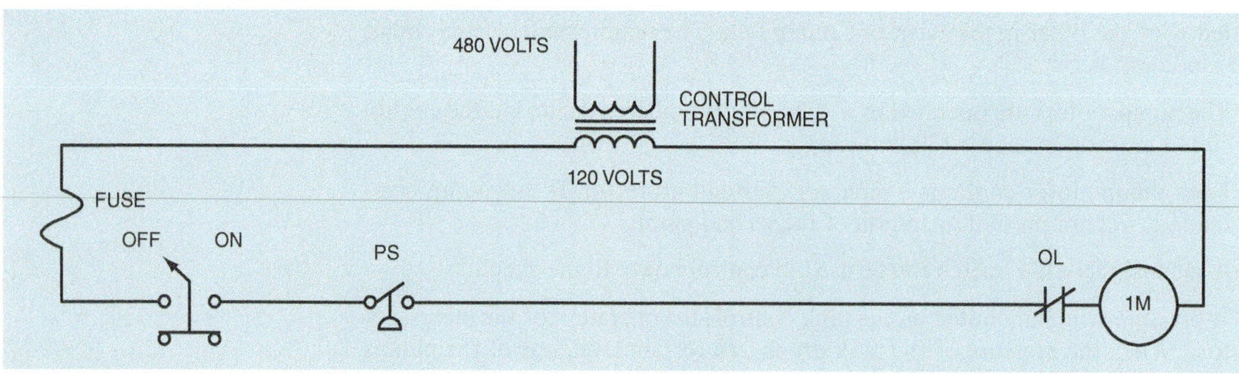

Figure 41–1 The pressure switch starts pump #1.

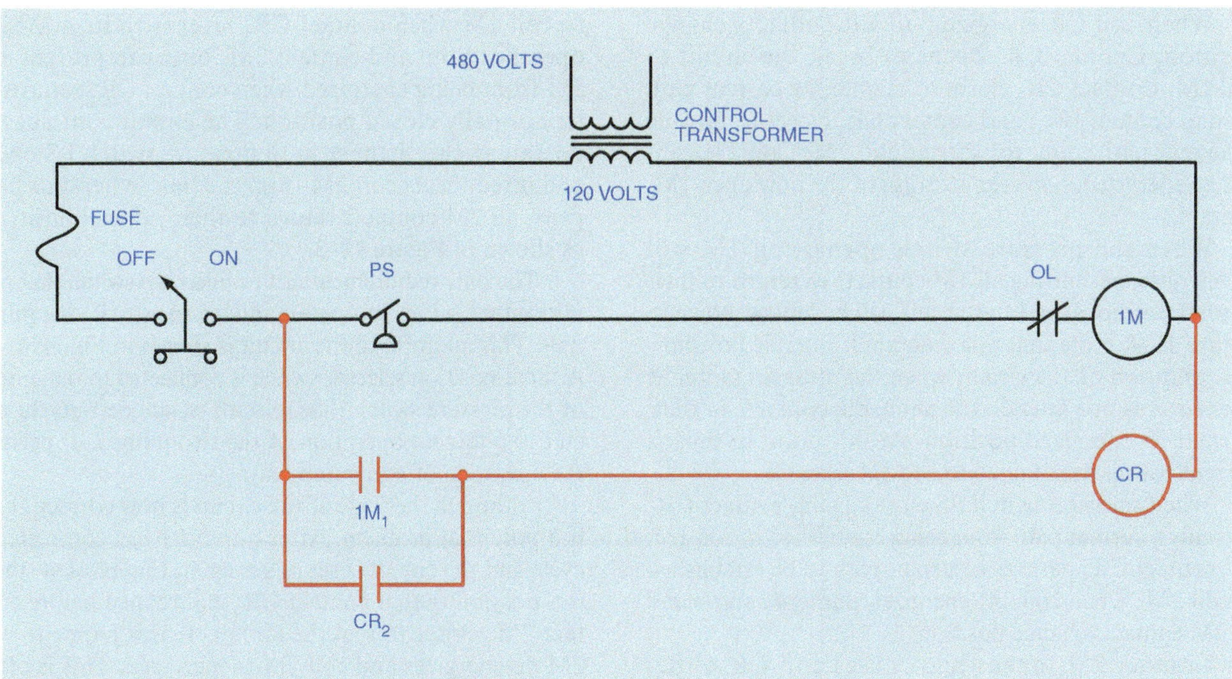

Figure 41–2 The control relay is used as a memory device.

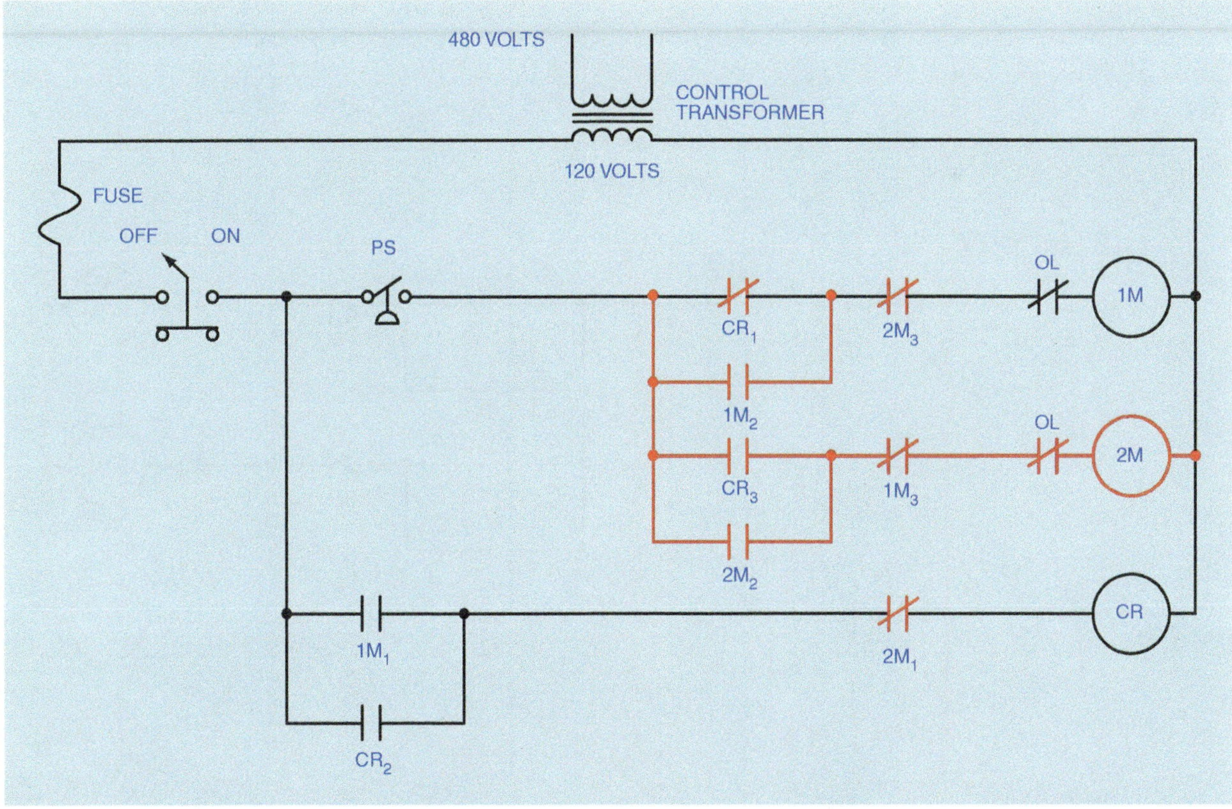

Figure 41–3 The addition of the second motor starter.

When coil CR energizes, all CR contacts change position. Contact CR_1 opens to break the circuit to coil 1M. Contact CR_2 closes to maintain a current path around contact $1M_1$, and contact CR_3 closes to provide a current path to motor starter coil 2M. Coil 2M cannot be energized, however, because of the now open $1M_3$ contact.

When the pressure switch opens, coil 1M will de-energize, permitting all 1M contacts to return to their normal positions, and the circuit will be left as shown in Figure 41–4. Note that this diagram is intended to show the condition of the circuit when the pressure switch is opened; it is not intended to show the contacts in their normal de-energized position. At this point in time, a current path is maintained to control relay CR.

When pressure switch PS closes again, contact CR_1 prevents a current path from being established to coil 1M, but contact CR_3 permits a current path to be established to coil 2M. When coil 2M energizes, pump #2 starts and all 2M contacts change position.

Contact $2M_1$ opens and causes coil CR to de-energize. Contact $2M_2$ closes to maintain a circuit to coil 2M when contact CR_3 returns to its normally open position, and contact $2M_3$ opens to prevent coil 1M from being energized when contact CR_1 returns to its normally closed position. The circuit continues to operate in this manner until pressure switch PS opens and disconnects coil 2M from the line. When this happens, all 2M contacts return to their normal positions as shown in Figure 41–3.

The only requirement not fulfilled is a switch that permits either pump to operate independently if one pump fails. This addition to the circuit is shown in Figure 41–5. A three-position selector switch is connected to the output of the pressure switch. The selector switch permits the circuit to alternate operation of the two pumps, or permits the operation of one pump only.

Although the logic of the circuit is now correct, there is a potential problem. After pump #1 has completed a cycle and the circuit is set as shown in Figure 41–4, there is a possibility that contact CR_3 will reopen before contact $2M_2$ closes to seal the circuit. If this happens, coil 2M de-energizes and coil 1M is energized. This is often referred to as a contact race. To prevent this problem, an

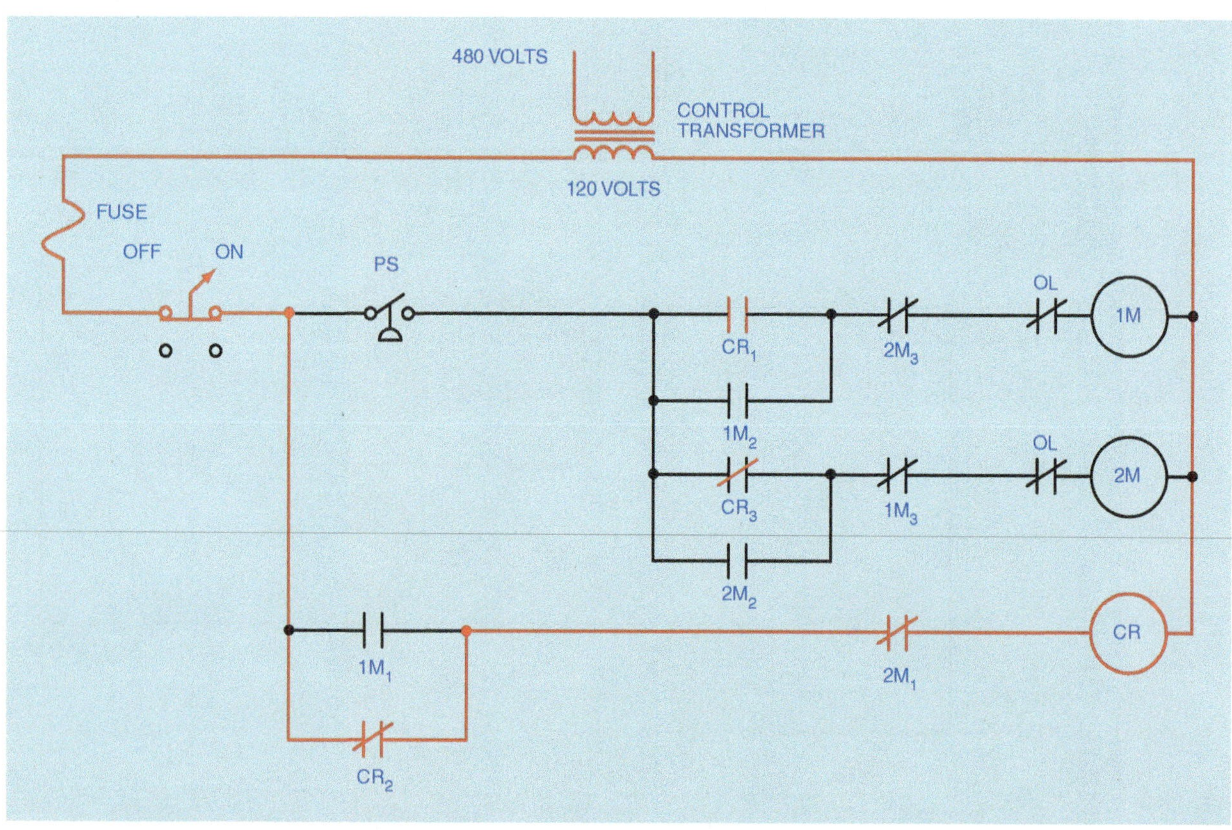

Figure 41–4 Coil CR remembers which pump operated last.

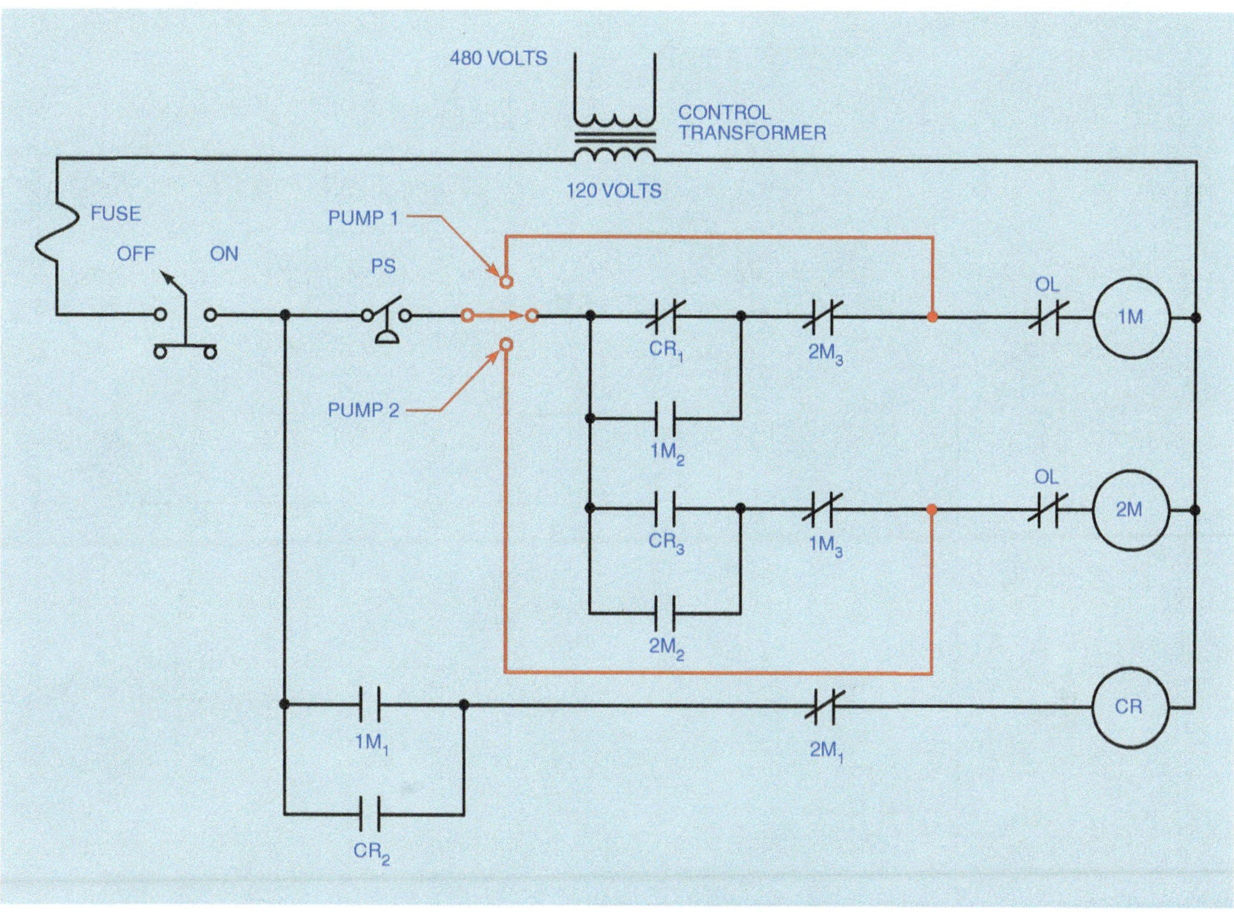

Figure 41–5 The basic logic of the circuit is complete.

off-delay timer will be added as shown in Figure 41–6. In this circuit, coil CR has been replaced by coil TR of the timer. When coil TR energizes, contact TR closes immediately, energizing coil CR. When coil TR de-energizes, contact TR remains closed for one second before reopening and permitting coil CR to de-energize. This short delay time ensures proper operation of the circuit.

Circuit #2: Speed Control of a Wound Rotor Induction Motor

The second circuit to be developed will control the speed of a wound rotor induction motor. The motor will have three steps of speed. Separate push buttons are used to select the speed of operation. The motor will accelerate automatically to the speed selected. For example, if second speed is selected, the motor must start in the first or lowest speed and then accelerate to second speed. If third speed is selected, the motor must start in first speed, accelerate to second speed, and then accelerate to third speed. The requirements of the circuit are as follows:

1. The motor is to operate on a 480-volt three phase power system, but the control system is to operate on 120 volts.
2. One STOP button can stop the motor regardless of which speed has been selected.
3. The motor will have overload protection.
4. Three separate push buttons will select first, second, or third speed.
5. There will be a three-second time delay between accelerating from one speed to another.
6. If the motor is in operation and a higher speed is desired, it can be obtained by pushing the proper button. If the motor is operating and a lower speed is desired, the STOP button must be pressed first.

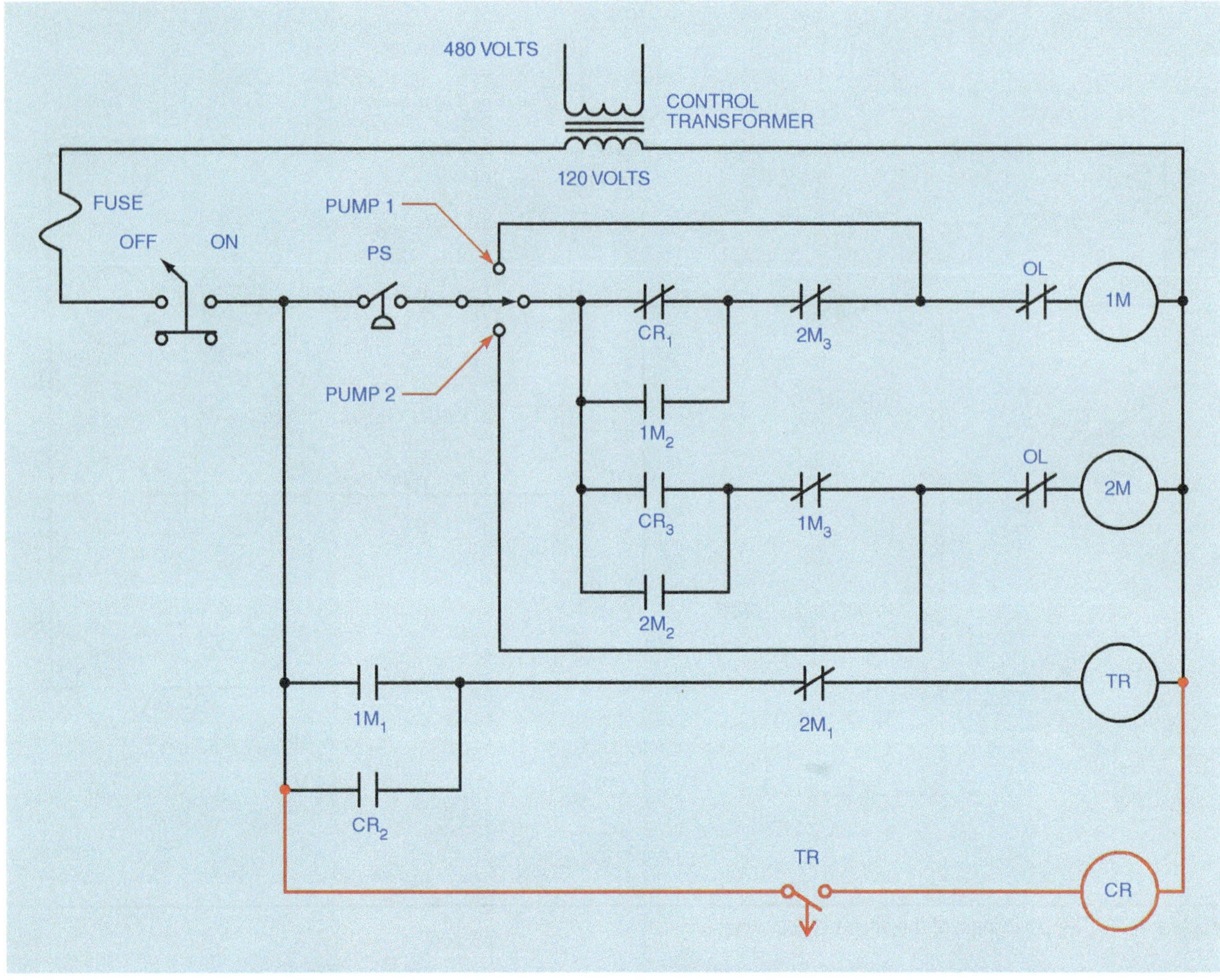

Figure 41–6 A timer is added to ensure proper operation.

Recall that speed control for a wound rotor motor is obtained by placing resistance in the secondary or rotor circuit as shown in Figure 41–7. In this circuit, load contacts 1M are used to connect the stator or primary of the motor to the power line. Two banks of three phase resistors have been connected to the rotor. When power is applied to the stator, all resistance is connected in the rotor circuit and the motor will operate in its lowest or first speed. Second speed is obtained by closing contacts 1S and shorting out the first three phase resistor bank. Third speed is obtained by closing contacts 2S. This shorts the rotor winding and the motor operates as a squirrel cage motor. A control transformer is connected to two of the three phase lines to provide power for the control system.

The first speed can be obtained by connecting the circuit shown in Figure 41–8. When the FIRST SPEED button is pressed, motor starter coil 1M will close and connect the stator of the motor to the power line. Because all the resistance is in the rotor circuit, the motor operates in its lowest speed. Auxiliary contact $1M_1$ is used as a holding contact. A normally closed overload contact is connected in series with coil 1M to provide overload protection. Notice that only one overload contact is shown, indicating the use of a three phase overload relay.

The second stage of the circuit can be seen in Figure 41–9. When the SECOND SPEED button is pressed, the coil of on-delay timer 1TR is energized. Because the motor must be started in the first speed position, instantaneous timer contact $1TR_1$ closes to energize coil 1M and connect the stator of the motor to the line. Contact $1TR_2$ is used as a holding contact to keep coil 1TR energized when the SECOND SPEED button is released. Contact $1TR_3$ is a timed contact. At the end of 3 seconds, it will close and energize contactor coil 1S,

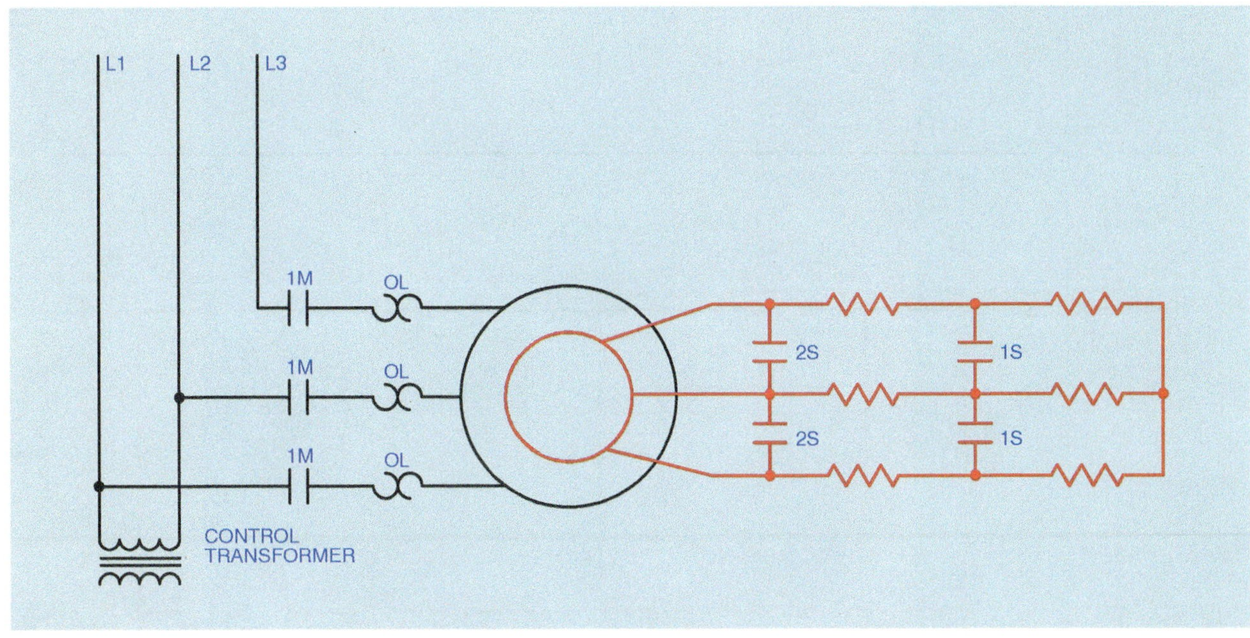

Figure 41–7 Speed is controlled by connecting resistance in the rotor circuit.

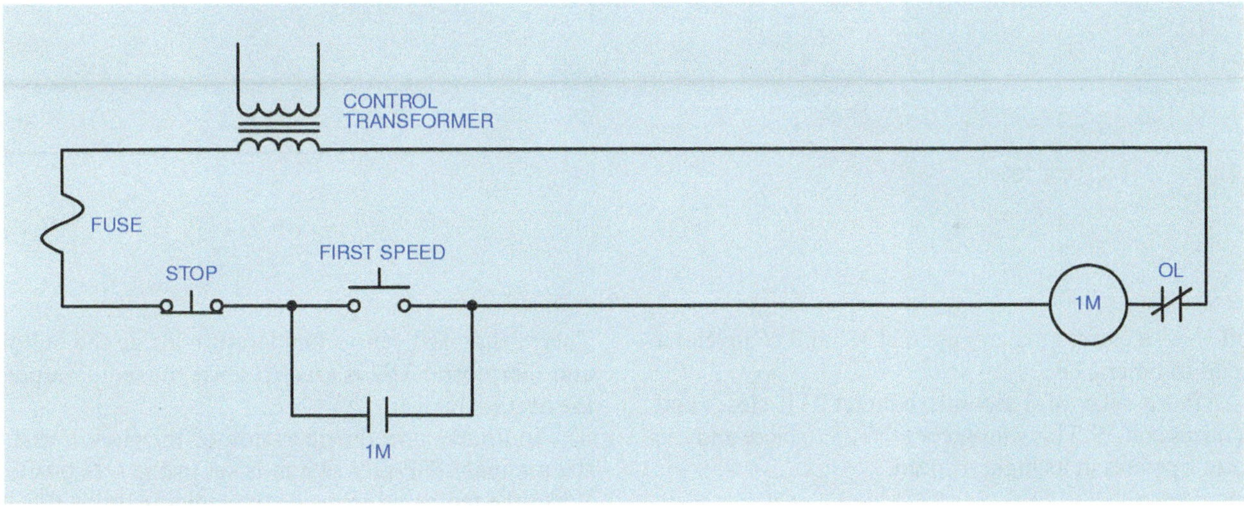

Figure 41–8 First speed.

causing all 1S contacts to close and shunt the first set of resistors. The motor now operates in second speed.

The final stage of the circuit is shown in Figure 41–10. The THIRD SPEED button is used to energize the coil of control relay 1CR. When coil 1CR is energized, all 1CR contacts change position. Contact $1CR_1$ closes to provide a current path to motor starter coil 1M, causing the motor to start in its lowest speed. Contact $1CR_2$ closes to provide a current path to timer 1TR. This permits timer 1TR to begin its timer operation. Contact $1CR_3$ maintains a current path to coil 1CR after the third speed button is opened, and contact $1CR_4$ permits a current path to be established to timer 2TR. This contact is also used to prevent a current path to coil 2TR when the motor is to be operated in the second speed.

After timer 1TR has been energized for a period of 3 seconds, contact $1TR_3$ closes and energizes coil 1S. This permits the motor to accelerate to the second speed.

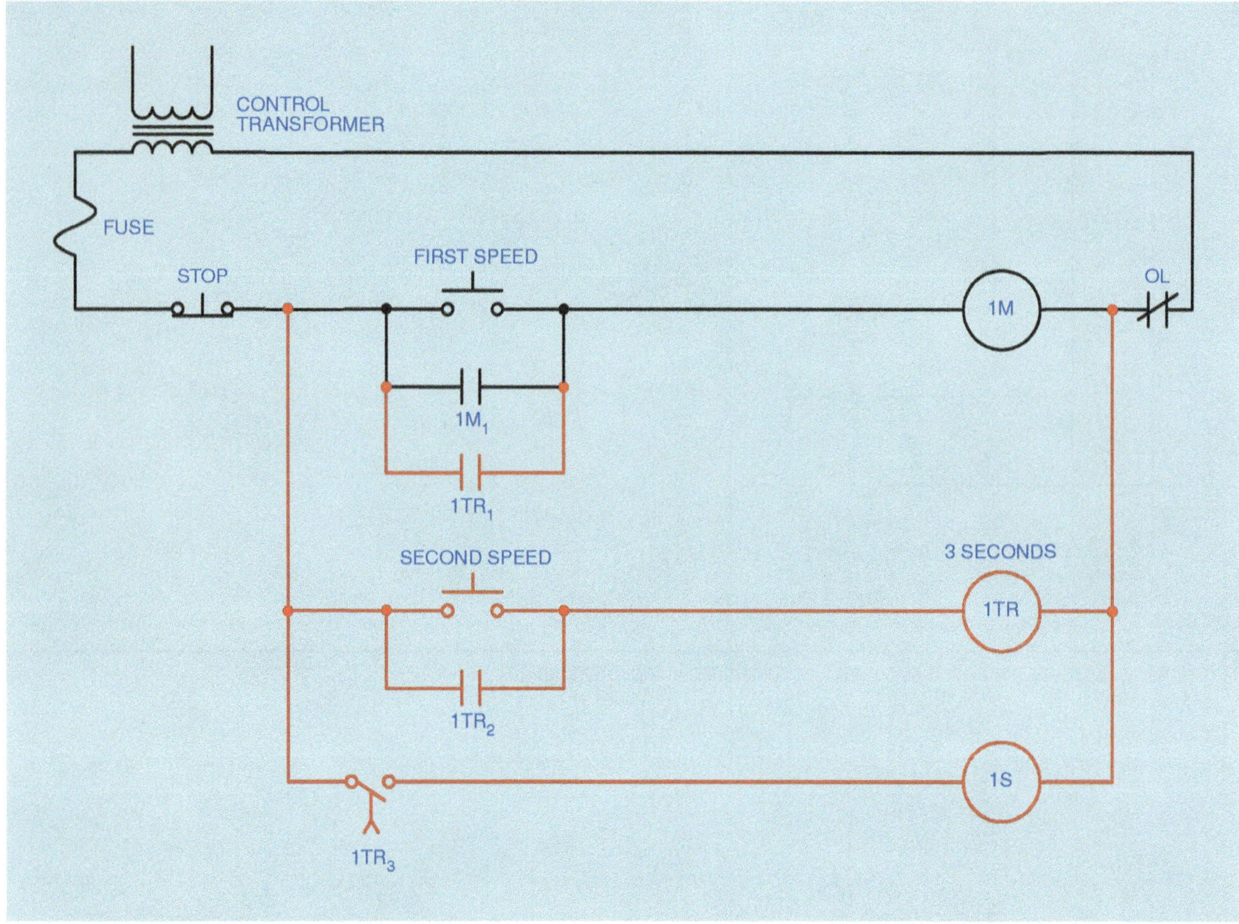

Figure 41–9 Second speed.

Coil 1S also closes auxiliary contact $1S_1$ and completes a circuit to timer 2TR.

After a delay of 3 seconds, contact 2TR closes and energizes coil 2S. This causes contacts 2S to close and the motor operates in its highest speed.

Circuit #3: An Oil Heating Unit

In the circuit shown in Figure 41–11, motor starter 1M controls a motor that operates a high-pressure pump. The pump is used to inject fuel oil into a combustion chamber where it is burned. Motor starter 2M operates an air induction blower that forces air into the combustion chamber when the oil is being burned. Motor starter 3M controls a squirrel cage blower that circulates air across a heat exchanger to heat a building. A control transformer is used to change the incoming voltage from 240 volts to 120 volts, and a separate OFF-ON switch can be used to disconnect power from the circuit. Thermostat TS1 senses temperature inside the building and thermostat TS2 is used to sense the temperature of the heat exchanger.

To understand the operation of the circuit, assume the manual OFF-ON switch is set in the ON position. When the temperature inside the building drops to a low enough level, thermostat TS1 closes and provides power to starters 1M and 2M. This permits the pump motor and air induction blower to start. When the temperature of the heat exchanger rises to a high enough level, thermostat TS2 closes and energizes starter 3M. The blower circulates the air inside the building across the heat exchanger and raises the temperature inside the building. When the building temperature rises to a high enough level, thermostat TS1 opens and disconnects the pump motor and air induction motor. The blower continues to operate until the heat exchanger has been cooled to a low enough temperature to permit thermostat TS2 to open its contact.

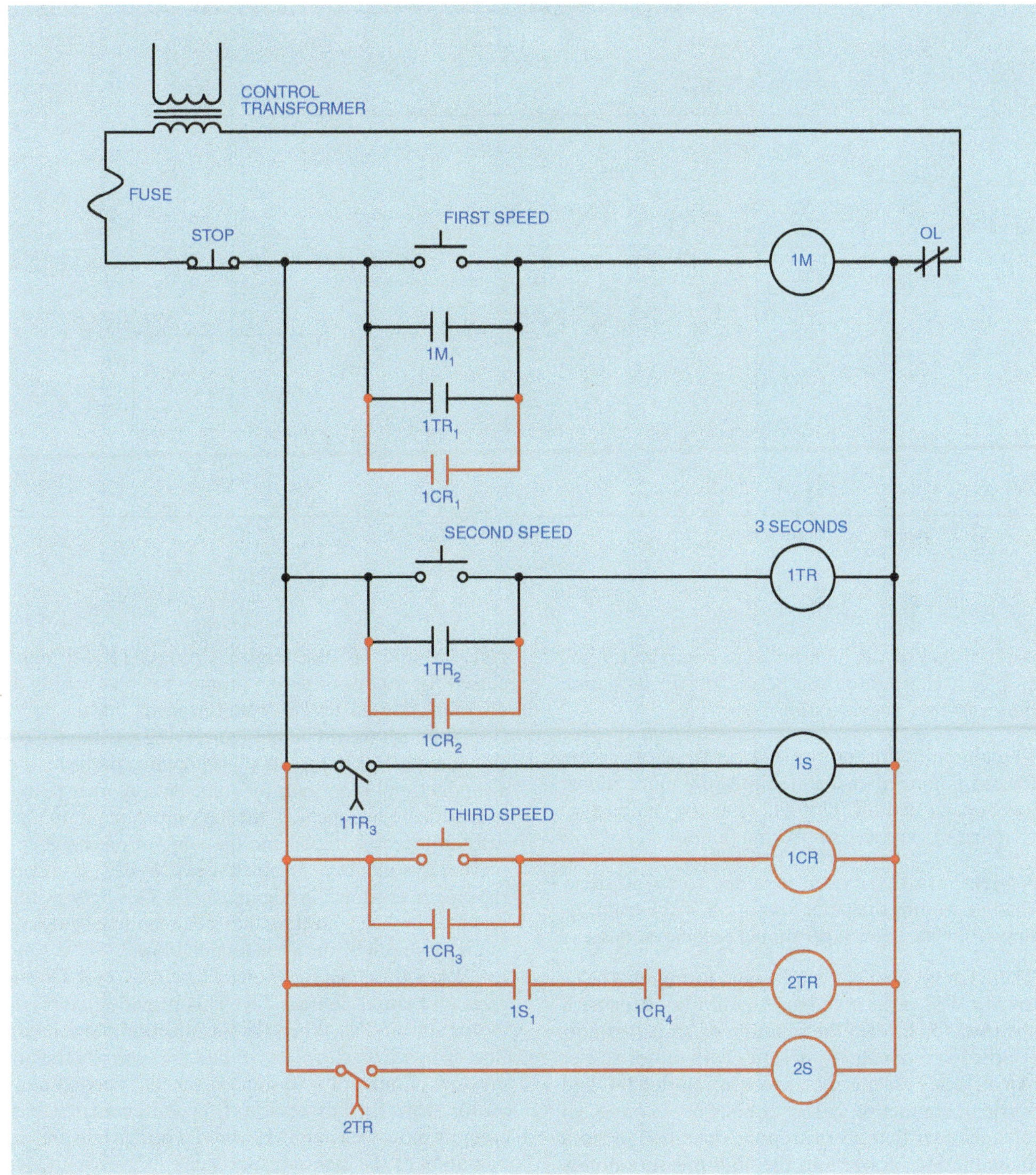

Figure 41–10 Third speed.

After a period of operation, it is discovered that the design of this circuit can lead to some serious safety hazards. If the overload contact connected to starter 2M should open, the high-pressure pump motor will continue to operate without sufficient air being injected into the combustion chamber. Also, there is no safety switch to turn the pump motor off if the blower motor fails to provide cooling air across the heat exchanger. It is recommended that the following changes be made to the circuit:

1. If an overload occurs to the air induction motor, it will stop operation of both the high-pressure pump motor and the air induction motor.

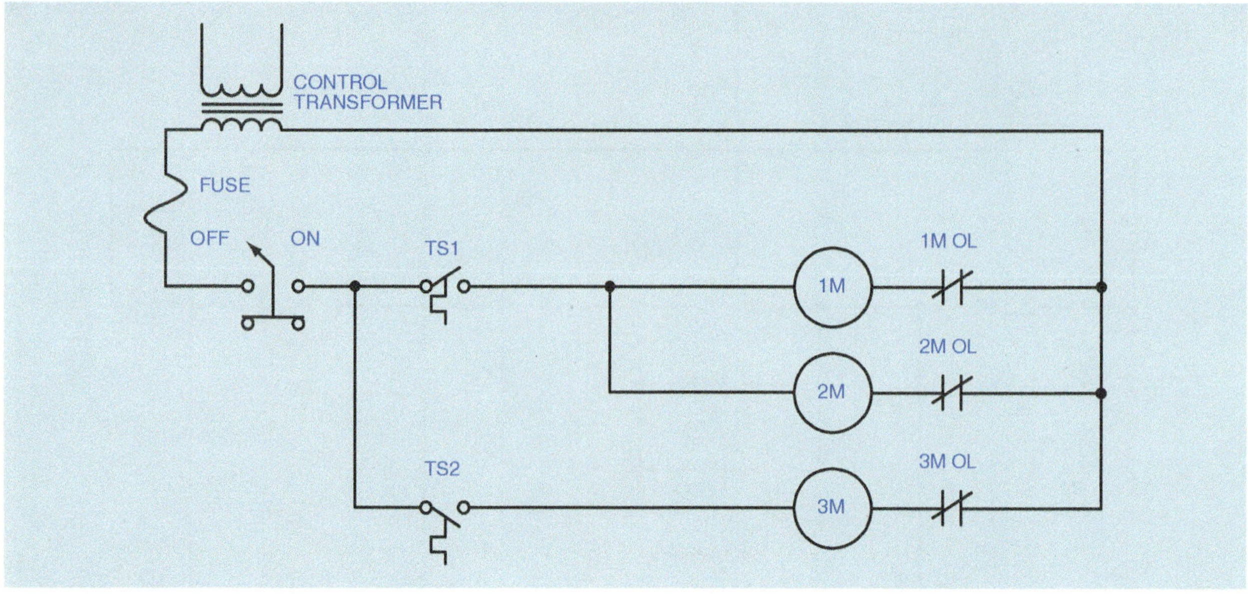

Figure 41–11 Heating system control.

2. An overload of the high-pressure pump motor will stop only that motor and permit the air induction motor to continue operation.

3. The air induction motor will continue operating for one minute after the high-pressure pump motor has been turned off. This will clear the combustion chamber of excessive smoke and fumes.

4. A high-limit thermostat is added to the heat exchanger to turn the pump motor off if the temperature of the heat exchanger should become excessive.

These circuit changes can be seen in Figure 41–12. Thermostat TS3 is the high limit thermostat. Because it is to be used to perform the function of stop, it is normally closed and connected in series with motor starter 1M. An off-delay timer is used to control starter 2M, and the overload contact of starter 2M has been connected in such a manner that it can stop the operation of both the air induction blower and the high-pressure pump. Notice, however, that if 1M overload contact opens, it will not stop the operation of the air induction blower motor. The air induction blower motor would continue to operate for a period of one minute before stopping.

The logic of the circuit is as follows: When thermostat TS1 closes its contact, coils 1M and TR are energized. Because timer TR is an off-delay timer, contact TR closes immediately, permitting motor starter 2M to energize. When thermostat TS1 is satisfied and reopens its contact, or if thermostat TS3 opens its contact, coils 1M and TR de-energize. Contact TR will remain closed for a period of one minute before opening and disconnecting starter 2M from the power line.

Although the circuit in Figure 41–12 satisfies the basic circuit requirement, there is still a potential problem. If the air induction blower fails for some reason other than the overload contact opening, the high-pressure pump motor will continue to inject oil into the combustion chamber. To prevent this situation, an air-flow switch, FL1, is added to the circuit as shown in Figure 41–13. This flow switch is mounted in such a position that it can sense the movement of air produced by the air induction blower.

When thermostat contact TS1 closes, coil TR energizes and closes contact TR. This provides a circuit to motor starter 2M. When the air injection blower starts, flow switch FL1 closes its contact and permits the high-pressure pump motor to start. If the air injection blower motor stops for any reason, flow switch FL1 will disconnect motor starter 1M from the power line and stop operation of the high-pressure pump.

Although the circuit now operates as desired, the owner of the building later decides the blower should circulate air inside the building when the heating system is not in use. To satisfy this request, an AUTO-MANUAL switch is added as shown in Figure 41–14. When the switch is set in the AUTO position, it permits the blower motor to be controlled by the thermostat TS2. When the switch is set in the MANUAL position, it connects the coil of starter 3M directly to the power line and permits the blower motor to operate independently of the heating system.

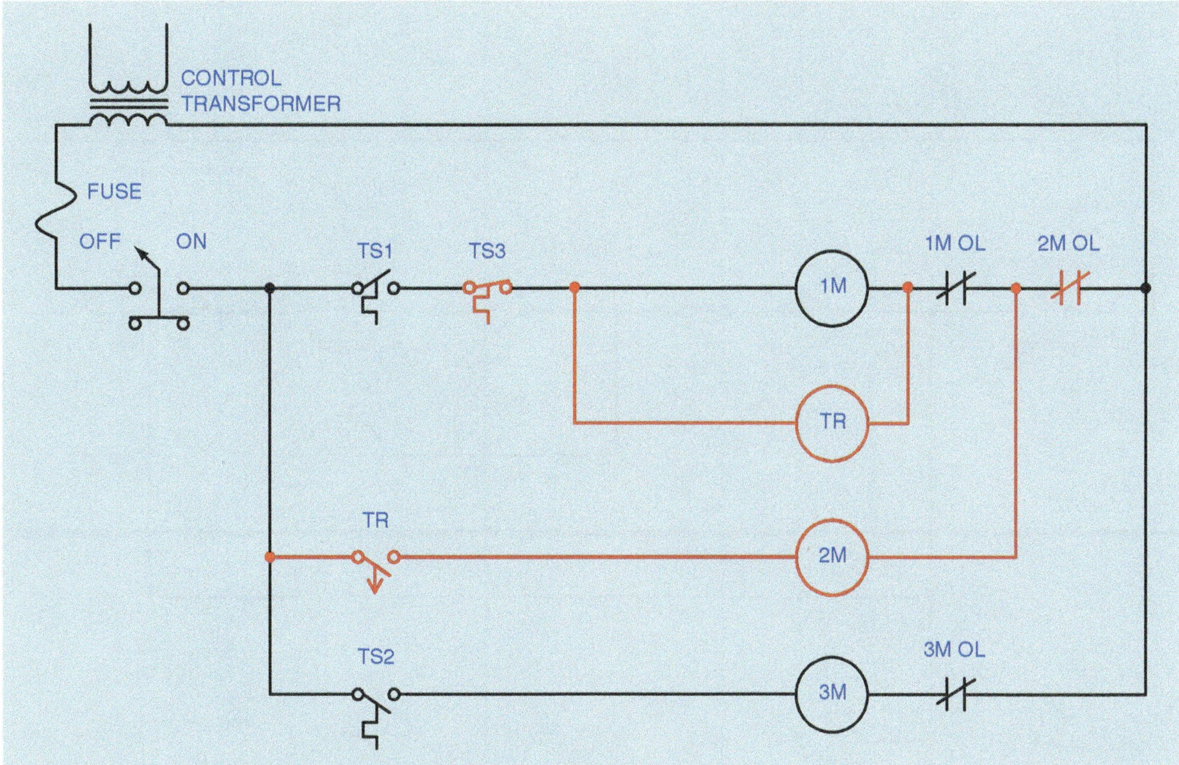

Figure 41–12 A timer is added to operate the air induction blower.

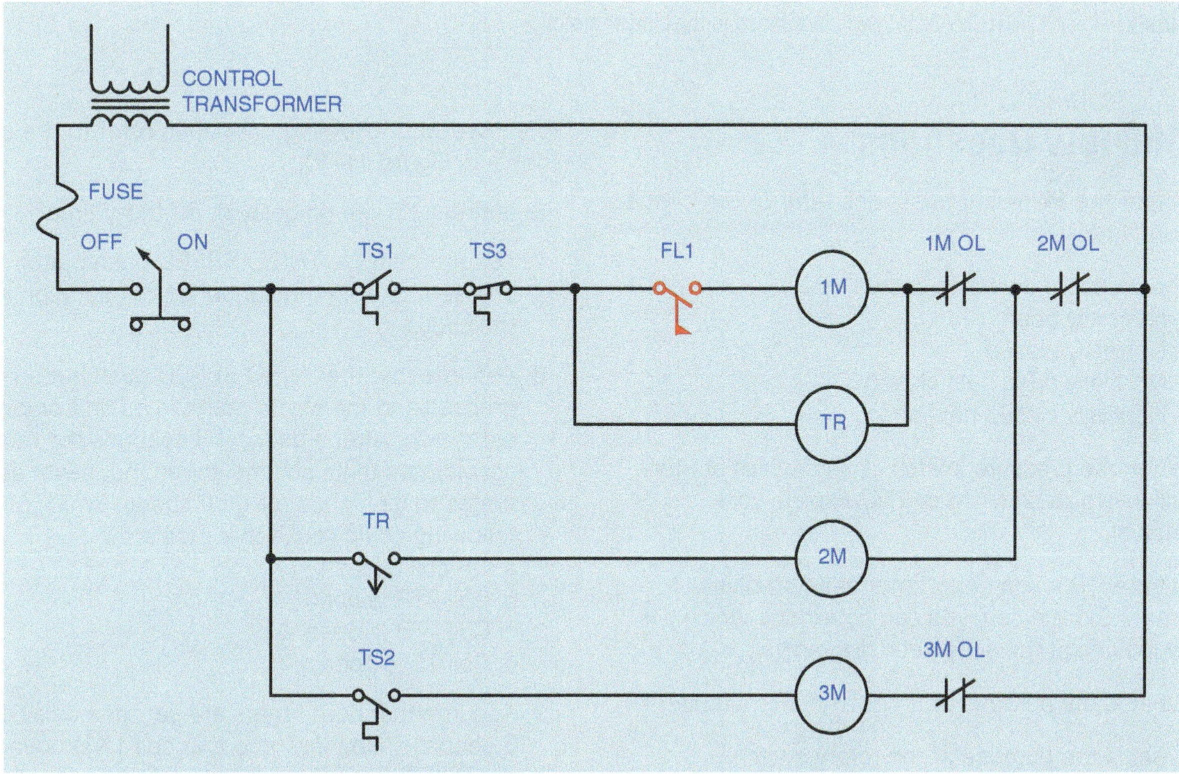

Figure 41–13 An air flow switch controls the operation of the high-pressure burner motor.

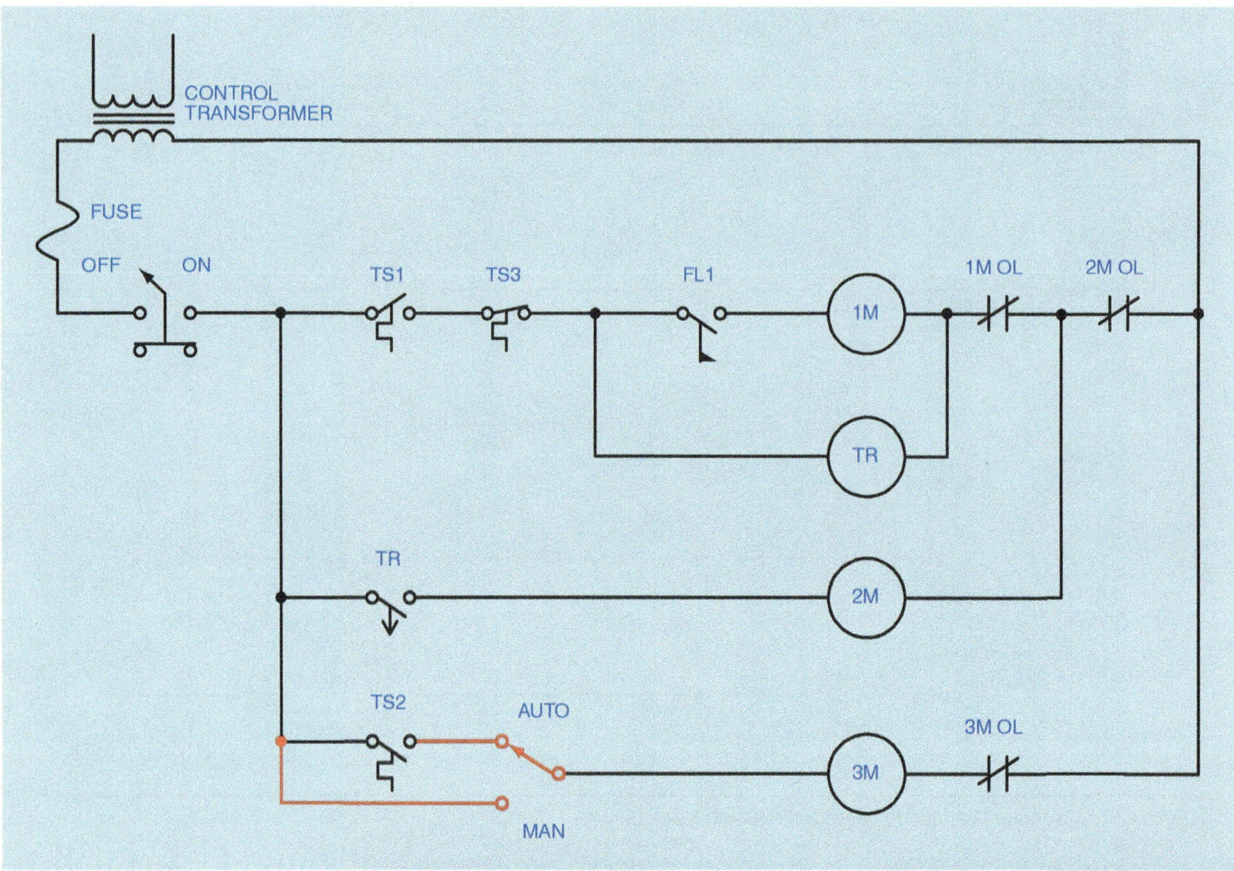

Figure 41–14 An AUTO-MANUAL switch is added to the blower motor.

Review Questions

To answer the following questions refer to the circuit in Figure 41–6.

1. The pressure switch is shown as:
   a. Normally open
   b. Normally closed
   c. Normally open held closed
   d. Normally closed held open
2. When the pressure switch closes, which starter will energize first, 1M or 2M? Explain your answer.
3. Is timer TR an on-delay timer or an off-delay timer? Explain how you can determine which it is by looking at the schematic diagram.
4. What is the purpose of timer TR in this circuit?
5. What is the purpose of the rotary switch connected after the pressure switch?

To answer the following questions refer to the circuit shown in Figure 41–10.

6. Is timer 1TR an on-delay or off-delay timer?
7. Assume that the THIRD SPEED push button is pressed. Explain the sequence of operation for the circuit.
8. Assume that the THIRD SPEED push button is pressed and the motor starts in its first or lowest speed. After a delay of 3 seconds the motor accelerates to its second speed, but never accelerates to its highest or third speed. Which of the following could cause this problem?
   a. CR coil is open.
   b. Coil 2TR is open.
   c. Coil 1TR is open.
   d. Coil 1S is open.
9. Assume that both timers are set for a delay of 3 seconds. Now assume that coil 1S is open. If the THIRD SPEED push button is pressed, will the motor accelerate to third speed after a delay of 6 seconds? Explain your answer.

10. Assume that timer 2TR is replaced with an off-delay timer, and that both timers are set for a delay of 3 seconds. Explain the operation of the circuit when the THIRD SPEED push button is pressed. Also explain the operation of the circuit when the STOP button is pressed.

To answer the following questions refer to the circuit shown in Figure 41–14.

11. Temperature switch TS1 is shown as:
    a. Normally open
    b. Normally closed
    c. Normally open held closed
    d. Normally closed held open
12. Temperature switch TS2 is shown as:
    a. Normally open
    b. Normally closed
    c. Normally open held closed
    d. Normally closed held open
13. Is timer TR an on-delay or off-delay timer?
14. Temperature switch TS3 is shown as:
    a. Normally open
    b. Normally closed
    c. Normally open held closed
    d. Normally closed held open
15. Assume that contact TS1 closes and the air injection blower motor starts operating, but the high-pressure pump motor does not start. What could cause this problem?
    a. Temperature switch TS3 is open.
    b. Coil 2M is open.
    c. Flow switch FL1 is defective and did not close.
    d. Coil TR is open.

Chapter 42

TROUBLESHOOTING

Objectives

After studying this chapter the student will be able to:

» Safely check a circuit to determine whether power is disconnected.

» Use a voltmeter to troubleshoot a control circuit.

» Use an ohmmeter to test for continuity.

» Use an ammeter to determine whether a motor is overloaded.

It is not a question of *if* a control circuit will eventually fail, but *when* it will fail. One of the main jobs of an industrial electrician is to troubleshoot and repair a control circuit when it fails. In order to repair or replace a faulty component, it is first necessary to determine which component is at fault. The three main instruments an electrician uses to troubleshoot a circuit are the voltmeter, ohmmeter, and ammeter. The voltmeter and ohmmeter are generally contained in the same meter (Figure 42–1). These meters are called *multimeters* because they can measure several different electrical quantities. Some electricians prefer to use a plunger type voltage tester because these testers are not susceptible to ghost voltages, Figure 42–2.High impedance voltmeters often give an indication of some amount of voltage, caused by

Courtesy of Amprobe

Figure 42–1 Digital multimeter.

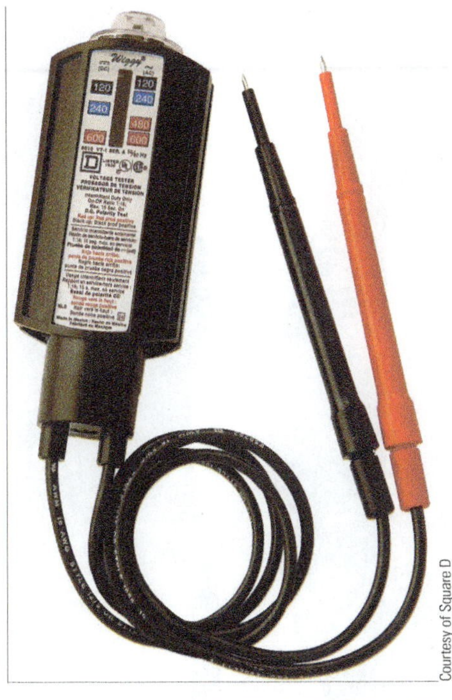

Courtesy of Square D

Figure 42–2 Low impedance voltage tester often called a Wiggy.

feedback and induction. Plunger type voltage testers are low impedance devices and require several milliamperes to operate. The disadvantage of plunger type voltage testers is that they cannot be used to test control systems that operate on low voltage, such as 24 volt systems.

Ammeters are generally clamp-on type (Figure 42–3). Both analog and digital meters are in common use. Clamp-on type ammeters have an advantage in that the circuit does not have to be broken to insert the meter in the line.

Figure 42–3 (A) Analog type clamp-on ammeter with vertical scale. (B) Analog type clamp-on ammeter with flat scale. (C) Clamp-on ammeter with digital scale.

Safety Precautions

It is often necessary to troubleshoot a circuit with power applied to the circuit. When this is the case, safety should be the first consideration. **When de-energizing or energizing a control cabinet or motor control center module, the electrician should be dressed in flame retardant clothing while wearing safety glasses, a face shield, and hard hat. Motor control centers employed throughout industry generally have the ability to release enough energy in an arc-fault situation to kill a person 30 feet away.** Another rule that should always be observed when energizing or de-energizing a circuit is to stand to the side of the control cabinet or module. Do not stand in front of the cabinet door when opening or closing the circuit. A direct short condition can cause the cabinet door to be blown off.

After the cabinet or module door has been opened, the power should be checked with a voltmeter to make certain the power is off. A procedure called *check, test, check*, should be used to make certain that the power is off.

1. Check the voltmeter on a known source of voltage to make certain the meter is operating properly.
2. Test the circuit voltage to make certain that it is off.
3. Check the voltmeter on a known source of voltage again to make certain that the meter is still working properly.

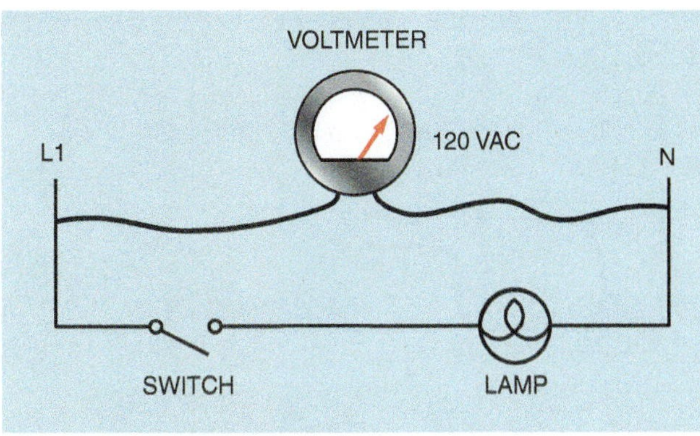

Figure 42–4 The voltmeter measures electrical pressure between two points.

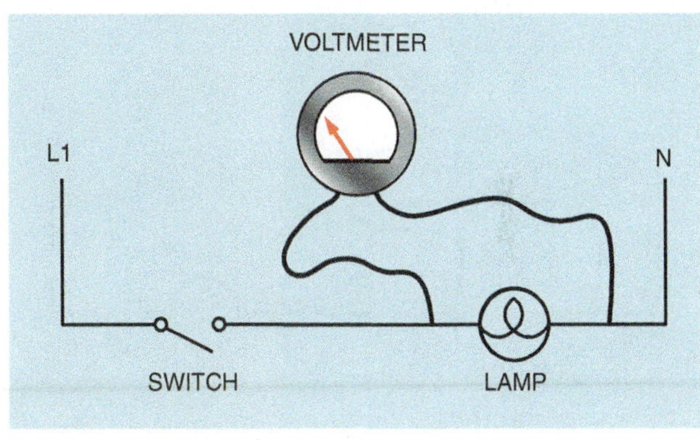

Figure 42–5 The voltmeter is connected across the lamp.

Voltmeter Basics

Recall that one definition of voltage is *electrical pressure*. The voltmeter indicates the amount of potential between two points in much the same way a pressure gauge indicates the pressure difference between two points. The circuit in Figure 42–4 assumes that a voltage of 120 volts exists between L1 and N. If the leads of a voltmeter were to be connected between L1 and N the meter would indicate 120 volts.

Now assume that the leads of the voltmeter are connected across the lamp (Figure 42–5).

> *Question 1:* Assuming that the lamp filament is good, would the voltmeter indicate 0 volts, 120 volts, or some value between 0 and 120 volts?
>
> *Answer:* The voltmeter would indicate 0 volts. In the circuit shown in Figure 42–5, the switch and lamp are connected in series. One of the basic rules for series circuits is that the voltage drop across all circuit components must equal the applied voltage. The amount of voltage drop across each component is proportional to the resistance of the components and the amount of current flow. Because the switch is open in this example, there is no current flow through the lamp filament and no voltage drop.
>
> *Question 2:* If the voltmeter was to be connected across the switch as shown in Figure 42–6, would it indicate 0 volts, 120 volts, or some value between 0 and 120 volts?
>
> *Answer:* The voltmeter would indicate 120 volts. Because the switch is an open circuit, the resistance is infinite at this point, which is millions of times greater than the resistance of the lamp filament. Recall that voltage is electrical pressure. The only current flow through this circuit is the current flowing through the voltmeter and the lamp filament (Figure 42–7).

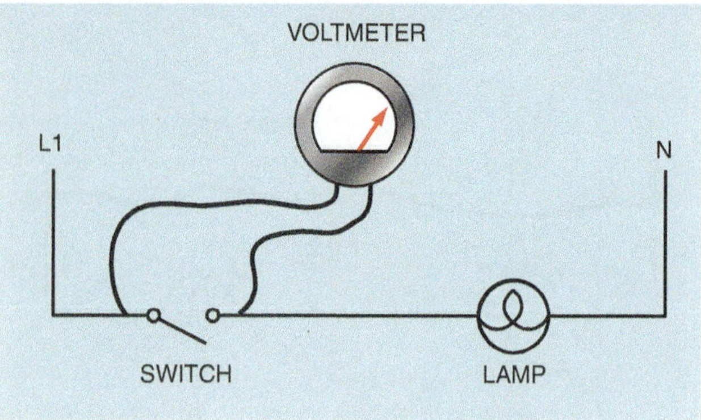

Figure 42–6 The voltmeter is connected across the switch.

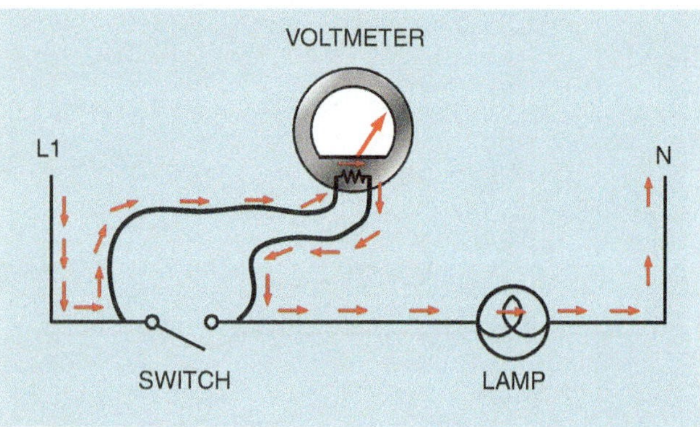

Figure 42–7 A current path exists through the voltmeter and lamp filament.

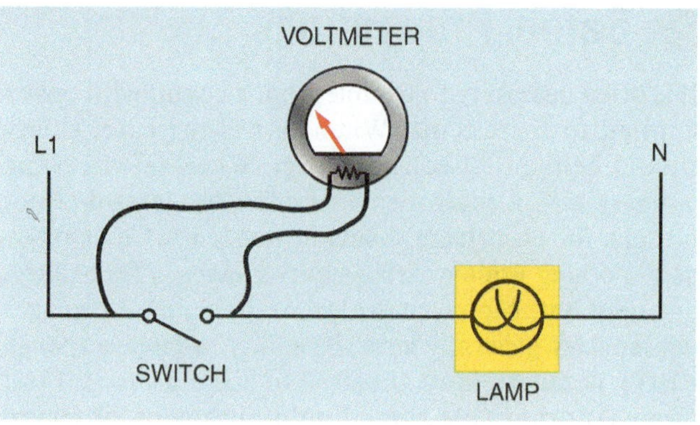

Figure 42–8 The lamp filament is burned open.

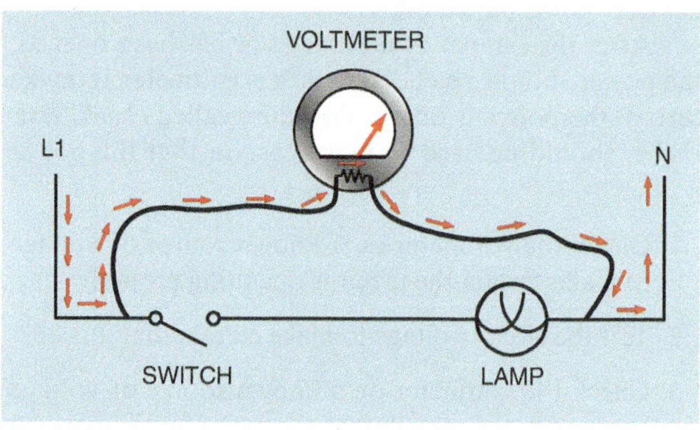

Figure 42–9 The voltmeter is connected across both components.

Question 3: If the total or applied voltage in a series circuit must equal the sum of the voltage drops across each component, why is all the voltage drop across the voltmeter resistor and none across the lamp filament?

Answer: There is some voltage drop across the lamp filament because the current of the voltmeter is flowing through it. The amount of voltage drop across the filament, however, is so small as compared to the voltage drop across the voltmeter it is generally considered to be zero. Assume the lamp filament to have a resistance of 50 ohms. Now assume that the voltmeter is a digital meter and has a resistance of 10,000,000 ohms. The total circuit resistance is 10,000,050 ohms. The total circuit current is 0.000,011,999 ampere (120/10,000,050) or about 12 microamps. The voltage drop across the lamp filament would be approximately 0.0006 volts or 0.6 millivolts (50 Ω × 12 μA).

Question 4: Now assume that the lamp filament is open or burned out. Would the voltmeter in Figure 42–8 indicate 0 volts, 120 volts, or some value between 0 and 120 volts?

Answer: The voltmeter would indicate 0 volts. If the lamp filament is open or burned out, a current path for the voltmeter does not exist and the voltmeter would indicate 0 volts. In order for the voltmeter to indicate voltage, it would have to be connected across both components so that a complete circuit would exist from L1 to N (Figure 42–9).

Question 5: Assume that the lamp filament is not open or burned out and that the switch has been closed or turned on. If the voltmeter is connected across the switch, would it indicate 0 volts, 120 volts, or some value between 0 and 120 volts (Figure 42–10)?

Answer: The voltmeter would indicate 0 volts. Now that the switch is closed, the contact resistance is

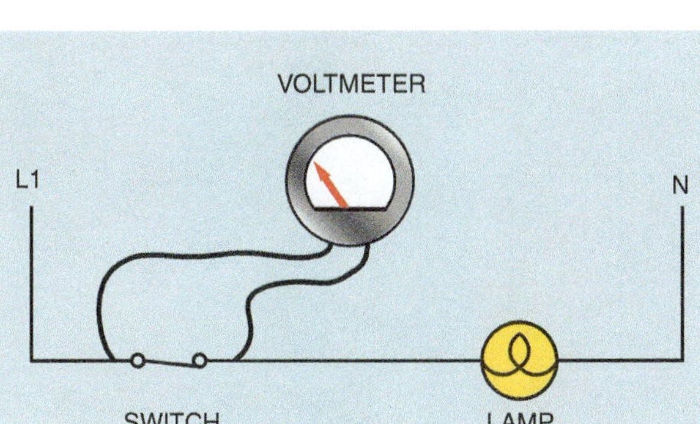

Figure 42–10 The switch is turned on or closed.

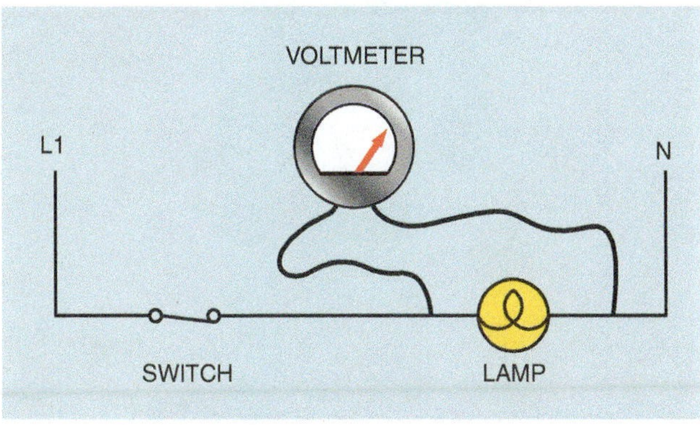

Figure 42–11 Practically all the voltage drop is across the lamp.

extremely small and the lamp filament exhibits a much higher resistance than the switch. Practically all the voltage drop will now appear across the lamp (Figure 42–11).

Test Procedure Example 1

The type of problem determines the procedure to be employed when troubleshooting a circuit. Assume that an overload relay has tripped several times. The first step is to determine what conditions could cause this problem. If the overload relay is a thermal type, a source of heat is the likely cause of the problem. Make mental notes of what could cause the overload relay to become overheated.

1. Excessive motor current.
2. High ambient temperature.
3. Loose connections.
4. Incorrect wire size.

If the motor has been operating without a problem for some period of time, incorrect wire size can probably be eliminated. If it is a new installation, it would be a factor to consider.

Because overload relays are intended to disconnect the motor from the power line in the event that the current draw becomes excessive, the motor should be checked for excessive current. The first step is to determine the normal full load current from the nameplate on the motor. The next step is to determine the percent of full load current setting for the overload relay.

EXAMPLE: A motor nameplate indicates the full load current of the motor is 46 amperes. The nameplate also indicates the motor has a service factor of 1.00. The *National Electrical Code* indicates the overload should be set to trip at 115 percent of the full load current. Overload heaters are generally selected from a manufacturers chart based on full load current and percentage as determined by service factor and/or temperature rise listed on the motor nameplate.

The next step is to check the running current of the motor with an ammeter. This is generally accomplished by measuring the motor current at the overload relay (Figure 42–12). The current in each phase should be measured. If the motor is operating properly, the

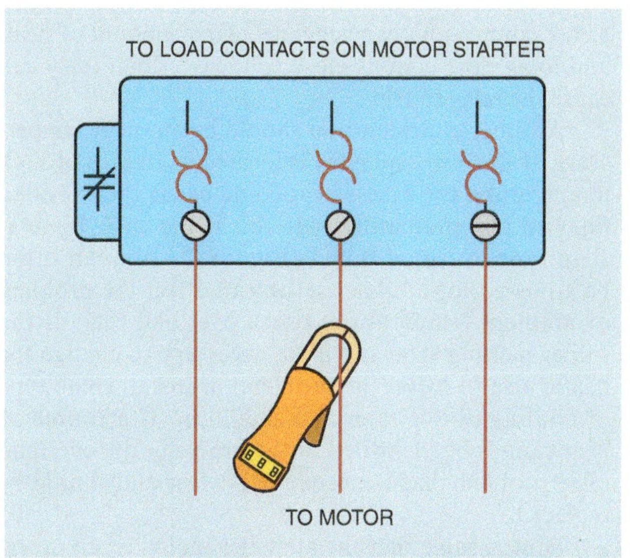

Figure 42–12 A clamp-on ammeter is used to check motor current.

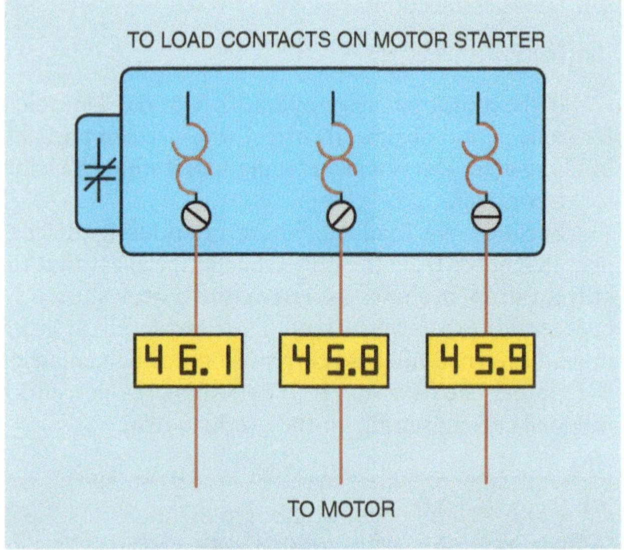

Figure 42–13 Ammeter readings indicate that the motor is operating normally.

Figure 42–14 Bimetal strip type overload relays can be set for a higher value of current.

readings may not be exactly the same, but they should be close to the full load current value if the motor is operating under load and relatively close to each other. In the example shown in Figure 42–13, phase 1 has a current flow of 46.1 amperes, phase 2 has a current flow of 45.8 amperes, and phase 3 has a current flow of 45.9 amperes. These values indicate that the motor is operating normally. Because the ammeter indicates that the motor is operating normally, other sources of heat should be considered. After turning off the power, check all connections to ensure that they are tight. Loose connections can generate a large amount of heat, and loose connections close to the overload relay can cause the relay to trip.

Another consideration should be ambient temperature. If the overload relay is located in an area of high temperature, the excess heat could cause the overload relay to trip prematurely. If this is the case, bimetal strip type overload relays (Figure 42–14), can often be adjusted for a higher setting to offset the problem of ambient temperature. If the overload relay is the solder melting type, it will be necessary to change the heater size to offset the problem, or install some type of cooling device such as a small fan. If a source of heat cannot be identified as the problem, the overload relay probably has a mechanical defect and should be replaced.

Now assume that the ammeter indicated an excessively high current reading on all three phases. In the example shown in Figure 42–15, phase 1 has a current flow of 58.1 amperes, phase 2 has a current flow of

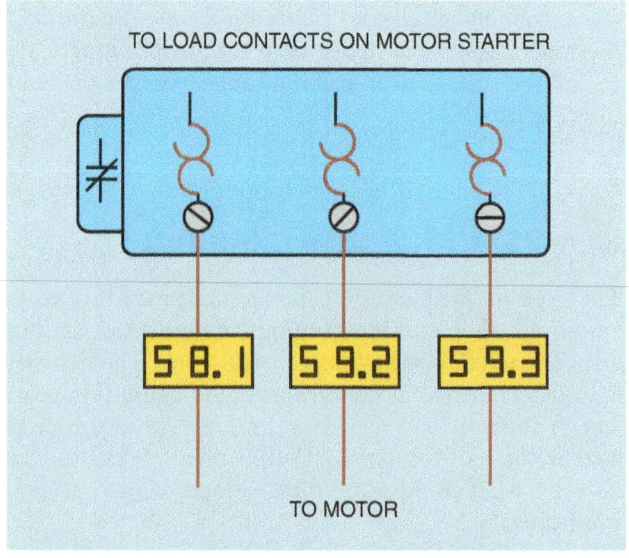

Figure 42–15 Ammeter readings indicate that the motor is overloaded.

59.2 amperes, and phase 3 has a current flow of 59.3 amperes. Recall that the full load nameplate current for this motor is 46 amperes. These values indicate that the motor is overloaded. The motor and load should be checked for some type of mechanical problem, such as a bad bearing or possibly a brake that has become engaged.

Now assume that the ammeter indicates one phase with normal current and two phases that have excessively high current. In the example shown in Figure 42–16, phase 1 has a current flow of 45.8 amperes, phase 2 has a current flow of 73.2 amperes, and phase 3 has a current flow of 74.3 amperes. Two phases with excessively high current indicate that the motor probably has a shorted winding. If two phases have a normal amount of current and one phase is excessively high, it is a good indication that one of the phases has become grounded to the case of the motor.

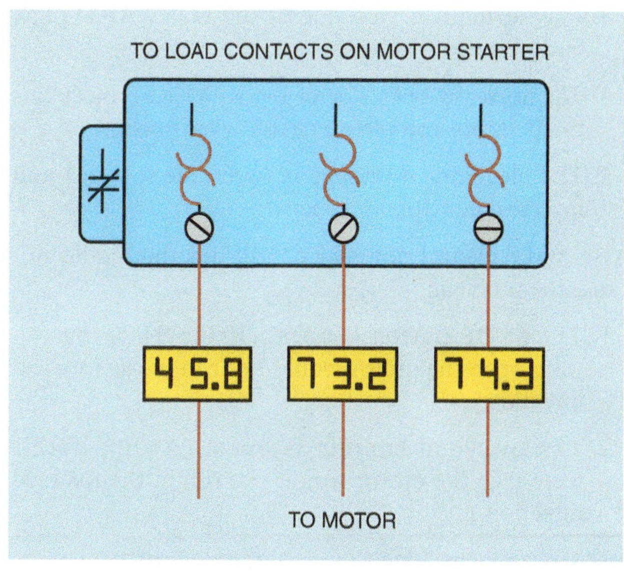

Figure 42–16 Ammeter readings indicate that the motor has a shorted winding.

Test Procedure Example 2

The circuit shown in Figure 42–17 is a reversing starter with electrical and mechanical interlocks. Note that double-acting push buttons are used to disconnect one contactor if the START button for the other contactor is pressed. Now assume that if the motor is operating in the forward direction, and the REVERSE push button is pressed, the forward contactor de-energizes but the reverse contactor does not. If the FORWARD push button is pressed, the motor will restart in the forward direction.

To begin troubleshooting this problem, make mental notes of problems that could cause this condition.

1. The reverse contactor coil is defective.
2. The normally closed F auxiliary contact is open.

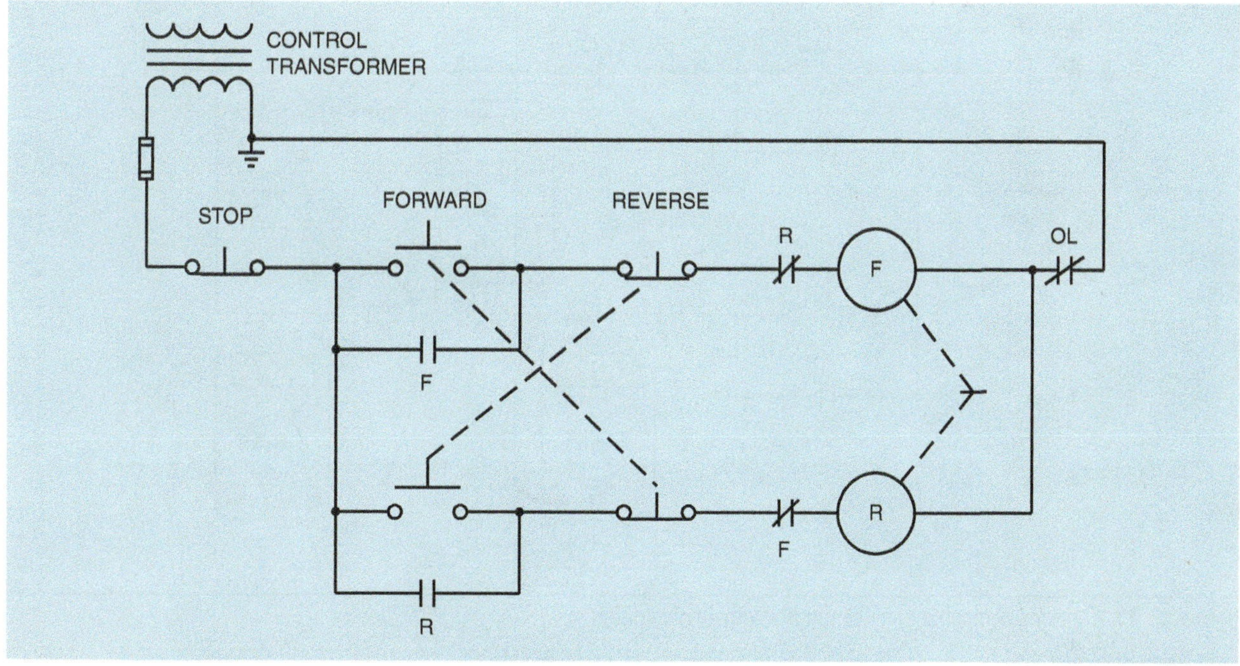

Figure 42–17 Reversing starter with interlocks.

3. The normally closed side of the FORWARD push button is open.
4. The normally open side of the REVERSE push button does not complete a circuit when pressed.
5. The mechanical linkage between the forward and reverse contactors is defective.

Also make mental notes of conditions that could not cause the problem.

1. The STOP button is open. (If the STOP button were open the motor would not run in the forward direction.)
2. The overload contact is open. (Again, if this were true the motor would not run in the forward direction.)

To begin checking this circuit, an ohmmeter can be used to determine whether a complete circuit path exists through certain components. **When using an ohmmeter, make certain that the power is disconnected from the circuit.** A good way to do this in most control circuits is to remove the control transformer fuse. The ohmmeter can be used to check the continuity of the reverse contactor coil, the normally closed F contact, the normally closed section of the FORWARD push button, and across the normally open REVERSE push button when it is pressed (Figure 42–18).

The ohmmeter can be used to test the starter coil for a complete circuit to determine whether the winding has been burned open, but it is generally not possible to determine whether the coil is shorted. To make a final determination, it is generally necessary to apply power to the circuit and check for voltage across the coil. Because the REVERSE push button must be closed to make this measurement, it is common practice to connect a fused jumper across the push button if there is no one to hold the button closed (Figure 42–19). A fused jumper is shown in Figure 42–20. When using a fused jumper, power should be disconnected when the jumper is connected across the component. After the jumper is in position, power can be restored to the circuit. If voltage appears across the coil, it is an indication that the coil is defective and should be replaced or that the mechanical interlock between the forward and reverse contactors is defective.

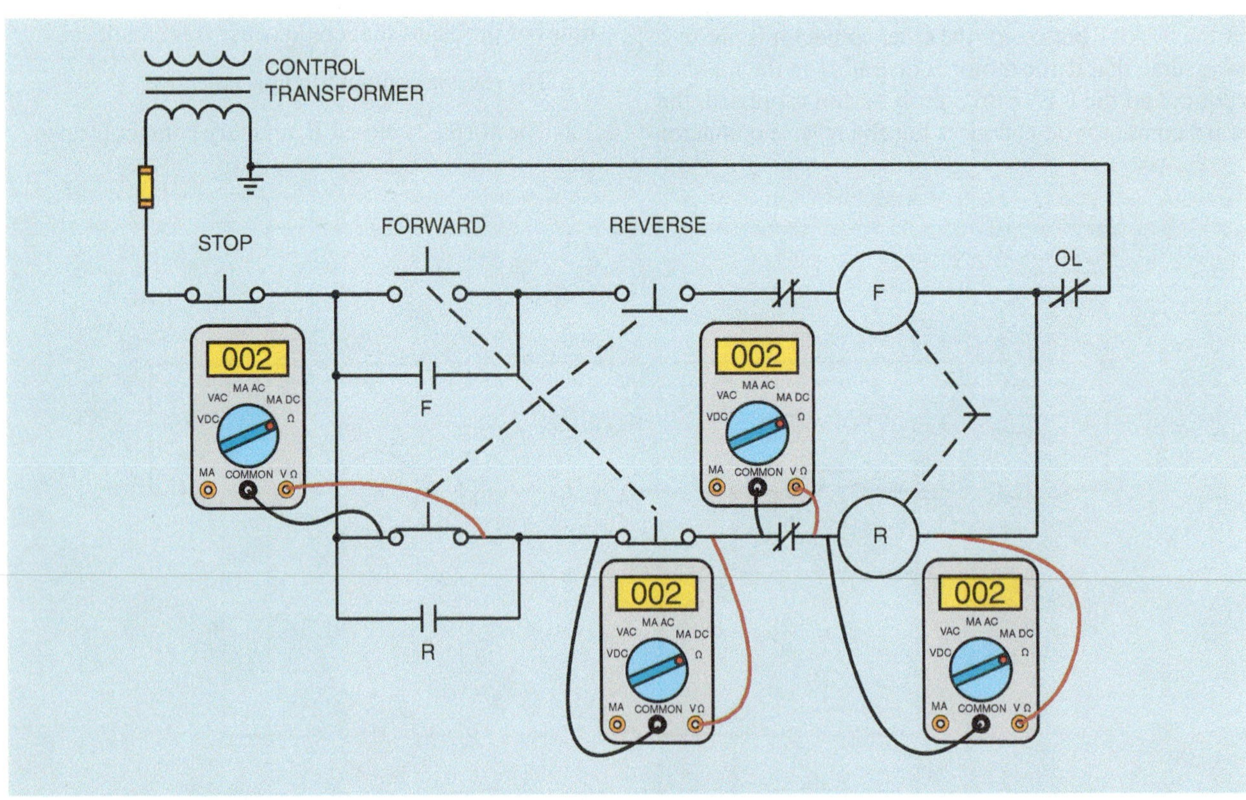

Figure 42–18 Checking components for continuity with an ohmmeter.

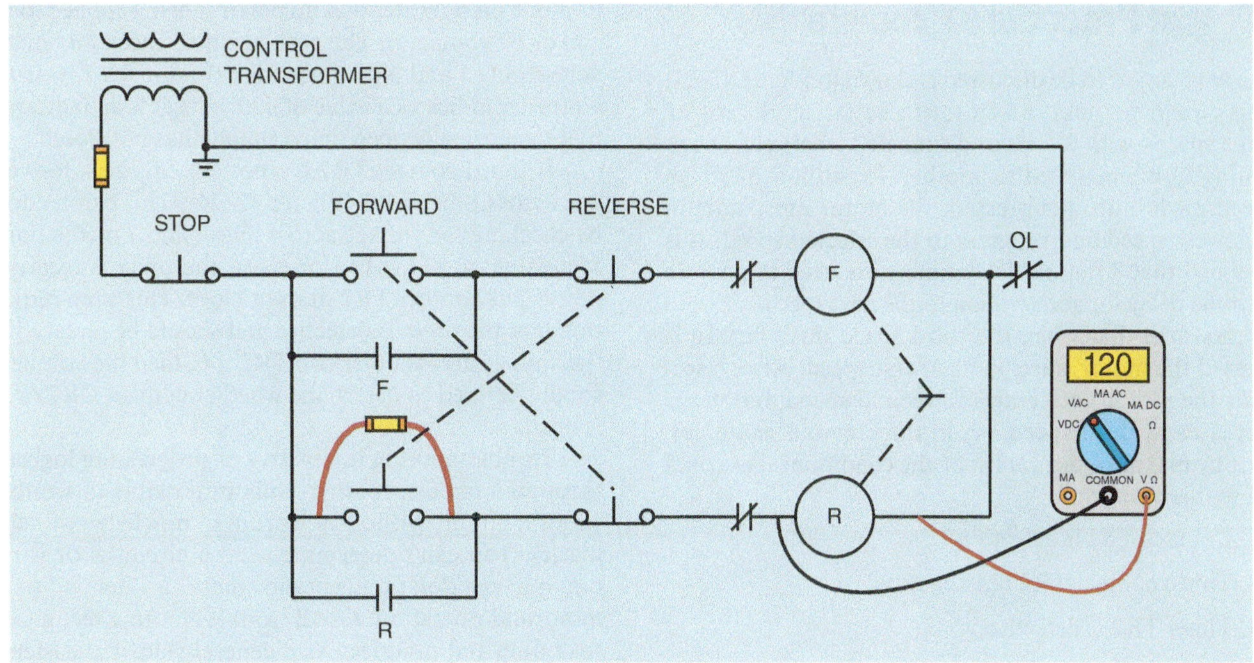

Figure 42–19 Testing to determine whether voltage is being applied to the coil.

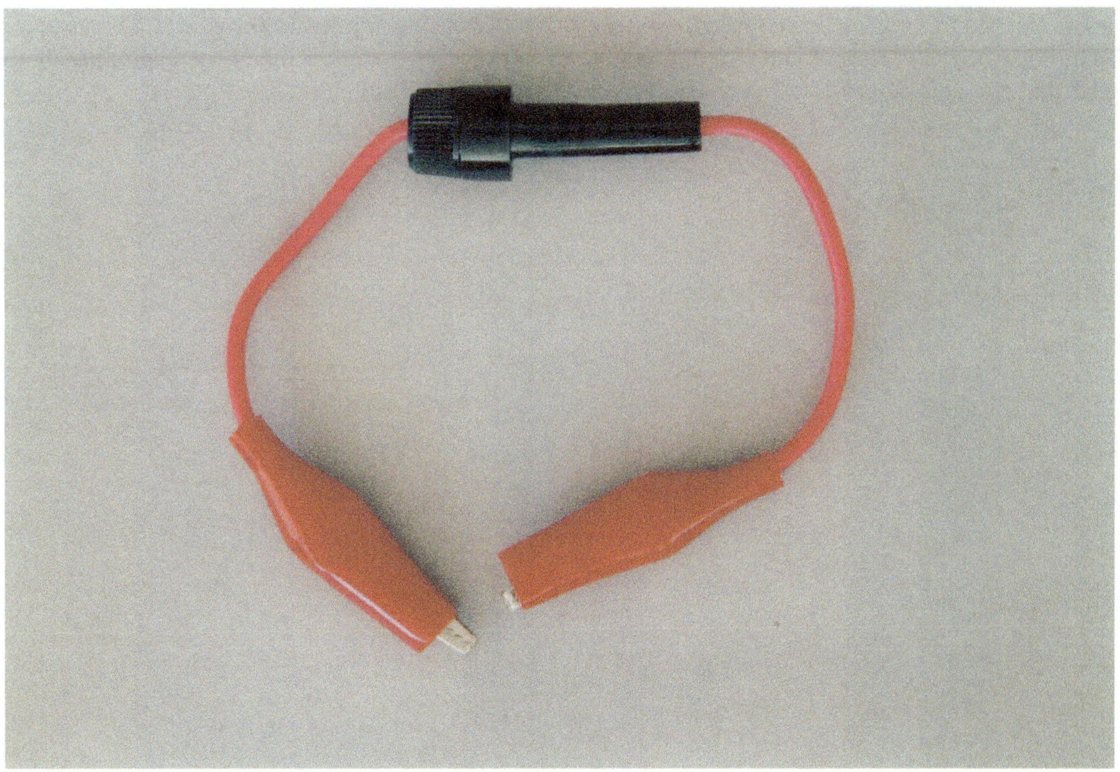

Figure 42–20 A fused jumper is often used to complete a circuit when troubleshooting.

Test Procedure Example 3

The next circuit to be discussed is shown in Figure 42–21. This circuit permits the motor to be started in any of three speeds with a 5-second time delay between accelerating from one speed to another. Regardless of which speed push button is pressed, the motor must start in its lowest speed and progress to the selected speed. It is assumed that 8 pin on-delay timers are used to provide the time delay for acceleration to the next speed.

Assume that when the third speed push button is pressed the motor starts in its lowest speed. After 5 seconds the motor accelerates to second speed, but never increases to third speed. As in the previous examples, start by making a mental list of the conditions that could cause this problem.

1. Contactor S2 is defective.
2. Timed contact TR2 did not close.
3. Timer TR2 is defective.
4. CR2 or S1 contacts connected in series with timer TR2 did not close.

Begin troubleshooting this circuit by pressing the THIRD SPEED push button and permit the motor to accelerate to second speed. Wait at least 5 seconds after the motor has reached third speed and connect a voltmeter across the coil of S2 contactor (Figure 42–22). It will be assumed that the voltmeter indicated a reading of 0 volts. This indicates that no power is being applied to the coil of S2 contactor. The next step is to check for voltage across pins 1 and 3 of timer TR2 (Figure 42–23). If the voltmeter indicates a value of 120 volts, it is an indication that the normally open timed contact has not closed.

If timed contact TR2 has not closed, check for voltage across timer TR2 (Figure 42–24). This can be done by checking for voltage across pins 2 and 7 of the timer. If a value of 120 volts is present, the timer is receiving power but contact TR2 did not close. This is an indication that the timer is defective and should be replaced. If the voltage across timer coil TR2 is 0, then the voltmeter should be used to determine whether contact CR2 or S1 is open.

Troubleshooting is a matter of progressing logically through a circuit. It is virtually impossible to troubleshoot a circuit without a working knowledge of schematics. You can't determine what a circuit is or is not doing if you don't understand what it is intended to do in normal operation. Good troubleshooting techniques take time and practice. As a general rule, it is easier to progress backward though the circuit until the problem is identified. For example, in this circuit, contactor S2 provided the last step of acceleration for the motor. By starting a contactor S2 and progressing backward until determining what component was responsible for no power being applied to the coil of S2 was much simpler and faster than starting at the beginning of the circuit and going through each component.

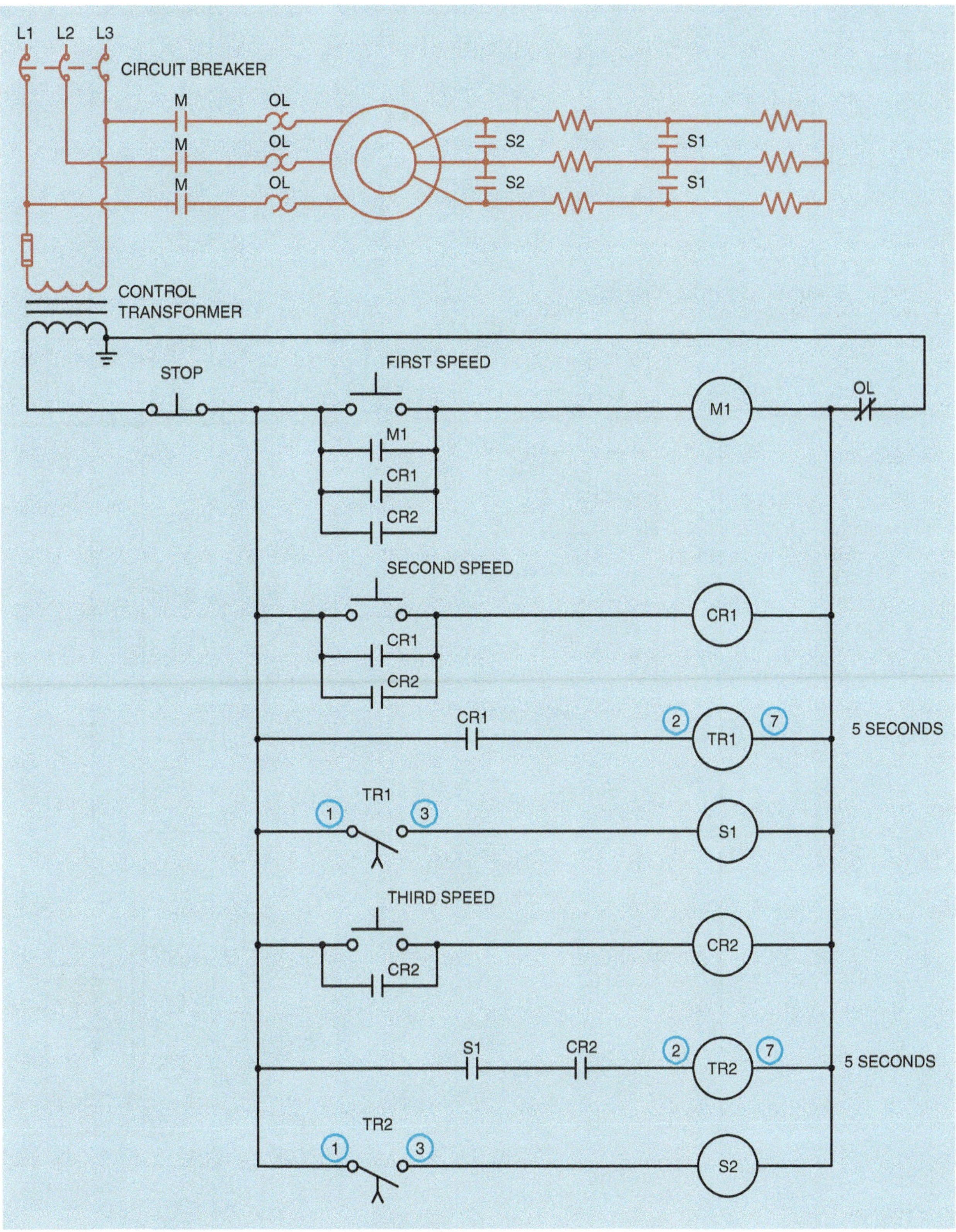

Figure 42–21 Three speed control for a wound rotor induction motor.

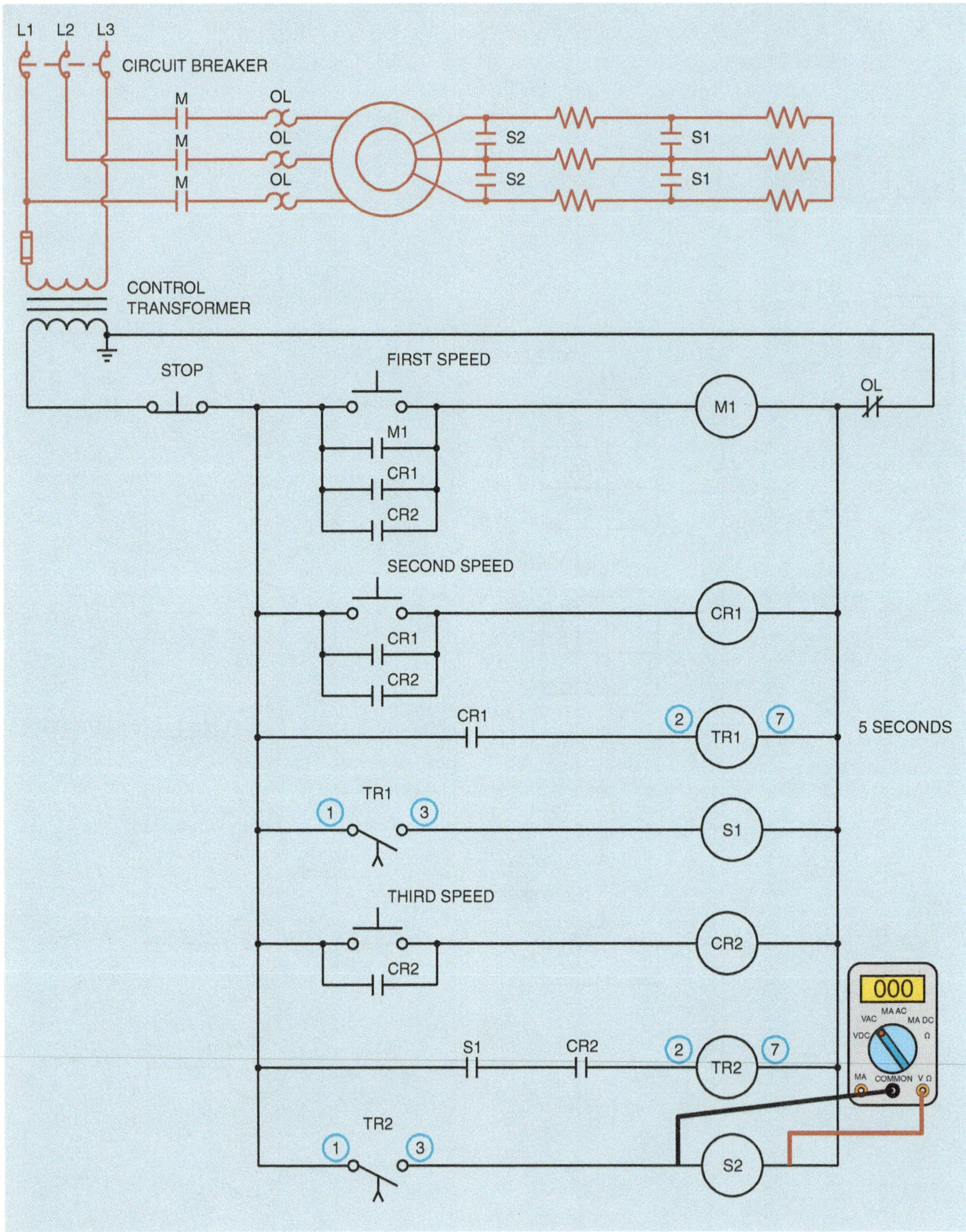

Figure 42–22 Checking for voltage across S2 coil.

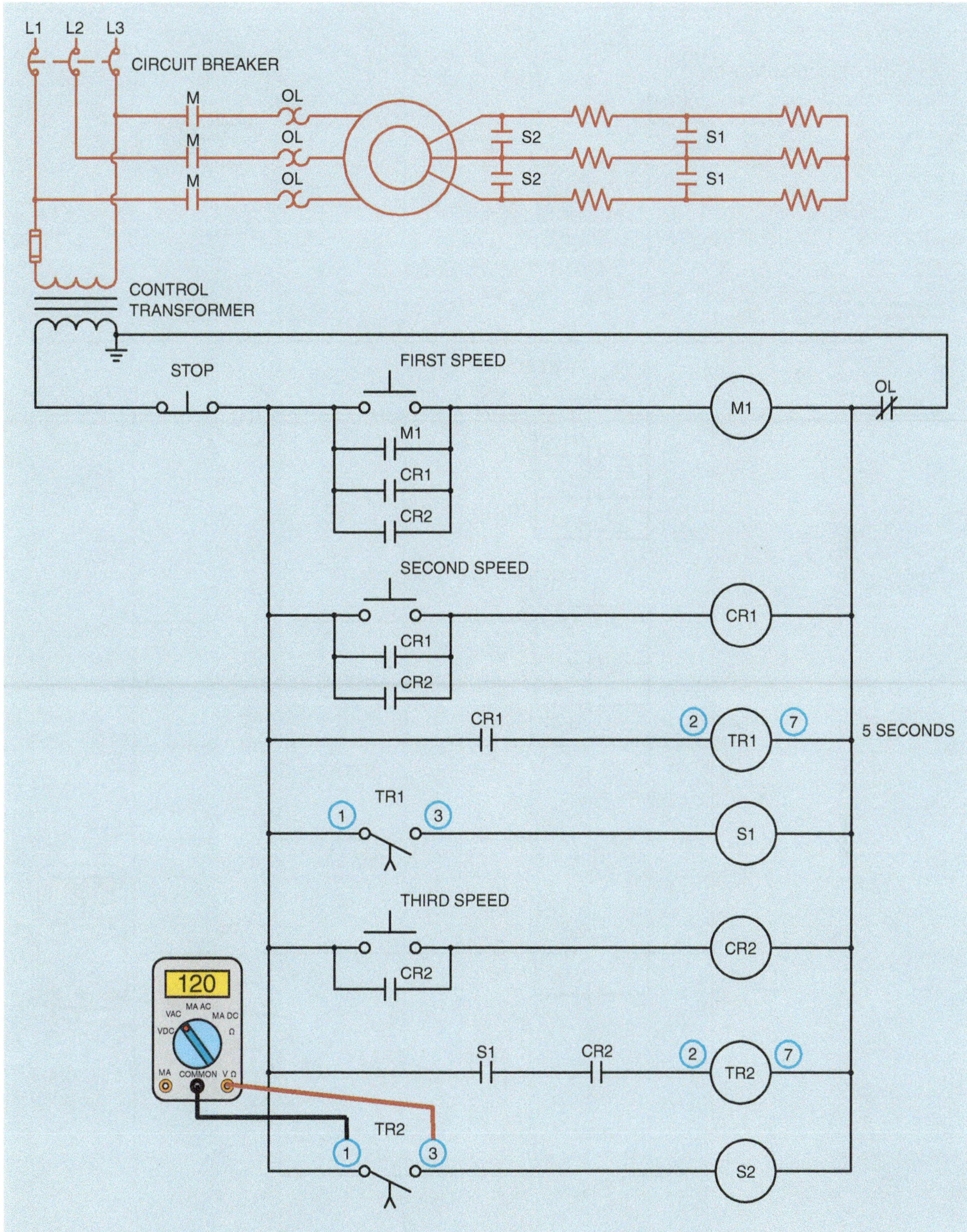

Figure 42–23 Checking for voltage across pins 1 and 3 of TR2 timer.

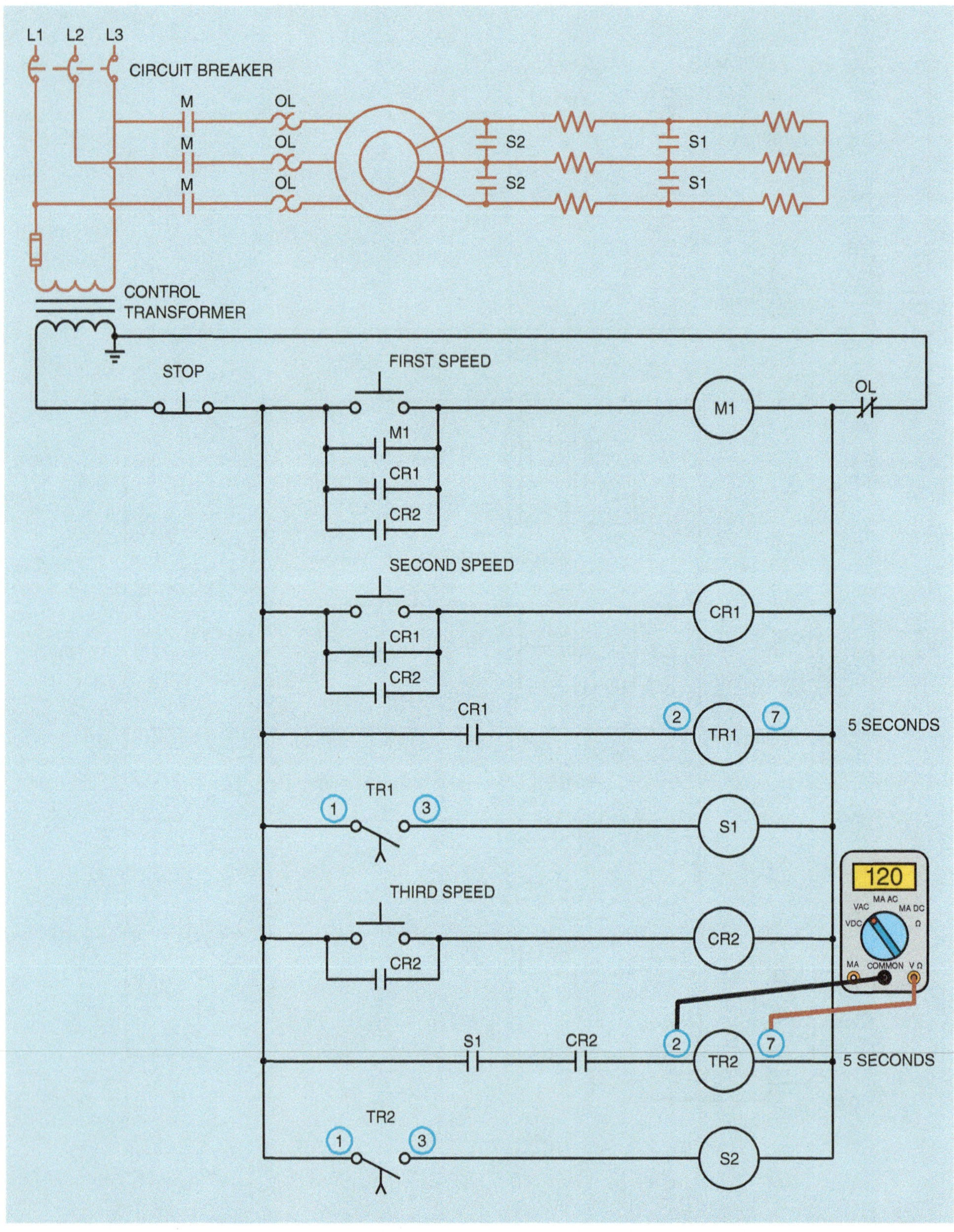

Figure 42–24 Checking for voltage across TR2 coil.

Motors

The control circuit is intended to control the operation of a motor or motors. It is sometimes necessary to determine if a problem exists with the motor itself. If the motor is in operation, the fastest way to check for a problem is to measure the current of each line supplying power to the motor. The running current should be within the nameplate full load current rating. Also, the currents should be within a few percent of each other. If the current is high on all phases, it could be an indication that there is a mechanical problem with the motor or the load. Mechanical problems can be caused by bad bearing in the motor or load, or by shorted or grounded motor windings. It is also possible that mechanical brakes are not adjusted properly and the brake lining is dragging when the motor is in operation. To check for a bad bearing it is generally necessary to disconnect the motor from the load. The rotor should turn freely. A bad bearing will often cause the rotor to drag and often times a grinding sound can be heard. In some instances it may be necessary to use a vibration sensor to determine if a bearing is bad.

Windings

If the currents measured are significantly different in value it can be an indication of a problem with the motor windings. Often a visual inspection will reveal a faulty winding. A strong acidic smell can also reveal a badly charred winding. If the winding is grounded, an ohmmeter can generally be used to check for continuity between the T leads and the case of the motor. If only one winding is grounded it will exhibit a lower resistance to the motor case than the other two, Figure 42–25.

Open windings can also be determined with an ohmmeter by measuring the resistance between each T lead. If one winding is open it will exhibit no continuity between it and the other two T leads, Figure 42–26.

Determining if a winding is shorted with an ohmmeter can be extremely difficult. In theory, if one winding is shorted it should have a lower resistance than the other two. Therefore, an ohmmeter should measure a lower resistance between the shorted winding and the other two T leads than it will between the two windings that are not shorted. In practice, the winding resistance of large horsepower motors is very low to begin with, sometimes less than 1 ohm. To determine if a motor winding is shorted it is generally necessary to measure the inductance of the windings. This can be accomplished with an inductance bridge if one is available. Another method that often works employs the use of an autotransformer to supply a low voltage to the windings and measure the current flow between two T leads at a time, Figure 42–27. Autotransformers that can supply a variable voltage

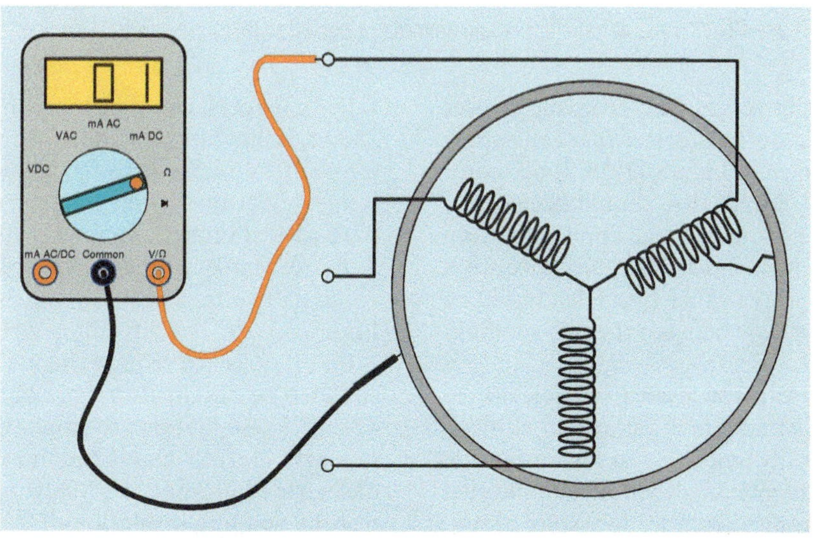

Figure 42–25 A grounded winding can generally be found with an ohmmeter.

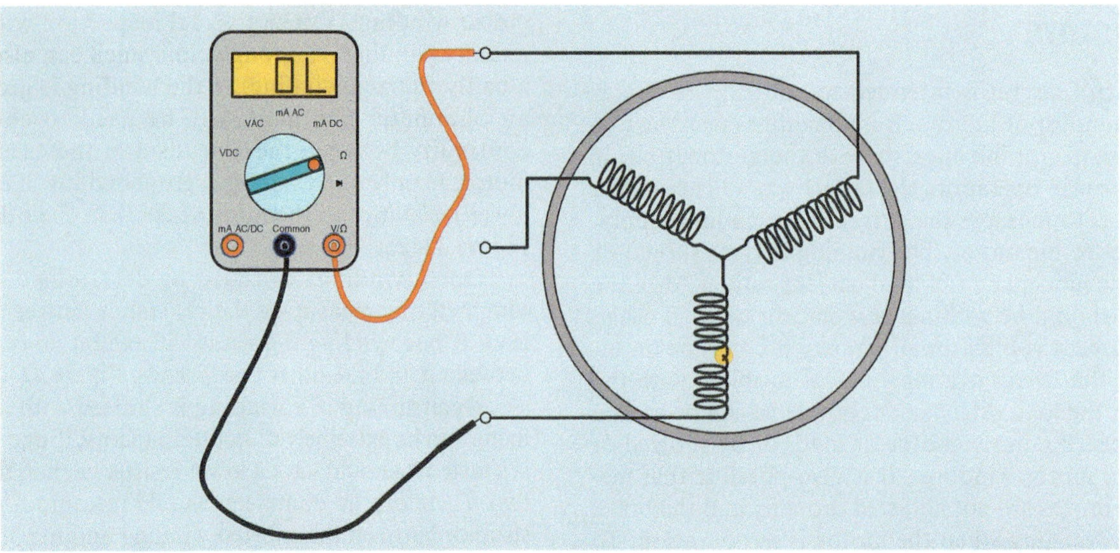

Figure 42–26 The ohmmeter will indicate no continuity for an open winding.

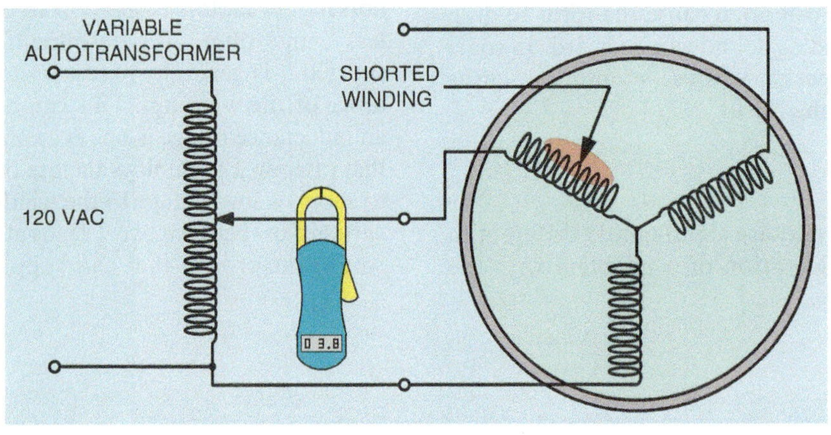

Figure 42–27 An autotransformer and ammeter are used to test for a shorted winding.

permit the voltage to be increased slowly to avoid an over current condition. An autotransformer that can supply a variable voltage is shown in Figure 42–28. If all motor windings are good, the current flow should be approximately the same with the same voltage applied between all T leads. If one winding is shorted the current between the shorted winding and the other two T leads will be greater than the current flow between the two windings that are not shorted.

Megohmmeters are often used to check motor windings. These meters measure resistance in millions of ohms and are generally employed to test insulation resistance. Common ohmmeters depend on an internal battery that generally ranges from 1.5 to 9 volts to supply the voltage for making a resistance measurement. This low voltage is often not high enough to detect a problem with wire insulation that may break down under a higher voltage. Megohmmeters, commonly called meggers, can develop voltages that range from 250 to 5000 volts depending on the model of the meter and the range setting. Many meggers contain a range selection switch that permits the meter to be used as a standard ohmmeter or as a megohmmeter. A hand-crank megger with voltage settings of 100 volts, 250 volts, 500 volts, and 1000 volts is shown in Figure 42–29. The advantage of a hand-crank megger is that it does not require the use of batteries. Meggers can also be obtained that are battery powered, Figure 42–30. These meters are small, light weight, and particularly useful when it becomes necessary to test the dielectric of a capacitor. Meggers are generally used to test the winding insulation of a motor to determine if it has become weak or damaged. To test the insulation, connect one of the meter leads to the case of the motor and the other to each of the motor T leads one at a time. Set the voltage range to a value that is close to

Figure 42–28 An autotransformer that can supply a variable voltage.

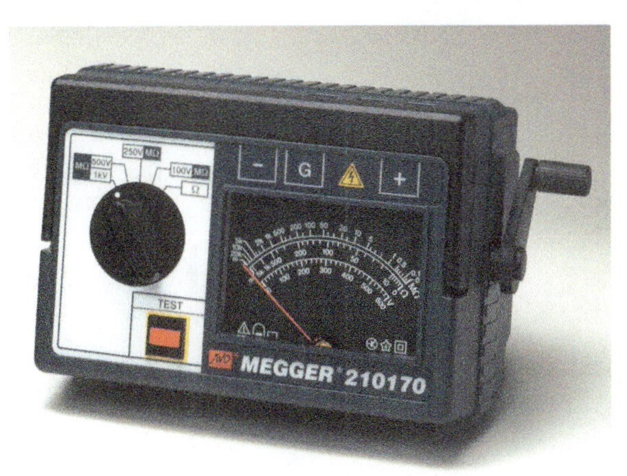

Courtesy of Megger

Figure 42–29 A hand-crank Megger.

Courtesy of Megger

Figure 42–30 Battery powered Megger.

or slightly higher than the rated motor voltage. The insulation should measure in the millions of ohms if it is good. Meggers cannot be used to determine if a winding is shorted.

Motor Overheating

A very common problem faced by many electricians is motor overheating. Motor overheating can be caused by several factors:

- Mechanical
- Poor Ventilation
- Improper Wire Size
- Overload Heaters Not Sized Correctly
- Motor Undersized for the Load
- Over and Under Voltage
- Unbalanced Voltages

Overheating can lead to poor efficiency and motor burnout. If the motor is overheating the cause should be determined and corrected.

Mechanical

The mechanical problems with bearings and windings discussed previously can cause the motor to overheat.

Poor Ventilation

Poor ventilation can cause an accumulation of dirt or other foreign matter to accumulate on the motor. Also, if the motor is located in an enclosed space there may not be enough air circulation to remove heat from the motor. The amount of heat removal is proportional to the surface area of the motor, the difference between motor temperature and ambient temperature, and the amount of air flow. Fins are often part of the motor case to increase surface area to aid in cooling, Figure 42–31. In areas of high ambient temperature it may be necessary to place a fan in a position that it will blow on the motor. The increased air flow will remove heat at a faster rate.

Improper Wire Size

If the wire supplying the motor is too small it will cause excessive voltage drop. The procedure for determining the proper wire size in accord with the *National Electrical Code* was presented in Chapter 37. The only other basic consideration is the length of the conductor. The wire size determined in accord with the NEC assumes that the wire length will not greatly affect the resistance and therefore the amount of voltage drop at the motor. As a general rule, for each 100 feet of wire length the size should be increased one standard size. Assume that a motor requires a 10 AWG copper conductor in accord with the NEC. Now assume that the motor is located 150 feet from the source. The conductor size should be

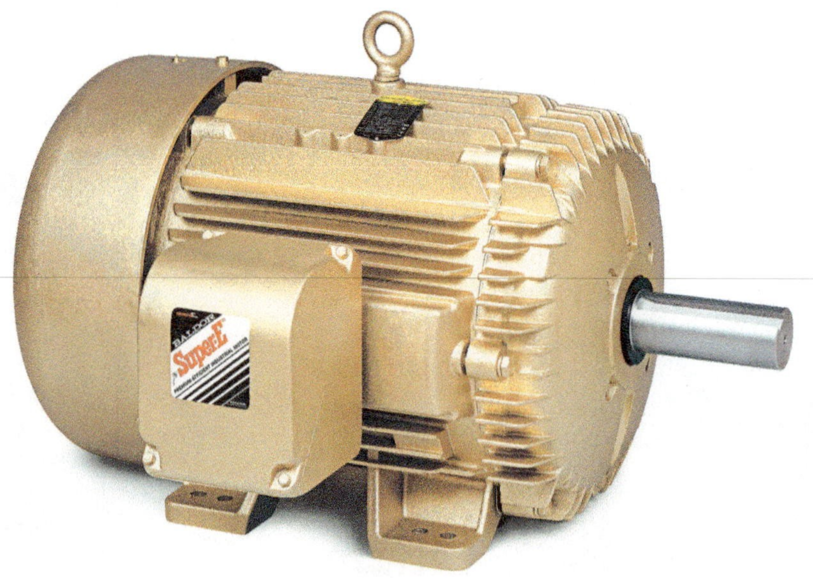

Courtesy of Baldor Electric Company

Figure 42–31 Fins increase the surface area to aid in cooling the motor.

increased to 8 AWG to reduce the resistance because of the length of the conductors.

Overload Heaters Not Sized Correctly

One of the leading causes of motor burn-out is improper sizing of the overloads. Overloads are designed to disconnect a motor from power in the event of excessive current draw. Thermal overloads depend on a heater connected in series with the motor to produce enough heat to disconnect the motor in the event the current becomes too great, Figure 42–32. Overload heaters should be sized in accord with the NEC as discussed in Chapter 37.

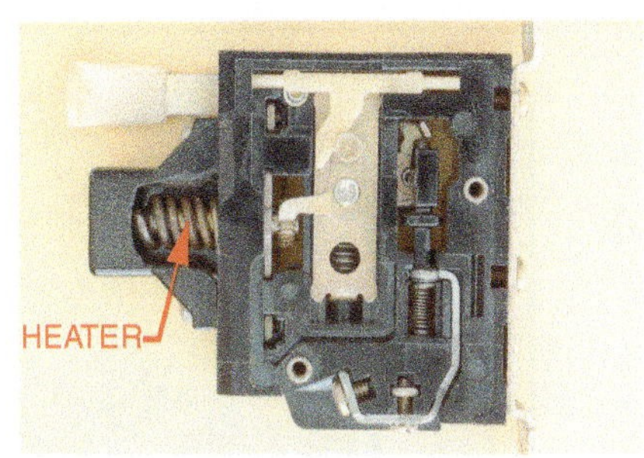

Figure 42–32 The overload heater is connected in series with the motor.

Motor Undersized for the Load

Another cause of motor overheating is the motor horsepower is not sufficient for the load. It is common practice to use the service factor rating of the motor to increase the horsepower. For example, assume that a load requires 22 horsepower. Often a 20 horsepower motor with a service factor of 1.15 will be used (20 HP × 1.15 = 23 HP) to supply the power. Although motors can run continually in their service factor rating, doing so will generally cause excessive heat in the winding and greatly shorten the life of the motor. A better solution is to install a motor with a larger horsepower than that required by the load.

Over and Under Voltage

Another cause of motor overheating and possibly burn out is improper voltage. NEMA rated motors are designed to operate at ± 10 percent of their rated voltage. The chart in Figure 42–33 shows the approximate change in motor current and temperature for typical squirrel cage electric motors operated over their rated voltage (110%) and under their rated voltage (90%). These values assume the motor is connected to a balanced three-phase system. The NEC generally limits voltage drop to 3 percent of the rated voltage.

Unbalanced Voltage

One of the worst operating conditions for a poly-phase motor is when it is connected to unbalanced voltages. Three-phase voltages generally become unbalanced when single-phase loads are added to the system. NEMA recommends that the voltages in a three-phase system should be maintained within 1 percent of each other. Unbalanced voltages can cause a much greater heating effect in the motor windings than over or under voltages in a balanced system. The most extreme example of an unbalanced voltage is when a three-phase motor loses one phase and operates as a single-phase motor, Figure 42–34. Single phasing causes a 73 percent increase in motor current in the two windings connected to power.

To determine the percent of voltage unbalance follow the steps shown:

- Step 1 - Take voltage measurements between all phases. In this example it will be assumed that the voltage between AB = 496 volts, BC = 460 volts, and AC = 472 volts.
- Step 2 - Determine the average voltage.

496 + 460 + 472 = 1428 volts
1428 / 3 = 476 volts (Average)

Voltage Variation	Full Load Current	Starting Current	Approximate Temperature Variation @ Full Load
110%	7% Decrease	10–12% Increase	3–4°C (37–39°F) Decrease
90%	11% Increase	10–12% Decrease	6–7°C (43–45°F) Increase

Figure 42–33 Voltage versus current table.

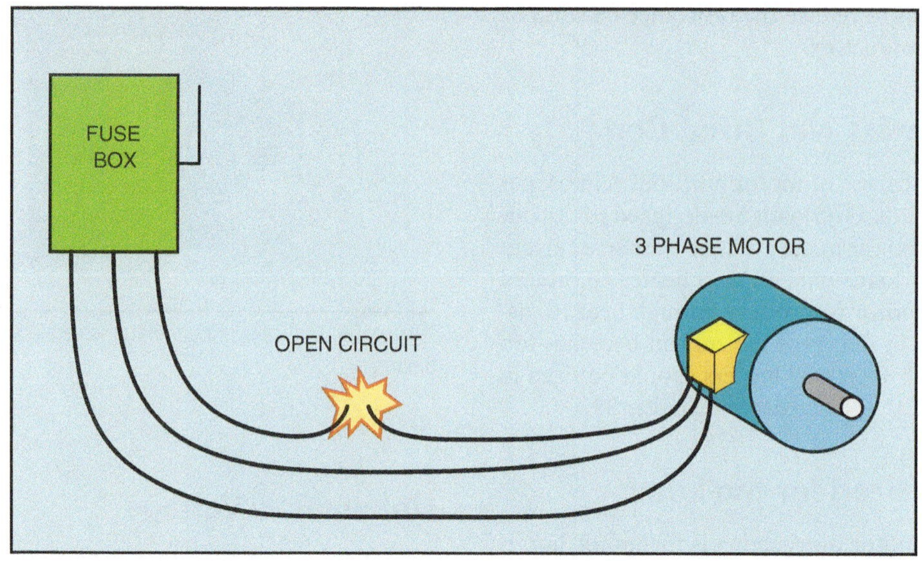

Figure 42–34 Single phasing occurs when a three-phase motor looses one phase.

- Step 3 - Subtract the average voltage from the reading that results in the greatest difference.

$$496 - 476 = 20 \text{ volts}$$

- Step 4 - Determine the percent difference.

The amount of heat rise in the winding with the highest current due to unbalanced voltages is equal to twice the percent squared.

$$\text{Heat rise} = 2 \times (\text{percent voltage unbalance})^2$$
$$2 \times 4.2 \times 4.2 = 35.28\%$$

In this example the winding with the highest current will experience a 35.28 percent increase in temperature.

Review Questions

1. What are the three main electrical test instruments used in troubleshooting?
2. What is the advantage of a plunger type voltage tester?
3. A motor is tripping out on overload. The motor nameplate reveals a full load current of 68 amperes. When the motor is operating under load, an ammeter indicates the following: Phase #1 = 106 amperes, Phase #2 = 104 amperes, and Phase #3 = 105 amperes. What is the most likely problem with this motor?
4. A motor is tripping out on overload. The motor nameplate reveals a full load current of 168 amperes. When the motor is operating under load, an ammeter indicates the following: Phase #1 = 166 amperes, Phase #2 = 164 amperes, and Phase #3 = 225 amperes. What is the most likely problem with this motor?
5. Refer to the circuit shown in Figure 42–17. The motor will not start in either the forward or reverse direction when the START push buttons are pressed. Which of the following could *not* cause this problem?
   a. F coil is open.
   b. The overload contact is open.
   c. The control transformer fuse is blown.
   d. The stop push button is not making a complete circuit.

6. Refer to the circuit shown in Figure 42–17. Assume that the motor is running in the forward direction. When the REVERSE push button is pressed, the motor continues to run in the forward direction. Which of the following could cause this problem?
   a. The normally open side of the reverse push button is not making a complete circuit when pressed.
   b. R contactor coil is open.
   c. The normally closed side of the reverse push button is not breaking the circuit when the reverse push button is pressed.
   d. There is nothing wrong with the circuit. The stop push button must be pressed before the motor will stop running in the forward direction and permit the motor to be reversed.

7. Refer to the circuit shown in Figure 42–21. When the THIRD SPEED push button is pressed, the motor starts in first speed but never accelerates to second or third speed. Which of the following could *not* cause this problem?
   a. Control relay CR1 is defective.
   b. Control relay CR2 is defective.
   c. Timer TR1 is defective.
   d. Contactor coil S1 is open.

8. Refer to the circuit shown in Figure 42–21. Assume that the THIRD SPEED push button is pressed. The motor starts in second speed, skipping first speed. After 5 seconds, the motor accelerates to third speed. Which of the following could cause this problem?
   a. S1 contactor coil is open.
   b. CR1 contactor coil is open.
   c. TR1 timer coil is open.
   d. S1 load contacts are shorted.

9. Refer to the circuit shown in Figure 42–17. If a voltmeter is connected across the normally open FORWARD push button, the meter should indicate a voltage value of:
   a. 0 volts
   b. 30 volts
   c. 60 volts
   d. 120 volts

10. Refer to the circuit shown in Figure 42–21. Assume that a fused jumper is connected across terminals 1 and 3 of TR2 timer. What would happen if the jumper were left in place and the FIRST SPEED push button pressed?
    a. The motor would start in its lowest speed and progress to second speed, but never increase to third speed.
    b. The motor would start operating immediately in third speed.
    c. The motor would not start.
    d. The motor would start in second speed and then increase to third speed.

11. The voltages supply a three-phase squirrel cage induction motor are as follows: A−B = 202 volts; A−C = 216 volts; B−C = 207 volts. What is the percent voltage unbalance?

12. Using the values in question 11, determine the percent of heat rise in the winding with the highest current.

Section 10

LABORATORY EXERCISES

EXERCISE 1
Basic Control

EXERCISE 2
START–STOP Push Button Control

EXERCISE 3
Multiple Push Button Stations

EXERCISE 4
Forward–Reverse Control

EXERCISE 5
Sequence Control

EXERCISE 6
Jogging Controls

EXERCISE 7
On-Delay Timers

EXERCISE 8
Off-Delay Timers

EXERCISE 9
Designing a Printing Press Circuit

EXERCISE 10
Sequence Starting and Stopping for Three Motors

EXERCISE 11
Hydraulic Press Control

EXERCISE 12
Design of Two Flashing Lights

EXERCISE 13
Design of Three Flashing Lights

EXERCISE 14
Control for Three Pumps

EXERCISE 15
Oil Pressure Pump Circuit for a Compressor

EXERCISE 16
Autotransformer Starter

LABORATORY EXERCISES

Foreword

The laboratory exercises presented in this book are connected to full voltage. Safety must be practiced at all times by students performing these exercises. It is recommended that the power supply be protected by fuses or circuit breakers. It is also recommended that ground fault protection be used when possible to reduce the hazard of electrocution. These exercises are written with the assumption that a 208-volt, three phase, four-wire power system is available. The instructor should be asked each time before power is applied to the circuit. Power should be turned off each time before making changes to the circuit.

Parts List for Laboratory Exercises

All of the exercises presented in this textbook can be constructed using the following components:

1 ea.—control transformer to step your laboratory line voltage down to 120 VAC

3 ea.—three phase motor starter that contains at least two normally open and one normally closed auxiliary contact

3 ea.—three phase contactors (no overload relays) containing at least one normally open and one normally closed auxiliary contact

3 ea.—three phase motors ⅓ to ¼ HP or simulated motor loads. Note: Assuming a 208-volt, three phase, four-wire system, a simulated motor load can be constructed by connecting three lamp sockets to form a wye connection (Figure F1). These lamps will have a voltage drop of 120 volts each. If a 240-volt three phase system in is use, it may be necessary to connect two lamps in series for each phase or obtain lamps rated for 240-volt operation. These three sets of series lamps can then be connected wye or delta (Figure F2).

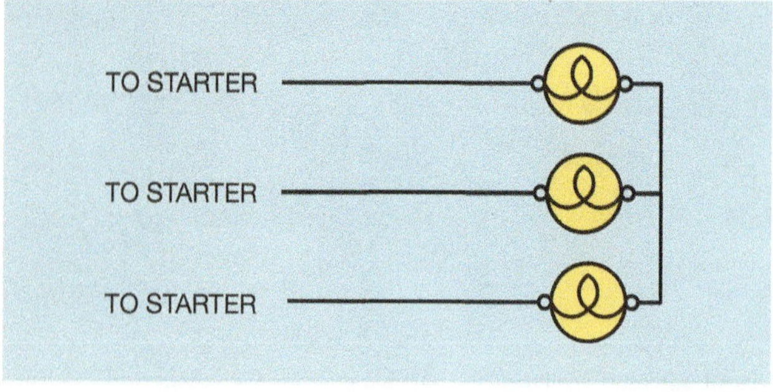

Figure F1 Simulated three phase motor load.

1 ea.—three phase overload relay or three single phase overload relays with the overload contacts connected in series

2 ea.—0.5 kVA control transformers 480/240-120

1 ea.—reversing started, or two three phase contactors that contain one normally open and one normally closed contact, and one three phase overload relay

4 ea.—double-acting push buttons

6 ea.—three-way toggle switches to simulate float switches or limit switches

4 ea.—electronic timers (Dayton model 6A855 recommended; available from Grainger) (Refer to Chapter 11 for an example of this timer.)

3 ea.—11 pin control relays (120-volt coil)

3 ea.—8 pin control relays (120-volt coil)

4 ea.—11 pin tube sockets

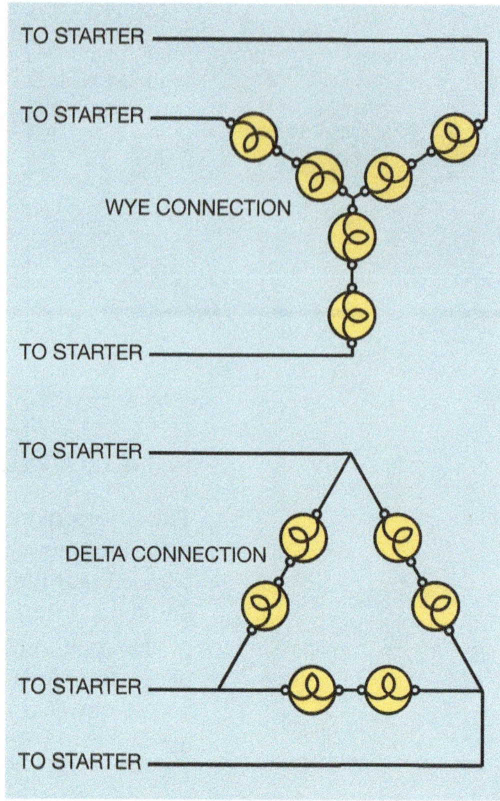

Figure F2 Lamps connected in series may be used for 240-volt systems and may be connected in wye or delta as a simulated motor load.

3 ea.—8 pin tube sockets

3 ea.—pilot light indicators

1 ea.—three phase power supply, 208-volt, four-wire system

Suppliers

Most of the parts listed can be obtained from Grainger Industrial Supply (http:/www.grainger.com). The Dayton model 6A855 timer is recommended because of its availability and price. Also, it is a multifunction timer and can be used as both an on- and off-delay device. It will also work as a one-shot timer and a pulse timer. Although the Dayton timer is recommended, any 11 pin electronic timer with the same pin configuration can be used. One such timer is available from Magnecraft (model TDR

SRXP-120, http:/www.magnecraft.com). This timer is also available from Mouser Electronics (http:/www.mouser.com). Other electronic timers can be employed, but if they have different pin configurations, the wiring connections shown in the textbook will have to be modified to accommodate the different timer.

The 8 and 11 pin control relays and sockets can be purchased from Grainger, Mouser Electronics, or Newark Electronics (http:/www.newark.com). The control transformer for use in the controls sections can be purchased from Mouser Electronics or Sola/Hevi-duty (http:/www.solaheviduty.com). Model E250JN is recommended because it has primary taps of 208/240/277 volts. The secondary winding is 120/24. It is also recommended that any control transformer used be fuse protected. Another control transformer that can be used is available from Grainger. It is rated at 150 VA and has a 208-volt primary and 120-volt secondary. The 0.5 kVA control transformers are available from Grainger or Newark Electronics. Transformers rated at 0.5 kVA are used because they permit the circuit to be loaded heavy enough to permit the use of clamp-on type ammeters.

Stackable banana plugs are available from both Grainger and Newark Electronics. The oil-filled capacitors listed are available from Grainger. Color-coded resistors can be obtained from Newark Electronics or Mouser Electronics.

Exercise 1

BASIC CONTROL

Objectives

After completing this exercise the student should be able to:

- Discuss the basic principles of control.
- Connect a relay circuit.
- Connect an 8 pin relay.
- Connect an 11 pin relay.
- Connect a three phase motor to a motor starter.

Materials Needed

- Three phase power supply
- Three phase squirrel cage induction motor or equivalent motor load
- Three phase motor starter
- Control transformer
- 8 pin control relay
- 8 pin tube socket
- 11 pin control relay
- 11 pin tube socket
- Single pole switch (Note: A three-way toggle switch can be used as a single pole switch)
- three pilot lamps

Basic control can be broken down to something as simple as a light and a switch (Figure Exp. 1–1). The switch is used to control the operation of the light. Now assume that the light is replaced with a single phase motor (Figure Exp. 1–2). A switch with the proper rating can control the operation of a single phase motor. Now assume that it is desirable to control the operation of a three phase motor. The single pole switch cannot control a three phase load. A switch with three separate poles must be used to provide control of a three phase load. A three pole switch could be employed (Figure Exp. 1–3). Now assume that the three phase motor is to be controlled by a float switch. Float switches do not contain three load contacts that can be used to connect a motor to the power line. Therefore, the float switch must control the operation of the three pole switch shown in Figure Exp. 1–3. If the three pole switch was controlled by a solenoid coil, the float switch could be used to control the switch (Figure Exp. 1–4). This is exactly the operation of a relay or contactor. Relays and contactors can provide control of multiple outputs from a single input. The outputs are the contacts, and the input is the solenoid coil.

Because the three phase load described in the circuit shown in Figure Exp. 1–4 is a motor, it should be overload protected. If an overload relay is added to the circuit (Figure Exp. 1–5), the three pole switch becomes a motor starter. Notice that the current sensing elements (overload heaters) are connected in series with the motor. The normally closed overload contact is connected in series with the solenoid coil.

Now that magnetic control for the motor has been achieved, any number of pilot devices, such as limit switches, pressure switches, proximity switches, and so on can be

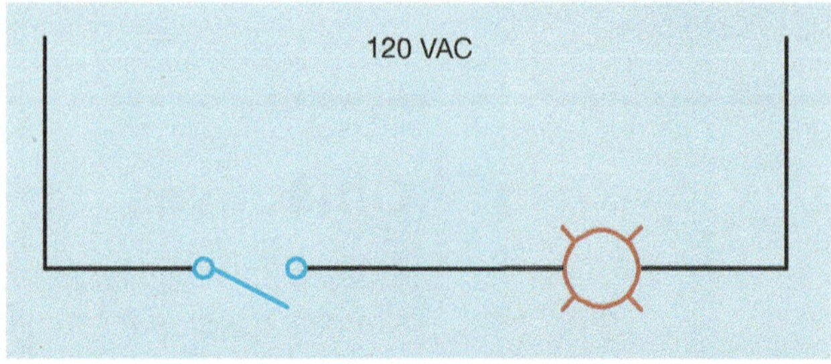

Figure Exp. 1–1 Basic control can be as simple as a light and a switch.

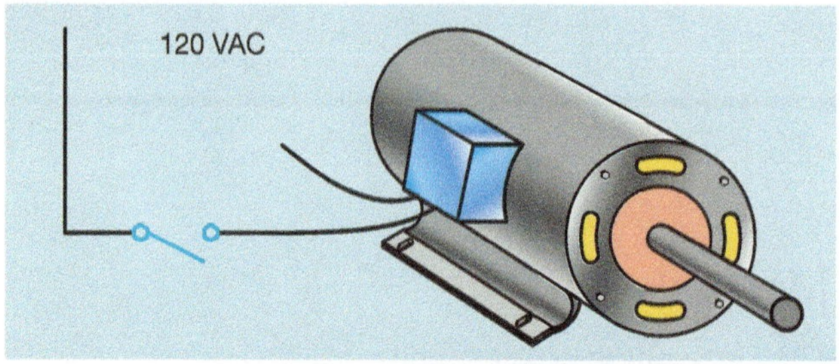

Figure Exp. 1–2 The light has been replaced by a single phase motor.

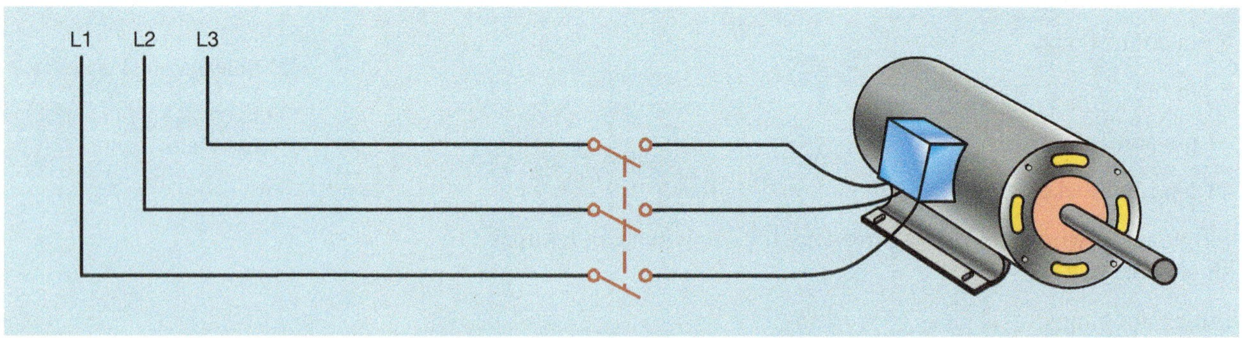

Figure Exp. 1–3 A three pole switch is required to control a three phase load.

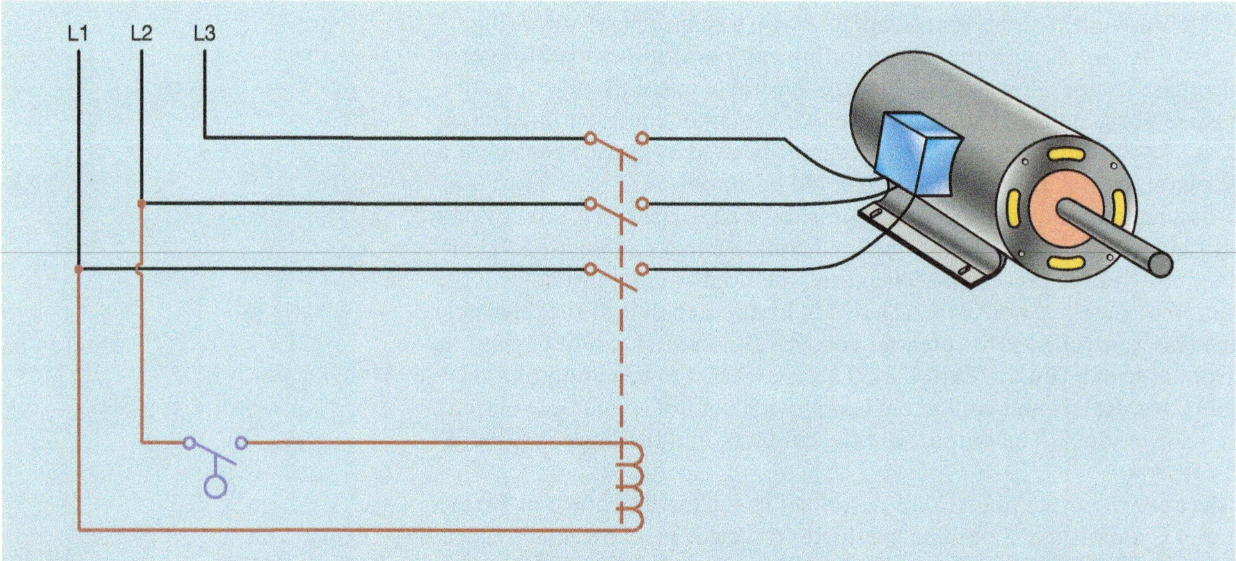

Figure Exp. 1–4 A float switch controls the operation of the motor.

added to control the operation of the motor (Figure Exp. 1–6). This is the basis of motor control. Basically, it is the expansion of a light and a switch.

Practical Exercise

In this exercise, an 8 pin control relay, 11 pin control relay, and motor starter will be controlled by a single pole switch. The pin diagram of an 8 pin control relay is shown in Figure Exp. 1–7. The relay will be used to control the operation of two pilot lights. A schematic diagram of this circuit is shown in Figure Exp. 1–8. When using 8 and 11

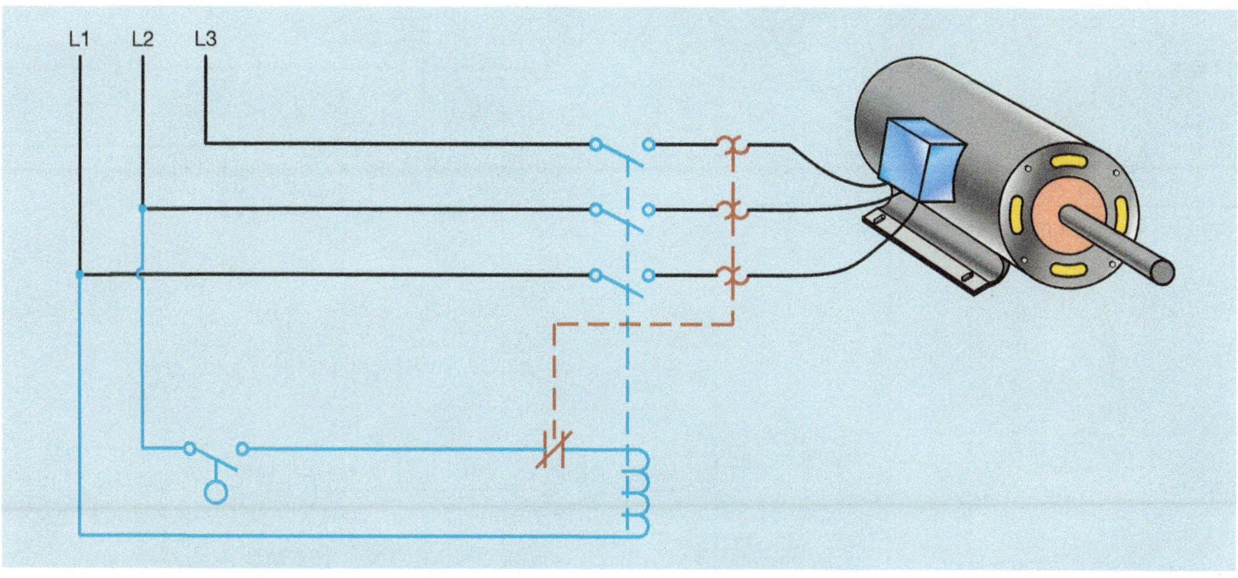

Figure Exp. 1–5 The addition of an overload relay changes the contactor into a motor starter.

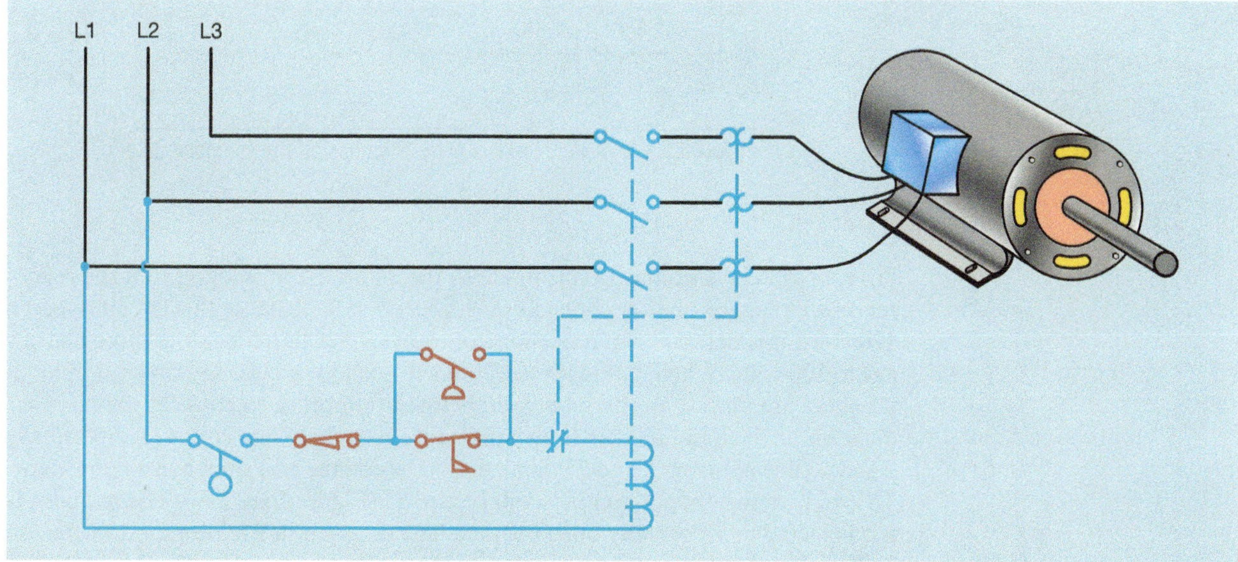

Figure Exp. 1–6 Any number of pilot devices can now be used to control the motor.

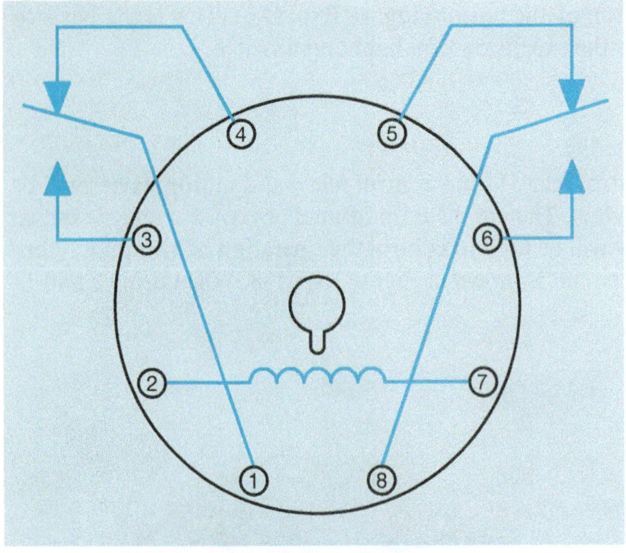

Figure Exp. 1–7 Diagram of an 8 pin relay.

Figure Exp. 1–8 Schematic diagram of Circuit 1.

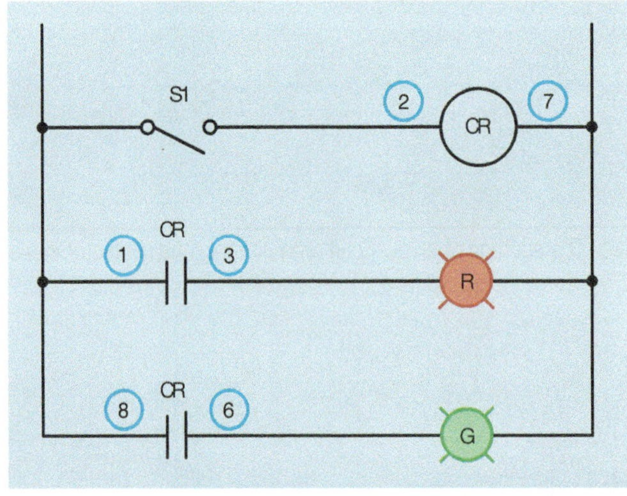

Figure Exp. 1–9 Pin numbers are placed beside the components.

pin control relays, connection is made to the 8 and 11 pin sockets, not the relay. The relays are plugged in after connection has been made. Because this circuit operates on 120 volts, it is necessary to use a control transformer to change the three phase voltage to 120 volts. When making connections to sockets, it becomes much simpler if pin numbers are placed on the schematic before attempting to connect the circuit. The diagram of the 8 pin relay is shown in Figure Exp. 1–7. The coil is connected to pins 2 and 7. The numbers 2 and 7 will be placed beside the coil shown in Figure Exp. 1–8. To avoid confusion, pin numbers will be circled (Figure Exp. 1–9). The red pilot lamp is connected to a normally open contact. The diagram in Figure Exp. 1–8 shows that an 8 pin relay has two normally open contacts. One is between pins 1 and 3, and the other is between pins 8 and 6. The numbers 1 and 3 will be placed beside the normally open contact that controls the red pilot lamp, and the numbers 8 and 6 will be placed beside the contact that controls the green pilot lamp. A diagram showing the components used to connect this circuit is shown in Figure Exp. 1–10. In the circuit shown in Figure Exp. 1–9, one side of the switch is connected to power. The other side of the switch is connected to terminal 2 of the 8 pin tube socket. Terminal 7 of the 8 pin socket is connected to the other side of the power supply. This connection is shown in Figure Exp. 1–11. The circuit shown in Figure Exp. 1–9 indicates that power is connected to terminals 1 and 8 of the relay. Terminal 3 is connected to one side of the red pilot light, and terminal 6 is connected to one side of the green pilot light. The other side of both pilot lights is connected back to the opposite power supply terminal (Figure Exp. 1–12).

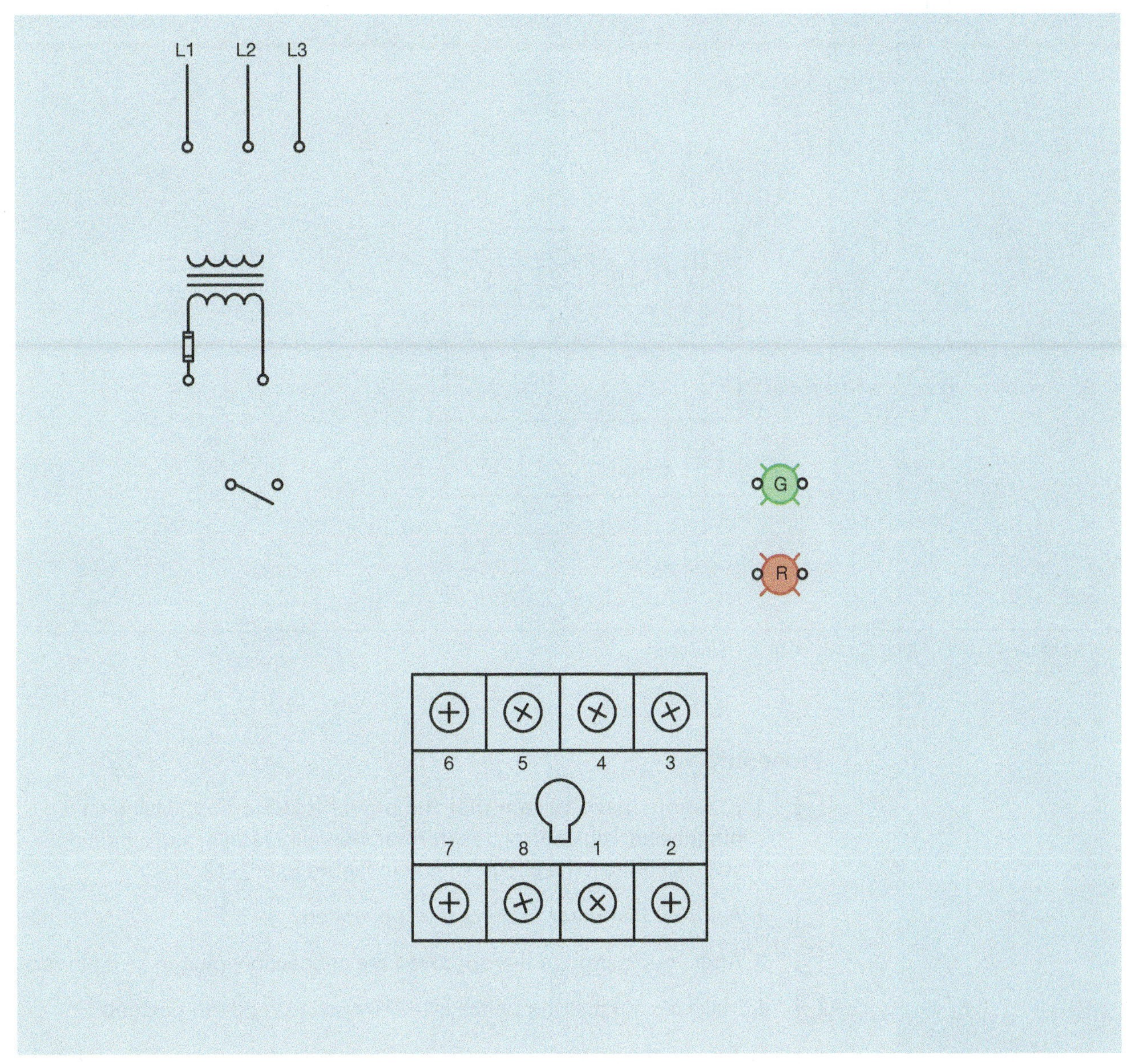

Figure Exp. 1–10 Components needed to connect the circuit in Figure Exp. 1–9.

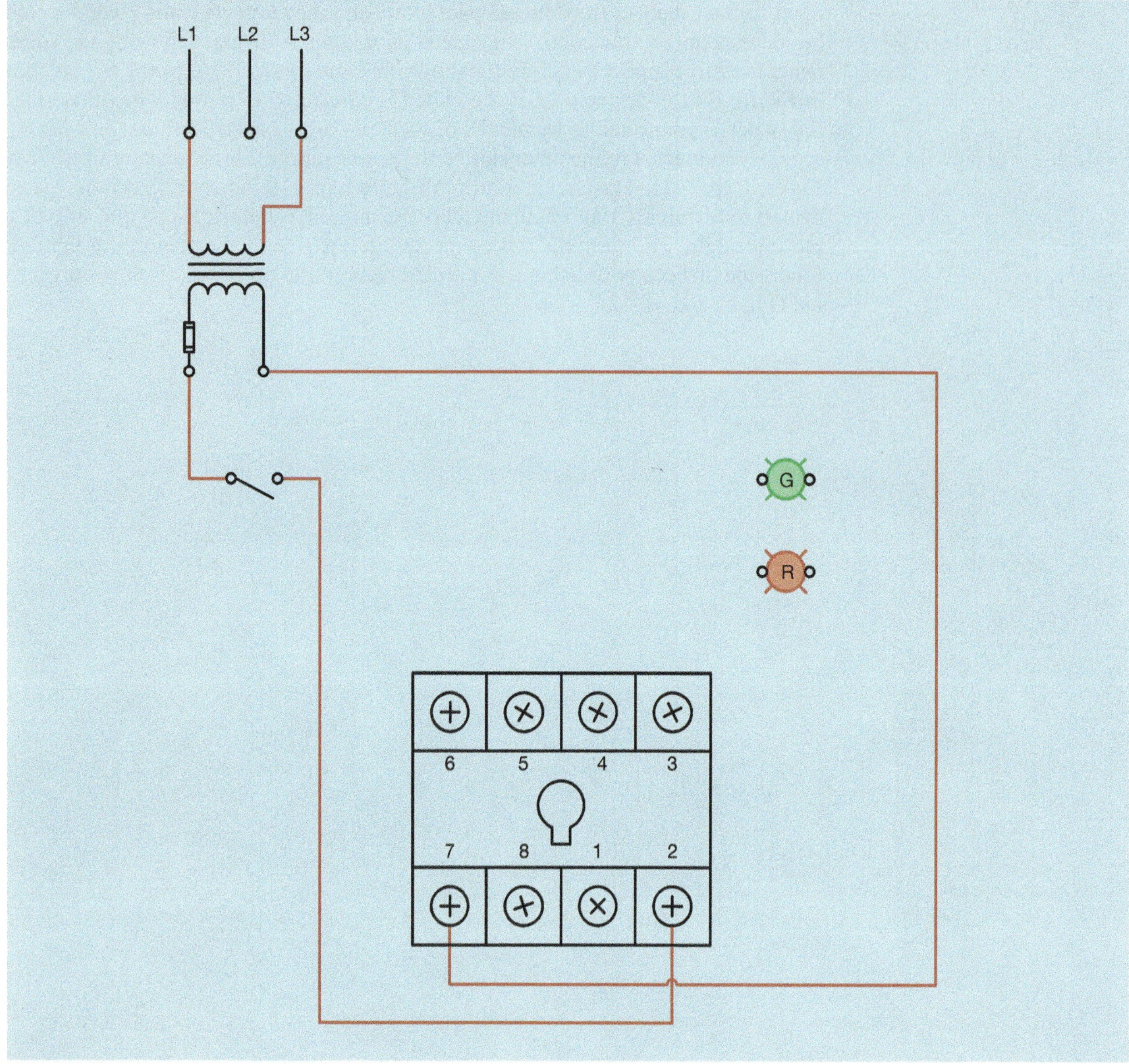

Figure Exp. 1–11 Connecting the switch to the relay coil.

Procedure 1

- [ ] 1. **(Caution: Make certain that the power is turned off.)** Using an 8 pin tube socket, control transformer, two pilot lamps, and single pole switch, connect the circuit shown in Figure Exp. 1–12.
- [ ] 2. Ask your instructor to check the connection.
- [ ] 3. After your instructor has approved the connection, plug in an 8 pin relay.
- [ ] 4. Make certain that the switch is in the open (turned off) position.
- [ ] 5. Turn on the power.
- [ ] 6. Are the pilot lamps on?

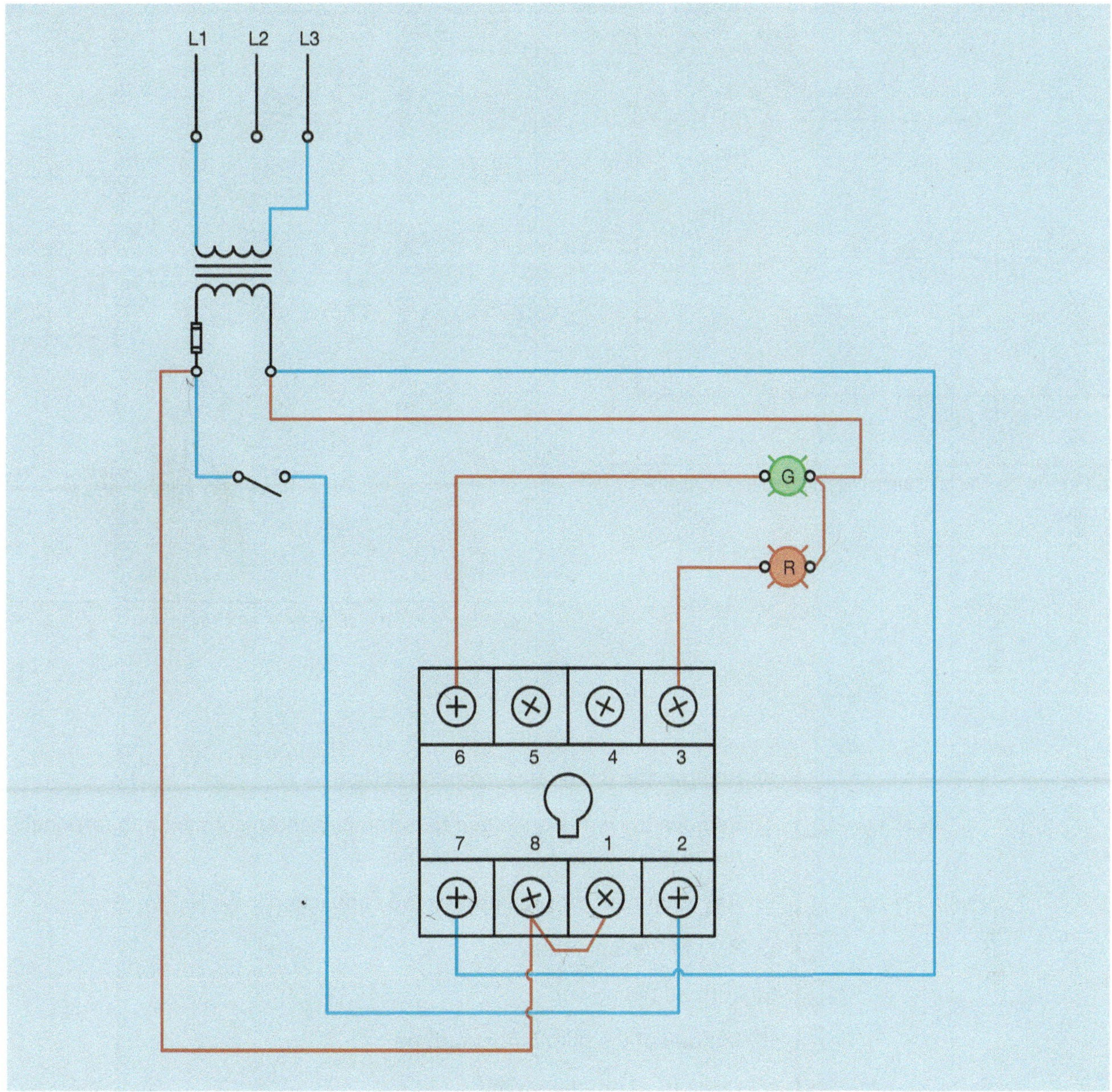

Figure Exp. 1–12 Connecting the lamps to the power supply.

- ❑ 7. Close (turn on) the switch.
- ❑ 8. Did the relay energize and turn on the two pilot lamps?
- ❑ 9. Open the switch and **turn off the power**.

In this circuit, both lamps are connected to normally open contacts. The circuit will now be changed so that one of the lamps is connected to a normally closed contact. The revised circuit is shown in Figure Exp. 1–13. The diagram of the 8 pin relay shown in Figure Exp. 1–7 indicates that a normally closed contact exists between pins 8 and 5. Notice that the schematic in Figure Exp. 1–13 shows that pin number 6 has been replaced with pin number 5.

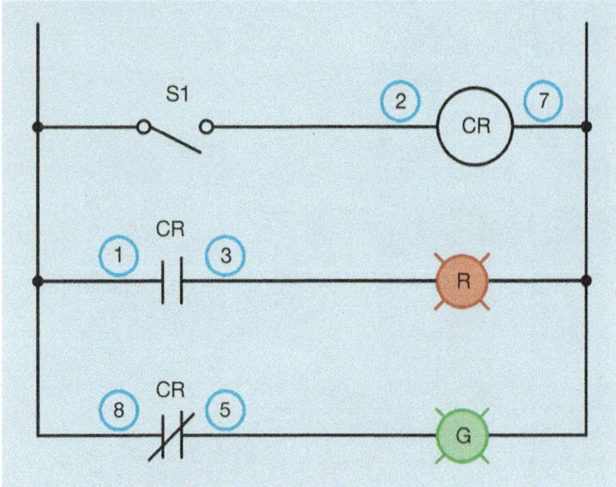

Figure Exp. 1–13 One pilot lamp is connected to a normally closed contact.

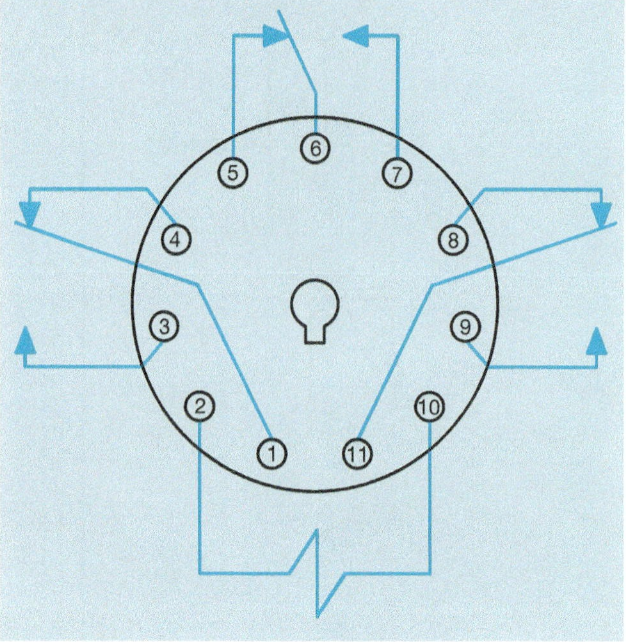

Figure Exp. 1–14 Pin diagram of an 11 pin relay.

- ❑ 10. Remove the 8 pin relay from the socket.
- ❑ 11. Remove the wire connected to terminal 6 and reconnect it to terminal number 5.
- ❑ 12. Ask your instructor to approve the connection.
- ❑ 13. Plug in the 8 pin relay.
- ❑ 14. Turn on the power.
- ❑ 15. Did one of the pilot lamps turn on?
- ❑ 16. Close (turn on) the switch.
- ❑ 17. Did the pilot lamp that was turned on turn off, and did the pilot lamp that was off turn on?
- ❑ 18. Open the switch and **turn off the power**.
- ❑ 19. Disconnect the circuit.

Procedure 2

In the second procedure, a single pole switch will be used to control three separate pilot lamps. An 11 pin relay will be used because it contains three separate sets of double-acting contacts (Figure Exp. 1–14). A schematic diagram of the circuit to be constructed is shown in Figure Exp. 1–15. In order to make connecting the circuit simpler, pin numbers will be placed on the schematic. The diagram in Figure Exp. 1–14 indicates that the coil is connected to pins 2 and 10. The red pilot lamp

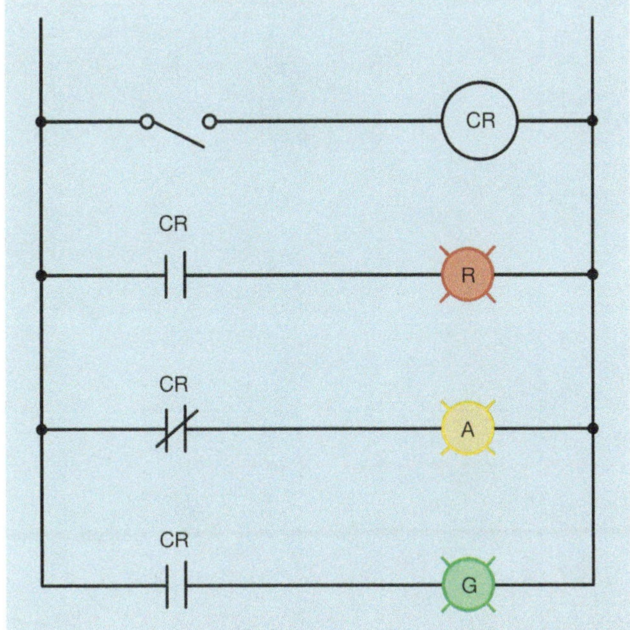

Figure Exp. 1–15 A single pole switch controls three output devices.

Figure Exp. 1–16 Pin numbers are added to the schematic.

is connected to a normally open contact. Pins 1 and 3 will be used for this connection. The amber pilot lamp is connected to a normally closed contact. Pins 6 and 5 will be used to control the amber light. The green pilot lamp is connected to a normally open contact. Pins 11 and 9 will be used for this connection (Figure Exp. 1–16).

As in the previous procedure, connection will be made to a socket instead of the actual relay. An 11 pin tube socket and connection diagram is shown in Figure Exp. 1–17.

- ❑ 20. Using an 11 pin tube socket, connect the circuit shown in Figure Exp. 1–17.
- ❑ 21. Have your instructor check the circuit for proper connection.
- ❑ 22. Make certain that the single pole switch is open (turned off).
- ❑ 23. Plug in the 11 pin relay and turn on the power.
- ❑ 24. List which lamps are turned on and which are turned off.
- ❑ 25. Close (turn on) the single pole switch.
- ❑ 26. List which lamps are turned on and which are turned off.
- ❑ 27. Open the single pole switch and **turn off the power**.
- ❑ 28. Disconnect the circuit.

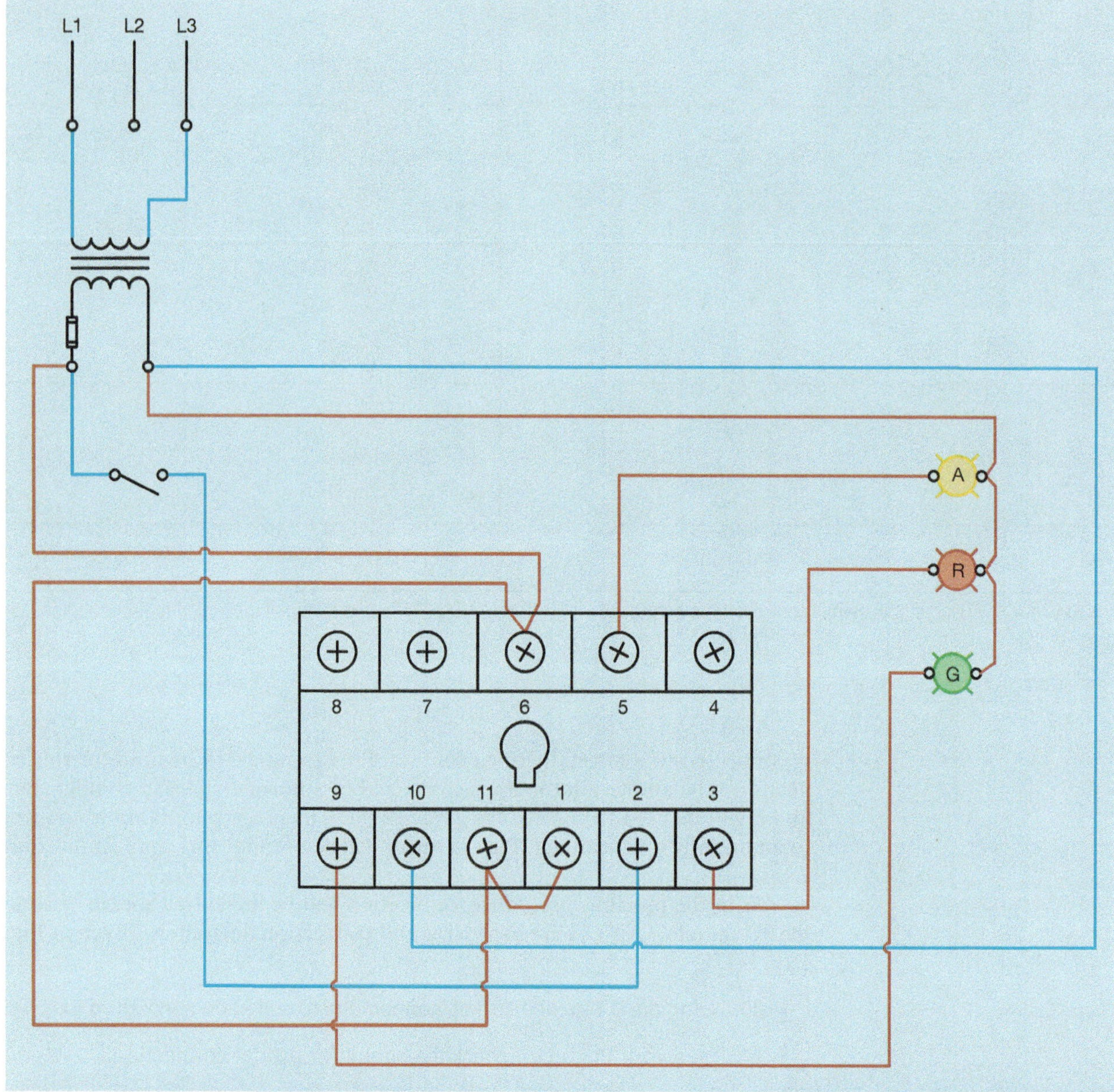

Figure Exp. 1–17 Connecting the 11 pin tube socket.

Procedure 3

In this procedure, the single pole switch will be used to control the operation of a three phase squirrel cage motor. The switch will control the operation of a motor starter by controlling the current flow to the coil. A basic schematic diagram of the connection is shown in Figure Exp. 1–18. The dashed lines indicate connections that are generally made internally on most motor starters and do not have to be connected by the electrician.

- ❑ 29. **Make sure that the power is turned off.**
- ❑ 30. Connect the circuit shown in Figure Exp. 1–18.

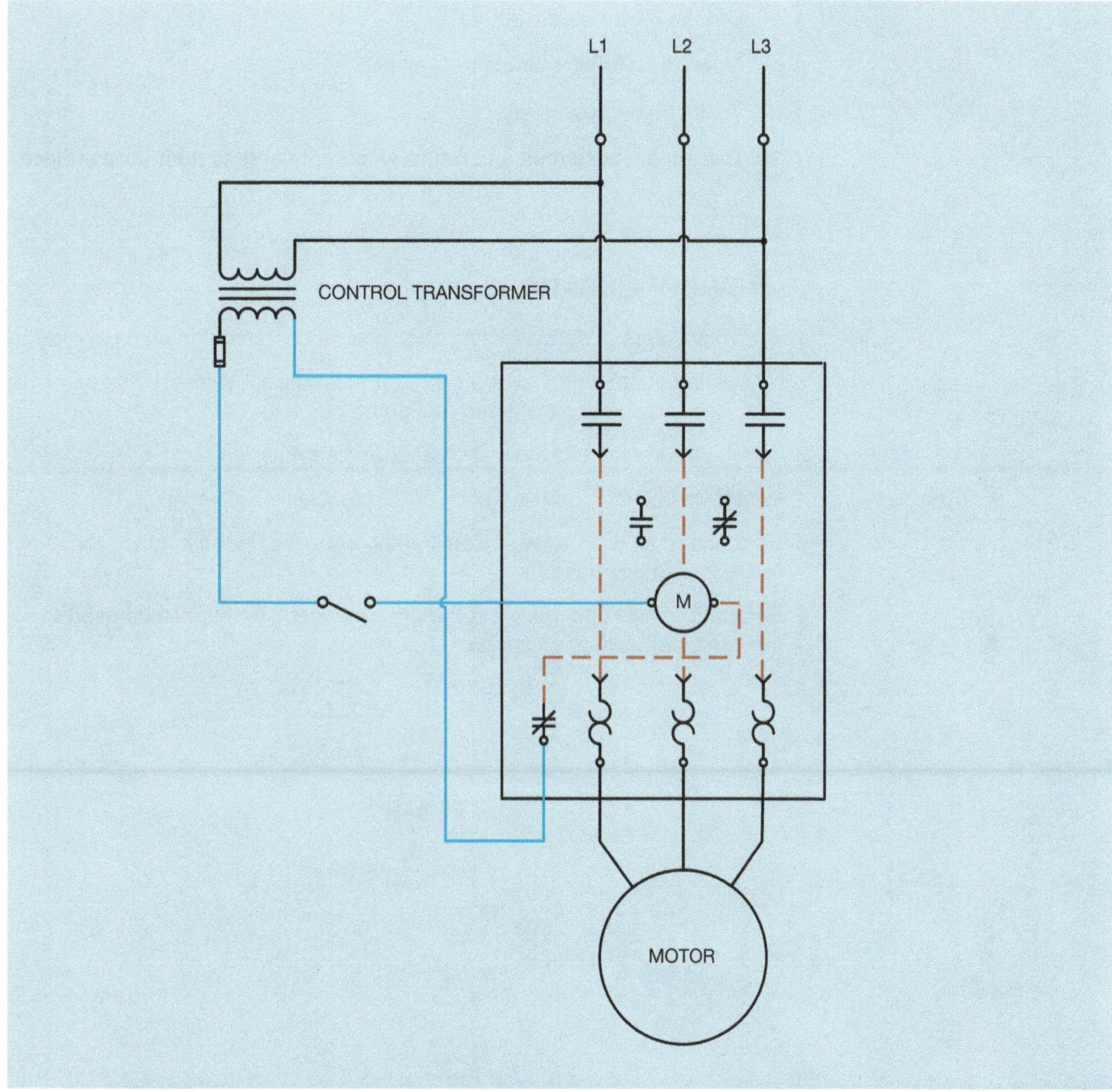

Figure Exp. 1–18 A single pole switch controls a three phase motor.

- ❑ 31. Ask your instructor to check the circuit for proper connection.
- ❑ 32. Make sure that the single pole switch is open (turned off).
- ❑ 33. Turn on the power.
- ❑ 34. Did the motor start running?
- ❑ 35. Close (turn on) the single pole switch.

- [ ] 36. Did the motor start running?
- [ ] 37. Open (turn off) the single pole switch.
- [ ] 38. **Turn off the power.**
- [ ] 39. Disconnect the circuit and return all components to their proper place.

Review Questions

1. How many sets of double-acting contacts are contained on an 8 pin relay?
2. Relays and contactors are single-input, multi-output devices. What part of a relay or contactor is considered the input?
3. Which pins connect to the coil on an 8 pin relay?
4. Which pins connect to the coil on an 11 pin relay?
5. Referring to an 8 pin relay, if pins 1 and 4 are used, is the contact normally open or normally closed?
6. Referring to an 11 pin relay, list the pin combinations that can be used to provide a normally open contact.

Exercise 2

START–STOP PUSH BUTTON CONTROL

Objectives

After completing this experiment the student should be able to:

- Place wire numbers on a schematic diagram.
- Place corresponding numbers on control components.
- Draw a wiring diagram from a schematic diagram.
- Define the difference between a schematic or ladder diagram and a wiring diagram.
- Connect a START–STOP push button control circuit.

Materials Needed

- Three phase power supply
- Three phase squirrel cage induction motor or simulated load
- Two double-acting push buttons (NO/NC on same button)
- Three phase motor starter or contactor with overload relay containing three load contacts and at least one normally open auxiliary contact
- Control transformer

In this exercise, a schematic diagram of a START–STOP push button control will be converted to a wiring diagram and then connected in the laboratory. *A schematic diagram shows components in their electrical sequence without regard for the physical location of any component* (Figure Exp. 2–1). *A wiring diagram is a pictorial representation of components with connecting wires.* The pictorial representation of the components is shown in Figure Exp. 2–2.

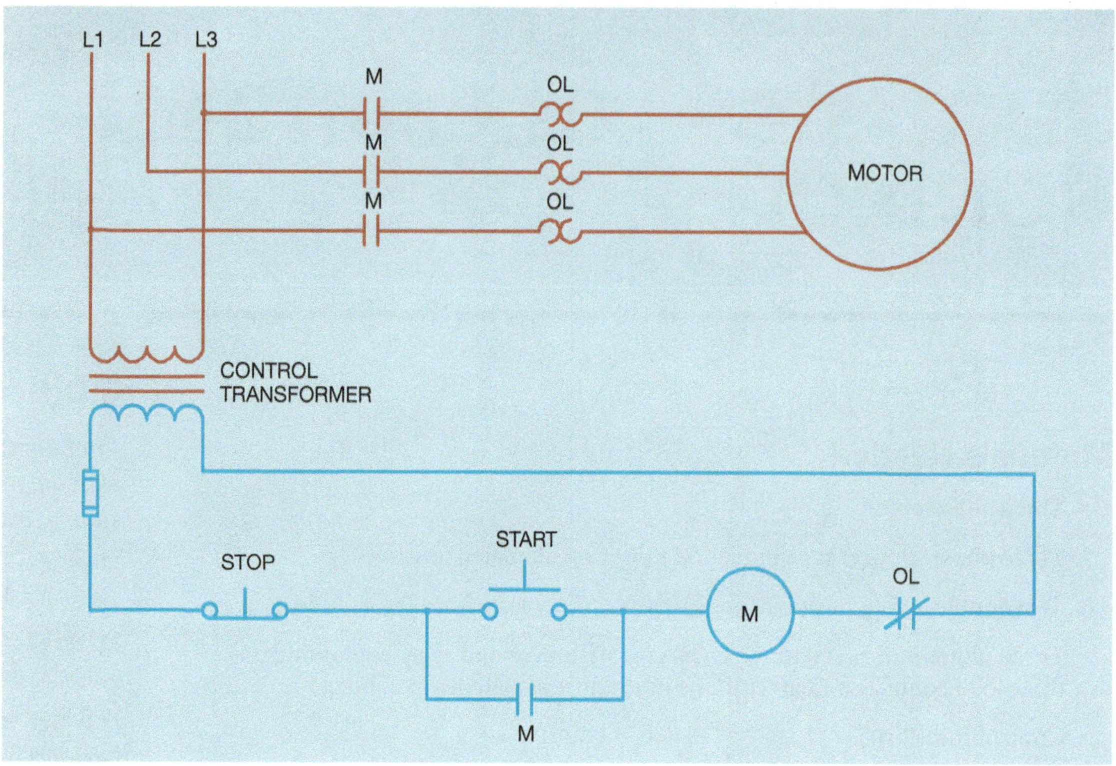

Figure Exp. 2–1 Schematic diagram of a basic START–STOP push button control circuit.

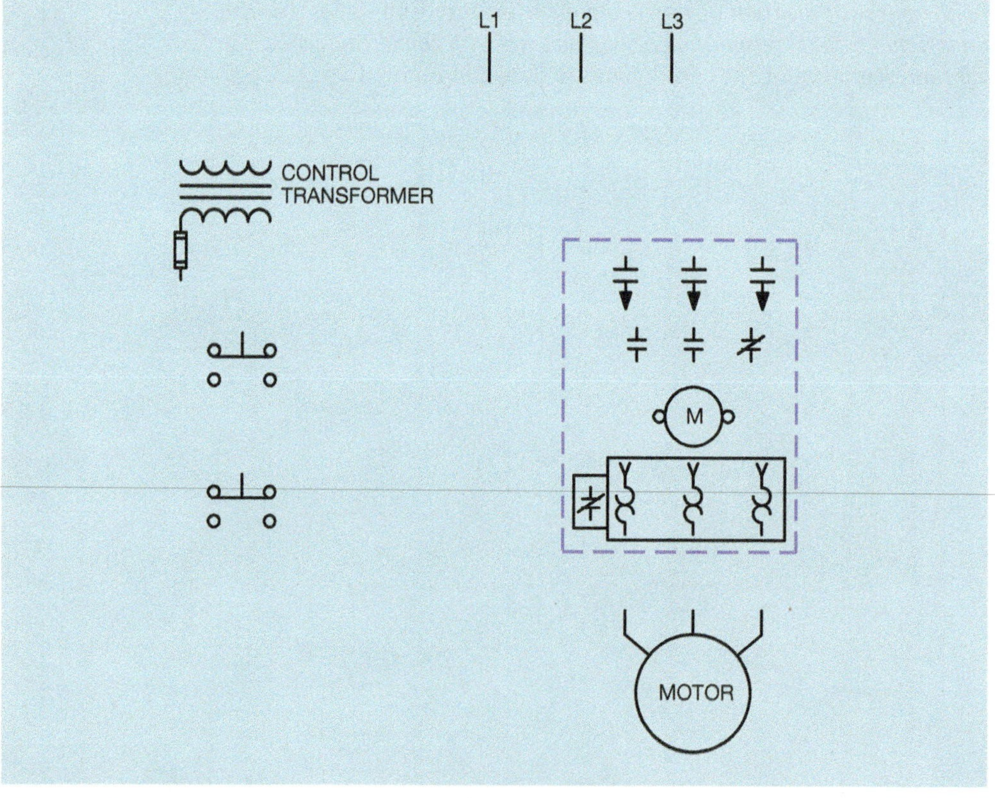

Figure Exp. 2–2 Components of the basic START–STOP control circuit.

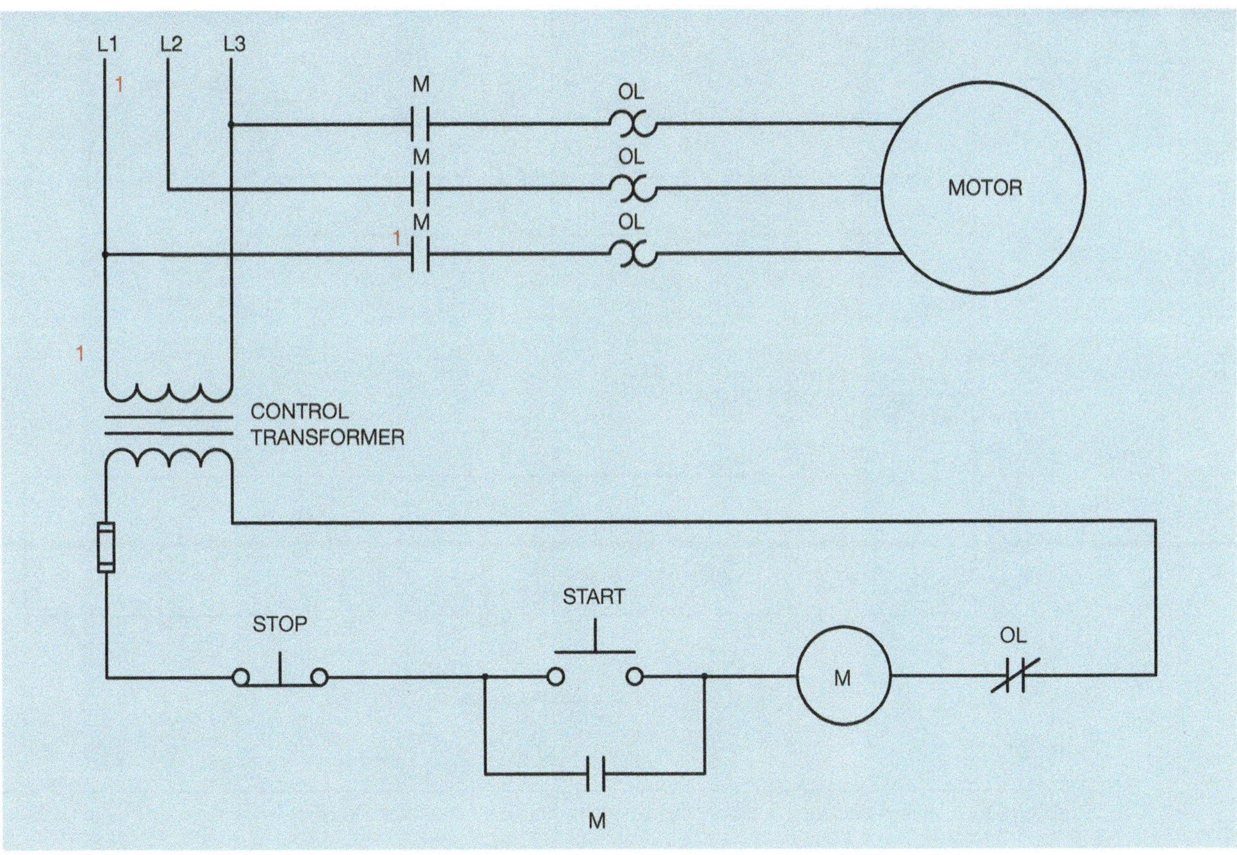

Figure Exp. 2–3 The number 1 is placed beside each component connected to L1.

To simplify the task of converting the schematic diagram into a wiring diagram, wire numbers will be added to the schematic diagram. These numbers will then be transferred to the control components shown in Figure Exp. 2–2. The rules for numbering a schematic diagram are as follows:

- ❑ 1. A set of numbers can be used only once.
- ❑ 2. Each time you go through a component the number set must change.
- ❑ 3. All components that are connected together will have the same number.

To begin the numbering procedure, begin at Line 1 (L1) with the number 1 and place a number 1 beside each component that is connected to L1 (Figure Exp. 2–3). The number 2 is placed beside each component connected to L2 (Figure Exp. 2–4), and a number 3 is placed beside each component connected to L3 (Figure Exp 2–5). The number 4 will be placed on the other side of the M load contact that already has a number 1 on one side and on one side of the overload heater (Figure Exp. 2–6). Number 5 is placed on the other side of the M

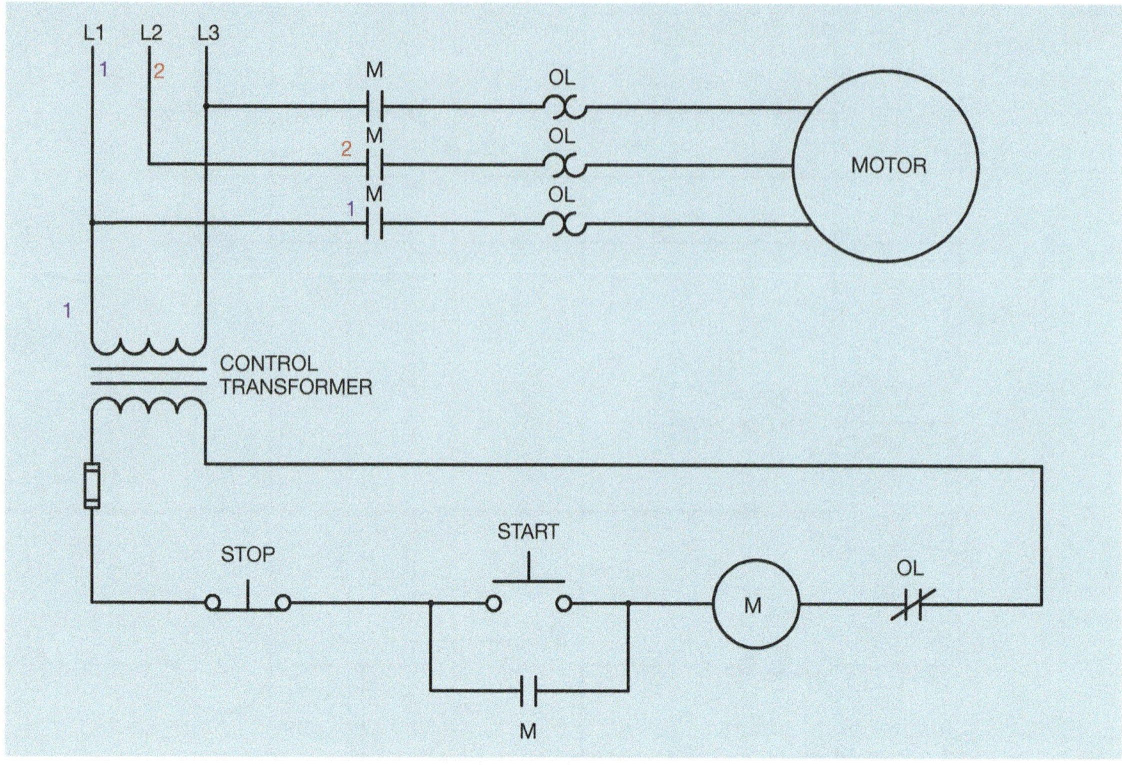

Figure Exp. 2–4 A number 2 is placed beside each component connected to L2.

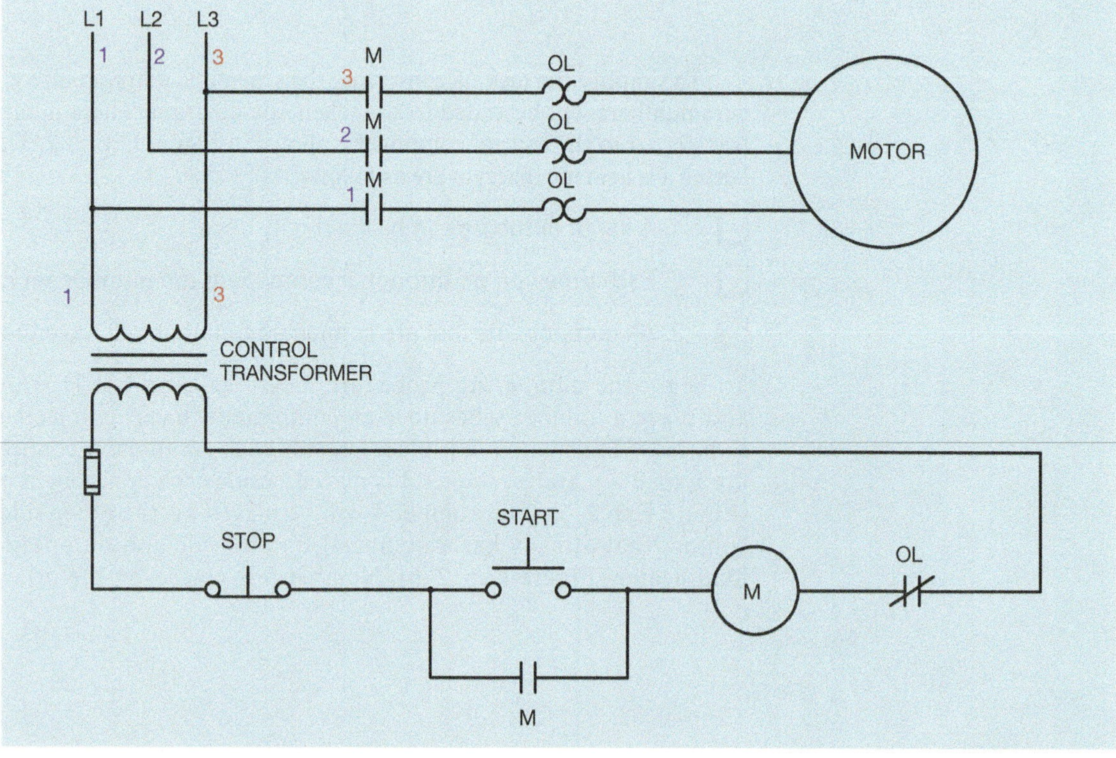

Figure Exp. 2–5 A number 3 is placed beside each component connected to L3.

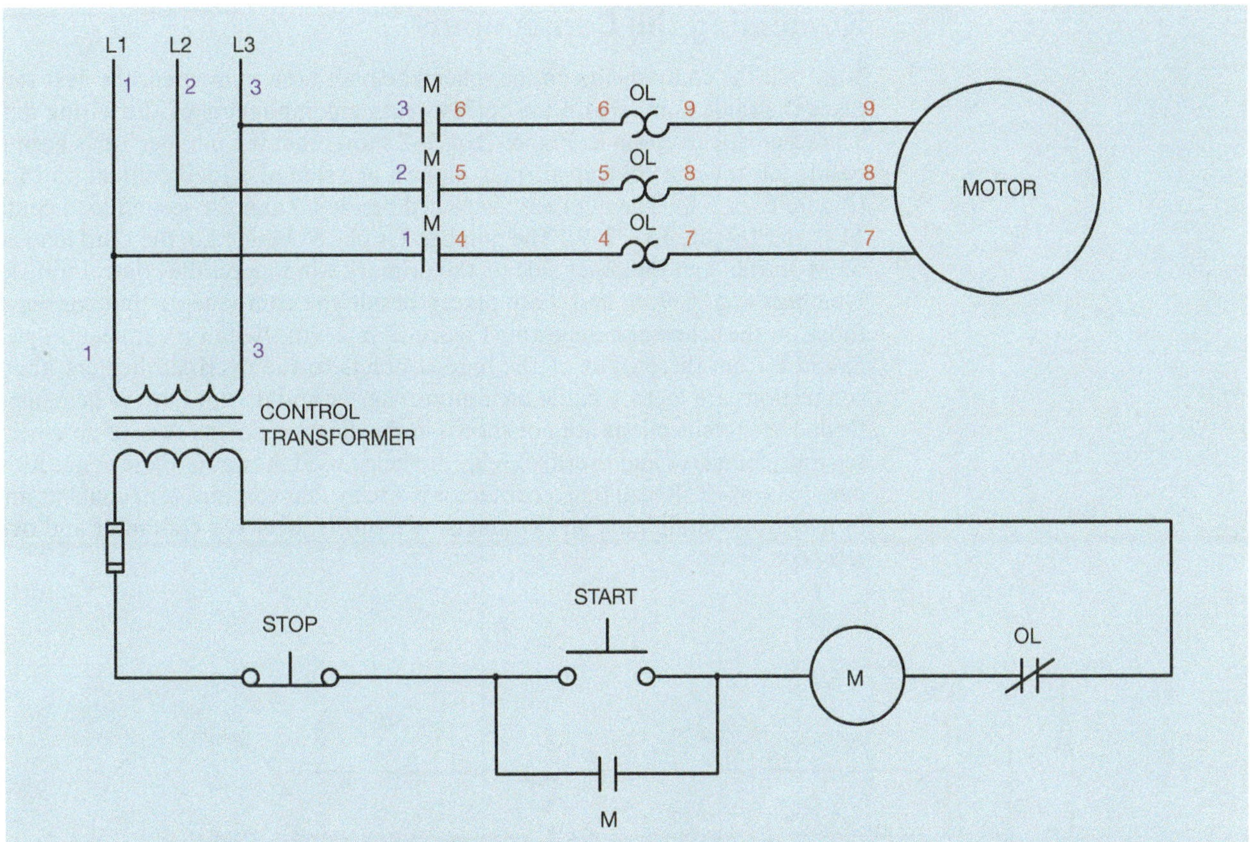

Figure Exp. 2–6 The number changes each time you proceed across a component.

load contact, which has one side numbered with a 2, and a 5 will be placed beside the second overload heater. The other side of the M load contact that has been numbered with a 3 will be numbered with a 6, and one side of the third overload heater will be labeled with a 6. Numbers 7, 8, and 9 are placed between the other side of the overload heaters and the motor T leads.

The number 10 will begin at one side of the control transformer secondary and go to one side of the normally closed STOP push button. Because the fuse is generally part of the control transformer, connection will begin after the fuse. The number 11 is placed on the other side of the STOP button and on one side of the normally open START push button and normally open M auxiliary contact. A number 12 is placed on the other side of the START button and M auxiliary contact, and on one side of M coil. Number 13 is placed on the other side of the coil to one side of the normally closed overload contact. Number 14 is placed on the other side of the normally closed overload contact and on the other side of the control transformer secondary winding (Figure Exp. 2–7).

Numbering the Components

Now that the components on the schematic have been numbered, the next step is to place the same numbers on the corresponding components of the wiring diagram. The schematic diagram in Figure Exp. 2–7 shows that the number 1 has been placed beside L1, the control transformer, and on one side of a load contact on M starter (Figure Exp. 2–8). The number 2 is placed beside L2 and the second load contact on M starter (Figure Exp. 2–9). The number 3 is placed beside L3, the third load contact on M starter, and the other side of the primary winding on the control transformer. Numbers 4, 5, 6, 7, 8, and 9 are placed beside the components that correspond to those on the schematic diagram (Figure Exp. 2–10). Note on connection points 4, 5, and 6 from the output of the load contacts to the overload heaters, that these connections are factory made on a motor starter and do not have to be made in the field. These connections are not shown in the diagram for the sake of simplicity. If a separate contactor and overload relay are being used, however, these connections will have to be made. Recall that a contactor is a relay that contains *load* contacts and may or may not contain auxiliary contacts. A motor starter is a contactor and overload relay combined.

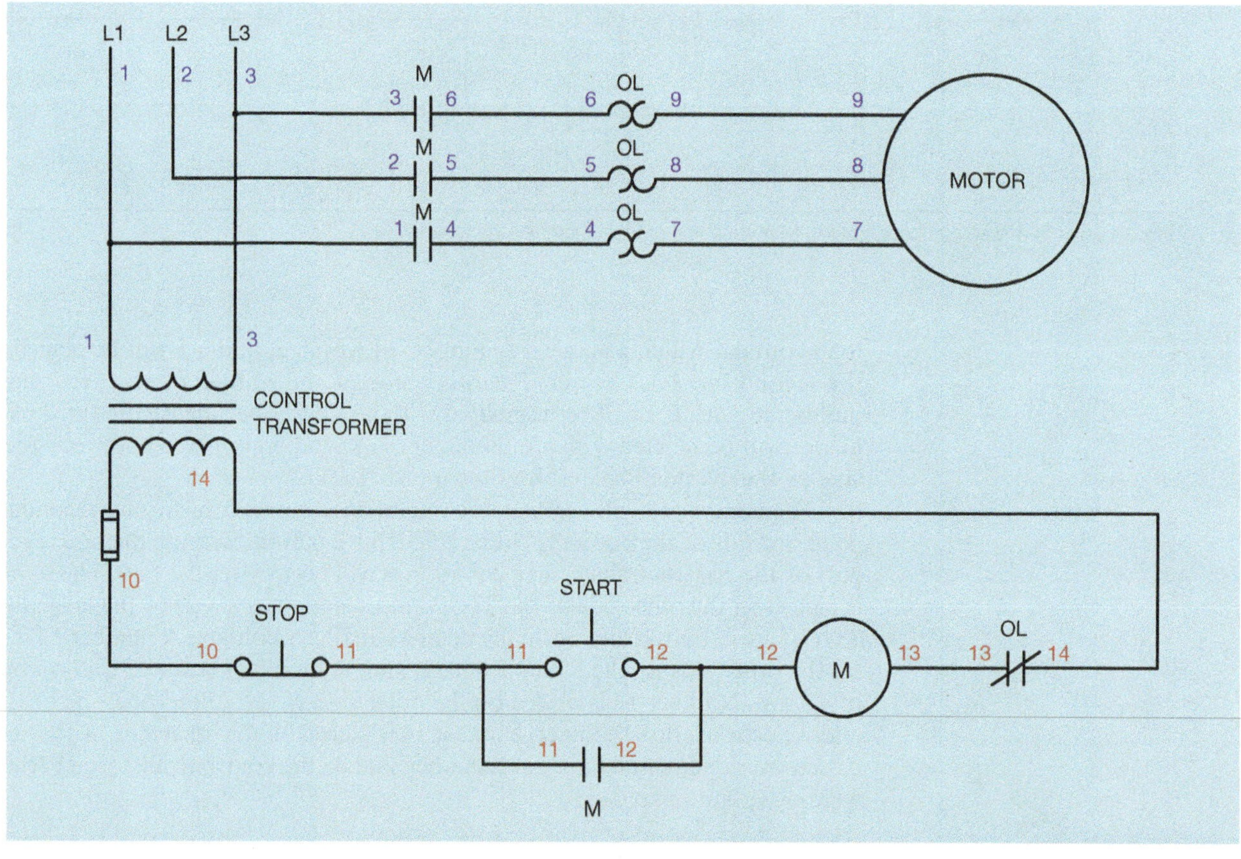

Figure Exp. 2–7 Numbers are placed beside all components.

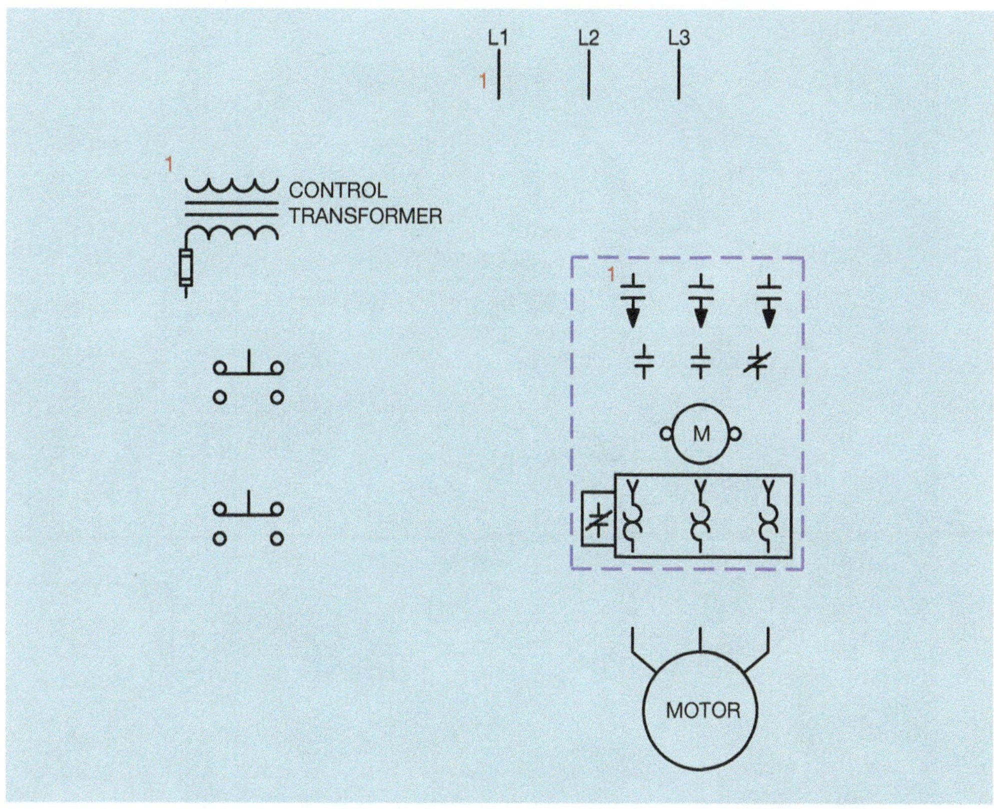

Figure Exp. 2–8 The number 1 is placed beside L1, the control transformer, and M load contact.

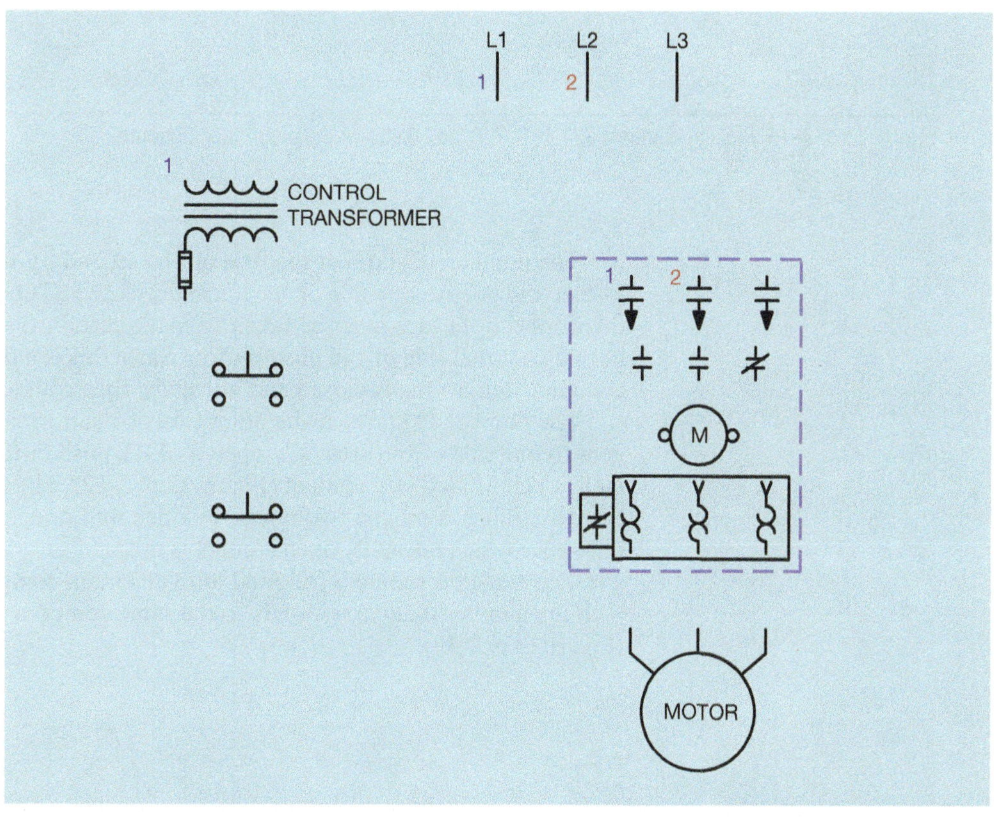

Figure Exp. 2–9 The number 2 is placed beside L2 and the second load contact on M starter.

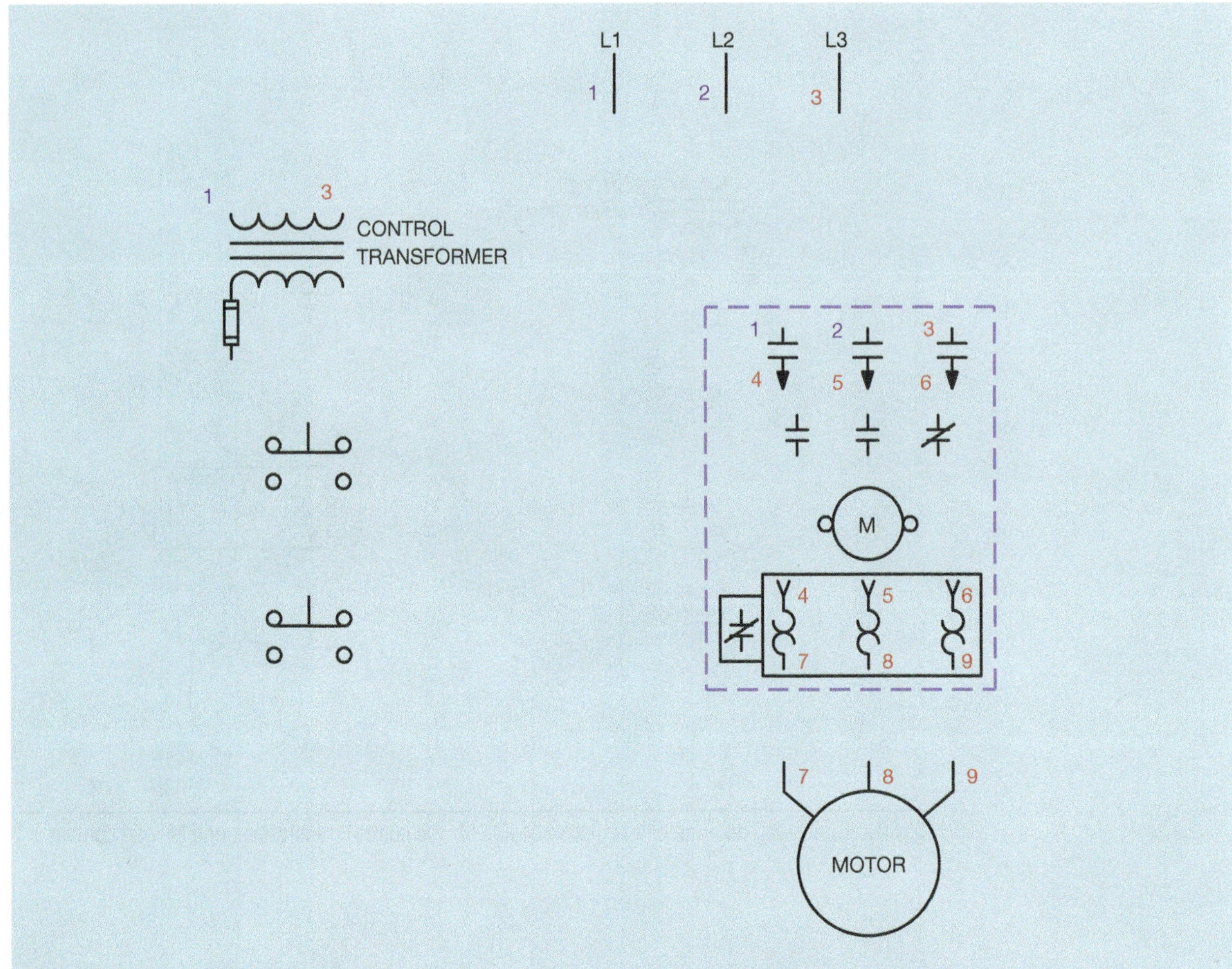

Figure Exp. 2–10 Placing numbers 3, 4, 5, 6, 7, 8, and 9 beside the proper components.

The number 10 starts at the fuse on the secondary winding of the control transformer and goes to one side of the normally closed STOP push button. When making this connection, care must be taken to make certain that connection is made to the normally closed side of the push button. Since this is a double-acting push button, it contains both normally closed and normally open contacts (Figure Exp. 2–11).

The number 11 starts at the other side of the normally closed STOP button and goes to one side of the normally open START push button and to one side of a normally open M auxiliary contact (Figure Exp. 2–12). The starter in this example shows three auxiliary contacts: two normally open and one normally closed. It makes no difference which normally open contact is used.

This same procedure is followed until all circuit components have been numbered with the number that corresponds to the same component on the schematic diagram (Figure Exp. 2–13).

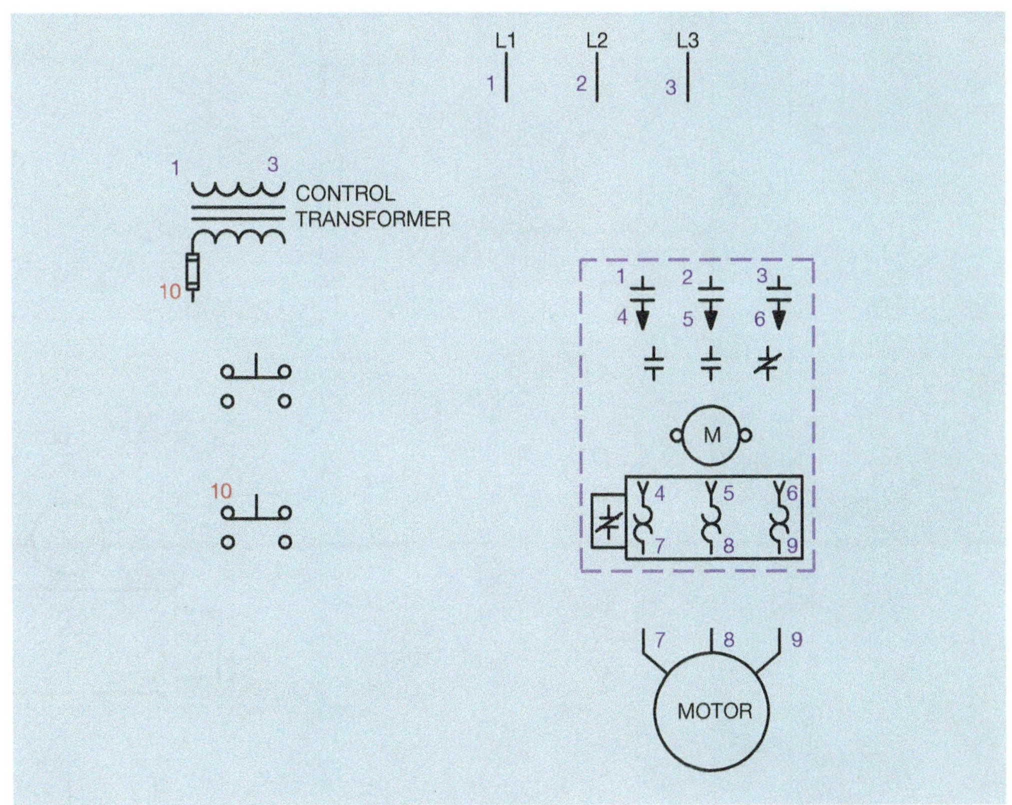

Figure Exp. 2–11 Wire number 10 connects from the transformer secondary to the STOP button.

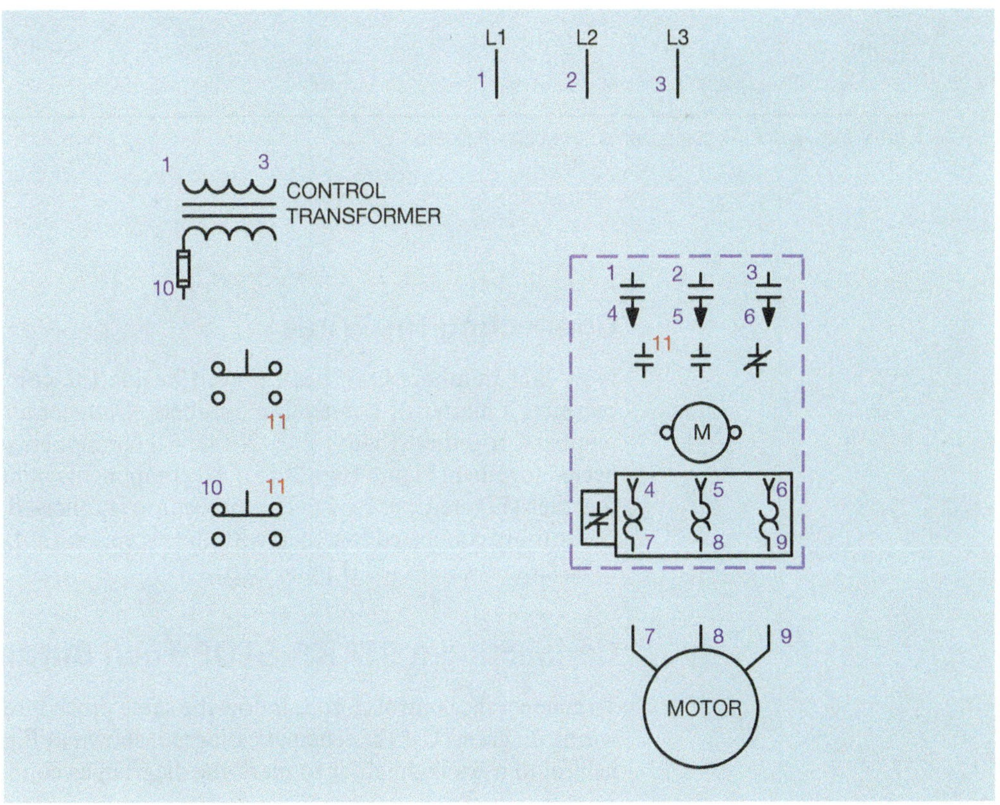

Figure Exp. 2–12 Number 11 connects to the STOP button, START button, and holding contact.

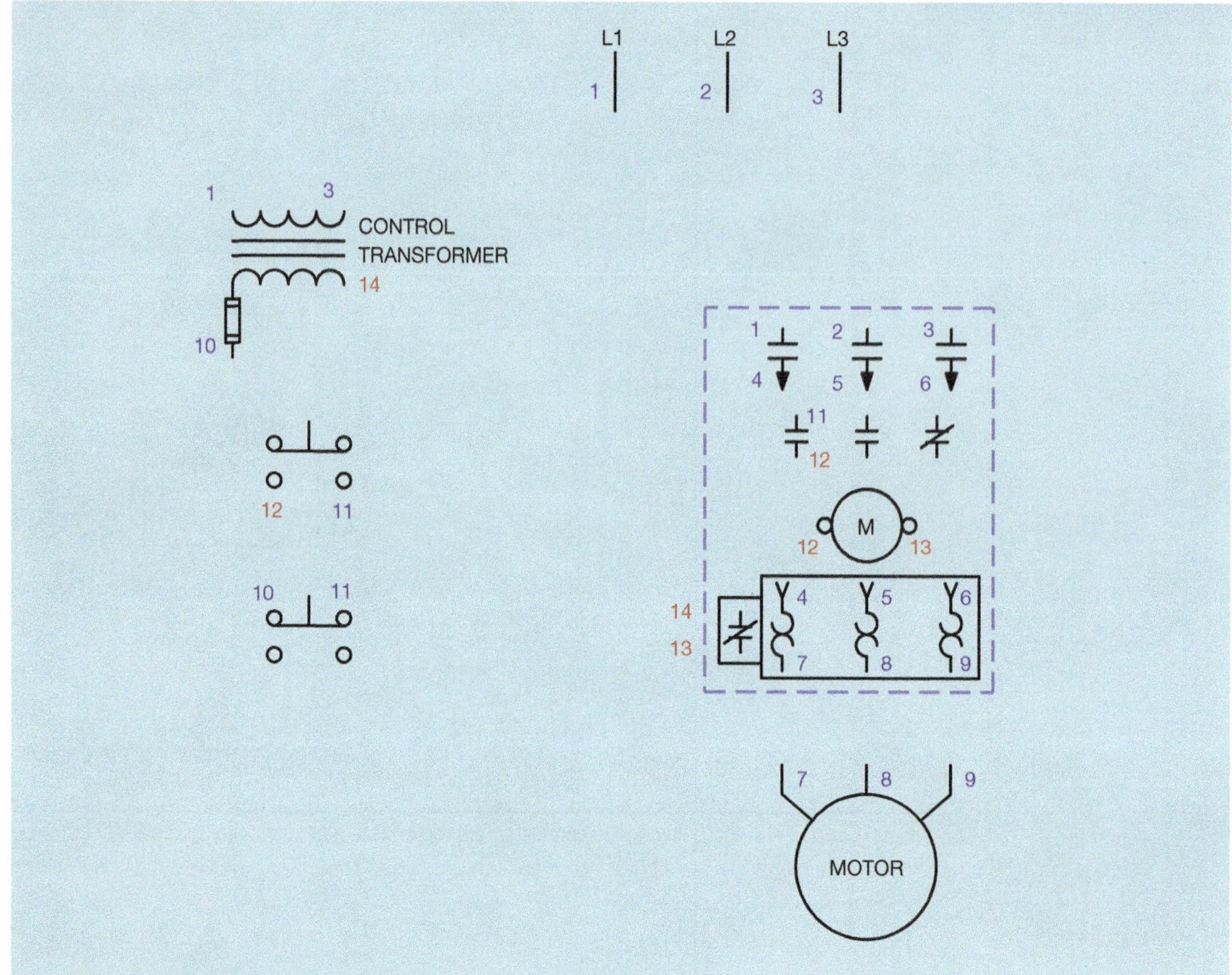

Figure Exp. 2–13 All components have been numbered.

Connecting the Wires

Now that numbers have been placed beside the components, wiring the circuit becomes a matter of connecting numbers. Connect all components labeled with a number 1 together (Figure Exp. 2–14). All components numbered with a 2 are connected together (Figure Exp. 2–15). All components numbered with a 3 are connected together (Figure Exp. 2–16). This procedure is followed until all the numbered components are connected together, with the exception of 4, 5, and 6, which are assumed to be factory connected (Figure Exp. 2–17).

Connecting a START–STOP Push Button Control Circuit

To connect the control circuit, follow the same procedure that was used to develop the wiring diagram. Use the schematic diagram shown in Figure Exp. 2–7. It is sometimes helpful to use a highlighter to mark the diagram as connections are made.

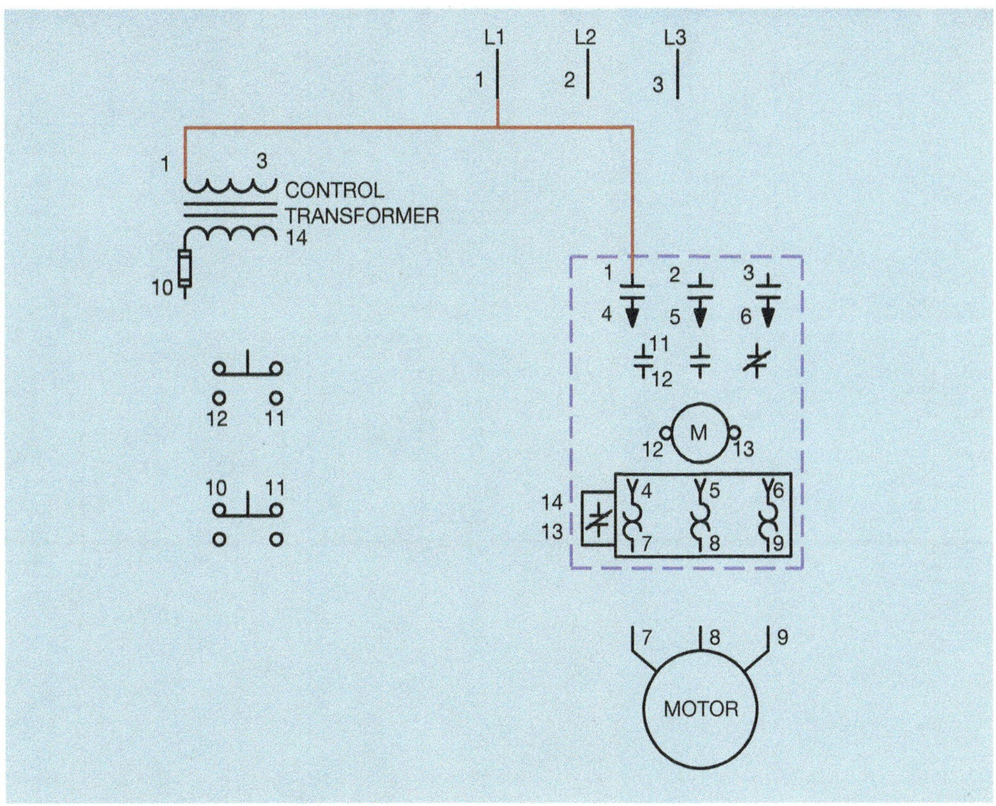

Figure Exp. 2–14 Connecting all the number 1s together.

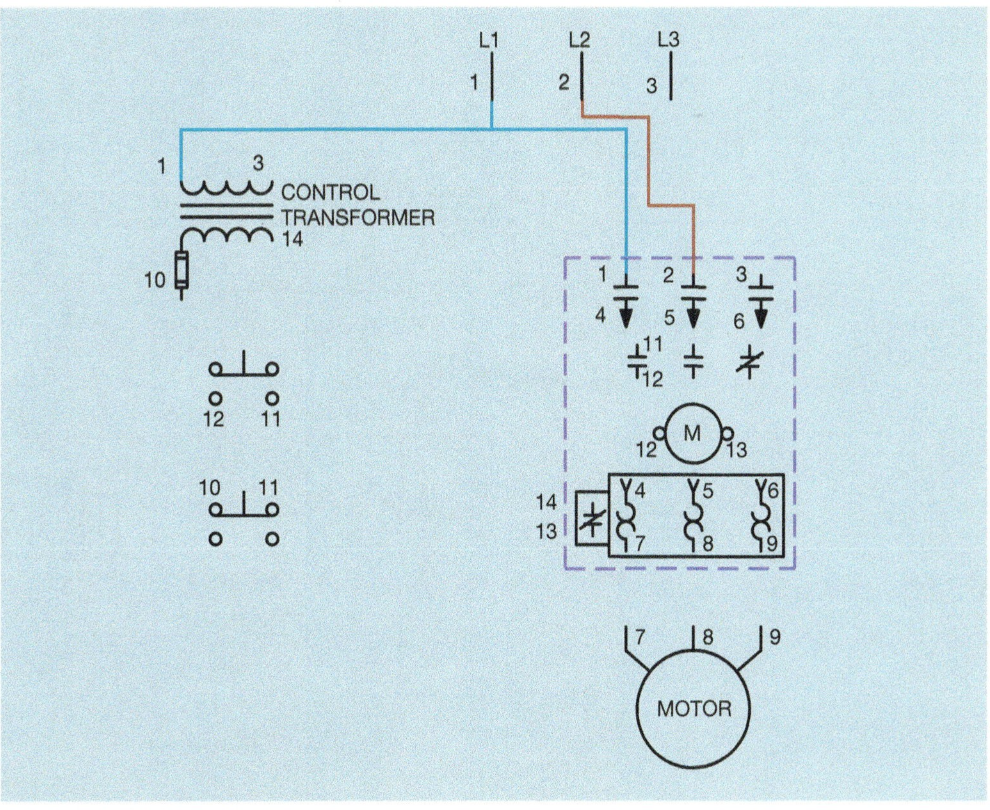

Figure Exp. 2–15 Connecting all the number 2s together.

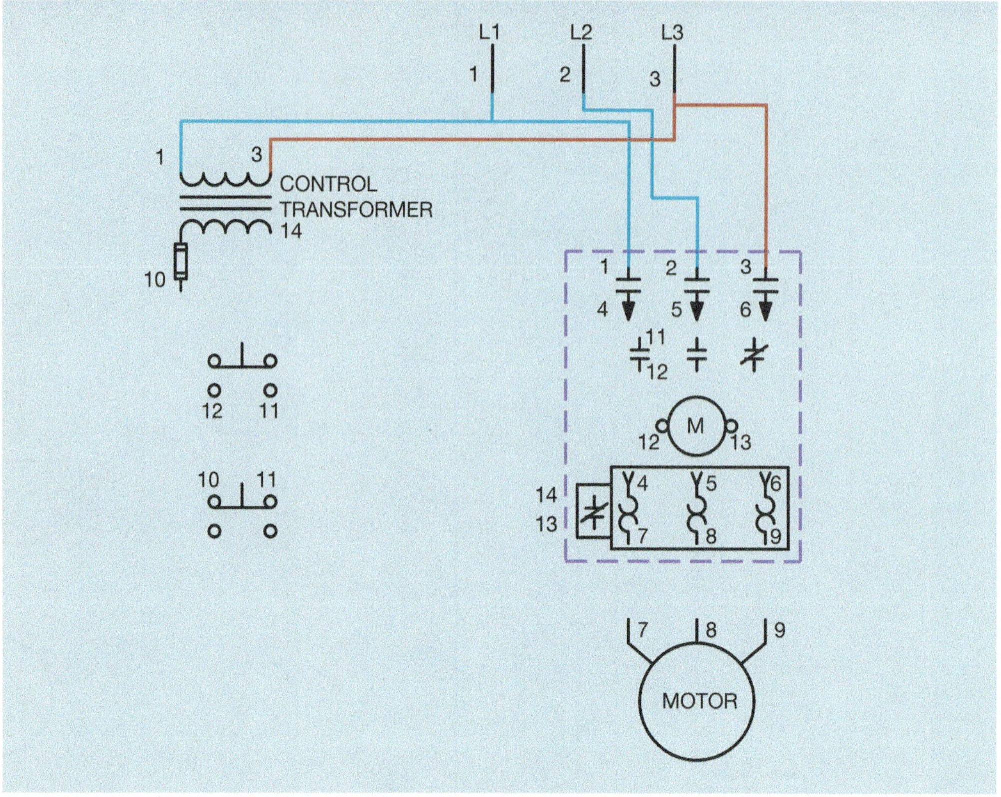

Figure Exp. 2–16 Connecting all the number 3s together.

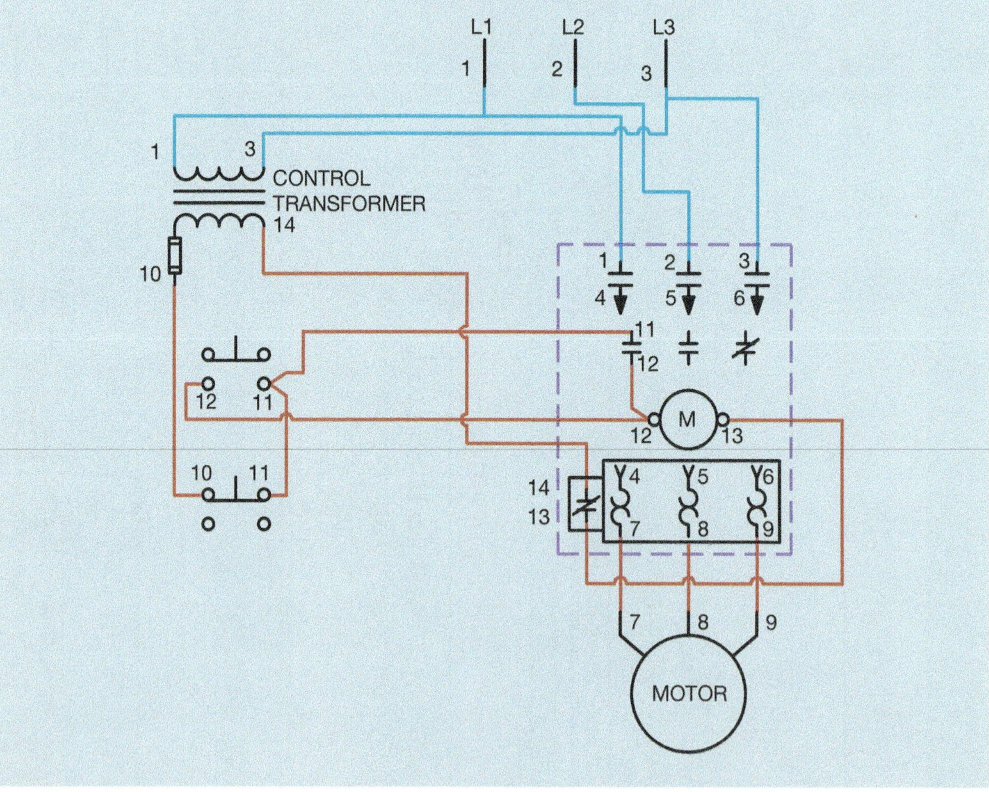

Figure Exp. 2–17 The circuit is wired by connecting all like numbered components together.

- ❑ 1. Connect all components that are labeled with a number 1. Make certain to connect to a *load* contact on the starter or contactor.
- ❑ 2. Connect all components labeled with a number 2. Again make sure to connect to a load contact on the starter or contactor.
- ❑ 3. Connect all components labeled with a 3.
- ❑ 4. Wire connections 4, 5, and 6 may or may not have to be made depending on whether you are using a starter or a contactor and separate overload relay.
- ❑ 5. Wires 7, 8, and 9 connect from the output of the heaters on the overload relays to the motor T leads. Your circuit may contain a single three phase overload relay or three separate overload relays if you are using a contactor and separate overload relays.
- ❑ 6. Wire number 10 connects from the secondary winding of the control transformer to one side of the normally closed push button used for the STOP button. If using a double-acting push button, make certain to connect to the closed side.
- ❑ 7. Wire number 11 connects from the other side of the normally closed push button to the normally open push button used for the START button. If a double-acting push-button is being used, make certain to connect to the open side. Wire number 11 also connects to a normally open auxiliary contact on M starter. Auxiliary contacts are smaller than the load contacts and are used as part of the control circuit. Make certain to connect to one side of an *open* contact.
- ❑ 8. Wire number 12 connects from the other side of the normally open START button to the other side of the normally open auxiliary contact and to one side of the coil on M starter.
- ❑ 9. Wire number 13 connects from the other side of the coil on M starter to one side of the normally open contact located on the overload relay. If a three phase motor starter is being used, or if a separate three phase overload relay is being used, there will be only one overload contact. Note the number of contacts on the overload relay. Some overload relays contain both normally open and normally closed contacts, and some do not. Make certain that connection is made to the normally closed contact if the relay contains more than one contact. If three separate single phase overload relays are being used, each overload relay contains an overload contact. These three contacts will have to be connected in series so that if one opens, the circuit will be broken.
- ❑ 10. Wire number 14 connects from the other side of the normally closed overload contact to the other side of the secondary winding on the control transformer.
- ❑ 11. Check with your instructor before turning on the power.
- ❑ 12. Test the circuit for proper operation.
- ❑ 13. If the circuit works properly, **turn off the power** and disconnect the circuit. Return the wires and components to their proper place.

Review Questions

1. Refer to the circuit shown in Figure Exp. 2–7. If wire number 11 were disconnected at the normally open auxiliary M contact, how would the circuit operate?
2. Assume that when the START button is pressed, M starter does not energize. List seven possible causes for this problem.
   1. ______________________________
   2. ______________________________
   3. ______________________________
   4. ______________________________
   5. ______________________________
   6. ______________________________
   7. ______________________________
3. Explain the difference between a motor starter and a contactor.
4. Refer to the schematic in Figure Exp. 2–7. Assume that when the START button is pressed, the control transformer fuse blows. What is the most likely cause of this trouble?
5. Explain the difference between load and auxiliary contacts.

Exercise 3

MULTIPLE PUSH BUTTON STATIONS

Objectives

After completing this exercise the student should be able to:

» Place wire numbers on a schematic diagram.

» Place corresponding numbers on control components.

» Draw a wiring diagram from a schematic diagram.

» Connect a control circuit using two STOP and two START push buttons.

Materials Needed

- Three phase power supply
- Three phase squirrel cage induction motor or simulated load
- Four double-acting push buttons (NO/NC on same button)
- Three phase motor starter or contactor with overload relay containing three load contacts and at least one normally open auxiliary contact
- Control transformer

There may be times when it is desirable to have more than one START–STOP push button station to control a motor. In this exercise, the basic START–STOP push button control circuit discussed in Exercise 1 will be modified to include a second STOP and START push button station.

When a component is used to perform the function of *stop* in a control circuit, it will generally be a normally closed component and be connected in series with the motor starter coil. In this example, a second STOP push button is to be added to an existing START–STOP control circuit. The second push button will be added to the control circuit by connecting it in series with the existing STOP push button (Figure Exp. 3–1).

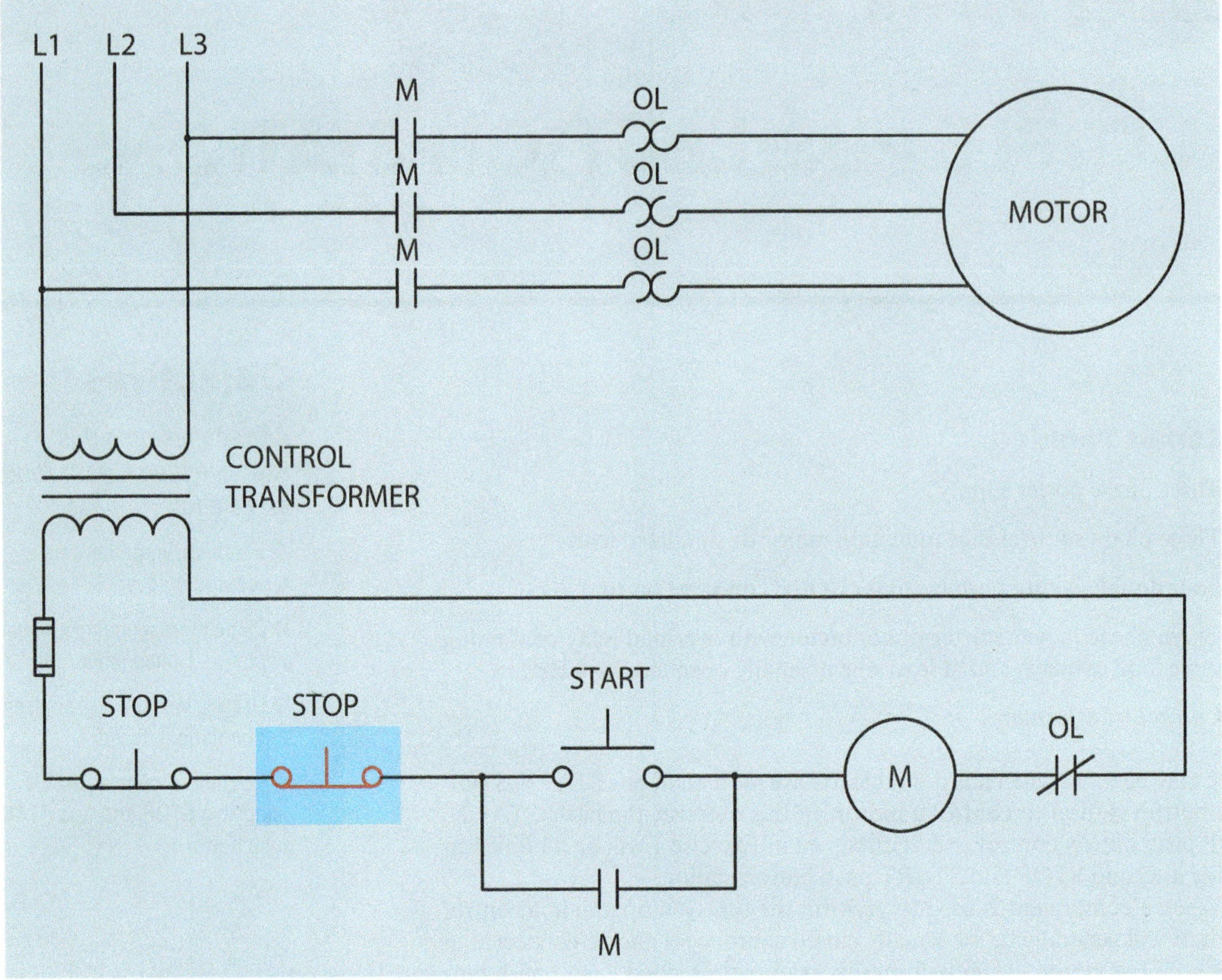

Figure Exp. 3–1 Adding a STOP button to the circuit.

When a component is used to perform the function of *start*, it is generally normally open and connected in parallel with the existing START button (Figure Exp. 3–2). If either START button is pressed, a circuit is completed to M coil. When M coil energizes, all M contacts change position. The three load contacts connected between the three phase power line and the motor close to connect the motor to the line. The normally open auxiliary contact connected in parallel with the two START buttons closes to maintain the circuit to M coil when the START button is released.

Developing the Wiring Diagram

Now that the circuit logic has been developed in the form of a schematic diagram, a wiring diagram will be drawn from the schematic. The components needed to connect this circuit are shown in Figure Exp. 3–3. Following the same procedure discussed in Exercise 1, wire numbers will be placed on the schematic diagram

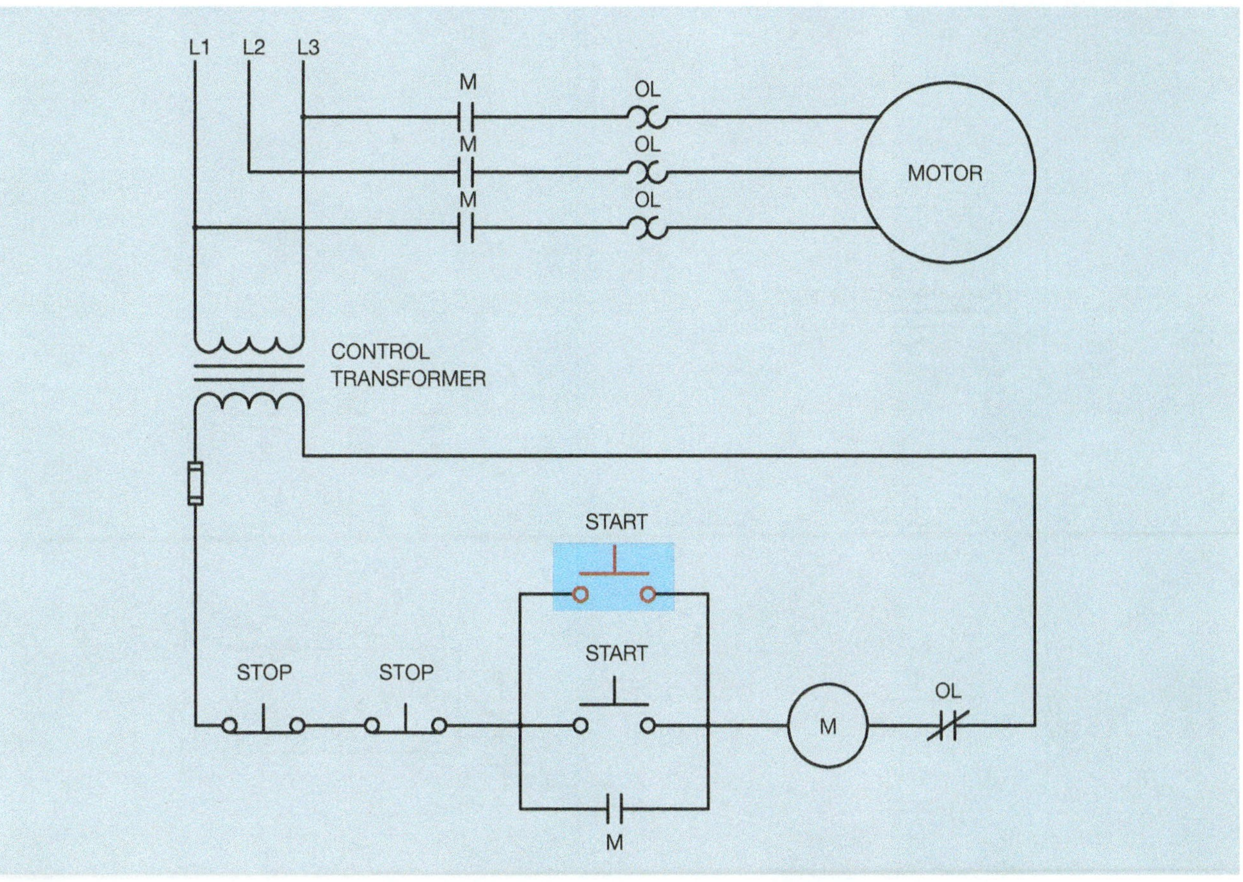

Figure Exp. 3–2 A second START button is added to the circuit.

(Figure Exp. 3–4). After wire numbers are placed on the schematic, corresponding numbers will be placed on the control components (Figure Exp. 3–5).

Connecting the Circuit

- ☐ 1. Using the schematic in Figure Exp. 3–4 or the diagram with numbered components in Figure Exp. 3–5, connect the circuit in the laboratory by connecting all like numbers together.
- ☐ 2. After the circuit has been connected, check with your instructor before turning on the power.
- ☐ 3. Turn on the power and test the circuit for proper operation.
- ☐ 4. If the circuit operates properly, **turn off the power** and disconnect the circuit. Return all components to their proper place.

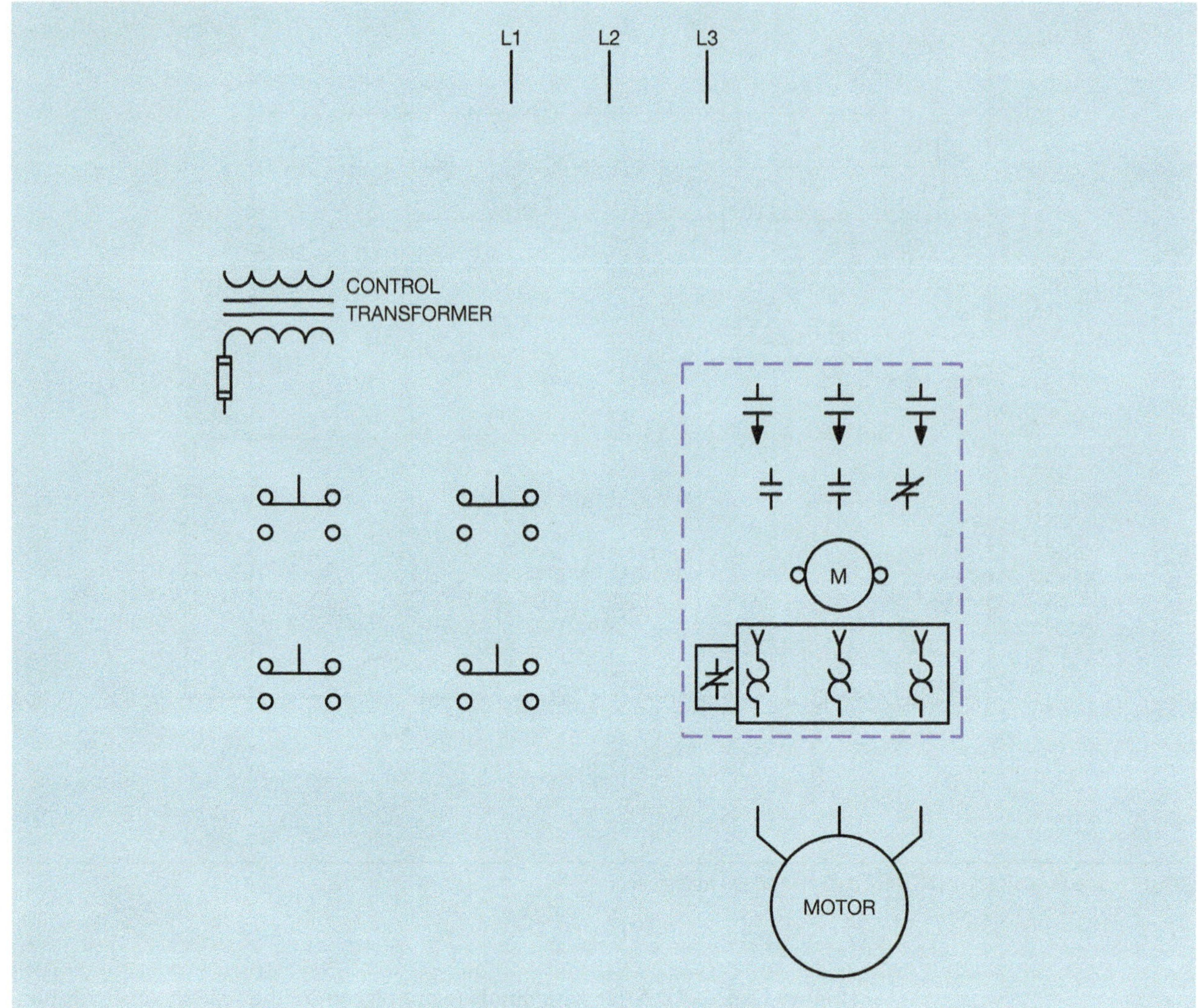

Figure Exp. 3–3 Components needed to produce a wiring diagram.

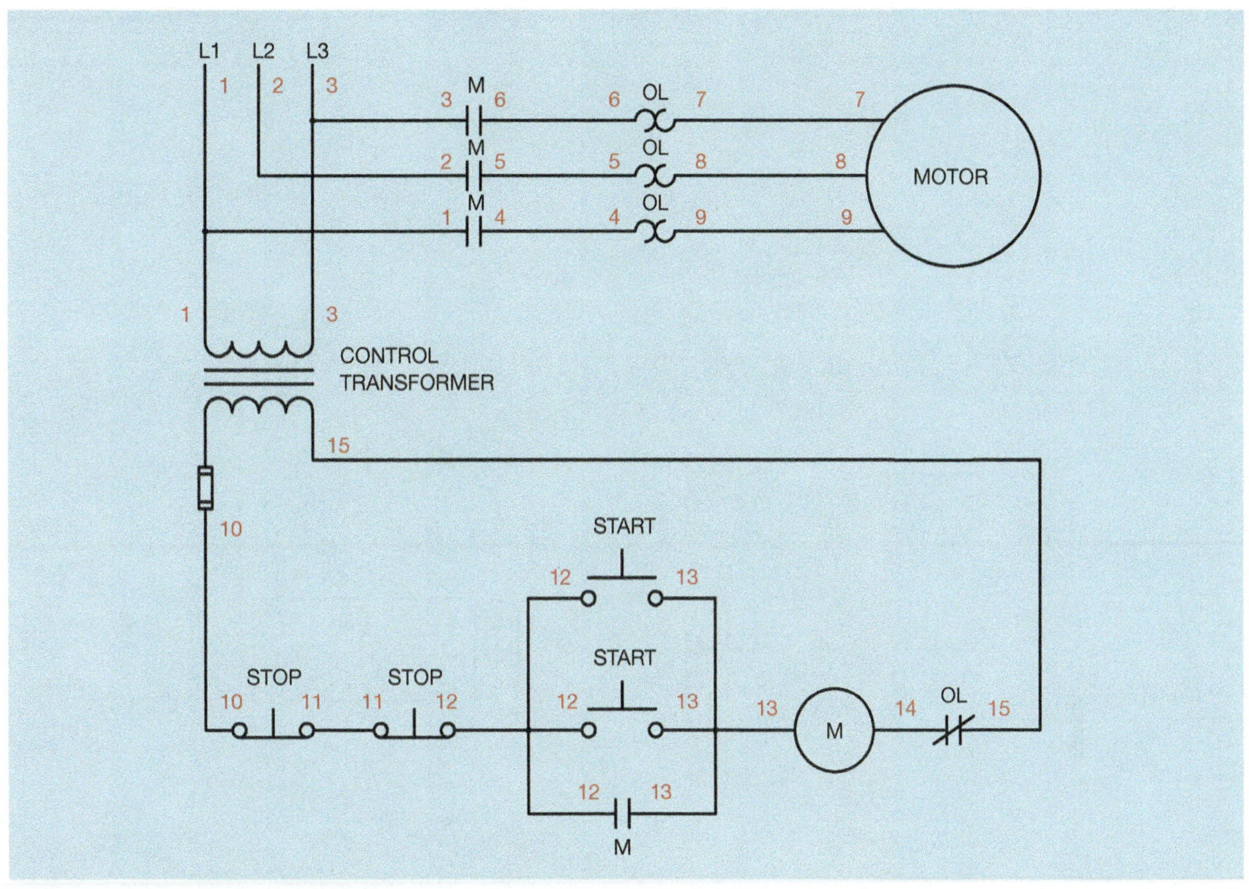

Figure Exp. 3–4 Numbering the schematic diagram.

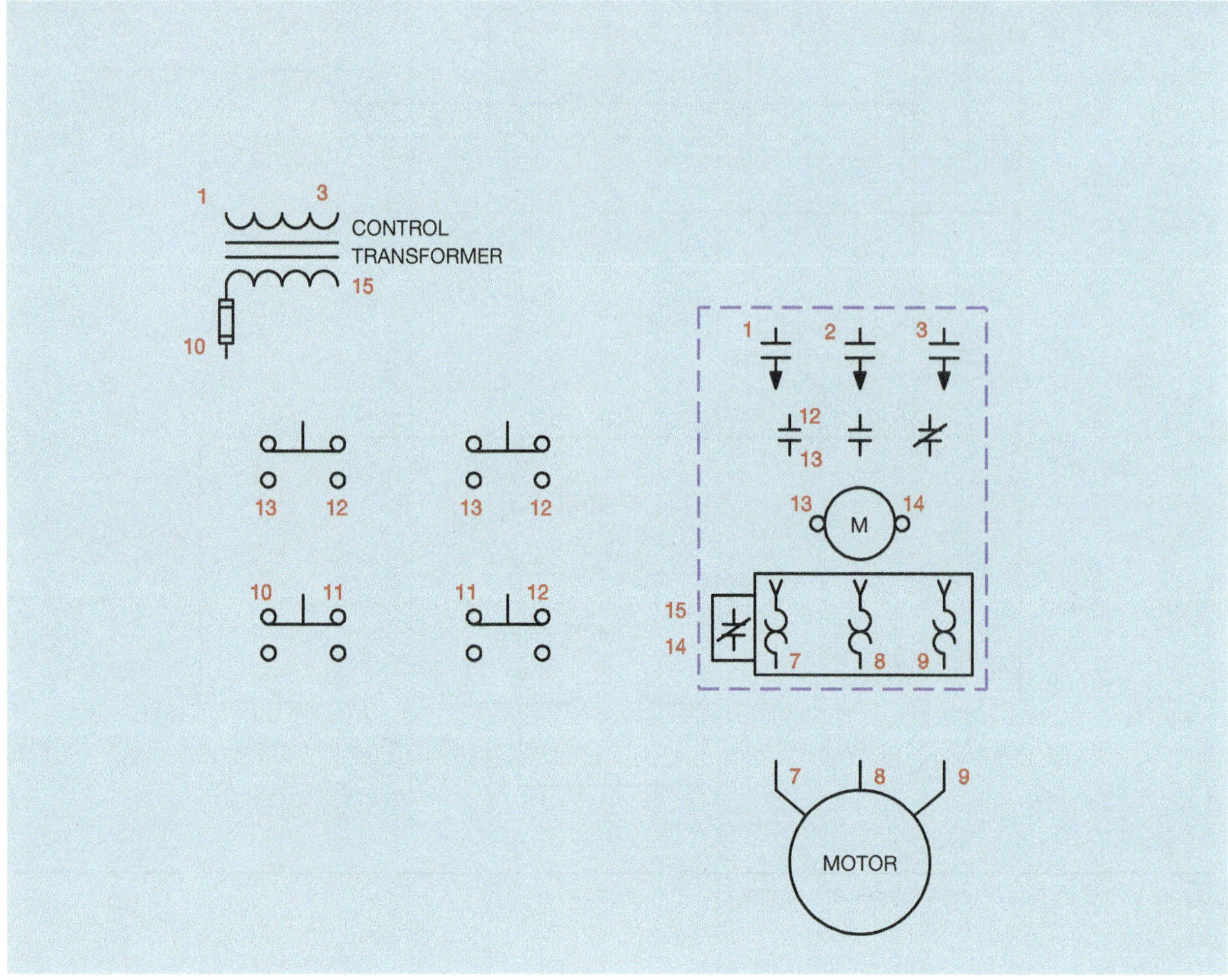

Figure Exp. 3–5 Numbering the components.

Review Questions

1. When a component is to be used for the function of start, is the component generally normally open or normally closed?
2. When a component is to be used for the function of stop, is the component generally normally open or normally closed?
3. The two STOP push buttons in Figure Exp. 3–2 are connected in series with each other. What would be the action of the circuit if they were to be connected in parallel as shown in Figure Exp. 3–6?
4. What would be the action of the circuit if both START buttons were to be connected in series as shown in Figure Exp. 3–7?
5. Following the procedure discussed in Exercise 1, place wire numbers on the schematic in Figure Exp. 3–7. Place corresponding wire numbers on the components shown in Figure Exp. 3–8.

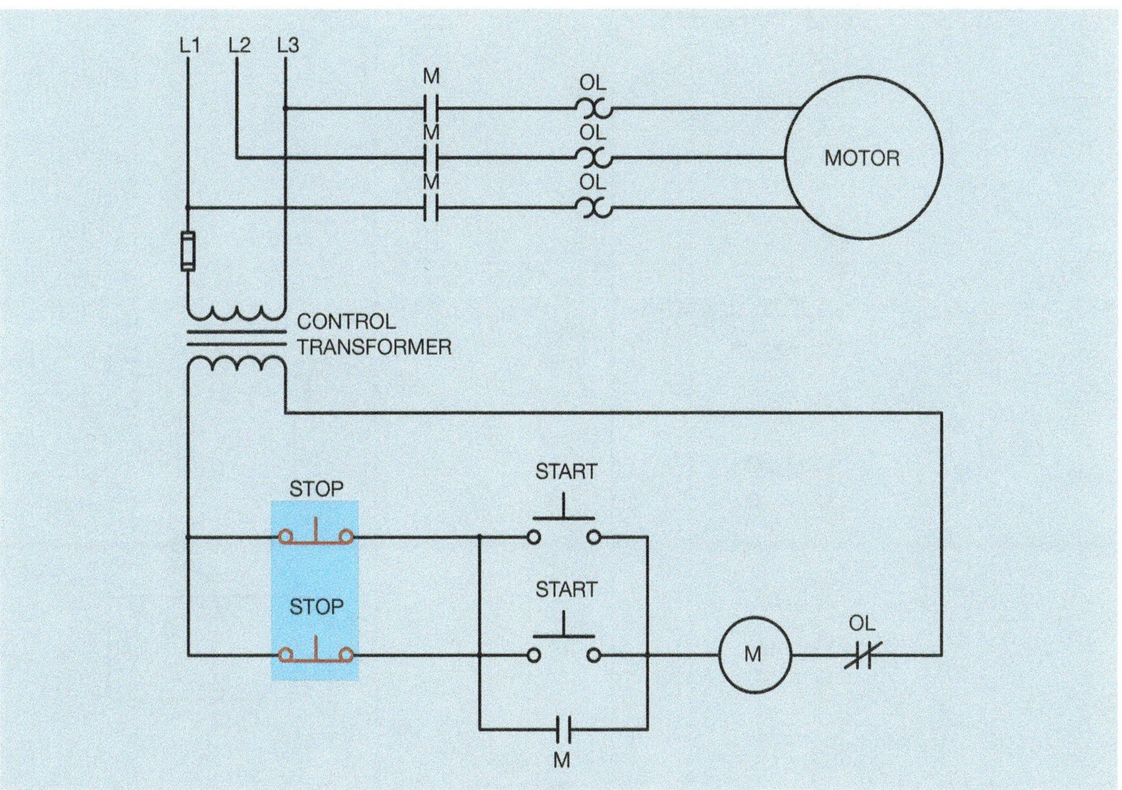

Figure Exp. 3–6 The STOP buttons have been connected in parallel.

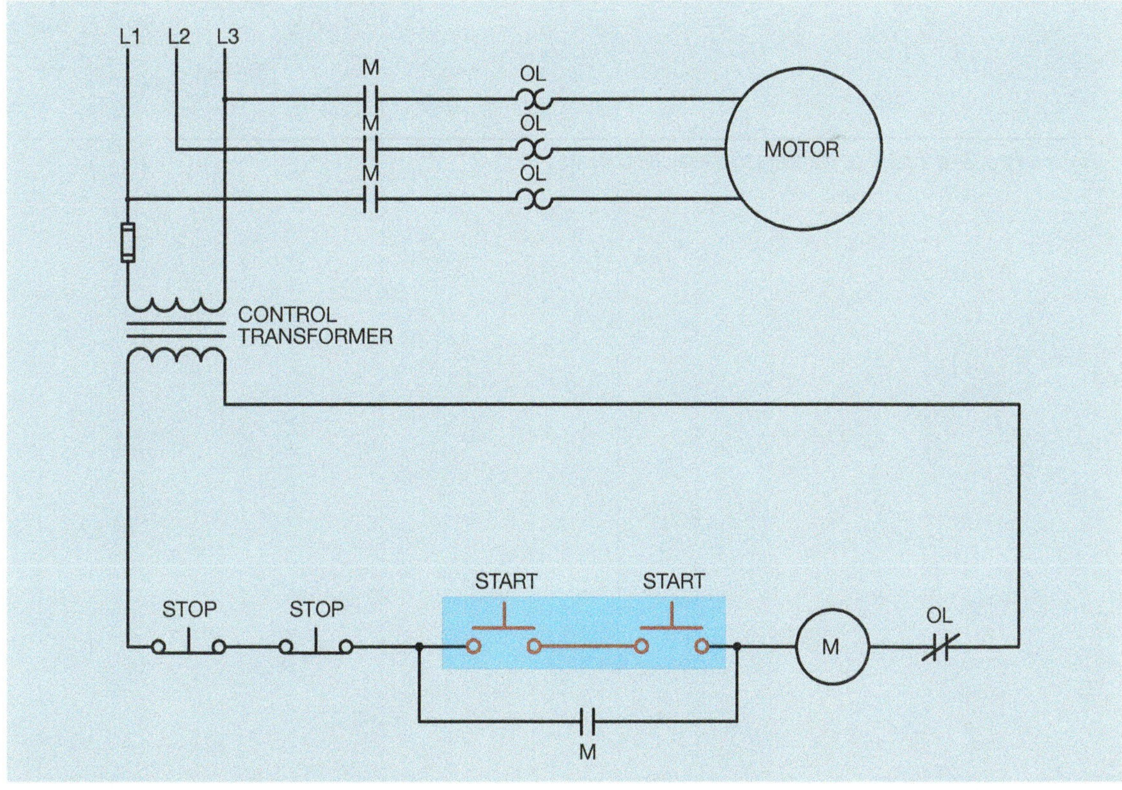

Figure Exp. 3–7 The START buttons are connected in series.

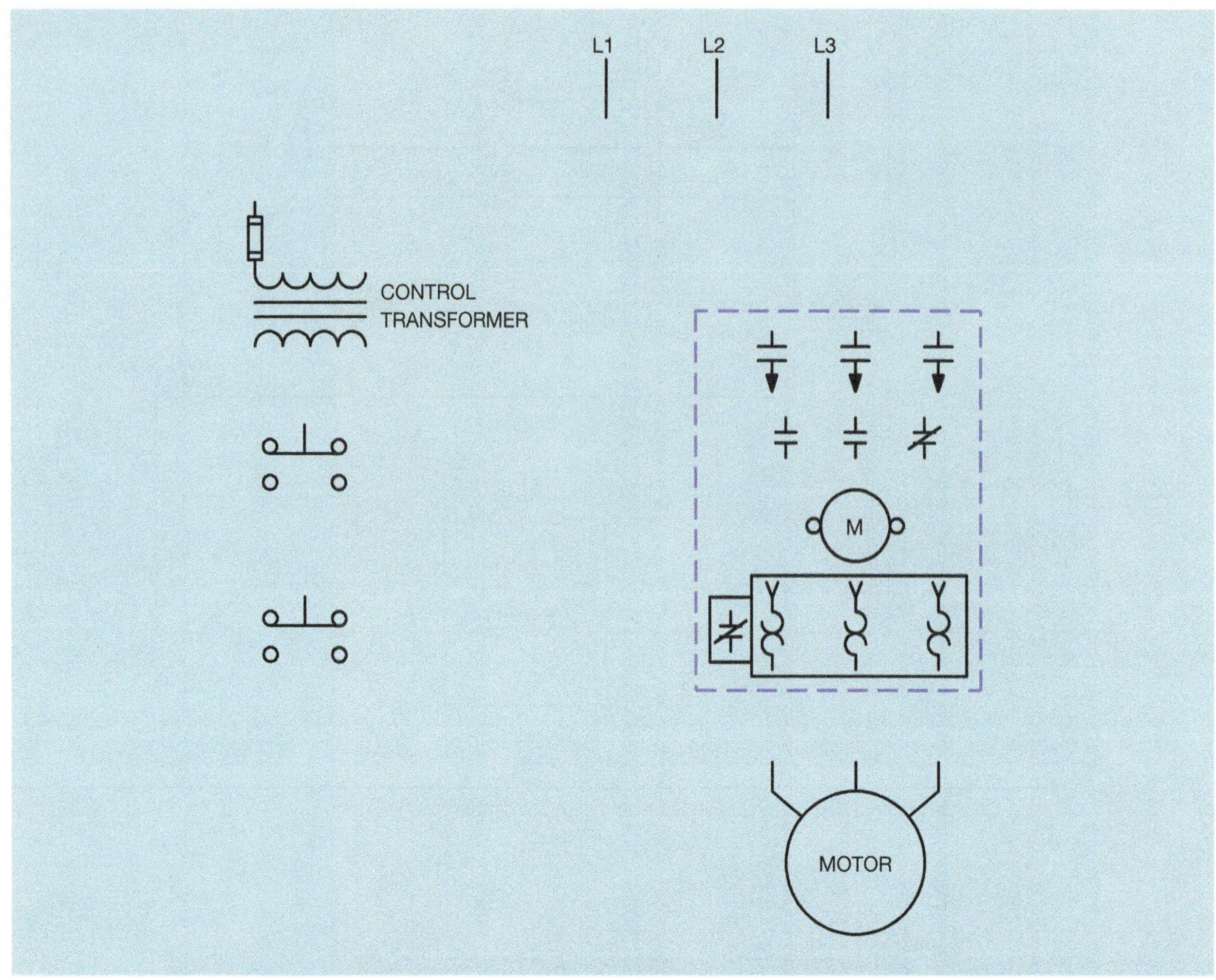

Figure Exp. 3–8 Add wire numbers to these components.

Exercise 4

FORWARD–REVERSE CONTROL

Objectives

After completing this exercise the student should be able to:

» Discuss cautions that must be observed in reversing circuits.

» Explain how to reverse a three phase motor.

» Discuss interlocking methods.

» Connect a forward–reverse motor control circuit.

Materials Needed

- Three phase power supply
- Control transformer
- One of the following:
    1. A three phase reversing starter
    2. Two three phase contactors with at least one normally open and one normally closed auxiliary contact on each contactor; one three phase overload relay or three single phase overload relays
- Three phase squirrel cage motor or simulated motor load
- Three double-acting push buttons (NO/NC on each button)

The direction of rotation of any three phase motor can be reversed by changing any two motor T leads. Because the motor is connected to the power line regardless of which direction it operates, a separate contactor is needed for each direction. Only one overload relay is needed, however, because the motor can operate in only one direction at a time. True reversing controllers contain two separate contactors and one overload relay built into one unit. Refer to Figures 9–3 and 9–4.

Interlocking

Interlocking prevents one action from taking place until another action has been performed. In the case of reversing starters, interlocking is used to prevent both contactors from being energized at the same time. Energizing both contactors would result in two of the three phase lines being shorted together. Interlocking forces one contactor to be de-energized before the other one can be energized.

Most reversing controllers contain mechanical interlocks as well as electrical interlocks. Mechanical interlocking is accomplished by using the contactors to operate a mechanical lever that prevents one contactor from closing while the other is energized.

Electrical interlocking is accomplished by connecting the normally closed auxiliary contacts on one contactor in series with the coil of the other contactor (Figure Exp. 4–1). Assume that the forward push button is pressed and F coil

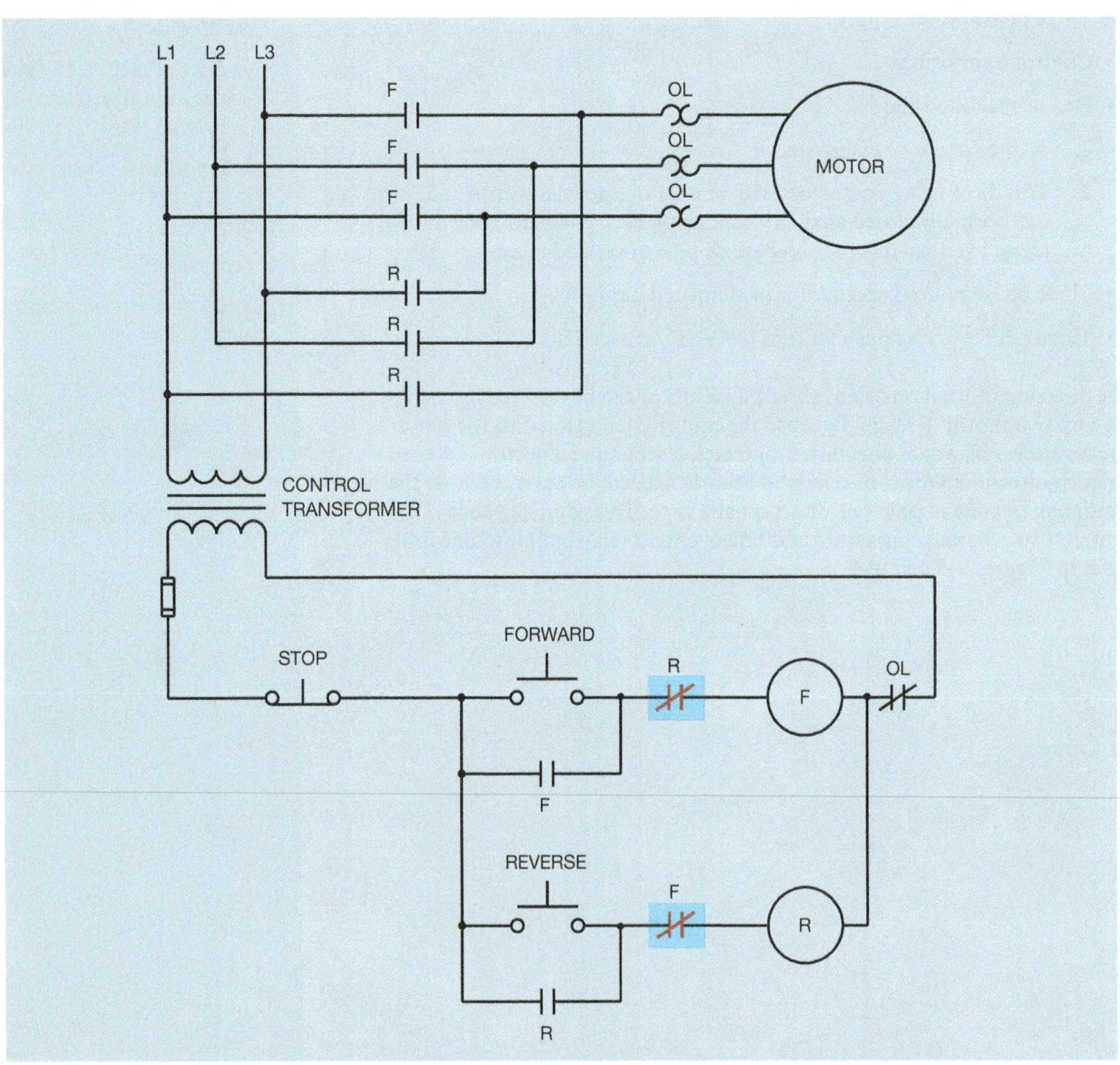

Figure Exp. 4–1 Forward–reverse control with interlock.

energizes. This causes all F contacts to change position. The three F load contacts close and connect the motor to the line. The normally open F auxiliary contact closes to maintain the circuit when the forward push button is released, and the normally closed F auxiliary contact connected in series with R coil opens (Figure Exp. 4–2). (Note: Figure Exp. 4–2 illustrates the circuit as it is when the forward starter has been energized.)

If the opposite direction of rotation is desired, the STOP button must be pressed first. If the reverse push button were to be pressed first, the now open F auxiliary contact connected in series with R coil would prevent a complete circuit from being established. Once the STOP button has been pressed, however, F coil de-energizes and all F contacts return to their normal position. The reverse push button can now

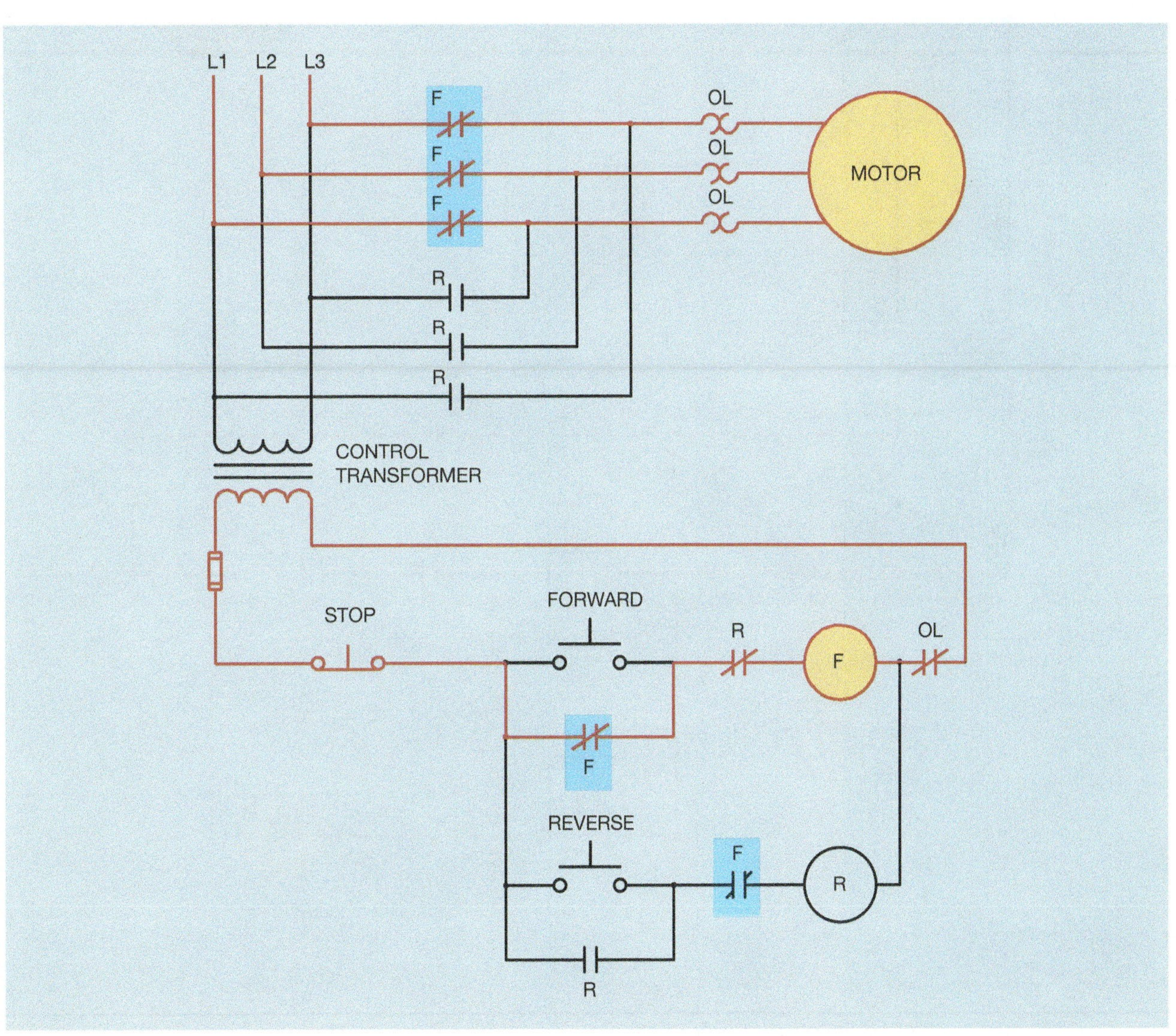

Figure Exp. 4–2 Motor operating in the forward direction.

be pressed to energize R coil (Figure Exp. 4–3). When R coil energizes, all R contacts change position. The three R load contacts close and connect the motor to the line. Notice, however, that two of the motor T leads are connected to different lines. The normally closed R auxiliary contact opens to prevent the possibility of F coil being energized until R coil is de-energized.

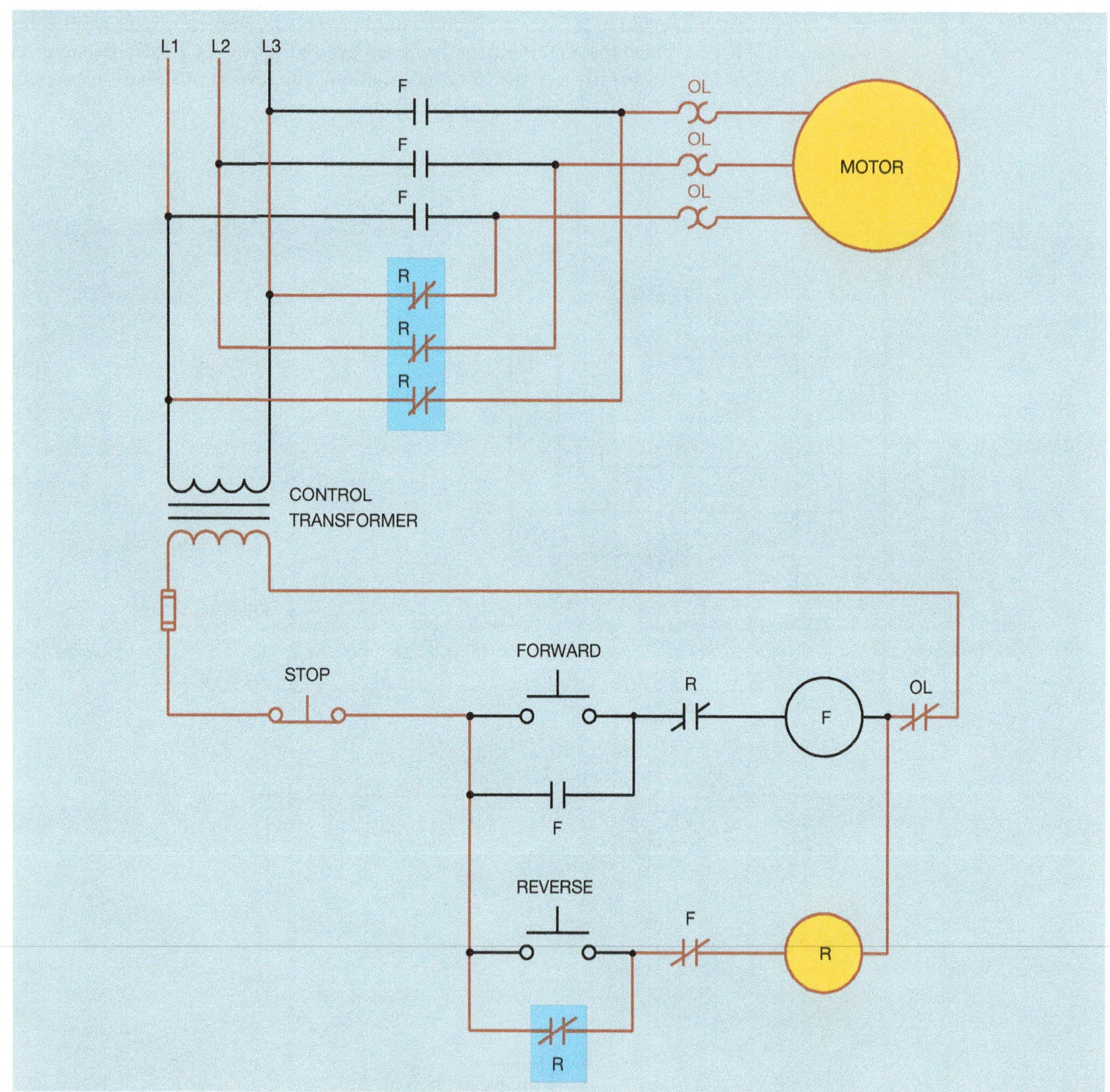

Figure Exp. 4–3 Motor operating in the reverse direction.

Developing a Wiring Diagram

The same basic procedure is used to develop a wiring diagram from the schematic as was followed in the previous exercises. The components needed to construct this circuit are shown in Figure Exp. 4–4. In this example it is assumed that two contactors and a separate three phase overload relay will be used.

The first step is to place wire numbers on the schematic diagram. A suggested numbering sequence is shown in Figure Exp. 4–5. The next step is to place the wire numbers beside the corresponding components of the wiring diagram (Figure Exp. 4–6).

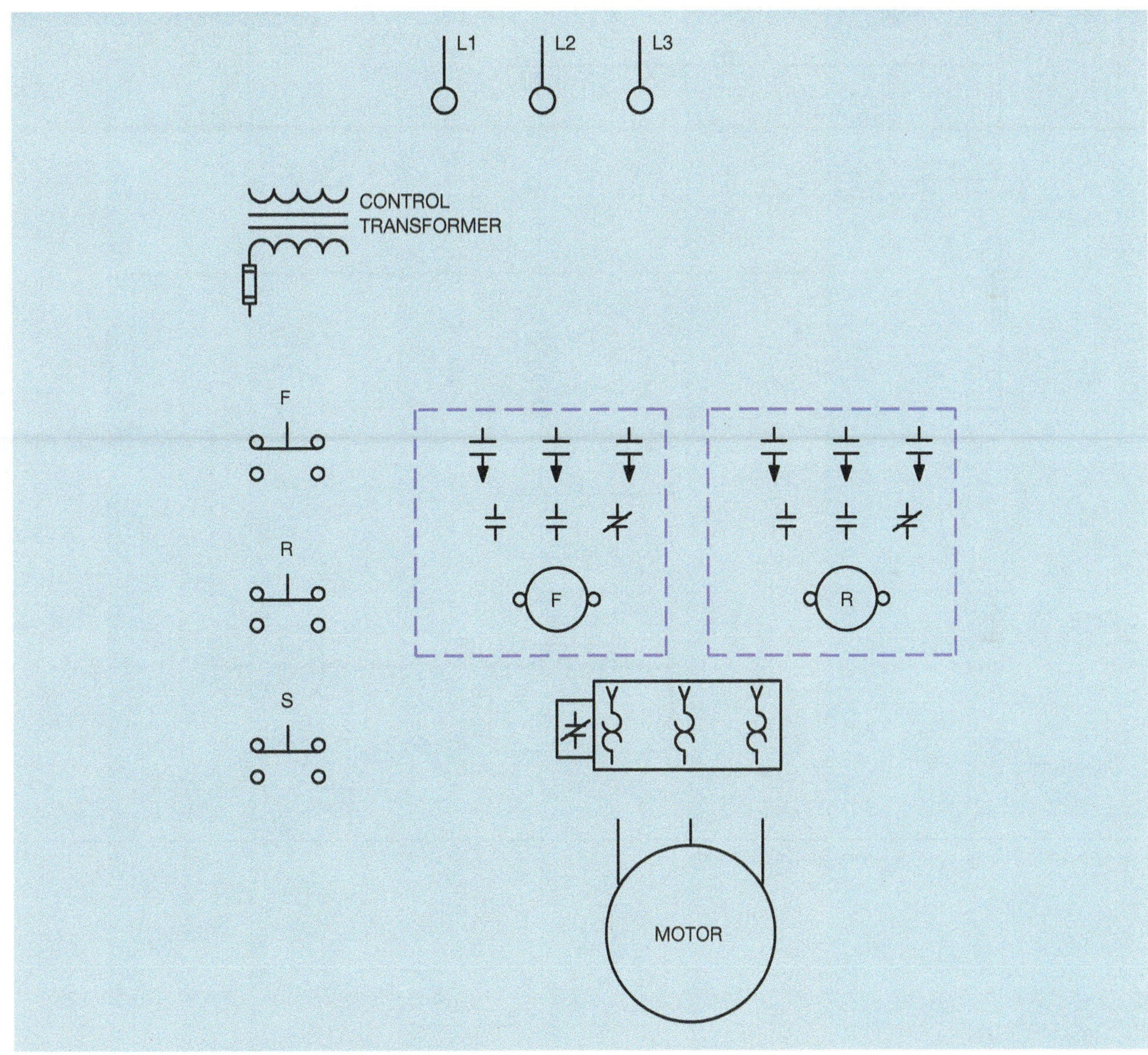

Figure Exp. 4–4 Components needed to construct a reversing control circuit.

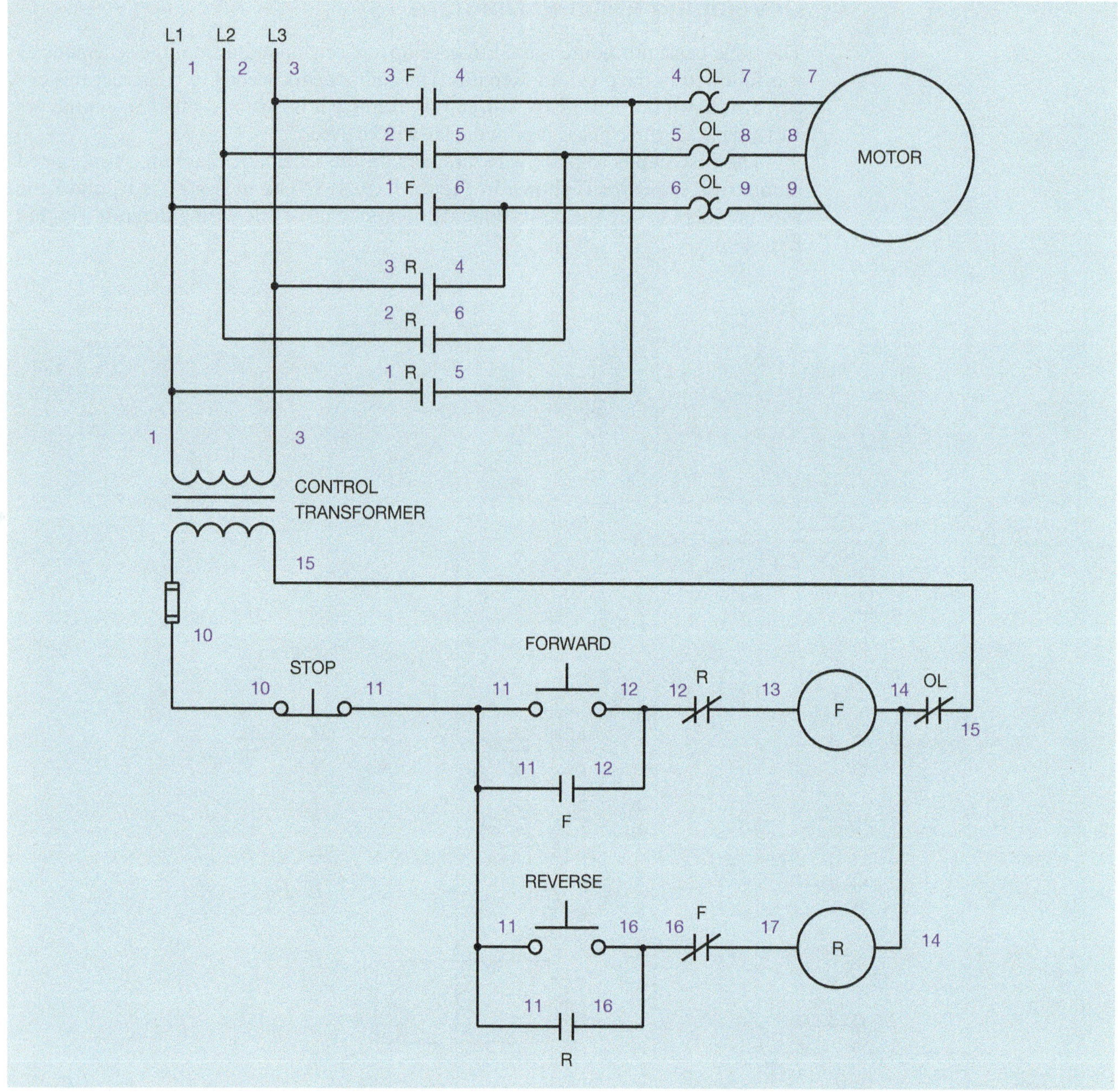

Figure Exp. 4–5 Placing wire numbers on the schematic.

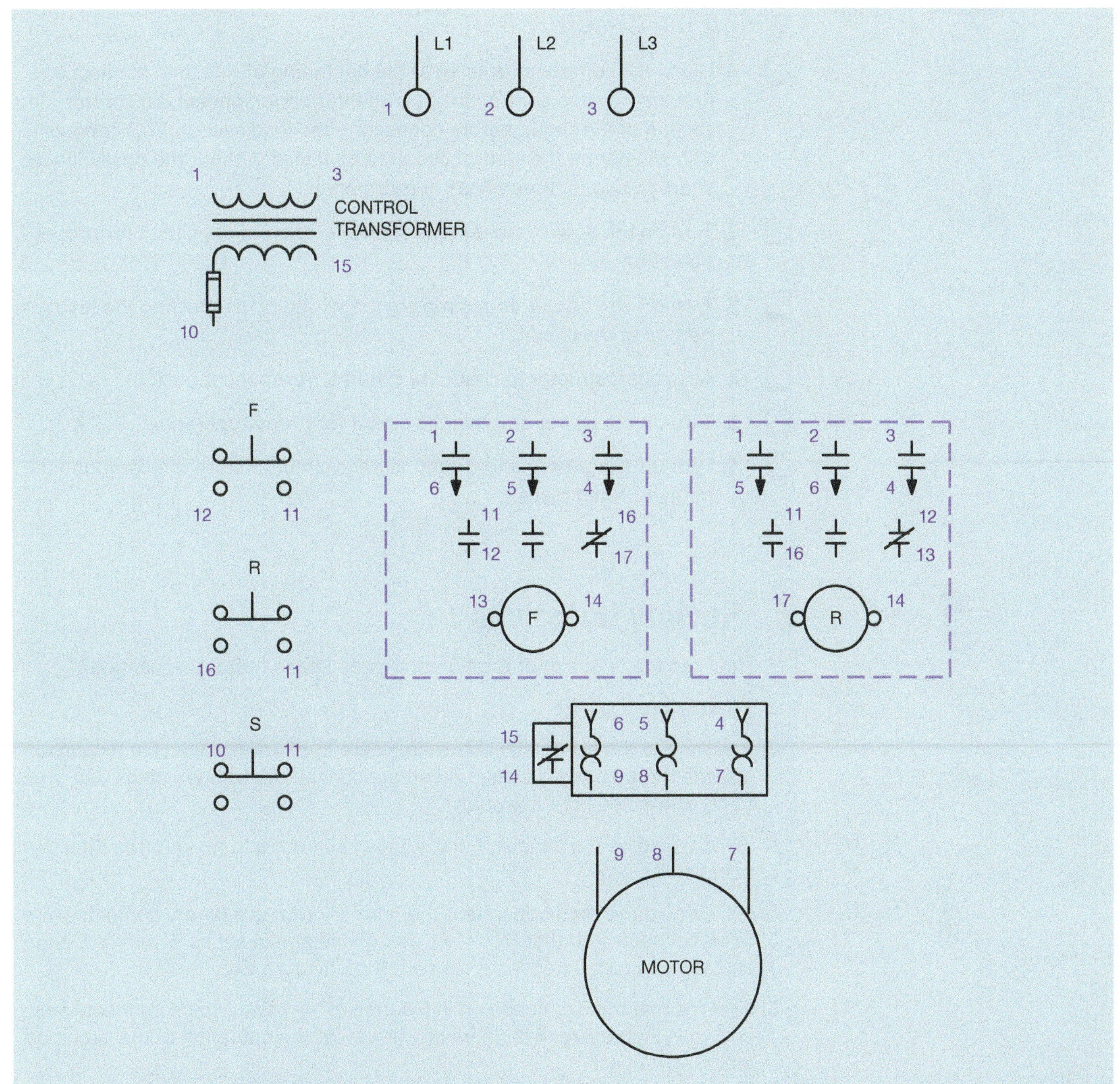

Figure Exp. 4–6 Placing corresponding wire numbers on the components.

Wiring the Circuit

- ☐ 1. Using the components listed at the beginning of this unit, connect a forward–reverse control circuit with interlocks. Connect the control section of the circuit before connecting the load section. This connection will permit the control circuit to be tested without the possibility of shorting two to three phase lines together.
- ☐ 2. Turn on the power and test the control section of the circuit for proper operation.
- ☐ 3. **Turn off the power** and complete the wiring by connecting the load portion of the circuit.
- ☐ 4. Ask your instructor to check the circuit for proper connection.
- ☐ 5. Turn on the power and test the circuit for proper operation.
- ☐ 6. **Turn off the power** and disconnect the circuit. Return the components to their proper place.

Review Questions

1. How can the direction of rotation of a three phase motor be changed?
2. What is interlocking?
3. Referring to the schematic shown in Figure Exp. 4–1, how would the circuit operate if the normally closed R contact connected in series with F coil were to be connected normally open?
4. What would be the danger, if any, if the circuit were to be wired as stated in question 3?
5. How would the circuit operate if the normally closed auxiliary contacts were to be connected so that F contact was connected in series with F coil, and R contact was connected in series with R coil (Figure Exp. 4–7)?
6. Assume that the circuit shown in Figure Exp. 4–1 were to be connected as shown in Figure Exp. 4–8. In what way would the operation of the circuit be different, if at all?

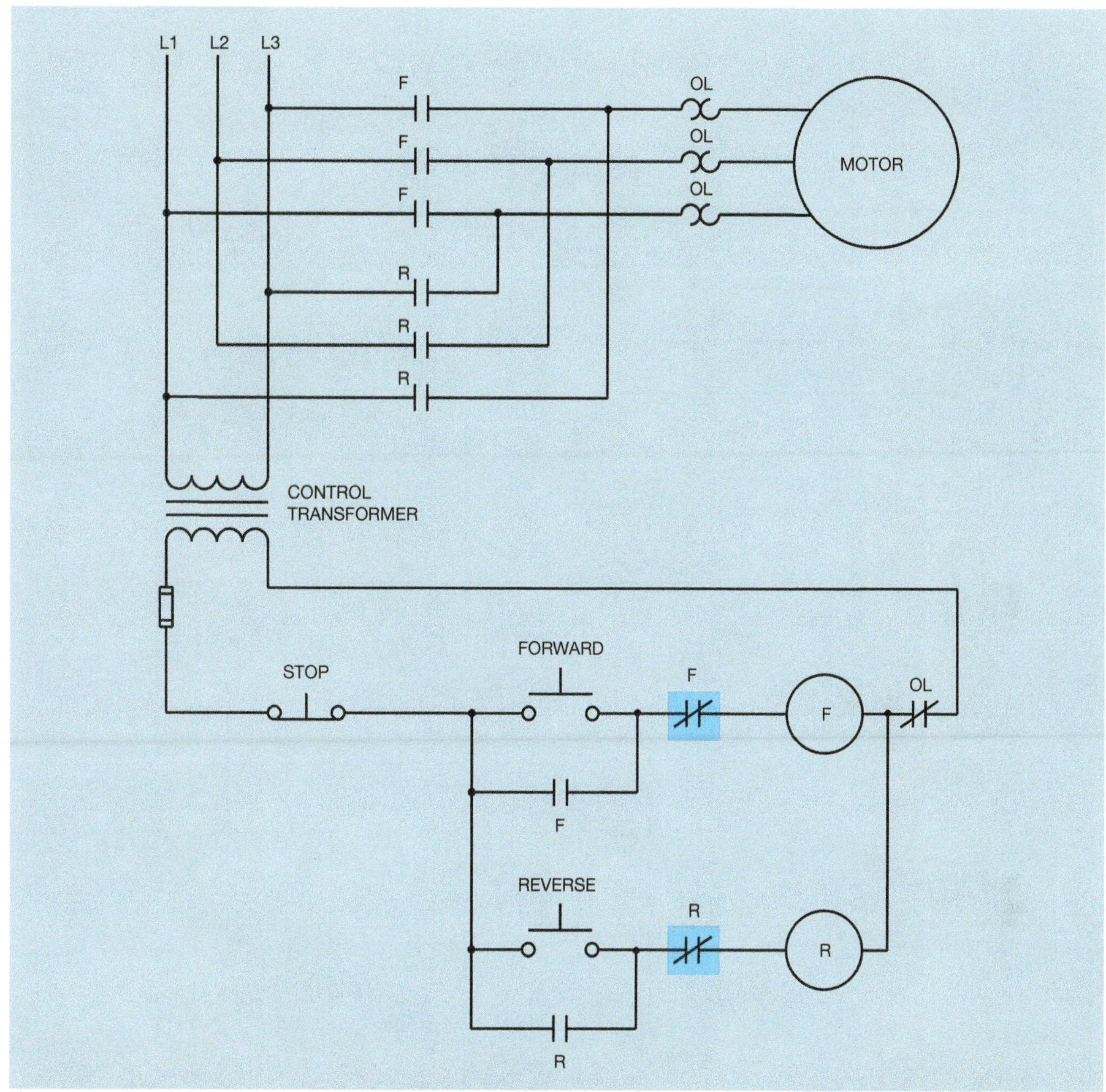

Figure Exp. 4–7 F and R normally closed auxiliary contacts are connected incorrectly.

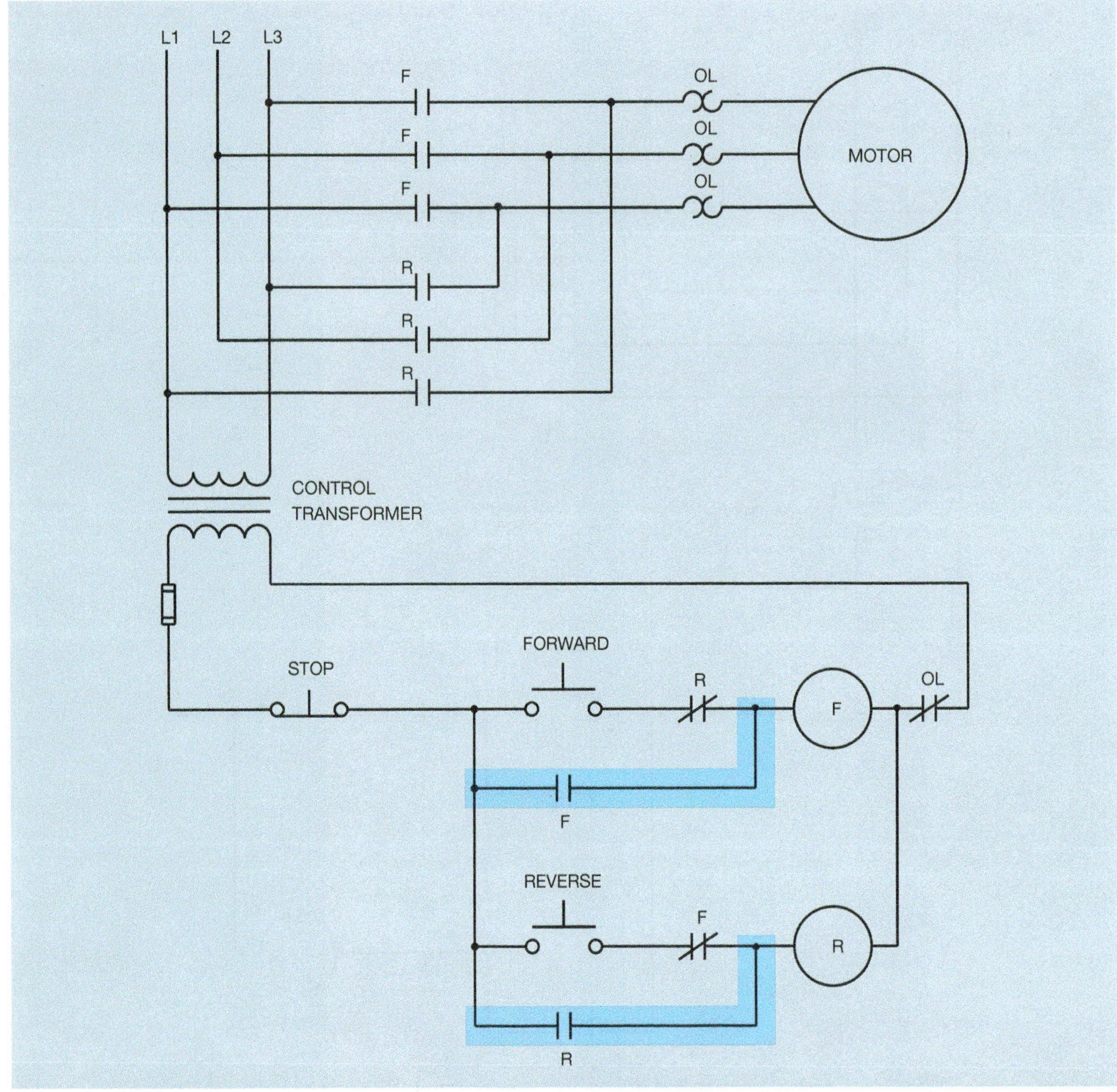

Figure Exp. 4–8 The position of the holding contacts has been changed.

Exercise 5

SEQUENCE CONTROL

Objectives

After completing this exercise the student should be able to:

» Define sequence control.

» Discuss methods of obtaining sequence control.

» Connect a control circuit for three motors that must be started in a predetermined sequence.

Materials Needed

- Three phase power supply
- Control transformer
- Three motor starters containing at least three load contacts and two normally open auxiliary contacts
- Three squirrel cage motors or three simulated motor loads
- Four double-acting push buttons (NO/NC on each button)

Sequence control forces a circuit to operate in a predetermined manner. In this exercise, three motors are to be started in sequence from motor 1 to 3. The requirements for the circuit are as follows:

- ☐ 1. The motors must start in sequence from 1 to 3. For example, motor 1 must be started before motor 2 can be started, and motor 2 must start before motor 3 can be started. Motor 2 cannot start before motor 1, and motor 3 cannot start before motor 2.
- ☐ 2. Each motor is started by a separate push button.
- ☐ 3. One STOP button will stop all motors.
- ☐ 4. An overload on any motor will stop all three motors.

As a general rule, there is more than one way to design a circuit that will meet the specified requirements, just as there is generally more than one road that can be taken to reach a destination. One design that will meet the requirements is shown in Figure Exp. 5–1. Because the logic of the circuit is of primary interest, the load contacts and motors are not shown. In this circuit, push button 1 must be pressed before power can be provided to push button 2. When motor starter 1 energizes, the normally open auxiliary contact 1M closes providing power to coil 1M and to push button 2. Motor starter 2 can now be started by pressing push button 2. Once motor starter 2 energizes,

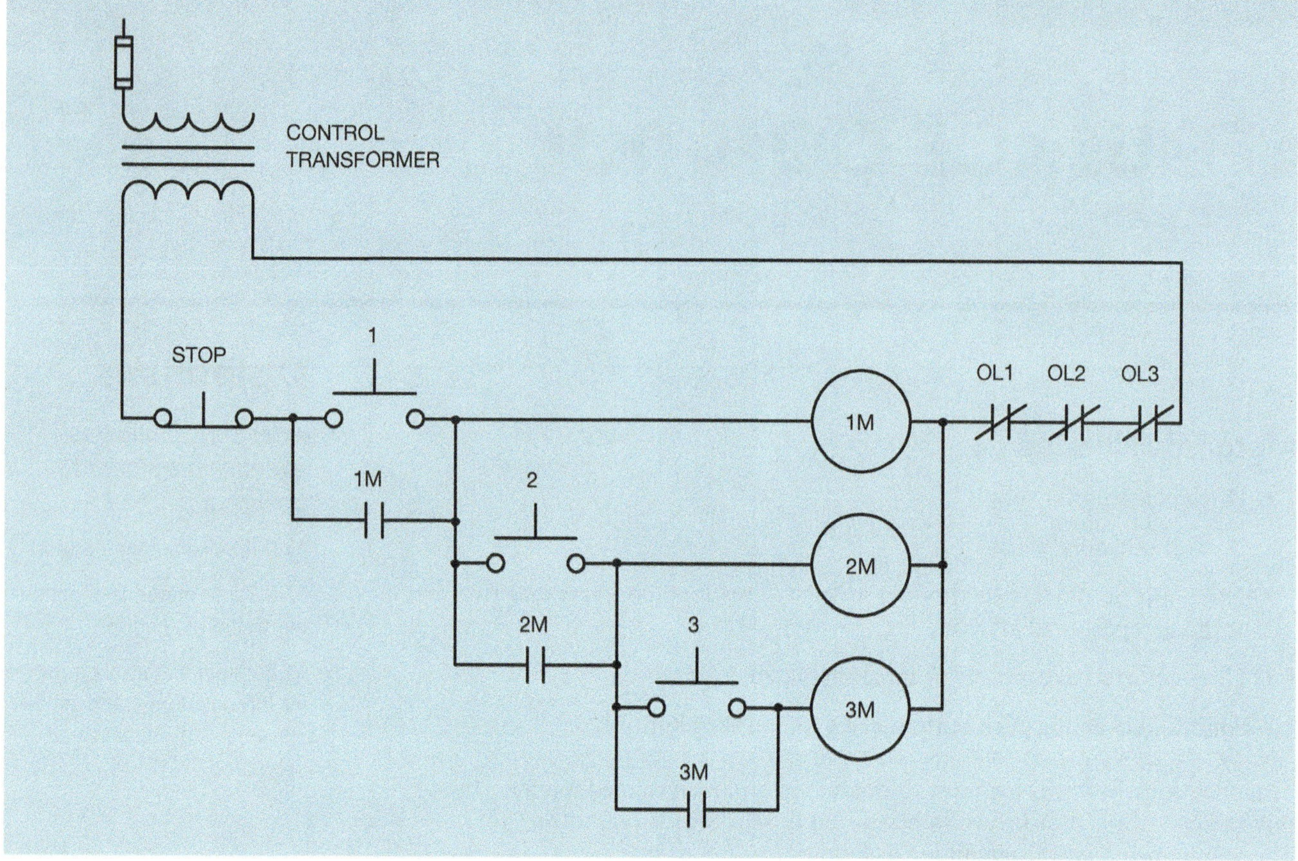

Figure Exp. 5–1 First example of starting three motors in sequence.

auxiliary contact 2M closes and provides power to coil 2M and push button 3. If the STOP button should be pressed or any overload contact opens, power is interrupted to all starters.

A Second Circuit for Sequence Control

A second method of providing sequence control is shown in Figure Exp. 5–2. In this circuit, normally open auxiliary contacts located on motor starters 1M and 2M are used to ensure that the three motors start in the proper sequence. A normally open 1M auxiliary contact connected in series with starter coil 2M prevents motor 2 from starting before motor 1, and a normally open 2M auxiliary contact connected in series with coil 3M prevents motor 3 from starting before motor 2. If the STOP button should be pressed or if any overload contact should open, power is interrupted to all starters.

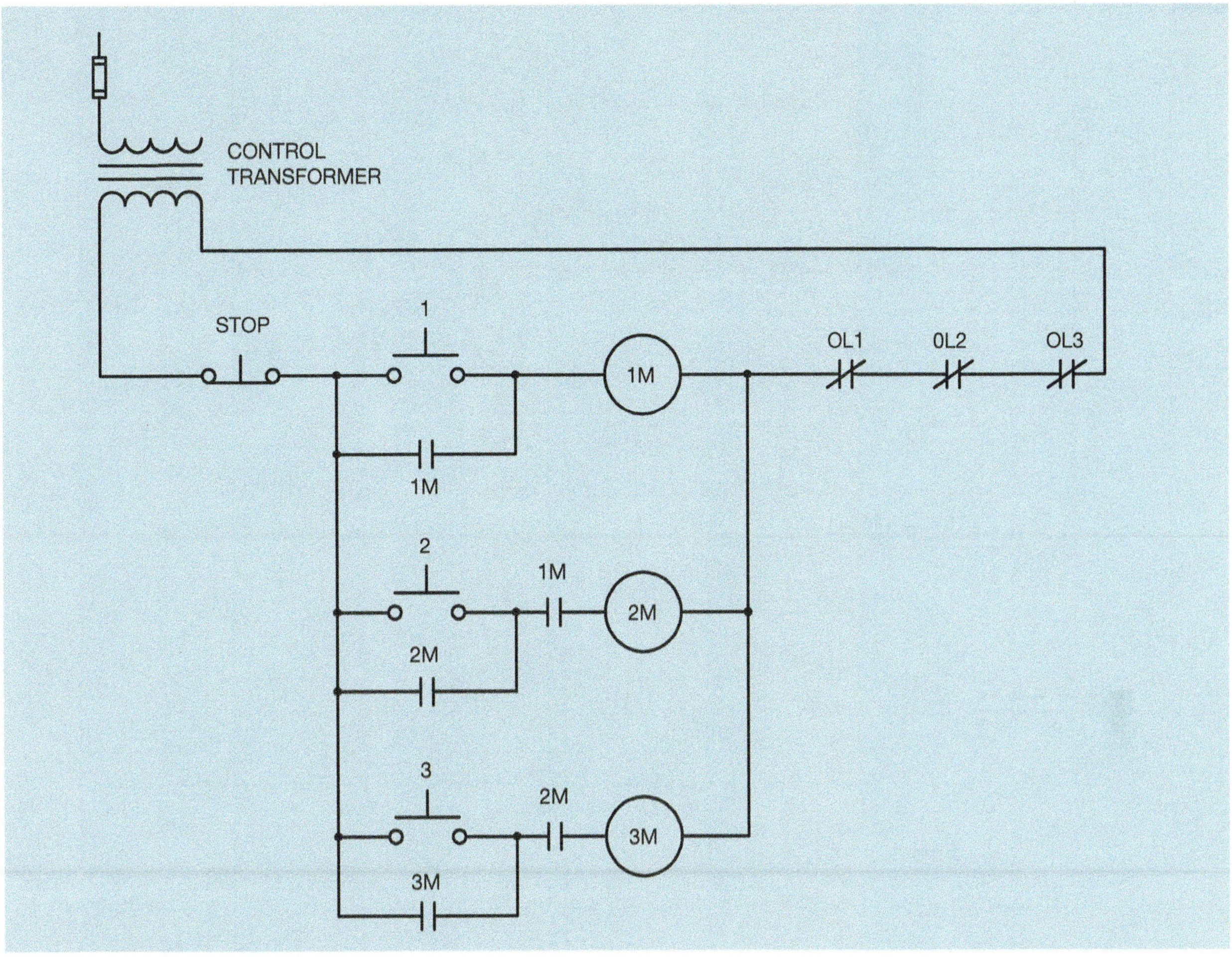

Figure Exp. 5–2 A second circuit for sequence control.

Developing a Wiring Diagram

The schematic shown in Figure Exp. 5–2 is shown with the motors in Figure Exp. 5–3. A drawing of the components needed to connect this circuit is shown in Figure Exp. 5–4. The schematic diagram shown in Figure Exp. 5–3 is shown with wire numbers in Figure Exp. 5–5. The components with corresponding wire numbers are shown in Figure Exp. 5–6 (on page 533).

Connecting the Circuit

❑ 1. Using the materials listed at the beginning of this exercise, connect the circuit shown in Figure Exp. 5–5. Follow the number sequence shown. After connection has been made, ask your instructor to check the circuit for proper connection.

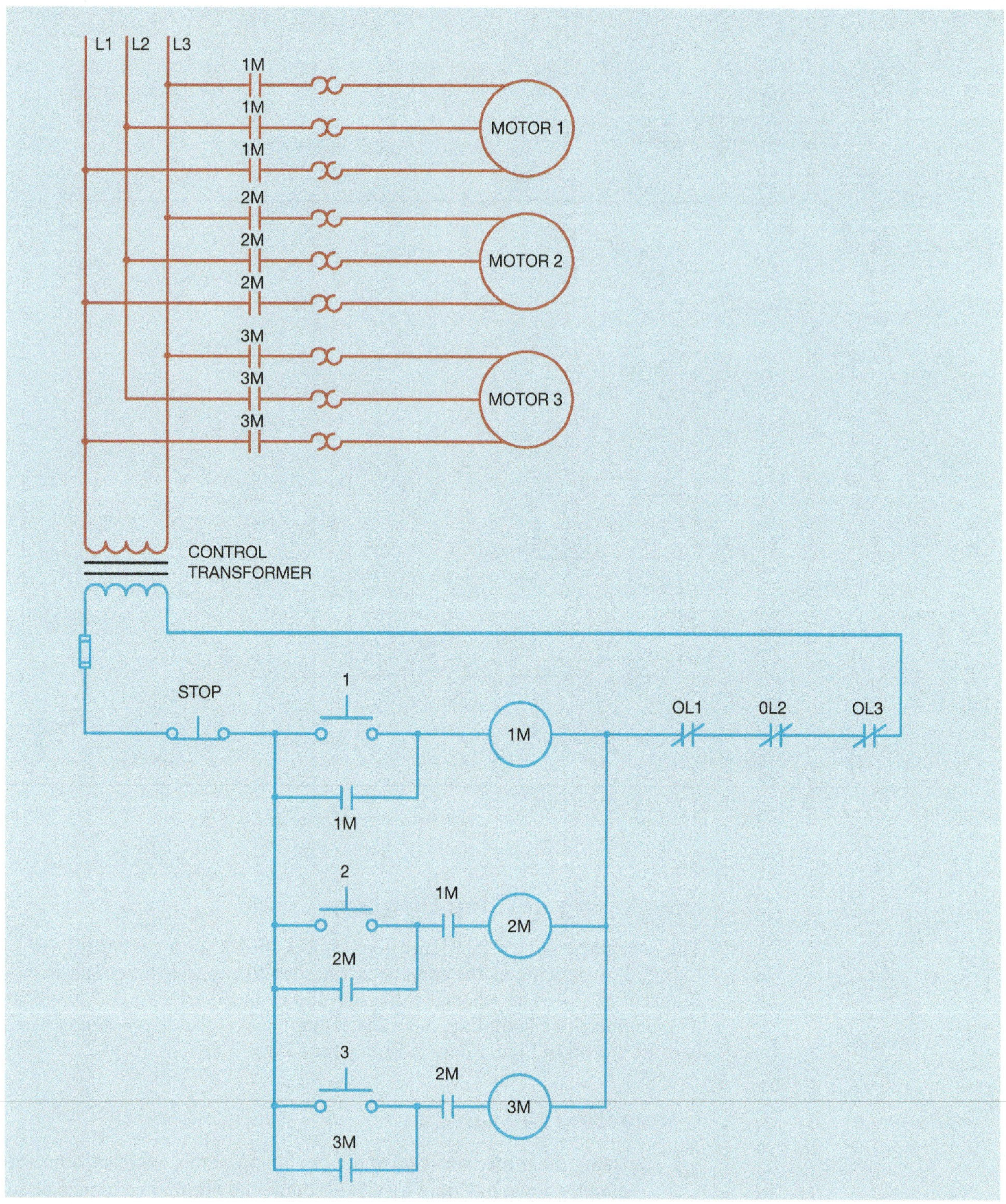

Figure Exp. 5–3 Sequence control with motors.

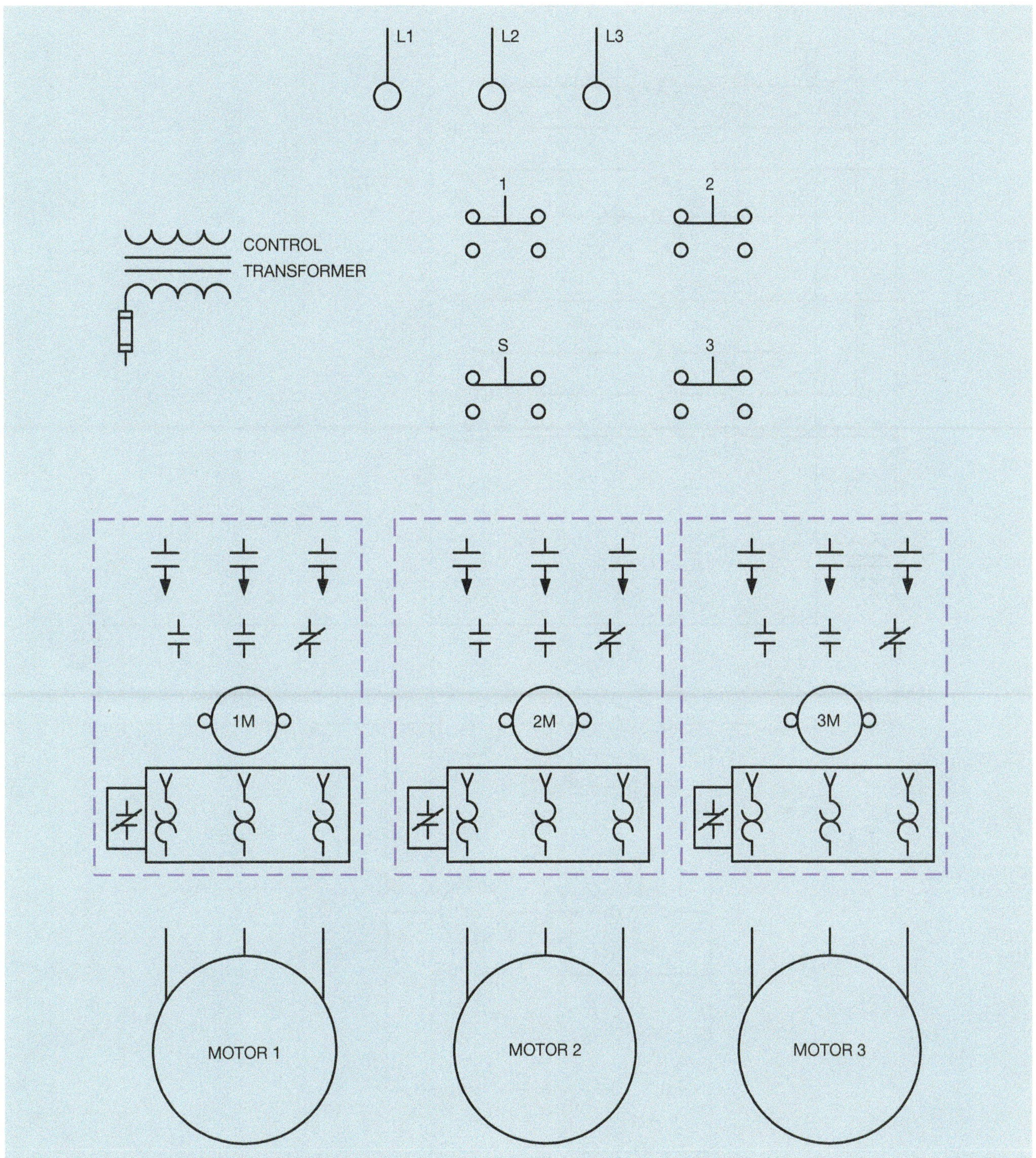

Figure Exp. 5–4 Components needed to connect the circuit.

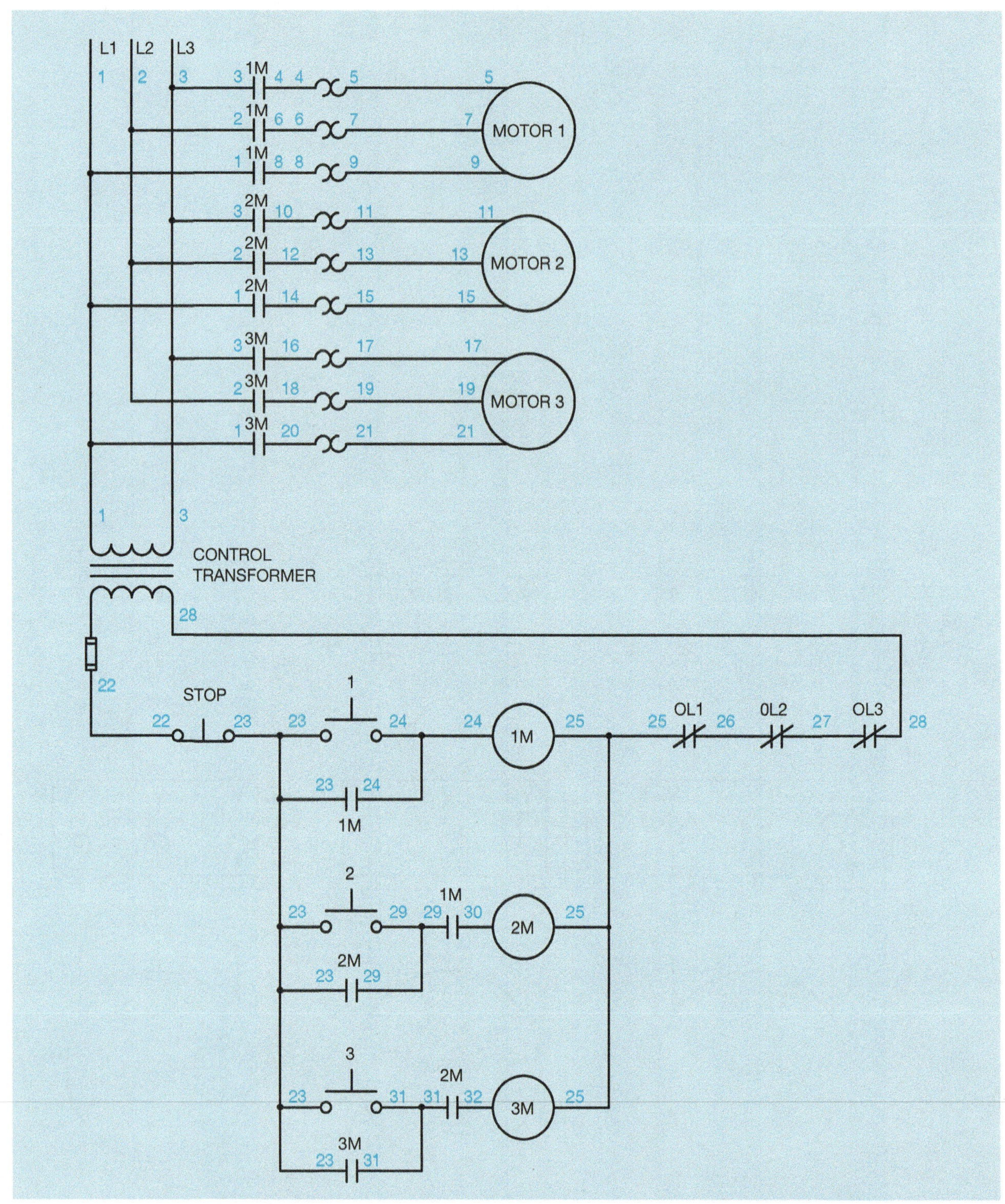

Figure Exp. 5–5 Numbering the schematic.

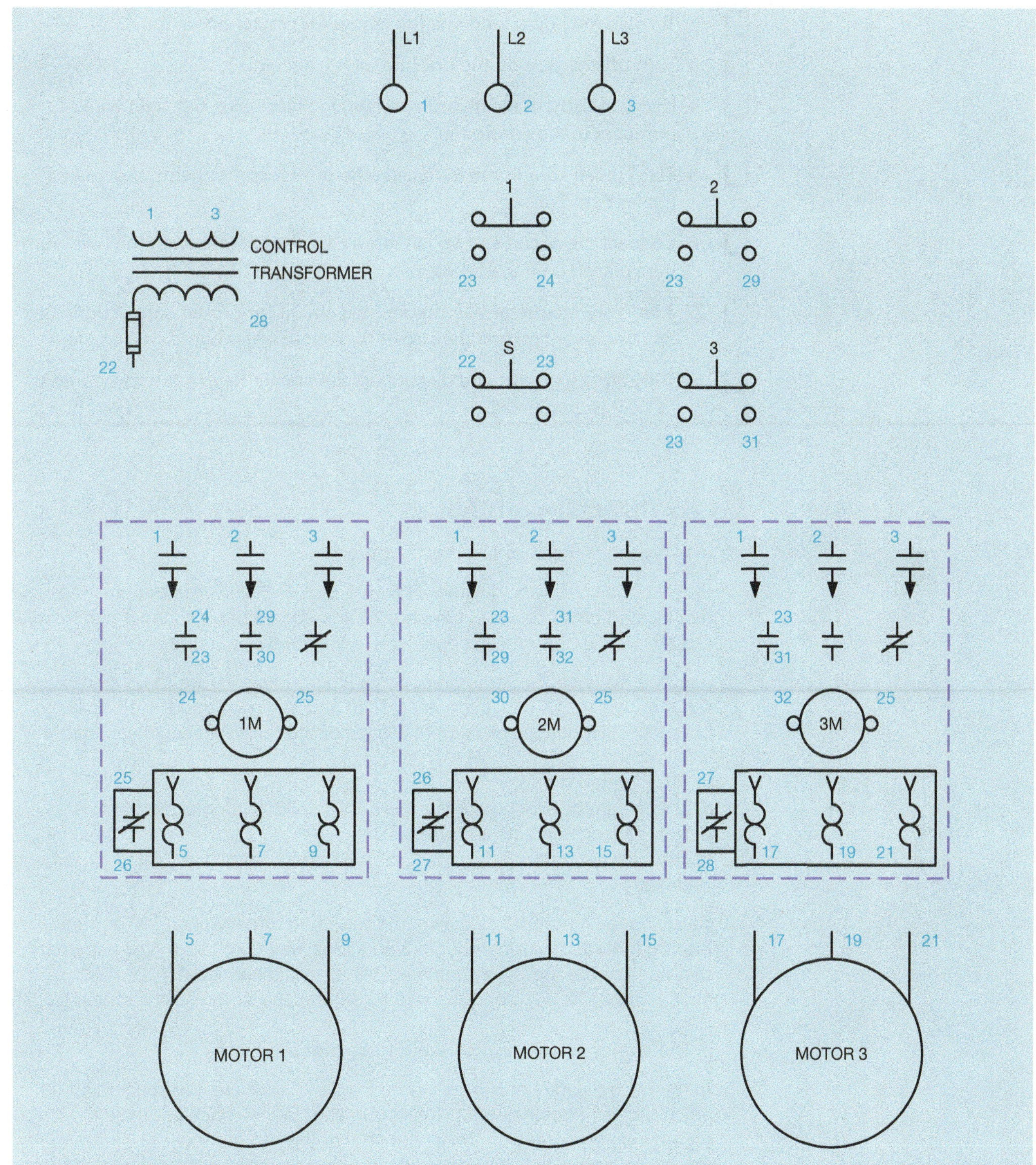

Figure Exp. 5–6 Numbering the components.

- ☐ 2. Turn on the power and test the circuit for proper operation.
- ☐ 3. **Turn off the power** and disconnect the circuit.
- ☐ 4. Using the schematic diagram shown in Figure Exp. 5–1, add wire numbers to the schematic.
- ☐ 5. Place these wire numbers beside the proper components shown in Figure Exp. 5–4.
- ☐ 6. Connect the circuit shown in Figure Exp. 5–1 by following the wire numbers placed on the schematic.
- ☐ 7. After your instructor has checked the circuit for proper connection, turn on the power and test the circuit for proper operation.
- ☐ 8. **Turn off the power** and disconnect the circuit. Return the components to their proper place.

Review Questions

1. What is the purpose of sequence control?
2. Refer to the schematic diagram in Figure Exp. 5–5. Assume that the 1M contact located between wire numbers 29 and 30 had been connected normally closed instead of normally open. How would this circuit operate?
3. Assume that all three motors shown in Figure Exp. 5–5 are running. Now assume that the STOP button is pressed and motors 1 and 2 stop running, but motor 3 continues to operate. Which of the following could cause this problem?
   a. STOP button is shorted.
   b. 2M contact between wire numbers 31 and 32 is hung closed.
   c. The 3M load contacts are welded shut.
   d. The normally open 3M contact between wire numbers 23 and 31 is hung closed.
4. Referring to Figure Exp. 5–5, assume that the normally open 2M contact located between wire numbers 23 and 29 is welded closed. Also assume that none of the motors are running. What would happen if:
   a. The number 2 push button were to be pressed before the number 1 push button?
   b. The number 1 push button were to be pressed first?
5. In the control circuit shown in Figure Exp. 5–2, if an overload occurs on any motor, all three motors will stop running. In the space provided in Figure Exp. 5–7, redesign the circuit so that the motors must still start in sequence from 1 to 3, but an overload on any motor will stop only that motor. If an overload should occur on motor 1, for example, motors 2 and 3 would continue to operate.

Figure Exp. 5–7 Circuit redesign.

Exercise 6

JOGGING CONTROLS

Objectives

After studying this exercise the student should be able to:

›› Describe the difference between inching and jogging circuits.

›› Discuss different jogging control circuits.

›› Draw a schematic diagram of a jogging circuit.

›› Discuss the connection of an 8 pin control relay.

›› Connect a jogging circuit in the laboratory using double-acting push buttons.

›› Connect a jogging circuit in the laboratory using an 8 pin control relay.

Materials Needed

- Three phase power supply
- Three phase motor starter
- One three phase motor or equivalent motor load
- Three double-acting push buttons. (NO/NC on each button)
- One 8 pin tube socket
- One 8 pin control relay
- One single pole switch
- Control transformer

Jogging or inching control is used to help position objects by permitting the motor to be momentarily connected to power. Jogging and inching are very similar and the terms are often used synonymously. Both involve starting a motor with short jabs of power. The difference between jogging and inching is that when a motor is jogged, it is started with short jabs of power at full voltage. When a motor is inched, it is started with short jabs at reduced power. Inching circuits require the use of two contactors, one to run the motor at full power and the other to start the motor at reduced power (Figure Exp. 6–1). The run contactor is generally a motor starter that contains an overload relay while the inching contactor does not. In the circuit shown in Figure Exp. 6–1, if the inch push button is pressed, a circuit is completed to S contactor coil, causing all S contacts to close. This connects the motor to the line through a set of series resistors used to reduce power to the motor. Note that there is

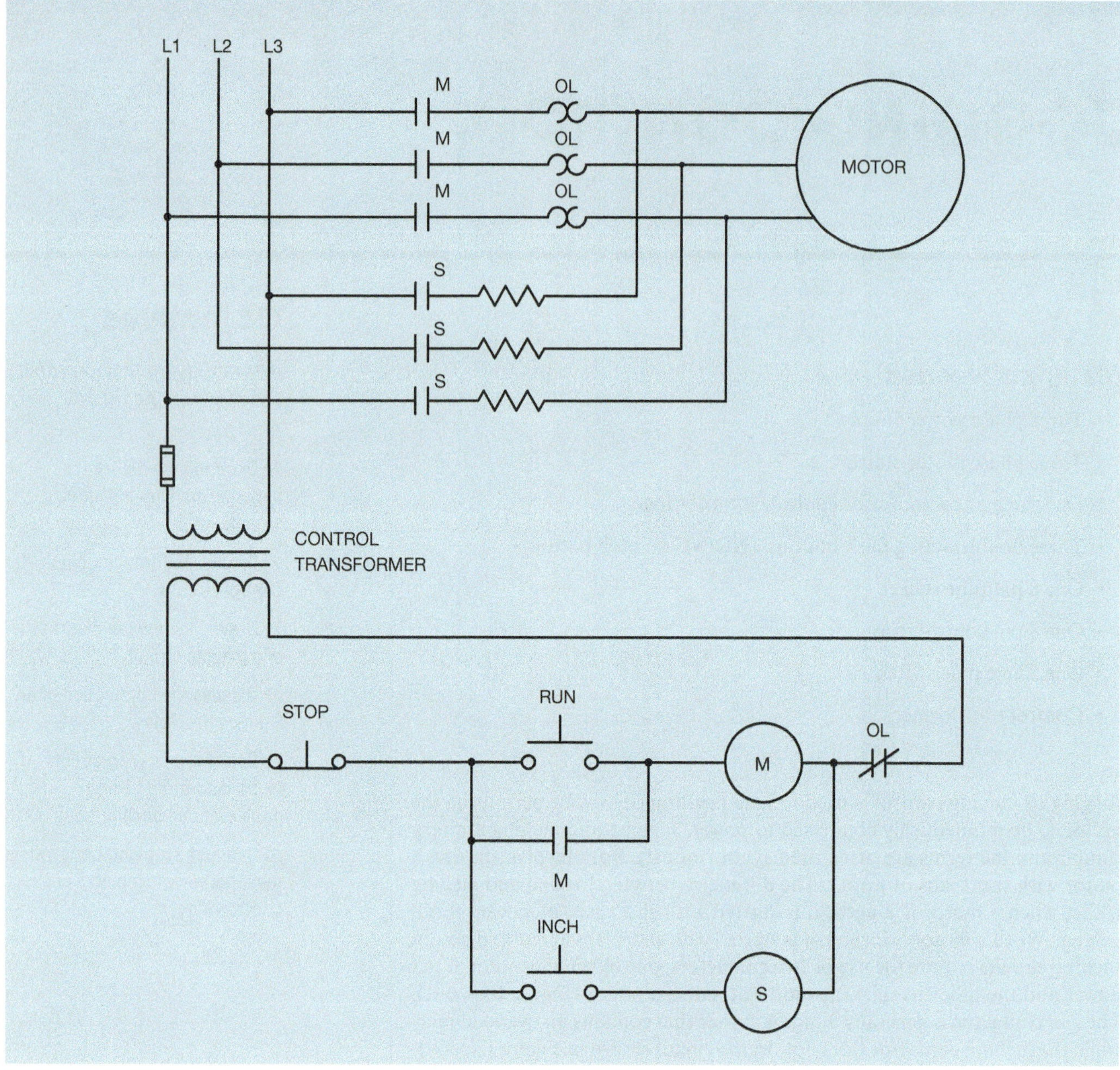

Figure Exp. 6–1 Inching control circuit.

no S holding contact in parallel with the inch push button. When the push button is released, S contactor de-energizes and all S contacts reopen and disconnect the motor from the power line. If the run push button is pressed, M contactor energizes and connects the motor directly to the power line. Note the normally open M auxiliary contact is connected in parallel with the run push button to maintain the circuit when the button is released.

Other Jogging Circuits

Like most control circuits, jog circuits can be connected in different ways. One method is shown in Figure Exp. 6–2. In this circuit, a simple single pole switch is inserted in series with the normally open M auxiliary contact connected in parallel with the START button. When the switch is open, it is in the *jog* position and prevents M holding contact from providing a complete path to M coil. When the START button is

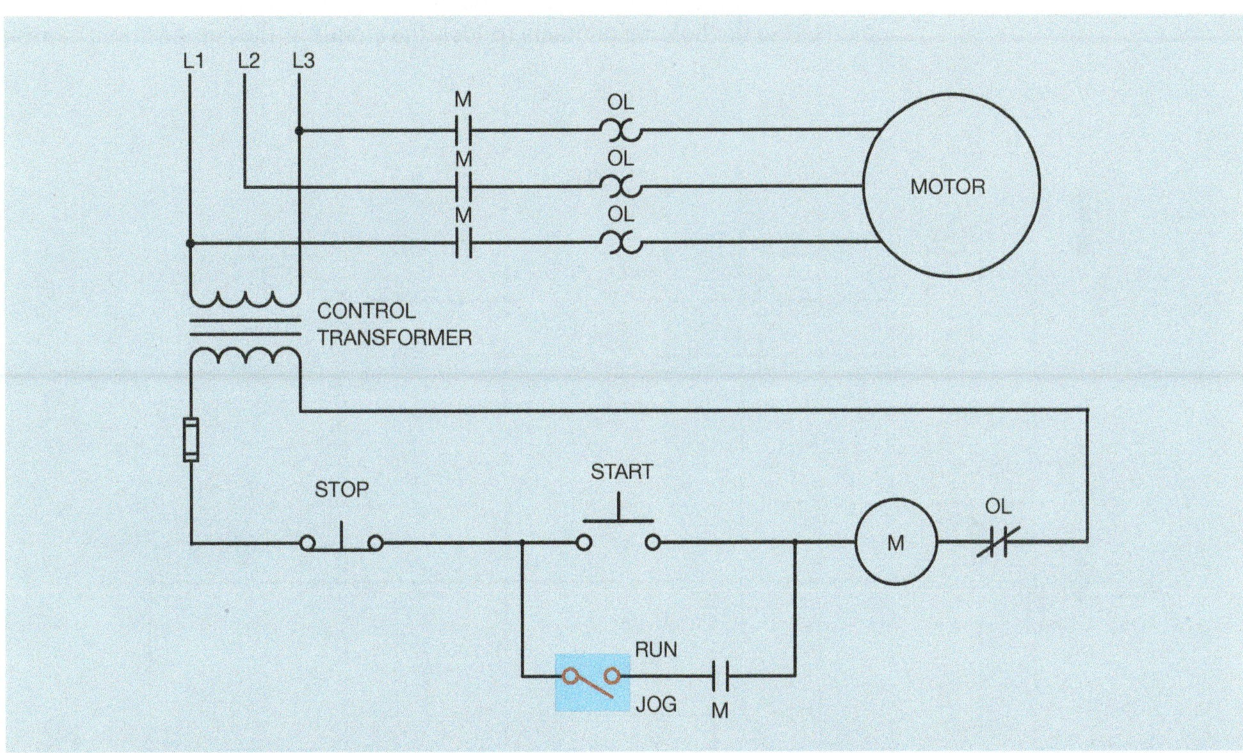

Figure Exp. 6–2 Run-jog control using a single pole switch.

pushed, M coil energizes and connects the motor to the power line. When the START button is released, M coil de-energizes and disconnects the motor from the line. If the switch is closed, it is in the *run* position and permits the holding contact to complete a circuit around the START button.

Another method of constructing a run-jog control is shown in Figure Exp. 6–3. This circuit employs a double-acting push button as the jog button. The normally closed section of the jog push button is connected in series with the normally open M auxiliary holding contact. If the jog button is pressed, the normally closed section of the button opens to disconnect the holding contacts before the normally open section of the button closes. Although M auxiliary contact closes when M coil energizes, the now open jog button prevents it from completing a circuit to the coil. When the jog button is released, the normally open section reopens and breaks contact before the normally closed section can reclose.

Although a double-acting push button can be used to construct a run-jog circuit, it is not generally done because there is a possibility that the normally closed section of the jog button could be reclosed before the normally open section reopens. This could cause the holding contacts to lock the circuit in the run position, causing an

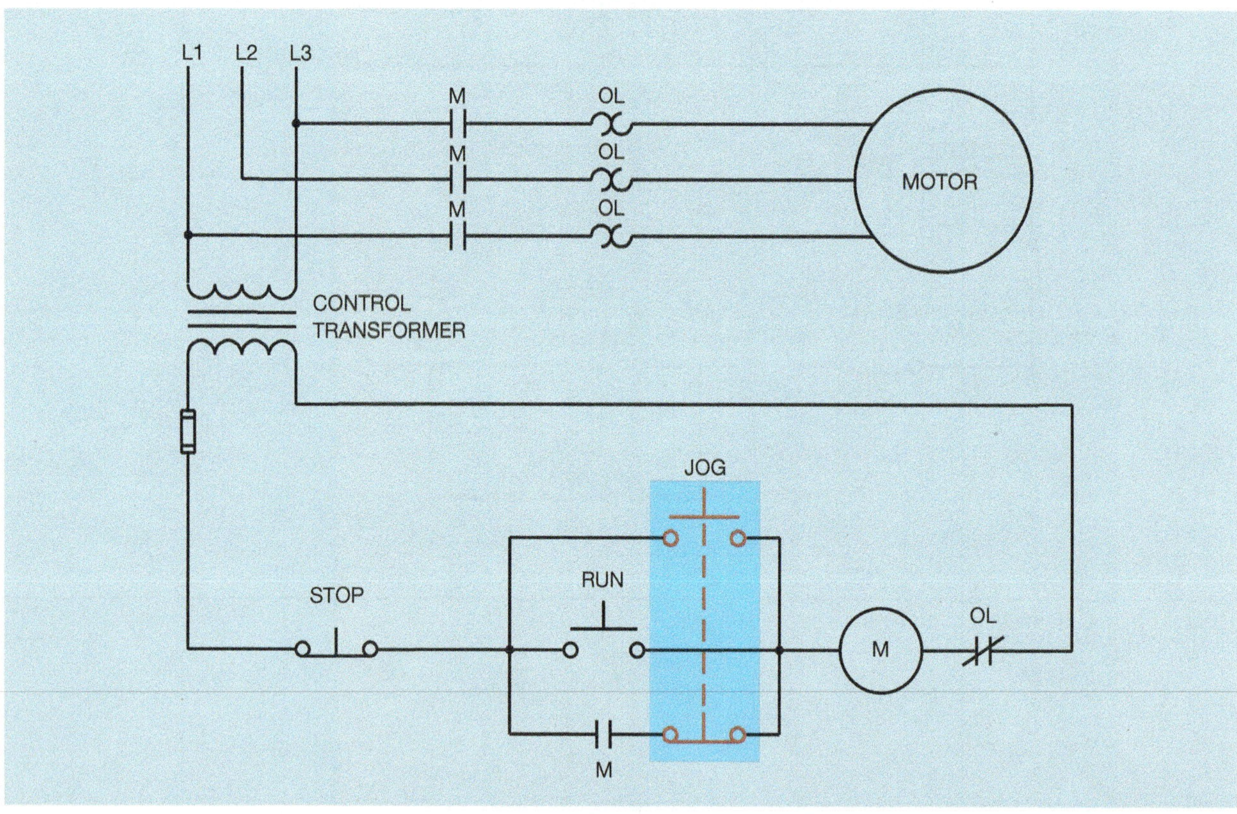

Figure Exp. 6–3 Jogging control using a double-acting push button.

accident. To prevent this possibility, a control relay is often employed (Figure Exp. 6–4). In the circuit shown in Figure Exp. 6–4, if the jog push button is pressed, M contactor energizes and connects the motor to the line. When the jog button is released, M coil de-energizes and disconnects the motor from the line.

When the run push button is pressed, CR relay energizes and closes both CR contacts. The CR contacts connected in parallel with the run button close to maintain the circuit to CR coil, and the CR contacts connected in parallel with the jog button close and complete a circuit to M coil.

Connecting Jogging Circuits

In this exercise, four different jog circuits will be connected in the laboratory. Three of these circuit are illustrated in Figures Exp. 6–2, 6–3, and 6–4. You will design the fourth circuit in accord with given circuit parameters.

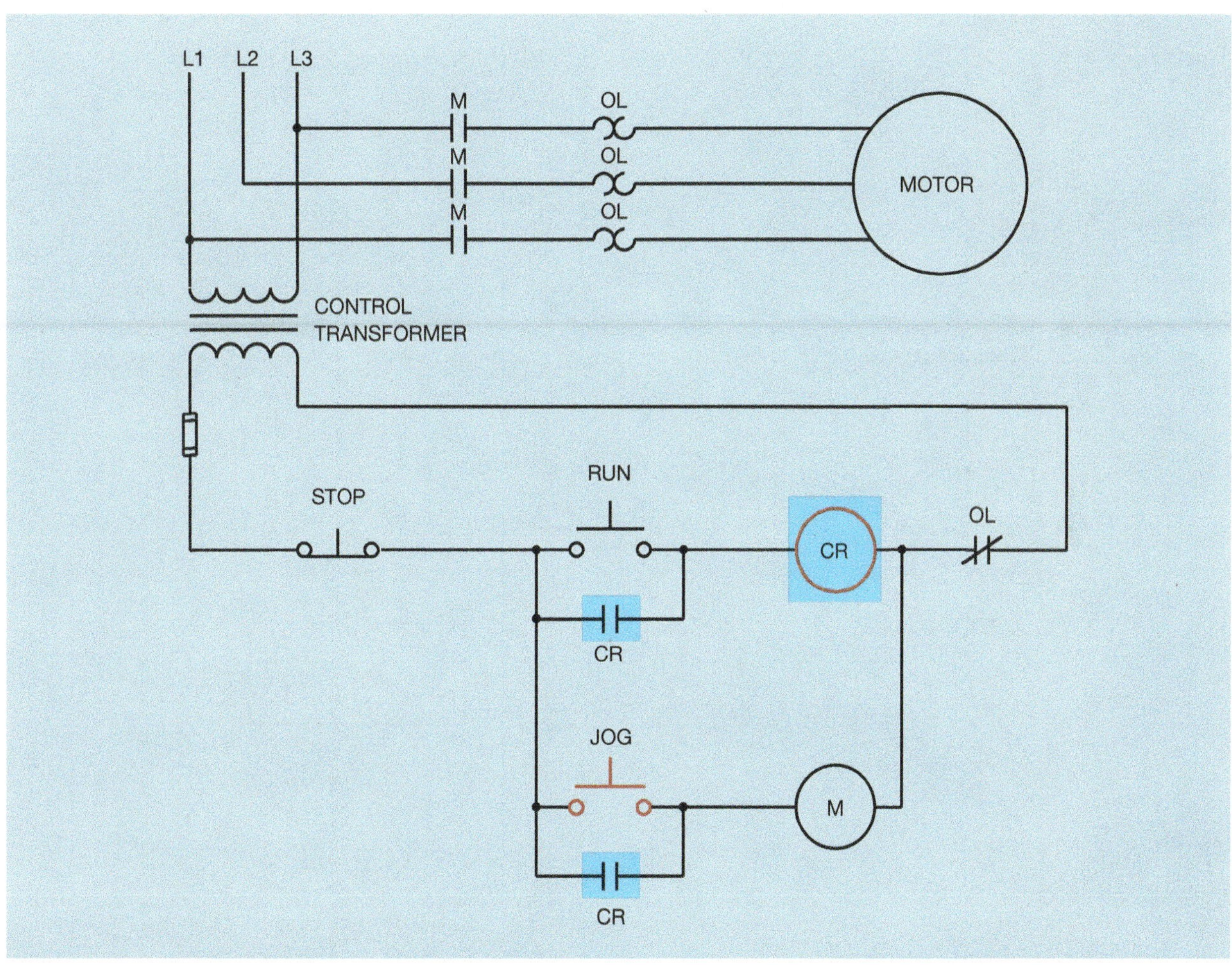

Figure Exp. 6–4 Run-jog control using a control relay.

Connecting Circuit 1

- ☐ 1. Refer to the schematic diagram in Figure Exp. 6–2. Place wire numbers beside the components following the procedure discussed in previous exercises.
- ☐ 2. Using the components shown in Figure Exp. 6–5, place corresponding wire numbers beside the components.
- ☐ 3. Connect the circuit by following the wire numbers in the schematic diagram in Figure Exp. 6–2.
- ☐ 4. After your instructor has checked the circuit for proper connection, turn on the power and test the circuit for proper operation. The motor should jog when the switch is open and run when the switch is closed.
- ☐ 5. **Turn off the power** and disconnect the circuit.

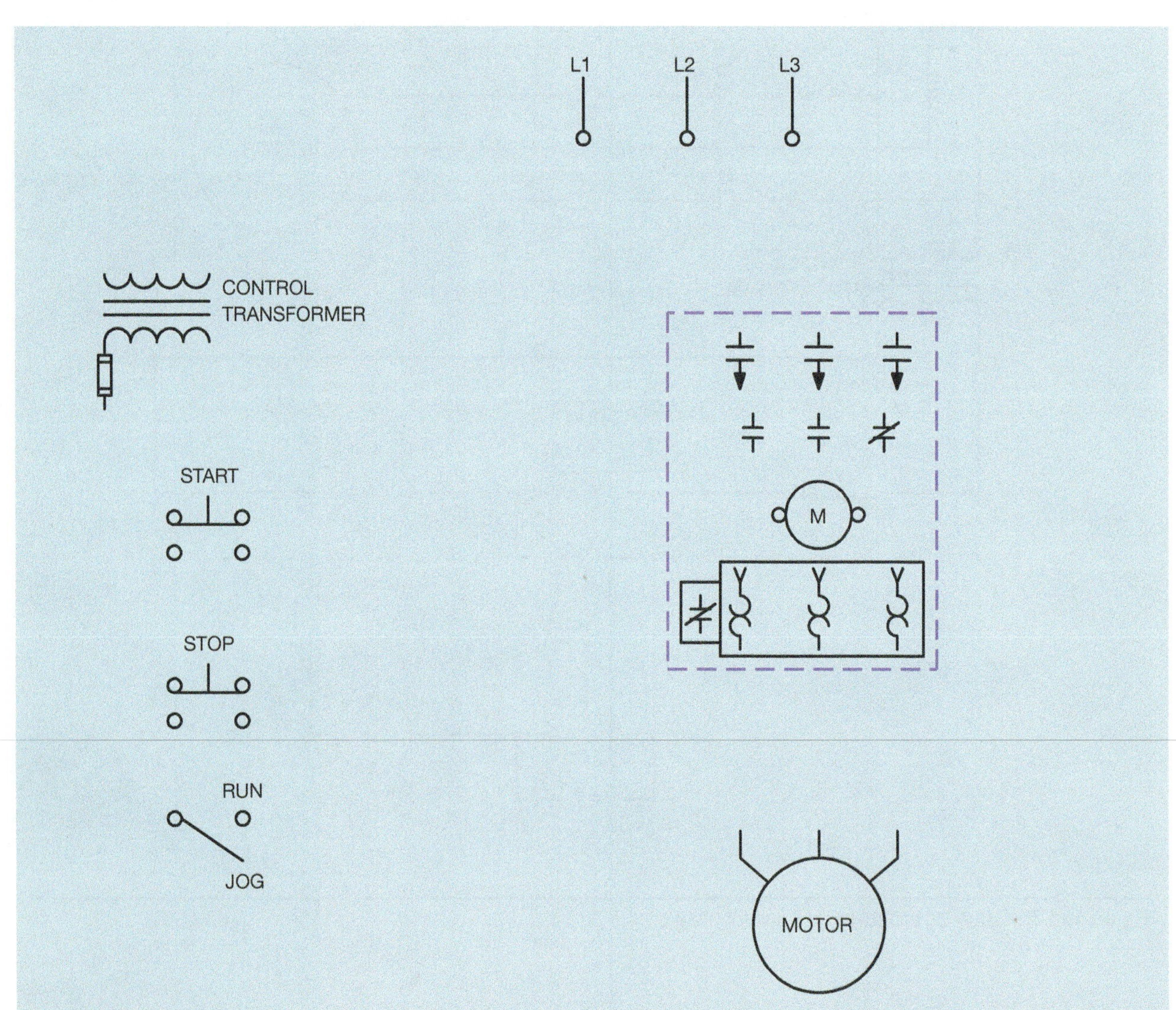

Figure Exp. 6–5 Components needed to connect circuit 1.

Connecting the Second Run-Jog Circuit

- ☐ 6. Using the schematic shown in Figure Exp. 6–3, place wire numbers beside the components.
- ☐ 7. Place corresponding wire numbers beside the components shown in Figure Exp. 6–6.
- ☐ 8. Connect the circuit using the schematic diagram in Figure Exp. 6–3.
- ☐ 9. After your instructor has checked the circuit for proper connection, turn on the power and test the circuit for proper operation.
- ☐ 10. **Turn off the power** and disconnect the circuit.

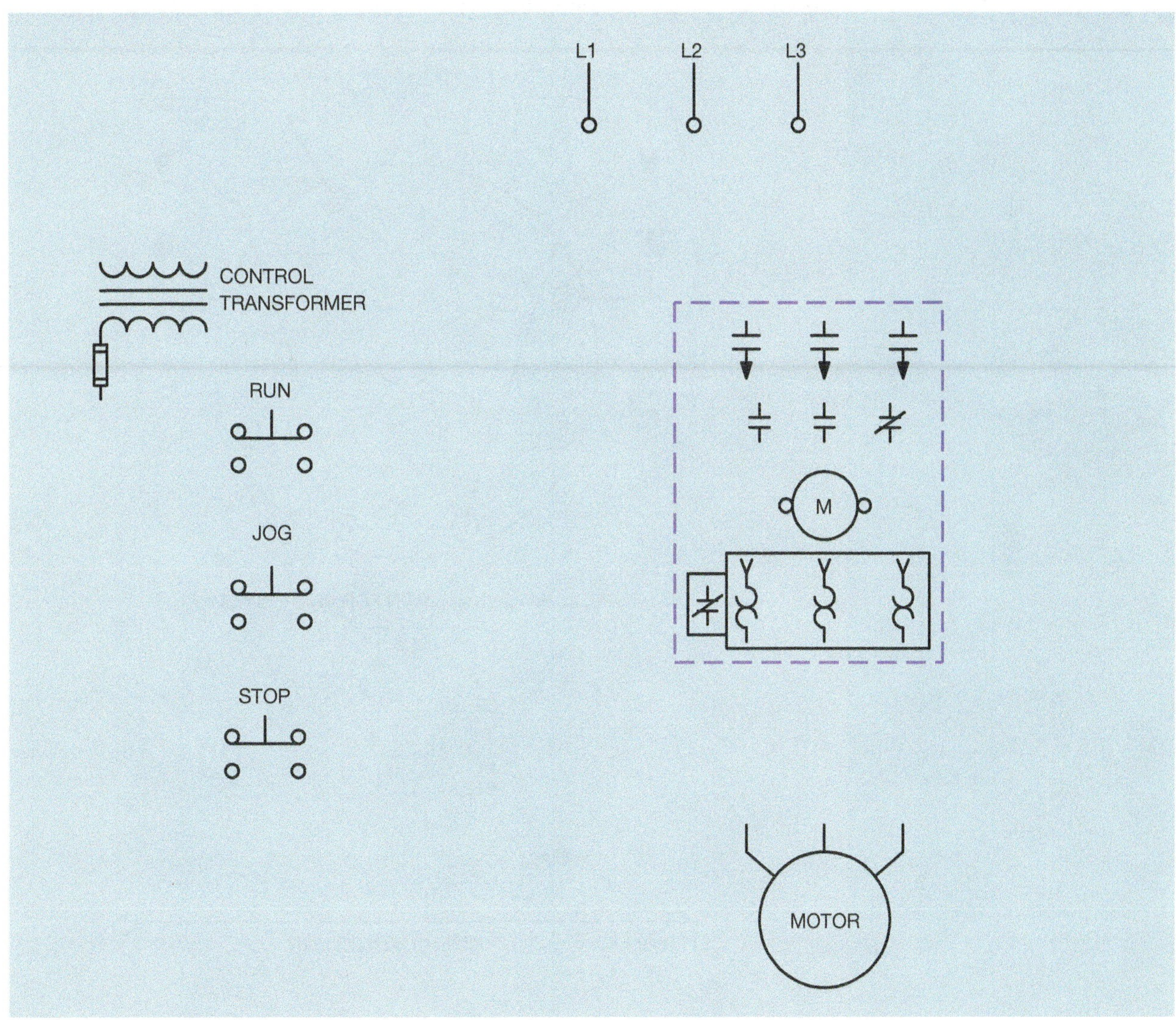

Figure Exp. 6–6 Components needed to connect the second run-jog circuit.

Connecting the Third Run-Jog Circuit

The third run-jog circuit involves the use of a control relay. In this circuit, an 8 pin control relay will be used. Eight pin relays are designed to fit into an 8 pin tube socket; therefore, the socket is the device to which connection is made, not the relay itself. Eight pin relays commonly have coils with different voltage ratings such as 12 VDC, 24 VDC, 24 VAC, and 120 VAC, so make certain that the coil of the relay you use is rated for the circuit control voltage. Most 8 pin relays contain two single pole, double throw contacts. A diagram showing the standard pin connection for 8 pin relays with two sets of contacts is shown in Figure Exp. 6–7.

Connecting the Tube Socket

When making connections to tube sockets, it is generally helpful to place the proper relay pin numbers beside the component on the schematic diagram. To distinguish pin numbers from wire numbers, pin numbers will be circled. The schematic in

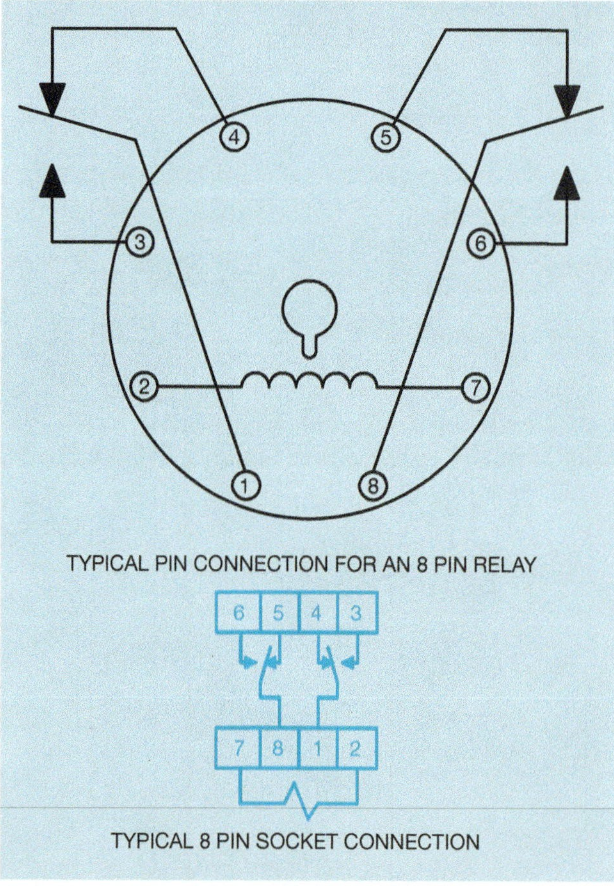

Figure Exp. 6–7 Pin diagram of an 8 pin relay.

Figure Exp. 6–4 is shown in Figure Exp. 6–8 with the addition of relay pin numbers. The connection diagram in Figure Exp. 6–7 shows that the relay coil is connected to pins 2 and 7. Note that CR relay coil in Figure Exp. 6–8 has a circled 2 and 7 placed beside it.

The connection diagram also indicates that the relay contains two sets of normally open contacts. One set is connected to pins 1 and 3, and the other set is connected to pins 8 and 6. Note in the schematic of Figure Exp. 6–8 that one of the normally open CR contacts has the circled numbers 1 and 3 beside it and the other normally open CR contact has the circled numbers 8 and 6 beside it.

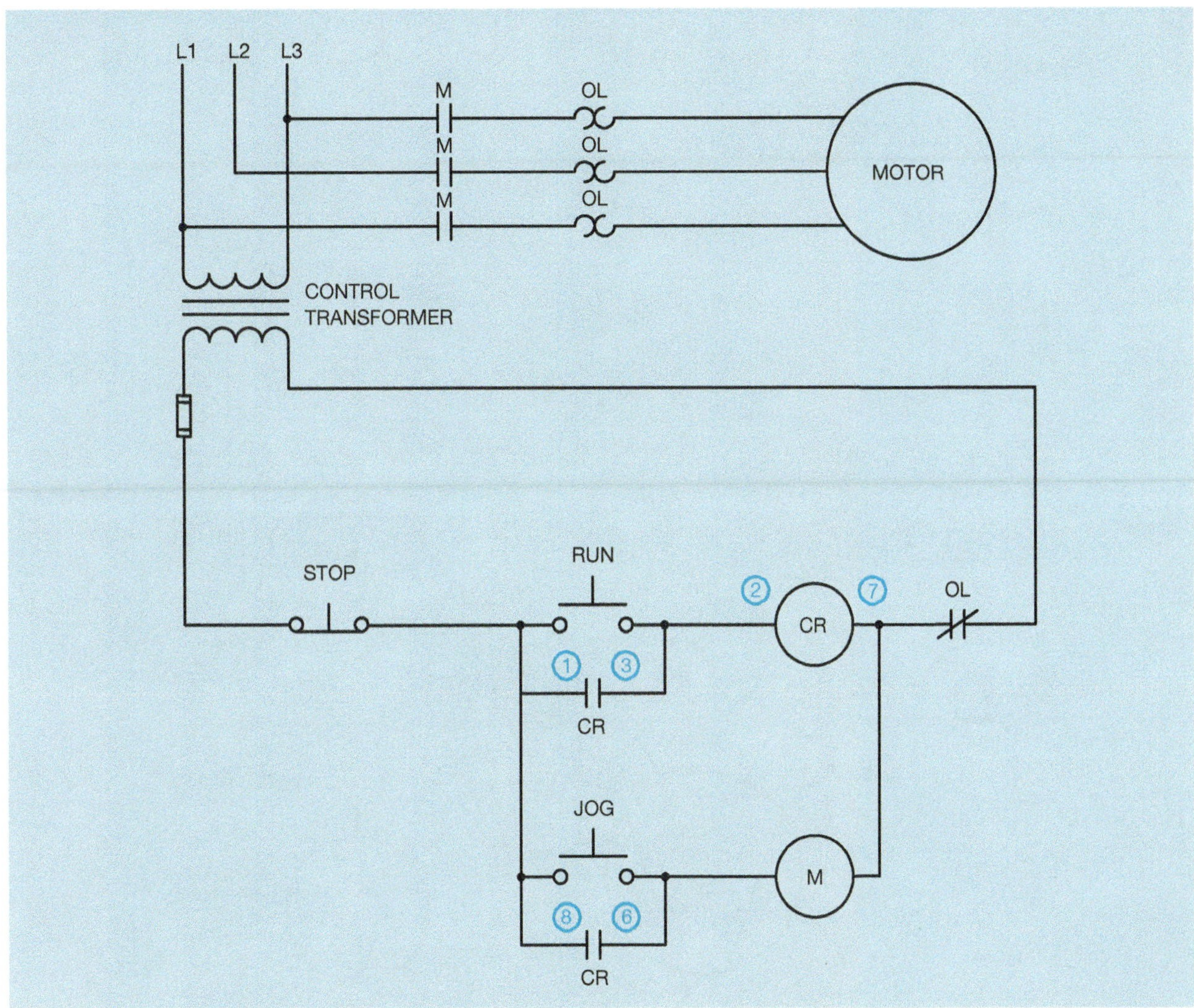

Figure Exp. 6–8 Adding pin numbers aids in connecting the circuit.

- ☐ 11. Using the drawing in Figure Exp. 6–8, place wire numbers on the schematic.
- ☐ 12. Using the wire numbers placed on the schematic diagram in Figure Exp. 6–8, place corresponding wire numbers beside the proper components shown in Figure Exp. 6–9.
- ☐ 13. Connect the circuit shown in Figure Exp. 6–8.

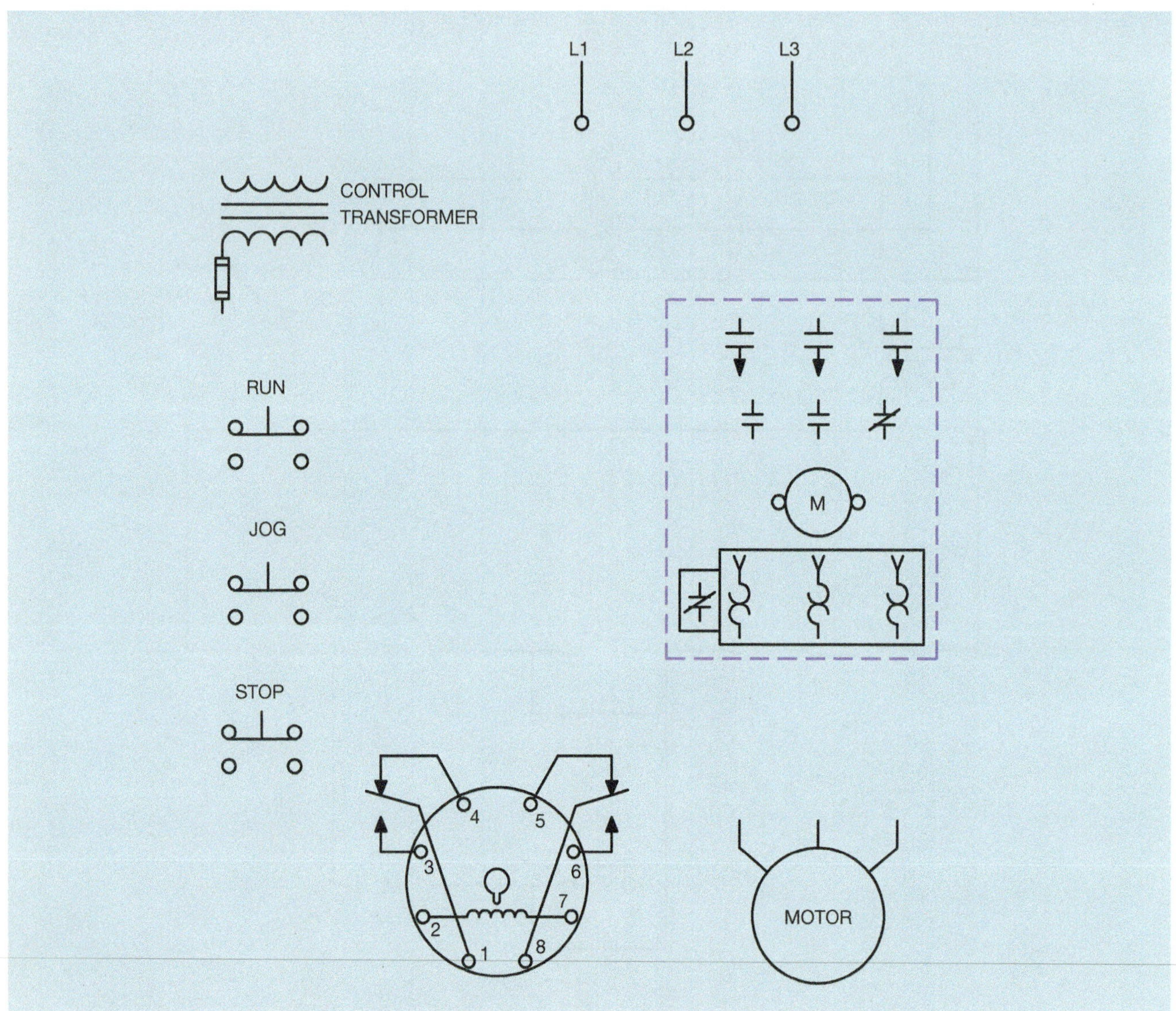

Figure Exp. 6–9 Components needed to connect circuit 3.

☐ 14. After your instructor has checked the circuit for proper connection, turn on the power and test the circuit for proper operation.

☐ 15. **Turn off the power** and disconnect the circuit.

Review Questions

1. Explain the difference between inching and jogging.
2. What is the main purpose of jogging?
3. Refer to the circuit shown in Figure Exp. 6–10. In this circuit, the jog button has been connected incorrectly. The normally closed section has been connected in parallel with the run push button and the normally open section has been connected in series with the holding contacts. Explain how this circuit operates.
4. Refer to the circuit shown in Figure Exp. 6–11. In this circuit the jog push button has again been connected incorrectly. The normally closed section of the button has been connected in series with the normally open run push button and the normally open section of the jog button is connected in parallel with the holding contacts. Explain how this circuit operates.

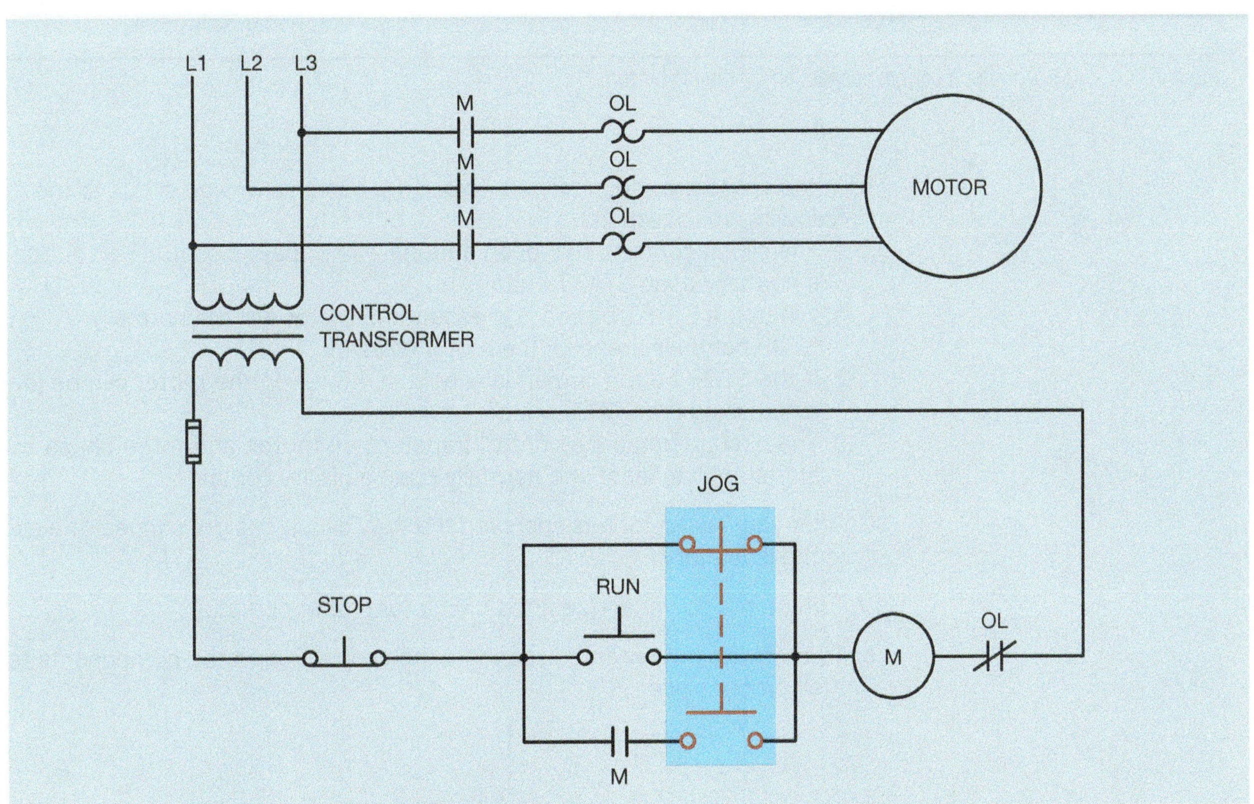

Figure Exp. 6–10 The jog button is connected incorrectly.

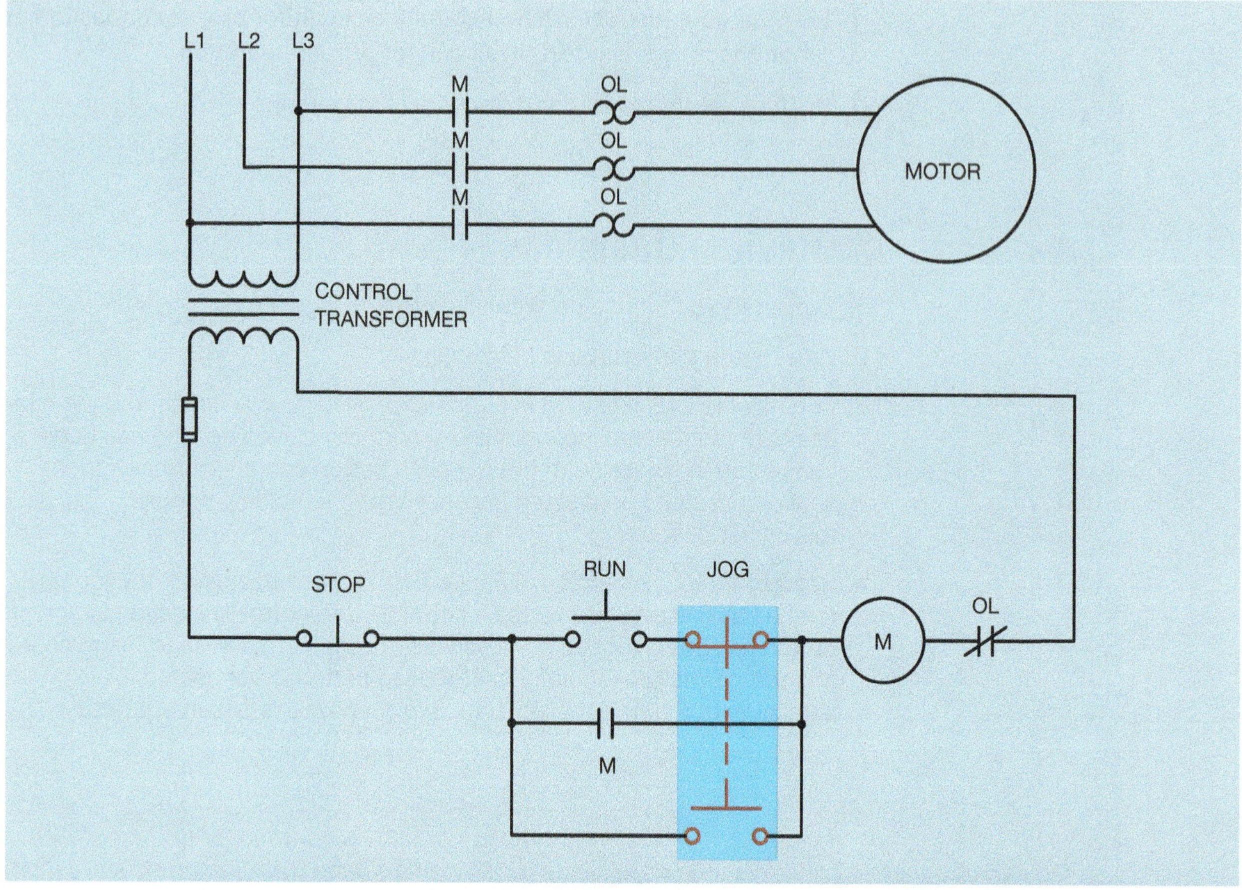

Figure Exp. 6–11 Another incorrect connection for the jog button.

5. In the space provided in Figure Exp. 6-12, design a run-jog circuit to the following specifications.
   a. The circuit contains two push buttons—a normally closed STOP button and a normally open START button.
   b. When the START button is pressed, the motor will run normally. When the STOP button is pressed, the motor will stop.
   c. If the STOP button is manually held in, however, the motor can be jogged by pressing the START button.
   d. The circuit contains a control transformer, motor, and three phase motor starter with at least one normally open auxiliary contact.

6. After your instructor has approved the new circuit design, connect the circuit in the laboratory.

7. Turn on the power and test the circuit for proper operation.

8. **Turn off the power** and disconnect the circuit. Return the components to their proper place.

Figure Exp. 6–12 Circuit design.

Exercise 7

ON-DELAY TIMERS

Objectives

After completing this exercise the student should be able to:

›› Discuss the operation of an on-delay timer.

›› Draw the NEMA contact symbols used to represent both normally open and normally closed on-delay contacts.

›› Discuss the difference in operation between pneumatic and electronic timers.

›› Connect a circuit in the laboratory employing an on-delay timer.

Materials Needed

- Three phase power supply
- Control transformer
- Two double-acting push buttons (NO/NC on each button)
- 2 ea.—three phase motor starter with at least one normally open auxiliary contact
- Dayton solid-state timer–(model 6A855 or equivalent) and 11 pin socket
- 8 pin control relay and 8 pin socket
- 2 ea.—three phase motors or equivalent motor loads

Timers can be divided into two basic types: on delay and off delay. Although there are other types such as one shot and interval, they are basically on- or off-delay timers. In this unit, the operation of on-delay timers will be discussed. The operating sequence of an on-delay timer is as follows.

When the coil is energized, the timed contacts will delay changing position for some period of time. When the coil is de-energized, the timed contacts will return to their normal position immediately. In this explanation, the word *timed* contacts is used. The reason is that some timers contain both timed and instantaneous contacts. When using a timer of this type care must be taken to connect to the proper set of contacts.

Timed Contacts

The timed contacts are controlled by the action of the timer, while the instantaneous contacts operate like any standard set of contacts on a control relay; when the coil energizes, the contacts change position immediately and when the coil de-energizes they change back to their normal position immediately.

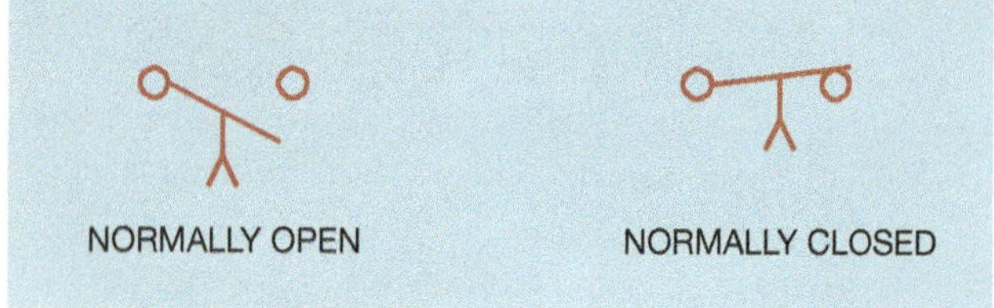

Figure Exp. 7–1 NEMA standard symbols for on-delay contacts.

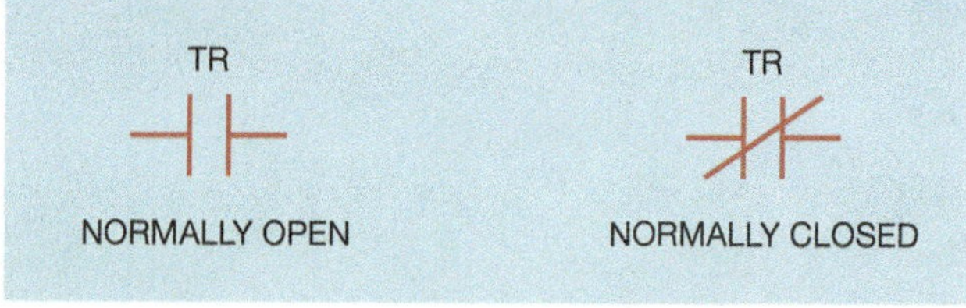

Figure Exp. 7–2 Instantaneous contact symbols.

The standard NEMA symbols used to represent on-delay contacts are shown in Figure Exp. 7–1. The arrow points in the direction the contact moves after the delay period. The normally open contact, for example, closes after the time delay period, and the normally closed contact opens after the time delay period.

Instantaneous Contacts

Instantaneous contacts are drawn in the same manner as standard relay contacts. Figure Exp. 7–2 illustrates a set of instantaneous contacts controlled by timer TR. The instantaneous contacts are often used as holding or sealing contacts in a control circuit. The control circuit shown in Figure Exp. 7–3 illustrates an on-delay timer used to delay the starting of a motor. When the START push button is pressed, TR coil energizes and the normally open instantaneous TR contacts close immediately to

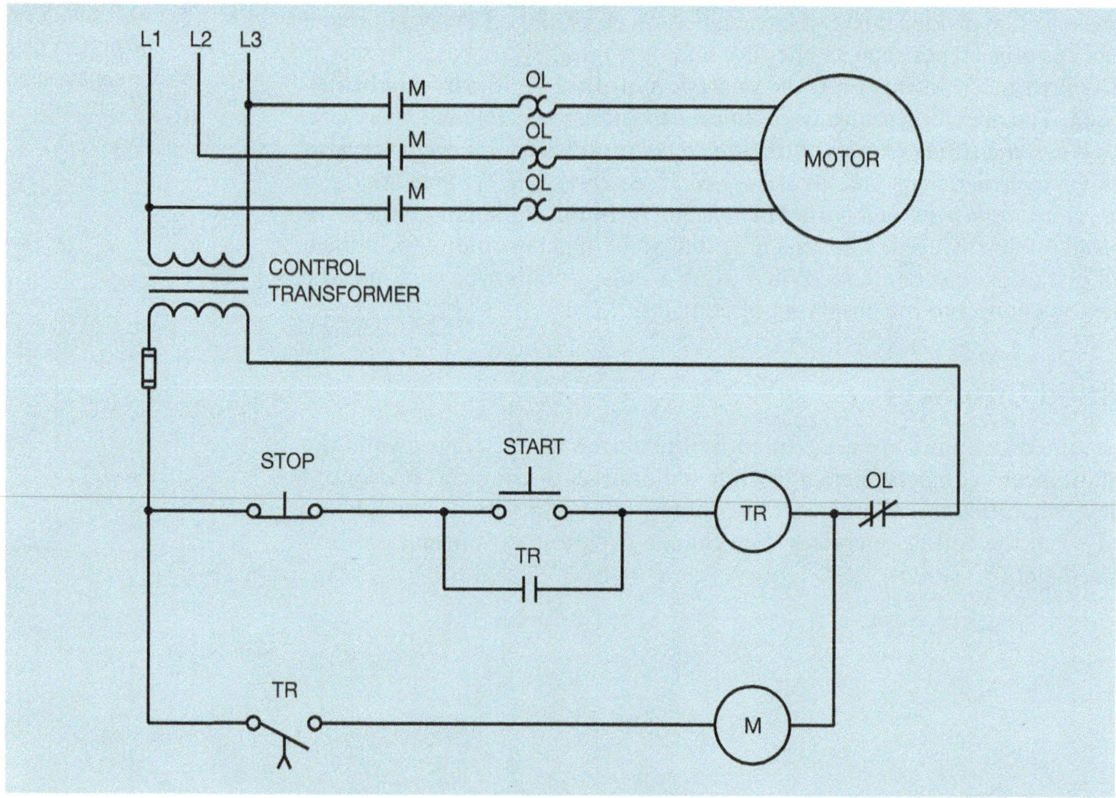

Figure Exp. 7–3 The motor starts after the START button is pressed.

hold the circuit. After the preset time period, the normally open TR timed contacts close and energize the coil of M starter, which connects the motor to the line.

When the STOP button is pressed and TR coil de-energizes, both TR contacts return to their normal position immediately. This de-energizes M coil and disconnects the motor from the line.

Control Relays Used with Timers

Not all timers contain instantaneous contacts. Most electronic timers, for example, do not. When an instantaneous contact is needed and the timer does not have one available, it is common practice to connect the coil of a control relay in parallel with the coil of the timer (Figure Exp. 7–4). In this way, the electronic timer will operate with the control relay. In the circuit shown in Figure Exp. 7–4, both coils TR and CR energize when the START button is pressed. This causes CR contact to close and seal the circuit.

The First Circuit

The first circuit to be connected is shown in Figure Exp. 7–4. In this circuit, it will be assumed that an 11 pin timer is being used and that the coil is connected to pins 2 and 10, and a set of normally open timed contacts are connected to pins 1 and 3. The coil of the 8 pin control relay is connected to pins 2 and 7, and a normally open contact is

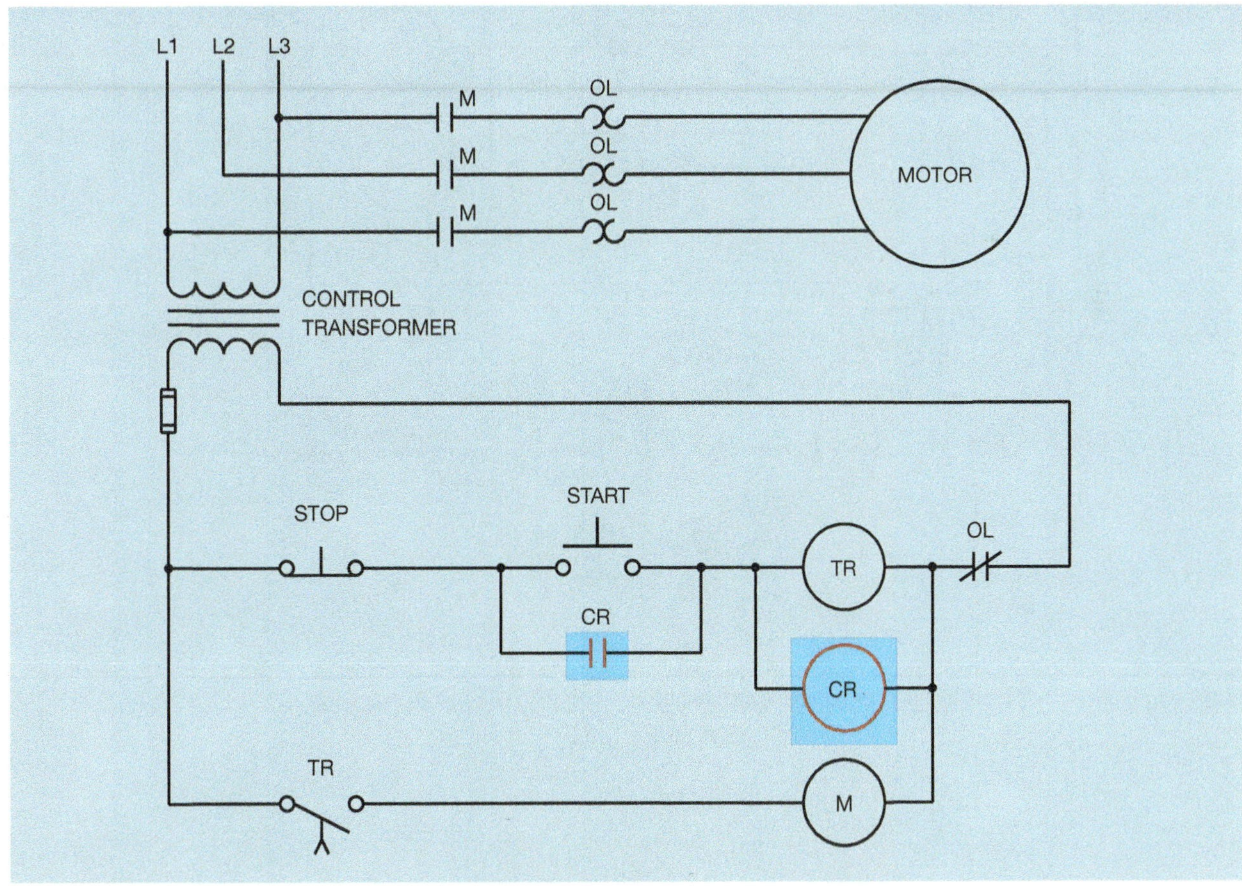

Figure Exp. 7–4 A control relay furnishes the instantaneous contact.

connected to pins 1 and 3. When using control devices that are connected with 8 and 11 pin sockets, it is generally helpful to place pin numbers beside the component. To prevent pin numbers from being confused with wire numbers, a circle will be drawn around the pin numbers (Figure Exp. 7–5). The pin diagram for the Dayton model 6A855 timer is also shown on this schematic.

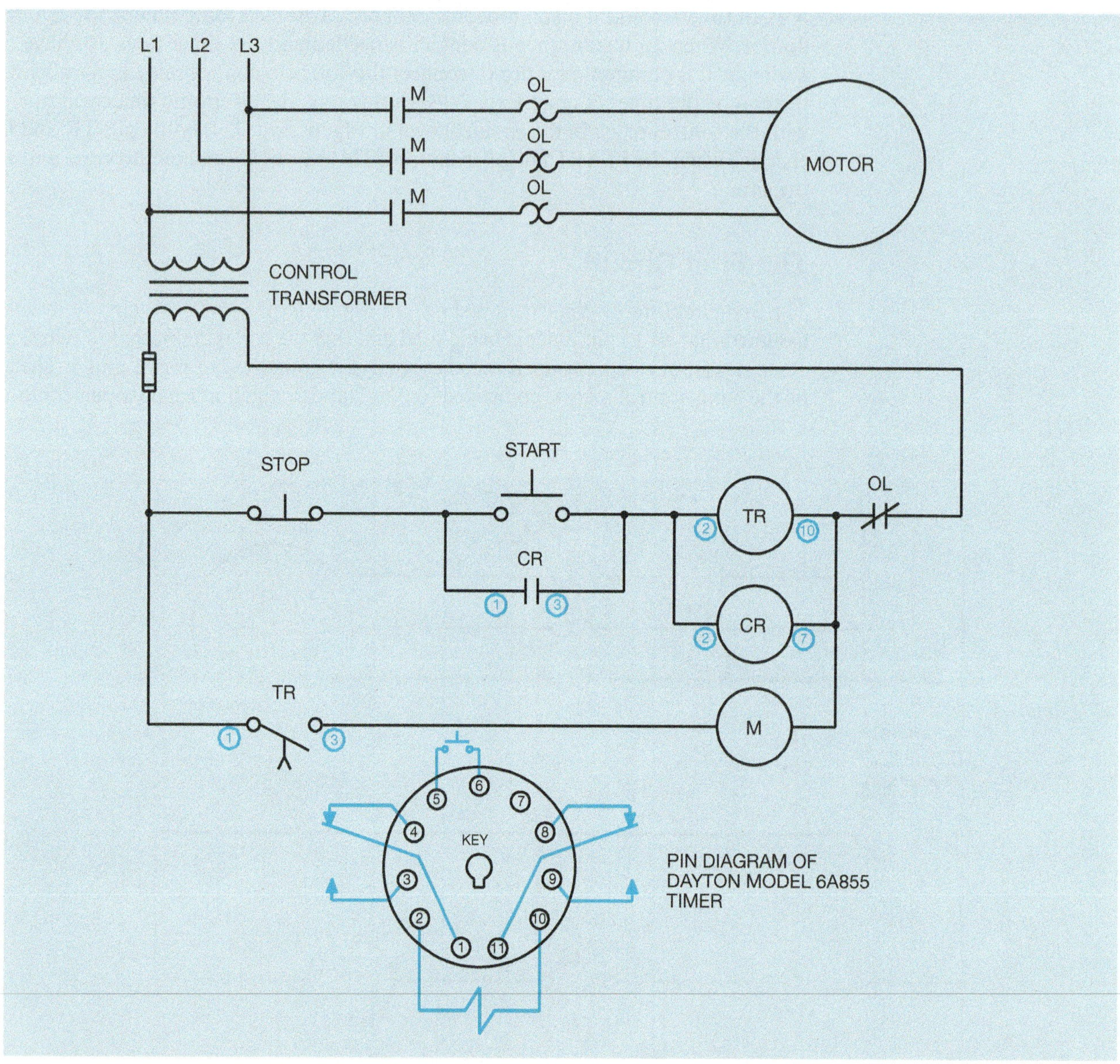

Figure Exp. 7–5 Placing pin numbers beside the components.

Connecting Circuit 1

- ❑ 1. Using the circuit shown in Figure Exp. 7–5, place wire numbers beside the components.
- ❑ 2. Connect the control part of the circuit by following the wire numbers placed beside the components. Note the pin numbers beside the coils and contacts of the timer and control relay. When connecting the timer and control relay, make connection to the 8 and 11 pin tube sockets.
- ❑ 3. Plug the timer and control relay into their appropriate sockets. Set the timer to operate as an on-delay timer and set the time period for 5 seconds.
- ❑ 4. Turn on the power and test the operation of the circuit.
- ❑ 5. **Turn off the power.**
- ❑ 6. If the control part of the circuit operated correctly, connect the motor or equivalent motor load.
- ❑ 7. Turn on the power and test the total circuit for proper operation.
- ❑ 8. **Turn off the power** and disconnect the circuit.

Discussing Circuit 2

In the next circuit, two motors are to be started with a 5-second time delay between the starting of the first motor and the second motor. In this circuit, a normally open auxiliary contact on starter 1M is used as the holding contact, making the use of the control relay unnecessary.

When the START button is pressed, coils 1M and TR energize immediately. This causes motor 1 to start operating and timer TR to begin timing. After 5 seconds, TR contacts close and connect motor 2 to the line. When the STOP button is pressed, or if an overload on either motor should occur, all coils will be de-energized and both motors will stop.

Connecting Circuit 2

- ❑ 1. Using the circuit shown in Figure Exp. 7–6, place pin numbers beside the timer coil and normally open contact.
- ❑ 2. Place wire numbers on the circuit in Figure Exp. 7–6.
- ❑ 3. Connect the control part of the circuit.
- ❑ 4. Turn on the power and test it for proper operation.
- ❑ 5. **Turn off the power.**
- ❑ 6. If the control part of the circuit operated properly, connect the motors or equivalent motor loads.
- ❑ 7. Turn on the power and test the circuit for proper operation.
- ❑ 8. **Turn off the power** and disconnect the circuit.

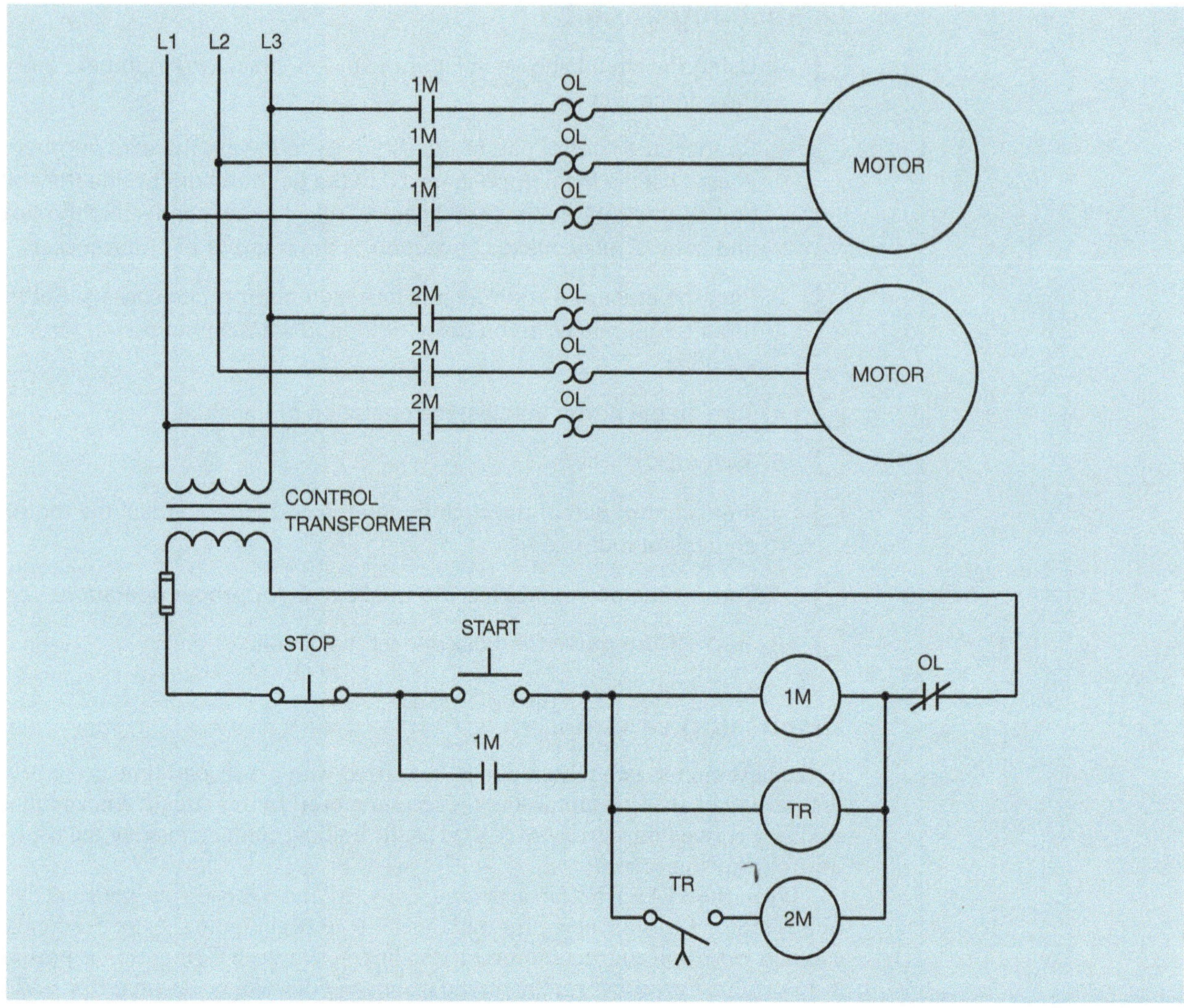

Figure Exp. 7–6 Motor 2 starts after motor 1.

Review Questions

1. Explain the operation of an on-delay timer.
2. Explain the difference between timed contacts and instantaneous contacts.
3. Refer to the circuit shown in Figure Exp. 7–3. If the timer has been set for a 10-second delay, explain the operation of the circuit when the START button is pressed.
4. In the circuit shown in Figure Exp. 7–3, is it necessary to hold the START button closed for a period of at least 10 seconds to ensure that the circuit will remain energized? Explain your answer.
5. Assume that the timer in Figure Exp. 7–3 is set for a delay of 10 seconds. Now assume that the START button is pressed, and after an 8-second delay the STOP button is pressed. Will the motor start 2 seconds after the STOP button was pressed?
6. What is generally done to compensate when a set of instantaneous timer contacts are needed and the timer does not contain them?
7. Refer to the circuit shown in Figure Exp. 7–6. Assume that it is necessary to stop the operation of both motors after the second motor has been operating for a period of 10 seconds. Using the space provided in Figure Exp. 7–7, redraw the circuit to turn off both motors after the second motor has been in operation for 10 seconds. (Note: It will be necessary to use a second timer.)
8. After your instructor has approved the design change, connect the new circuit in the laboratory and test it for proper operation.

Figure Exp. 7–7 Circuit redesign.

Exercise 8

OFF-DELAY TIMERS

Objectives

After completing this exercise the student should be able to:

›› Discuss the operation of an off-delay timer.

›› Draw the NEMA contact symbols used to represent both normally open and normally closed off-delay contacts.

›› Discuss the difference in operation between pneumatic and electronic timers.

›› Connect a circuit in the laboratory employing an off-delay timer.

Materials Needed

- Three phase power supply
- Control transformer
- Two double-acting push buttons (NO/NC on each button)
- 2 ea.—Three phase motor starter with at least one normally open auxiliary contact
- Dayton solid-state timer—MODEL 6A855 or equivalent
- An 11 pin control relay and two 11 pin sockets
- 2 ea.—Three phase motors or equivalent motor loads

The logic of an off-delay timer is as follows: *When the coil is energized, the timed contacts change position immediately. When the coil is de-energized the timed contacts remain in their energized position for some period before changing back to their normal position.* Figure Exp. 8–1 shows the standard NEMA contact symbols used to represent an off-delay timer. Notice that the arrow points in the direction the contact moves after the time delay period. The arrow indicates that the normally open contact will delay reopening, and that the normally closed contact will delay reclosing. Like on-delay timers, some off-delay timers contain instantaneous contacts as well as timed contacts, and some do not.

Example Circuit 1

The circuit shown in Figure Exp. 8–2 illustrates the logic of an off-delay timer. It is assumed that the timer has been set for a 5-second delay. When switch S1 closes, TR coil energizes. This causes the normally open TR contacts to close immediately and turn on the lamp. When switch S1 opens, TR coil de-energizes, but the TR contacts remain closed for 5 seconds before they reopen. Notice that the time delay period does not start until the coil is de-energized.

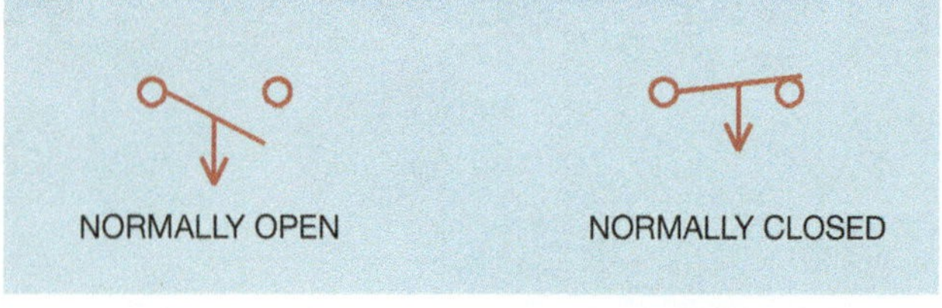

Figure Exp. 8–1 NEMA standard symbols for off-delay contacts.

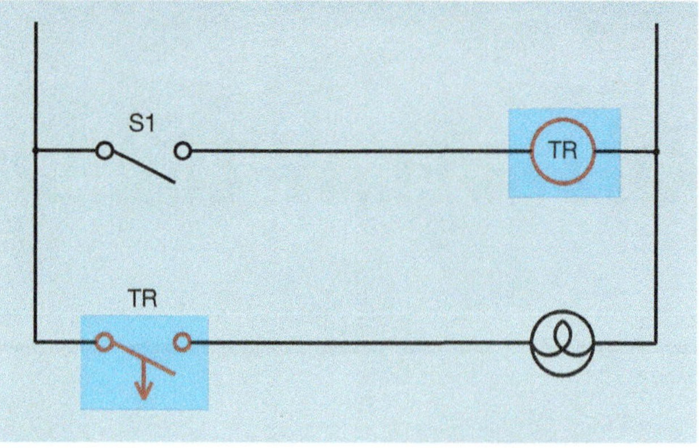

Figure Exp. 8–2 Basic operation of an off-delay timer.

Example Circuit 2

In this example, it is assumed that timer TR has been set for a 10-second delay. Two motors start when the START button is pressed. When the STOP button is pressed, motor 1 stops operating immediately, but motor 2 continues to run for 10 seconds (Figure Exp. 8–3). In this circuit, the coil of the off-delay timer has been placed in parallel with motor starter 1M permitting the action of the timer to be controlled by the first motor starter.

Example Circuit 3

Now assume that the logic of the previous circuit is to be changed so that when the START button is pressed, both motors still start at the same time, but when the STOP button is pressed, motor 2 must stop operating immediately and motor 1 continues to run for 10 seconds. In this circuit, the action of the timer must be controlled by the operation of starter 2M instead of starter 1M (Figure Exp. 8–4). In the circuit shown in Figure Exp. 8–4, a control relay is used to energize both motor starters at the same time. Notice that timer coil TR energizes at the same time as starter 2M, causing the normally open TR contacts to close around the CR contact connected in series with coil 1M.

When the STOP button is pressed, coil CR de-energizes and all CR contacts open. Power is maintained to starter 1M, however, by the now closed TR contacts. When the CR contact connected in series with coils 2M and TR opens, these coils de-energize, causing motor 2 to stop operating and starting the time sequence for the off-delay timer. After a 10-second delay, TR contacts reopen and de-energize coil 1M, stopping the operation of motor 1.

Using Electronic Timers

In the circuits shown in Figures Exp. 8–3 and Exp. 8–4, it was assumed that the off-delay timers were pneumatic. It is common practice to develop circuit logic assuming that the timers are pneumatic. The reason is that the action of a pneumatic timer is controlled by the coil being energized or de-energized. The action of the timer depends on air pressure, not an electric circuit. This, however, is generally not the case when using solid-state time delay relays. Solid-state timers that can be used as off-delay timers are generally designed to be plugged into an 11 pin relay socket. The pin connection for a Dayton model 6A855 timer is shown in Figure Exp. 8–5. Although this is by no means the only type of electronic timer available, it is typical of many.

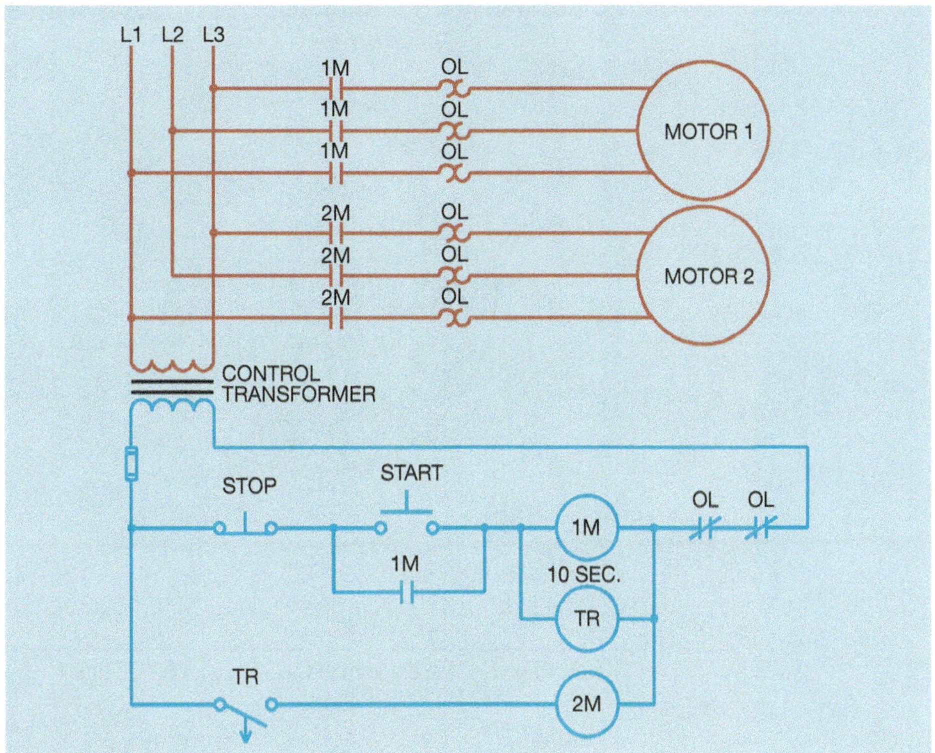

Figure Exp. 8–3 Off-delay motor circuit using pneumatic timer.

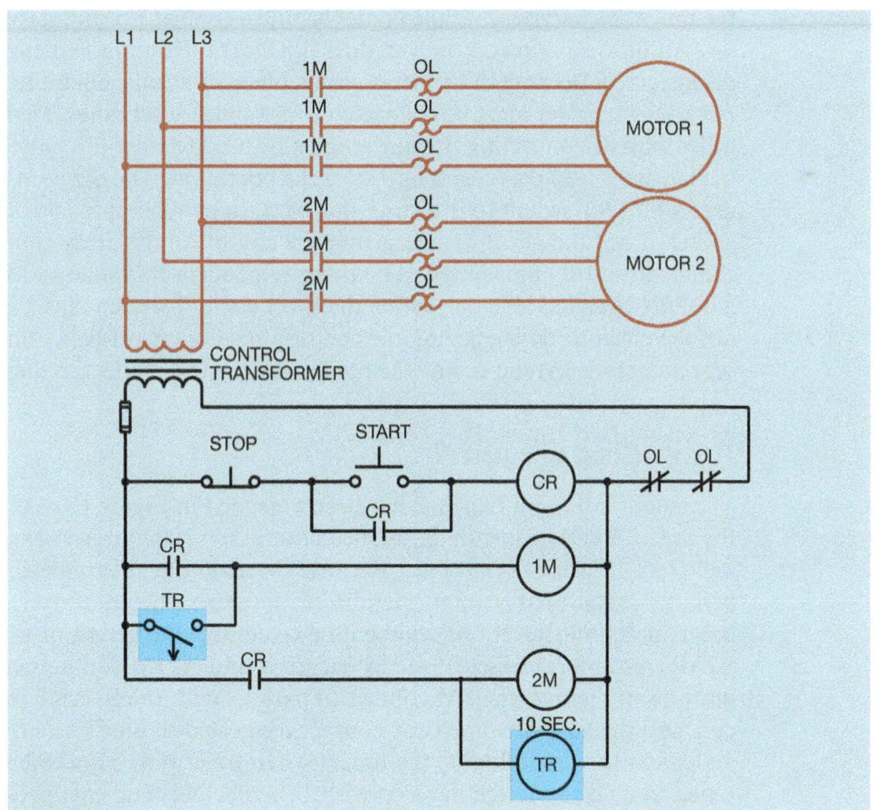

Figure Exp. 8–4 Motor 1 stops after motor 2.

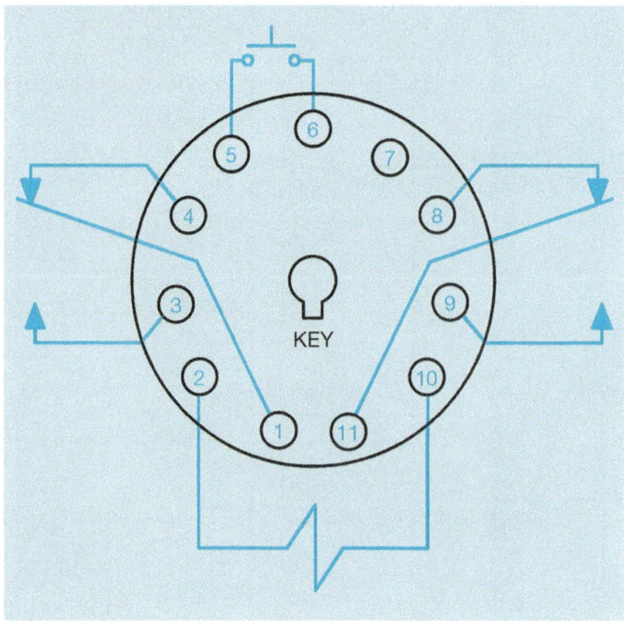

Figure Exp. 8–5 Pin diagram of Dayton model 6A855 timer.

Notice in Figure Exp. 8–5 that power is connected to pins 2 and 10. When this timer is used in the on-delay mode, there is no problem with the application of power because the time sequence starts when the timer is energized. When power is removed the timer de-energizes and the contacts return to their normal state immediately.

An off-delay timer, however, does not start the timing sequence until the timer is de-energized. Because this timer depends on an electronic circuit to operate the timing mechanism, power must be connected to the timer at all times. Therefore, some means other than disconnecting the power must be used to start the timing circuit. This particular timer uses pins 5 and 6 to start the operation. The diagram in Figure Exp. 8–5 uses a START switch to illustrate this operation. When pins 5 and 6 are shorted together, it has the effect of energizing the coil of an off-delay timer and all contacts change position immediately. The timer remains in this state as long as pins 5 and 6 are short-circuited together. When the short circuit between pins 5 and 6 is removed, it has the effect of de-energizing the coil of a pneumatic off-delay timer, and the timing sequence starts. At the end of the period, the contacts return to their normal position.

Amending Circuit 1

The circuit in Figure Exp. 8–3 has been amended in Figure Exp. 8–6 to accommodate the use of an electronic timer. Notice in this circuit that power is connected to pins 2 and 10 of the timer at all times. Because the action of the timer in the original circuit is that the coil of the timer operates at the same time as starter coil 1M, an auxiliary contact on starter 1M will be used to control the action of timer TR. When the START button is pressed, coil 1M energizes and all 1M contacts close. This connects motor 1 to the line, the 1M contact in parallel with the START button seals the circuit, and the normally open 1M contact connected to pins 5 and 6 of the timer closes and starts the operation of the timer. When timer pins 5 and 6 become shorted, the timed contact connected in series with 2M coil closes and energizes starter 2M.

When the STOP button is pressed, coil 1M de-energizes and all 1M contacts return to their normal position, stopping the operation of motor 1. When the 1M contacts connected to timer pins 5 and 6 reopen, it starts the timing sequence of the

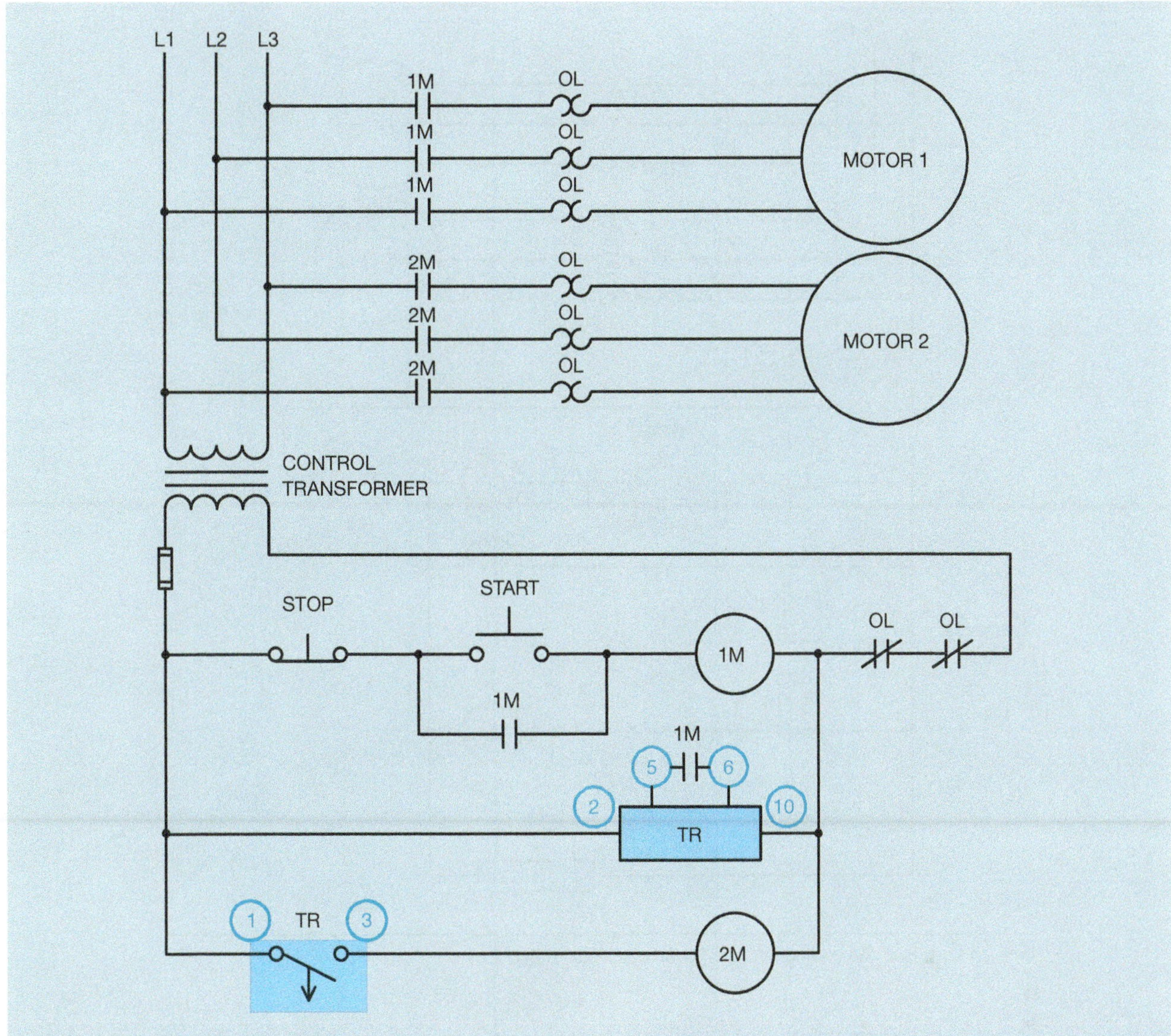

Figure Exp. 8–6 Amending the first circuit for an electronic timer.

timer. After a 10-second delay, timed contact TR reopens and disconnects starter coil 2M from the circuit. This stops the operation of motor 2.

Amending Circuit 2

Circuit 2 will be amended in much the same way as circuit 1. The timer must have power connected to it at all times (Figure Exp. 8–7). Notice in this circuit that the action of the timer is controlled by starter 2M instead of 1M. When coil 2M energizes, a set of normally open 2M contacts close and short pins 5 and 6 of the timer. When coil 2M de-energizes, the 2M auxiliary contacts reopen and start the time sequence of timer TR.

Circuit 2 assumes the use of an 11 pin control relay instead of an 8 pin. An 11 pin control relay contains three sets of contacts instead of two. Figure Exp. 8–8 shows the connection diagram for most 11 pin control relays. Notice that normally open contacts are located on pins 1 and 3, 6 and 7, and 9 and 11. The coil pins are 2 and 10. Pin numbers have been placed beside the components in Figure Exp. 8–7.

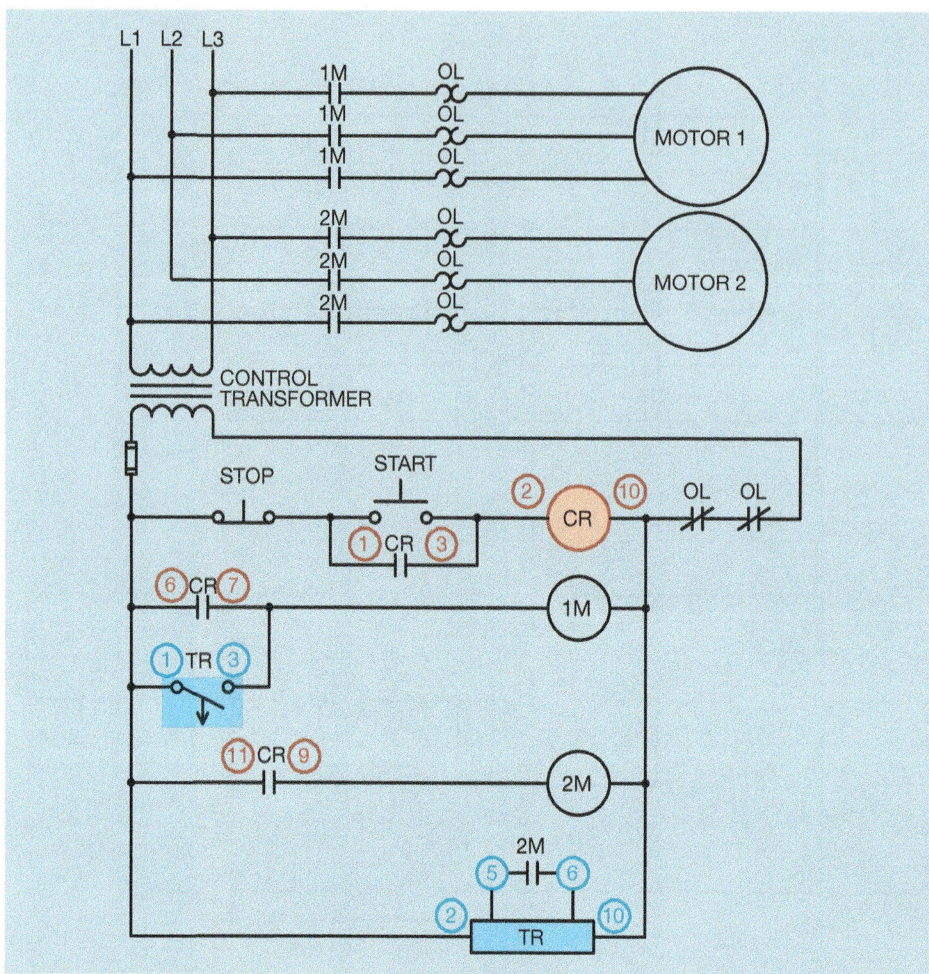

Figure Exp. 8–7 Amending circuit 2 for an electronic timer.

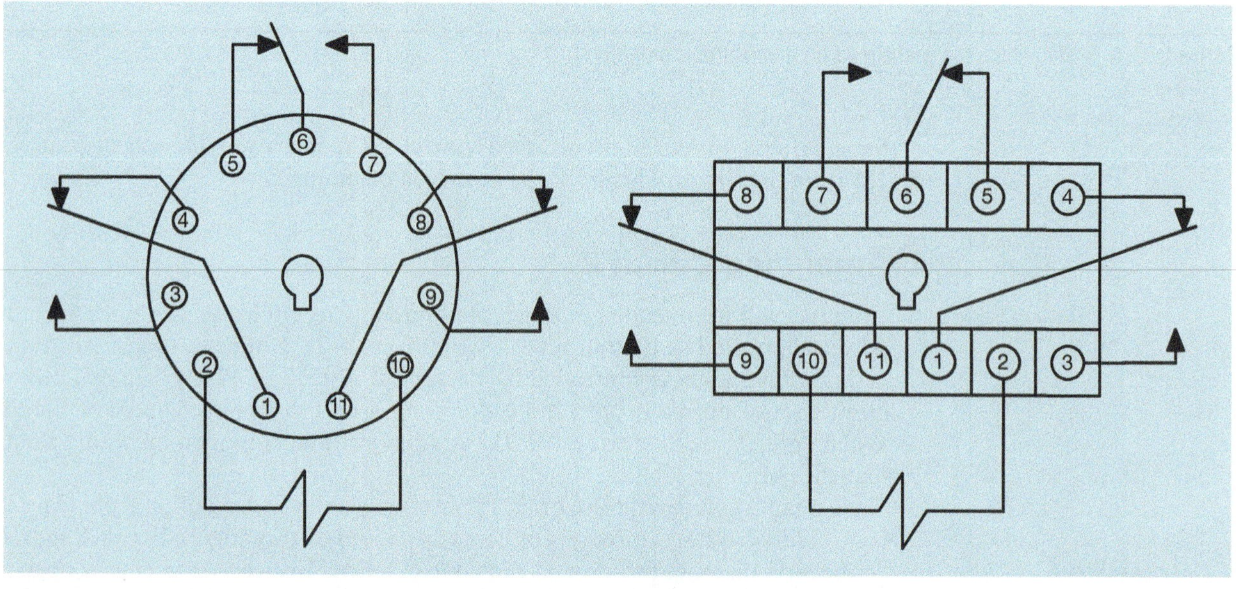

Figure Exp. 8–8 Connection diagram of an 11 pin control relay and socket.

Connecting the First Circuit

- ❑ 1. Place wire numbers on the schematic shown in Figure Exp. 8–6.
- ❑ 2. Using an 11 pin tube socket, connect the control part of the circuit in Figure Exp. 8–6.
- ❑ 3. Set the electronic timer to operate as an off-delay timer and set the time delay for 10 seconds.
- ❑ 4. Plug the timer into the tube socket and turn on the power.
- ❑ 5. Test the control part of the circuit for proper operation.
- ❑ 6. If the control portion of the circuit operated properly, connect the motors or equivalent motor loads and test the entire circuit for proper operation.
- ❑ 7. **Turn off the power** and disconnect the circuit.

Connecting the Second Circuit

- ❑ 8. Place wire numbers on the schematic diagram shown in Figure Exp. 8–7.
- ❑ 9. Using two 11 pin tube sockets, connect the control part of the circuit.
- ❑ 10. Set the electronic timer to operate as an off-delay timer and set the time delay for 10 seconds.
- ❑ 11. Plug the timer and control relay into the tube sockets and turn on the power.
- ❑ 12. Test the control part of the circuit for proper operation.
- ❑ 13. If the control portion of the circuit operated properly, connect the motors or equivalent motor loads and test the entire circuit for proper operation.
- ❑ 14. **Turn off the power** and disconnect the circuit.
- ❑ 15. Return the components to their proper location.

Review Questions

1. Describe the operation of an off-delay timer.
2. Why is it common practice to develop circuit logic assuming all timers are pneumatic?
3. Refer to the schematic diagram shown in Figure Exp. 8–6. Assume that starter coil 2M is open. Describe the action of the circuit when the START button is pressed and when the STOP button is pressed.
4. Refer to the circuit shown in Figure Exp. 8–7. Assume that when the START button is pressed, motor 1 starts operating immediately, but motor 2 does not start. When the STOP button is pressed, motor 1 stops operating immediately. Which of the following could cause this condition?
   a. 1M coil is open.
   b. 2M coil is open.
   c. Timer TR is not operating.
   d. CR coil is open.

5. Refer to the circuit shown in Figure Exp. 8–7. When the START button is pressed, both motors 1 and 2 start operating immediately. When the STOP button is pressed, motor 2 stops operating immediately, but motor 1 remains running and does not turn off after the time delay period has expired. Which of the following could cause this condition?
   a. CR contacts are shorted together.
   b. 2M auxiliary contacts connected to pins 5 and 6 of the timer did not close.
   c. 2M auxiliary contacts connected to pins 5 and 6 of the timer are shorted.
   d. The STOP button is shorted.
6. Refer to the circuit shown in Figure Exp. 8–7. Assume that timer TR is set for a 10-second delay. Now assume that timer TR is changed from an off-delay timer to an on-delay timer. Explain the operation of the circuit.
7. Using the space provided in Figure Exp. 8–9, modify the circuit in Figure Exp. 8–7 to operate as follows:
   a. When the START button is pressed, motor 1 starts running immediately. After a 10-second delay, motor 2 begins running. Both motors remain operating until the STOP button is pressed or an overload occurs.
   b. When the STOP button is pressed, motor 2 stops operating immediately, but motor 1 continues to operate for a period of 10 seconds before stopping.
   c. An overload on either motor will stop both motors immediately.
   d. Assume the use of electronic timers in final design.
8. After your instructor has approved the modifications connect your circuit in the laboratory.
9. Turn on the power and test the circuit for proper operation.
10. **Turn off the power** and disconnect the circuit. Return the components to their proper location.

Figure Exp. 8–9 Redesigning the circuit.

Exercise 9

DESIGNING A PRINTING PRESS CIRCUIT

Objectives

After completing this exercise the student should be able to:

- Describe a step-by-step procedure for designing a motor control circuit.
- Design a basic control circuit.
- Connect the completed circuit in the laboratory.

Materials Needed

- Three-phase power supply
- Three-phase motor starter
- 8 or 11 pin on-delay relay with appropriate socket
- Three-phase motor or equivalent motor load
- Pilot light
- Buzzer or simulated load
- Control transformer
- 8 pin control relay and 8 pin socket

In this exercise, a circuit for a large printing press will be designed in a step-by-step procedure. The owner of a printing company has the following concern when starting a large printing press:

> The printing press is very large and the surrounding noise level is high. There is a danger that when the press starts, a person unseen by the operator may have their hands in the press. To prevent an accident, I would like to install a circuit that sounds an alarm and flashes a light for 10 seconds before the press actually starts. This would give the person time to get clear of the machine before it starts.

To begin the design procedure, list the requirements of the circuit. List not only the concerns of the owner, but also any electrical or safety requirements that the owner may not be aware of. Understand that the owner is probably not an electrical technician and does not know all the electrical requirements of a motor control circuit.

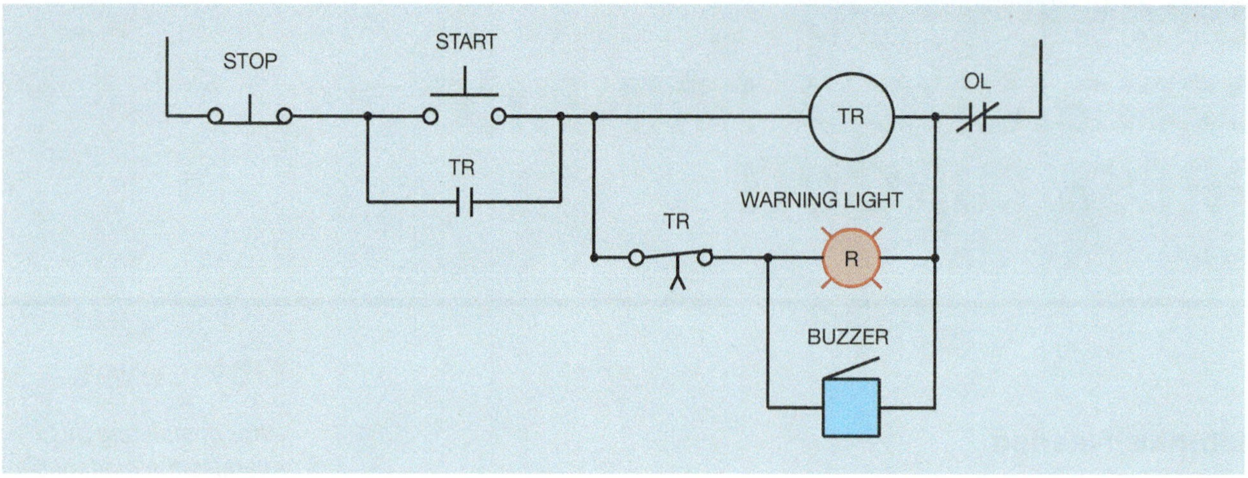

Figure Exp. 9–1 First step in the circuit design.

- ☐ 1. There must be a START and STOP push button control.
- ☐ 2. When the START button is pressed, a warning light and buzzer turn on.
- ☐ 3. After a 10-second delay, the warning light and buzzer turn off and the press motor starts.
- ☐ 4. The press motor should be overload protected.
- ☐ 5. When the STOP push button is pressed, the circuit de-energizes even if the motor has not started.

To begin design of the circuit, fulfill the first requirement of the logic, "When the START button is pressed, a warning light and buzzer turn on for a period of 10 seconds." This first part of the circuit can be satisfied with the circuit shown in Figure Exp. 9–1. In this example, a timer is used because the warning light and buzzer are to remain on for only 10 seconds. Because the warning light and buzzer are to turn on immediately when the START button is pressed, a normally closed timed contact is used. This circuit also assumes that the timer contains an instantaneous contact that is used to hold the circuit in after the START button is released.

The next part of the logic states that after a 10-second delay, the warning light and buzzer are to turn off and the press motor is to start. As shown in Figure Exp. 9–1 in the present circuit, when the START button is pressed, TR coil energizes. This causes the normally open instantaneous TR contacts to close and hold TR coil in the circuit when the START button is released. At the same time, timer TR starts its timing sequence. After a 10-second delay, the normally closed TR timed contact connected in series with the warning light and buzzer opens and disconnects them from the circuit.

The only remaining circuit logic is to start the motor after the warning light and buzzer have turned off. This can be accomplished with a normally open timed contact controlled by timer TR (Figure Exp. 9–2). At the end of the timing sequence, the normally closed TR contact opens and disconnects the warning light and buzzer. At

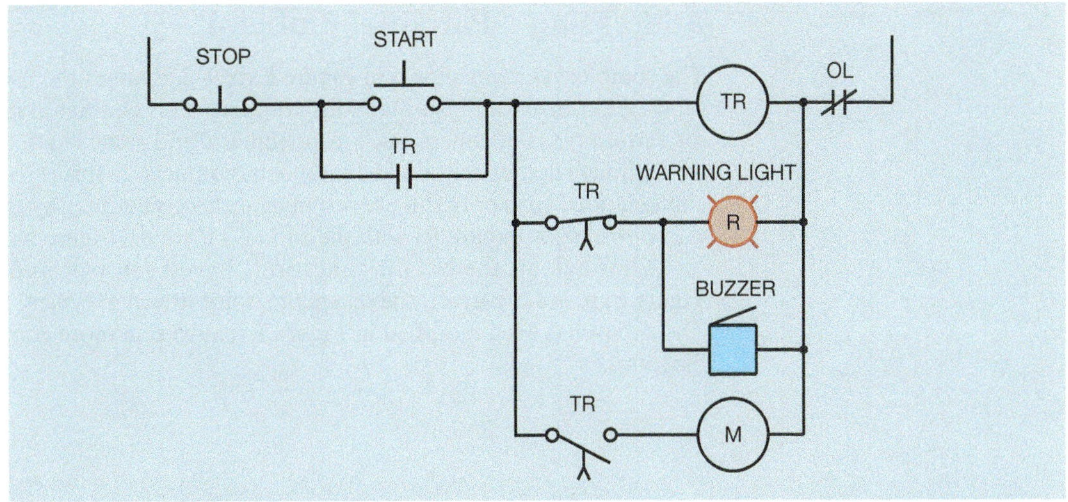

Figure Exp. 9–2 Completing the circuit logic.

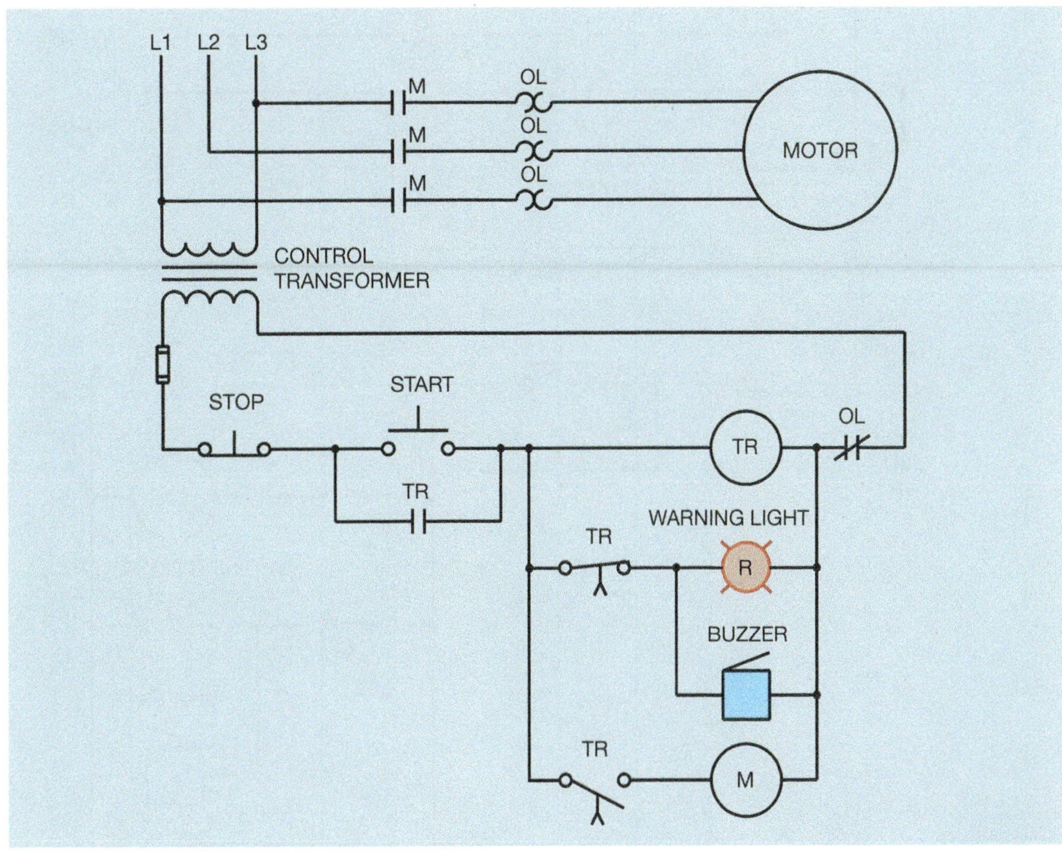

Figure Exp. 9–3 The complete circuit.

the same time, the normally open TR timed contact closes and energizes the coil of M starter. The normally closed overload contact connected in series with the rest of the circuit de-energizes the entire circuit in the event of motor overload.

Now that the logic of the control circuit has been completed, the motor load can be added as shown in Figure Exp. 9–3.

Addressing a Potential Problem

The completed circuit shown in Figure Exp. 9–3 assumes the use of a timer that contains both timed and instantaneous contacts. This contact arrangement is common for certain types of timers, such as pneumatic and some clock timers, but most electronic timers do not contain instantaneous contacts. If this is the case, a control relay can be added to supply the needed instantaneous contact by connecting the coil of the control relay in parallel with the coil of TR timer (Figure Exp. 9–4).

Although all the circuit conditions have been met and the circuit logic in Figure Exp. 9–4 is correct, the schematic is not drawn in typical ladder diagram form. The circuit has been modified in Figure Exp. 9–5 to a more common form for ladder diagrams.

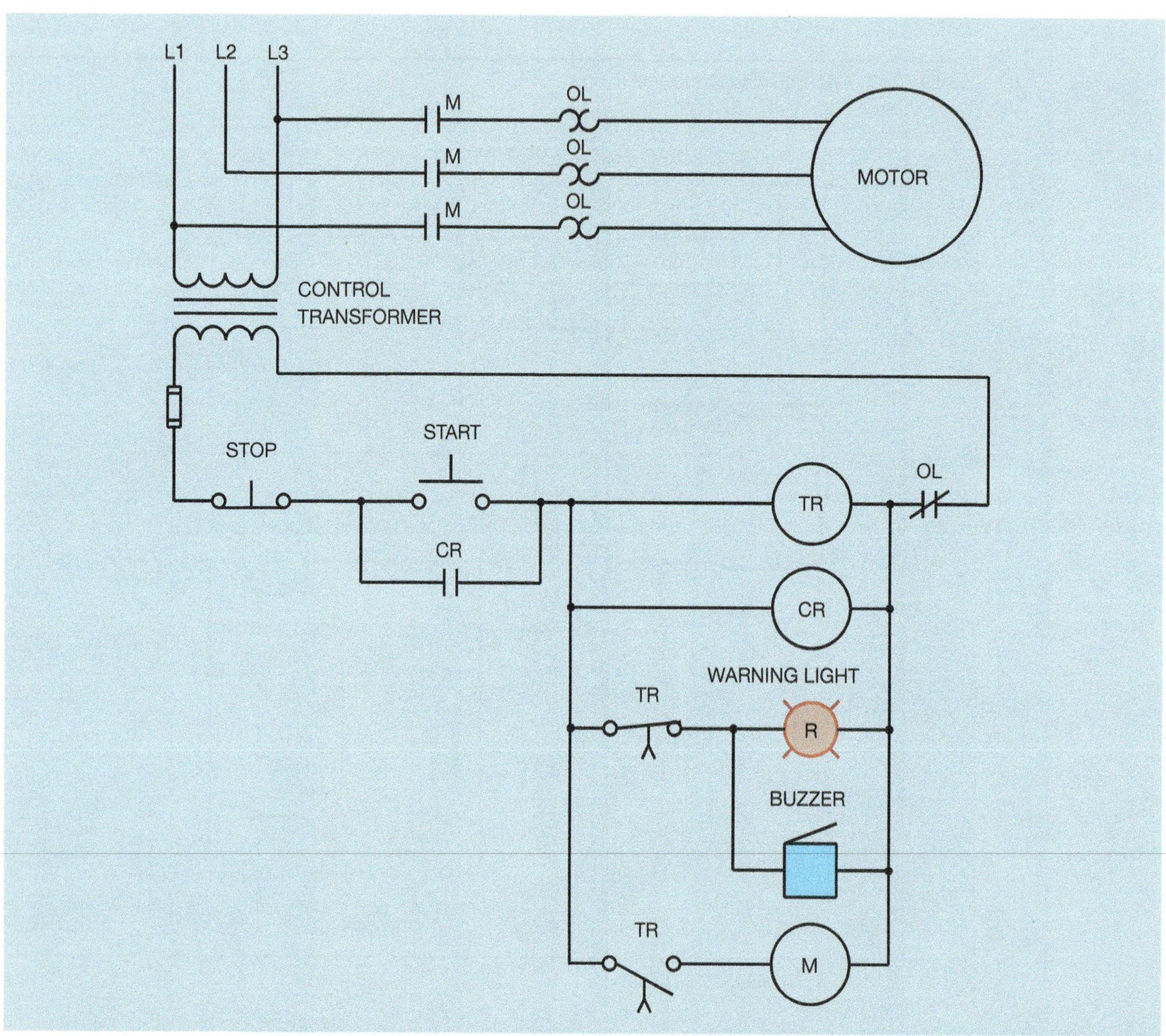

Figure Exp. 9–4 Adding a control relay.

Connecting the Circuit

- [ ] 1. It is assumed that the timer in this circuit is the electronic type. Therefore, it is assumed that a control relay will be used to provide the normally open holding contacts. Assuming the use of an electronic on-delay timer and an 8 pin control relay, place pin numbers beside the components of the timer and control relay shown in Figure Exp. 9–5. Circle the numbers to distinguish them from wire numbers.
- [ ] 2. Place wire numbers beside the components in Figure Exp. 9–5.
- [ ] 3. Connect the control portion of the circuit. (Note: It may be necessary to use a pilot light for the buzzer if one is not available.)

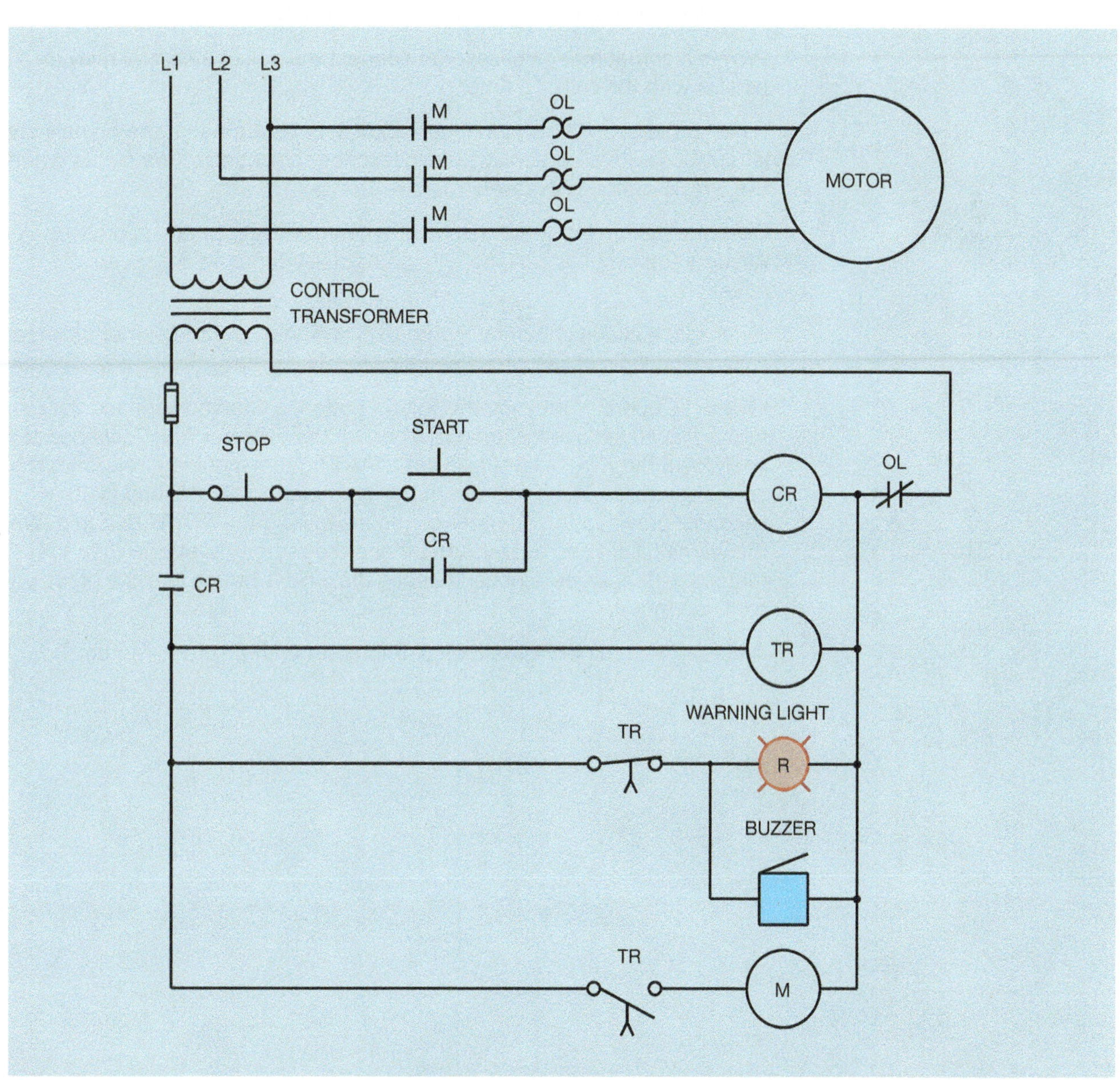

Figure Exp. 9–5 The circuit has been modified to a more common form.

- [ ] 4. Turn on the power and test the control part of the circuit for proper operation.
- [ ] 5. **Turn off the power.**
- [ ] 6. If the control part of the circuit operated properly, connect the motor or simulated motor load to the circuit.
- [ ] 7. Turn on the power and test the entire circuit for proper operation.
- [ ] 8. **Turn off the power** and return the components to their proper location.

Review Questions

1. What should be the first step when beginning the design of a control circuit?
2. Why is it sometimes necessary to connect the coil of a control relay in parallel with the coil of a timer?
3. Refer to the circuit shown in Figure Exp. 9–3. Assume that the on-delay timer is replaced with an off-delay timer. Describe the action of the circuit when the START button is pressed.
4. Describe the operation of the circuit when the STOP button is pressed. Assume the circuit is running with an off-delay timer as described in question 3.
5. Refer to the circuit shown in Figure Exp. 9–5. Assume the owner decides to change the logic of the circuit as follows:

   When the operator presses the START button, a warning light and buzzer turn on for 10 seconds. During this period, the operator must continue to hold down the START button. If the START button should be released, the timing sequence stops and the motor will not start. At the end of 10 seconds, provided the operator continues to hold the START button down, the warning light and buzzer turn off and the motor will start. When the motor starts, the operator can release the START button and the press will continue to run.

   Amend the circuit in Figure Exp. 9–5 to meet with this requirement.

Exercise 10

SEQUENCE STARTING AND STOPPING FOR THREE MOTORS

Objectives

After completing this experiment the student should be able to:

- Discuss the step-by-step procedure for designing a circuit.
- Change a circuit designed with pneumatic timers into a circuit that uses electronic timers.
- Connect the circuit in the laboratory.
- Troubleshoot the circuit.

Materials Needed

- Three-phase power supply
- Control transformer
- 2 ea.—8 pin control relays and 8 pin sockets
- 3 ea.—three-phase motor starters
- 4 ea.—electronic timers (Dayton model 6A855 or equivalent) and 11 pin sockets
- 3 ea.—three-phase motors or equivalent motor loads

In this exercise, a circuit will be designed and connected. The requirements of the circuit are as follows:

- ☐ 1. Three motors are to start in sequence from motor 1 to 3.
- ☐ 2. There is to be a 3-second delay between the starting of each motor.
- ☐ 3. When the STOP button is pressed, the motors are to stop in sequence from motor 3 to 1.
- ☐ 4. There is to be 3-second delay between the stopping of each motor.
- ☐ 5. An overload on any motor will stop all motors.

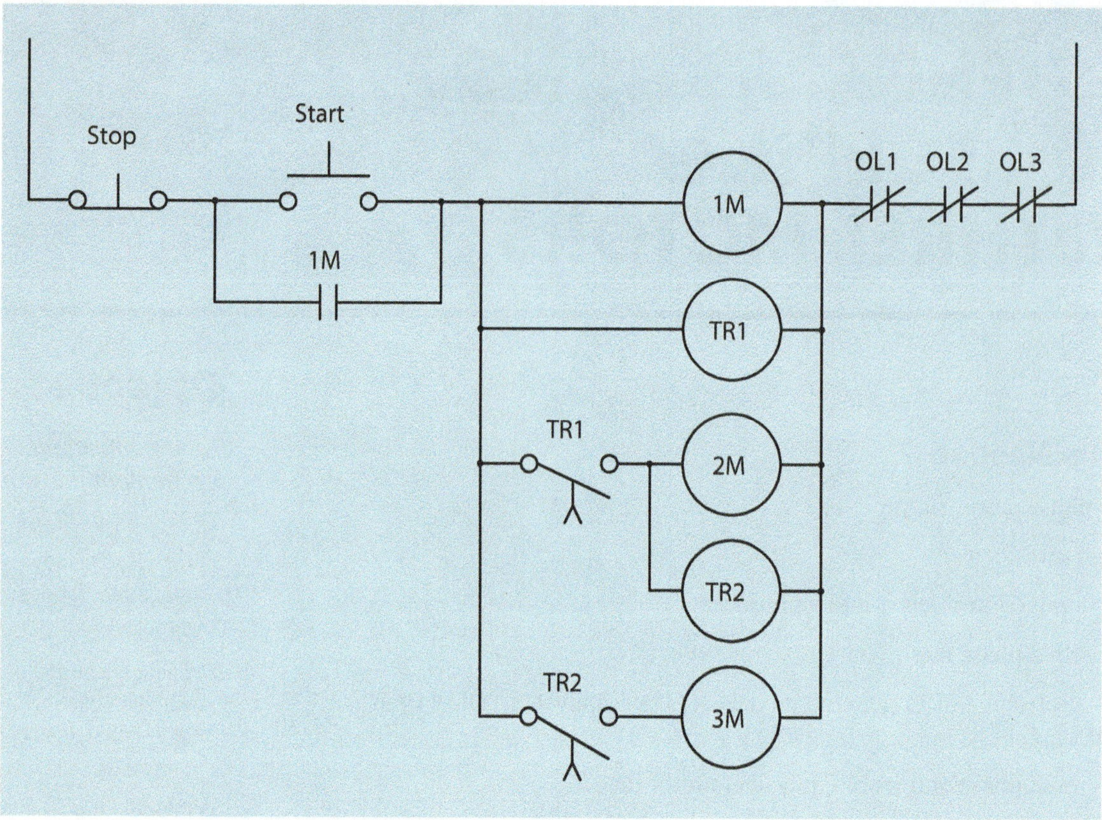

Figure Exp. 10–1 The motors start in sequence from 1 to 3.

When designing a control circuit, satisfy one requirement at a time. This may, at times, lead to an unforeseen dead-end, but don't let these dead-ends concern you. When they happen, back up, and redesign around them. In this example, the first part of the circuit is to start three motors in sequence from motor 1 to 3 with a 3-second delay between the starting of each motor. This is also the time to satisfy the requirement that an overload on any motor will stop all motors. The first part of the circuit can be satisfied by the circuit shown in Figure Exp. 10–1. (Note: In this exercise the motor connections will not be shown because of space limitations. It is assumed that the motor starters are controlling three-phase motors. It is also assumed that all timers are set for a 3-second delay.)

When the START button is pressed, coils 1M and TR1 energize. Starter 1M starts motor 1 immediately, and timer TR1 starts its time sequence of 3 seconds. After a 3-second delay, timed contact TR1 closes and energizes coils 2M and TR2. Starter 2M starts motor 2 and timer TR2 begins its 3-second timing sequence. After a 3-second delay, timed contact TR2 closes and energizes motor 3. The motors have been started in sequence from 1 to 3 with a 3-second delay between the starting of each motor. This satisfies the first part of the circuit logic.

The next requirement is that the circuit stop in sequence from motor 3 to 1. To fulfill this requirement, power must be maintained to starters 2M and 1M after the STOP button has been pushed. In the circuit shown in Figure Exp. 10–1, this is not possible. Because all coils are connected after the M auxiliary holding contact, power will be disconnected from all coils when the STOP button is pressed and the holding contact opens. This circuit has proven to be a dead-end. There is no way to fulfill the second requirement with the circuit connected in this manner. Therefore, the circuit must be amended in such a manner that it will not only start in sequence from motor

1 to 3 with a 3-second delay between the starting of each motor, but also be able to maintain power after the START button is pressed. This amendment is shown in Figure Exp. 10–2.

To modify the circuit so that power can be maintained to coils 2M and 1M, a control relay has been added to the circuit. Contact $1CR_2$ prevents power from being applied to coils 1M and TR1 until the START button is pressed.

Designing the Second Part of the Circuit

The second part of the circuit states that the motors must stop in sequence from motor 3 to 1. Do not try to solve all the logic at once. Solve each problem as it arises. The first problem is to stop motor 3. In the circuit shown in Figure Exp. 10–2, when the STOP button is pressed, coil 1CR de-energizes. This causes contact $1CR_2$ to open and de-energize coils 1M and TR1. Contact TR1 opens immediately and de-energizes coils 2M and TR2, causing contact TR2 to open immediately and de-energize coil 3M. Notice that coil 3M does de-energize when the STOP button is pressed, but so does everything else. The circuit requirement states that there is to be a 3-second time delay between the stopping of motor 3 and 2. Therefore, an off-delay timer will be added to maintain connection to coil 2M after coil 3M has de-energized (Figure Exp. 10–3).

The same basic problem exists with motor 1. In the present circuit, motor 1 turns off immediately when the STOP button is pressed. To help satisfy the second part of the problem, another off-delay relay must be added to maintain a circuit to motor 1 for a period of 3 seconds after motor 2 has turned off. This addition is shown in Figure Exp. 10–4.

Motors 2 and 1 will now continue to operate after the STOP button is pressed, but so will motor 3. In the present design, none of the motors turn off when the

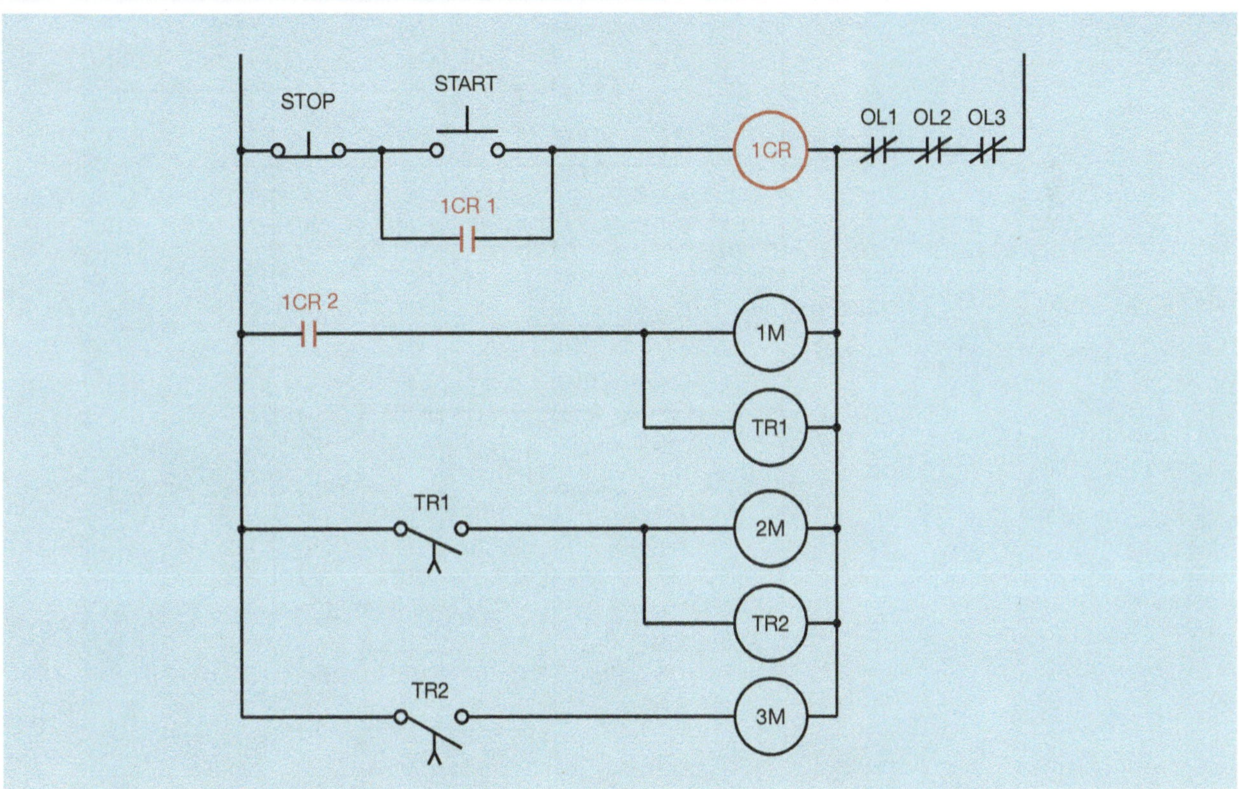

Figure Exp. 10–2 A control relay is added to the circuit.

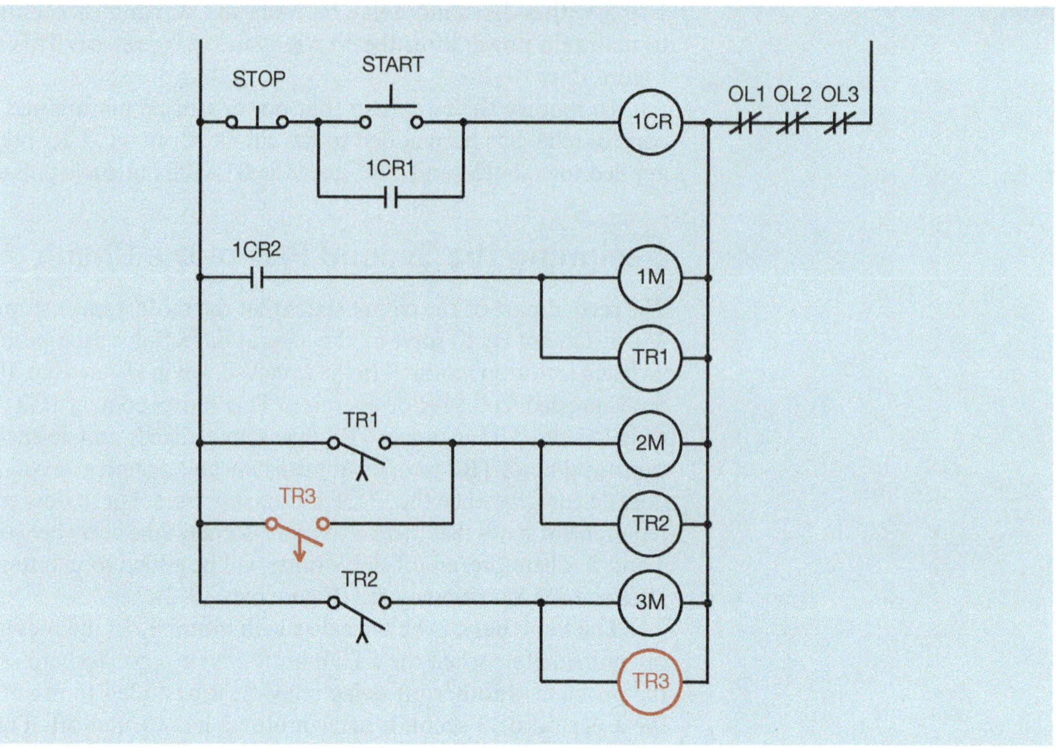

Figure Exp. 10–3 Timer TR3 prevents motor 2 from stopping.

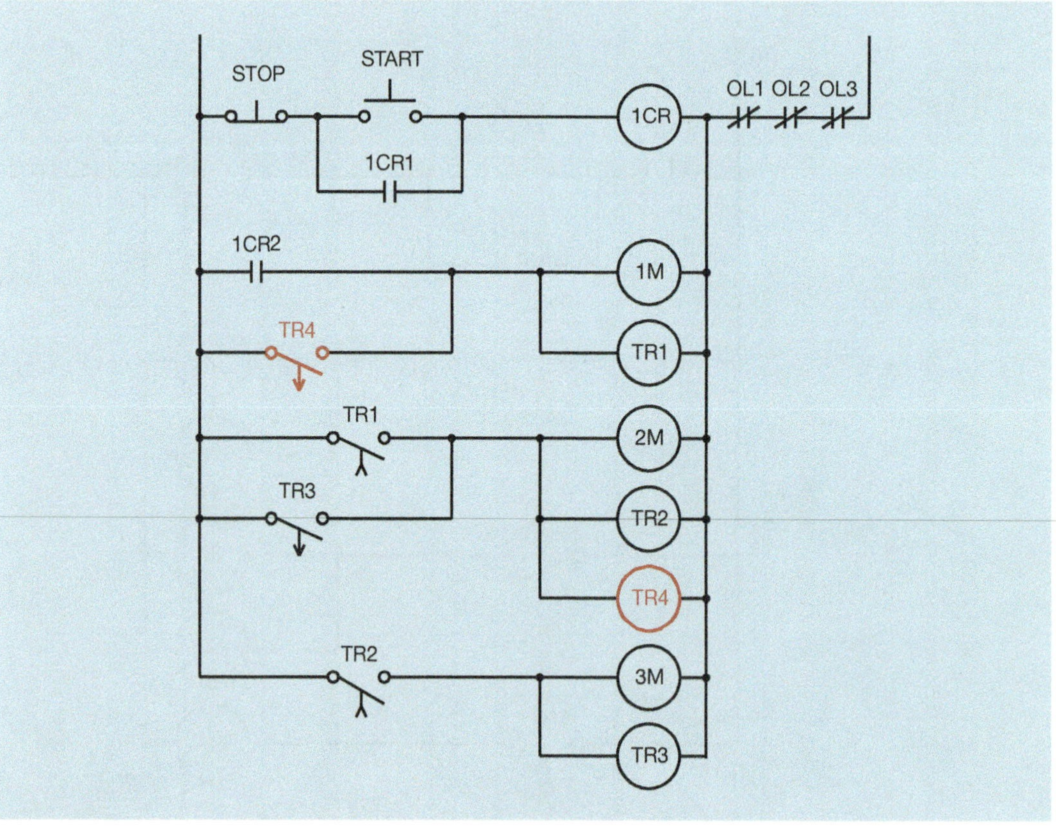

Figure Exp. 10–4 Off-delay timer TR4 prevents motor 1 from stopping.

STOP button is pressed. To understand this condition, trace the logic step by step. When the START button is pressed, coil 1CR energizes and closes all 1CR contacts. When contact $1CR_2$ closes, coils 1M and TR1 energize. After a period of 3 seconds, timed contact TR1 closes and energizes coils 2M, TR2, and TR4. Timed contact TR4 closes immediately to bypass contact $1CR_2$. After a 3-second delay, timed contact TR2 closes and energizes coils 3M and TR3. Timed contact TR3 closes immediately and bypasses contact TR1. When the STOP button is pressed, coil 1CR de-energizes and all 1CR contacts open, but a circuit is maintained to coils 1M and TR1 by contact TR4. This prevents timed contact TR1 from opening to de-energize coils 2M, TR2, and TR4, which in turn prevents timed contact TR2 from opening to de-energize coils 3M and TR3. To overcome this problem, two more contacts controlled by control relay 1CR will be added to the circuit (Figure Exp. 10–5). The circuit will now operate in accord with all the stated requirements.

Modifying the Circuit

The circuit in Figure Exp. 10–5 was designed with the assumption that all the timers are pneumatic. When this circuit is connected in the laboratory, 8 pin control relays and electronic timers will be used. The circuit will be amended to accommodate these components. The first change to be made concerns the control relays.

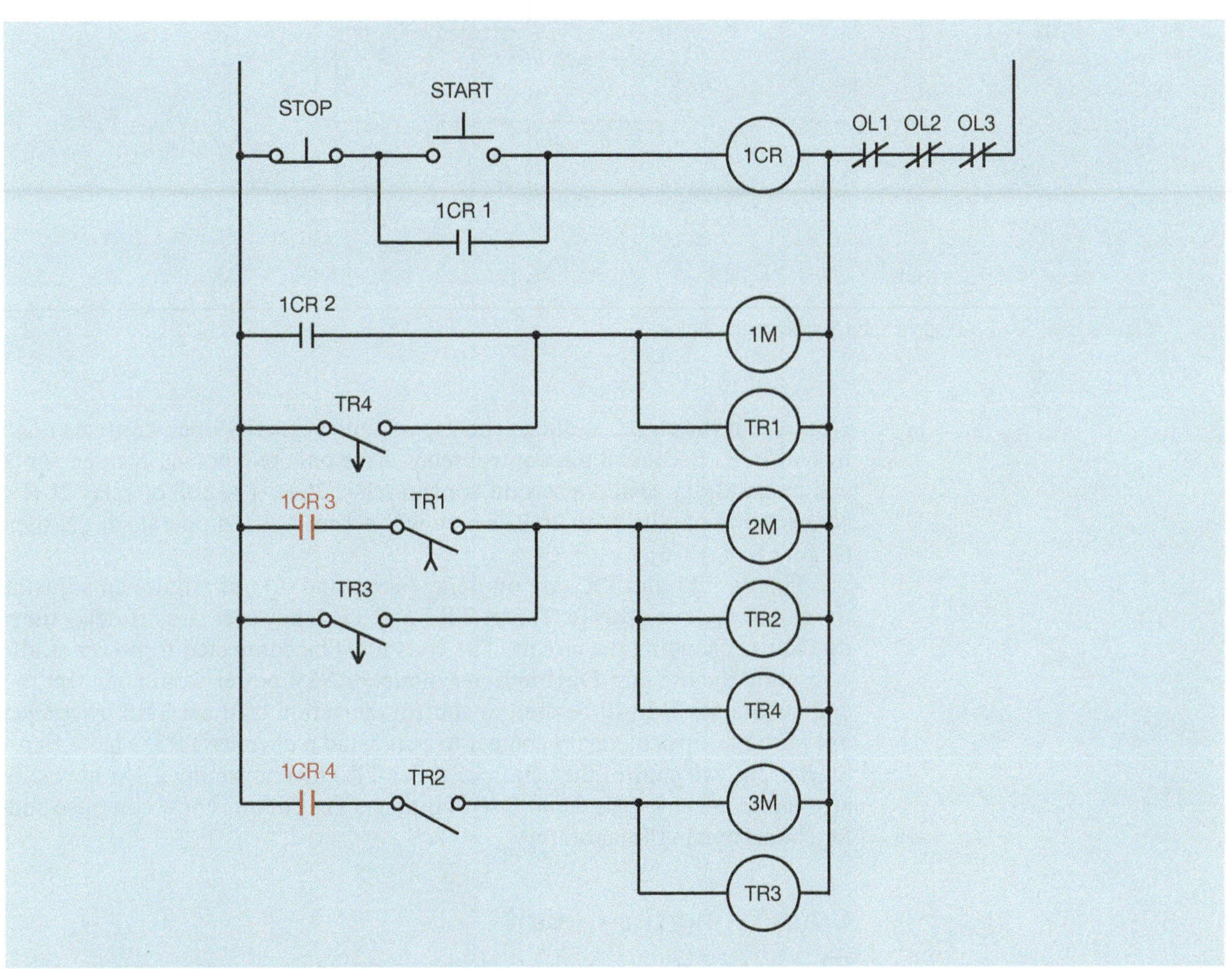

Figure Exp. 10–5 Control relay contacts are added to permit the circuit to turn off.

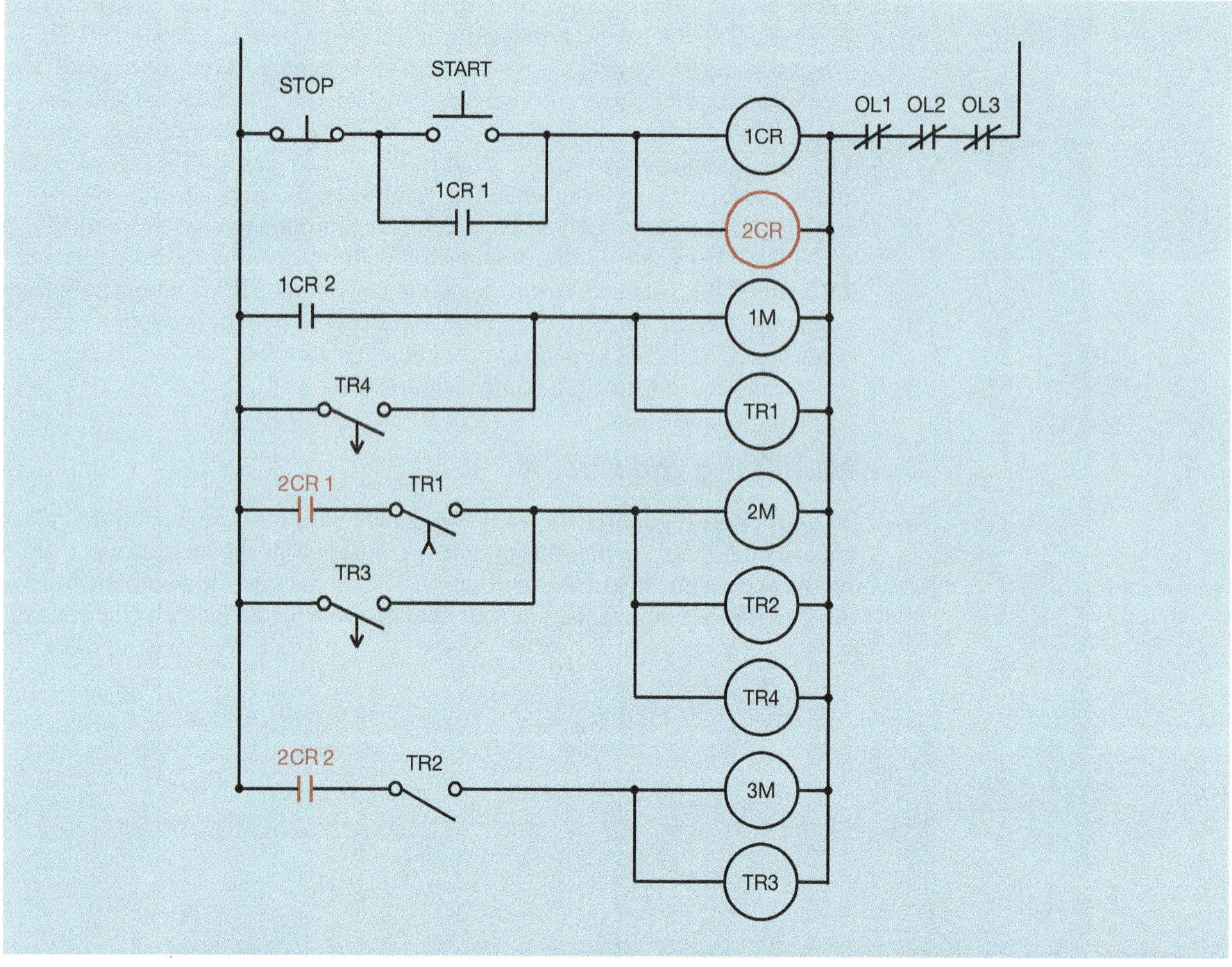

Figure Exp. 10–6 Adding a control relay to the circuit.

Notice that the circuit requires the use of four normally open contacts controlled by coil 1CR. Because 8 pin control relays have only two normally open contacts, it will be necessary to add a second control relay, 2CR. The coil of relay 2CR will be connected in parallel with 1CR, which will permit both to operate at the same time (Figure Exp. 10–6).

Timers TR1 and TR2 are on-delay timers and do not require an adjustment in the circuit logic to operate. Timers TR3 and TR4, however, are off-delay timers and do require changing the circuit. The coils must be connected to power at all times. Assuming the use of a Dayton timer model 6A855, power would connect to pins 2 and 10. Starter 3M will be used to control the action of timer TR3 by connecting a 3M normally open auxiliary contact to pins 5 and 6 of timer TR3 (Figure Exp. 10–7). Starter 2M will control the action of timer TR4 by connecting a 2M normally open auxiliary contact to pins 5 and 6 of that timer. The circuit is now complete and ready for connection in the laboratory.

Connecting the Circuit

❑ 1. Using the circuit shown in Figure Exp. 10–7, place pin numbers beside the proper components. Circle the pin numbers to distinguish them from wire numbers.

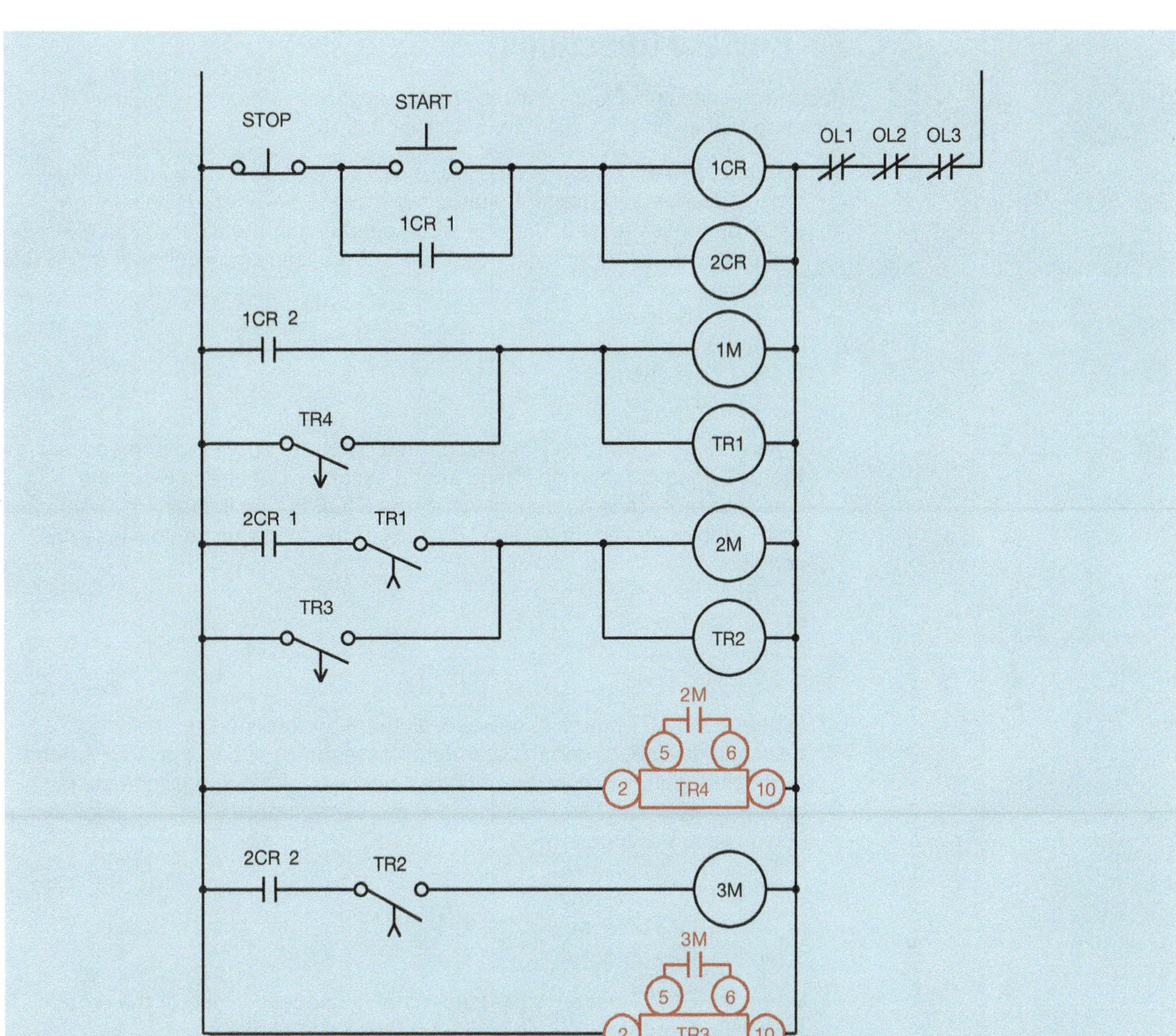

Figure Exp. 10–7 Changing pneumatic timers for electronic timers.

- ❑ 2. Place wire numbers on the schematic.
- ❑ 3. Connect the control circuit in the laboratory.
- ❑ 4. **Turn on the power** and test the circuit for proper operation.
- ❑ 5. Turn off the power and connect the motor loads to starters 1M, 2M, and 3M.
- ❑ 6. Turn on the power and test the complete circuit.
- ❑ 7. **Turn off the power.**
- ❑ 8. Disconnect the circuit and return the components to their proper place.

Review Questions

Refer to the circuit in Figure Exp. 10–7 to answer the following questions. It is assumed that all timers are set for a 3-second delay.

1. When the START button is pressed, motor 1 starts operating immediately. Three seconds later, motor 2 starts, but motor 3 never starts. When the STOP button is pressed, motor 2 stops operating immediately. After a 3-second delay, motor 1 stops running. Which of the following could *not* cause this condition?
   a. TR3 coil is open.
   b. 3M coil is open.
   c. TR2 coil is open.
   d. 2CR coil is open.

2. When the START button is pressed, motor 1 starts operating immediately. Motor 2 does not start operating after 3 seconds, but after a 6-second delay motor 3 starts operating. When the STOP button is pushed, motors 3 and 1 stop operating immediately. Which of the following could cause this condition?
   a. 2CR coil is open.
   b. TR1 coil is open.
   c. TR3 coil is open.
   d. 2M coil is open.

3. When the START button is pressed, all three motors start normally with a 3-second delay between the starting of each motor. When the STOP button is pressed, motor 3 stops operating immediately. After a 3-second delay, both motors 2 and 1 stop operating at the same time. Which of the following could cause this problem?
   a. Timer TR1 is defective.
   b. Timer TR2 is defective.
   c. Timer TR3 is defective.
   d. Timer TR4 is defective.

4. When the START button is pressed, nothing happens. None of the motors start. Which of the following could *not* cause this problem?
   a. Overload contact OL1 is open.
   b. 1CR relay coil is open.
   c. 2CR relay coil is open.
   d. The STOP button is open.

5. When the START button is pressed, motor 1 does not start, but after a 3-second delay motor 2 starts, and 3 seconds later motor 3 starts. When the STOP button is pressed, motor 3 stops running immediately and after a 3-second delay motor 2 stops running. Which of the following could cause this problem?
   a. Starter coil 1M is open.
   b. TR1 timer coil is open.
   c. Timer TR4 is defective.
   d. 1CR coil is open.

Exercise 11

HYDRAULIC PRESS CONTROL

Objectives

After completing this experiment the student should be able to:

- Discuss the operation of this hydraulic press control circuit.
- Connect the circuit in the laboratory.
- Operate the circuit using toggle switches to simulate limit and pressure switches.

Materials Needed

- Three-phase power supply
- Control transformer
- Three-phase motor starter with at least two normally open auxiliary contacts
- 5 ea.—double-acting push buttons (NO/NC on each button)
- Pilot light
- 3 ea.—toggle switches that can be used to simulate two limit switches and one pressure switch
- One three-phase motor or equivalent motor load
- 2 ea.—solenoid coils or lamps to simulate solenoid coils
- 3 ea.—control relays with three sets of contacts (11 pin) and 11 pin sockets
- 3 ea.—control relays with two sets of contacts (8 pin) and 8 pin sockets

The next circuit to be discussed is a control for a large hydraulic press (Figure Exp. 11–1). In this circuit, a hydraulic pump must be started before the press can operate. Pressure switch PS closes when there is sufficient hydraulic pressure to operate the press. If switch PS opens, it stops the operation of the circuit. A green pilot light is used to tell the operator that there is enough pressure to operate the press.

Two run push buttons are located far enough apart so that both of the operator's hands must be used to cause the press to cycle. This is to prevent the operator from getting his or her hands in the press when it is operating. Limit switches UPLS and DNLS are used to determine when the press is at the bottom of its downstroke and when it is at the top of its upstroke. In the event that one or both of the run push buttons are released during the cycle, a reset button can be used to reset the press to its top position. The up solenoid causes the press to travel upward when it is energized, and the down solenoid causes the press to travel downward when it is energized.

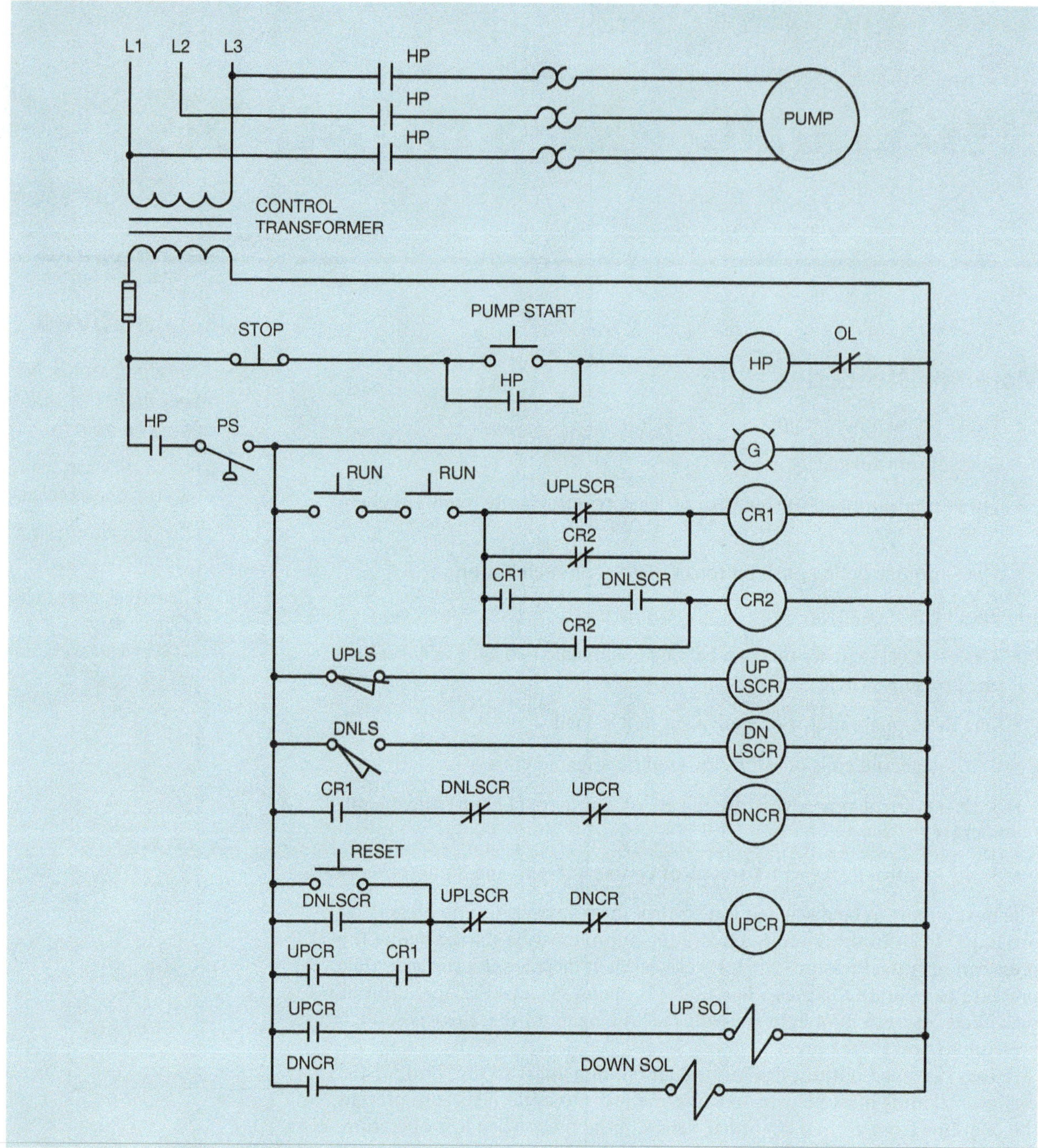

Figure Exp. 11–1 Hydraulic press.

To understand the operation of this circuit, assume that the press is in the up position. Notice that limit switch UPLS is shown normally open held closed. This limit switch is connected normally open, but when the press is in the up position, it is being held closed. Now assume that the hydraulic pump is started and that the pressure switch closes. When pressure switch PS closes, the green pilot light turns on and UPLSCR (up limit switch control relay) energizes, changing all UPLSCR contacts (Figure Exp. 11–2).

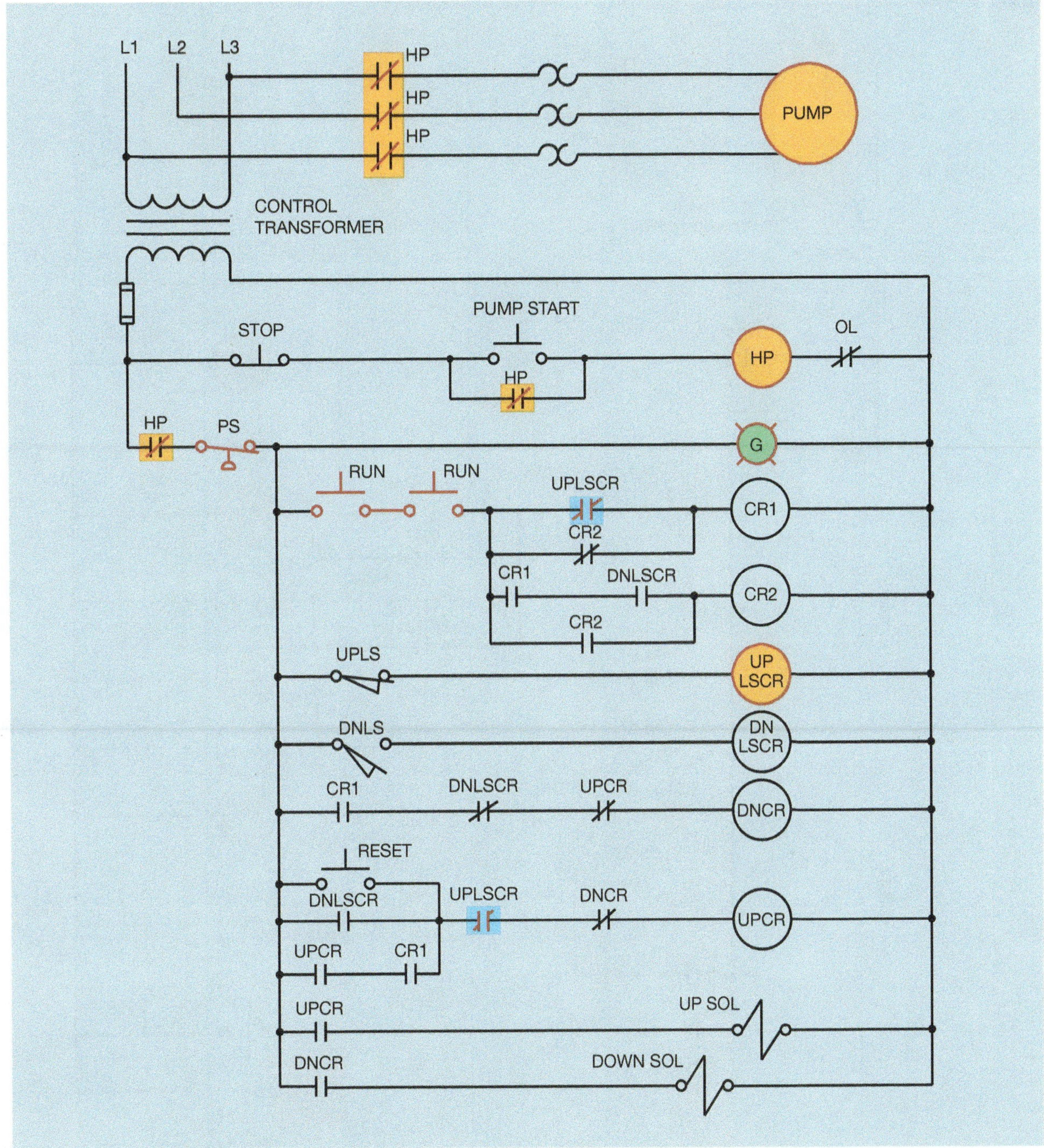

Figure Exp. 11–2 The circuit with pump operating.

When both run push buttons are held down, a circuit is completed to CR1 relay, causing all CR1 contacts to change position (Figure Exp. 11–3). The CR1 contact connected in series with the coil of DNCR closes and energizes the relay, causing all DNCR contacts to change position. The DNCR contact connected in series with the down solenoid coil closes and energizes the down solenoid.

As the press begins to move downward, limit switch UPLS opens and de-energizes coil UPLSCR, returning all UPLSCR contacts to their normal position (Figure Exp. 11–4).

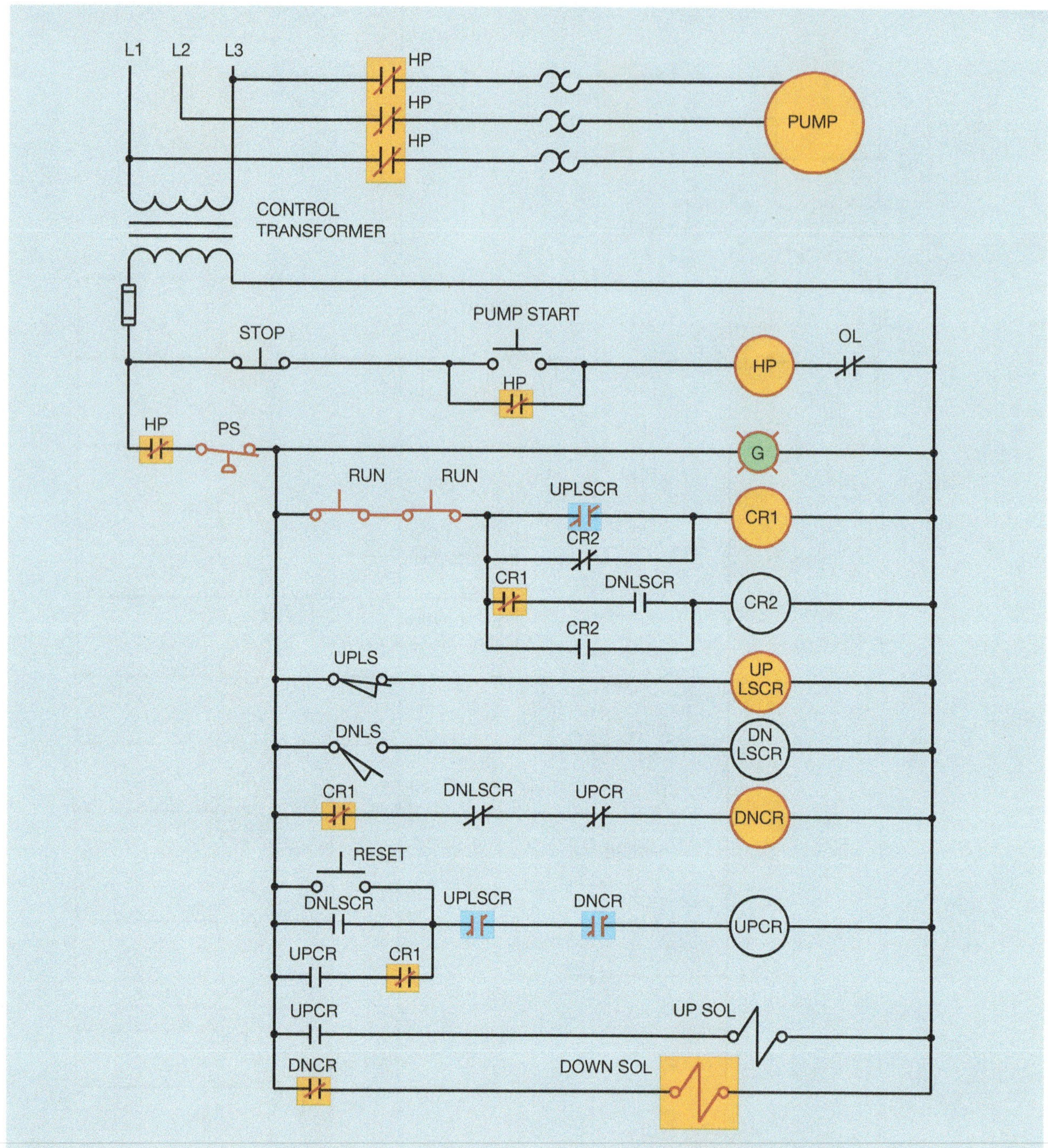

Figure Exp. 11–3 The circuit is started.

When the press reaches the bottom of its stroke, it closes down limit switch DNLS. This energizes the coil of the down limit switch control relay, DNLSCR, causing all DNLSCR contacts to change position (Figure Exp. 11–5). The normally open DNLSCR contact connected in series with the coil of CR2 closes and energizes that relay, causing all CR2 contacts to change position. The normally closed DNLSCR contact connected in series with DNCR coil opens and de-energizes that relay. All

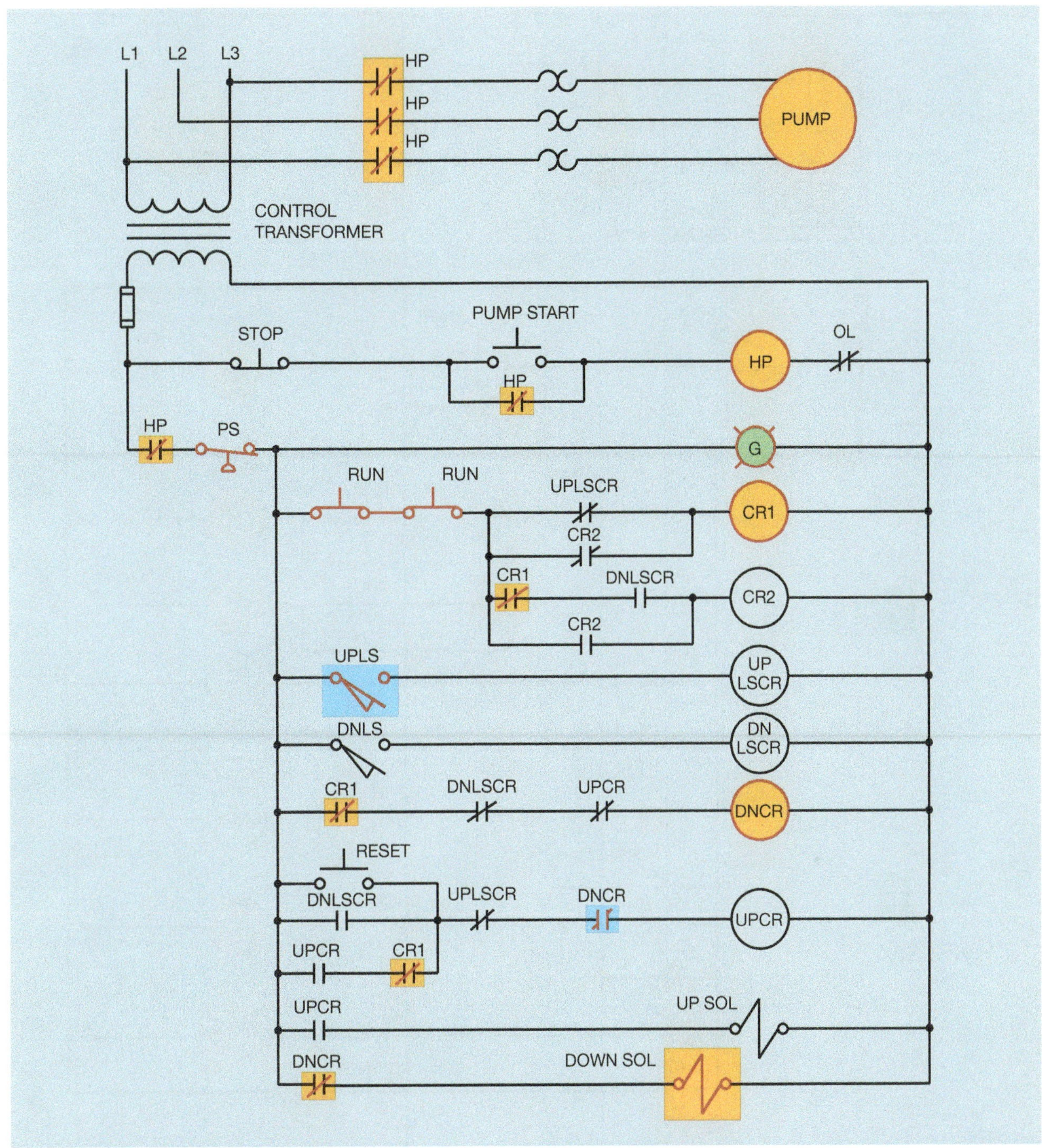

Figure Exp. 11–4 The up limit switch opens.

DNCR contacts return to their normal position. The normally open contact connected in series with the down solenoid coil opens and de-energizes the solenoid. The normally closed DNCR contacts connected in series with UPCR coil reclose to provide a current path to that relay.

The UPCR contact connected in series with coil DNCR opens to prevent coil DNCR from re-energizing when coil DNLSCR de-energizes. The normally

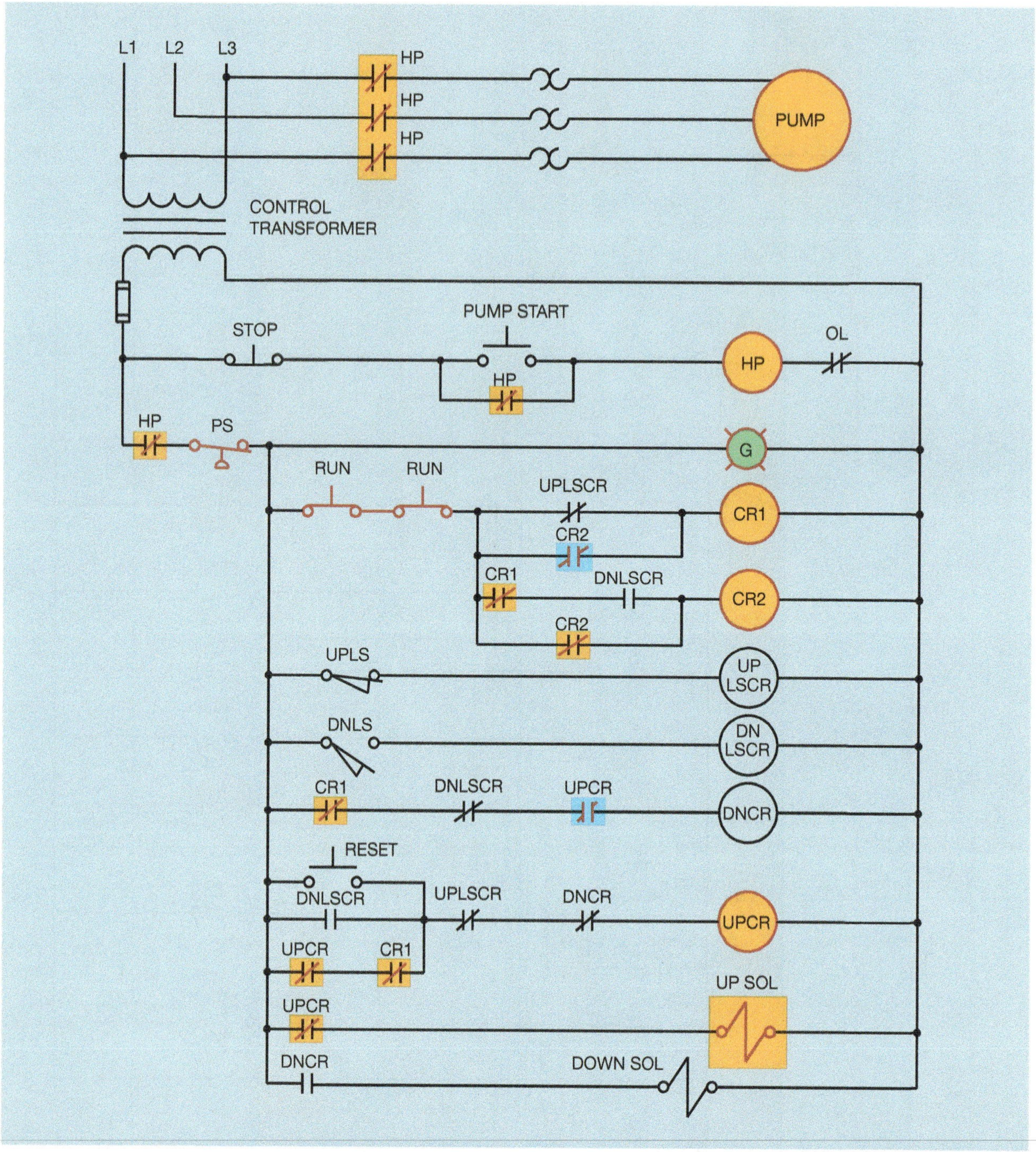

Figure Exp. 11–5 DNLSCR and CR2 relays energize.

open UPCR contact connected in series with the up solenoid closes and provides a current path to the up solenoid. When the press starts upward, limit switch DNLS reopens and de-energizes coil DNLSCR. A circuit is maintained to UPCR coil by the now closed UPCR contact connected in series with the CR1 contact (Figure Exp. 11–6).

The press continues to travel upward until it reaches its upper limit and closes limit switch UPLS, energizing coil UPLSCR (Figure Exp. 11–7). This causes both

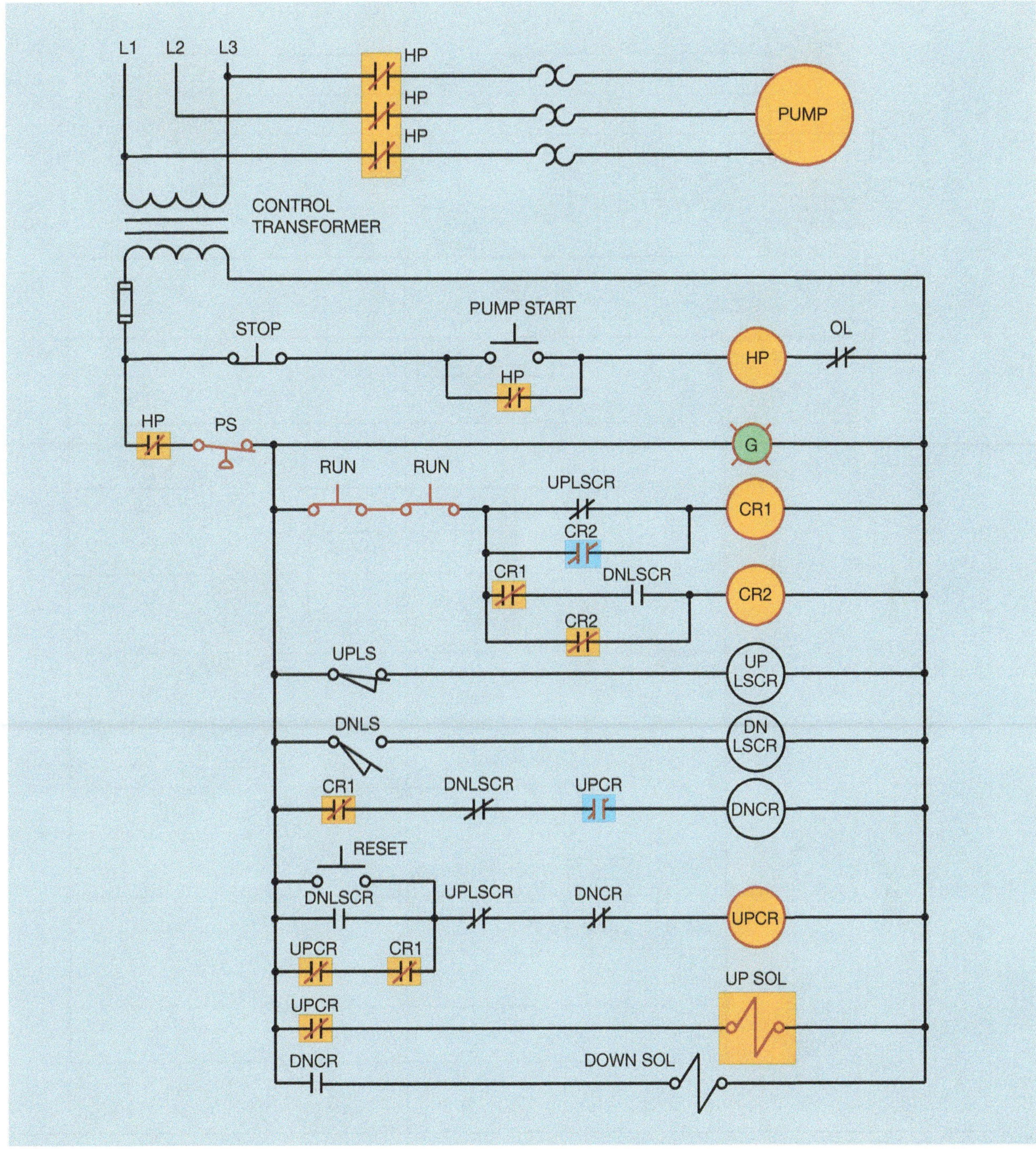

Figure Exp. 11–6 Limit switch DNLS reopens.

UPLSCR contacts to change position. The UPLSCR contact connected in series with coil UPCR opens and de-energizes the up solenoid. Notice that control relay CR2 is still energized. Before the press can be recycled, one or both of the run buttons must be released to break the circuit to the control relays. This permits the circuit to reset to the state shown in Figure Exp. 11–2. If for some reason the press should be stopped during a cycle, the reset button can be used to return the press to the starting position.

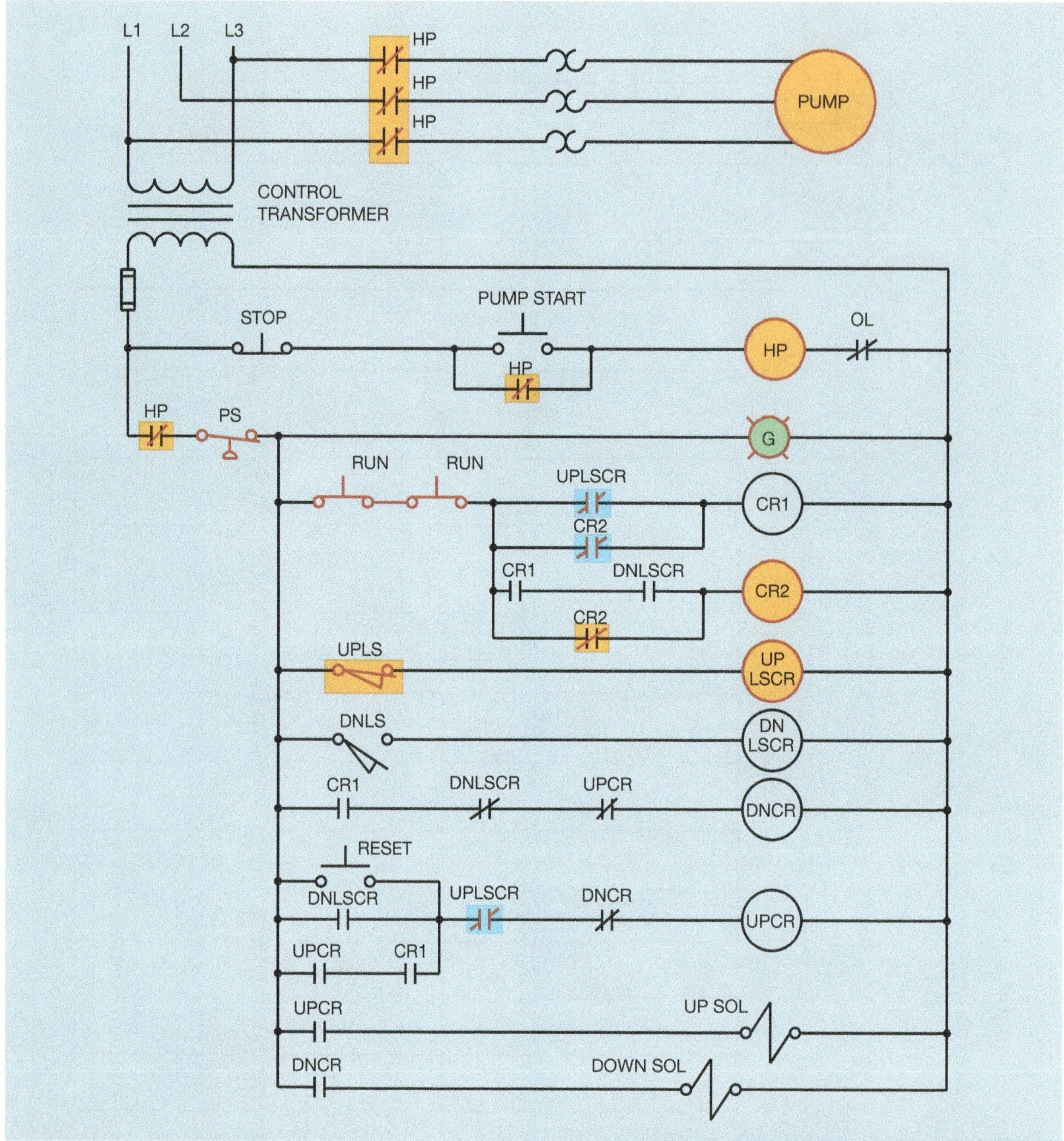

Figure Exp. 11–7 The press completes the cycle.

Connecting the Circuit

In this exercise, toggle switches will be used to simulate the action of the pressure switch and the two limit switches. Lights may also be substituted for the up and down solenoid coils.

❑ 1. Refer to the circuit shown in Figure Exp. 11–1. Count the number of contacts controlled by each of the control relays to determine which should be 11 pin and which should be 8 pin. Relays that need three contacts will have to be 11 pin, and relays that need two contacts may be 8 pin.

- ☐ 2. After determining whether a relay is to be 11 pin or 8 pin, identify the relay with some type of marker that can be removed later. Identifying the relays as CR1, CR2, and so on, can make connection much simpler.
- ☐ 3. Place the pin numbers on the schematic in Figure Exp. 11–1 to correspond with the contacts and coils of the control relays. Circle the numbers to distinguish them from wire numbers.
- ☐ 4. Place wire numbers beside each component on the schematic.
- ☐ 5. Connect the circuit. (Note: When connecting the two run push buttons, connect them close enough together to permit both to be held closed with one hand.)

Testing the Circuit

To test the circuit for proper operation:

- ☐ 1. Set the toggle switches used to simulate the pressure and down limit switch in the open (OFF) position. Set the toggle used to simulate the up limit switch in the closed (ON) position.
- ☐ 2. Press the "PUMP START" button and the motor or simulated motor load should start operating.
- ☐ 3. Close the pressure switch. The pilot light and UPLSCR relay should energize.
- ☐ 4. Press and hold down both of the run push buttons. Relays CR1 and DNCR should energize. The down solenoid should also turn on.
- ☐ 5. The press is now traveling in the down direction. Open the up limit switch. This should cause UPLSCR to de-energize. The down solenoid should remain turned on.
- ☐ 6. Close the down limit switch to simulate the press reaching the bottom of its stroke. DNLSCR, CR2, and UPCR should energize. The press is now starting to travel upward.
- ☐ 7. Open the down limit switch. DNLSCR should de-energize, but the UPCR should remain energized.
- ☐ 8. Close the up limit switch to simulate the press reaching the top of its stroke. The up solenoid should turn off. Control relay CR2 should remain on as long as the two run buttons are held closed.
- ☐ 9. To restart the cycle, release the run buttons and reclose them.

Review Questions

1. Assume that the hydraulic pump is running and the pilot light is turned on indicating that there is sufficient pressure to operate the press. Now assume that the up limit switch is not closed. What will be the action of the circuit if both run buttons are pressed?
2. Assume that the press is in the middle of its downstroke when the operator releases the two run push buttons. Explain the action of the circuit.

3. Referring to the condition of the circuit as stated in question 2, what would happen if the two run push buttons are pressed and held closed? Explain your answer.
4. Referring to the condition of the circuit as stated in question 2, what would happen if the reset button was pressed and held closed? Explain your answer.
5. Assume that the press traveled to the bottom of its stroke and then started back up. When it reached the middle of its stroke, the power was interrupted. After the power has been restored, if the two run buttons are pressed, will the press continue to travel upward to complete its stroke, or will it start moving downward?

Exercise 12

DESIGN OF TWO FLASHING LIGHTS

Objectives

After completing this exercise the student should be able to:

- Design a circuit from a written statement of requirements.
- Connect the circuit in the laboratory after the design has been approved.

Materials Needed

- Depends on the circuit design

In the space provided in Figure Exp. 12–1, draw a schematic diagram of a circuit that will fulfill the following requirements. Use two separate timers. Do not use an electronic timer set in the repeat mode. Remember that there is generally more than one way to design any circuit. Try to keep the design as simple as possible. The fewer components a circuit has, the less it is likely to fail.

- ☐ 1. An ON-OFF toggle switch is used to connect power to the circuit.
- ☐ 2. When the switch is turned on, two lights will alternately flash on and off. Light 1 will be turned on when light 2 is turned off. When light 1 turns off, light 2 will turn on.
- ☐ 3. The lights are to flash at a rate of on for 1 second and off for 1 second.

When the circuit design is completed, have your instructor approve it. After the design has been approved, connect it in the laboratory.

Figure Exp. 12–1 Design of two flashing lights.

Review Questions

1. When designing a control circuit that requires the use of a timing relay, what type of timer is generally used during the design?
2. Should schematic diagrams be drawn to assume that the circuit is energized or de-energized?
3. Explain the difference between a schematic and a wiring diagram.
4. In a forward–reverse control circuit, a normally closed F contact is connected in series with the R starter coil, and a normally closed R contact is connected in series with the F starter coil. What is the purpose of doing this and what is the contact arrangement called?
5. What type of overload relay is not sensitive to changes in ambient temperature?

Exercise 13

DESIGN OF THREE FLASHING LIGHTS

Objectives

After completing this exercise the student should be able to:

» Design a motor control circuit using timers.

» Discuss the operation of this circuit.

» Connect this circuit in the laboratory.

Materials Needed

- Depends on the design of the circuit

The design of this circuit will be somewhat similar to the circuit in Exercise 12. This circuit, however, contains three lights that turn on and off in sequence. Use the space provided in Figure Exp. 13–1 to design this circuit. The requirements of the circuit are as follows:

- ❑ 1. A toggle switch is used to connect power to the circuit. When the power is turned on, light 1 will turn on.
- ❑ 2. After a 1-second delay, light 1 will turn off and light 2 will turn on.
- ❑ 3. After a 1-second delay, light 2 will turn off and light 3 will turn on.
- ❑ 4. After a 1-second delay, light 3 will turn off and light 1 will turn back on.
- ❑ 5. The lights will repeat these actions until the toggle switch is opened.

Figure Exp. 13–1 Design of three lights that turn on and off in sequence.

Procedure

- ❑ 1. After the design of your circuit has been approved by your instructor, connect the circuit in the laboratory.
- ❑ 2. Turn on the power and test the circuit for proper operation.
- ❑ 3. **Turn off the power** and disconnect the circuit. Return the components to their proper location.

Review Questions

1. A 60 HP, three-phase squirrel cage induction motor is to be connected to a 480-volt line. What size NEMA starter should be used to make this connection?
2. An electrician is given a NEMA size 2 starter to connect a 30 HP, three-phase squirrel cage motor to a 575-volt line. Should this starter be used to operate this motor?
3. Assume that the motor in question 2 has a design code B. What standard size inverse time circuit breaker should be used to connect the motor?
4. The motor described in questions 2 and 3 is to be connected with copper conductors with type THHN insulation. What size conductors should be used? The termination temperature rating is not known.
5. Assume that the motor in question 2 has a nameplate current rating of 28 amperes and a marked service factor of 1. What size overload heater should be used for this motor?

Exercise 14

CONTROL FOR THREE PUMPS

Objectives

After completing this exercise the student should be able to:

- Analyze a motor control circuit.
- List the steps of operation in a control circuit.
- Connect this circuit in the laboratory.

Materials Needed

- Three-phase power supply
- Control transformer
- 3 ea.—three-phase motor starters with normally open auxiliary contacts
- 6 ea.—toggle switches to simulate auto-off-man switches and float switches
- 8 pin control relay and 8 pin socket
- 3 ea.—three-phase motors or equivalent motor loads
- One normally open and one normally closed push button

One of the primary duties of an industrial electrician is to troubleshoot existing control circuits. To troubleshoot a circuit, the electrician must understand what the circuit is designed to do and how it accomplishes it. To analyze a control circuit, start by listing the major components. Next, determine the basic function of each component, and finally determine what occurs during the circuit operation.

To illustrate this procedure, the circuit previously discussed in Exercise 11 will be analyzed. The hydraulic press circuit is shown in Figure Exp. 14–1. In order to facilitate circuit analysis, wire numbers have been placed beside the components. The first step will be to list the major components in the circuit.

- ☐ 1. Normally closed STOP push button
- ☐ 2. Normally open push button used to start the hydraulic pump
- ☐ 3. 2 ea.—normally open push buttons used as run buttons
- ☐ 4. Normally open push button used for the reset button

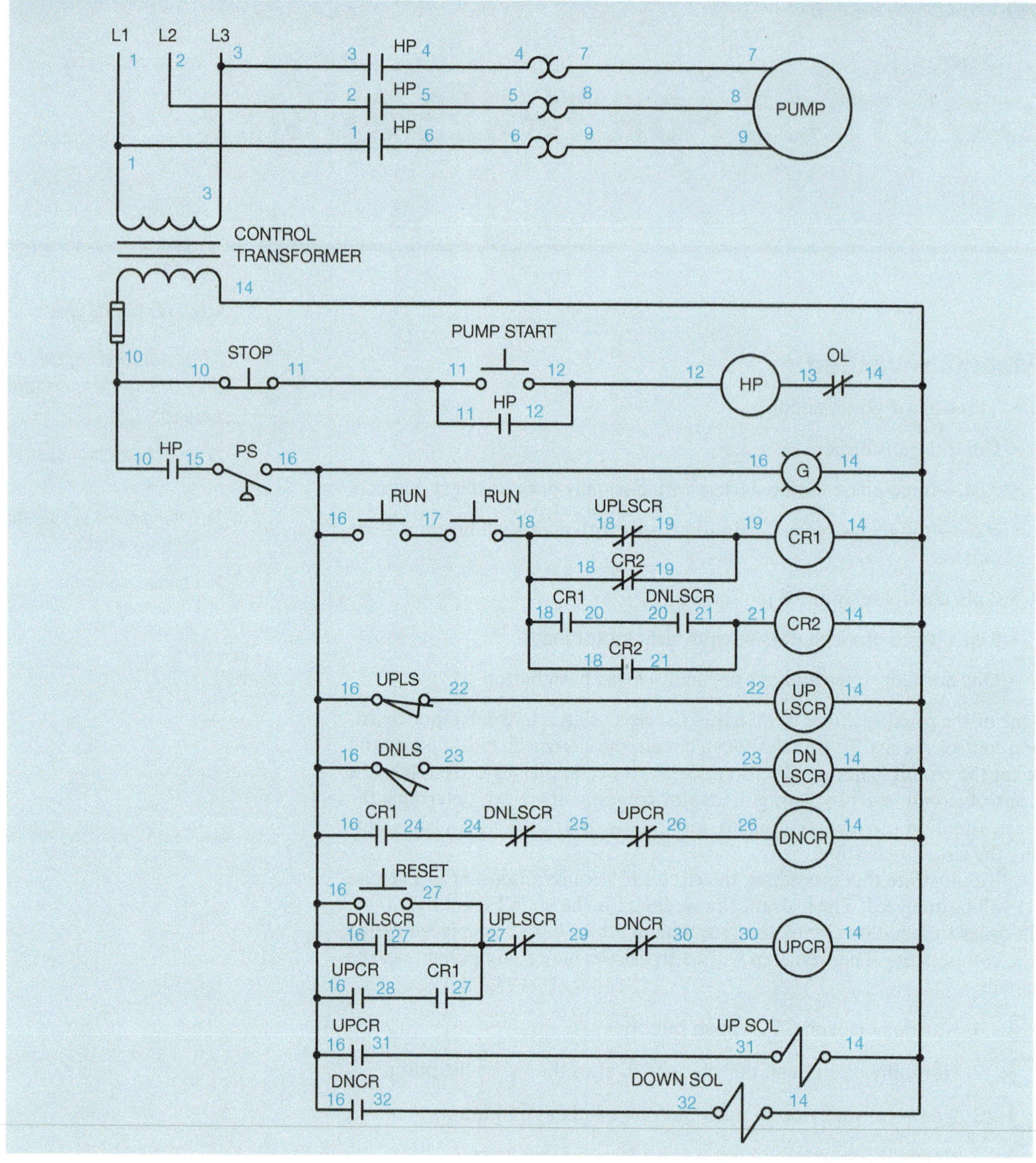

Figure Exp. 14–1 Analyzing the circuit.

- ❑ 5. Normally open pressure switch
- ❑ 6. 2 ea.—normally open limit switches
- ❑ 7. 2 ea.—solenoid valves
- ❑ 8. 3 ea.—8 pin control relays (CR2, UPLSCR, and DNCR)

❑ 9. 3 ea.—11 pin control relays (CR1, DNLSCR, and UPCR)

❑ 10. Control transformer

❑ 11. Green pilot light

The next step in the process is to give a brief description of the function of each listed component.

❑ 1. (Normally closed STOP push button) Used to stop the operation of the hydraulic pump motor.

❑ 2. (Normally open push button used to start the hydraulic pump) Starts the hydraulic pump.

❑ 3. (2 ea.—normally open push buttons used as run buttons) Both push buttons must be held down to start the action of the press.

❑ 4. (Normally open push button used for the reset button) Resets the press to the top-most position.

❑ 5. (Normally open pressure switch) Determines whether or not there is enough hydraulic pressure to operate the press.

❑ 6. (2 ea.—normally open limit switches) Determine when the press is at the top of its stroke or at the bottom of its stroke.

❑ 7. (2 ea.—solenoid valves) The up solenoid valve opens on energize to permit hydraulic fluid to move the press upward. The down solenoid valve opens on energize to permit hydraulic fluid to move the press downward.

❑ 8. (3 ea.—8 pin control relays, CR2, UPLSCR, and DNCR) Part of the control circuit.

❑ 9. (3 ea.—11 pin control relays, CR1, DNLSCR, and UPCR) Part of the control circuit.

❑ 10. (Control transformer) Reduces the value of the line voltage to the voltage needed to operate the control circuit.

❑ 11. (Green pilot light) Indicates there is enough hydraulic pressure to operate the pump.

The final step is to analyze the operation of the circuit. To analyze circuit operation, trace the current paths each time a change is made in the circuit. Start by pressing the pump START button.

❑ 1. When the pump START button is pressed, a circuit is completed to the coil of starter HP.

❑ 2. When coil HP energizes, all HP contacts change position. The three load contacts close to connect the pump motor to the line. The HP auxiliary contact located between wire points 11 and 12 closes to maintain the circuit after the pump START button is released, and the HP auxiliary contact located between wire numbers 10 and 15 closes to provide power to the rest of the circuit.

❑ 3. After the hydraulic pump starts, the hydraulic pressure in the system increases and closes the pressure switch.

- ☐ 4. When the pressure switch closes, a current path is provided to the green pilot light to indicate that there is sufficient hydraulic pressure to operate the press. A current path also exists through the normally open held closed up limit switch to control relay coil UPLSCR.
- ☐ 5. When UPLSCR energizes, both UPLSCR contacts open. The UPLSCR contact located between wire numbers 18 and 19 opens to break a current path to CR1 coil. UPLSCR contact located between wire numbers 27 and 29 opens to break the current path to coil UPCR.
- ☐ 6. Both run push buttons must be held down to provide a current path through the normally closed CR2 contact located between wire numbers 18 and 19 to the coil of CR1 relay.
- ☐ 7. When CR1 relay coil energizes, the CR1 contact located between wire numbers 18 and 20 closes to provide a path to CR2 coil in the event that the DNLSCR contact should close. The CR1 contact located between wire numbers 16 and 24 closes to provide a current path to the down control relay (DNCR). The CR1 contact located between wire numbers 28 and 27 closes to provide an eventual current path to the up control relay (UPCR).
- ☐ 8. When DNCR coil energizes, the DNCR contact located between wire numbers 29 and 30 opens to provide interlock with the UPCR. The DNCR contact between wire numbers 16 and 32 closes and provides a current path to the down solenoid valve.
- ☐ 9. When the down solenoid valve energizes, the press begins its downward stroke. This causes the normally open held closed up limit switch to open and de-energize the coil of the up limit switch control relay (UPLSCR).
- ☐ 10. Both UPLSCR contacts reclose.
- ☐ 11. When the press reaches the bottom of its stroke, the down limit switch located between wire numbers 16 and 23 closes to provide a current path to the coil of the down limit switch control relay (DNLSCR).
- ☐ 12. All DNLSCR contacts change position. The DNLSCR contact located between wire numbers 20 and 21 closes to provide a current path through the now closed CR1 contact to the coil of CR2 relay. The DNLSCR contact located between wire numbers 24 and 25 opens and breaks the current path to DNCR relay. The DNLSCR contact located between wire numbers 16 and 27 closes to provide a current path to UPCR relay when the DNCR contact located between 29 and 30 recloses.
- ☐ 13. When CR2 coil energizes, the normally closed CR2 contact located between wires 18 and 19 opens to prevent a maintained current path to CR1 when the UPLSCR contact reopens. The normally open CR2 contact located between 18 and 21 closes to maintain a current path to the coil of CR2 in the event that CR1 or DNLSCR contacts should open.
- ☐ 14. When the DNCR relay coil de-energizes, the DNCR contact located between wires 29 and 30 recloses to permit coil UPCR to be energized. The DNCR contact located between 16 and 32 reopens to break the current path to the down solenoid valve.

- ❑ 15. When the UPCR coil energizes, the normally closed UPCR contact located between wires 25 and 26 opens to provide interlock with the DNCR relay coil. The UPCR contact located between 16 and 28 closes to maintain a circuit through the now closed CR1 contact to the coil of UPCR. The UPCR contact located between 16 and 31 closes and provides a current path to the up solenoid valve.
- ❑ 16. When the up solenoid valve opens, hydraulic fluid causes the press to begin its upward stroke.
- ❑ 17. When the press starts upward, the down limit switch reopens and de-energizes the coil of DNLSCR relay.
- ❑ 18. When coil DNLSCR de-energizes, the DNLSCR contact located between wires 20 and 21 reopens, but a current path is maintained by the now closed CR2 contact. The DNLSCR contact located between 24 and 25 recloses, but the current path to DNCR coil remains broken by the UPCR contact located between 25 and 26. The DNLSCR contact located between wires 16 and 27 reopens, but a current path is maintained by the now closed UPCR and CR1 contacts.
- ❑ 19. When the press reaches the top of its stroke, the up limit switch again closes and provides a current path to the coil of UPLSCR relay.
- ❑ 20. The UPLSCR contact located between wires 18 and 19 opens to break the current path to CR1 coil. The UPLSCR contact located between wires 27 and 29 opens to break the current path to the coil of UPCR.
- ❑ 21. When CR1 coil de-energizes, all CR1 contacts return to their normal position. The CR1 contact between wires 18 and 20 reopens, CR1 contact between wires 16 and 24 reopens to prevent a current path from being established to the DNCR relay coil, and CR1 contact between wires 27 and 28 reopens.
- ❑ 22. When coil UPCR de-energizes, its contacts return to their normal position. The UPCR contact located between wires 16 and 28 reopens, and the UPCR contact located between wires 16 and 31 reopens to break the circuit to the up solenoid.
- ❑ 23. Before the circuit can be restarted, the current path to relay CR2 must be broken by releasing one or both of the run push buttons. This will return all contacts back to their original state.
- ❑ 24. In the event the press should be stopped in the middle of its stroke, the up limit switch will be open and coil UPLSCR will be de-energized. The DNCR coil will also be de-energized. If the reset button is pressed and held, a circuit will be completed through the normally closed DNLSCR and DNCR contacts to the coil of UPCR. This will cause the up solenoid valve to energize and return the press to its up position.

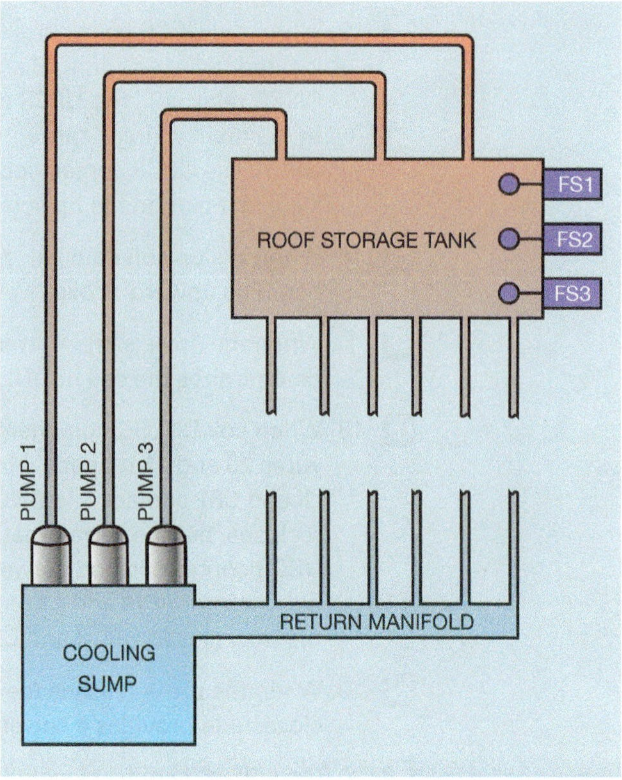

Figure Exp. 14–2 Roof-mounted tank for plant cooling system.

Determining What the Circuit Does

The circuit in this experiment is intended to operate three pumps. The pumps are used to pump water from a sump to a roof storage tank. The water in the storage tank is used for cooling throughout the plant. After the water has been used for cooling, it returns to the sump to be recooled. Three float switches are used to detect the water level in the storage tank. As the water is drained out of the tank, the level drops and the float switches turn on the pumps to pump water from the sump back to the storage tank (Figure Exp. 14–2).

List the Components

In the space provided, list the major components in the control circuit shown in Figure Exp. 14–3.

1. ______________________________

2. ______________________________

3. ______________________________

4. ______________________________

5. ______________________________

6. ______________________________

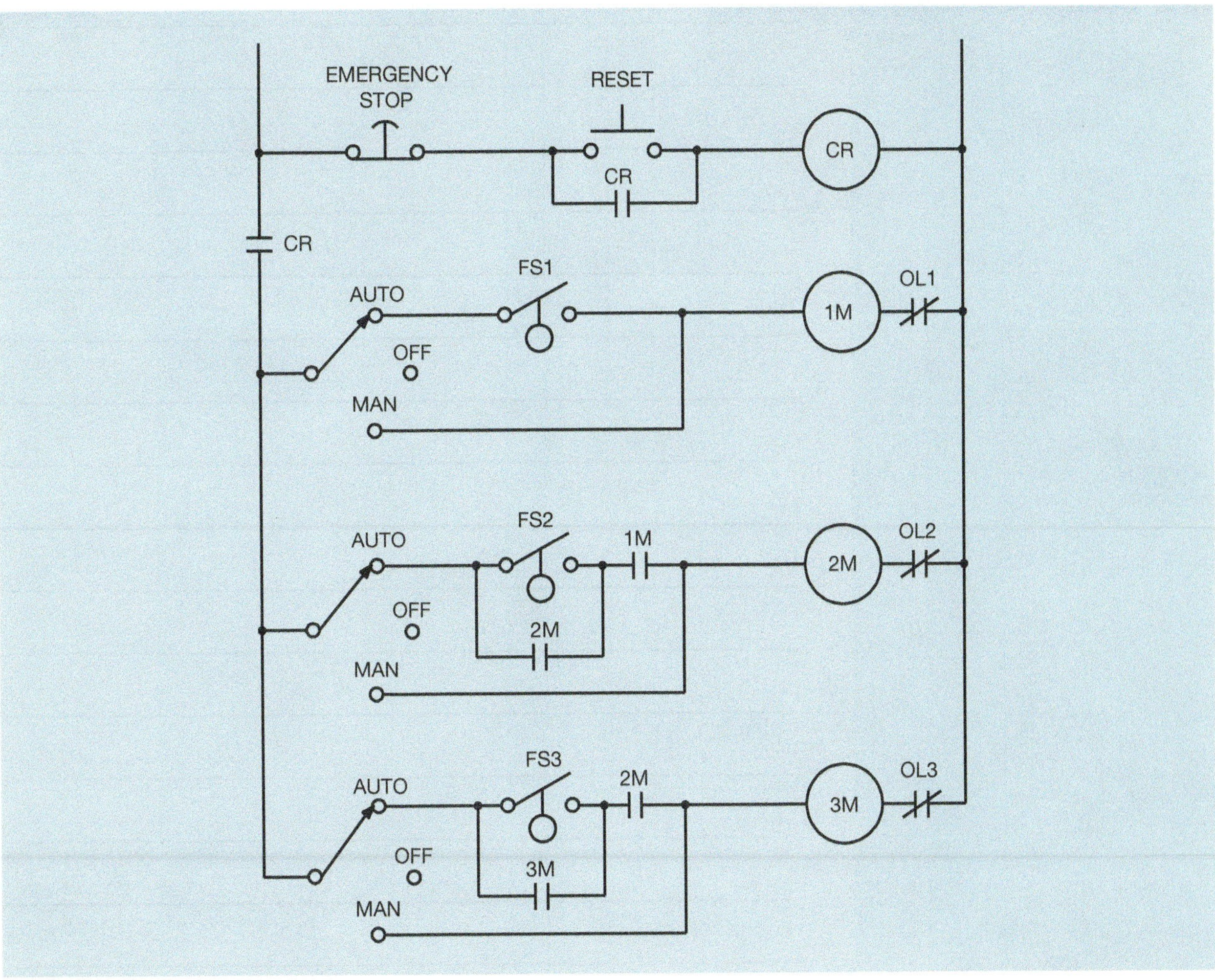

Figure Exp. 14–3 Control circuit for three pumps.

7. ______________________________

8. ______________________________

9. ______________________________

10. ______________________________

Describe the Components

In the space provided below give a brief description of the function of the components in this circuit.

1. ______________________________

____

____

2. ______________________________

____

____

3. ______________________________

____

____

4. ______________________________

____

____

5. ______________________________

____

____

6. ______________________________

____

____

7. ______________________________

____

____

8. ______________________________

____

____

9. ______________________________

____

____

10. ______________________________

____

____

Describing the Circuit Operation

In the space provided below, describe the operation of the circuit. Assume that in the normal state the roof storage tank is filled with water, and all the auto-off-man switches are set in the auto position. Also assume that the three motor starters control the operation of the three pumps, although the pumps are not shown on the schematic.

1. ____

2. ____

3. ____

4. ____

5. ____

6. ____

7. ____

8. ____

____

____

9. ____

____

____

10. ____

____

____

11. ____

____

____

12. ____

____

____

13. ____

____

____

14. ____

____

____

15. ____

____

____

Review Questions

To answer the following questions, refer to the circuit shown in Figure Exp. 14–3.

1. Assume that all three pumps are operating. What would be the action of the circuit if the auto-off-man switch of pump 2 were to be switched to the off position?
2. Assume that the auto-off-man switch of pump 3 is set in the manual position. What will be the operation of the circuit if float switch FS1 closes?
3. Assume that the roof storage tank empties completely, but none of the pumps have started. Which of the following could *not* cause this condition?
   a. The emergency STOP button has been pushed and the control relay is de-energized.
   b. The auto-off-man switch of pump 1 has been set in the off position.
   c. The auto-off-man switch of pump 1 has been set in the manual position.
   d. 1M coil is open.
4. Assume that all three pumps are in operation and OL3 contact opens. Will this affect the operation of the other two pumps?
5. Assume that FS2 float switch is defective. If the water level drops enough to close float switch FS3, will pump 3 start running?

Exercise 15

OIL PRESSURE PUMP CIRCUIT FOR A COMPRESSOR

Objectives

After completing this exercise the student should be able to:

- Analyze a motor control circuit.
- List the steps of operation in a control circuit.
- Connect this circuit in the laboratory.

Materials Needed

- Three-phase power supply
- 2 ea.—three-phase motor starters
- Control transformer
- 2 ea.—electronic timers (Dayton model 6A855) and 11 pin tube sockets
- 2 ea.—pilot lights
- 2 ea.—double-acting push buttons

In the circuit shown in Figure Exp. 15–1, the oil pump must start for some time before the compressor is started. When the START button is pressed, the oil pump should continue to run for some time after the compressor stops operating.

Listing the Components

In the space provided below, list the circuit components.

1. ____

2. ____

3. ____

4. ____

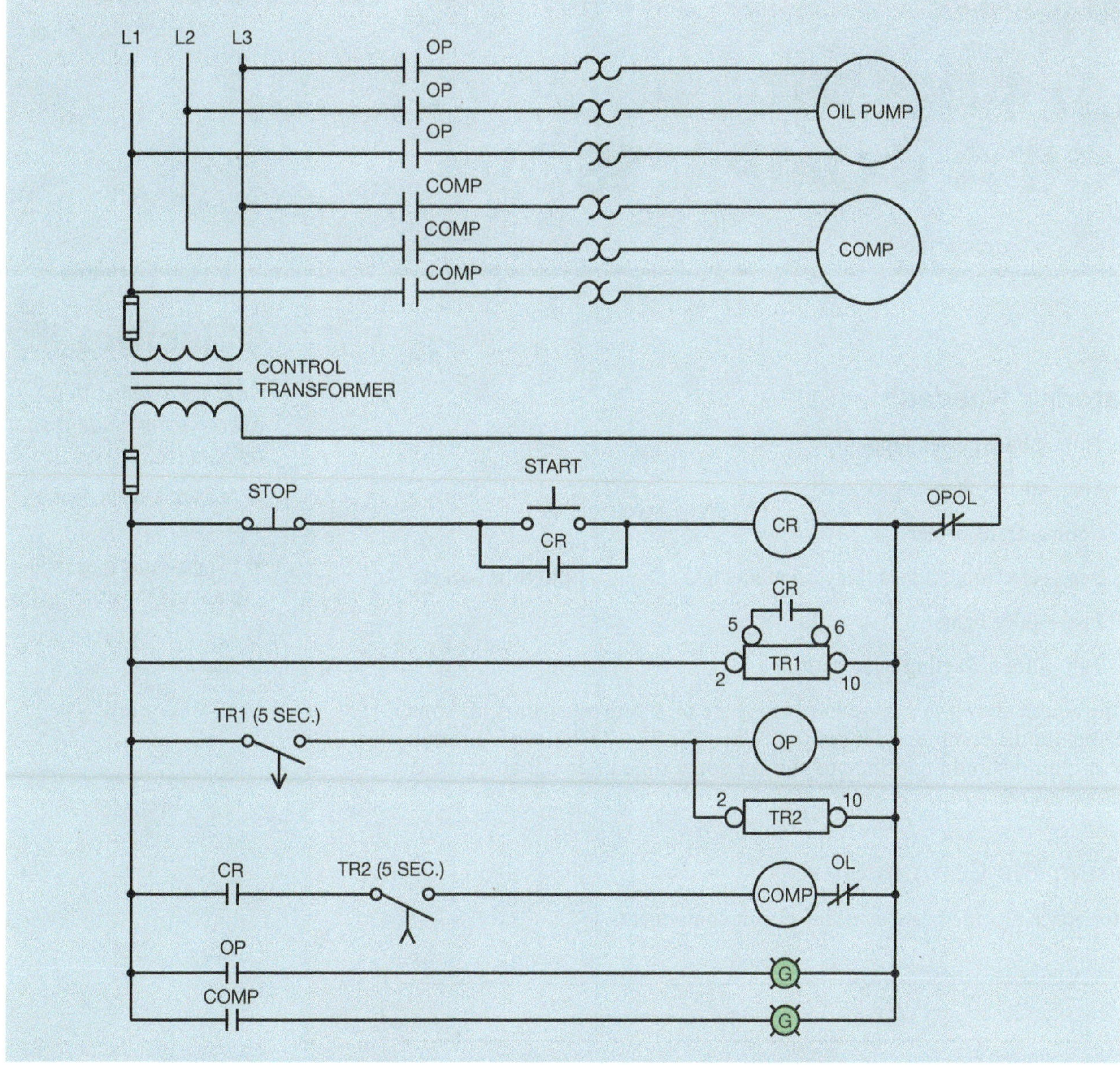

Figure Exp. 15–1 Compressor oil pump circuit.

5. ______________________________

6. ______________________________

7. ______________________________

8. ______________________________

9. ______________________________

10. ______________________________

Describe the Components

In the space provided below, give a brief description of what function each component performs.

1. ____

____

2. ____

____

3. ____

____

4. ____

____

5. ____

____

6. ____

____

7. ____

____

8. ____

____

9. ____

____

10. ____

____

Circuit Operation

In the space provided below, describe the operation of the circuit in a step-by-step sequence.

1. ____

____

2. ____

____

3. ____

____

4. ____

____

5. ____

____

6. ____

____

7. ____

____

8. ____

____

9. ____

____

10. ____

____

11. ____

____

12. ____

____

13. ____

____

14. ____

____

15. ____

____

Connecting the Circuit

- ☐ 1. Connect the circuit shown in Figure Exp. 15–1.
- ☐ 2. Turn on the power and test the circuit for proper operation.
- ☐ 3. **Turn off the power** and disconnect the circuit. Return the components to their proper location.

Review Questions

To answer the following questions, refer to the circuit shown in Figure Exp. 15–1.

1. Assume that the START button is pressed and the oil pump starts operating. After a 5-second delay, the COMP pilot light turns on, but the compressor motor does not start. Which of the following could cause this condition?
   a. TR2 timer is defective.
   b. COMP starter coil is defective.
   c. The compressor motor is defective.
   d. All the above.
2. Assume that the circuit is in operation. When the STOP button is pressed, both the compressor and oil pump stop operating immediately. Which of the following could cause this condition?
   a. CR relay is defective.
   b. TR1 timer is defective.
   c. OP starter is defective.
   d. Timer TR2 is defective.
3. When the START button is pressed, the oil pump starts operating immediately. After a 5-second delay, the oil pump motor turns off. An electrician finds that the control transformer fuse is blown. Which of the following could cause this condition?
   a. TR1 coil is shorted.
   b. OP coil is shorted.
   c. TR2 coil is shorted.
   d. COMP coil is shorted.

4. When the START button is pressed, the oil pump motor starts operating immediately. After a long time delay, it is determined that the compressor motor will not start. Which of the following could *not* cause this condition?
   a. OP coil is defective.
   b. TR2 coil is defective.
   c. COMP coil is defective.
   d. The compressor overload contact is open.

5. When the START button is pressed, the oil pump motor starts operating immediately. When the START button is released, however, the oil pump motor turns off. The operator then presses the START button and holds it down for 10 seconds. This time the oil pump motor starts operating immediately, but the compressor motor never starts. When the START button is released, the oil pump motor again immediately turns off. Which of the following could cause this condition?
   a. CR coil is defective.
   b. TR1 coil is defective.
   c. TR2 coil is defective.
   d. COMP coil is defective.

Exercise 16

AUTOTRANSFORMER STARTER

Objectives

After completing this experiment the student should be able to:

- Discuss the operation of an autotransformer starter.
- Explain the operation of an autotransformer starter.
- Connect an autotransformer starter in the laboratory.

Materials Needed

- Three-phase power supply
- Control transformer
- 3 ea.—three-phase contactors with at least one normally open and one normally closed auxiliary contact
- 2 ea.—0.5 kVA control transformer (480/240–120)
- Three-phase motor or equivalent motor load
- On-delay timer—Dayton model 6A855 or equivalent and 11 pin tube socket
- 8 pin control relay and 8 pin tube socket
- 2 ea.—double-acting push buttons (NO/NC on each button)
- Three-phase overload relay or three single-phase overload relays with the overload contacts connected in series

Autotransformer starters are used to reduce the amount of in-rush current when starting a large motor. The autotransformer starter accomplishes this by reducing the voltage applied to the motor during the starting period. If the voltage is reduced by one-half, the current will be reduced by one-half, and the torque will be reduced to one-fourth of normal.

There are several different ways to construct an autotransformer starter. Some use three transformers, and others use two transformers. In this exercise, two transformers connected as an open delta will be used. Two 0.5 kVA control transformers will be employed. Because these transformers are to be used as autotransformers, only the high-voltage windings will be connected. The low-voltage windings (X1 and X2) will not be used in this experiment. The high-voltage windings can be identified by the markings on the terminal leads of H1 through H4. These high-voltage windings are to be connected in series by connecting a jumper between terminals H2 and H3. This jumpered point provides a center tap for the entire winding.

Obtaining Enough Contacts

A schematic diagram of this connection is shown in Figure Exp. 16–1. Notice that a total of five S contacts are needed during the starting period. Contactors that contain five load contacts can be purchased, but they are difficult to obtain and expensive. For this reason, two three phase contactors will be employed to provide the needed load contacts. This can be accomplished by connecting the coils of S1 and S2 contactors in parallel with each other.

Circuit Operation

When the START button is pressed, coils CR, TR, S1, and S2 energize. When the S1 and S2 load contacts close, the motor is connected to the center tap of the

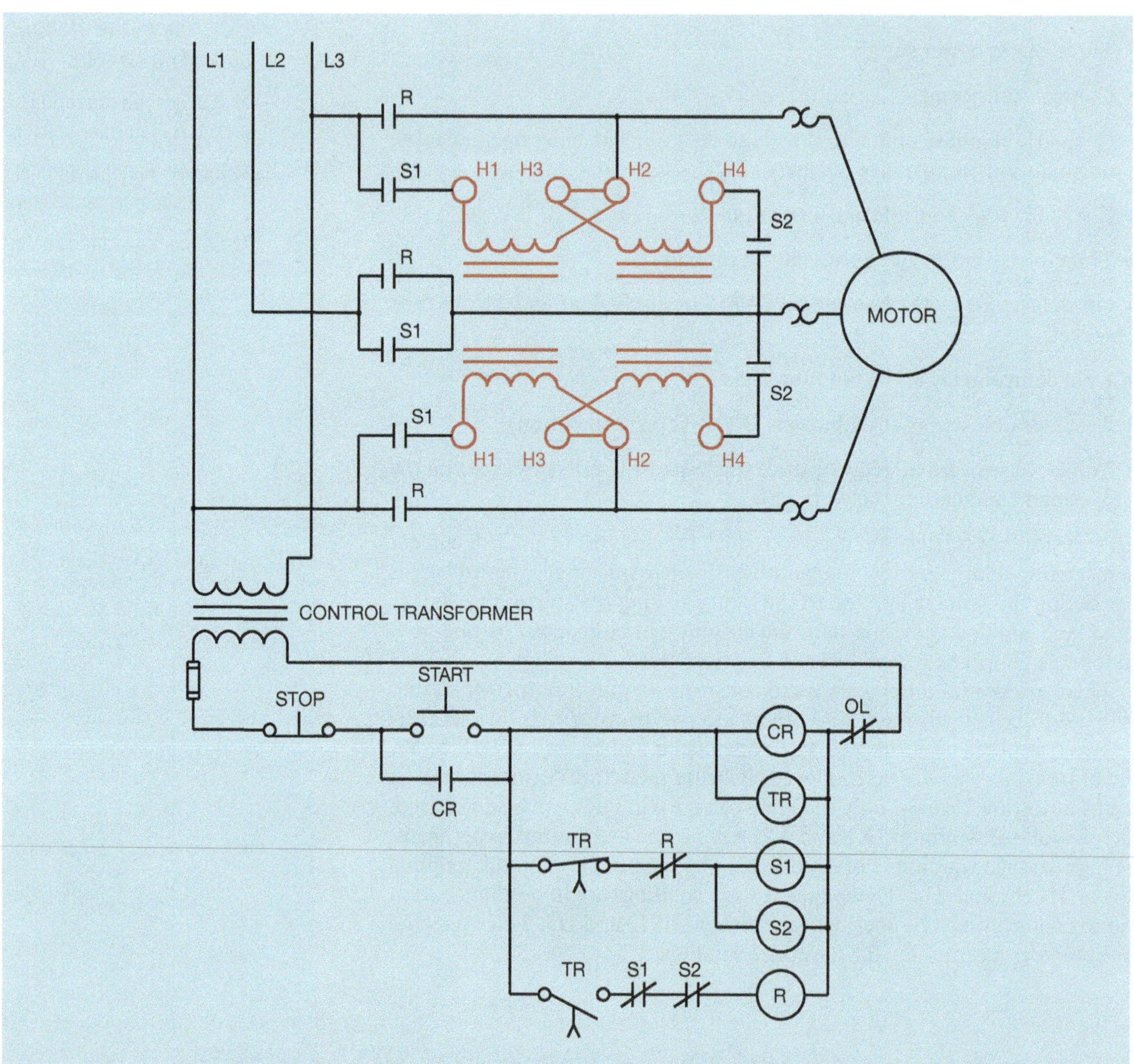

Figure Exp. 16–1 Autotransformer starter.

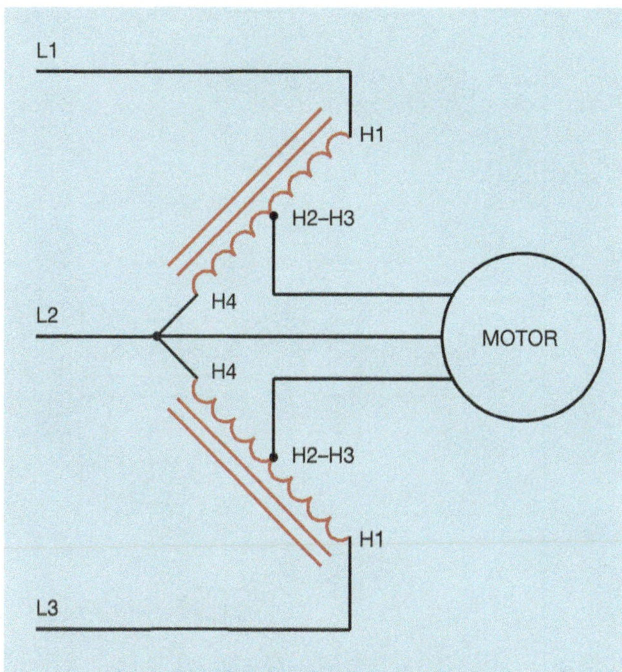

Figure Exp. 16–2 Schematic diagram of basic autotransformer connection.

open delta autotransformer. Because the transformers have been center tapped, the motor is connected to one-half of the line voltage. A basic schematic diagram of this connection is shown in Figure Exp. 16–2. The normally closed S1 and S2 auxiliary contacts connected in series with R coil open to provide interlock and prevent R contactor from energizing as long as S1 or S2 is energized.

After a period of time, TR timer reaches the end of its timing sequence and the two timed TR contacts change position. The normally closed TR contact connected in series with coils S1 and S2 opens and de-energizes these contactors. This causes all S1 and S2 load contacts to open and disconnect the autotransformer from the line. The normally closed S1 and S2 auxiliary contacts connected in series with R coil are reclosed.

When the normally open TR contact connected in series with R coil closes, R contactor energizes and closes all R load contacts. This connects the motor directly to the power line. The normally closed R auxiliary contact connected in series with coils S1 and S2 opens to provide interlock. The motor will continue to run until the STOP button is pressed or an overload occurs.

Connecting the Circuit

❑ 1. Assuming that relay CR is an 8 pin control relay, and that timer TR is a Dayton model 6A855, place pin numbers beside the components of CR and TR in Figure Exp. 16–1. Circle the pin numbers to distinguish them from wire numbers.

❑ 2. Place wire numbers beside all circuit components in Figure Exp. 16–1.

❑ 3. Place corresponding wire numbers beside the components shown in Figure Exp. 16–3. Make certain to make the connection between H2 and H3 on the high voltage side of the control transformers.

❑ 4. Connect the control section of the circuit shown in Figure Exp. 16–1.

❑ 5. Set the timing relay for a 5-second delay.

❑ 6. After your instructor has inspected the circuit for proper operation, turn on the power and test the control section of the circuit for proper operation.

❑ 7. **Turn off the power.**

❑ 8. Connect the load section of the circuit.

❑ 9. Turn on the power and test the circuit for proper operation. (Note: Connect a voltmeter across the motor or equivalent motor load terminals and monitor the voltage. When the circuit is first energized, the voltage applied to the motor should be one-half the full line value. After a 5-second delay, the voltage should increase to full value.)

❑ 10. **Turn off the power** and disconnect the circuit. Return the components to their proper place.

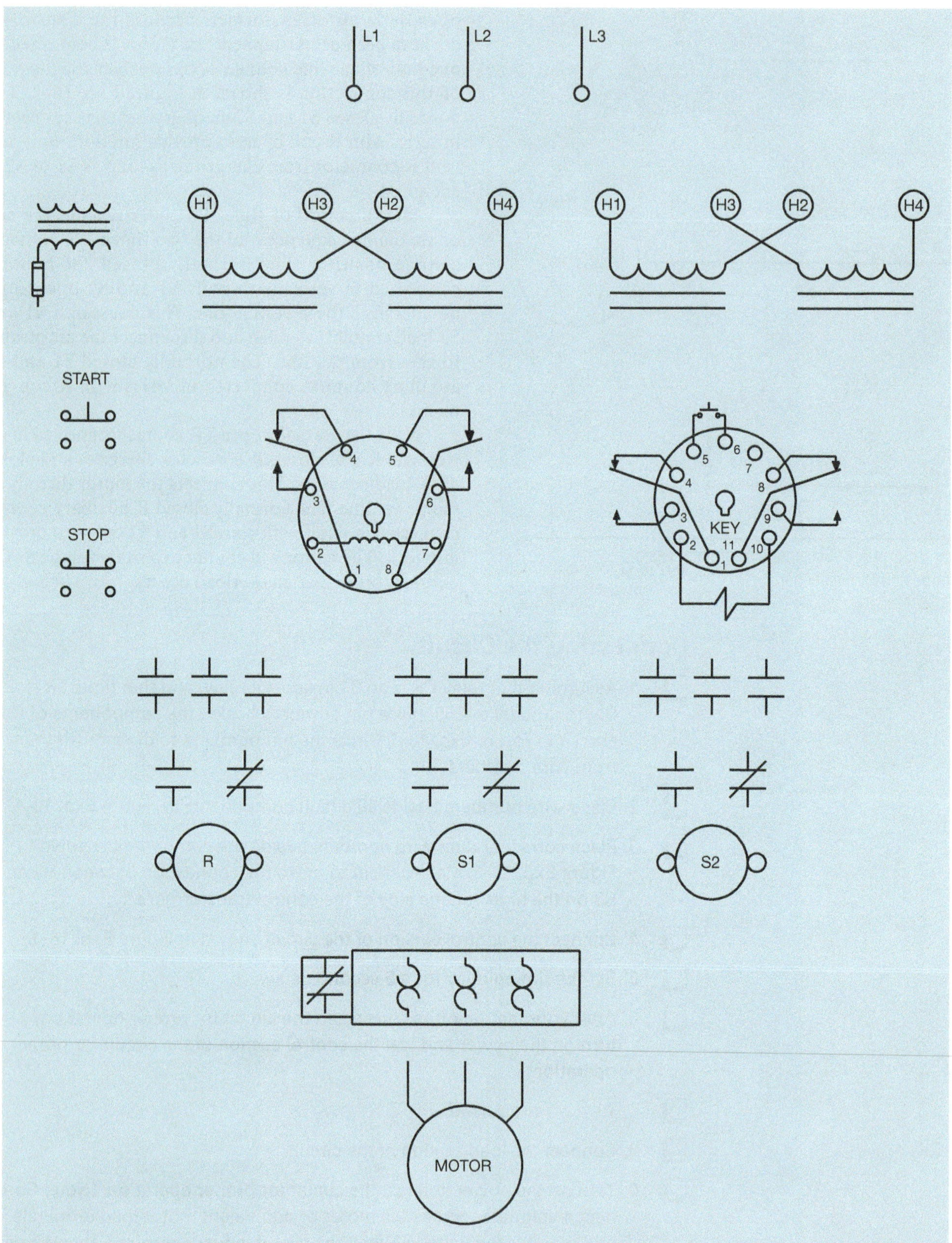

Figure Exp. 16–3 Developing a wiring diagram.

Review Questions

1. How does the autotransformer reduce the amount of starting current to a motor?
2. Is the autotransformer used in this exercise connected as a wye, delta, or open delta?
3. What is the advantage, if any, of using an open delta connection as opposed to a closed delta or wye?
4. Assume that the line-to-line voltage in Figure Exp. 16–1 is 480 volts. Also assume that when the START button is pressed, the motor starts with 240 volts applied to the motor. When the START button is released, however, the motor stops running. Which of the following could cause this problem?
   a. S1 coil is open.
   b. CR coil is open.
   c. TR coil is open.
   d. The STOP push button is open.
5. Refer to the circuit shown in Figure Exp. 16–1. When the START button is pressed, nothing happens for a period of 5 seconds. After 5 seconds, the motor suddenly starts with full voltage connected to it. Which of the following could cause this problem?
   a. CR coil is open.
   b. TR coil is open.
   c. R coil is open.
   d. R normally closed auxiliary contact is open.

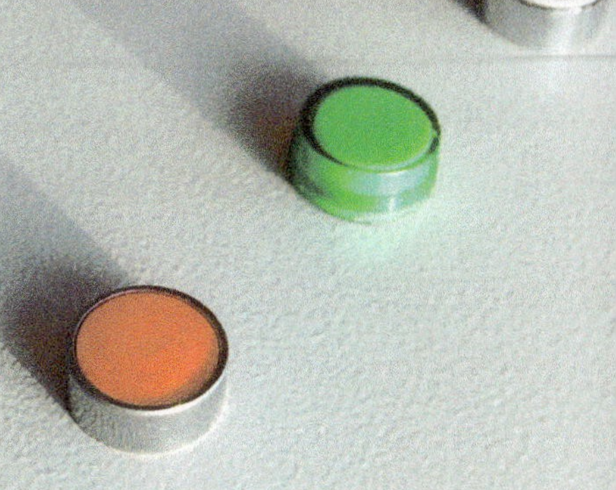

APPENDIX

Identifying the Leads of a Three Phase, Wye-Connected, Dual-Voltage Motor

The terminal markings of a three-phase motor are standardized and used to connect the motor for operation on 240 or 480 volts. Figure A–1 shows these terminal markings and their relationship to the other motor windings. If the motor is to be connected to a 240-volt line, the motor windings are connected parallel to each other as shown in Figure A–2. If the motor is to be operated on a 480-volt line, the motor windings are connected in series as shown in Figure A–3.

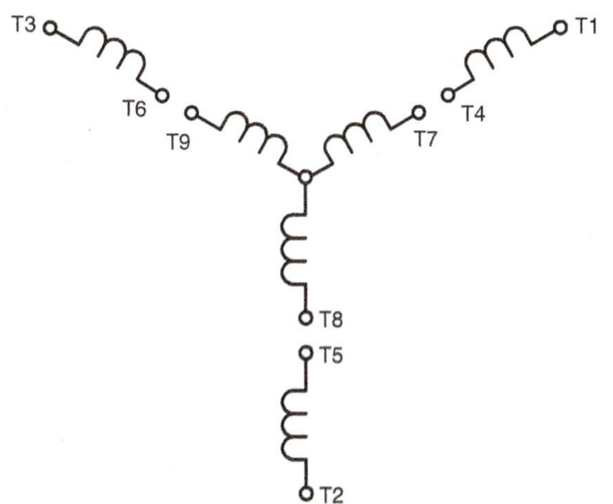

Figure A–1 Standard terminal markings for a three-phase motor.

As long as these motor windings remain marked with the proper numbers, connecting the motor for operation on a 240- or 480-volt power line is relatively simple. If these numbers are removed or damaged, however, the leads must be reidentified before the motor can be connected. The following procedure can be used to identify the proper relationship of the motor windings.

1. Using an ohmmeter, divide the motor windings into four separate circuits. One circuit will have continuity to three leads, and the other three circuits will have continuity between only two leads (Figure A–1). *Caution: the circuits that exhibit continuity between two leads must be identified as pairs, but do not let the ends of the leads touch anything.*
2. Mark the three leads that have continuity with each other as T7, T8, and T9. Connect these three leads to a 240-volt, three-phase power source (Figure A–4). (Note: Since these windings are rated at 240 volts each, the motor can be safely operated on one set of windings as long as it is not connected to a load.)
3. With the power turned off, connect one end of one of the paired leads to the terminal marked T7. Turn the power on, and using an AC voltmeter set for a range not less than 480 volts, measure the voltage from the unconnected end of the paired lead to terminals T8 and T9 (Figure A–5). If the measured voltages are unequal, the wrong paired lead is connected to terminal T7. Turn the power off, and connect another paired lead to T7. When the correct set of paired leads is connected to T7, the voltage readings to T8 and T9 will be equal.

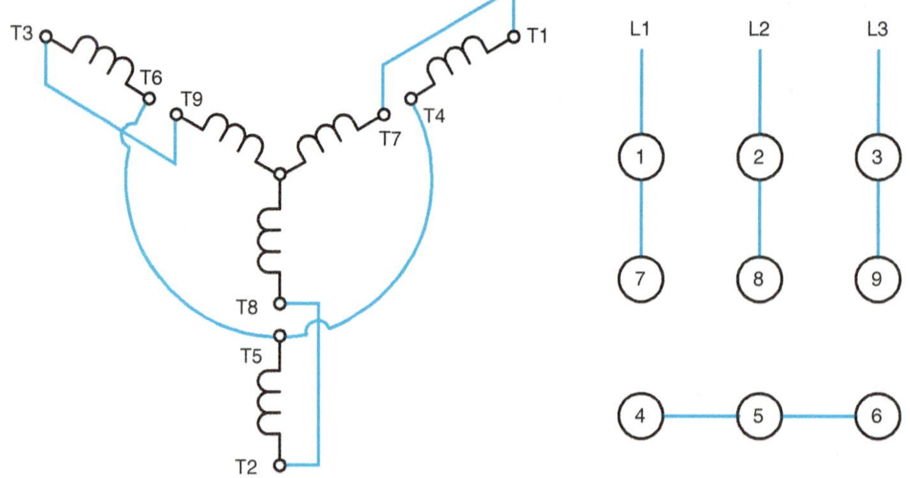

Figure A–2 Low voltage connection.

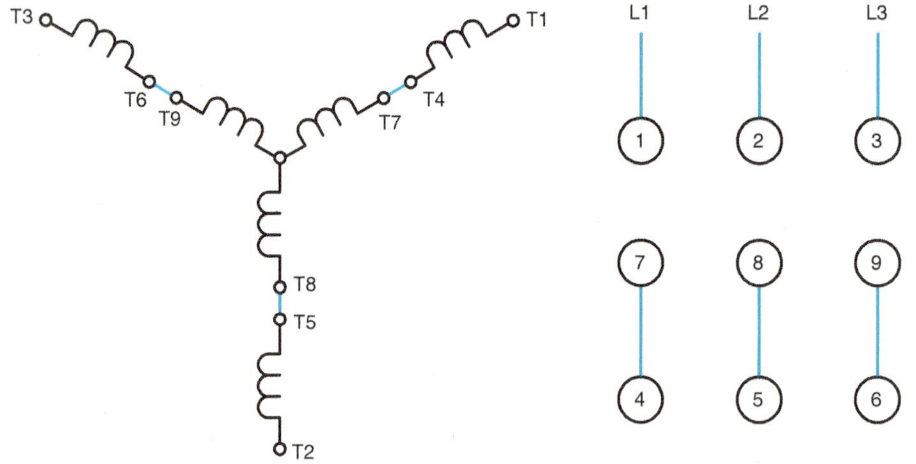

Figure A–3 High voltage connection.

4. After finding the correct pair of leads, a decision must be made as to which lead should be labeled T4 and which should be labeled T1. Because an induction motor is basically a transformer, the phase windings act very similar to a multiwinding autotransformer. If terminal T1 is connected to terminal T7, it will operate similar to a transformer with its windings connected to form subtractive polarity. If an AC voltmeter is connected to T4, a voltage of about 140 volts should be seen between T4 and T8 or T4 and T9 (Figure A–6).

   If terminal T4 is connected to T7, the winding will operate similar to a transformer with its windings connected for additive polarity. If an AC voltmeter is connected to T1, a voltage of about 360 volts will be indicated when the other lead of the voltmeter is connected to T8 or T9 (Figure A–7).

   Label leads T1 and T4 using the preceding procedure to determine which lead is correct. Then disconnect and separate T1 and T4.

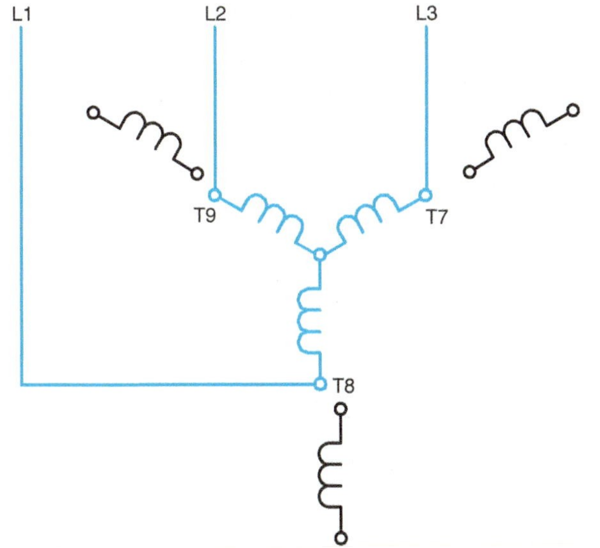

Figure A–4 T7, T8, and T9 connected to a three phase, 240-volt line.

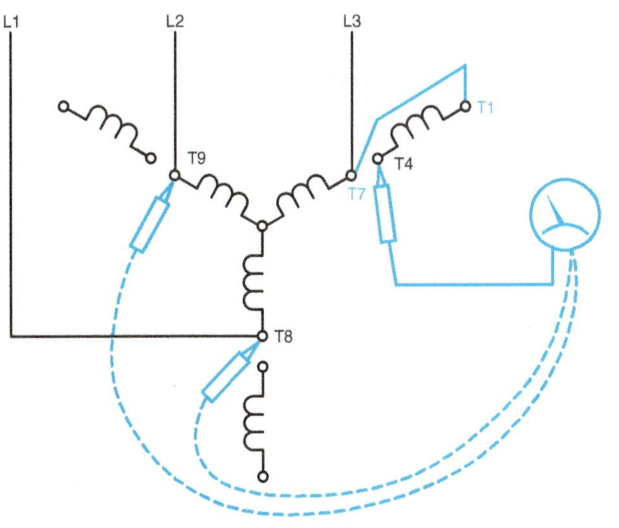

Figure A–6 T1 connected to T7.

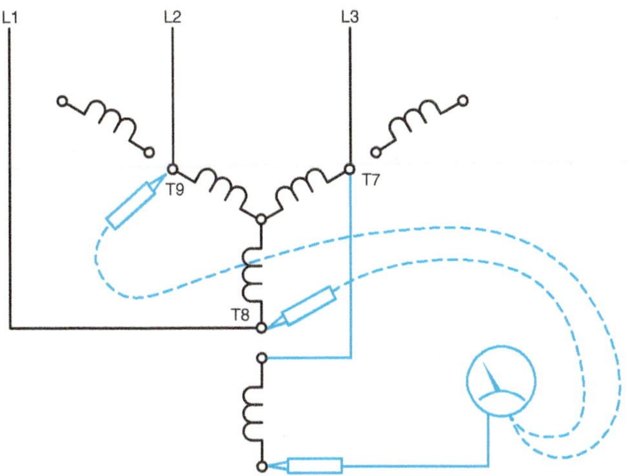

Figure A–5 Measure voltage from unconnected paired lead to T8 and T9.

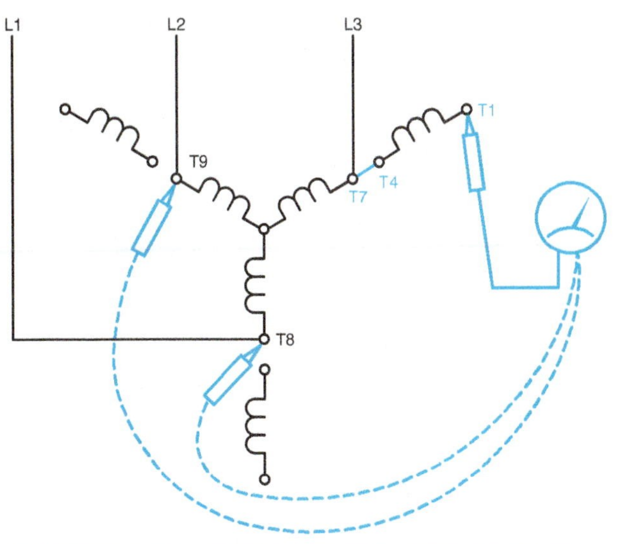

Figure A–7 T4 connected to T7.

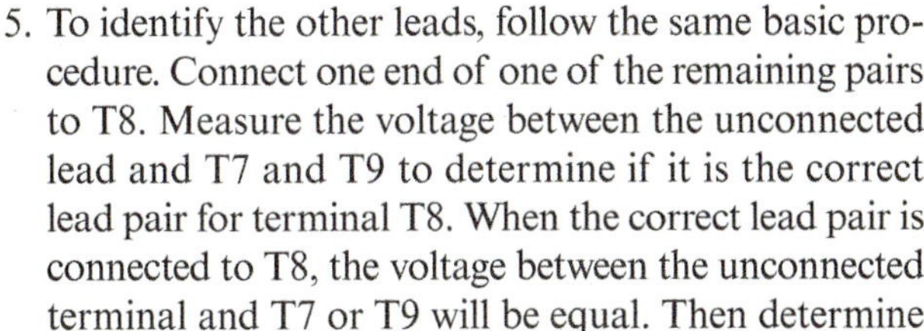

5. To identify the other leads, follow the same basic procedure. Connect one end of one of the remaining pairs to T8. Measure the voltage between the unconnected lead and T7 and T9 to determine if it is the correct lead pair for terminal T8. When the correct lead pair is connected to T8, the voltage between the unconnected terminal and T7 or T9 will be equal. Then determine which is T5 or T2 by measuring for a high or low voltage. When T5 is connected to T8, about 360 volts can be measured between T2 and T7 or T2 and T9.

6. The remaining pair can be identified as T3 or T6. When T6 is connected to T9, a voltage of about 360 volts can be measured between T3 and T7 or T3 and T8.

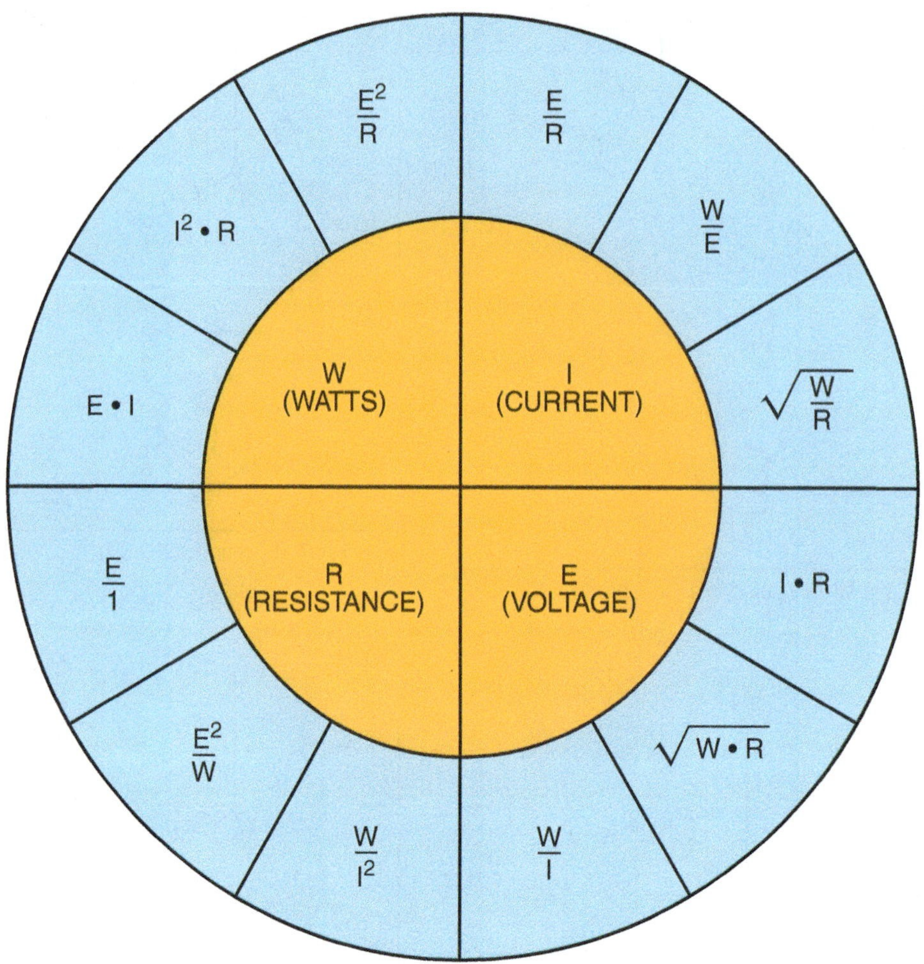

Figure A–8 Ohm's Law formulas.

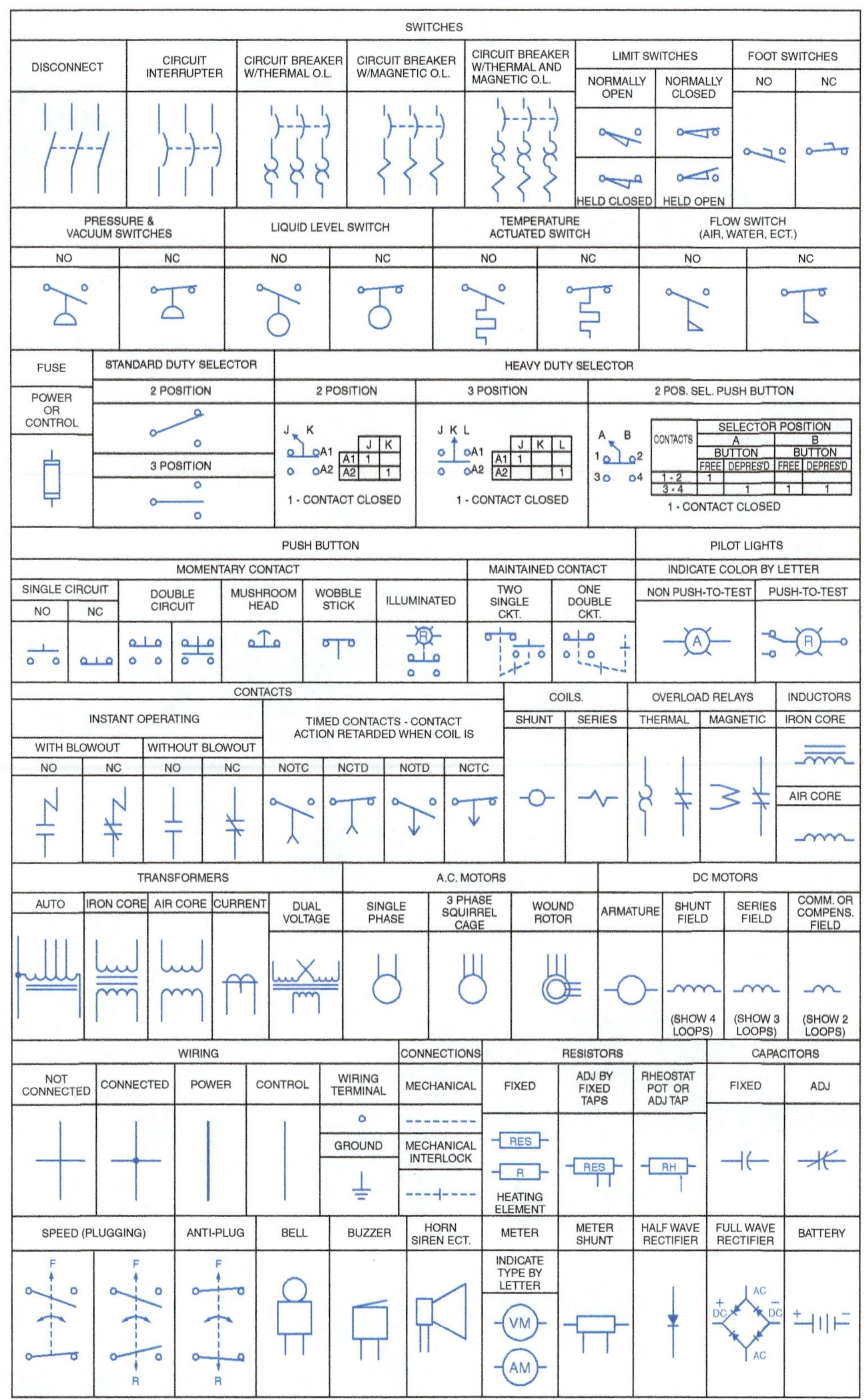

Figure A–9 Standard wiring diagram symbols.

GLOSSARY

AC Alternating current.

Across-the-line Method of motor starting which connects the motor directly to the supply line on starting or running. (Also called Full Voltage Control.)

Automatic Self-acting, operating by its own mechanism when actuated by some triggering signal such as a change in current strength, pressure, temperature, or mechanical configuration.

Automatic Control Employs the use of pilot sensing devices but does not require an operator to initiate each change of action.

Bimetal Strip A strip made by bonding two unlike metals together that, when heated, expand at different rates. This causes a bending or warping action.

Blowout Coil Electromagnetic coil used in contactors and starters to deflect an arc when a circuit is interrupted.

Cad Cell A device that changes its resistance with a change of light intensity.

Cam Timers Also known as sequence timers. Used to provide a continuous timing operation by the setting of adjustable cams.

Capacitor A device made with two conductive plates separated by an insulator or dielectric.

Cardiopulmonary Resuscitation (CPR) A method of alternating closed heart massage and artificial respiration for a person that is not breathing and has no pulse.

Circuit Breaker Automatic device that opens under abnormal current in carrying circuit; circuit breaker is not damaged on current interruption; device is ampere, volt, and horsepower rated.

Clock Timer A time-delay device that uses an electric clock to measure the delay period.

Coil Clearing Contacts Contacts used to prevent power being continuously supplied to the coil of a latching type relay.

Collector The semiconductor region of a transistor which must be connected to the same polarity as the base.

Confined Spaces Spaces that have very limited entrance and exit.

Contactor A device that repeatedly establishes or interrupts an electric power circuit.

Current Limit A setting that limits the amount of current a controller can supply.

Current Relay A relay that functions at a predetermined value of current. A current relay may be either an overcurrent relay or an undercurrent relay.

Dashpot Consists of a piston moving inside a cylinder filled with air, oil, mercury, silicon, or other fluid. Time delay is caused by allowing the air or fluid to escape through a small orifice in the piston. Moving contacts actuated by the piston close the electrical circuit.

DC Direct current.

De-energized Circuit A circuit that has no power applied to it.

Delta Connection A circuit formed by connecting three electrical devices in series to form a closed loop. Most often used in three-phase connections.

Diode A solid state device that performs the function of an electric check valve by permitting current to flow through it in only one direction.

Diodes A two-element device that permits current to flow through it in only one direction.

Direct Current (DC) A type of current that does not reverse polarity or direction of current flow.

Dynamic Braking A method of braking a motor using magnetic fields.

Eddy Currents Circular induced currents contrary to the main currents; a loss of energy that shows up in the form of heat.

Electronic Timers Timers that employ electronic components to produce a time delay.

Emitter The semiconductor region of a transistor which must be connected to a polarity different than the base.

Energized Circuit A circuit that has power applied to it.

Fire-Retardant Clothing Clothing that is treated with chemicals that reduce its ability to burn.

Floating System A control system that does not have one side grounded.

Fuse An overcurrent protective device with a fusible member, which is heated directly and destroyed by the current passing through it to open a circuit.

Ghost Voltages Voltages that can appear on high impedance voltmeters that are caused by induction from surrounding magnetic fields.

Grounded System A control system that has one side connected to ground.

Holding contact A contact used to maintain the circuit when the start push button is released.

Horsepower Measure of the time rate of doing work (working rate).

Hysteresis Loss A type of power loss created by a magnetic field changing polarity at regular intervals.

Idiot Proofing Designing a piece of equipment so that it cannot be misused or connected in an incorrect manner.

IEC International Electrotechnical Commission. A European organization that sets standards for electrical equipment.

Inching (jogging) Momentary operations; the quickly repeated closure of the circuit to start a motor from rest for the purpose of accomplishing small movements of the driven machine.

Interlocking A method of preventing actions from taking place at the same time.

Internal relays Relays that exist only in the program of a programmable logic controller.

International Electrotechnical Commission (IEC) A European organization that sets standards for electrical equipment.

Inverter Rated A motor that is designed to be used with variable frequency controls.

Jogging (inching) Momentary operations; the quickly repeated closure of the circuit to start a motor from rest for the purpose of accomplishing small movements of the driven machine.

Kick-Back Diode A diode used to eliminate the voltage spike induced in a coil by the collapse of a magnetic field.

Latch Coil The coil of a latching type contactor or relay that causes the contacts to close and lock in place.

Latching Contactors and Relays A type of contactor or relay that mechanically locks the contacts in the open or closed position after the power has been removed.

Leakage current The current that flows through an unintended path.

Lenz's Law A basic law concerning magnetism that basically states that induced voltages and currents oppose the force that produces them.

Locked Rotor Torque (of a motor) The minimum torque that a motor will develop at rest for all angular positions of the rotor with the rated voltage applied at a rated frequency. (ASA)

Lockout and Tagout The act of placing a lock and warning tag on the disconnect of a piece of equipment that is not to be energized.

Magnet Braking Friction brake controlled by electromagnetic means.

Magnetic Field The space in which a magnetic force exists.

Maintaining contact See holding contact.

Manual Control Characterized by the fact that an operator must go to the location of the controller to initiate any change in the control circuit.

Material Safety Data Sheets (MSDS) Written material supplied by various manufacturers that list safety information about a particular product.

Maximum Hertz The highest frequency a controller will produce.

Meter A device for measuring some quantity such as voltage, current, resistance, etc.

Milliamperes (mA) 1/1000 of an ampere (0.001 ampere).

Minimum Hertz The lowest frequency a controller will produce.

Molecular Friction Heat produced by molecules continually changing direction due to alternating magnetic fields.

MOV (Metal Oxide Varistor) An electronic component that changes its resistance in accord with an amount of voltage.

National Electrical Code (NEC) A recognized authority on the safe installation of electrical wiring. Generally used by federal, state, and local authorities in establishing their own laws and ordinances concerning the installation of electrical wiring.

NEMA National Electrical Manufacturers Association.

Normally Closed A switch in the closed or on position when the circuit is de-energized.

Normally Closed Held Open A switch connected normally closed but the contacts are being held open when the circuit is de-energized.

Normally Open A switch in the open or off position when the circuit is de-energized.

Normally Open and Normally Closed When applied to a magnetically-operated switching device, such as a contactor or relay, or to the contacts of these devices, these terms signify the position taken when the operating magnet is de-energized. The terms apply only to nonlatching types of devices.

Normally Open Held Closed A switch that is connected normally open but the contacts are held closed when the control circuit is de-energized.

Off-Delay A timer in which the contacts change position immediately when the coil or circuit is energized, but delay returning to their normal positions when the coil or circuit is de-energized.

On-Delay A timer in which the contacts delay changing position when the coil or circuit is energized, but change back immediately to their normal positions when the coil or circuit is de-energized.

Op Amp Operational amplifier.

Photodetectors Devices used to detect the presence or absence of light.

Photodiode A diode that conducts in the presence of light, but not in darkness.

Plugging The act of starting a motor with short jabs of power.

Pneumatic Timer A device that uses the displacement of air in a bellows or diaphragm to produce a time delay.

Poles The output or input terminal of a switch.

Potentiometer A variable resistor with a sliding contact, which is used as a voltage divider.

Pushbutton A switch with open or closed contacts that change position when a button is pressed.

Ramping Accelerating or decelerating a motor over some period of time.

Rectifier A device used to changes alternating current into direct current.

Relays Operated by a change in one electrical circuit to control a device in the same circuit or another circuit; rated in amperes; used in control circuits.

Resistance Start Induction Run Motor One type of split-phase motor that uses the resistance of the start winding to produce a phase shift between the current in the start winding and the current in the run winding.

Resistor A device used primarily because it possesses the property of electrical resistance. A resistor is used in electrical circuits for purposes of operation, protection, or control; commonly consists of an aggregation of units.

Scaffolds Portable structures that can be assembled to provide a high working platform

Schematic Diagram A control diagram that shown components in their electrical sequence without regard for physical location.

SCR Silicon Controlled Rectifier.

Sealing Contacts See holding contacts.

Semi-Automatic Control Employs the use of pilot sensing devices. An operator is required to initiate changes in the action of the control circuit.

Service Factor (of a general-purpose motor) An allowable overload; the amount of allowable overload is indicated by a multiplier which, when applied to a normal horsepower rating, indicates the permissible loading.

Shaded-Pole Motors Single-phase induction motors provided with an auxiliary short-circuited winding or windings displaced in magnetic position from the main winding. (NEMA)

Slip The difference in speed between the rotating magnetic field and the rotor of a motor.

Solder Pot See Eutectic Alloy.

Solenoid An electromagnetic with a movable metal core the changes electrical energy into linear mechanical motion.

Solid-State Devices Electronic components that control electron flow through solid materials such as crystals; e.g., transistors, diodes, integrated circuits.

Solid-State Relay A relay that controls current flow with a solid state device instead of mechanical contacts.

Split phase A type of single-phase motor that produces a separate phase to create a rotating magnetic field.

Stealer Transistor A transistor used in such a manner as to force some other component to remain in the off state by shunting its current to electrical ground.

Step-Down Transformer A transformer that produces a lower voltage at its secondary winding than is applied to its primary winding.

Switch A mechanical or electrical devices that completes or breaks an electric circuit, or sends it on a different path.

Synchronous Speed The speed of the rotating magnetic field of an AC induction motor.

Thermistor An electronic device that changes it resistance in accord with temperature.

Thyristor An electronic component that has only two states of operation, on and off.

Transformer An electromagnetic device that converts voltages for use in power transmission and operation of control devices.

Transistor A solid-state device made by combining three layers of semiconductor material. A small amount of current flow through the base-emitter can control a larger amount of current flow through the collector-emitter.

Triac An electronic device used to control alternating current.

Unijunction Transistor (UJT) A special transistor that is a member of the thyristor family of devices and operates like a voltage-controlled switch.

Unlatch Coil The coil of a latching type contactor or relay that causes the contacts to open.

Volt/Voltage An electrical measure of potential difference, electromotive force, or electrical pressure.

Volts per Hertz A control that regulates the output voltage in accord with the frequency.

Watchdog Timer A time used to prevent a programmable controller from becoming locked in a loop.

Wiring Diagram A diagram that shows a pictorial representation of components with connecting wires.

Wye Connection A connection of three components made in such a manner that one end of each component is connected. This connection generally connects devices to a three-phase power system.

INDEX

D

E

F

G

H

I

J

K

L

N

O

P

R

S

W

Z